Telecommunications Technology Handbook

For a complete listing of the *Artech House Telecommunications Library*,
turn to the back of this book . . .

Telecommunications Technology Handbook

Daniel Minoli

Bell Communications Research, Inc.
Red Bank, New Jersey

New York University

Artech House
Boston • London

Library of Congress Cataloging-in-Publication Data

Minoli, Daniel, 1952-
Telecommunications technology handbook / Daniel Minoli.
p. cm.
Includes bibliographical references and index.
ISBN 0-89006-425-3
1. Telecommunication. 2. Data transmission systems. I. Title.
TK5101.M54 1991 90-26023
621.382--dc20 CIP

British Library Cataloguing in Publication Data

Minoli, Daniel
Telecommunications technology handbook
1. Telecommunication
I. Title
621.38

ISBN 0-89006-425-3

685 Canton Street
Norwood, MA 02062

International Standard Book Number: 0-89006-425-3
Library of Congress Catalog Card Number: 90-26023

10 9 8

For
Gino, Angela, Anna,
Emmanuelle, Emile, and Gabrielle

Contents

Preface

The objective of this text is to expose managers to contemporary issues in the telecommunication and data communication fields. Both technical and managerial aspects are addressed. The text provides a comprehensive guide to the telecommunication field, with emphasis on issues affecting the industry in the 1990s.

The subject matter is written by a practitioner, for practitioners. We often take an end-user perspective; we may not follow the historical development of a subject, but we take advantage of our insight from the current vantage point.

We examine the major components of a network, including the transmission and switching segments. We survey microwave, fiber, infrared, and satellite transmission systems, and discuss private branch exchanges (PBXs). We examine basic data communication technologies, including evolving international standards and the Open Systems Interconnection (OSI) reference model. We cover management of long-haul and local area networks, and address issues pertaining to security.

This text takes an advanced view of the field, and is not primarily intended as an introductory, fully pedagogical treatment of elementary issues in communication, for which scores of other excellent texts are available. A reader with some exposure to telecommunication who wants to become aware of contemporary technical issues affecting the industry should find the text of value.

The telecommunication and data communication fields are fast-moving industries. A hypothetical practitioner who last designed a network in 1985 and then went on to do something different, would, upon returning in 1990, find a virtually new environment—the technology is moving that fast. In some areas, the half-life of the information is down to six months! This book aims at familiarizing the reader with these important evolving communication technologies and ensuing opportunities.

What makes this book different from other textbooks?

1. We present the subject from a contemporary perspective: issues and trends of the early 1990s. These issues are certainly not the same as those of the mid- or even late 1980s. Some topics and emphases are:

Emphasis on high-speed communication, including the new digital hierarchies: fractional T1 (DS1), T1 (DS1), T3 (DS3), SONET, ATM;
Coverage of FDDI issues and standards, DQDB MAN systems, IEEE 802.9 integrated voice-data LAN interfaces;
Heavy OSI reference model perspective and an emphasis on upper layer standards;
Emphasis on LANs, MANs, and interconnection issues;
Coverage of network security and the ISO security standards;
Emphasis on network management and the ISO standards;
Emphasis on fiber optics technology;
Coverage of ISDN and broadband ISDN applications, including frame relay;
VSAT systems and issues and discussion of WARC-92; and
Emphasis on the new digital nature of communication (more than 80% of the *Fortune* 500 companies now have high-speed digital networks).

2. Covers topics not often treated in other texts (or at least not all covered in one text). Examples are:

Wireless data communication, including FM subcarrier, cellular radio, packet radio, and "cordless telephone 2";
Free-space infrared communication technology;
Detailed user-to-network and user-to-user signaling issues, including CCSS7 information;
Information on high-speed digital equipment such as T1 multiplexers and digital cross-connect systems; coverage of ADPCM technology, transcoders, and testing and monitoring of highly integrated corporate networks;
Coverage of SONET and ATM;
PBX network design, including practical information on SMDRs and computer-to-PBX links;
Detailed information on fiber optics technology, trends, and issues;
Coverage of high-speed dialup V.32 and V.42 modems;
Detailed discussion of LAN and WAN internetworking, including OSI methods, TCP/IP, bridges, and routers;
LAN security (IEEE 802.10) and network security in general; and
Description of the standards-making process and 24 of the most important standards-setting bodies.

All of these factors and technologies complicate the network design process. Understanding the new design requirements and acquiring the appropriate tools is a mandatory effort in establishing cost-effective, reliable, and flexible corporate networks. The responsibilities of a communication manager are becoming more demanding under the thrust of industry competition and the ensuing multiplicity of available communication options; they are further complicated by the increased obsolescence rate of the technology. In designing a network, we must analyze multiple options and take into account the issue of hidden costs associated with these options. This richness of communication options offers the telecommunication manager major opportunities, not only in terms of network design, but also in terms of network management, grade of service, and reliability.

While some chapters are formal (for example, Chapter 4 on ISDN and Chapter 13 on the OSIRM), others are practical (for example, Chapter 5 on radio technology, Chapter 10 on PBX networking, and Chapter 14 on LAN management).

3. This text aims at giving a balanced, yet comprehensive view of telecommunication and data communication, from a contemporary perspective. The topics treated are ones that people are discussing in the 1990s. The text can serve as a one-or two-semester graduate course in telecommunication, assuming some previous exposure to the field or a good treatment by the instructor. The text is being used by the author at New York University and has evolved from lecture notes developed over a period of seven years of teaching.

While conveying fast-developing technical information through a book medium is becoming increasingly difficult, this presentation is aimed at describing the foundation on which the developments likely to occur over the next decade are based. The book is current as of 1990, and it is designed to cover a three- to four-year window of the industry.

Acknowledgments

This text draws on the expertise of hundreds of specialists identified in the references listed at the end of each chapter.

The following individuals have made direct contributions in reviewing the manuscript and providing valuable comments: K. Tesink, E. W. Geer, A. L. Lang, T. R. Farese, C. F. Newman, A. A. Knapp, R. G. Spusta, D. Spears, R. Sinha, T. M. Bauman, M. Ward-Callan, J. A. Ross, D. L. Alt, S. Wainberg, R. M. Ephraim, M. Romeiser, C. Lin, R. Goldberg, S. F. Knapp, R. A. Graff, T. Peak, B. A. Mordecai, L. J. Lang, D. Tow, G. Tom, E. W. Soueid, and M. Seeley. D. Kenney, President of Bank Data Bank Consultants, made substantial contributions to the microwave technology section of this work.

The following individuals are thanked for their moral support: Dick Vigilante, New York University; Lester Sitzes, DataPro Research Corporation; John J. Amoss and James E. Holcomb, Bell Communications Research; and Ben Occhiogrosso, DVI Communications.

This book does not reflect *any* policy, position, or posture of Bell Communications Research (Bellcore); the work was not funded or supported financially by Bellcore or Bell Operating Company (BOC) resources. All ideas expressed are strictly those of the author. Data pertaining to the public switched network are based on open literature and were not reviewed by the BOCs. Bellcore did support the project in spirit and is hereby thanked.

This book contains hundreds of thousands of words, and I alone take responsibility for any incomplete, uneven, or unintuitive treatment of the subject matter.

Dan Minoli
November 1990

Chapter 1
The Telecommunication Environment

1.1 INTRODUCTION

We are now in what is called the "Information Age." Information has become a commodity not only to the business community, but to all of society. In fact, the whole economic well-being of a nation may well depend on the telecommunication infrastructure in place in that nation, and the reach of that infrastructure to other nations. Just as there are techniques on how best to handle physical commodities, so there are techniques on how best to manage information, and how to transmit it in a timely fashion to where it is needed. This text addresses issues pertaining to and affecting the field of communication of information in the 1990s.

The word communication derives from the Latin word *communicare*: to impart, participate. The term "information" can be viewed as a primitive (axiomatic) concept, requiring no further definition. The science of "communication" is the study of all information transfer processes.

Given two entities A and B, communication is the action of transferring information from A to B, and *vice versa.* A and B are communicating when A is suitably able to code a message of information and relay it to B through an appropriate medium, while B is suitably equipped to receive the message by using or interpreting it in some fashion. A set of entities A, B, C, . . . , is in mutual communication when the pairs (A,B), (A,C), (B,C), *et cetera* can, at appropriate times, communicate with each other. Protocols are agreed conventions between communicating entities on how to carry out the mechanics of the communication process.

Telecommunication entails disciplines, means, and methodologies to communicate over distances, in effect, to transmit voice, video, facsimile, and computer data. *Data communication* entails disciplines, means, and methodologies particular to transmission of computer data, possibly over a specially engineered network, and typically from a protocol perspective. The data communication field is a subset of the telecommunication field.

A formal theory of *communication and information* has evolved over the past 40 years, heralded by Claude Shannon in 1948 [1.1]. In this framework, a simplified model is described, as illustrated in Figure 1.1. An information source generates a message, which in turn is submitted through a transmitter and coder as a signal to a medium (or channel). The medium, however, is not a perfect conveyer, so that medium-specific pathologies may be added to the message. Noise from external sources can also affect the message. The message collected by the receiver module of the destination (or sink) is decoded by a decoder, and sent along to its destination for further usage. The model of this process is illustrated in Figure 1.1

Information theory provides a mathematical treatment of the fundamental limits on the reliability and efficiency of a communication system. This formal theory permits the derivation of certain theoretical results for the channel and the coder-decoder, based on the type of influences affecting the channel. One of these results is discussed later in the chapter.

1.1.1 An Historical View

Electrical communication started in the nineteenth century. The relationship between magnetism and electricity was discovered by Oersted in 1819. Further work was undertaken by Faraday and Ampere in the 1820s. Laboratory systems able to transmit information via the use of rotating magnetized needles became available in that decade. In the early 1830s, Gauss and Weber developed a small-scale telegraph system operating in the city of Göttingen. In 1844, Morse set up a 40-mile telegraph line between Washington, DC, and Baltimore; the system employed dots and dashes [1.2]. In 1849, the first slow telegraph printer link was installed. By the 1860s, the transmission speed was up to 15 bits (binary digits) per second. In 1874, Baudot invented the first "carrier" system able to multiplex up to six telegraph channels onto a single copper wire pair. Telephone lines started to appear

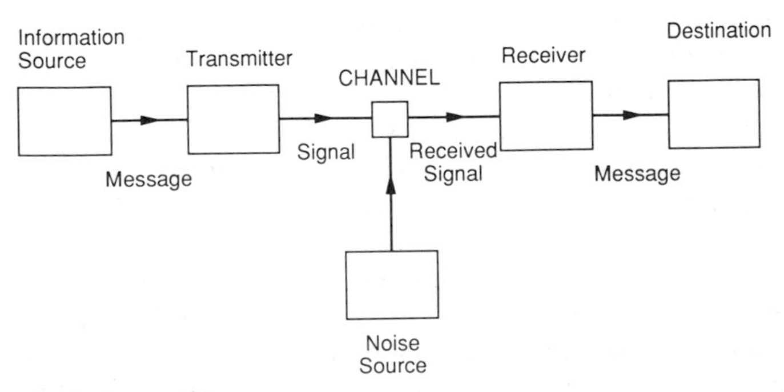

Figure 1.1 Communication model.

in the 1890s. Around 1920, several voice channels were being multiplexed over one wire pair.

Today, thousands of information channels can be carried by a single underlying medium. In the 1940s, 600 voice channels could be carried over a coaxial cable, with a composite carrying capacity of 1800 channels over a system of multiple tube pairs (known as L1, 1946 vintage); in digital equivalent this corresponds to 1.15×10^8 bits per second (b/s). By the early 1950s, 1860 voice channels could be carried over a coaxial cable, with a composite carrying capacity of 9300 channels over a system of multiple tube pairs (known as L3, 1953 vintage); in digital equivalent this corresponds to 5.95×10^8 b/s. By the 1960s, 3600 voice channels could be carried over a coaxial cable, with a composite carrying capacity of 32,400 channels over a system of multiple tube pairs (known as L4, 1967 vintage); in digital equivalent this corresponds to 2×10^9 b/s. By the 1970s, some 10,800 voice channels could be carried over a coaxial cable, with a composite carrying capacity of 108,000 channels over a system of multiple tube pairs (known as L5, 1974 vintage); in digital equivalent this corresponds to 6×10^9 b/s. By the early 1980s, some 13,200 voice channels could be carried over a coaxial cable, with a composite carrying capacity of 132,000 channels over a system of multiple tube pairs (known as L5E, 1978 vintage); in digital equivalent this corresponds to 8×10^9 b/s [1.3]. As of 1990, a fiber optic link can carry multiple gigabit data rates, in the 10^{10} b/s range.

Microwave-based radio systems also achieved similar bandwidth growth, with 2400 channels in 1950 (TD-2 systems); 6000 channels in 1959 (TD-2); 9000 channels in 1967 (TD-3); 12,000 channels in 1968 (TD-3); 16,500 channels in 1973 (TD-3A); 19,800 channels in 1979 (TD-3D); and 42,000 channels in 1981 (AR6A) [1.3].

The evolution of key telecommunication services over time, beginning in 1847, is shown in Figure 1.2. We discuss many of these services in the following chapters.

1.1.2 The Future

An empirical review of the bandwidth that can be carried over transmission systems, as discussed in the previous section, shows that the bandwidth has increased by an order of magnitude every 20 years: 1950s: 10^8 b/s; 1970s: 10^9 b/s; 1990s: 10^{10} b/s. Transmission bandwidth may increase even more in the future, particularly considering the potential of fiber optic technology. If the above empirical rule were to hold, the data rate of the next few decades will be:

Year 2010	Year 2030
10^{11} b/s	10^{12} b/s

Considering that the bible, when coded with an 8-bit character set code, results in approximately 40,000,000 bits, we could transmit the content of 250 bibles in 1

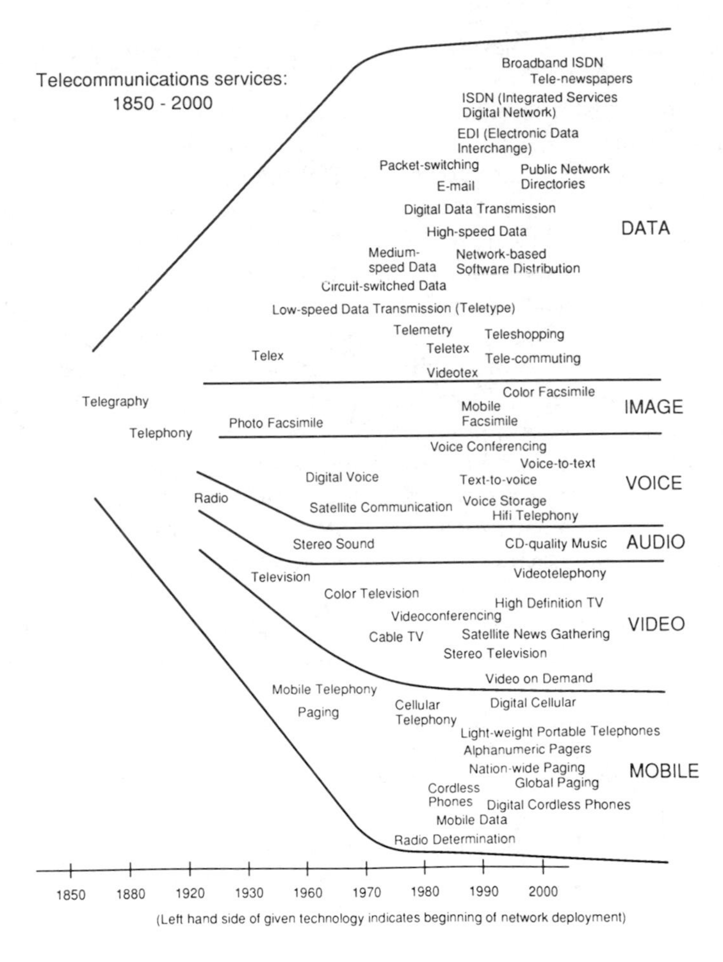

Figure 1.2 Evolution of telecommunication services.

second in 1990; 2500 bibles in 1 second in the year 2010; and 25,000 bibles per second in the year 2030.

Figure 1.3 depicts the approximate retail cost of bandwidth in the early 1990s. As we can see from the figure, the cost can be as little as 0.2 cents per b/s per month for a 1000-mile link; wholesale bandwidth available to carriers is even cheaper. Superficially, we might be inclined to believe that the cost of managing information, including communication, is high. In reality, the cost of information in constant money has been decreasing for centuries. Four major phases in this process have been: (1) the introduction of papyrus and then paper in place of earlier types of "hard" media (first to sixth century); (2) the invention of the movable-character press (fifteenth century); (3) electronic communication of all

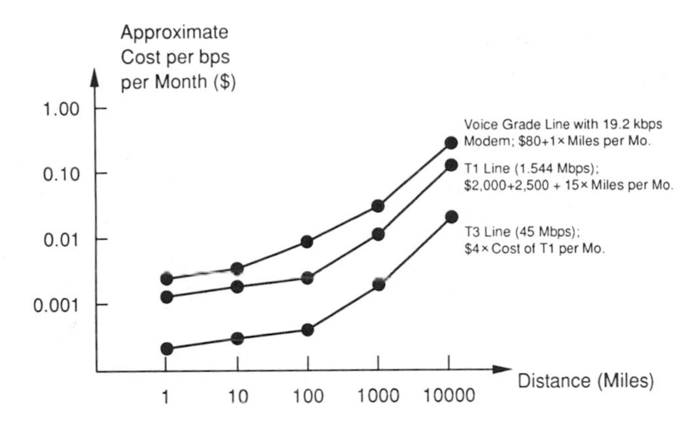

Figure 1.3 Approximate cost of bandwidth, in b/s per month.

types (beginning in the nineteenth century); and (4) the recent proliferation of computers. What has been increasing is the appetite for how much information needs to be moved. Whereas, until 30 years ago, a telecommunication manager was (probably) glad to have a 50 b/s channel between two sites, now managers aspire to DS1 facilities (1.5 million bits per second—one bible in 30 seconds), or even DS3 facilities (45 million bits per second, the equivalent of one bible per second).

1.1.3 The Course of Investigation

The first chapter contains a summary of key concepts in a traditional telecommunication and data communication environment, which are utilized in the rest of the book. Almost every one of the topics covered in Chapter 1 could be the subject of a lengthy book; only the most basic introductory treatment is provided here. Subsequent chapters concentrate on evolving issues that are likely to be of key importance in the 1990s. Our goal is to present a three- to five-year window on the future of the industry. Again, the topics covered in a given chapter could be expanded into a book or sets of books. Consequently, we will not address every possible aspect, pathology, minute exception, or obscure twist of a given topic in this book.

The presentation generally follows the following formula, with amplification of pertinent topics:

- Telecommunication;
  - Transmission;
  - Switching;

- Data Communication;
  - Transmission; and
  - Switching.

See Table 1.1; numerous secondary evolving issues and technologies are covered under each broad topic identified by the chapter heading. The table is only meant to assist the reader in a general appreciation of how the various topics loosely fit the overall structure of the subject matter. By classifying local area networks (LANs) and metropolitan area networks (MANs) in the *switching* category, we employ the property of being able to reach any other network customer as a

Table 1.1
Subject Synthesis as Undertaken in this Text

TELCOMMUNICATION	
Chapter 2:	The criticality of standardization
Transmission	
Chapter 3:	Digital transmission issues: High-capacity systems
Chapter 4:	Digital transmission in the loop: Integrated services digital network
Chapter 5:	Transmission media and radio services
Chapter 6:	Transmission systems: Satellite and microwave
Chapter 7:	Transmission systems: Fiber
Chapter 8:	Transmission systems: Free-space infrared
Switching	
Chapter 9:	Switching and signaling
Chapter 10:	Private branch exchange technology and signaling
DATA COMMUNICATION	
Transmission	
Chapter 11:	Data communication: Issues affecting the 1990s
Chapter 12:	Computer channel-to-channel communication
Chapter 13:	Standards: Fundamental building block to achieve end-to-end connectivity
Switching	
Chapter 14:	Local area networks
Chapter 15:	Metropolitan area networks
Managing Data Communication	
Chapter 16:	Network management and evolving standards
Chapter 17:	Network security and evolving standards

characteristic of switching; in reality LANs and MANs also clearly involve transmission (in addition, switching is done in a distributed fashion, rather than in a centralized fashion, as is more traditional in a telecommunication environment).

1.2 COMPONENTS OF A TELECOMMUNICATION NETWORK

In this section, we examine various types of networks. A telecommunication network can be viewed as an ensemble of a number of links, such as those shown in Figure 1.1. Telecommunication networks consist of three general categories of equipment: *termination equipment*, *transmission equipment*, and *switching equipment*. Each of these three categories, in turn, comprises a number of subcategories or technologies. Figure 1.4 depicts a typical telecommunication network of the 1990s. As we can see, these networks can be very complex and may employ a variety of technologies.

Networks can be *public* or *private*. Public networks can be used by everyone, similar to public transportation. Private networks are typically restricted to one organization (or a group of organizations); the general public is not allowed access to a private network, just as one cannot employ (without permission or by arrangement) somebody else's private automobile. Both types of networks basically have the same functionality and capabilities, although different capabilities may be accentuated in one or the other. Public networks are provided by *common carriers*.

Public networks usually utilize carrier-provided switches, also known as *exchanges*. Prior to the Bell System divestiture, the distinction between exchange access and interexchange communication was not a matter of major regulatory importance, except possibly in terms of the rates and tariffs. After divestiture, terms such as "local exchange" and "interexchange" acquired an additional legal distinction; local exchange access service and interexchange service (more specifically, inter-LATA—local access and transport area) must be provided by separate entities, and the old Bell companies are currently precluded from offering domestic interexchange service (with minor exceptions in some major corridors, for example, Northern New Jersey and New York city).

The sections that follow address technical issues pertaining to traditional circuit-switched voice networks, while the latter part of the chapter addresses data networks. A strong trend toward integration of voice, data, and video is apparent. The activities that are taking place under the auspices of ISDN (integrated services digital network), discussed in detail throughout this text, are an indication of this trend. The separate treatment undertaken herewith of the voice and data disciplines is not meant to detract in any way from this desirable and attainable goal. The separate treatment is dictated solely by the presentation's logistics.

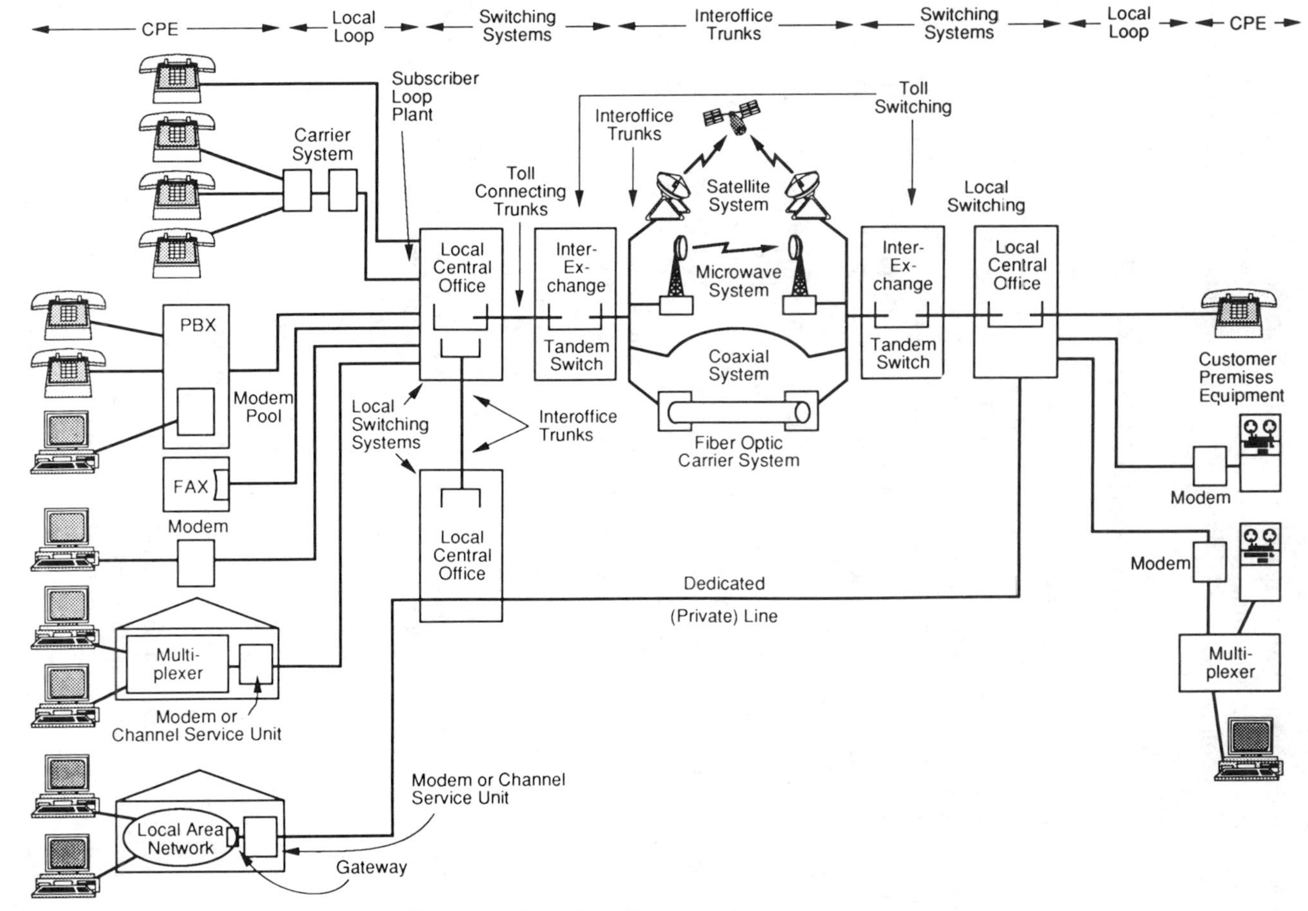

Figure 1.4 A modern telecommunication network.

1.2.1 Voice Networks

The traditional public switched telephone network was originally developed to service voice traffic. Data can also be carried by the same network when a *modem* (modulator-demodulator) is employed by users at each end of the link. In effect, the modem transforms the data into an acoustical signal that fits into the nominal 4-kHz bandwidth of a standard telephone channel. This method of carrying data is called *voiceband* or *circuit-mode* data. Improved network facilities more suited to carrying data in their native digital mode are now beginning to emerge, as discussed in Chapters 3 and 4.

1.2.1.1 Customer-Premises Equipment

Customer-premises equipment (CPE) is equipment that is owned and maintained by the user. It includes the signal-entry termination equipment (called *station* in the voice environment), concentration equipment, in-building or in-campus wiring, or even an entire subnetwork. In the latter case, the user may own the communication network up to the demarcation point (the point between the public network and the user's network), or in some cases the user can also own the long-haul transmission and switching equipment.

Typical CPE in the voice area includes:

- Telephone sets (particularly after Computer Inquiry II, 1980, and divestiture, 1984);
- Key telephone equipment, including telephone sets, wiring, and other components;
- PBXs;
- Inside wiring (including wire closets, jacks, and connectors);
- Recording, answering, and voice mail equipment.

CPE *termination equipment* (as compared to other types of CPE) accepts the user's voice, data, or video signals and encodes them so that they can be transmitted over a telecommunication network. Examples of termination CPE include voice station sets, CRTs and other data terminals, facsimile machines, and videoconferencing cameras.

Analog termination devices encode the user signal (i.e., speech) into an electrical signal that replicates the energy content of the original signal. This electrical signal is then transmitted to the desired remote location. Intuitively, *digital termination* devices "measure" the height of the signal at frequent intervals, and then represent that height with a binary coded number (more sophisticated methods perform signal analysis rather than simple energy measurement). For a computer terminal, the signal is already in digital form. The digital representation of the signal is then transmitted to the desired remote location, where it is utilized by the receiving CPE in digital form, or is converted into analog form. As an alternative, the user's CPE delivers an analog signal to the network; the network transforms this signal into a digital stream for more reliable transmission, and then reconverts it into an analog signal for delivery to the intended destination.

Binary numbers, employed in digital transmission, are composed of combinations of 0s and 1s. The individual 0s and 1s are called *bits*. A collection of eight bits is called an *octet* (a less precise term is *byte*). Sometimes octets and bytes are also called words, although the term is more appropriate to describe a collection of octets. For example a 32-bit computer is said to have 32-bit words (four octets). Digital communication channels are measured in terms of their information carrying capacity in b/s.

Initially, the public telephone plant was an analog network, optimized for analog voice transmission. This included analog transmission and analog switching facilities. Beginning in the early 1960s many interoffice trunks (links between switches) began to be replaced with digital links. In the mid-1970s, switches also began to handle digitized voice directly (i.e., without multiple conversions between analog and digital). Today digital links are common and prevalent. A new network architecture, ISDN, aims at providing end-to-end digital circuits to the customer. All major telecommunication carriers in the United States and abroad have stated that ISDN is the strategic direction of their networks. Digital circuits are more suited to data transmission applications than are analog circuits.

Voice Digitization Schemes

The telephone instrument performs the function of coding the user's voice into a signal suitable for subsequent transmission over the network. For public switched networks, an evolution is taking place in this area, tracking (sometimes leading) a similar evolution in private networks. The evolution is as follows. Until the 1960s, the telephone set generated an analog signal, which was transmitted through the network in an analog fashion, end-to-end. Beginning in the 1960s and continuing through the 1980s, while the set still generated analog signals, the voice could be digitized in the transmission or in the switching components of the network [first in time in trunks between central offices (COs), 1962; then in the loops between the user and the CO, 1973; then in the proximity of the switch itself, so that a digital switch could terminate undemultiplexed loop or trunk carrier systems, 1976]. With ISDN in the 1990s, the telephone set will be allowed to generate a digital signal representing the user's voice, for end-to-end transmission in digital form. (In a number of PBXs, this digitization at the telephone set is already taking place).

Historically, digitization techniques have been identified with activities performed in the network, particularly in reference to trunk carrier systems, digital loop carriers, and digital switches, as described in the previous paragraph. Because the future belongs to ISDN (or, at least, to digital communication), voice digitization techniques are here properly seen from a CPE perspective (the continued introduction of CPE high-speed digital multiplexers also supports this perspective). The digitization techniques do not change with this change in perspective; the perspective only determines in what context the subject is treated. Remember, however, in spite of the present perspective, that as of 1990, a very large percentage of the voice digitization still occurs within the network proper.

To digitize the voice means to represent it with a stream of numbers coded in binary representation. Two classes of methods are used to digitize voice: *waveform coding* and *vocoding*. In waveform coding, we attempt to code and then reproduce the analog voice curve by modeling its physical shape. The number of b/s to represent the voice with this method is high: 64, 32, 16, or at least 9.6 kb/s, depending on the technology. Vocoding attempts to reproduce the analog voice curve by performing a mathematical analysis (fast Fourier transform) that "identifies" abstractly the type of curve; what is transmitted is a small set of parameters describing the nature of the curve. The number of kb/s to represent the voice with this method is low: 9.6, 4.8, 2.4, and even 1200 b/s, depending on the technology. Voice quality is increasingly degraded as the digitization rate becomes smaller. An extensive body of research on vocoding methods has evolved in the past 15 years (at last count more than 700 technical papers have been written).

Digital speech quality through a network is also degraded by the accumulation of quantization noise introduced at signal conversion points; conversion can occur several times in a network, in a partially digital environment. In a nearly totally digital environment, only one analog-to-digital conversion close to the source and one digital-to-analog conversion close to the destination are required. In a totally digital environment, the conversion takes place right at the source, and not in the network. Voice quality will improve substantially in these environments, particularly with the high-quality coding discussed later.

Pulse Code Modulation

The simplest waveform method to convert analog speech to a digital stream is a process called *pulse code modulation* (PCM). PCM was invented in the 1930s, but only became prevalent in the 1960s when transistors and integrated circuits became available.

Nyquist theory specifies that to code properly an analog signal of bandwidth W with basic PCM techniques, we need $2W$ samples per second. For voice, band-limited to a nominal 4000-Hz bandwidth, we need 8000 samples per second (the actual telephony frequency range used is 300 to 3400 Hz). The dynamic range of the signal [and ultimately the signal-to-noise ratio (S/N)] dictates the number of quantizing levels required. For telephonic voice, 256 levels suffice, based on psychoacoustic studies conducted in the 1950s and early 1960s; it follows that 8 bits are needed to represent this many levels uniquely. This, in turn, implies that we need 64,000 b/s to encode telephonic human speech in digital form. Figure 1.5 depicts the steps involved in PCM at a high level. PCM does not require sophisticated signal processing techniques and related circuitry; hence, it was the first method to be employed, and is the prevalent method used today in telephone plant. PCM provides excellent quality. This is the method used in modern *compact disc* (CD) music recording technology (although the sampling rate is higher and the coding words are longer, to guarantee a frequency response to 22 kHz). The problem with PCM is that it requires a fairly high bandwidth (64 kb/s) to represent a voice signal. PCM is specified by the CCITT's (Consultative Committee on International Telephone and Telegraph) Recommendation G.711. The CCITT is a standards-making body described in more detail in Chapter 2. Two "laws" (recommended standards) describe voice compression in PCM: in the United States, the μ-law is used; in Europe, the A-law is employed. The reason to follow specific PCM standards is that we want to be able to install equipment from different manufacturers and still retain system integrity and compatibility. (Although PCM can be mathematically treated as a type of modulation—which we will discuss later—many people today view it as an example of signal processing; this is the perspective we will use herewith.)

One key issue is the spacing in the signal amplitude postulated by the sampling codec (also known as quantizer) to establish the boundaries where the different levels are declared. If we divide the maximum amplitude in 256 equal intervals, voice, which normally has numerous low-level signal components, would not be coded adequately. Instead, the amplitude space is subdivided with logarithmic spacing with respect to the signal origin; this affords a stable S/N ratio over a wide

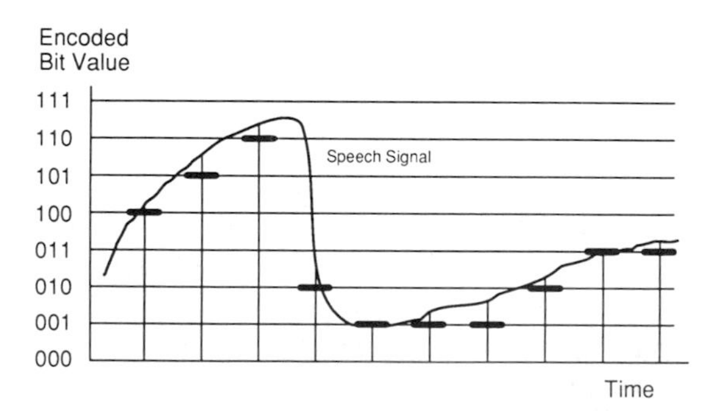

Figure 1.5 PCM coding (simplified).

range of voice levels. Note, as implied in Figure 1.5, that if the input signal amplitude exceeds the maximum quantizer level, the result is clipping distortion. A quantizer must be designed to avoid frequent clipping; hence, the quantizer's maximum level is determined by the power of the strongest signal that the quantizer must handle [1.4]. A signal-to-distortion ratio (S/D) of around 35 decibels is desired for a wide range of input levels.

Newer Coding Schemes

PCM has been around for a quarter century, and new technologies are beginning to demand attention. Sophisticated voice coding methods have become available in the past decade due to the evolution of VLSI technology; 64 kb/s PCM is no longer the only available technique. Coding rates of 32,000 b/s, 16,000 b/s, and even "vocoder" methods requiring 4800 b/s, 2400 b/s, and even less, have evolved (intelligibility, but not speaker recognition, can still be obtained at 800 b/s [1.5]).

Some interest exists in pursuing these new coding schemes because the implication is that we can double or quadruple the voice carrying capacity of the network in place without the introduction of new transmission equipment. Of all available schemes emerging from the laboratory the *adaptive differential pulse code modulation* (ADPCM) scheme is the most promising at this time. It effectively provides "toll quality" voice with minimal degradation at 32 kb/s. The CCITT studied this algorithm and a recommendation (G.721) followed in 1988. A problem with this method has been that of "passing" data at various speeds under these coding methods; a number of widely deployed U.S. modems (in particular, the Bell 202 type, at 1200 b/s in half-duplex mode) fail to transmit through a digital carrier system equipped with the 32 kb/s line cards. Algorithmic refinements to deal with the problem involve fine-tuning some of the parameters that characterize the coding scheme. Performance of the coding scheme revolves around the following parameters: frequency response and tracking, idle circuit noise, transient response and warmup period, single frequency distortion, intermodulation distortion, and S/D. A standard for a 16 kb/s coding scheme and a proposal for an 8 kb/s scheme has been studied by CCITT Study Group XV. A brief description of ADPCM follows.

Differential PCM

If a signal has a high correlation (exceeding 0.5) between adjacent samples, as is the case for speech sampled at the Nyquist rate, the variance of the difference between adjacent samples is smaller than the variance of the original signal. If this difference is coded, rather than the original signal, significant gains in S/D performance can be achieved (conversely, a given S/D can be achieved with fewer quantizer bits) [1.4]. This implies that, for the same desired accuracy, fewer bits are needed to describe the change value from one sample to the next than would be needed to describe the absolute value of both samples. This is the idea behind differential PCM (DPCM). DPCM systems are based primarily on a 1952 patent by Cutler.

In a typical DPCM system, the input signal is band-limited, and an estimate of the previous sample (or a prediction of the current signal value) is subtracted from the input. The difference is then sampled and coded. In the simplest case, the estimate of the previous sample is formed by taking the sum of the decoded values of all the past differences (which ideally differ from the previous sample only by a quantizing error). DPCM exhibits the greatest improvement over PCM when the signal spectrum is peaked at the lower frequencies and rolls off toward the higher frequencies.

The problem with this voice coding method is that if the input analog signal varies rapidly between samples, the DPCM technique is not able to represent with sufficient accuracy the incoming signal. Just as in the PCM technique, clipping can occur when the input to the quantizer is too large; in this case, the input signal is the change in signal from the previous sample. The resulting distortion is known as *slope-overload distortion*.

Adaptive DPCM

In *adaptive DPCM*, the coder can be made to adapt to slope overload by increasing the range represented by the encoded bits, which here number 4. In principle, the range implicit in the 4 bits can be increased or decreased to match different situations. This will reduce the quantizing noise for large signals, but will increase noise for normal signals; so, when the volume drops, the range covered by the 4-bit signal drops accordingly. These adaptive aspects of the algorithm give rise to its name. ADPCM transmits 4 bits per sample for 8000 samples per second, for a bandwidth of 32,000 b/s.

In practice, the ADPCM coding device accepts the PCM coded signal and then applies a special algorithm to reduce the 8-bit samples to 4-bit words using only 15 quantizing levels. These 4-bit words no longer represent sample amplitudes; instead, they contain only enough information to reconstruct the amplitude at the distant end. The adaptive predictor predicts the value of the next signal on the level of the previously sampled signal. A feedback loop ensures that voice variations are followed with minimal deviation. The deviation of the predicted value measured against the actual signal tends to be small and can be encoded with 4 bits. In the event that successive samples vary widely, the algorithm adapts by increasing the range represented by the 4 bits through a slight increase in the noise level over normal signals [1.6].

Lower Rate Voice Digitization

Some CPE equipment (for example, T1 multiplexers discussed in Chapter 3) now use *continuously variable slope delta* (CVSD) to achieve voice digitization rates below 32,000 b/s. To understand CVSD, we consider a form of DPCM where the length of the digital word per sample is a single bit. With such a small digital word, more samples compared to PCM-DPCM can be sent in the same bandwidth. Clearly, 1-bit words cannot measure loudness; hence, rather than sending the change in height of the analog signal curve, the 1-bit CVSD data refer to a change in slope (steepness) of the analog signal curve. At the sending end, CVSD compares the input analog voltage with a reference voltage: if the signal is greater than the reference, a "1" is sent and at the same time the slope of the reference is increased; if the input is less than the reference, a "0" is sent and at the same time the slope of the reference is reduced. CVSD attempts to bring the reference signal in line with the incoming analog signal. The steeper the slope (positively or negatively), the larger the output changes between samples; the CVSD algorithm increases the size of the step taken between samples each time the slope change continues in the same direction. This is similar in concept to the adaptive nature of ADPCM: a series of 1s produce a progressively larger increase in the output.

The receiver will reconstruct the sender's reference voltage, which, after being filtered, should be a replica of the original input. While one normally employs the CVSD at 32 kb/s, the actual digitization rate can be selected by the user in some systems (with ensuing voice quality implications). CVSD can operate from 64 to 9.6 kb/s. The quality deteriorates as the bandwidth decreases; this is the method employed by systems that provide 16 kb/s voice rates (the speaker is still recognizable at 16 kb/s, and speech is still intelligible at 9.6 kb/s).

A typical equipment line card converting four analog voice channels into four 16 kb/s digitized speech streams and multiplexing them onto a 64 kb/s channel costs around $4000 in 1990 [1.7].

High-Quality Telephony

Recently, the CCITT recommended an international standard (G.722/G.725) for coding wideband speech and music (50 Hz to 7 kHz at 3-dB attenuation) at 64 kb/s. This frequency range, extended at both the high end and the low end, considerably improves telephonic voice quality over the existing norm, approaching the quality of a typical car's FM radio. Extending the cutoff frequency from 300 to 50 Hz improves the naturalness of the audio signal. Some applications for these coders are in high-grade telephones in ISDN and for teleconferencing applications. In audiovisual conferencing applications, we would like to approach the quality of face-to-face communication.

The codec performance requirements for voiceband data are substantially different from those of voice signals. If the codec is required to encode voiceband data signals as well, its cost and complexity increase. If the codec is not required to carry data, it can be optimized for best performance on speech signals [1.5]. Subband coding techniques separate the signal into components occupying contiguous frequency bands, and encode the components separately. With the audio signal subdivided into two 4-kHz bands, a high S/N in the lower band becomes perceptually more important than in the higher band. An advantage of a design that uses two equally wide subbands is that each component can be subsampled to 8 kHz and the total transmission rate may be reduced in 8 kb/s steps by reducing the number of bits assigned to samples in one or the other band. A typical wideband coder accepts a 16-kHz sampled input signal and splits it into two 4-kHz (8-kHz sampled) bands using a quadrature mirror filter. The two 8-kHz sample rate subband signals are then encoded using an ADPCM coder. The upper band is encoded using 2 bits, while the lower band is allocated various bits, depending on the desired overall rate: 6 bits are used for 64 kb/s, 5 bits for 56 kb/s, and 4 bits for 48 kb/s. The two lower rates allow simultaneous transmission of an 8 kb/s (6.4 kb/s for data and 1.6 kb/s service channel) and 16 kb/s (14.4 kb/s for data and 1.6 kb/s service channel) data stream, respectively, in addition to the voice, on a single 64 kb/s (ISDN) channel.

Voice Digitization Summary

Figure 1.6 summarizes the quality *versus* digitization rate relationship associated with various voice coding schemes (see [1.8] for details).

1.2.1.2 Transmission Equipment

The process of moving information from one point to another is called *transmission*. To undertake transmission, one needs a variety of facilities. Transmission equip-

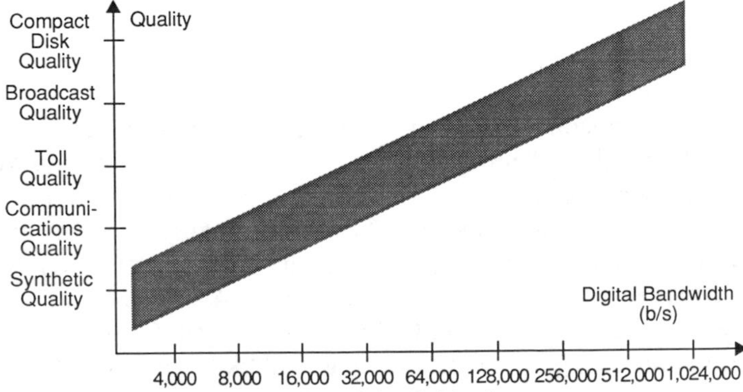

Figure 1.6 Digital voice quality as a function of the digitization rate.

ment typically includes terminal equipment (not to be confused with CPE termination equipment, discussed earlier), which accepts the user's signal and changes it appropriately, and an interconnection medium, such as copper wire or coaxial cable among others. To superimpose the user signal onto the medium, the transmission equipment needs to modulate a carrier signal. Additionally, the equipment may multiplex a large number of users over the same physical medium. (Multiplexers can also be CPE, if desired or appropriate.)

Multiplexing Schemes

A number of multiplexing schemes are available to place multiple calls in a standardized fashion on one medium. In Chapter 3, we will discuss the hierarchies of U.S. telephone plant multiplex schemes. The basic multiplexing schemes are:

- Frequency division multiplexing (FDM). This is typical of analog coaxial, microwave, and radio systems.
- Time division multiplexing (TDM). This is typical of digital transmission; it lends itself well to computer interfaces. In the traditional telephone network TDM has been used in conjunction with PCM coded signals.
- Space division multiplexing. An example is the frequency reuse in a cellular system or satellite.
- Code division multiplexing. Systems where the multiplexing is achieved by employing different data-stream coding methods. Used principally by military communication systems.
- Random access techniques. A method used in conjunction with TDM in which the multiplexing occurs via statistical bid and assignment of the channel. This is typically employed in LANs that do not use token disciplines.
- Demand assignment techniques. Bandwidth reservation-based systems; used in conjunction with random access. This approach is typical of satellite systems.

Modulation

At the functional level, modulation is the process of imparting an intelligent signal onto an underlying carrier signal so that it can then be transmitted over a distance. The carrier signal depends on the media at hand (copper, microwave, fiber, *et cetera*). Modulation is very common: radio and television, to mention only two obvious communication systems, employ modulation. Modulation functions come into play across the network for all types of transmission systems. The function of the modulator is to match the encoder output to the transmission channel.

Figure 1.7 depicts the three characteristics of an electrical sinusoidal carrier, typical of media such as copper, coaxial, and radio: the *amplitude*, the *frequency*,

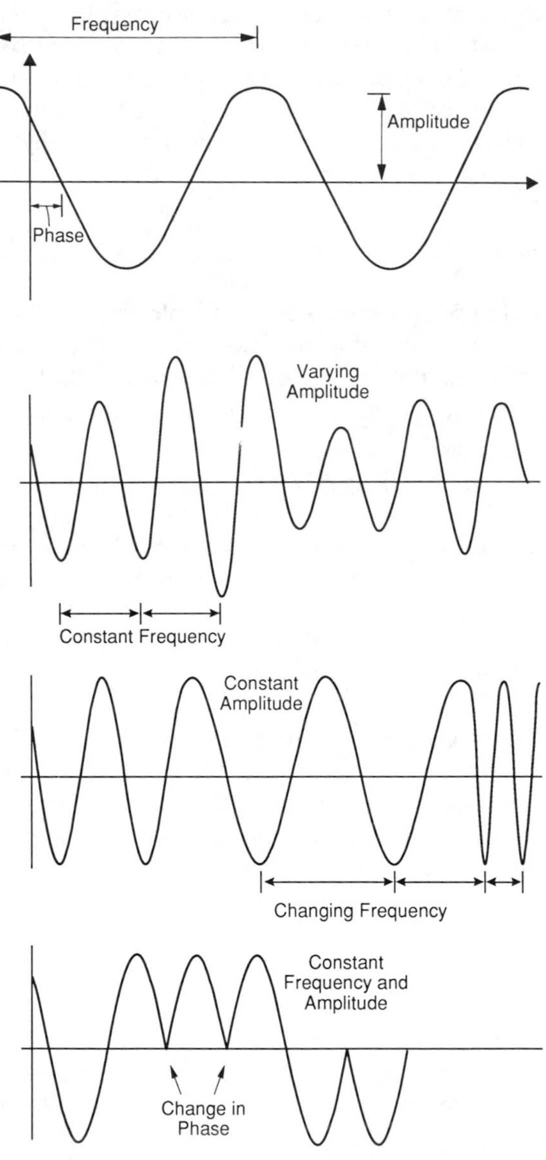
Frequency
Amplitude
Phase
Varying
Amplitude
Constant Frequency
Constant
Amplitude
Changing Frequency
Constant
Frequency and
Amplitude
Change in
Phase
Top: Three characteristics of a carrier signal
Second from top: Amplitude Modulation (simplified picture)
Third from Top: Frequency Modulation (simplified picture)
Bottom: Phase Modulation (simplified picture)

Figure 1.7 Modulation of the carrier.

and the *phase*. All three of these factors can be affected in order to achieve modulation. The three types of modulation are: amplitude modulation, frequency modulation, and phase modulation. The change to the carrier under each of the three methods is also shown in Figure 1.7. The modulating signal (the intelligence to carry) can be analog or digital.

Amplitude Modulation. A carrier's amplitude represents the instantaneous strength of the signal, as depicted in Figure 1.7. A carrier's amplitude, when amplitude modulated (AM), varies according to the amplitude of the modulating signal, while keeping the frequency constant. The modulation process produces a power spectrum that is symmetrical with respect to the frequency of the carrier: when viewed in the frequency domain, the modulated signal will have power spectral lines at sums and differences of the carrier frequency with the frequencies of the modulating signal. Although a major portion of the transmitter's power remains at the carrier frequency, amplitude modulation shifts energy into the sideband frequencies. This energy in the sidebands is what allows the remote end to demodulate the original intelligent signal.

Single-sideband (SSB) modulation, an improvement of AM, concentrates most of the energy of the transmitter into the intelligence-bearing portion of the signal, enhancing the receiver signal. Because the upper and lower sidebands in AM modulation contain redundant information, one of the bands can be suppressed, after the modulation stage, giving rise to a "single sideband." This differs from AM, which uses transmitter energy to feed the carrier frequency and adds little to the intelligence received at the far end, as indicated above. SSB modulation results in a signal that requires reduced transmission bandwidth, in effect, allowing more intelligence to be transmitted over the same channel. The carrier signal can also be partially suppressed so as to use less power [1.9]. This technique is often employed in microwave systems.

Frequency Modulation. Frequency modulation (FM) was developed in the 1930s as an improvement over AM to provide high-quality music broadcasting. With FM, intelligence is added to the carrier wave by varying the frequency of the carrier in step with the frequency of the intelligence signal, while holding the output power of the carrier constant. FM is more immune to major sources of noise than AM because the most common type of noise tends to affect the amplitude. However, a frequency-modulated signal needs more bandwidth as compared to AM; even narrowband FM requires nearly twice the transmitter bandwidth of AM. FM is still used extensively today for analog microwave transmission systems.

Phase Modulation. Phase modulation (PM) is similar in some respects to FM. The phase of the transmitter is increased or decreased in accordance with the modulating intelligence signal. However, because small changes in phase are difficult to detect,

PM is not generally used for analog applications. PM is used more commonly in digital modulation, for data transmission applications, as discussed later. Figure 1.8 depicts the modulation process from a hardware components perspective.

1.2.1.3 Subscriber Loop Plant

CPE equipment normally is connected to remote locations through a telephone company's CO. The subscriber loop is the physical link by which customers are connected with the telecommunication network. In the United States, local loops are usually provided by *local exchange carriers* (LECs). A star topology is employed, with loops emanating from the central point, the CO, to all local users of the network. This not only facilitates management and troubleshooting of the loops,

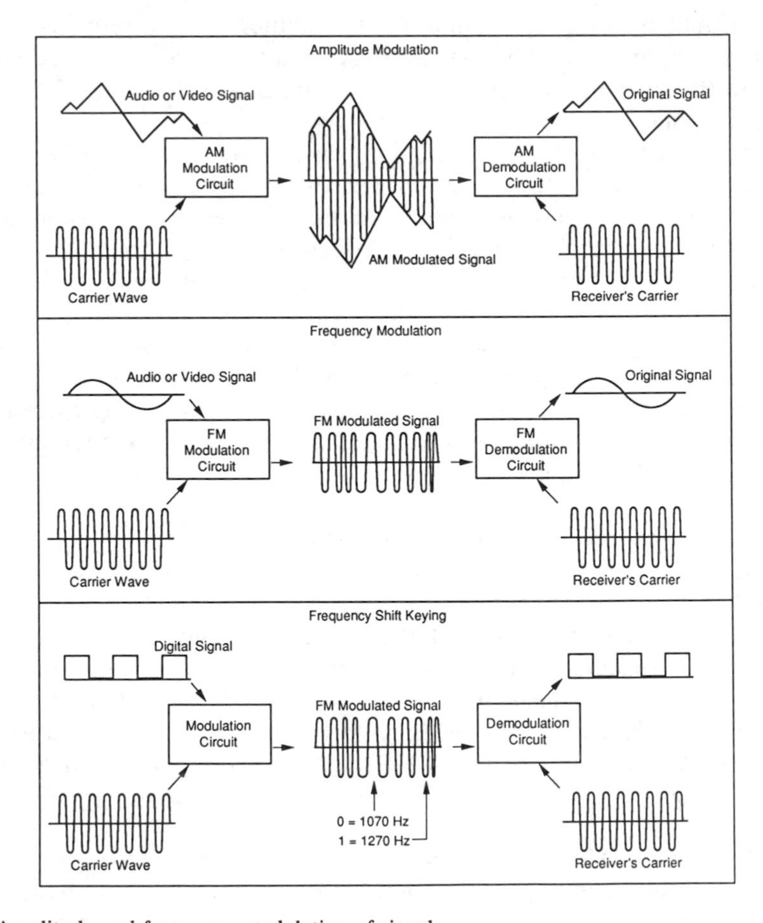

Figure 1.8 Amplitude and frequency modulation of signals.

but also allows central switching. Loops may be discrete two-wire copper facilities from the CO to the user, or may be partially multiplexed over a few miles of common medium. (Local loops for special services, such as data communication, may be four-wire.) In this context, transmission equipment consists of the underlying physical channel, such as twisted pair, coaxial cable, optical fiber, microwave, and others, and multiplexing equipment such as digital loop carriers, which we will discuss in Chapter 3. To gain a perspective, we note that long-haul circuit mileage is only around 10% of the total U.S. telephone plant circuit mileage; 90% is the local loop, where the LECs install nearly 200 million miles of copper wire a year [1.10].

The physical infrastructure supporting local loops is known as *outside plant.* The outside plant includes conduit, poles, cable protection devices, terminals, aerial drop wire, feeder cable, and distribution cable. The cable may be strung from poles, buried, or placed in conduit. The cable is often buried underground without conduit because this approach is usually less expensive than housing the cable in conduit. The local loop is the part of the circuit that traditionally has been most susceptible to transmission impairments; therefore, design and testing are both very important to maintaining quality. In designing a loop cable's characteristics, loop resistance and capacitance of the pair must all be carefully considered by the telephone company's engineers. Today's telephone networks are built around *carrier serving areas* (CSAs), and are served by a combination of digital or analog transmission and switching equipment. The radius of CSAs was originally defined by the bandwidth-distance performance characteristics of copper; typically, this radius is 12,000 feet [1.11]. Fiber systems may eventually change the CSA concept.

The local loop infrastructure is made up of two components: the *feeder plant* and the *distribution plant.* The feeder plant provides (but not in every case) what are called "carrier facilities"—multiplexing-transmission facilities in which a number of customers may be multiplexed onto a single transmission medium, as shown in Figure 1.9. (The term "carrier," in this context, is only loosely related to the concept of carrier discussed in the modulation section.) The number of multiplexed channels can vary from 24 to 96, or even to 672 or more. The LEC can combine many user channels with the multiplexing equipment, and deliver the signal to the CO where it is demultiplexed and fed to the switch for further handling. (In newer switches, it can be fed directly to the switch without demultiplexing equipment.) The advantage of multiplexing is that construction of the physical plant is typically the most expensive component of the telecommunication network (often more expensive than the electronic equipment itself); thus, one of the objectives is to minimize the number of discrete physical channels needed. Two methods of multiplexing in the feeder plant are frequency division multiplexing (analog) and time division multiplexing (digital). Time division techniques are becoming prevalent because of the increased deployment of digital networks and the lower cost to achieve multiplexing compared to analog techniques. The equipment at the remote

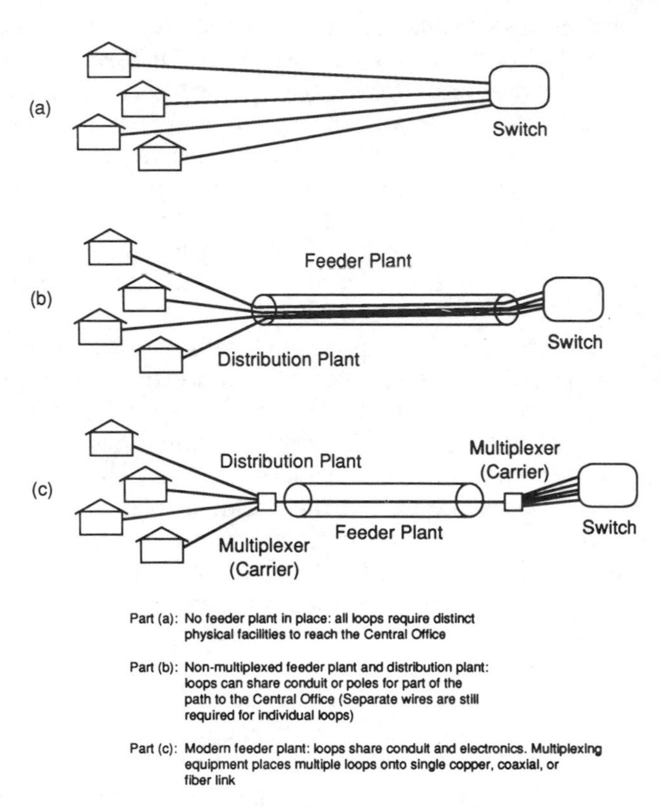
(a)
Switch
Feeder Plant
(b)
Switch
Distribution Plant
Distribution Plant
Multiplexer
(Carrier)
(c)
Feeder Plant
Switch
Multiplexer
(Carrier)
Part (a): No feeder plant in place: all loops require distinct physical facilities to reach the Central Office
Part (b): Non-multiplexed feeder plant and distribution plant: loops can share conduit or poles for part of the path to the Central Office (Separate wires are still required for individual loops)
Part (c): Modern feeder plant: loops share conduit and electronics. Multiplexing equipment places multiple loops onto single copper, coaxial, or fiber link

Figure 1.9 The feeder and distribution plants.

end of the feeder plant also typically provides the voice digitization function for the group of stations being served.

As Figure 1.9 shows, the early examples of the feeder plant (through the 1960s) did not necessarily involve the use of multiplexing equipment. The only shared facilities in this case would be underground conduit or the poles. The distribution plant would be put in place once, to reflect anticipated growth in the sector served; the resulting junction between feeder and distribution plant occurred at a pedestal box or pole-mounted cross-connect box. In the late 1950s and 1960s, carrier systems grew in importance, due to the increased cost-effectiveness of remote electronics; an objective was set to achieve 30% of the feeder plant using carrier. At the interface, several lines can be integrated onto a single link (fiber or copper carrier) through multiplexing. The economies of scale established in the feeder plant reduce wire and cable costs that outweigh those of multiplexing. In the contemporary context, the feeder plant is largely implemented using carrier systems, and an increasing proportion of all loops are multiplexed.

The distribution plant is that portion of the network between the feeder termination (normally, some type of channel bank or remote switching module—a satellite CO) and the customer. The distribution plant is generally copper-based to interoperate with the existing generation of telephone sets. Existing telephone sets require a current to operate the bell; the equipment at the feeder termination typically provides the appropriate signaling to the stations over the distribution plant. Two-wire twisted-pair cable (sometimes four-wire) is used in the distribution plant. Ideally the distance between the users and the feeder plant interface should be optimized to reduce the total length of wire required.

Fiber facilities are being increasingly introduced in the feeder plant. The only interfaces of the feeder plant are the (digital) switch at one end, and the carrier electronics at the other end. The feeder plant is thus amenable to fiber replacement without requiring additional changes to the network. Cable ducts supporting the feeder plant are often overcrowded, and therefore offer ideal opportunities for fiber. End-to-end fiber-based local loops are a prospect for the future (see below). On the other hand, feeder lines are becoming the prime area for introduction of fiber at this time, particularly because this will facilitate the introduction of broadband ISDN, and also narrowband ISDN to an extent. About three-quarters of all new fiber installation projects are now in the feeder plant. Numerically, the feeder market is larger than the long-haul (about 45% of the total by circuit mile). Eventually, fiber will also be deployed in the distribution plant, for a general customer, providing end-to-end fiber connectivity. At this time, many commercial customers are already being equipped with fiber local loops—the large office buildings, computer centers, *et cetera*. The deployment of fiber in the local loop (distribution plant portion) for small businesses and residential customers will probably have to wait until the mid-1990s, though numerous experimental trials have already been undertaken [1.12, 1.13, 1.14]. Residential fiber will open up opportunities for new services, including video-on-demand, high-definition television (HDTV), and other graphics services (HDTV provides large-screen projection with 35-mm slide quality).

As for advanced local loops, some trends can perhaps be extrapolated from the French experiment at Biarritz in the mid-1980s [1.10] . The Biarritz city testbed offered 5000 homes a host of "futuristic" services on the all-fiber network, in addition to cable TV. These services were videophone (two-way video), sensing of facilities for remote home management, video databanks (movie libraries), HDTV, and computer services (the last three were originally scheduled for a future time). Initially, fiber loops for 1.4 million homes in France were to be installed by the end of 1987, and 7 million homes were to be equipped by 1992 (estimated project cost: $10 billion). The plant was to be used for multiservice broadband applications, notably cable TV (which at the beginning of the trial period in 1982 was nonexistent in France). Instead of the traditional branch-and-tree architecture, which is typical of one-way cable TV distribution applications, the plan called for a star configuration to allow full-function two-way interactive services. Because of political, technical, cost, and user-acceptance problems, that target has now been scaled down: only half a million homes will be completed by the early phase. The fiber demultiplexing equipment turned out to be more expensive than anticipated. Only 30% of the houses passed subscribed to the service; videophone in particular has not been very successful [1.13]. The government promoters of the system are reevaluating the cost-effectiveness of the entire concept, and of fiber in particular. Hybrid coaxial systems are now being considered in an effort to reduce cost.

Experts believc that the installed hardware cost per subscriber must be in the $1500 to $2000 range (preferably, the lower figure) if fiber loop technology is to proliferate. The French fiber system requires household connection hardware that with current technology costs between $2000 and $3000. The cost of a domicile fiber loop depends not only on "fixed" elements such as CO equipment (which may amount to 5% of the total cost), CPE (possibly 35% of the total), remote terminal equipment (possibly 30% of the total for this multiplexing function), and optoelectronics (typically 10% of the total cost), but also on the distance of the house from the CO. For typical distances (2 to 4 miles), the fiber cable would cost $400 (20% of the total). For longer distances, the cost of the cable would increase linearly [1.14].

A number of trials for fiber deployment in the local loop have been undertaken in the past couple of years, as discussed in more detail in Chapter 7.

1.2.1.4 Switching Systems

From a user's perspective, the primary function of a voice telephone switch is to connect, on demand, telephone instruments or other properly configured CPE. Because it would be impossible (and unrealistically expensive) to have a direct line between every possible user pairing, switching systems were developed to achieve the needed interconnectivity at an affordable price. This switched network allows a user to connect with any other subscriber by dialing the subscriber's address, and it eliminates the need for point-to-point wiring. An important aspect of any CO is the dc voltage supply that is applied to all loops. No current flows in any of the loops until the switchhook contacts of a telephone set are closed. The dc voltage permits the telephone set to signal the CO by merely closing or opening contacts with the switchhook or rotary dial.

Switches are connected by interoffice trunks and must communicate with one another to set up an interoffice call. Originally, this conversation was undertaken in-band, namely, within the same channel as the user's conversation; however, this process was prone to fraud and was slow and inefficient. Now many switches can communicate with one another in an out-of-band fashion using a separate supervisory network called *common channel signaling*. During the 1990s, the deployment of this signaling network will become more extensive.

Modern switches contain a common control section, which manages the call connection process, and a switching matrix, which actually makes the connection possible. Four generations of switching systems have occurred: (1) step-by-step technology, (2) common control switching systems, (3) analog electronic switching systems, and (4) digital switching systems. Modern switches are in reality computers: *stored program control* (SPC) implies the ability to program the switch using software instead of having to add discrete hardware modules (as was the case in older switches). Digital switches are becoming prevalent, although many analog switches are still embedded in the telephone network, as we will discuss in Chapter 9.

A network may contain a hierarchy of switches, beginning with local switching systems, which are closest to the user, and then going higher via tandem switches to a regional or national switch. A five-level hierarchical infrastructure was used in the United States prior to divestiture, but other topologies have now emerged.

Common Control Offices

Early switches had to dedicate the facilities required to control a call to each active call for its entire duration. This was inefficient and also uneconomical; if a call was blocked by equipment in use, the system was incapable of rerouting the call. A control method that eliminated these problems was required. Equipment introduced after 1940 makes use of a limited number of specialized shared equipment units to control the process. Much of the control function to direct the path that the call takes through the system is concentrated in a small number of pieces of equipment, and these are used repeatedly. A unit of control equipment performs its function on a call and then becomes immediately available to perform the same function on another call. This mode of operation is known as *common control.*

Computer-Controlled Switching Systems

The common control hardware can be electromechanical or hard-wired electronic (as was the case until the early 1960s) or can be computerized (which was introduced in 1965). Beginning in the late 1950s, designers realized that if the hard-wired common control was replaced with a programmable computer, options could be supplied to users more easily. Additionally, new services could be provided at the switch. As mentioned earlier, these systems are referred to as stored program control. With SPC, the control of switching functions is achieved by instructions stored in a memory and new service features may be added by changing the contents of the machine's memory. Examples of such services include speed calling, three-way calling, call-waiting, call forwarding, and others.

The advantages of computer SPC include:

1. Labor saving as a result of simpler administration (for changing subscribers' information), and reduced maintenance effort;
2. Higher traffic capacity;
3. Space saving: we can replace an existing switch with a smaller one having a larger capacity;
4. Power saving;
5. Cost reductions due to continued VLSI cost improvements;
6. Flexibility to changes over the life cycle of the switch (which could be as long as 40 years); and
7. Economical offering of new advanced services to the subscriber.

The Bell System's first SPC switch was the Number 1 Electronic Switching System (No. 1 ESS™ AT&T), which went into service in 1965. Number 1 ESS was designed to handle the heavy traffic loads and high density of telephone customers in metropolitan areas. The same basic principles were used in the No. 2 ESS, introduced in 1970. The No. 2 ESS was intended for communities with local switching offices serving around 10,000 telephone lines. Three basic elements of this type of switch in its early manifestations were: (1) a switching matrix using high-speed electromechanical ferrite switches; (2) a control unit, which directs the switching operations and maintenance in the system; and (3) two memories: a temporary memory (call store) for storing information such as the availability of circuits, called number, calling number, and type of call; and a semipermanent memory (program store) containing all the information that the control unit needs to process the call and make a connection. The program store is semipermanent because it does not have to be changed as calls are processed by the system. Two control units and a maintenance frame make up the control complex; switches are designed with two control units with lock-step full duplex processing so that no service degradation is experienced for active calls (or calls being processed) should a single processor fail. The various subunits that form a control unit include the program control, input-output control (of peripheral units such as switching networks), the call store, and the program store [1.15].

Digital Switching Offices

Initially, switching was achieved by activating electromagnetic relays that would close to achieve a continuous metallic circuit end-to-end. This type of switching, also called analog, has a number of drawbacks, including unreliability due to mechanical components, noise added to signals due to the opening and closing of relays, large size, inconsistency with digital transmission systems, and other drawbacks. Digital switching technology accepts digital signals and switches these to the desired destination by redistributing the signal electronically.

Voice digitization techniques have been widely applied, as indicated earlier, and their use is growing. Digitization is valuable in transmission because of its ability to protect the signal against the corruptive influences that degrade analog transmission. The use of corresponding digital carrier systems is increasing at such a rate (more recently because of the use of fiber) that to switch the digital signal directly is more advantageous than converting it to analog for space-division switching, then re-encoding it for transmission [1.16]. We will discuss digital switches in Chapter 9.

Interoffice Trunks

Telephone switches are interconnected by a group of circuits called interoffice trunks. Typically these trunks are pooled, meaning that they can be seized by switches at either end and put on line as needed. The size of the pool is calculated in such a fashion that at the busy hour in the busy season no more than 1% (or some other small number) of the calls are unable to be routed to the next stage of the switching over the trunk pool system. Trunks are usually derived from carrier systems connecting the switching offices. If two COs are connected directly via a system of trunks, they are said to have direct trunks between them (see Figure 1.10).

Tandem Switching Offices

In telephone networks with a large number of COs, having trunks between every pair of switches would not be practical. In these cases, the local telephone exchanges are connected to trunks that can provide access to an intermediate switching center, known as tandem switching offices, to which all switches are in turn connected. Connectivity between two local COs that are not directly connected by direct trunks must go through the tandem switch. Today, access to toll facilities is typically (but not always) via an interexchange tandem (see Figure 1.11).

Interexchange Trunks

As a product of divestiture, long-haul communication is handled in almost all situations by a carrier other than the one providing the local loops and the local switching system. A tandem arrangement may be employed to reach one or more interexchange carriers (see Figure 1.12).

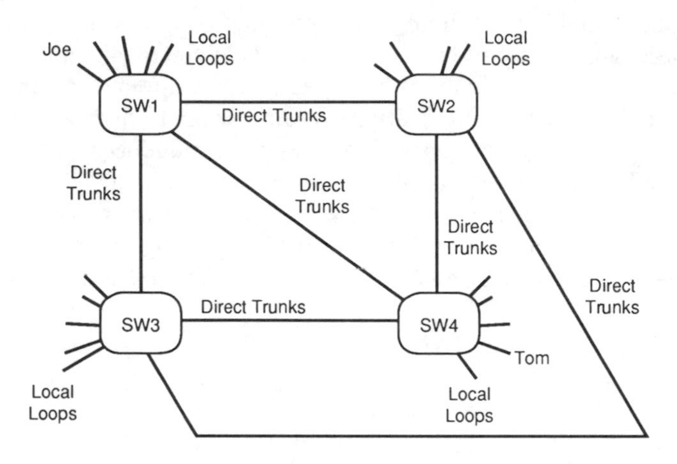

Figure 1.10 Direct trunks.

Traditional Switching Hierarchy

A traditional tree-based hierarchy of switches ensures that there is a path from each switching office in the network to any other switching office in the network. It is an architecture which had been employed by the Bell System for decades. A characteristic of the hierarchical structure of switching offices is that each office is connected to an office at a higher level except, of course, for those at the highest level; these top level offices are completely interconnected. The traditional network structure for the United States and Canada is divided into 12 regions; each region has one switching office at the highest level. The trunk group that connects a switching office to the next highest level switching office within a region is called a final group. Additional trunk groups supplementing the tree structure are permissible; in fact, they are desirable where sufficient traffic exists between switching offices not directly connected by the tree structure. These trunk groups, which are not part of the tree structure, are called high-usage groups. If one group is busy, a call will be rerouted to a different group until the final group is reached.

Since divestiture, AT&T has utilized nonhierarchical dynamic routing, schematically shown in Figure 1.13.

Evolution of the Telephone Company Plant

Figure 1.14 depicts the five-stage evolution of the telephone company plant over the years. Stage 1 (1890s to 1950s) involved an all-analog plant. Stage 2 (1960s

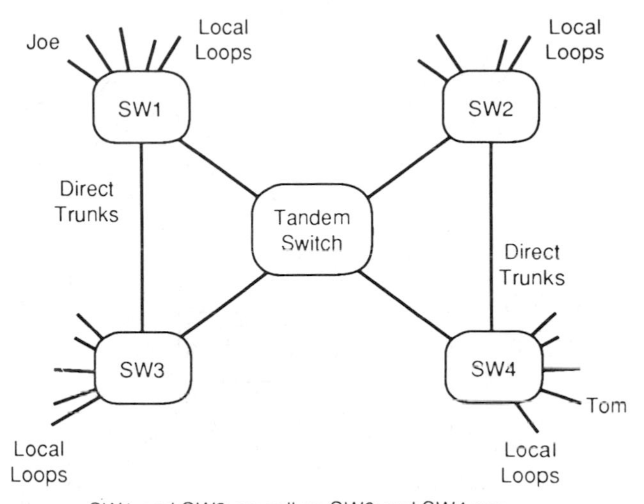
Joe
Local
Loops
Local
Loops
SW1
SW2
Direct
Trunks
Tandem
Switch
Direct
Trunks
SW3
SW4
Tom
Local
Loops
Local
Loops
SW1 and SW3 as well as SW2 and SW4 are
connected with direct trunks because of high traffic
Other connections are made via the tandem switch
Joe talks to Tom over trunks to the tandem switch,
the tandem itself, and then over trunks to SW4

Figure 1.11 Tandem arrangement.

and early 1970s) saw the emergence of digital transmission. Stage 3 (mid-1970s and 1980s) saw the introduction of digital switching. Stage 4 (early 1990s) is experiencing the introduction of ISDN for true end-to-end digital connectivity. All these stages have involved voiceband bandwidths (4000 Hz or 64,000 b/s); in stage 5 (late 1990s), we will see the introduction of end-to-end broadband digital communication.

North American Numbering Plan

Numbering schemes are mandatory for orderly identification of network subscribers. In traditional telecommunication networks, numbering plans are also used to accomplish routing. The North American numbering plan is briefly discussed here.

A domestic toll voice call (with presubscription to an interexchange carrier) can be placed using an address such as 1-NXX-NXX-XXXX, where the first three digits after the 1 represent the area code, the middle three digits the exchange, and the last four digits the specific station connected to the switch specified by the NXX. A total of 160 area code combinations is possible under the current numbering plan, which requires the first digit to be a number between 2 and 9 (an "N"), the second digit to be a 0 or a 1, and the third digit to be between 0 and 9

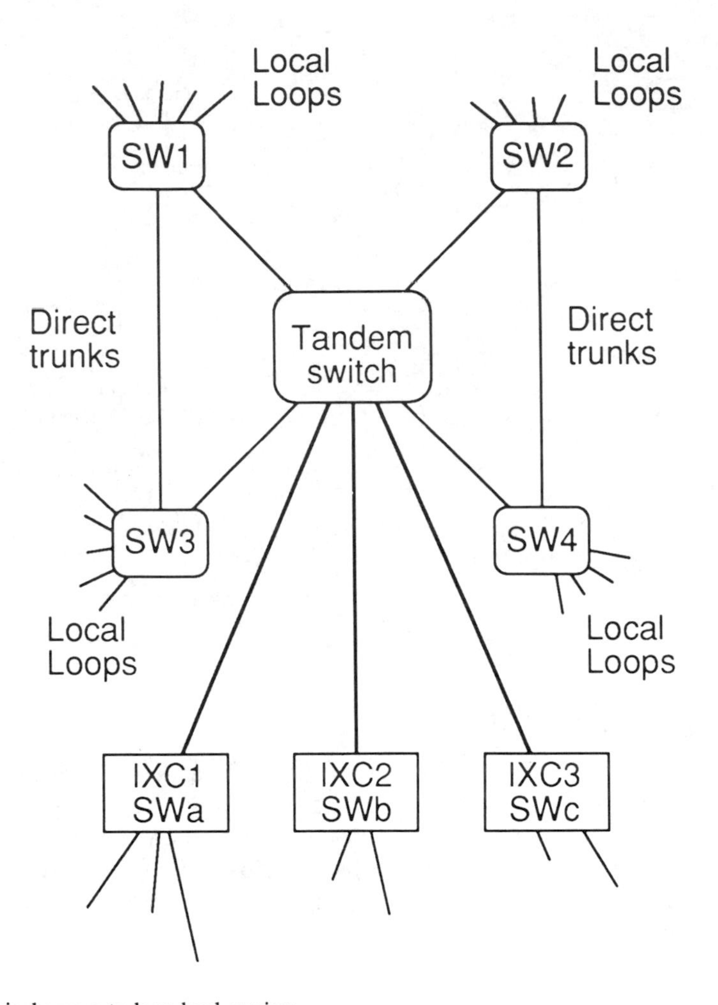

Figure 1.12 Typical access to long-haul carriers.

(an "X"). Code combinations 211, 311, 411, 511, 611, 711, 811, and 911, as well as 200, 300, 400, 500, 600, 700, 800, and 900 are reserved for special applications. When AT&T first assigned the initial area codes for the United States, Canada, and parts of the Caribbean in 1947, engineers projected that they would last 100 years. With a potential one billion phone numbers now assigned, the 100-year projection will fall short by more than 50 years. All currently available area codes will be assigned by 1995. The overall growth in telephone number usage is running at about 7 to 9% a year. The growth is fueled by growth in population, multiloop households, cellular telephony, paging, and facsimile machines.

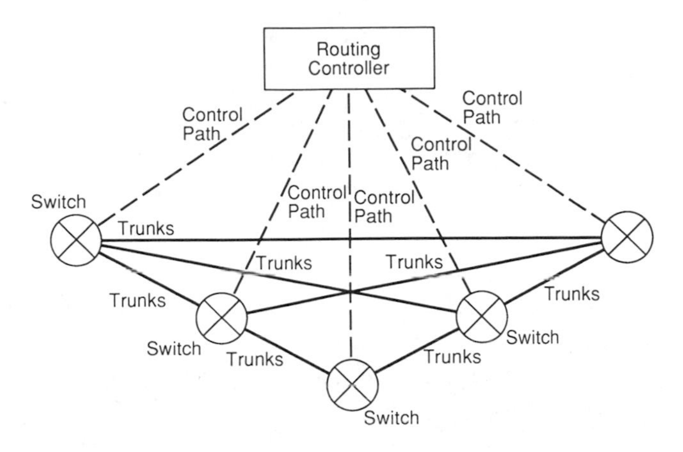

Figure 1.13 Nonhierarchical routing.

Two recently introduced area codes are 708 in Illinois and 908 in New Jersey. Area codes that have been reserved include 903 (Dallas, late 1990); 510 (San Francisco, late 1991); and 310 (Los Angeles, 1992). The remaining unassigned codes under the original numbering plan are: 210, 410, 706, 810, 905, 909, 910, and 917 [1.17].

During the 1960s, a plan was developed to provide for the time when the existing area codes would be exhausted. The plan removes the requirement that the second digit of each area code be a 0 or a 1; eliminating this restriction makes 640 additional codes available. Implementation of the plan, however, is a major undertaking for all LECs. Each of the 792 COs in an area code has a maximum of 10,000 telephone numbers associated with it, which means that each area code has approximately 7.9 million numbers than can potentially be assigned to customers.

Traffic Engineering

The traffic offered to a switch is a function of two factors: the average rate of arrival of new call attempts and the average holding time of a call (assuming that the variance is small enough to be safely ignored). The averaging period for the origination rate is the busy hour, a one-hour period chosen to typify for a given CO the annually recurring hour during which the offered traffic load is a maximum. *Peak busy hour calls* is the unit used for expressing the processing capacity of a switching machine's control. The offered traffic load is expressed in hundred call seconds ("CCS" where the first C is from *centum* = hundred), and is the product of the number of calls and the average holding time, or the sum of the holding times of all calls under consideration. By convention, the units of CCS are often

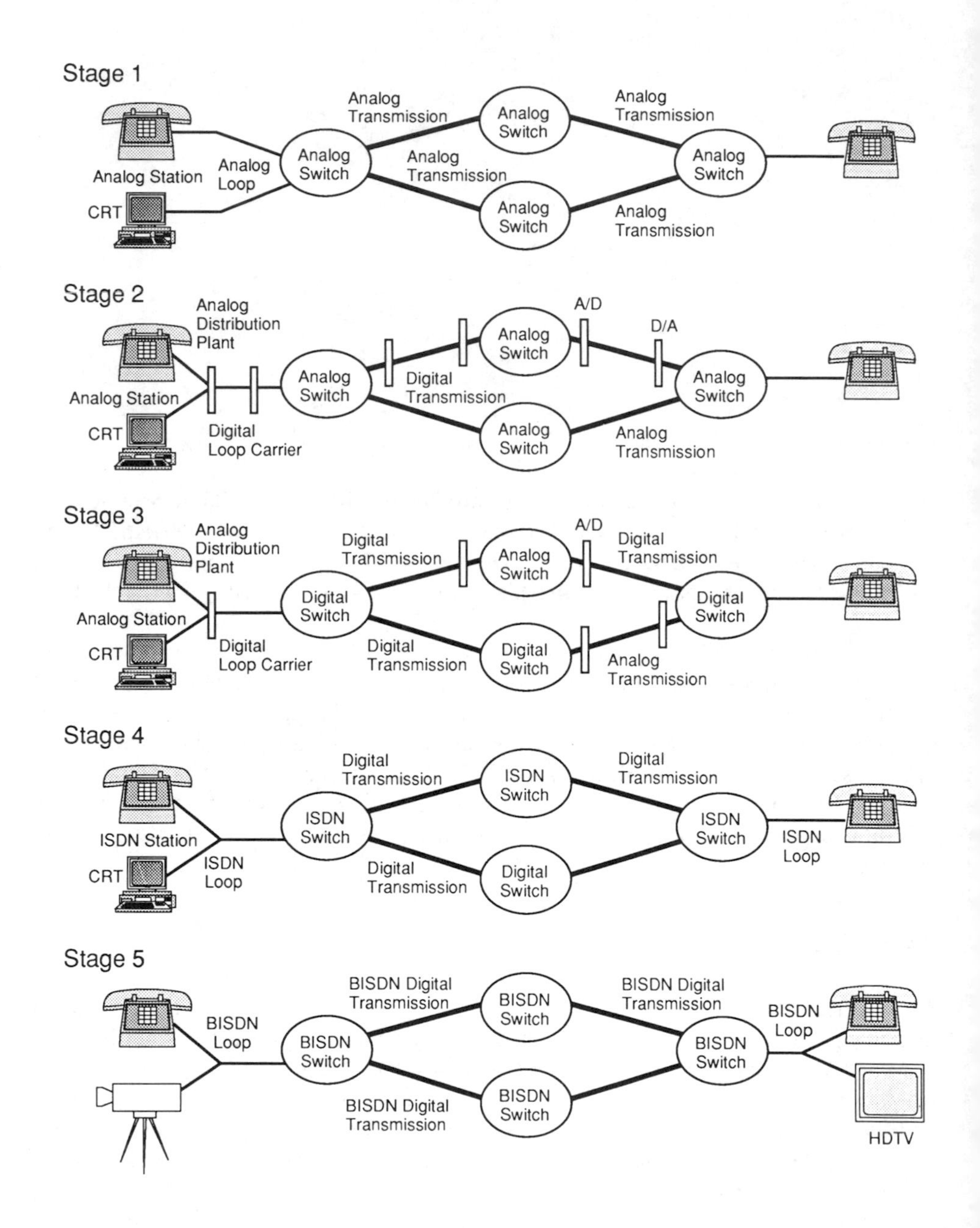

Figure 1.14 Evolution of telco plant.

used to mean CCS per hour. The average holding time multiplied by the number of calls placed per unit time is a measure of the traffic intensity and is often expressed in erlangs. The erlang, named after an early contributor to traffic theory, is the traffic intensity equivalent to one call held for an entire hour, and is therefore equal to 36 CCS per hour (it is also equivalent, for example, to 36 calls held for 100 seconds each in a one-hour period, or a circuit occupied 100% for a full hour by any number of calls). Put differently, the traffic intensity in erlangs is the number of channels that would be sufficient to serve a given offered load in a one-hour period, if the load timing could be rearranged so that all the channels would be continuously busy. It thus constitutes a lower bound on the number of channels to carry that traffic intensity.

1.2.2 Data Networks

While data networks differ in many ways from voice networks, the components identified above (CPE, switching, and transmission) are all still present. CPE may consist of CRTs (cathode ray tubes), computers, remote peripherals, *et cetera*. High-speed private networks, such as LANs, usually do not include an explicit switching function because the switching function is achieved via message acceptance or rejection based on the address posted on the message (the message is broadcast to all users of the network). Lower speed networks may involve circuit-switched or packet-switched facilities in the network, or switching in the host computer or associated front-end hardware.

1.2.2.1 CPE

CPE in the data environment includes:

- CRTs, printers, plotters, computers;
- Local area networks (LANs);
- Private wide-area networks (WANs);
- Modems, multiplexers, channel service units; and
- Front-end processors.

DTE and DCE

A data communication system consists of *data terminal equipment* (DTE), data circuit-terminating equipment (DCE)—colloquially known as *data communication equipment*—and the transmission circuit, also variously known as channel, line, link, or trunk. The DTE is a device, such as a terminal or a computer (mainframe, minicomputer, or microcomputer). The DTE supports end-user applications, for example, data entry, inquiry or response, and database management functions. The DCE provides the connection of the user DTE into the communication circuit. Notice that both DTE and DCE can be CPE (although some DCE may also be part of a public network). See Figure 1.15.

A *physical level specification* (also known as a *physical level interface*), such as EIA 232-C, defines the following attributes of a data communication system [1.18]:

1. The wiring connection between devices (when wires are used);
2. The electrical, electromagnetic, or optical characteristics of thc signal between communicating devices;
3. The provision for mechanical connectors (dimension, number of pins, *et cetera*);
4. The agreement on the type of clocking signals that will enable the devices to synchronize onto each other's signal;
5. The provision for electrical grounding (if needed).

For example, RS-232-C describes a standardized interface between DTE and DCE employing serial binary data interchange. RS-449 describes a general-purpose 37-position and 9-position interface for DTE and DCE employing serial binary data interchange.

ASCII and EBCDIC Character Coding

One of the issues in data communication is the representation of the data, particularly when dealing with several computer systems. The applications in different systems may wish to represent data structures in different ways. Yet a common exchange structure is needed. Many computers use the 7-bit ASCII (American Standard Code for Information Interchange) code; other systems use the 8-bit EBCDIC (Extended Binary Coded Decimal Interchange Code). We will discuss a more sophisticated method using the Abstract Syntax Notation One and basic encoding rules in Chapter 13. Table 1.2 depicts the ASCII bit coding assignment.

1.2.2.2 Transmission

Figure 1.16 depicts some of the components found in a typical data communication environment. The transmission link can be analog or digital. It can also be point-

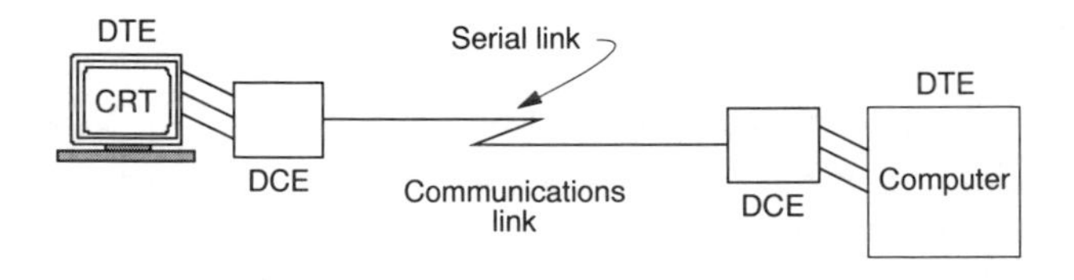

Figure 1.15 DTE and DCE.

Table 1.2
ASCII Coding

$b_7\ b_6\ b_5$ / $b_4\ b_3\ b_2\ b_1$	0 0 0	0 0 1	0 1 0	0 1 1	1 0 0	1 0 1	1 1 0	1 1 1
0 0 0 0	NUL	DLE	SP	0	@	P	`	p
0 0 0 1	SOH	DC1	!	1	A	Q	a	q
0 0 1 0	STX	DC2	"	2	B	R	b	r
0 0 1 1	ETX	DC3	#	3	C	S	c	s
0 1 0 0	EOT	DC4	$	4	D	T	d	t
0 1 0 1	ENQ	NAK	%	5	E	U	e	u
0 1 1 0	ACK	SYN	&	6	F	V	f	v
0 1 1 1	BEL	ETB	'	7	G	W	g	w
1 0 0 0	BS	CAN	(	8	H	X	h	x
1 0 0 1	HT	EM	)	9	I	Y	i	y
1 0 1 0	LF	SUB	*	:	J	Z	j	z
1 0 1 1	VT	ESC	+	;	K	[	k	{
1 1 0 0	FF	FS	,	<	L	\	l	|
1 1 0 1	CR	GS	−	=	M	]	m	}
1 1 1 0	SO	RS	.	>	N	^	n	~
1 1 1 1	SI	US	/	?	O	_	o	DEL

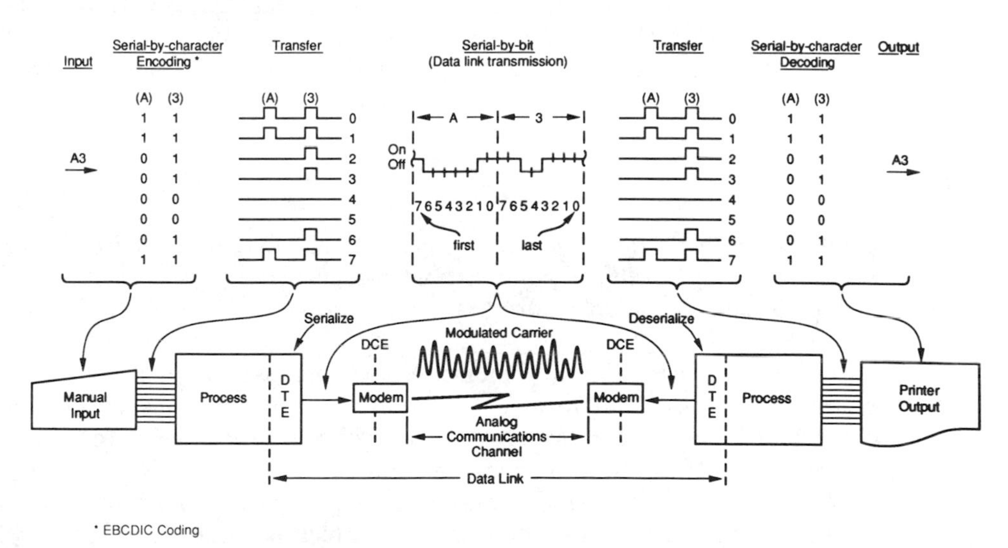

Figure 1.16 A data communication system.

to-point or multipoint. Additionally, the link can also be characterized as being two-wire or four-wire, and half-duplex or full-duplex. These concepts are summarized below.

Analog Transmission Methods

Analog signals vary in time in terms of amplitude and frequency. As indicated earlier, at the current time, analog network interfaces are still very common. To send data over an analog line we use a modem. The modem takes a digital signal and produces a signal suitable for transmission over the analog network.

The analog signal generated by a modem to transmit data consists of a carrier frequency, plus sidebands that change as the data bit pattern varies. These sidebands must fit within the attenuation limits of the voice-grade channel. If the channel distorts the signal by attenuating the sidebands unequally, the data bit timing will be distorted when reproduced by the distant modem. To minimize the effects of channel attenuation, the modems may include or require conditioning filters to produce an attenuation curve which complements that of the channel. Modems are discussed in more detail below.

Digital Transmission Methods

New transmission facilities, specifically designed for data, accept a digital signal directly, although a network termination device is still required to connect the CPE to the public network. The internal transmission method, however, is still based on analog media (typically, the alternate mark inversion method described in Chapter 3).

Types of Circuits

Point-to-Point and Multipoint Data Circuits. A data circuit can also be classified as *point-to-point* or *multipoint.* A point-to-point circuit connects two devices only and a multipoint circuit connects more than two devices.

Two-wire versus *Four-wire.* The communication channel on the line side of the DCE is usually described as "two-wire" or "four-wire." These terms are derived from copper-based telephone nomenclature, in which two or four wires (or equivalent) are employed to transmit information. Four-wire affords more flexibility, but requires more transmission resources (i.e., cable). Typically, one pair of wires is used for transmission of information in each direction, and higher effective bandwidth is achievable. With two wires, more sophisticated electronics is required to achieve intelligible simultaneous two-way data transmission. The majority of traditional local loops for voice use employ two wires to minimize the investment in copper, as discussed earlier. Long-distance voice circuits employ four wires, so that a separate transmission path is used for each direction. The interface between the two configurations is provided by a device called a "hybrid."

Data circuits that need to carry 9.6 kb/s or higher on a sustained basis have traditionally consisted of dedicated transmission facilities (also known as private lines), which are four-wire equivalent end-to-end. Dedicated four-wire circuits are particularly common in multipoint applications. Two-wire, full-duplex, echo-cancelling, error-correcting, high-throughput modems for use with switched lines are increasingly becoming available.

Half-Duplex versus *Full-Duplex*. These terms describe how data are transmitted across the channel, regardless of the physical configuration of the channel. Half-duplex defines a transmission in which data are transmitted in both directions but not at the same time. The DCE alternates between transmitting and receiving the data between the devices. Full-duplex describes the simultaneous transmission of data in both directions. The term "two-way alternate" is also used to describe a half-duplex data flow, and "two-way simultaneous" to describe a full-duplex data flow.

The ability to operate in a full-duplex mode depends on: (1) the channel (i.e., if the echo suppressor found in a long-haul dialup circuit can be disengaged); (2) the modem; and (3) the protocol used by the communicating entities. If any one of these three components fails to operate in a full-duplex mode, then the overall transmission becomes (effectively) half-duplex. Today, the majority of the data communication systems are full-duplex.

Line Turn-Around

When the DCEs use half-duplex schemes, a time interval is required for the devices to stabilize and adjust to the signal before transmission in the other direction. This stabilization period is called "training time," and the process of reversing the signal is called "line turn-around." Turn-around times of 100 to 200 ms are not uncommon. Because of this delay, most multipoint systems keep the master device's carrier on constantly (constant carrier) with the slaves configured for switched carrier operation (switched carrier). This approach eliminates half the turn-around delay. Some DCEs have used split channel techniques to eliminate the switched carrier delay completely.

Asynchronous and Synchronous Transmission

Two methods for formatting and transmitting user data through the channel exist: asynchronous and synchronous. The asynchronous approach is an older technique; however, due to its simplicity and relative low cost, it is still a common method and is found in many modern systems such as personal computers. Asynchronous transmission is characterized by the use of timing bits (start and stop bits) enveloping each transmitted character. The purpose of the start and stop bits is to provide for character synchronization and timing between the transmitting device and the receiving device. The start bit notifies the receiving device that a character is being transmitted on the channel. The stop bit indicates that all the bits have arrived and provides for each other's timing functions. In asynchronous transmission, each character is framed by start and stop bits.

In the synchronous approach, all characters are directly blocked together and are transmitted without the intervening start and stop bits. Framing codes (called syncs or flags) are placed in front and behind the full data unit (usually called a frame) to indicate to the receiver where the user data begins and ends.

Synchronous transmission can provide timing signals by one of three techniques [1.18]:

- A separate clocking line. A separate clocking line is a technique used for short-distance nontelecommunication connections: in addition to the data line, another line transmits an associated timing signal, which is used to clock the data into the receiver. For example, the EIA-232-D and CCITT V.24 specifications provide several options for synchronous transmission and clocking. Separate clocking channels are not practical for longer distances because the installation of a separate link or wire is expensive. Longer distance also increases the probability that the clocking line will lose its synchronization with the data line because each line has its own unique transmission characteristics. The telephone network does not provide clocking lines.
- Embedding the clocking signal in the data stream with the data acting as a clock to a simple receiver circuit. To embed the clocking signal, the data bits are encoded at the transmitter to provide frequent transitions of the channel.
- Embedding the clocking signal in the data stream and using it to synchronize a receiver clock. The line transitions in the incoming bit stream keep the receiver clock aligned (synchronized) onto each bit in the data block.

Digital Modulation

The principles of digital modulation are the same as those of analog modulation discussed earlier, except that the modulating signal is digital. In a digital environment, the modulator accepts binary encoded symbols and produces waveforms appropriate to the physical transmission medium at hand, which is always analog with today's technology [1.19]. Because these signals involve discrete jumps from one state to another, the modulating action is generally described as *keying*.

The simplest technique, *amplitude-shift keying* (ASK), modulates the carrier with the binary signal to produce an AM signal. Binary 0 is represented by no amplitude of the carrier and binary 1 with full amplitude of the carrier. The off-on keying is simple, but ASK makes inefficient use of transmission power. Amplitude modulation is not often used by itself because of transmission line power problems and sensitivity to line errors. However, it is commonly used with phase modulation to yield a method superior to either FM or AM.

Frequency-shift keying (FSK) is also common. This simple FM technique uses the binary signal to switch between two frequencies. In its simplest form, it is characterized by abrupt phase changes at the transition between one state and the other, which distorts the spectral energy and reduces spectral efficiency.

Phase-shift keying (PSK) is also being increasingly used because of higher efficiencies achievable with this technique (packing more data onto the carrier, or "more bits per baud," in engineering jargon). Simple PSK uses the binary signal to alternate the phase of a carrier between 0 degrees and 180 degrees. A variation, known as differential PSK, compares the phase of the previous bit with the present bit to determine if a transition has occurred.

Efficiencies of around 1 b/Hz of bandwidth are achieved with basic PSK, FSK, and ASK. Two bits per hertz can be obtained through the use of 4-phase PSK and *quadrature amplitude modulation* (QAM) (QAM is discussed below). More complex schemes, such as 16-phase PSK and 16-state QAM, are needed for 3 b/s per hertz. Four bits per hertz is achieved with 64 QAM. However, as the complexity of the modulation techniques increases, a better S/N is required to achieve the same *bit error rate* (BER) grade of service. (To be more precise, we mean bits per baud, which are changes in signal state; for example, the V.32 modem operating at 9.6 kb/s has a modulation rate of 2400 baud and encodes 4 bits/baud.)

The Modem

As indicated earlier, the modem is responsible for providing the required translation and interface between the digital environment and an analog communication link. Modems are DCEs, and have become CPE in the United States. The modem provides a digital-domain to voice-domain conversion; modems should not be confused with the analog-to-digital process discussed earlier under the PCM technique. Modems are designed around the use of an analog carrier frequency in the voiceband domain. The carrier is modulated with the DTE's data stream. The carrier signal is changed back to digital at the receiver DCE modem by the process of demodulation. Modems can use amplitude, frequency, and phase modulation, or a combination thereof to impart the data signal over the analog carrier. Each method impresses the data on a carrier signal, which is altered to carry the properties of the digital data stream.

Amplitude modulation modems alter the carrier signal in accordance with the modulating digital bit stream, while the frequency and phase of the carrier are held constant (the carrier is turned on or off, or at least changed in magnitude).

Frequency modulation modems alter the frequency of the carrier in accordance with the digital bit stream. The amplitude is held constant. In its simplest form, a binary 1 is represented by a specified frequency and a binary 0 by another. The most common FSK modems use four frequencies within the usable bandwidth of a standard telephone circuit (nominally, the bandwidth is 4000 Hz). A typical full-duplex two-wire FSK modem transmits 1070- and 1270-Hz signals to represent a binary 0 (space) and binary 1 (mark), respectively. It receives 2025- and 2225-Hz signals as a binary 0 and binary 1. FSK has been a widely used technique for low-speed modems (up to 1200 b/s). It is relatively inexpensive and simple; many PCs use FSK modems. The most sophisticated modems employ multiple carriers, instead of two carriers.

Phase modulation modems alter the phase of the signal to represent a 1 or 0. A common approach to implement PSK is to compare the phase of the current signal state to the previous signal state. PSK techniques use bandwidth more efficiently than FSK, but they require more elaborate equipment for signal generation and data representation. Phase modulation techniques are almost exclusively used today on high-speed modems. Phase shifting can be used to provide multilevel modulation. Table 1.3 depicts a 4-phase PSK and an 8-phase PSK modulation scheme. In the former we can map 2 bits per baud (one 360-degree carrier signal rotation); in the latter, 3 bits per baud. Figure 1.17 depicts the resulting *signal constellation*.

An extension of multiphase PSK modulation is *quadrature amplitude modulation* (QAM), in which the sine and cosine of the modem carrier frequency are amplitude modulated with two or more amplitudes. QAM techniques are widely used in high-speed modems. As can be seen from Figure 1.17, PSK puts the signal points over a circular periphery, which produces a crowding of the points and ensuing difficulties for demodulating the signal in the presence of noise. QAM is

Table 1.3
Quadrature Modulation

4-PSK		*8-PSK*	
Bits to Code	*Phase Change in Signal (deg)*	*Bits to Code*	*Phase Change in Signal (deg)*
11	45°	111	22.5°
10	135°	110	67.5°
01	225°	101	112.5°
00	315°	100	157.5°
		011	202.5°
		010	247.5°
		001	292.5°
		000	337.5°

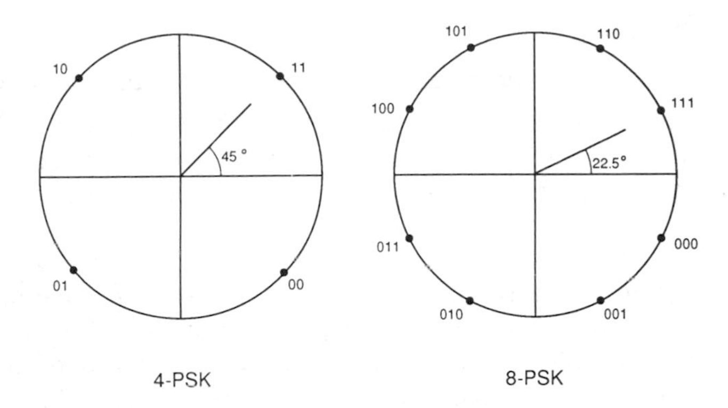

Figure 1.17 PSK constellations.

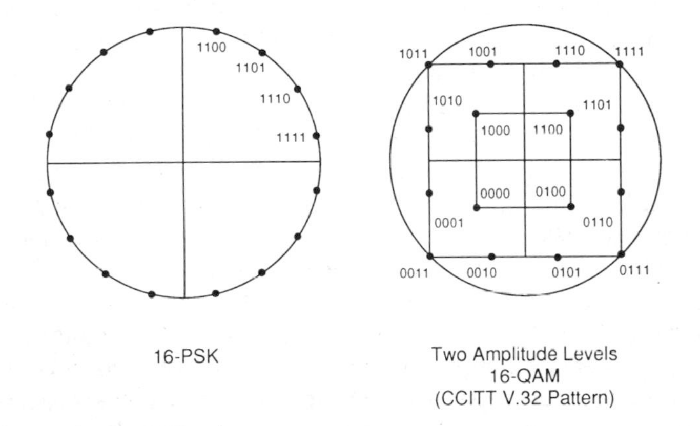

Figure 1.18 Comparison between PSK and QAM.

a combination of PSK with an amplitude differentiation for a set of signal points, as seen in Figure 1.18.

Most modems use scrambling techniques to ensure a proper number of state transitions for accurate timing recovery at the receiver modem. The scrambling is usually done by the DCEs. Scrambling provides modem (DCE-to-DCE) synchronization. However, synchronization of synchronous transmission between user devices and modems (DTE-to-DCE) must be performed with a separate timing (or clocking) circuit.

Most modems that operate with speeds up to 4.8 kb/s employ fixed equalizers; these circuits are designed to compensate for the average conditions on a circuit. However, the fixed equalizers are being replaced with dynamic (or automatic) equalization: the modem analyzes the line conditions and adjusts its equalization accordingly. The adjustments occur very rapidly, on the order of 2400 times a second for a 9.6 kb/s modem. More information on equalization follows.

History of Modems

The 70-year history of voiceband modems falls into four phases [1.20]:

1. From 1919 to the mid-1950s, work arose out of the need to transmit telegraphic information over the voice network. Research focused primarily on the basic properties of copper lines and on basic theories of data communication. The maximum data rate was around 100 b/s.
2. Starting in the mid-1950s, growing military requirements, and nascent commercial interest in transmitting large amounts of data, led to efforts to achieve greater transmission speeds. The technical investigations concentrated on modulation techniques, telephone line characteristics, and receiver design. Methods to deal with marginal phase distortion in additive noise channels, equalization, intersymbol interference, channel amplitude distortion, and delay distortion were developed. This resulted in an increase in the speed from 100 b/s at the beginning of this period to 9600 b/s in the late 1960s. By the late 1950s, AT&T introduced the Bell 103 (300 b/s) and Bell 202 (1200 b/s) modems; these employed FSK principles. In the early 1960s, the application of 4-phase PSK modulation, resulted in the Bell 201, which provided 2400 b/s over conditioned lines. Commercial products in the late 1960s provided reliable higher speed bandwidths; notable in retrospect were Milgo, which achieved 4800 b/s with 8-phase PSK, and Codex, which achieved 9600 b/s with QAM techniques with a 16-point signal constellation (i.e., 4 b/Hz).
3. During the 1970s, the speed on commercially available modems remained around 9600 b/s, but major design improvements led to significant reductions in size and power. Techniques implemented in this period, including LSI and VLSI, timing recovery, adaptive filtering, and digital signal processing, would establish the basis for the advancements of the next phase. Some improvements in speed (to 14,400 b/s) were obtained by using more advanced equalization techniques.
4. During the 1980s, error-correcting modems, advanced signal processing, and higher speeds were introduced—19,200 b/s is now routinely possible on dedicated lines. Speeds as high as 38.4 kb/s are now achieved using data compression. Because telephone lines use fiber for a large portion of their total span, they are less prone to some of the traditional problems, including phase jitter; S/N of better than 28 dB are achievable. QAM signal constellations with 64 points (6 b/Hz) become practical. The 19,200 b/s modems use orthogonal multiplexing (transmission of several noninterfering subsignals over the common channel), or multidimensional trellis-coded modulation, which we will discuss later. In band-limited channels an increase in transmission rate requires an increase in the number of coding points in the constellation; this, however, can run into marginal-performance areas (in terms of signal quality) of the channel. An approach to dealing with possible mutilation of

some constellation points is to use error-correction bits. Until the early 1980s, however, it was thought that the increased speed would wash out compared to the overhead needed to provide the needed error correction. This turns out not to be true. Trellis-coded modulation can improve the performance of a modem by 3 to 6 dB.

In spite of repeated predictions over the past decade that modems would soon be eliminated by end-to-end digital networks, the modem industry continues to prosper. While digital backbones are becoming popular, a large portion of data communication in 1990 is still carried by voiceband modems over the analog telephone network [1.20, 1.21]. Modems are now available on one or two chips. Effective bandwidth has increased approximately 30-fold during the past decade, and the cost per bit per second has decreased 20-fold. In the early 1980s, a 1200 b/s modem cost $1000, providing 1.2 b/s per dollar; in 1987 one could obtain 5 b/s per dollar ($2,000 for a 9.6 kb/s V.32 modem); in 1990 one can obtain 16–18 b/s per dollar ($2,000 for a top of the line 38.4 kb/s V.32 modem with error correction, or a 9.6 kb/s V.32 modem for $600). Both the bandwidth and the cost per b/s have improved for the dialup modems as well as for the private line modems.

Table 1.4 provides an assessment of both the actual growth of data, and of digital facilities in particular [1.22], and may be used to approximate the total data rate in the United States. The long-haul data carried on analog private lines in 1989 is estimated to be 3 billion b/s; that number decreases to 0.3 billion b/s by 1993. (Remember, however, that a lot of data are and will continue to be carried on dial-up facilities.) The long-haul data carried on digital private lines in 1989 are estimated to be 23 billion b/s; that number increases to 120 billion b/s by 1993.

Table 1.4
Demand for Telecommunications Services

Transmission Service	*Number of Private Lines*			
	1989 IXC[f]	*1989 LEC*[g]	*1993 IXC*[f]	*1993 LEC*[g]
Analog dedicated circuit	303,000	1,035,000	36,000	100,000[c]
Digital dedicated circuits	47,000[a]	88,000	310,000[b]	600,000[c]
Fractional T1 circuits	1,600	([d])	21,000	([e])
DS1 circuits	8,200	65,000	16,000	110,000[c]
DS3 circuits	200	2,000[c]	1,200	10,000[c]

[a] 50% are 56 kb/s lines.
[b] 90% are 64 kb/s lines.
[c] Estimated.
[d] Not generally tariffed.
[e] Unknown if tariffed.
[f] Interexchange Carrier.
[g] Local Exchange Carrier.

Compatibility Issues Pertaining to Modems

Given the plethora of techniques available to the modem to encode the digital signal for transmission over an analog circuit, we need to ascertain that the modems at both ends of the circuit are compatible, preferably following some established standard. In theory, two modems designed to the same standard should interoperate. In practice, people generally buy the two modems needed on a link from the same vendor; this is particularly true for higher speed modems that may be using proprietary encoding, modulation, error correction, and data compression schemes.

CCITT has published a series of recommendations to bring some standardization to the equipment. For example, CCITT Recommendation V.22 describes standardized 1200 b/s full-duplex modems for use on the general switched telephone network. V.29 describes standardized 9600 b/s modems for use on point-to-point leased circuits. Table 1.5 depicts some key modem families.

In 1982 CCITT started to develop recommendations for full-duplex two-wire modems operating at 9600 b/s over the public switched network. This was the first time that a modem standard was developed prior to a commercial product; in the past, all modem recommendations had been developed based on successful commercial products. An eight-dimensional error-correcting code is included in Recommendation V.32 (two-wire switched line, 9600 b/s) and in V.33 (four-wire dedicated line, 14,400 b/s). V.32bis (possibly a standard by the end of 1991) is designed to let a V.32-standard 9600 b/s modem operate at 14.4, 12.0, 9.6, 7.2, and 4.8 kb/s in full duplex over a dialup link, with on-line rate negotiation.

Table 1.5
Some Key Families of Modems

Modem Type	*Speed and Characteristics*	*1990 Price*[a]
Bell 103A	300 b/s full-duplex operation on dial-up line	$20–50
Bell 212A	1200 b/s full-duplex operation on dial-up line	$50–100
V.22 bis	2400, 1200 b/s full-duplex operation on dial-up line	$100
V.22 bis	19,200 b/s full-duplex operation on dial-up line, with compression	$700
V.29	9600, 4800 b/s half-duplex (2-wire) and full-duplex (4-wire) leased line	$800
V.32	9600, 4800 b/s full-duplex operation on dial-up line with echo cancellation	$600
V.32	9.6 kb/s to 28.8 kb/s to full-duplex operation on dial-up line with MNP error correction	$1200–2000
V.42	38.4 kb/s–19.2 kb/s with error correction and compression	$1500[b]

[a] See, for example, [1.22].
[b] See, for example, [1.23].

The new V.42 standard, primarily concerned with error correction, was formally adopted by CCITT late in 1988. The protocol includes both *Microcom's networking protocol* (MNP) and CCITT's LAPM (*link access procedure for modems*). LAPM uses *cyclic redundancy checking* (CRC) to detect errors in transmission and recover with a retransmission. Because an installed base of a half million error-correcting modems using MNP already exists, V.42 included both options. MNP is a *de facto* standard for error correction that has evolved through several "classes." The basic classes provide error detection and correction; the more advanced classes include compression techniques. Microcom released the first four classes of MNP to the industry and they license Class 5 and Class 6 to vendors that desire to include the protocol in their own products (the higher classes were still proprietary in 1990). Currently, only a couple of vendors have a V.42 product.

V.42bis aims at providing data compression. Compression ratios of 2-to-1, or even 4-to-1 may be possible. This means that a 9.6 kb/s modem can provide an effective throughput of 38.4 kb/s. The compression is based on an algorithm known as Lempel-Ziv. As of early 1990, only one modem on the market in the United States utilized this technique (the Telebit Trailblazer™ modem, which operates at 19.2 kb/s, first used an earlier version of a compression algorithm, but now is fully compliant with V.42bis [1.23]). Formal approval of the V.42bis was planned for 1990 [1.24].

The multitude of modem standards often leads to interoperability problems. Many modem manufacturers now include multiple transmission schemes into a single modem. Such modems allow users to choose an appropriate scheme for the application at hand, and facilitate interoperation with various existing modems. As of early 1990, more than half a dozen companies provided these "multimodems." One such product can, for example, implement nine distinct modulation techniques.

Capacity of a Data Link

The maximum digital capacity of an analog communication channel, in b/s, is given by Shannon's equation [1.1, 1.25]:

$$C = W \log_2 (1 + S/N)$$

where

C = channel capacity in b/s,
W = channel bandwidth in hertz,
S = signal power in watts,
N = noise power in watts.

[If the S/N is expressed as x decibels, then S/N = $10^{(x/10)}$.]

Application of Shannon's equation to a voice-grade line with a S/N of 30 dB (which is typical given the amount of power that can be applied to a copper medium, and the effects of noise and impairments—39 dB being the theoretical maximum) leads to a maximum bandwidth of 30,000 b/s for a 3000-Hz telephonic voice-grade channel [obtained as $3000 \log_2 (1+10^3)$]. When a modem has an effective throughput of 38.4 kb/s, it is achieved with data compression techniques along with efficient modulation schemes (producing perhaps an uncompressed throughput of 19,200 b/s; the 2-to-1 compression algorithm would produce an effective rate of 38.4 kb/s). This does not contradict Shannon's equation. A higher data rate could also be carried over an unloaded copper wire with no filters because the bandwidth W is higher. A 22-gauge copper wire can, in theory, carry up to 5 MHz, allowing higher digital throughput, as, for example, in the case of twisted-pair LANs (see Chapter 15), digital loop carrier systems (see Chapter 3), and ISDN local loops (see Chapter 4).

Sources of Noise in Data Circuits

Many physical factors inhibit successful data transmission. Some of the factors applicable to a traditional telephone channel are shown in Table 1.6. A brief description of some of these traditional problems affecting data transmission over an analog network follows. This discussion should clarify the need for digital

Table 1.6
Impairments on Traditional Copper-Based Telephone Channels

Attenuation as a function of frequency
Attenuation as a function of the signal level
Crosstalk
Echo
Transients
Impulse noise
Gain hits
Phase hits
Dropouts
Thermal noise
Intermodulation distortion
Delay as a function of frequency
Phase and amplitude jitter
Frequency errors

networks, fiber-based networks, and networks designed specifically for data communication.

Attenuation

Attenuation (A) is loss of energy in the signal as it is transmitted through the medium. It is described in decibels and is defined as:

$$A = -10 \log_{10} P_{in}/P_{out} \quad \text{dB}$$

where P_{in} is the input power at one end of the channel and P_{out} is the output power at the other end of the channel. Without amplification, the power out will be less than the power in. For example, a loss of 50% of the power corresponds to a 3-dB loss. A gain of 3 dB is a doubling of power. The measure "dBW" is often used to indicate the gain relative to 1 W of base power. Attenuation affects all types of transmission systems including coaxial-based systems, microwave systems, and fiber optic systems.

In a band-limited channel, such as a traditional long-distance voice circuit, designers customarily design for minimum attenuation in the middle of the spectrum. The loss is usually measured with reference to the 1000-Hz level. The useful bandwidth of a copper circuit is expressed as the frequency difference between points on the attenuation curve that represent 10 dB of loss with respect to the level of a 1000-Hz reference signal.

Attenuation *versus* signal level refers to dynamic loss in the channel, such as that caused by a compander operating at a syllabic rate. A compander is a device installed in the voice channel that amplifies weak signals more than strong signals at the source, and amplifies strong signals more than the weak signals at the destination, to restore the original characteristics. It is fast acting, following the syllables of speech for maximum effect. The companding technique permits the voice signal to be transmitted at a higher average level than would otherwise be the case and results in a consequent improvement of the S/N at the receiver. In data transmission, companding has the undesirable effects of decreasing the apparent modulation percentage while producing spurious frequencies that show up as noise. Line conditioning can ameliorate the situation, as discussed later.

Crosstalk

Inductive, capacitive, or conductive coupling between adjacent circuits results in having an undesired signal "leak" across and appear as a weak background signal added to the desired signal. Crosstalk degrades the S/N and increases the error rate. A limit is placed on the amount of noise that is allowed on the circuit. If the noise cannot be attenuated more than 24 dB below the signal, the circuit may need to be repaired or changed.

Echo

When an electrical circuit such as a telephone line is not properly terminated, some of the data signal energy is absorbed by the receiver and some energy, called echo, is returned to the sender. A portion of the echo may be reflected by an improper line termination at the sender and reach the receiver as noise. Like other kinds of noise, it increases the error rate, unless the signal level of the data is much greater than the signal level of the echo. The ability of a circuit to reduce reflections is called *echo return loss* and is measured in dB. The greater the return loss, the better is the termination. A ratio of 24 dB or more is desired.

Transients

Transients include impulse noise, gain hits, phase hits, and dropouts. *Impulse noise* is defined as any noise that exceeds the root mean squared level of the background noise by 12 dB for more than 10 ms. This kind of noise can be expected on most metallic facilities. It is the result of coupling from nearby circuits that carry electrical current surges originating in switching equipment, line faults, or from lightning surges. The desired data signal usually exceeds the background noise by at least 26 dB. A correlation exists between the frequency of occurrence of noise impulses and the number of electromechanical switching offices through which the data circuit must pass. Reduction of these switches in the 1980s has had the effect of decreasing the importance of switches as a source of impulse noise.

Telephone circuits are rated according to the average number of "hits" of impulse noise that exceed threshold over a period of time. When the statistical average is exceeded, an attempt ought to be made to find the cause and to make the necessary repairs. A typical rating is 15 counts in 15 minutes. Impulse noise hits tend to be clustered with long periods between clusters; also, a typical noise impulse is much larger than the data signal, and mutilates several successive data bits. Impulse noise affects any kind of modem and any modulation scheme. It will affect more data in a high data rate stream than in a slow one because of the disruptive effect of a large hit on the modem. The terminal equipment user should be prepared to cope with various patterns of impulse noise by providing data streams that can be acknowledged within the average interval of noise hits. If data blocks are too long, the probability is high that each one will receive a noise hit and require retransmission (with the chance of another hit). On the other hand, if blocks are short, the overhead incurred by the synchronizing sequence, header, and acknowledgments will reduce transmission efficiency.

Thermal Noise

Thermal noise (also called broadband or white noise) arises from the thermal agitation of electrons in resistors and in semiconductors. The greater the absolute temperature of the noise source, the greater the noise level. In telephone systems, this type of noise appears as a background hiss. Its strong appearance usually indicates a faulty component. The usual effect of thermal noise is to cause infrequent and random single-bit or double-bit errors, until a modem signal-to-noise threshold is reached, after which the error rate becomes catastrophic.

Intermodulation Distortion

Nonlinear components in the network will produce distortion of the data signal. The distortion becomes harmful if the time constant of the nonlinearity is short compared to the frequencies that compose the data signal. Intermodulation results in the production of spurious frequencies by a nonlinear medium. These frequencies are the sum and difference of fundamental frequencies present in the undistorted analog data signal, as well as the sum and difference between their harmonic frequencies, and between fundamentals and harmonics. The level of intermodulation distortion is typically less than 5% of the desired signal level and appears as correlated background noise that can cause characteristic distortion of the data signal (i.e., weak bit patterns).

Delay Distortion

In the process of transmission, all frequencies may not arrive at their destinations with the same relative phase relationships that existed at the transmission point. The effect is that some portions of the data signal appear to be delayed in time, with reference to other portions of the data signal. A modulated signal has a fundamental frequency plus harmonic frequencies. Each harmonic wave has a fixed-phase relationship with the fundamental wave that must be preserved if the data signal is to be reproduced by the receiver. When the harmonics are advanced or retarded in phase, they appear to arrive early or late. Some signal components overlap the time domains of adjacent signal elements, causing intersymbol interference. The solution to the delay distortion problem is called *delay equalization*.

Incidental FM

This phenomenon is peculiar to frequency-based carrier equipment, for example, in a link with analog microwave equipment. Incidental FM usually arises from the influence of varying power supply voltages on the oscillators used in the equipment. If the oscillators are not stabilized, hum frequencies and lower frequency ringing voltages can modulate the oscillators, both in frequency (or phase) and in amplitude. The amplitude effect is usually of lesser consequence. The effect of phase jitter on a data signal is to decrease the certainty of detection of data bits in the receiving modem. High-speed modems that use PSK are most vulnerable. The pool of analog carrier systems is becoming smaller in the United States, as fiber is increasingly deployed, diminishing the importance of this impairment.

Frequency Error

This phenomenon, like incidental FM, is endogenous to carrier frequency equipment. It results from the fact that the heterodyning oscillators in location A can differ in frequency by several hertz from those in location B, with the result that the data carrier and wideband frequencies are offset. This is not noticeable to a telephone listener, but modem performance can be degraded slightly, and a poorly designed modem will fail to receive data correctly.

Line Conditioning

Two options are available to reduce the effects of attenuation and delay distortion in a traditional data circuit: *conditioning* and *equalization*. Line conditioning may be used when data transmission on dedicated voice-grade lines occurs at speeds of more than 4800 b/s. Line conditioning will ensure that the circuit conforms to transmission specifications and is less susceptible to errors. Line conditioning is provided on leased lines by carriers at a monthly fee. The carrier may add special equipment to the circuit; conditioning provides a method to diminish the problems of attenuation and delay, but it does not totally eliminate the impairments; it provides for more consistency across the bandwidth. For attenuation, the carrier adds equipment that attenuates the frequencies in the signal that tend to remain at a higher level than others. Thus, attenuation still occurs but is more evenly distributed across the channel. Conditioning is not available over the switched telephone network.

Two types of line conditioning are available in the United States: *C-conditioning* and *D-conditioning*. C-conditioning is used to deal with attenuation distortion, which occurs when the relative amplitudes of the different frequency components change during transmission. This can occur because of uneven attenuation or uneven amplification. D-conditioning is used to deal with harmonic distortion, which is caused by the presence of unwanted harmonics from the input signal in the output signal. D-conditioning also improves the S/N. The standard specification requires a S/N of not less than 24 dB, a second harmonic distortion of not more than 25 dB, and a third harmonic distortion of not more than 30 dB. With D-conditioning, the carrier gives a S/N ratio specification of 28 dB, a signal to second harmonic distortion ratio of 35 dB, and a signal to third harmonic distortion of 40 dB. D-conditioning is accomplished by eliminating or avoiding noisy facilities for the given circuit.

An attenuation equalizer adds loss to the power frequencies of the signal because these frequencies decay less than the higher frequencies in the band. The signal loss is then consistent throughout the transmitted signal. After equalization is applied, amplifiers restore the signal back to its original level. A delay equalizer compensates for total signal delay. The higher frequencies thus may reach the receiver ahead of the lower frequencies. Consequently, the equalizer introduces more delay to these frequencies to make the entire signal propagate into the receiver at the same time.

With the recent development of sophisticated modems able to cope with problems on dialup lines (and thus, by extension, in private lines), and with the continued introduction of digital lines, the importance of conditioning, as currently defined, may diminish in the 1990s.

Data Error Detection and Correction

Data communication systems require error control mechanisms to deal more explicitly with the problems listed in the previous sections than is possible with line conditioning. In a metropolitan environment with considerable man-made electromagnetic noise and multipath problems, the need for error correction is imperative, especially for data carried on copper loops or radio systems.

The original work on error-control coding was undertaken in the 1940s and 1950s by Shannon, Hamming, and Golay. Coding is now a mature branch of communication, with a strong mathematical foundation. These advances in theory combined with the emergence of inexpensive VLSI provide cost-effective means to achieve efficient and highly reliable communication.

In terrestrial links (where the signal propagation delay is small), the commonly used method of error control is the automatic repeat request (ARQ). Once an error is detected, the receiving system asks for retransmission from the sender (automatic refers to the fact that no user intervention occurs). Two types of ARQ systems exist: (a) stop-and-wait ARQ and (b) continuous ARQ. In the stop-and-wait ARQ, the sender system waits for an acknowledgment from the receiver system on the status of the transmitted block. If the acknowledgment is positive, then the sender system transmits the next block; if it is negative, it repeats the transmission of the blocks that have errors. This approach is unsuitable in a satellite environment where a round-trip delay of half a second is required for the reception of the acknowledgment. A more acceptable form of ARQ is the continuous ARQ.

With this approach, the sender sends the blocks and receives the acknowledgment continuously. Once a negative acknowledgment is received, the transmitter sends either the block with an error and all blocks that follow it, or sends only the block that has the error.

Both of these ARQ approaches require a duplex channel, so that handshake information can be sent back and forth. If the channel is simplex, as is the case with the FM-subcarrier discussed in Chapter 5 (or the propagation time is so long that the channel is effectively simplex—for example, communicating with an interplanetary spaceship), then a totally different method is required. The form of error control commonly used in satellite systems is the *forward error correction* (FEC). In this system, extra bits of data are added to the blocks for error checking and correction. If an error is detected, enough redundant information is carried along, which permits the receiving end to fix the incoming message; the receiver need not go back to the sending party to obtain a retransmission as in ARQ.

In FEC, one wants to be able to remove all redundancy from the source of the information, so that the amount of data to be transmitted is minimized. The channel encoder performs all the digital operations needed to prepare the source data for modulation. The encoder accepts information at rate R_s, and adds its own redundancy, producing an encoded data stream at a higher rate R_c. There are two types of FEC methods, the block codes and the convolutional codes [1.19, 1.26, 1.28].

When using a block code, the encoder accepts information in sequential k-bit blocks and, for each k bit, generates a block of n bits, with $n \geq k$; the n-bit block is called a *code block* or *codeword*. The ratio k/n is called the *rate of the code*. Thus, the stream of data is broken into k symbols of information and $n-k$ redundant symbols for error control, where n is the total length of the code word. The resultant system is referred to as (n,k) block code, of which many forms exist; the most popular ones are the cyclic codes, which can simply be formed from cyclic shifting of bits of data in a block, thus creating a code word. The two most common coding techniques are the BCH (Bose-Chaudhuri-Hocquenghem) code and the Golay code.

Convolutional codes are another class of FEC methods. For encoding with a convolutional code, the encoder accepts information bits as a continuous stream and generates a continuous stream at a higher rate. The information stream is fed to the encoder b bits at the time (b typically ranges from 1 to 6). The encoder operates on the current b-bits and some number k (called constraint length) of immediately preceding b-bit inputs to produce B output bits, with $B > b$. Here the code rate is b/B. The encoder for the convolutional code might be thought as a form of digital filter with memory extending $k - 1$ symbols in the past. A typical binary convolutional code will have $b = 1$, $B = 2$ or 3, $k = 4$ or 5 or 6 or 7 (in special situations k can be as high as 70) [1.19, 1.26]. The channel decoder undertakes the conversion of the demodulator output into symbol decisions, which reproduce as accurately as possible the data that were encoded by the channel encoder. The most widely used convolutional code is the Viterbi code. Viterbi coders that transmit 256 kb/s are now available for less than $90 in quantities [1.28].

Trellis code modulation (TCM) is a relatively new technique now available in high-speed modems. The method is such that the signal (derived and coded from the user data bit stream) is allowed to assume only certain characteristics (states). User bits are interpreted such that only certain of the states are allowed to exist from prior states. The transmitting device accepts a series of user bits and develops additional (restricted) bit patterns from these bits. Moreover, a previous user bit pattern (a state) is used to determine the current bit patterns (states). Certain other states are not allowed and are never transmitted. The transmitter and receiver are programmed to understand the allowable states and the permissible state transitions. If the receiver states and state transitions differ from redefined conventions, we assume that an error has occurred in the circuit. By convention, the transmitter and the receiver know the transmission states and the permissible state transitions; the receiver analyzes the receiver signal and makes a "best guess" as to what state the signal should assume. It analyzes current states, compares them to previous states, and makes decisions as to the most relevant state. In effect, the receiver uses a path history to reconstruct damaged bits. Trellis coding is an error-correction code with a memory. It increases the BER performance on a line by two to three orders of magnitude [1.18].

Parity Checking

Parity checking is a simple but relatively unreliable method for basic error detection. It was primarily used in the late 1960s and early 1970s. Parity check schemes are simple examples of block codes. Here the encoder accepts k information bits and appends a set of r parity check bits, derived from the information bits according to some predefined algorithm. The information and parity bits are transmitted as a block of $n = k + r$ bits. A typical parity code is (8,7). Single parity check codes lack sufficient power to provide reliable communication. Hamming codes provide more powerful block codes and are also used for error control in computer memories and other mass storage systems.

Two versions exist: an odd-parity and an even-parity. In even-parity, the number of 1s in the 7-bit ASCII representation of each character being sent are counted. If that number is even, a 0-bit is concatenated to the 7-bit code; if the number is odd, a 1-bit is concatenated to the 7-bit code, thus giving an even number of bits (odd-parity does the reverse). For example, the code 1111111 becomes 11111111. The receiving end will count the number of ones; if there are an odd number of ones, the end system concludes that an error must have occurred in transmission.

The problem with this method is that line hits and dropouts tend to occur in bursts, affecting several contiguous bits, particularly at high speed. If more than one bit is affected, the method may fail to detect an error. For example, if 11111111 becomes 00000011, the receiving end will not be able to detect the problem.

Cyclic Redundancy Checking

Cyclic redundancy checking is the prevailing method used in conjunction with ARQ to detect errors in long blocks of data. The process of generating a CRC for a message involves dividing the message by a polynomial, producing a quotient and a remainder. The remainder, which usually is two characters (16 bits) in length, is appended to the message and transmitted. The receiver performs the same operation on the received messages and compares its calculated remainder. If the two CRCs fail to match, the protocol causes the block to be discarded, and a retransmission is requested. In contrast with the parity method discussed above, and other methods used in the 1970s (vertical and horizontal error checking), the undetectable error rate for CRC-protected data is extremely small (the undetectable error rate depends on the length of the CRC and the length of the data block). We will discuss the CRC method in more detail in Chapter 11.

1.2.2.3 Switching

Switching is as important for data communication as it is for voice communication. Six types of switching applicable (but not exclusively) to data are described below.

Circuit Switching

In a circuit switched connection, the end-to-end path of a fixed bandwidth exists only for the duration of the session. The destination is identified by an address; the network receives the address from the sender and sets up a path, typically within seconds, to the destination. At the completion of the call, the path is taken down. Circuit switching is not only suitable for voice transmission, which employs this method almost exclusively, but also for data transmission.

Circuit switched connections are economical if the end-to-end session is short (on the order of minutes), or the geographic area to be covered is wide. However, if the sessions are long and traffic to a specific destination is heavy, the cost of using circuit switching may exceed the cost incurred by other methods (for example, a dedicated line). If calls are very short, the setup and teardown overhead may be excessive; other methods (such as packet switching discussed below) may be better.

Channel Switching

Channel switching refers to a service that allows the user to establish a channel that can stay in place for hours or days, and can be reterminated (typically in minutes) as needed. The nomenclature used to describe this service varies, and the terms "reserved," "semipermanent," and "permanent" have all been used [1.29, 1.30, 1.31]. Channel switching can be considered an extension of circuit switching with the following four variations: (1) the call setup time is in minutes rather than in seconds; (2) the duration of the call is in hours or days rather than in minutes; (3) it employs "slow" switches in the CO, such as a digital cross-connect system, rather than a traditional circuit switch; and (4) it is cheaper for the user, compared to straight circuit switching, for long sessions. Channel switching may be provided under ISDN, although manifestations appeared in the early 1980s using digital cross-connect systems. It is a service positioned between circuit switching and dedicated lines (no switching).

No Switching

Many data applications use dedicated lines (also known as a "private line") that do not include carrier-provided switching. The line is leased at a fixed monthly rate, and the charge is independent of usage. Once installed, these facilities can stay in place for years. T1 circuits discussed in Chapter 3, now commonly used as backbone data networks, are examples of dedicated channels. Systems Network Architecture (SNA), discussed in Chapter 11, predominantly employs dedicated lines, although a packet switched interface is also available.

Packet Switching

Early approaches to data communication were based on techniques for voice communication such as circuit switching. One soon discovered that the dynamic allocation of bandwidth would allow more efficient utilization of available network resources for interactive data communication. *Packet switching* has emerged as an important approach in data networks. In packet switching, information is exchanged as blocks of limited size or packets. At the source, long messages are divided into several packets that are transmitted across the network and then reassembled at the destination to reconstitute the original message. Many users can share network resources, although efficient use of transmission resources increases the network complexity. Data buffers are needed at each node; however, the storing is typically transient in nature, and should be of the order of tens or hundreds of milliseconds. Packet switching can be viewed as a case of message switching, except that the very formal procedures are used (absent in traditional message switching), and the period of nodal storage is small, as indicated.

Packet switching comes in two types: connection-oriented (such as in traditional X.25 wide-area networks), and connectionless (such as in LANs and MANs). We will explore these concepts in more detail in Chapters 4 and 11.

Message Switching

Message switching refers to a method of storing a message at intermediary nodes in the network for nontrivial amounts of time (i.e., more than a couple of minutes). This method was commonly used in telegraphy and telex networks. It almost disappeared, but may now be reemerging in store-and-forward E-mail, particularly with the message handling system discussed in Chapter 13.

Distributed Switching

LANs, MANs, and IBM's SNA, among other systems, employ a form of a distributed "self-switching."* In this environment, the data are labeled with an address and are broadcast to every user connected to the particular transmission system at hand (for example, in a LAN this would be every user on the bus or ring). Each user's equipment receiving the message examines the address on it and is trusted not to display or throw away the data unless the message is addressed to that user. This technique is becoming prevalent, and it has the advantage of being efficient because it relies on distributed user-provided intelligence to carry out the message sorting task; the sorting and switching task would otherwise have to be done centrally, by carrier-provided resources. Additionally, it can be more reliable, if properly designed, because failure of a single component does not necessarily affect the entire communication system.

1.2.2.4 Layered Protocols

Motivation

As we discussed at the beginning of this chapter, communication involves by definition two (remote) entities, also called end systems. The two entities should be *peers,* namely, enjoying the same set of communication privileges, although this is a relatively new approach (throughout the 1960s and 1970s, a master-slave approach was much more typical). To undertake communication, a fairly large

*In this formulation, packet-switching and distributed switching are not seen as mutually exclusive.

number of functions must be carried out. In addition, agreements must exist between the two end systems on how to undertake functions that have remote importance. These agreements are now known as *protocols,* and publicly agreed protocols are referred to as *standards.*

A *layer* is a defined set of related communication functions. Protocols describe ways in which remote peers can utilize functions within a layer. Layering (or modularization) provides the following benefits, among others [1.32]:

1. Easier understanding of the communication process by working with a small number of logical groupings;
2. Collecting related functions in the same groupings minimizes the number of interactions between layers and simplifies the interfaces;
3. Layers can be implemented differently and changed to take advantage of new developments without affecting the other layers;
4. Simple layer boundaries can be created with at most only two neighbors.

(Layering by itself does not necessarily imply peer-to-peer communication capabilities: a number of vendor-specific architectures of the 1970s employed the layering concept (i.e., IBM's SNA, DEC's DECNET); however, the open-layered architectures that have evolved in the 1980s can provide peer-to-peer capabilities.)

Open Systems Interconnection Reference Model

To facilitate interconnection, standards for open systems have been developed by the International Organization for Standardization. Seven major layers have been defined in what is now known as the *Open Systems Interconnection reference model* (OSIRM), which has been available since 1984, as follows: application (7), presentation (6), session (5), transport (4), network (3), data link (2), and physical (1) layers. See Table 1.7 for a listing of some of the key functions of the OSIRM layers. (The application layer should properly have been named "application support layer" because the ultimate user application utilizing communication facilities to interact with partners in other systems resided above the application layer.) This model is described in Specification ISO 7498 and also in CCITT X.200 [1.33]. The term upper layers (or higher layers) refers to layers (4) through (7); downward, or lower, layers are in the direction of the layer (3) through (1).

All contemporary descriptions of data communication (and telecommunication, for that matter), including network management, security, addressing, and internetworking, employ the framework defined in the OSIRM. The higher adjacent layer is called the user; the lower one is called the provider (the term here does not refer to the ultimate end-user). User and provider describe, respectively, the relationship between the consumer and the producer of a layer service. As

Table 1.7
Layers of the Open Systems Interconnection Reference Model

Layer	*Function**
Application	Support of user functions such as file transfer, transaction processing, *et cetera*
Presentation	Transfer syntaxes (character coding)
Session	Coordination services, dialogue, synchronization
Transport	Reliable end-to-end communication
Network	Delivery within a single subnetwork; end-to-end aspects, such as addressing and internetworking
Data Link	Delivery of blocks of data between two points
Physical	Bit transmission

*Refer to ISO 7498 or X.200 for a more formal and complete description of these functions.

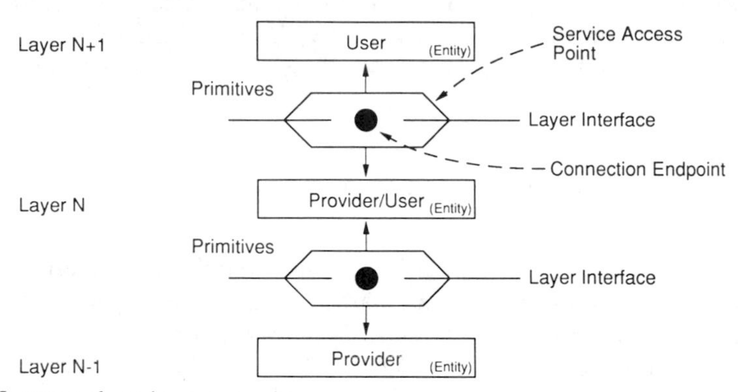

Figure 1.19 Layers and service access points.

one moves through layers, users become providers and *vice versa* as shown in Figure 1.19.

Entities exist in each layer. An *entity* is an active element within a layer that carries out the layer's prescribed functions. A layer may contain multiple entities for different functions. Entities in the same layer, but in different systems, that must exchange information to achieve a common objective are called "peer" entities. Peer defines an equivalent layer entity in another system within the open environment. A peer may be of a different hardware or software environment, but behaves consistently in all cases. Entities in adjacent layers interact through their common boundary.

The *application layer* can be viewed as an extension of local operating system supervisory functions to another system. These supervisory functions include the following: (1) identifying the intended partner and activating authentication procedures; (2) agreeing on quality of service, security, payment, *et cetera*; (3) supplying services to control the modification of shared data; (4) determining that required resources are available; and (5) specifying more detailed application requirements.

The upper layers—application, presentation, session, and transport—are generally, although not always, independent of the telecommunication network; the reason for the exception is that some carriers may offer functionality above the network layer, for example, E-mail. In general, however, these layers are components of the end-user systems and are insulated from networking operations. The interworking layers create computer-system-to-computer-system services operating across any combination of subnetworks. The OSIRM layering is applicable both when communicating over a carrier's network and when communicating over a private (CPE) network.

Communication with a remote peer, at the same layer, involves a protocol, as depicted in Figure 1.20 (only three layers are shown for simplicity). Adjacent entities communicate by exchanging primitives with each other via the *service access point* (SAP). The SAP is a conceptual delivery point, and as such it can be addressed. Note, however, that the relationship between SAPs and entities is not one to one.

(N)-layer Formalism

Some formalism related to important concepts associated with a layer, already described above in more intuitive form, follows (refer to Figure 1.21) [1.34]. This

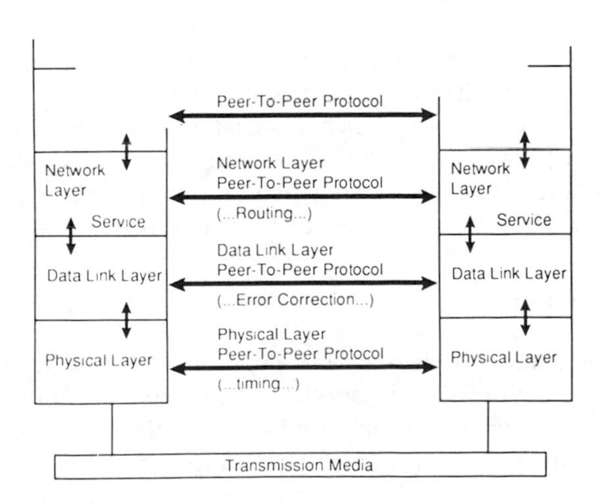

Figure 1.20 Peer-to-peer protocols.

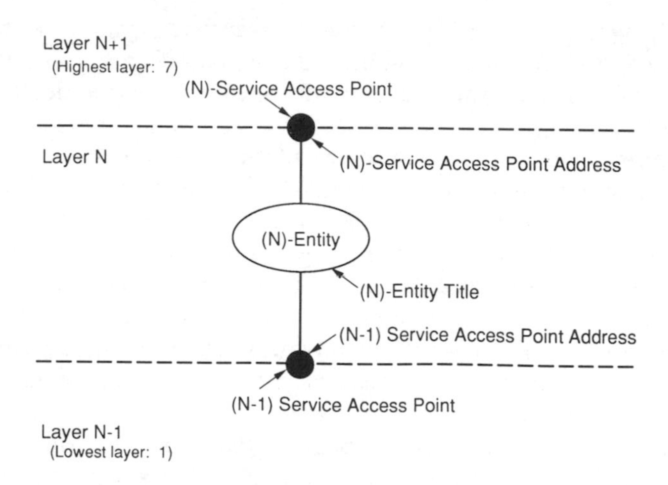

Figure 1.21 Formal view of OSI layers.

formalism will be required when discussing name and addressing issues in Chapter 13. The functionality of a generic (N)-layer is provided by one or more (N)-entities. These (N)-entities must be identified and located to perform communication. An (N)-title is a name that uniquely identifies a particular (N)-entity throughout the (N)-layer in an OSI environment; an (N)-title is independent of the location of an (N)-entity. (N)-service access points represent the logical interfaces between (N)-entities and ($N+1$)-entities. An (N)-SAP address is a name that identifies a set of one or more (N)-SAPs to which an ($N+1$)-entity is attached. An (N)-SAP address relates to an (N)-SAP, and not directly to an ($N+1$)-entity. Hence, an (N)-entity is identified by an (N)-title, independently of the (one or more) ($N-1$)-SAPs to which the (N)-entity may be bound. An (N)-entity is located by specifying the ($N-1$)-SAP address of the ($N-1$)-SAPs to which the (N)-entity is bound. A mechanism exists at each layer which associates (N)-entities with their ($N-1$)-SAPs; this is the (N)-directory. The (N)-directory maps (N)-titles of (N)-entities onto the ($N-1$)-SAP address through which they communicate. As we will discuss in Chapter 13, an OSI address comprises nested (N)-SAP addresses.

Primitives and Services

A *primitive* is the smallest unit of action that can be specified. More precisely, it is a conceptual instruction in a layer service. The primitives represent, in an abstract way, the logical exchange of information between a layer and the adjacent layers; they do not specify or constrain implementation. Examples of primitives are: establishing communication with a remote peer, sending data, and inserting a synchronization point. Most communication activities require a number of primitives to complete their tasks. A service primitive consists of a name and one or more parameters passed in the direction of the service primitive. The name of the service primitive contains three elements: (1) a type indicating the direction of the service primitive, (2) a name, which specifies the action to be performed, and (3) an initial (or initials), which specifies the layer (or sublayer) providing the service.

Two kinds of services are available: *confirmed* and *unconfirmed.* A confirmed service produces information from the remote peer entity on the outcome of the service request (this may be needed when additional action is contingent on a successful outcome). An unconfirmed service only passes a request along; this is a faster interaction because no overhead is involved with the response.

Four generic types of service primitives for confirmed service are [1.35]:

1. Request: a service request from a higher layer to a lower layer (more formally: a primitive issued by a service user to invoke a service element);
2. Indication: a notification from a lower layer to a higher layer that a significant event has occurred (more formally: a primitive issued by a service provider to advise that a service element has been invoked by the service user at the peer service access point or by the service provider);
3. Response: the response to a request (more formally: a primitive issued by the service user to complete at a particular service access point, some service element whose invocation has been previously indicated at that service access point);
4. Confirm: message passed from a lower layer to a higher layer to indicate the results of a previous service request (more formally: a primitive issued by a service provider to complete, at a particular service access point, some service element previously invoked by a request at that service access point).

The unconfirmed service only employs the request and indication primitives. A pictorial view of a primitive action sequence is provided in Figure 1.22.

Peers exchange *protocol data units* (PDUs) containing (1) protocol control information (PCI) and (2) data. A user initiates activity by issuing a service request across the SAP. The entity receives the service request and constructs a PDU, the type and values of which are determined by the request and locally available information. The PDU is delivered to the remote peer partner using the services of the underlying layers (the PDU will be enclosed as data in a subsequent service request to a lower layer, as shown in Figure 1.23). When the remote entity receives the PDU, it generates a primitive that it passes upward via the SAP to the user.

More details on the OSIRM and related standards are provided in Chapter 13; however, a basic understanding is required at this juncture for use in the intervening chapters. ISDN, broadband ISDN, signaling, LANs, MANs, network management, and security all require an understanding of the OSIRM and related recommendations. We describe the first three layers of the OSIRM below. The entire model is described in Chapter 13; a reader may wish to go directly to that

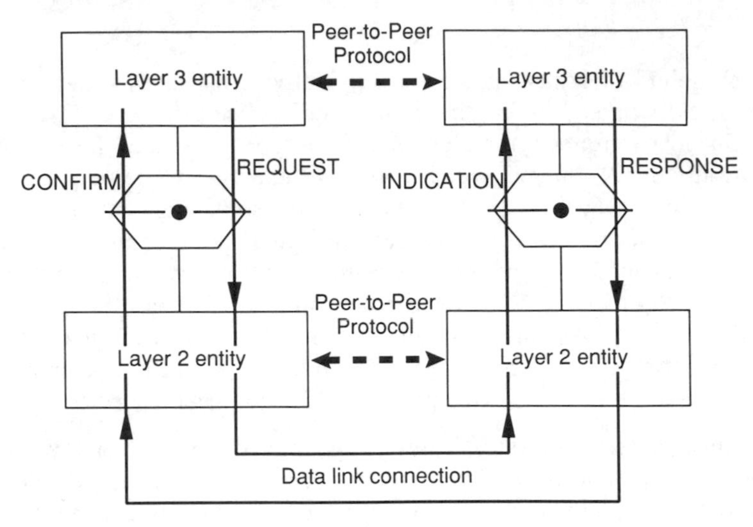

Figure 1.22 Sequence of primitives (layer 3 example).

User Application							Information
Application						A	Information
Presentation					P	A	Information
Session				S	P	A	Information
Transport			T	S	P	A	Information
Network		N	T	S	P	A	Information
Data Link	D	N	T	S	P	A	Information

Figure 1.23 PDU encapsulation.

chapter at the conclusion of Chapter 1, although that path is not required to read the intervening chapters.

Layer 1—Physical

This layer deals with the physical connection of the circuits. Recommendations at this layer standardize pin connections between DTEs and DCEs and are also concerned with the transmission of bits between machines. Examples of protocol standards for this layer are RS-232-C and RS-449, noted earlier for connection between terminal devices and modems, and I.430 for ISDN.

Layer 2—Data Link

This layer deals with data transmission over a single link. The protocol in this layer detects and corrects errors in the transmission. This layer creates frames for the data to be transmitted, including a CRC code. The functionality includes sending acknowledgments to the sender end system. It will signal for retransmission if a frame is out of sequence or is mutilated. This layer uses flags and headers in the frame, so that the receiver end system can recognize the start and end of a frame. Typical standards for this layer are: IBM's *Synchronous Data Link Control* (SDLC), ISO's *High-level Data Link Control* (HDLC), and ISO's *Link Access Procedure B* (LAP-B).

Layer 3—Network

This layer receives messages from the higher layer, segments them into packets, and sends them to the receiver through the data link and physical layers. At the receiving end, functionality in this layer reassembles the messages into its original form. It enables routing, multiplexing, and flow control. Typical standards for this layer are: CCITT's X.25 Packet Level Protocol and ISO's 8473, which deals with internetworking.

1.3 TELECOMMUNICATION CARRIERS

This section describes the commercial providers of telecommunication services. In the United States, a common carrier is an agency that provides telecommunication services to the general public. A common carrier typically provides a spectrum of services. Descriptions of services and rates are called tariffs, and are filed with the Federal Communications Commission (FCC) or with the local state public utility commissions (PUCs). Common carriers can be grouped as follows:

1. Exchange-access carriers (intra-LATA), such as the Bell Operating Companies (BOCs) and independent telephone companies; these are collectively known as LECs.
2. Interexchange carriers (IXCs), such as AT&T, US Sprint, and MCI; and
3. Specialized common carriers (SCCs), such as Telenet; SCCs are like other common carriers, but specialize in providing specific services (for example data or video communication); these are also called value-added carriers, or value-added networks (VANs).

While 1500 or so local telephone companies provide exchange access common carrier services, 85% of the business is provided by the BOCs. Approximately 5% of the LEC's customers account for 50% of the revenues [1.36]. Approximately 200 IXCs exist in the United States.

Other types of carriers also exist. A *private carrier* offers point-to-point links, has no switched service, negotiates with customers on an individual basis, and has a majority of leases that are long-term, with a limited and stable customer base.

A company that offers only nonswitched point-to-point service to business customers, and does not offer any services to the general public, can be considered a private carrier. These carriers are also referred to as "noncommon carriers." A nondominant fiber optic carrier is a carrier that provides principally or solely end-to-end fiber-based communication links to another carrier or to end-users, generally in an inter-LATA arrangement. Teleport is, formally, an access facility to a satellite or other long-haul telecommunication medium that incorporates a distribution network (usually fiber-based) serving the greater regional community, and is associated with comprehensive related real estate or other economic development. In practice, however, most teleports are not usually involved in major real estate or local economic development. About 20 teleports operate or plan to operate in the United States.

1.4 NETWORK DESIGN PHILOSOPHIES

The remainder of this book describes dozens of network architectures and configurations. In reading this material, keep in mind that, depending on the objective, one may be led to design and use different network topologies, technologies, and architectures. For example, some objectives could be to:

- Minimize cost (build the cheapest network);
- Maximize the cost-performance ratio (get the most network for the dollar);
- Maximize profit (build a network that allows the firm to be very aggressive, reaching new markets, *et cetera*);
- Maximize profit rate (especially for a utility);
- Minimize risk of loss (military network);
- Maximize safety (network designed for police or fire departments);
- Maximize quality of service (for a given investment);
- Maximize growth opportunity for the firm (build a network that can easily grow in the future—this may be the course of action of a carrier);
- Maximize prestige of the firm (i.e., buy the newest equipment to impress investors, competitors, public);
- Many other possible objectives.

Clearly, these criteria are not all compatible with each other: a network built to minimize the cost will probably not maximize reliability and quality of service. Hence, the telecommunication manager must be familiar with the dynamics of the design synthesis. Complexity in design (and corresponding responsibility for the planners) continues to increase inexorably. Using 1985 as an arbitrary reference for discussion and ignoring the effects of divestiture, the planner must now deal with the following new technical issues and options (some of which were only embryonic at that time):

- Availability of high-speed dialup modems operating up to 38.4 kb/s, and cellular telephony modems operating up to 16.8 kb/s, which make the optimization process compared to dedicated lines and ISDN an interesting one;
- Proliferation of personal computers and workstations in the office and on the factory floor, and their need to communicate, probably through internetworked LANs. High-speed backbone networks operating at 16 and 100 Mb/s are also becoming available. In addition to coaxial-based solutions, the planner can choose unshielded twisted-pair and multimode or single-mode fiber (see Chapter 14);
- Fourth-generation PBXs that not only allow integration of data, but also include LANs, resource servers (file servers, communication servers, protocol gateways, etc.), ISDN access, and advanced internal algorithms such as "dynamic bandwidth allocation." Also, we have witnessed the substantial improvement of CENTREX services, making it more competitive with PBXs, and rendering the decision-making process in choosing a solution more difficult (see Chapter 10);
- Availability (in the United States) of high-speed digital interfaces to the telephone plant, to another networked PBX, or to a mainframe computer via a CPE T1 multiplexer. T1 multiplexers are being introduced in private networks at an ever-increasing rate for the purpose of corporate-wide voice, data, and video integration. Fractional T1 service is also becoming important, and there is talk of fractional T3 (see Chapter 3);
- The introduction of high-speed channel-to-channel communication among mainframe computers, and between mainframes and peripherals (Chapter 12);
- The ongoing introduction of ISDN with its user-to-user and user-to-network signaling capability, and advanced services, including network-based *automatic call distributor* (ACD), and *automatic number identification* (ANI), which facilitates integrated voice and data applications. Switched Multimegabit Data Service (SMDS), with data rates in the 1.5 to 45 Mb/s range, and broadband ISDN (BISDN), with data rates in the 150 to 600 Mb/s range, are also on the horizon (see Chapters 4 and 15);
- The possibility of interconnecting dispersed LANs via MANs and potentially using fiber optics facilities (see Chapters 7 and 15);
- The emergence of very small aperture terminal (VSAT) satellite networks, allowing cost-effective two-way communication over four-foot dishes, and other radio services for data usage (see Chapters 5 and 6);
- The increased need for network management: networks can cost millions of dollars every year; optimal usage is critical (see Chapter 16);
- Security has become a major area of concern as more and more computers are connected to networks, making them accessible and susceptible to attacks (see Chapter 17).

All of these factors and technologies complicate the design process; understanding the new design requirements and acquiring the appropriate tools is a mandatory effort in establishing cost-effective, reliable, and flexible corporate networks. The responsibilities of a data communication manager are becoming more demanding as industry competition increases along with the ensuing multiplicity of available communication options and the increased obsolescence rate of this technology. In designing a network, multiple options must be analyzed and the issue of hidden costs associated with these options must be taken into account. This richness of communication options offers the data communication manager major opportunities, not only in terms of optimal network design, but also in terms of network management, grade of service, and reliability. However, analytical methods are required to facilitate the decision-making.

The multiplicity of options makes the task more difficult in the decade of the 1990s. Additionally, deployment of the least expensive network may not always be the best course of action if the network constricts the company's ability to be competitive, deliver products and services quickly, and grow smoothly. Figure 1.24 depicts some of the factors impacting the decision-making process. As we can see, the problem is not a trivial one, due to the high multidimensionality of the solution space. The purpose of this book is to familiarize the reader with these important evolving communication technologies and ensuing opportunities.

REFERENCES

[1.1] C.E. Shannon, "A Mathematical Theory of Communication," *Bell System Technical Journal*, July and October 1948, Vol. 27, pp. 379–423 and 623–656.

[1.2] F.T. Andrews, "The Heritage of Telegraphy," *IEEE Communications Magazine*, August 1989, pp. 12–18.

[1.3] R.F. Rey, *Engineering and Operations in the Bell System*, 2nd Ed., AT&T, Morristown, NJ, 1983.

[1.4] Bell Telephone Laboratories, *Transmission Systems for Communications*, 5th Ed., Holmdel, NJ, 1982.

[1.5] P. Mermelstein, "G.722, A New CCITT Coding Standard for Digital Transmission of Wideband Audio Signals," *IEEE Communications Magazine*, January 1988, pp. 8 ff.

[1.6] N.J. Muller, "ADPCM Offers Practical Method for Doubling T1 Capacity," *Data Communications*, February 1987, pp. 213 ff.

[1.7] *Communications Week*, May 14, 1990, p. 62.

[1.8] N. Kitawaki and H. Nagabuchi, "Quality Assessment of Speech Coding and Speech Synthesis Systems," *IEEE Communications Magazine*, October 1988, pp. 36 ff.

[1.9] E.B. Carne, *Modern Telecommunication*, Plenum Publishing, New York, 1985.

[1.10] G. Friesen, "Optical Fiber for Subscribers," *CO Magazine*, March 1986.

[1.11] Bellcore TR-TSY-000303, Issue 1, September 1986.

[1.12] *AT&T Lightguide Digest*, Fall 1986.

[1.13] *Lightwave*, November 1986.

[1.14] *Fiberoptic Marketing Intelligence*, Kessler Publisher, Newport, RI, December 1986.

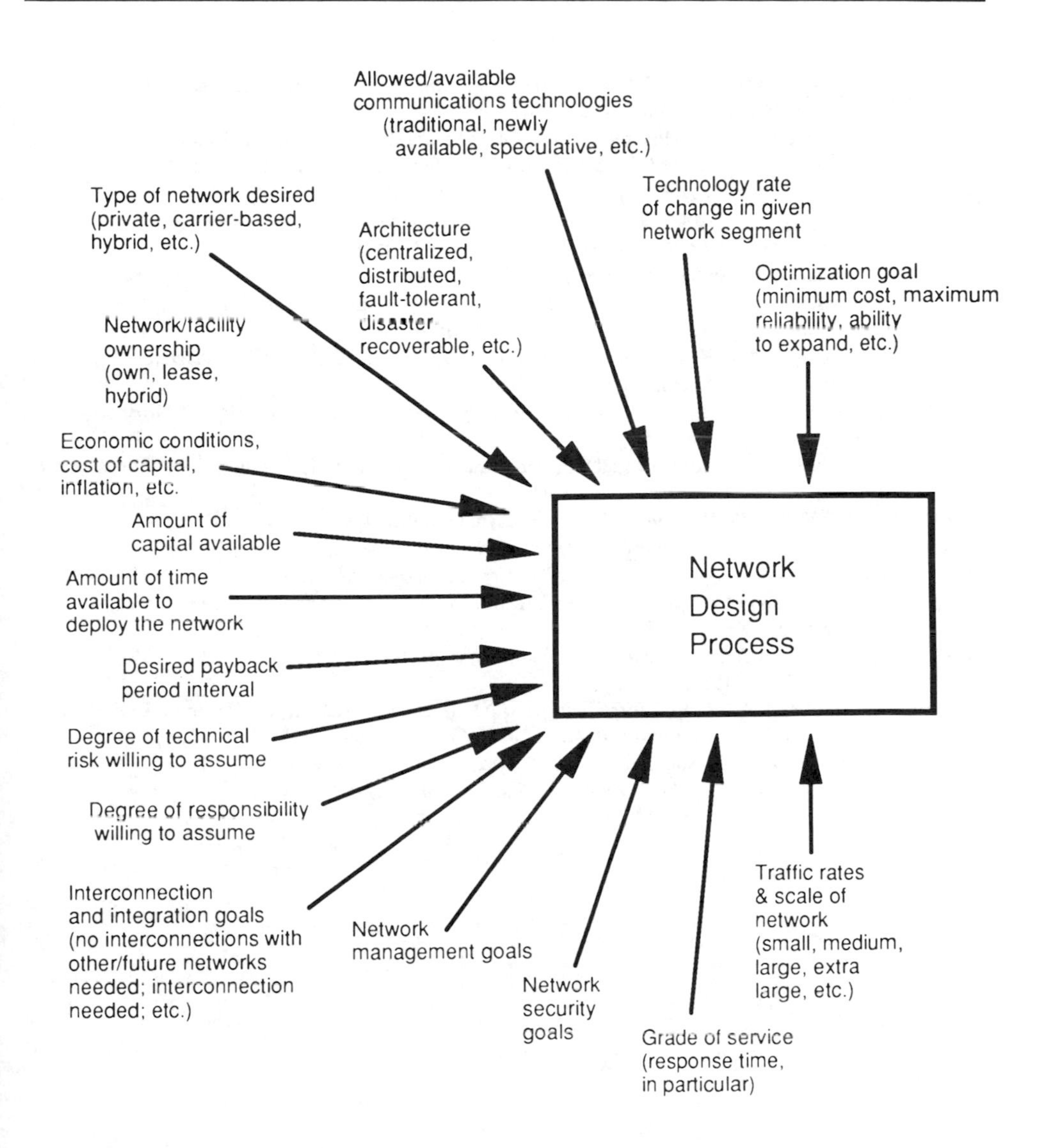

Figure 1.24 Some of the factors affecting the decision-making process.

[1.15] P.C. Richards and J.A. Herndon, "No. 2 ESS: An Electronic Switching System for the Suburban Community," *Bell Labs Record*, May 1973, Vol. 51, pp. 130–135.

[1.16] B.B. Briely, *Introduction to Telephone Switching*, Addison-Wesley, Reading, MA, 1983.

[1.17] *Bellcore News*, January 17, 1990, Vol. 7, No.1.

[1.18] U. Black, *Physical Level Interfaces and Protocols*, IEEE Computer Society Press, New York, 1988.

[1.19] A.M. Michelson and A.H. Levesque, *Error-Control Techniques for Digital Communication*, John Wiley and Sons, New York, 1985.

[1.20] K. Pahlavan and J. K. Holsinger, "Voice-band Data Communications Modems—A Historical Review: 1919–1988," *IEEE Communications Magazine*, January 1988, pp. 16 ff.

[1.21] D. Minoli, "The Future of Modems," *Teleconnect*, August 1986, pp. 51–52.

[1.22] Vertical Systems Group, T1 Network Industry Analysis, Dedham, MA. 1989.

[1.23] *OSI Products and Equipment News*, April 26, 1990, p. 12.

[1.24] *Communication Systems Worldwide*, May 1989, p. 10.

[1.25] J. Massey, "Information Theory: The Copernican System of Communications," *IEEE Communications Magazine*, pp. 26–28, December 1984.

[1.26] I. Csiszar and J. Korner, *Information Theory Coding Theorems for Discrete Memoryless Systems,* Academic Press, Orlando, 1981.

[1.27] D. Wiggert, *Codes for Error Control and Synchronization*, Artech House, Norwood, MA, 1988.

[1.28] J. Madsen, "A Lower Cost on High Performance," *Satellite Communications*, February 1990, p. 27.

[1.29] CCITT I.335, *ISDN Routing Principles*, Blue Book, 1988/89.

[1.30] CCITT I.340, *ISDN Connection Types*, Blue Book, 1988/89.

[1.31] CCITT I.140, *Telecommunication Network and Service Attributes*, Blue Book, 1988–89.

[1.32] "OSI and SNA Compared: The Integration of Network Standards," *Communication Solutions*, San Jose, CA, 1987.

[1.33] *Open Systems Interconnection Model and Notation*, X.200, CCITT Blue Book, Volume VIII, Fascicle VIII.4, 1989.

[1.34] A. Patel and, V. Ryan, "Introduction to Names, Addresses and Routes in an OSI Environment," *Computer Communication*, January–February 1990, Vol. 13, No.1, pp. 27 ff.

[1.35] American National Standard ANSI T1.602.1989, ANSI, New York.

[1.36] *Lightwave*, "Cable TV's Warning to Telcos," August 1989, p. 3.

Chapter 2
Standards-setting Bodies for Telecommunication

2.1 THE NEED FOR STANDARDS

2.1.1 Telecommunication

In the previous chapter, we indicated that many types of equipment from many manufacturers exist that must interwork in order to intelligibly achieve end-to-end communication. Standards are critically needed because they can reduce costs and allow users to choose from among a large selection of products. Industry and government have widely recognized that consensus standards and procedures are needed to preserve the integrity of nationwide telecommunication and facilitate the interconnection and interoperability of carrier services [2.1]. International standardization in the telecommunication field began in the 1860s with the ITU (International Telecommunication Union), and in the electrotechnical field in the 1910s.

Prior to divestiture, AT&T set many standards for compatibility and interoperability of all devices connected to the U.S. national telecommunication network. While these standards may have been for internal use, many became *de facto* standards. The corporate restructuring that resulted from the Second Computer Inquiry and the AT&T divestiture has changed the standards-setting process in the United States, and has initiated a new era of broadly based industry participation. In the new divested environment, not only has a new standards-setting process been required, but, more importantly, the need for standards has become accentuated.

Without standards, confusion would result. For example, manufacturers of equipment would not have uniform physical or electrical performance parameters as guidelines to design their products to meet business needs. Users of nonstandard equipment would have to provide such features as physical communication interfaces, mountings, power requirements, and performance norms to accommodate each manufacturer's design. Test equipment supplied by different manufacturers would not give compatible or comparable results. Connection cables supplied by different manufacturers and designed to their individual standards would lead to a proliferation of designs [2.2].

In the mid-1970s, international standardization activity began to accelerate in response to user demands, advances in technology, and growth in international trade. The 1970s and the early 1980s produced about 70% of the approximately 7500 international standards now on the books [2.3]. Carriers and manufacturers adopt and apply standards on a voluntary basis. Voluntary, consensus standards are becoming an essential ingredient of heterogeneous voice and data networks.

2.1.2 Data Communication

In contrast to the more homogeneous nature of the networks and the equipment suppliers for the traditional telecommunication industry, the data communication industry is supported by a large variety of vendors with revenues ranging from a few million to several billion dollars. More types of equipment are also needed to build a data communication network. A link between a terminal and the computer can involve facilities and equipment from a dozen vendors (without even considering the equipment intrinsic within a carrier's network). A typical financial services company can have as many as 200 suppliers and providers of data communication products and services. *Data Sources*, *Datacom*, 1989 version listed approximately 10,000 data communication products, in about 1000 categories, from about 2000 companies. Products include: network processors, network management systems, LAN systems, modems, multiplexers, emulation and protocol conversion equipment, security equipment, networking software, terminal emulation software, and PC communication utilities, among many others.

The advancement of distributed processing in general, and data communication in particular, is dependent on the capabilities of new sophisticated networks. Distributed data processing is a common trend for large information-based industries. The objective is to place more of the data closer to the users who need it, modify it, or are responsible for it; this approach is endogenous with the strategic business goal of increasing the quality of the service rendered to the ultimate user. The development of standardized interfaces has become essential to prevent the proliferation of unique and incompatible interfaces between end systems and networks, and between end systems.

Automation continues to be a powerful method for reducing costs and, in turn, results in an increased base of terminals and terminal types. Terminals are also becoming more intelligent. PCs and powerful workstations—capable of executing 10–15 million instructions per second (MIPS) and equipped with 80- to

200-Mbyte disk drives—are better suited to advanced office and factory floor applications, which may involve graphics, color, multiwindows, and even animation. These terminals typically involve file transmission and remote procedure calls as the means of communication. Appropriate standards are needed to support interconnection of these terminals. Increased productivity is also achieved through massive networking of previously stand-alone applications. Distributed data processing must be supported by high-connectivity networking, which, in turn, requires that standards be available for internetworking the various systems (or modules of systems), both for real-time links (transaction processing), as well as bulk data transfer links.

To facilitate interconnection of this plethora of multivendor hardware and software, standards must be available. One of the problems typically encountered now is that at least three generations of systems are usually in place: yesterday's systems ($n-1$st generation), present-day equipment (nth generation), and new-technology equipment ($n+1$st generation). This environment mandates the need for standardization.

As of 1990, a number of key data communication upper layer functions are supported by standardized protocols. These functions include: file transfer, electronic mail, electronic directories, common management information exchange for administrative tasks, and virtual terminals. Other functions are well on their way to being standardized, including network management (i.e., specific functions and management information), security, and transaction processing. This standardization has been achieved by arduous work on the part of many standards bodies, and their increased cooperation. Commercial products implementing these standards should become widely available by 1993. Other standards have already reached a level of maturity, such as X.25 and TCP/IP.

The major standards-setting bodies are identified and briefly described in this chapter. Each standards organization has a different set of clients who comprise spheres of influence, and this is the reason why so many standards organizations currently exist. References [2.4], [2.5], and [2.6] are excellent books on this topic, providing additional detailed information.

2.2 TYPES OF STANDARDS

Standards can be grouped in two categories. Regulatory standards are imposed by various regulatory agencies for safety, health, and environmental reasons. Voluntary, consensus standards are employed by users and providers as a market-enhancing method of doing business. The market usually decides the viability of any voluntary standard.

The two types of voluntary standards are *de facto* and *de jure*. The *de facto* standard is a standard that is offered initially by a single provider and subsequently becomes accepted by the market. Eventually a number of vendors may adopt the standard, particularly if the original provider is a dominant vendor. Sometimes, but not always, *de facto* standards have detrimental effects on the overall market. During the 1960s and 1970s, the presence of *de facto* standards forced users into vendor dependencies and limited product sources. A driver of the current focus of standards is the elimination (or at least diminution) of these dependencies through the promotion of interoperability, if nothing else, at the interface level. The abuse of the *de facto* standard is one of the reasons for such preeminent emergence of industry standards today; another force is the European community and the elimination of trade barriers in 1992. Interoperability is one of the key thrusts in Europe, and proprietary *de facto* standards are not generally acceptable there.

The *de jure* standard is not the product of a single provider, but represents the collective consensus of the industry, including users. A *de jure* standard is developed by a body of interested parties and organizations. In the competitive environment of the late 1980s and early 1990s, no single group can operate independently and create standards for the whole industry to follow, including the largest computer or communication companies. If one group of providers creates a standard that it unilaterally deems optimal, but for which no user demand exists, it will be unable to sell the incremental costs of the products, and another provider may make the product or service available at a discount, without the overhead of the standard. Mutual interdependence is the key to success in a consensus standard. A system of checks and balances is built into the standards development process. On the other hand, influential intraindustry user groups (for example MAP/TOP—see Chapter 13) can still drive standards development. However, these groups tend to rely on and apply existing *de jure* standards, rather than developing totally new ones.

2.3 THE STANDARDS DEVELOPMENT CYCLE

In the past, standards were drawn from existing technical solutions; this is somewhat less true today. Many standards are agreed to before they are implemented. Formal standards development can be divided into four phases:

1. Conceptualization;
2. Discussion;
3. Exposition; and
4. Implementation.

After it becomes evident that a standard is needed, the search for a technical solution and a means for developing the standard begin. During this conceptualization phase, a standards-developing organization with the appropriate jurisdiction is identified. An industry call for participation is published. For example, in the

United States, such a notification may be issued in the American National Standards Institute's *ANSI Standards Action*, which is a newsletter to the general industry [2.7].

People from interested manufacturers, carriers, users, special interest groups, *et cetera*, may respond to the invitation, participating on a voluntary basis. Once a committee is formed, the discussions begin. The committee defines its mission, its resources, and its structure. This phase can take from six months to two years, and some committees dissolve without reaching any agreements. Once the discussion phase is complete, the committee can proceed with the business of creating a standard. The standard creation process itself usually takes from one year to several years. Frequent meetings are held at which positions are proposed, debated, amended, or sent back for further technical clarification or study. Eventually a consensus agreement will emerge. Once the standard is technically firm, it is drafted into a document. This may be called a draft (international) standard. The draft standard is then released for public review. The actual mechanics of this review depend on the standards body. In the case of ANSI, the *ANSI Standards Action* announces the public review period; this period can vary from six months to a year or sometimes more, depending in part on how many levels of review are required. Public comments may be incorporated in a revision of the standard that may result from such review. This review process is iterated until a consensus is reached [2.5].

2.4 MAJOR BODIES

A listing of major standards-making bodies follows.

2.4.1 International Telecommunication Union (ITU)

The ITU is now a United Nations charter organization. The ITU is the major international organization for developing global agreements on the use of the radio spectrum, for example, for satellite transmission. For regulatory purposes, the ITU divides the world into three regions:

Region 1: Europe and Africa;
Region 2: North and South America; and
Region 3: The Soviet Union, Asia, and Australia.

The purpose of the ITU, according to its charter, is:

1. To maintain and extend international cooperation for the improvement and rational use of telecommunication of all kinds;
2. To promote the development of technical facilities and their most efficient operation with a view to improve the efficiency of telecommunication services, increase their usefulness, and make them, as far as possible, generally available to the public; and

3. To harmonize the action of nations in the attainment of those ends.

Particular responsibilities include [2.8]:

- Allocating the radio-frequency spectrum and registration of radio-frequency assignments in order to avoid harmful interference between radio stations of different countries;
- Eliminating harmful interference between radio stations of different countries and improvements of the use made of the radio spectrum;
- Harmonizing the development of telecommunication facilities, notably those using space technologies, with a view to full advantage being taken of their possibilities;
- Fostering collaboration among its members with a view to the establishment of rates at levels as low as possible, consistent within an efficient service in taking into account the necessity for maintaining independent financial administration of telecommunication on a sound basis;
- Fostering the creation, development, and improvement of telecommunication equipment and networks in developing countries by every means at its disposal; especially its participation in appropriate programs of the United Nations; and
- Promoting the adoption of measures for ensuring the safety of life to the cooperation of telecommunication services; continuing to undertake studies, make regulations, adopt resolutions, formulate recommendations and opinions, and collect and publish information concerning telecommunication matters.

Because the ITU is a UN organization, only a U.S. government body can vote at the ITU meetings. The Department of State officially represents the U.S. position at these meetings (the U.S. State Department relies, in turn, on ECSA/T1 input, and on other standards bodies discussed below) [2.9]. The ITU structure, function, and authority are governed by the International Telecommunication Constitution and Convention, a multilateral treaty that resulted from the Atlantic City Radio Conference of 1947, and subsequent revisions. These rules specify the rights and obligations of the members. In addition to the ITU Convention, two companion multilateral treaty documents exist that are of relevance: the *ITU Telegraph and Telephone Regulations* and the *ITU Radio Regulations*. These regulations are comprised of substantive technical and procedural rules governing communication. The Radio Regulations, for example, contain specific technical rules governing the use of the spectrum, including the definition of specific radio services, frequency allocation, frequency coordination procedures, and others. The Radio Regulations have been periodically updated since their original adoption in 1947 at World Administrative Radio Conferences (the next of which is to be held in 1992).

The United States is bound to enforce both the general terms of the convention and the technical requirements included in the Radio Regulations. Under

U.S. law, any revisions to the convention or regulations must be ratified by the Senate and signed by the president before they are binding on the United States as a treaty obligation. (Although the FCC abides by the provisions of any revised regulations or convention terms as soon as they are adopted, they do not become legally binding on the United States until entered into force as a treaty obligation) [2.10]. A nation may formulate a reservation to a treaty before ratifying the treaty and will thereby exempt itself from being bound by a specific clause in question. The United States has exercised this option in a number of instances related to specific provisions of the Radio Regulations that it finds incompatible with U.S. interests.

CCITT and its sister organization CCIR (Consultative Committee on International Radio) are subentities of the ITU. CCITT and CCIR are permanent operating units that develop standards for the ITU. The duties of CCITT are "to study technical and operating questions relating to telegraphy or telephony"; the duties of CCIR are "to study technical and operating questions pertaining to radiocommunications." Although the standards developed by CCITT and CCIR are not binding as treaty obligations, they have generally been accepted and used by the ITU [2.10]. The recommendations and standards developed by these technical committees are regularly published by the ITU.

Plenipotentiary Conferences are the supreme managing bodies of the ITU and are the only bodies empowered to adopt revisions to the ITU convention. In the interval between Plenipotentiary Conferences, ITU policy-making is conducted by administrative conferences, which are held periodically to consider specific telecommunication matters. In addition to the World Administrative Radio Conferences (WARCs), the ITU authorizes Regional Administrative Radio Conferences (RARCs). RARCs may be organized to address communication issues of a specific geographic region. For example, in 1983, a RARC was held for deployment of direct broadcast satellites in region 2 (North and South America). A WARC may adopt changes to ITU regulations; a RARC is not empowered to adopt any revised regulation but may only recommend that such changes be adopted by a WARC [2.10].

2.4.2 The CCITT

The CCITT (Consultative Committee on International Telephone and Telegraph or, more precisely, Comité Consultatif International Télègraphique et Télèphonique) is an association of international carriers, mostly government owned, as well as vendors. It generates functional and electrical recommendations for telecommunication and data communication. Countries are not bound legally by an international treaty to apply CCITT recommendations, but they are encouraged to do so. CCITT is one of seven ITU organizations. CCITT maintains liaison with carriers, ISO, and ANSI, among others. Many standards are harmonized with ISO. Most communication systems in the world operate through a quasigovernment monopoly such as a PTT (Postal, Telephone, and Telegraph Administration), and typically it is the PTT who participates in the CCITT to represent its own national telecommunication industry's interests. Generally, each CCITT study group has three plenary meetings that involve the entire study group within the four-year study period between plenary assemblies. Plenary meetings take place at the beginning, middle, and end of a study period. Every CCITT study group is subdivided into working parties, each led by a chairman and responsible for a subset of related items selected from the questions for study assigned to the study group by the plenary assembly. Often, two or more working parties meet simultaneously. The working parties may be further subdivided into subworking parties and "rapporteur" groups to study specific topics. Some of the key study groups are: Group II (telephone operation and quality of service), Study Group VII (data communication networks), Group XI (telephone switching and signaling), Group XVII (data communication over the telephone network), and Group XVIII (digital networks). There were 700 CCITT standards in 1968; that number has grown to more than 1700 in 1988, and the rate of growth is increasing [2.4, 2.11].

The two classes of CCITT-studied subscriber services are [2.12]:

1. End-to-end services, where CCITT provides standards (corresponding to the seven layers of the OSIRM, discussed in Chapter 1 and more in Chapter 13). This encompasses the networks supporting the services and the terminals at both ends (i.e., the public switched telephone service and messaging services).
2. Interface services, where CCITT standardizes the basic information transport functions (corresponding to the first three layers of the OSIRM). Examples include public (packet) switched data networks, leased lines, modem interfaces, and ISDN. In this case CCITT leaves the definition of the terminal to the customer.

End-to-End Services

The following series of recommendations applies to public switched telephone networks: Q Series pertains to signaling; M and O Series pertain to maintenance; P Series applies to telephone transmission performance; E Series pertains to service features and operating rules; D Series applies to tariff principle and international accounting. Other series include G and Z. Each series typically is associated with a specific study group. Some additional information on these series follows.

D Series. Includes tariff principles, rules on how to establish accounting procedures for international calls, how to undertake metering, and how to devise reduced rates during periods of low traffic (rate fixing is not a CCITT function).

E Series. Includes service definitions for the telephone service, international numbering plan, routing arrangements, and traffic engineering.

F Series. Includes recommendations on operations and quality of service for telegraph, mobile communication, data transmission, and teleconferencing services.

G Series. Includes recommendations on digital transmission systems, coaxial cables, and fiber optic systems.

I Series. Covers ISDN network and broadband aspects of ISDN.

M and O Series. Describe maintenance of transmission systems and switching systems, including the specification of measurement equipment.

P Series. Covers transmission quality, sending reference, and receiving reference equivalence of station sets.

Q Series. Describes telephone signaling systems (principally #4, #5, #6, and #7).

Z Series. Describes, among others, a (1) specification and description language (a formalized method of presenting a functional specification for software required in SPC) and (2) a human-machine interface for installing, testing, operating, and maintaining SPC switches.

Interfaces

These specifications are designed for customers who own CPE devices and only require network transport services. The X Series of standards applies to public data networks and services. X.1 through X.39 refer to user interface standards; X.40 through X.59 apply to transmission over an international circuit; X.60 through X.80 represent switch standards; X.81 through X.180 represent various numbering plan and performance characteristics specs (the upper numbers represent protocol family boundaries—standards corresponding to these numbers may not exist yet). Some specific examples of interface standards are: X.21 "Interface between data terminal equipment and data circuit-terminating equipment for synchronous operation over a public data network"; X.25 "Interface between data terminal equipment and data circuit-terminating equipment for terminals operating in the packet mode in public data networks"; X.75 "Terminal and transit call control procedures and data transfer system on international circuits between packet-switched data networks." Table 2.1 shows some of the key CCITT standardization activities in both areas over the years.

Table 2.1
Key Activities of CCITT Over the Years

CCITT Books	*Date*	*Major items*
Red Book	1960	Emerging specifications for coaxial cables and related FDM carrier system
Blue Book	1964	Small capacity FDM carrier systems; carrier systems for submarine cables
White Book	1968	Small and medium capacity FDM carrier systems
Green Book	1972	Emerging digital transmission technology; digital encoding of analog signals; specifications for Common Channel Signaling System 6 (CCSS6)
Orange Book	1976	Pulse code modulation (PCM) systems and multiplexers; interworking between CCITT's CCSS6 and national systems; packet switching interface X.25
Yellow Book	1980	PCM frame structures for use with digital COs; Signaling over PCM links; X.25 revisions
Red Book	1984	Message handling system (MHS); integrated serviced digital network (ISDN); Common Channel Signaling System 7 (CCSS7)
Blue Book	1988	Refinements to MHS; Abstract Syntax Notation One; telecommunication management network, directory, office document architecture; broadband ISDN (B-ISDN)
White Book	1992	Broadband ISDN

Accelerated Procedures

Until 1988, CCITT published standards at four-year intervals. At the November 1988 Plenary Assembly in Melbourne, Australia, attendees decided to emphasize accelerated procedures for the future. The Report of Committee A, Fascicle 1.1, states that "a general consensus has been reached that the four-year interval is too long and it has been suggested that the CCITT look at the working methods of other international organizations." A revised Resolution 2, adopted in Melbourne, describes the expedited approach. While the accelerated procedures were available in the past, the standards, even if available early, were not officially sanctioned until the end of the four-year study period. The new process would allow the official publication of standards as soon as they are available.

The United States Organization for the International Telegraph and Telephone Consultative Committee (U.S. CCITT)

The participation of the United States in the work of the CCITT is channeled through the national preparatory organization officially known as "The United

States Organization for the International Telegraph and Telephone Consultative Committee." This "U.S. CCITT," as it is popularly called, is headed by the Office of International Communications Policy, Bureau of Economic and Business Affairs, U.S. Department of State. The reason for this form of representation is because full voting membership in the United Nations-related ITU, of which the CCITT is a permanent organization, is composed of national governments, who, alone, may assume international treaty obligations [2.13].

The U.S. CCITT Charter of 1977, in delineating the purposes of the organization, states that the U.S. CCITT will:

1. Promote the best interests of the United States in CCITT activities;
2. Provide advice to the Department of State on matters of policy and positions in preparation for CCITT plenary assemblies and meetings of the international CCITT study groups;
3. Provide advice to the Department of State on the disposition of proposed contributions (documents) to the international CCITT; and
4. Assist in the resolution of administrative and procedural problems pertaining to U.S. CCITT activities.

The U.S. CCITT also provides the pool of informed public and private-sector personnel that is drawn on to staff official U.S. delegations to the international CCITT study group meetings and plenary assemblies. Broad-based participation in the activities of the U.S. CCITT is encouraged by the Department of State. Such participation gives delegates to international CCITT meetings a clearer understanding of the U.S. position on specific telecommunication issues. This broad participation is of particular importance in the formulation of positions because the United States is one of the very few nations in which the responsibility for and provision of telecommunication services are in the hands of common carriers and other private industries, with only a limited degree of governmental control and regulation. The National Committee of the U.S. CCITT constitutes a steering body and has purview over the agenda and nontechnical activity of the study groups. The major work of the U.S. study groups is the review of proposed contributions from the United States to international CCITT study group meetings. The "working paper" contributions that are reviewed by the U.S. CCITT study groups become part of the preparatory work done in the international CCITT study groups. The CCITT discussions are based on the technical contributions received from the member nations. These contributions are the materials from which the CCITT recommendations are eventually drafted [2.13]. Figure 2.1 shows the CCITT contribution approval process for U.S. contributions [2.14].

2.4.3 Consultative Committee on International Radio (CCIR)

The CCIR is the permanent organization of the ITU for the "study of technical and operating questions in radiocommunications and to issue recommendations"

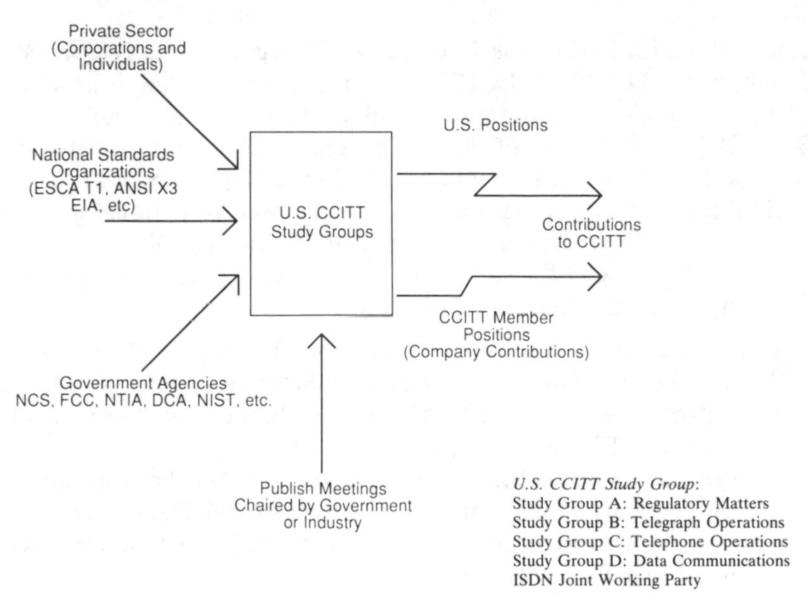

Figure 2.1 U.S. contributions to CCITT.

[2.15]. CCIR has recently been involved with issues pertaining to direct broadcast satellites, and even more recently with HDTV. CCIR was initially established in 1927 to study the questions that arose in preparation for international radio conferences on the allocation and regulation of the frequency spectrum. The structure of the CCIR was modified in 1947. Since then, CCIR activities have focused on two areas: (1) the technical aspects of radio spectrum use and (2) performance criteria and system characteristics for compatible interworking. Areas of coverage include spectrum utilization and monitoring, propagation in nonionized media, ionospheric propagation, space research and radioastronomy, fixed satellite services, mobile satellite services, sound broadcasting, and television broadcasting. About 200 CCIR standards existed in 1968; that number has grown slowly to about 300 in 1988 [2.11]. The importance of CCIR has declined relative to CCITT. To some extent, this can be attributed to different rates of technological changes in the two areas of activity. Much radio technology is relatively stable. In the view of some, however, the CCIR is poorly structured to come up with timely and relevant recommendations, which is a shibboleth for any standards-generating body. Some areas of technical movement include mobile communication and HDTV.

CCIR recommendations and reports provide the main technical basis for holding an ITU administrative radio conference to allocate and regulate the use of frequency bands. The results are used for the international coordination and planning of frequency assignments. System characteristics are provided for most domains of radio communication especially in the area of international connections (sound and television broadcasting), maritime and land mobile services, satellite and microwave links, and standard time and frequency.

2.4.4 International Organization for Standardization (ISO)

The ISO (the familiar acronym derives from its former name, the International Standards Organization) is an international organization dedicated to the writing and dissemination of technical standards for industry and trade. All member nations provide input to ISO; however, it is not a treaty organization and therefore does not require official governmental representation and approval of contributions. The activity of ISO started in 1946. The scope of ISO includes areas such as agriculture, petrochemicals, minerals, and clean air and environment, among numerous others. Of interest in this context is the Information Processing Systems study groups. ISO operates on a voluntary basis. Experts from the producer, consumer, governmental, and general interest communities contribute to national positions. National positions are represented by a member body, which is a given country's representative in a specified field. Although ISO is a nongovernmental organization, more than 70% of ISO member bodies are governmental standards institutions or organizations incorporated by public law. The U.S. member body is ANSI. Almost 400 international organizations have formal liaisons with ISO, including the IEC, CCITT, and all UN specialized agencies working in similar fields. The central secretariat of ISO is located in Geneva, Switzerland. The general assembly, which meets every three years, makes the basic decisions. The council meets yearly and administers the ISO operation. The technical work of ISO is undertaken by technical committees (TCs) and their working subcommittees (SCs) and working groups (WGs), all subject to the general auspices of the council. Every member body may choose to be represented on any TC or SC. Member bodies may take an active role in the committee work (designated as P for participant), or an observatory role (designated O).

The most important technical committee relevant to telecommunication is ISO/IEC JTC 1 (a joint organization of the former ISO TC97 and the IEC, discussed later). ISO/IEC JTC 1 is responsible for one of the most dynamic areas in international standardization. Its work takes place in an environment that is constantly changing and expanding in scope and criticality. Other relevant committees are: TC46 (automation and library science), TC68 (banking procedures), TC154 (documents and data elements in administration, commerce, and industry), TC159 (ergonomics), and TC184 (information processing systems as related to industrial automation). TC97 was formed in 1961 with ANSI as its secretariat. This committee dealt with computers and associated systems, peripherals, and media. In 1981, TC97 was merged with TC95, because office machine technology was becoming closely related to computer technology. Since 1984, TC97 has restructured itself into three groupings of subcommittees, as follows:

1. Applications Elements (SC1, SC7, and SC14);
2. Equipment and Media (SC10, SC11, SC13, SC15, SC17, SC19, and SC23); and
3. Systems (SC2, SC6, SC18, SC20, SC21, and SC22).

One of the more well-known accomplishments of ISO in this arena is the standardization of the OSIRM described in Chapter 1.

The development of an ISO international standard from the first proposal of the idea, to the publication of the standard involves a number of steps [2.3, 2.5]:

1. The new work item is included in the program work of an appropriate TC. The initial document ("working draft") is circulated among appropriate TCs, SCs, or WGs, in an effort to write a *draft proposal* (DP). This process usually takes several months. The DP must achieve a substantial level of support from the participating members of the technical committee.
2. After checking for conformity with ISO directives, the secretariat registers the DP as a *draft international standard* (DIS).
3. The DIS must be approved by member bodies of the ISO within six months. The DIS must receive a majority approval by the technical committee members, and 75% of all voting members.
4. The approved DIS and revisions are accepted by the ISO Council as an *international standard* (IS).
5. The IS is published by ISO. The member bodies are responsible for distribution of the ISO standards within their own countries. In the United States, the *American National Standards Institute* (ANSI) is the source of these publications.

One of the challenges of ISO is to move through these steps with technical expertise and in a timely manner.

2.4.5 American National Standards Institute (ANSI)

ANSI is not a standards-developing organization. It neither develops standards nor undertakes the technical work required to create a standard. Instead, it develops and maintains the consensus process for all U.S. standards-developing organizations by accrediting such organizations. ANSI represents the U.S. government in international standards negotiations, at ISO and at the International Electrotechnical Commission (IEC), in particular. ANSI is primarily concerned with functional and procedural standards. Currently more than 250 standards groups and 1000 companies belong to ANSI. When a standards organization is needed for a new work item, ANSI publishes a request for a group willing to sponsor a secretariat. The secretariat is the administrative function of the standards-developing organization, providing support, legal, and financial assistance. The

Computer and Business Equipment Manufacturers Association (CBEMA) supports the secretariat of X3, the primary standards-developing organization for information processing in the United States. The Exchange Carriers Standards Association is the secretariat for T1, the primary telecommunication standards organization in the United States. Both of these entities are accredited standards committees—bodies with recognized interest in creating standards in defined areas of expertise.

X3 undertakes standardization in the areas of computers and information processing and peripheral equipment, devices, and media. In addition, it provides standardization of the functional characteristics of office machines, particularly in the areas that influence the operators of such machines. X3 was created in 1960. The organization was revised three times (1969, 1980, and 1988) to improve its ability to meet current needs. X3 is responsible to the CBEMA for its finances and to ANSI for its technical standards activities. X3 includes about 40 members, which provide more than 3200 industry volunteers to support the standards-making activities. The actual work of X3 is done by its technical committees. These committees fall in seven groups:

X3A: Recognition;
X3B: Media;
X3H&J: Languages;
X3K: Documentation;
X3L: Data representation;
X3S: Communications; and
X3T&V: Systems technology.

The technical committees are further identified by a qualifier. For example, X3T1 deals with data encryption, X3T5 with OSI, and X3T9 with input-output interfaces.

ANSI had published more than 8000 standards by the end of 1989. It provides the machinery for creating voluntary standards. It serves to eliminate duplication of standards activities and to weld conflicting standards into single, nationally accepted standards under the designation "American national standard." Each standard represents general agreement among maker, seller, and user groups as to the best current practice with regard to some specific problem. Thus, the completed standards cut across the whole fabric of production, distribution, and consumption of goods and services. American national standards are used widely by industry and commerce and often by municipal, state, and federal governments.

ANSI, under whose auspices this work is being done, is the U.S. clearinghouse and coordinating body for voluntary standards activity on the national level. It is a federation of trade associations, technical societies, professional groups, and consumer organizations. More than 1000 companies are affiliated with the institute as company members. Through the ANSI channels, U.S. standards interests make their positions felt on the international level.

2.4.6 Exchange Carriers Standards Association (ECSA)

The ECSA is comprised of the telephone carriers in the United States. It was formed at the time of divestiture as a forum for the establishment of standards for the operation of the telephone system in the United States. The ECSA is both a trade association of exchange carriers and the sponsor of an independent standards committee, T-1. The name T-1, not to be confused with the well-known T1 digital transmission system (discussed in Chapter 3), is attributed to ANSI's coding of standards committees: T for telecommunication and 1 for the first such ANSI activity [2.1]. ECSA arose out of an FCC request for public comments regarding the vehicle for post-divestiture standards making. In August 1983, ECSA recommended to the FCC that a public standards committee be established that would be open in membership and use the same ANSI procedures used by many other groups; ECSA advised the FCC that it was prepared to sponsor an ANSI-affiliated committee and provide secretariat and administrative functions in support of the committee's work. In March 1985, the FCC formally endorsed T-1 as the vehicle for post-divestiture standards making [2.16]. Incorporated as a nonprofit organization in 1983, the ECSA is an association established voluntarily by members of the wireline exchange carrier industry to address exchange access interconnection standards and other technical issues.

T-1 provides a public forum for developing interconnection standards for the national telecommunication system. Before January 1, 1984, AT&T as the major telecommunication carrier in the United States set many *de facto* standards. These standards were shared with the independent telephone companies and manufacturers in various Bell System–United States Telephone Association (USTA) forums. Divestiture created a need to develop industry-wide telecommunication standards. The responsibility for the secretariat of T-1 was given to ECSA. Although T-1 is primarily concerned with domestic matters, the committee also provides a national position to the U.S. Department of State for international purposes. Figure 2.2 shows the activity flow of the T-1 committee.

The T-1 committee has a number of subcommittees. T1E1 covers carrier-to-customer installations (digital interfaces, ISDN primary rate access—T1E1.4); T1M1 covers internetwork operations, administration, maintenance, and provisioning (interfaces between network monitoring devices and related operation systems); T1Q1 deals with performance (recent work has been in television transmission issues); T1S1 deals with services, architectures, and signaling (particularly for ISDN and data communication networks—T1S1.1, T1S1.2, T1S1.3, T1S1.4, and T1S1.5); T1X1 covers digital hierarchy and synchronization (particularly for new optical networks); T1Y1 deals with specialized subjects (two recent topics: ADPCM algorithms and voice compression from PCM to ADPCM—a process known as "transcoding"). See Table 2.2 for some key working groups.

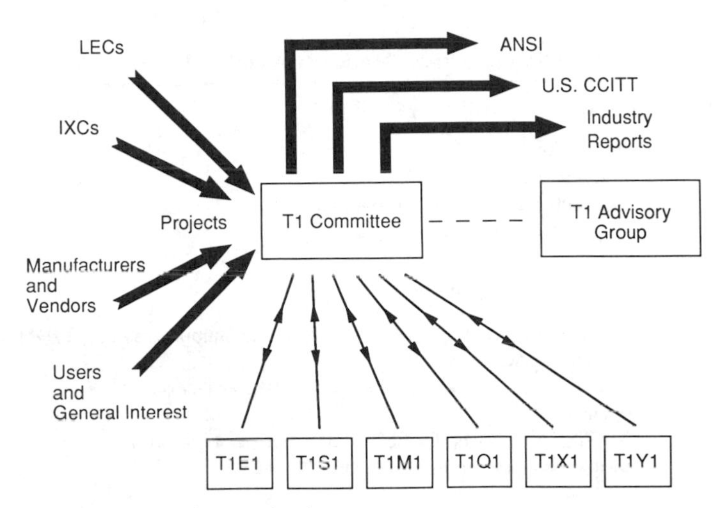

Figure 2.2 The T-1 committee project mechanism.

To maintain balance within T-1, members are grouped into four categories: exchange carriers, interexchange carriers and telecommunication resellers, manufacturers and vendors, and users. Currently, T-1 has more than 90 voting members and 75 observers. T-1 is one of the fastest growing accredited standards committees in the United States. Membership includes exchange carriers, interexchange carriers, resellers, manufacturers, vendors, government agencies, user groups, consultants, and liaisons.

2.4.7 Electronics Industries Association

The Electronics Industries Association (EIA) represents manufacturers and is concerned primarily with physical and electrical standards and interfaces. It submits international contributions through ANSI. Its more well-known efforts have been in the RS-232-C interface standard (first issued in the early 1960s through the TR-30 committee), and more recently in the EIA-232-D and EIA-449 interfaces (also known as RS-232-D and RS-449, respectively). The interface standards group, TR-30, coordinates its effort through ANSI X3S3.

2.4.8 Institute of Electrical and Electronics Engineers (IEEE)

The IEEE is a professional organization of electrical engineers. Through the years, the IEEE has produced a number of standards. Their standards activities deal mostly with LANs and interface of measurement instrumentation. IEEE is one of the largest professional organizations in electrical and electronics engineering; it

Table 2.2
Key ECSA Committees, Working Groups, and Work Items

T1E1 Network Interfaces	
T1E1.1	Analog access
T1E1.2	Wideband access
T1E1.3	Connector and wiring arrangements
T1E1.4	Digital subscriber loop access
T1S1 Services, Architectures and Signaling	
T1S1.1	Internetworking and rate adaption, packet-mode services, ISDN networking, intelligent networks
T1S1.2	Circuit-mode, packet-mode, supplementary services, management and maintenance
T1S1.3	Non-ISDN services, CCSS7, subscriber-network interfaces
T1S1.4	Switched access, numbering, broadband
T1M1 Operations Technology Standards	
T1M1.1	Coding standards, exchange carrier names, credit card issues
T1M1.2	Maintenance of CCSS7 networks, fiber optic maintenance and SONET, tones and announcements
T1M1.3	In-service measurements, digital measurements, analog measurement standards
T1M1.5	Operations systems-to-network element protocols, upper layers, functions, lower layers, user system language
T1Q1 Performance	
T1Q1.1	Voice and voiceband data
T1Q1.3	Digital packet and ISDNs
T1Q1.4	Digital circuit
T1Q1.5	Television and program audio
T1X1 Digital Hierarchy and Synchronization	
T1X1.3	Synchronization interfaces
T1X1.4	Metallic hierarchical interfaces administration
T1X1.5	Optical hierarchical interfaces
T1X1.6	Tributary analysis interfaces
T1Y1 Specialized Subjects	
T1Y1.1	Specialized video and audio services
T1Y1.2	Specialized voice and data processing
T1Y1.4	Environmental standards for exchange and interexchange carrier networks

develops standards that relate to those disciplines. IEEE taps the expertise of its membership to undertake standards development efforts. The Computer Society produces the majority of the information systems standards, and began work in 1980. The IEEE Communications Society deals with the development of communication-related standards. The Communications Society has 17 technical committees. Some of the more active committees in recent years have been the following: Data Communications Systems Committee, Transmission Systems Committee, Telecommunications Switching Committee, and the Computer Communication Committee.

The IEEE groups coordinate their activities with X3 and T1. IEEE is an ANSI-accredited standards-developing organization. ANSI procedures are followed closely, and the IEEE standards are submitted to ANSI for consideration as national standards. In particular, Project IEEE 802 was responsible for developing what are now well-known and well-established LAN and MAN standards (see Chapters 14 and 15).

2.4.9 International Electrotechnical Commission (IEC)

The IEC is concerned with user protection from undesirable side effects of radiation and electrical shock. National committees represent their country's electrical and electronics interests; the committees are composed of manufacturers, users, trade associations, government bodies, scholars, and engineers. The U.S. committee is closely related to the EIA. The organization of IEC dates back to the turn of the century. Forty-three countries now hold membership in IEC, represented by their national committees. More than 200 technical committees, subcommittees, and working groups exist. Typical work includes electronic components, test methods, electrical apparatus, wire, cable and connectors, measuring instruments, and even terminology and graphic symbols.

The IEC/ISO JTC 1 (Joint Technical Committee 1) is the first joint technical committee formed by two principal international standards organizations: the IEC and ISO. It was established in 1987. The scope of JTC 1 is: "standardization in the field of information technology." Its original components were ISO TC97 (information technology and all its subcommittees), IEC TC83 (information technology equipment), and IEC SC47B (microprocessor systems). Membership in JTC 1 is comprised of national bodies, which are countries that are eligible for membership through their existing membership in ISO or as national committees of IEC.

2.4.10 Computer and Business Equipment Manufacturers Association (CBEMA)

CBEMA, while not generating standards, sponsors standards activities on behalf of its members and provides legal, secretarial, and clerical services for standards agencies and the representatives of its member companies.

2.4.11 National Institute of Standards and Technology (NIST)

NIST (formerly the National Bureau of Standards) has a unique responsibility for computer systems technology within the federal government because it publishes Federal Information Processing Standards (FIPS) (which may be copies of ANSI standards). It maintains liaison with ANSI, EIA, and government agencies. It is based in Gaithersburg, Maryland. NIST sponsors many workshops, including the OSI Implementors Workshop, which provides a forum for continued work on implementation agreements for a variety of OSI special interest groups (SIGs), discussed below.

NIST was established (under its former name) by an act of Congress on March 3, 1901. The institute's overall goal is to strengthen and advance the nation's science and technology and to facilitate their effective application for public benefit. To this end, the institute conducts research to assure international competitiveness and leadership for U.S. industry, science, and technology, in support of continually improving U.S. productivity, product quality and reliability, innovation, and the underlying science and engineering. The institute's technical work is performed by the National Measurement Laboratory, the National Engineering Laboratory, the National Computer Systems Laboratory (NCSL), and the Institute for Materials Science and Engineering.

The NCSL develops standards and guidelines, provides technical assistance, and conducts research for computers and related telecommunication systems to achieve more effective utilization of federal information technology resources. NCSL's responsibilities include development of technical, management, physical, and administrative standards and guidelines for the cost-effective security and privacy of sensitive unclassified information processed in federal computers. NCSL assists agencies in developing security plans and in improving computer security awareness training.

The NCSL conducts research and provides scientific and technical services to aid federal agencies in the selection, acquisition, application, and use of computer technology to improve effectiveness and economy in government operations in accordance with Public Law 89-306 (40 U.S.C. 759), relevant executive orders, and other directives. NCSL carries out this mission by managing the FIPS program, developing federal ADP standards guidelines, and managing federal participation in ADP voluntary standardization activities. It also provides scientific and technological advisory services and assistance to federal agencies and provides the technical foundation for the computer-related policies of the federal government [2.17]. The laboratory consists of the following divisions:

- Information Systems Engineering;
- Systems and Software Technology;
- Computer Security;
- Systems and Network Architecture; and
- Advanced Systems.

NIST Workshop for Implementors of OSI

In February 1983, at the request of industry, NIST organized the NIST Workshop for Implementors of OSI to bring together future users and potential suppliers of OSI protocols. The workshop accepts as input the specifications of emerging standards for protocols and produces as output agreements on the implementation and testing particulars of these protocols. This process is expected to expedite the development of OSI protocols and promote interoperability of independently manufactured data communication equipment [2.17].

The workshop organizes its work through SIGs that prepare technical documentation. An executive committee of SIG chairpersons, led by the overall workshop chairperson, administers the workshop. NIST invites qualified technical leaders from participating organizations to assume leadership roles in the SIGs. The SIGs are encouraged to coordinate with standards organizations and user groups, and to seek widespread technical consensus on implementation agreements through international discussions and liaison activities.

The workshop meets four times a year and each SIG is required to convene its meeting. In addition, a plenary assembly of all workshop delegates is convened for consideration of SIG motions and other workshop business.

The workshop is an open public forum. The workshops are held for those organizations expressing an interest in implementing or procuring OSI protocols and open systems. However, no corporate commitment regarding the implementation of proposals associated with workshop participation is needed. Areas of interest to the OSI workshop are shown in Table 2.3.

2.4.12 European Standards-setting Bodies

This section provides a brief survey of European standards bodies. Key bodies include [2.18]: Comité Europeén de Normalisation (CEN); Comité Europeén de

Table 2.3
Areas Covered by OSI Workshop

Areas
Local Area Networks (LANs)
Wide Area Networks (WANs)
Integrated Services Digital Networks (ISDN)
Network Layer Protocols
Transport Layer Protocols
Session Layer Protocols
Abstract Syntax Notation One Encoding Rules
X.400
File Transfer, Access, and Management Phase 2
Directory Services Protocols
Virtual Terminal Protocol
Transaction Processing
Office Document Architecture

Normalisation Élèctrotechnique (CENELEC), the Conference Europeén des Administrations des Postes et des Télècommunications (CEPT); the Normés Européénes Télècommunications (NETs); the Standards Promotion and Application Group (SPAG); the European Workshop on Open Systems (EWOS); the European Computer Manufacturers Association (ECMA), and the newly instituted European Telecommunications Standards Institute (ETSI) [2.19].

2.4.12.1 Conference Europeén des Administrations des Postes et des Télècommunications (CEPT)

CEPT brings together the PTT administrations (not the governments) of 26 European countries, including Yugoslavia, but, as of 1990, excluding the Eastern European countries. CEPT, which is independent from any political or economic organization, was established in 1959 through the determination of PTT administrations who were aware of the need to tighten and institutionalize their links. The effort was aimed at harmonizing and improving postal and telecommunication relations among European countries in order to form a homogeneous, coherent, and efficient unit on a continental scale. The CEPT has two committees: the Post Committee and the Telecommunications Committee. The activities of the Telecommunications Committee concern the operation and planning of intercontinental links in the North Atlantic, policy concerning the use of regional satellites, new services, and new tariffs, among others.

CEPT is the sister organization to CEN-CENELEC, and participates in relevant aspects of the CEN-CENELEC program. CEPT has a number of working groups on subjects including terminal equipment signaling, protocols and switching, and network aspects, and ISDN. CEPT is based in Berne, Switzerland. CEPT is not a standardization organization *per se*, but does set standards. Its traditional purpose has been to harmonize and standardize services within Europe (for example, to define interfaces between national networks). All the administrations that belong to it are members of ITU, and partake of the CCITT and CCIR standards work. CEPT intervenes by consulting within specialized work groups to delineate a proposal that will be acceptable to everyone.

CEPT was originally responsible for the standards program that produces the European telecommunication standards, called "norms" or NETs. The NETs program is managed by the CEPT Technical Recommendations Applications Committee. The technical work for the NETs program is in the process of being transferred to ETSI, discussed below. NETs 1–10 are termed "access" NETs and specify the characteristics of terminal equipment at a network interface. Directive 86/361/EEC, which relates to telecommunication in the European community, effectively gave rise to the NETs program, with the objective of providing technical standardization in telecommunication to support a unified European market [2.18].

2.4.12.2 European Telecommunications Standards Institute

The ETSI is replacing CEPT as the regional standards body (CEPT will continue to exist, but its role will diminish). ETSI was formed by the CEPT in 1988 to implement the European community's policy of separation between the operational and regulatory functions in telecommunication. The ETSI will undertake the technical standards-writing activities of CEPT; ETSI is more accessible to non-PTT suppliers and end-users. ETSI is an autonomous body within CEPT and has a permanent secretariat located in Sophia Antipolis, France.

ETSI is governed by a general assembly, while the technical work is managed by the technical assembly and the detail drafting work is undertaken by technical committees that have a similar structure and membership to the CEPT working groups. Membership in ETSI is open and in late 1989, ETSI had 135 member organizations including national administrations, public network operators, manufacturers, users, private service providers, and research bodies. Seven initial project teams were as follows [2.20]:

PT1: ISDN basic access;
PT2: PSTN basic access;
PT3: terminal adaptor;
PT4: modems;
PT5: Facsimile Group 4 on ISDN;
PT6: X.32 access; and
PT7: PBX interconnection.

Other ETSI project teams scheduled for 1990 and beyond include:

PT8: measurement for equipment for private mobile radio service;
PT9: ISDN supplementary services;
PT10: digital cordless telephone; and
PT11: digital paging systems.

ETSI covers both private and public network standards, as the list above reveals. (Historically, private network standards have originated with ECMA and been approved by CENELEC; ETSI will now fill that role.)

2.4.12.3 CEN and CENELEC

CEN and CENELEC are the two official European standards organizations responsible for standardization in the field of information technology. CENELEC is responsible for standardization in the electrotechnical field, while CEN covers other subjects. CEN and CENELEC are, effectively, the European subsets of the members of ISO and the IEC, respectively. One of their main objectives is the preparation of standards that can be referenced in European directives. While in other areas of work they function separately, in the area of information technology, CEN and CENELEC have a common interest and have combined as the Joint European Standards Institution. The central secretariat of CEN-CENELEC is located in Brussels, Belgium, and their membership is drawn from the European national standards organizations [2.18].

Although a set of protocols may be available, the series of protocols applicable to a particular function must be further defined and then any options that may exist in those protocols must be specified. This definition is called a "functional standard," "functional profile," or "protocol stack," and is a definition that can be implemented as a product. Coordination of profile development is important because the profiles are being designed by regional groups, and small differences will lead to an inability to interwork. The main goal of CEN-CENELEC's work is the production of functional standards for OSI and related communication technologies. Current areas of work include OSI layers 1 through 4, message handling systems, and file transfer.

The CEN-CENELEC output documents are European standards, ENs (European Norms). In draft form, ENs have the designation ENV's and are described as "European Prestandards"; the objective is to provide an opportunity for industry and users to plan for their adoption and implementation. CEN-CENELEC intends for the ENVs to be converted into ENs within a two-year time frame. CEN-CENELEC retains the procedural and legal framework for the approval and publication of the ENs [2.18].

2.4.12.4 European Computer Manufacturers Association

The ECMA is an association of manufacturers resembling CBEMA, EIA, and ANSI. It provides input to ISO, and also publishes its own standards. ECMA is not a trade organization (as the name might imply) but a technical and standards review group. Its participation includes several North American companies. ECMA was established in 1961. ECMA works rapidly to produce voluntary standards and technical reports. The ECMA secretariat is in Geneva, Switzerland. ECMA is a liaison member of ISO/IEC JTC1. The ECMA technical committee responsible for OSI is TC 32 (communications, networks, and systems interconnection), which was formed in 1985. TC32 is responsible for coordinating the ECMA view for input into ISO and CCITT. TC32 is concerned with interface standards for connection of equipment to private switched networks, and has close liaisons with CCITT, ISO, IEC, CEPT, and CEN/CENELEC. TC32 has several specialist task groups on the following subjects [2.18]:

TG2: distributed interactive processing;
TG4: OSI management;
TG5: distributed office applications;

TG6: private switching networks (PSN);
TG9: security in open systems; and
TG10: OSI lower four layers and LANs.

2.4.13 Other Standards Organizations

This section identifies a few of the hundreds of other organizations involved in the standards process. While these organizations do not conform to the rigid procedural format of the agencies discussed above, their inputs are nonetheless important in some key areas of communication.

2.4.13.1 Bell Communications Research, Inc. (Bellcore)

An August 1982 court judgment, to which AT&T and the Department of Justice consented, resulted in the termination of a seven-year antitrust suit against AT&T. That judgment, known as the *Modification of Final Judgment* (MFJ), required AT&T to divest itself of the 22 BOCs. It mandated the formation of a single point of contact for coordination among the operating companies to meet the requirements for national security and emergency preparedness. This agreement also allowed the operating companies to "support and share the costs of a centralized organization for the provision of engineering, administrative and other services which can most efficiently be provided on a centralized basis." Bellcore was established to provide these functions [2.21].

Bellcore provides technical and other services to the BOCs to support their provision of exchange telecommunication and exchange access services. Seven regional companies through their subsidiaries share ownership of Bellcore. These are NYNEX Corporation, Bell Atlantic Corporation, BellSouth Corporation, Ameritech, Southwestern Bell Corporation, U S WEST, and Pacific Telesis Group. Together, these regional companies account for more than 80% of all local telecommunication service in the United States. To support these companies and to provide the technical assistance necessary for them to maintain these complex networks, Bellcore performs major work in the areas of technology support, network and operations architecture planning support, network service capabilities support, and network operations capabilities support [2.22]. The company provides the means for utilizing the newest technology in the evolution of operating company networks and in the planning of new services. In addition, Bellcore participates extensively in national and international standards forums that help to maintain the integrity of the operating telephone company communication networks.

Bellcore is also responsible for the North American numbering plan, described in Chapter 1, which facilitates the access, routing, and billing of calls. Bellcore administers the plan on behalf of the operating companies and the telecommunication industry. The primary mechanism for translating this research into supplier-provided products is through the publication of generic technical requirements, called *Technical References* (TRs). TRs describe the interfaces and the functional, environmental, documentation, and reliability requirements for network equipment. We will discuss several of these in Chapters 3 and 4. In the first five years of its existence, Bellcore published more than 1000 technical advisories and TRs [2.22]. Bellcore also conducts Technology Requirements Industry Forums (TRIFs) to provide equipment suppliers with an opportunity to understand the nature of the companies' needs and to influence the generic technical requirements relative to those needs while the requirements are still in a preliminary stage.

2.4.13.2 U.S. Government Activities

In addition to the U.S government's interest in the commercial side of standards, as evinced by the NIST activities, the government also has a strong military interest. In recognition of the extreme importance of standardization within the federal government, Congress passed the Federal Cataloging and Standardization Act in 1952. Immediately thereafter, the Department of Defense (DoD) established the Defense Standardization Program, with the purpose of meeting the intent of Congress within the DoD.

The Federal Telecommunication Standards Program (FTSP) was initiated in 1972 to provide for the development and coordination of federal telecommunication standards for national communication system (NCS) interoperability and computer-communication interface. Pursuant to the request of the director of telecommunication policy of the Executive Office of the President, the administrator of the General Services Administration on August 14, 1972, assigned the responsibility for the development of certain federal standards to the executive agent of NCS, who delegated this responsibility to the manager of NCS. As a result, the FTSP was established in October 1972. Within the Defense Communications Agency (DCA), the Protocol Standards Steering Group acts as executive agent for DoD for use of OSI protocols and TCP/IP (see below). One key area of interest to government agencies is communication security; this topic is discussed in Chapter 17.

2.4.13.3 Corporation for Open Systems (COS)

COS, while not generating standards, operates to facilitate equipment interoperability by certification of various (upper layer) products with respect to conformance to the applicable international standards. COS's main purposes are: (1) to select from among existing standards a small number that will support user applications in many industries; (2) to endorse specifications for implementing the selected standards; and (3) to provide conformance testing: to demonstrate that tested products or services operate in accordance with the endorsed implementation specifications. Both COS and MAP-TOP focus on standards developed by international standards bodies, but MAP-TOP supplements these with emerging standards to provide complete functionality in the short term. Because their long-term objectives are identical, COS and MAP-TOP have agreed to cooperate in specific areas. COS efforts are directed at accelerating the introduction of products and services conforming to its endorsed standards and specifications. COS formally commenced operations on March 6, 1986. More than 70 companies were members of COS in 1989. Possibly, COS activities might be subsumed under NIST.

2.4.13.4 The Internet Activities Board

In the late 1960s, the U.S. Defense Advanced Research Projects Agency (DARPA) initiated an effort to develop packet switching technology as an alternative transmission discipline in the telephone, radio, and satellite environment. This led to the development of the first packet-switching network, known as ARPANET. In the early 1970s, the Xerox Palo Alto Research Center began an exploration of packet switching on coaxial cable that ultimately led to the development of Ethernet LANs. The successful implementation of packet radio and packet satellite technology raised the possibility of interconnecting the ARPANET with other types of packet networks. The solution to this challenge was developed as part of a research program in internetting sponsored by DARPA and resulted in a collection of protocols based on the original Transmission Control Protocol (TCP) and its lower layer counterpart, Internet Protocol (IP). Together these *de facto* protocols, along with others developed during the years, are known as the TCP/IP or Internet protocol suites, and are now very common in the LAN environment, as discussed in Chapter 14 (this suite is not OSI-based except at layers 1 and 2).

In the early stages of the Internet research program, only a few researchers worked to develop and test versions of the Internet protocols. Over time, the size of this activity increased until, in 1979, an informal committee was formed to guide the technical evolution of the TCP/IP suite. This group was called the Internet Configuration Control Board (ICCB). In 1983, the DCA, then responsible for the operation of the ARPANET, declared the TCP/IP to be the standard for ARPANET and the Defense Data Network (DDN); all systems on the network converted from the earlier Network Control Program to TCP/IP. Late that year, the ICCB was reorganized around a series of task forces considering different technical aspects of internetting. The reorganized group was named the Internet Activities Board (IAB) [2.23].

The IAB is now the coordinating committee for Internet design, engineering, and management (Internet is a collection of more than 2000 packet switched networks located principally in the United States, but also includes systems in other parts of the world; interconnection is via the TCP/IP). The IAB is an independent committee of researchers and professionals with a technical interest in the evolution of the Internet system in general, and TCP/IP in particular. Membership changes with time to adjust to the needs of the Internet system and the concerns of the U.S. government, universities, and industrial sponsors. IAB members are committed to making the Internet function effectively. All IAB members are required to have at least one other major role in the Internet community, in addition to their IAB membership. New members are appointed by the chair of the IAB, with the advice and consent of the remaining members.

All decisions of the IAB are made public. The principal vehicle by which the decisions are propagated is the Request For Comment (RFC) note series. These describe experimental protocols. A Standard RFC starts out as a proposed standard and may be promoted as a draft standard, and finally standard, after suitable review, comment, implementation, and testing. Some of the key areas being addressed by the IAB as of the beginning of 1990 include internetworking with full OSI-based networks, security, and user services [2.23]. Each RFC has two attributes: a state, which indicates the document's level of standardization, and a status, which indicates the level of support the Internet community must accord the document. States are Internet standard, draft standard, proposed standard, experimental protocol, and historical protocol. Statuses are "required" (all hosts in the system must implement it), "recommended" (all hosts in the system are encouraged to implement it), "elective" (hosts in the system may decide whether to implement it or not), and "not recommended" [2.5].

2.5 ISSUES PERTAINING TO THE DELIVERY OF STANDARDS—A CRITICAL ASSESSMENT

2.5.1 The Environment

To meet the challenges of the 1990s, telecommunication standards must be developed with a much shorter life cycle compared to the life cycle of the past. Standards may still take up to 10 years to complete (examples: FDDI and IEEE 802.6 discussed in Chapter 15, among many others). Given this development cycle, the standards may be 10 years behind the clock in terms of current technology. For example, file transfer, e-mail, and transaction processing were being done in the 1970s and 1980s; yet only in the late 1980s did standards start to appear. Work on ISDN standards started in 1972 [2.24].* HDTV work started in the late 1960s; studies of DPCM digital transmission of television signals started even earlier, in 1958 [2.25]. However, a concept such as ISDN may require the development of 50 or 60 standards; this may in part explain the latency in the process, at least for some of the more sophisticated standards.

Statements made by some enthusiasts that "standards now lead technology" simply do not hold up to scrutiny. To mention just one example, all standards related to fiber technology (FDDI, Chapter 15; SONET, Chapter 4; BISDN, Chapter 4) are simply trying to catch up in the packaging—one way or another—of the enormous Gb/s-range bandwidth already made available by fiber for several years.

Besides the consensus issue intrinsic with agreement among dozens and even hundreds of parties, too many agencies are pursuing standards—at times in a fragmented, duplicative, and even conflicting fashion. Several hundred agencies deal with communication standards on a global basis.

A streamlined systematization and professionalization of the standards development process is needed. "We must operate standards activities as we would operate a business, with a keen eye to the bottom line" [2.26]. Industry experts are calling for process control in the standards bodies. Some believe that the slow process of delivering standards is delaying the deployment of equipment and services [2.19]. An influential reviewer of this book put it this way: "We need more 'officer'-level commitment to key deadlines, instead of technical purists arguing over bits/bytes for years."

2.5.2 The Embedded Base Push-Pull

Standards are not immediately needed as a new technology breaks new grounds and totally new products enter the commercial market. However, after a set of vendor products has been on the market for a number of years, users perceive the need to interwork these products. A *de facto* standard may have developed, but some segments of the market may be unwilling to accept it. A standards body willing to address the problem may emerge. A new work item is formulated, and the process begins to unfold as described in Sections 2.3 and 2.4.4. Figure 2.3

*"... In June 1971 ... Japan proposed adding the word 'Digital' to 'Integrated Services Network'. ... A key event in ISDN history in the CCITT began with the famous circular Letter No. 199, submitted on August 1972. This letter suggested a joint meeting of selected CCITT group representatives. The purpose ... was to make proposals for the Vth Plenary Assembly to determine how studies of digital networks ... should be conducted during the period 1973–1976. ... The letter mentioned three possibilities for the studies: (1) Service integrated digital network, (2) ..." [2.24]. By 1976, the ISDN term officially appeared in the list of terms contained in the Orange Book issued at the Sixth Plenary Session Assembly.

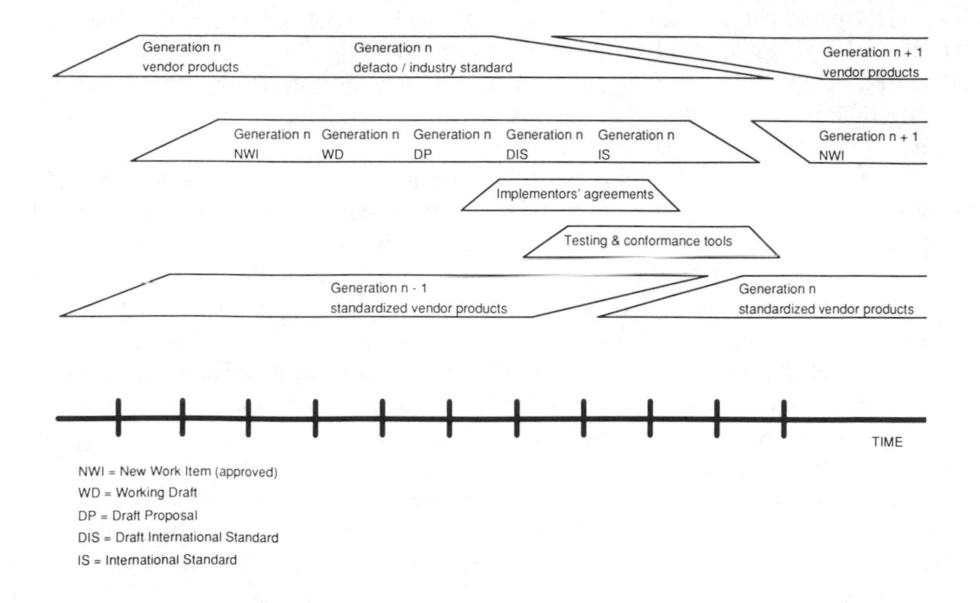

Figure 2.3 Bringing products to market.

depicts the other aspects involved in bringing a product to the market. Implementors' agreements are needed to fill in gaps left by the original body, perhaps for regional definition, or because a range of values is given, rather than a specific value. Ambiguities or errors in the standard may also exist that need to be resolved.

Testing is required both by a specific vendor to develop a conformant product, and by groups of vendors to "ensure" interworking of products. Testing determines whether an implementation conforms to the relevant protocol and associated implementors' agreements. These testing procedures must also be developed and agreed to. The functional capabilities of the implementation, as well as the dynamic behavior, need to be tested. This will increase the probability that different implementations are able to interoperate.

2.5.3 Final Standards

In many situations, the internationally published standard is not sufficient to begin development of a product. National clarifications, and even industry clarifications, are required. For example, CCITT Q.931 specifies network layer capabilities of the user-network interface for ISDN. ECSA T1.603-1989 provides North American clarifications. Bellcore's TR-TSY-000268 (1989 edition) specifies requirements the BOCs may utilize to fulfill their needs. This process of having to clarify or qualify adds additional time in the deployment of products.

2.5.4 Some Approaches

The challenge of the increasing urgency in the need for standards can be satisfied only by increasing the responsiveness of some of the key standards bodies. Both timely delivery of acceptable standards and the minimization of redundant application of scarce technical resources are being advocated. Today, standards bodies function in a bottom-up mode, in which direction is set by the individual committees, committee members, and some of the stronger contributors. For example, contributions may be driven by company business strategy or parochial interests. A more top-down approach is now being advocated by some; see Table 2.4 [2.26]. Interest is aimed at finding ways to speed up the process [2.19], and the CCITT's Accelerated Procedures is a major step in the right direction.

Table 2.4
Eight Steps for Expediting the Standards-Creation Process

1. Set work priorities.
2. Set timetables and rigorously adhere to them.
3. Restrict scope of the standards activities.
4. Set sequence of work activities.
5. Recognize the need to avoid duplication of efforts in public standards bodies.
6. Automate and mechanize the distribution of drafts, proposals, *et cetera,* to reduce travel.
7. Obtain executive-level involvement in standards.
8. Ensure competent and informed leadership in standards.

REFERENCES

[2.1] I. Lifchus, "Standards Committee T1—Telecommunications," *IEEE Communications Magazine*, January 1985, Vol. 23, No. 1.

[2.2] E.J. Cohen and W.B. Wilkens, "The IEEE role in Telecommunications Standards," *IEEE Communications Magazine*, January 1985, Vol. 23, No. 1.

[2.3] E. Loshe, "The role of ISO in Telecommunications and Information Systems Standardization," *IEEE Communications Magazine*, January 1985, pp. 18–20.

[2.4] G.W. Wallenstein, *Setting Global Telecommunications Standards: The Stakes, the Players, and the Process*, Artech House, Norwood, MA, 1990.

[2.5] M.T. Rose, *The Open Book, A Practical Perspective on OSI*, Prentice Hall, Englewood Cliffs, NJ, 1989.

[2.6] A.C. Macpherson, *International Telecommunication Organizations and Standards*, Artech House, Norwood, MA, 1990.

[2.7] C.F. Cargill, "Standards and Standards Organizations," *Journal of Data and Computer Communications*, Winter 1989.

[2.8] 1976 ITU Convention, Article 4, Paragraph 2.

[2.9] J. Jipguep, "The ITU and the Regulatory Revolution," *Telecommunications*, April 1989, pp. 88–89.

[2.10] C.A. Meagher, *Satellite Regulatory Compendium*, Phillips Publishing, Potomac, MD, 1985.

[2.11] G. Finne, "The Future of CCIR," *Telecommunications*, April 1989, p. 90.

[2.12] E. Hummel, "The CCITT," *IEEE Communications Magazine*, January 1985, Vol. 23, No. 1.

[2.13] D.M. Cerni, "The United States Organization for CCITT," *IEEE Communications Magazine*, January 1985, Vol. 23, No. 1.

[2.14] Accredited Standards Committee T1 Telecommunications Procedures Manual, February 1989.

[2.15] R.C. Kirby, "International Standards in Radio Communication," *IEEE Communications Magazine*, January 1985, Vol. 23, No. 1.

[2.16] M.N. Woinsky, "National Performance Standards for Telecommunication Services," *IEEE Communications Magazine*, October 1988, p. 70 ff.

[2.17] Stable Implementation Agreements for Open Systems Interconnection Protocols, Version 2, Edition 4, September 1989.

[2.18] R. Gibbons, "Surveying Europe's Standardization Scene," *Communications International*, November 1989, p. 28.

[2.19] R. Mosher, "Standards Topics," *IEEE Communications Magazine*, September 1989, p. 58.

[2.20] "ETSI Taking Over from CEPT," *European Telecommunications*, September 15, 1988, p. 1.

[2.21] Bellcore, *A New National Resource*, Bellcore, Livingston, NJ.

[2.22] *Bellcore Digest*, December 1988, Vol. 5, No. 9.

[2.23] Corporation for National Research Initiatives, IAB Secretariat, Reston, VA.

[2.24] K. Habara, "ISDN: A Look at the Future Through the Past," *IEEE Communications Magazine*, November 1988, p. 25 ff.

[2.25] J.B. O'Neal, "Predictive Quantizing Systems (Differential Pulse Code Modulation) for Transmission of Television Signals," *Bell System Technical Journal*, May–June 1966, pp. 689 ff.

[2.26] L. Hohmann, "Process Control in Standards Bodies," *IEEE Communications Magazine*, September 1989, pp. 59–61.

Chapter 3
Digital Transmission Systems

This chapter provides a basic description of digital transmission systems, commonly known as *digital carrier systems*. High-capacity digital systems are becoming increasingly important to both telecommunication carriers and end users. Transmission technology and termination equipment are discussed.

Digital carrier technology was initially deployed to support trunks between metropolitan COs (1962), and then in the feeder plant connecting the user to the CO (1973). Since the early 1980s, digital multiplex technology has become directly available to end users in the form of "T1" private lines, and lately as "fractional T1" private lines. ISDN can be considered a companion digital technology applicable to the distribution plant (we will discuss ISDN in Chapter 4). Figure 3.1 depicts some components of digital carrier systems, particularly as seen from a user's perspective.

Initially, carrier technology was intimately connected with digital voice transmission and PCM methods. With newer non-PCM voice digitization schemes, and the extensive commercial applications of T1 in the high-speed data environment, a treatment of digital carrier systems no longer needs to be subordinated as a mere footnote of PCM, as might have been the case 30 years ago. A top-down system perspective is taken herewith. However, the peculiarities these digital facilities inherited from the voice ancestry cannot be avoided or overlooked.

We discuss the following issues in this chapter: the multiplexing schemes employed in T1 (original and newer method); restrictions on the user signal imposed by these schemes (such as the zero-density problem); typical CPE and carrier equipment; schemes for higher digital rates; and evolving transmission approaches for broadband ISDN.

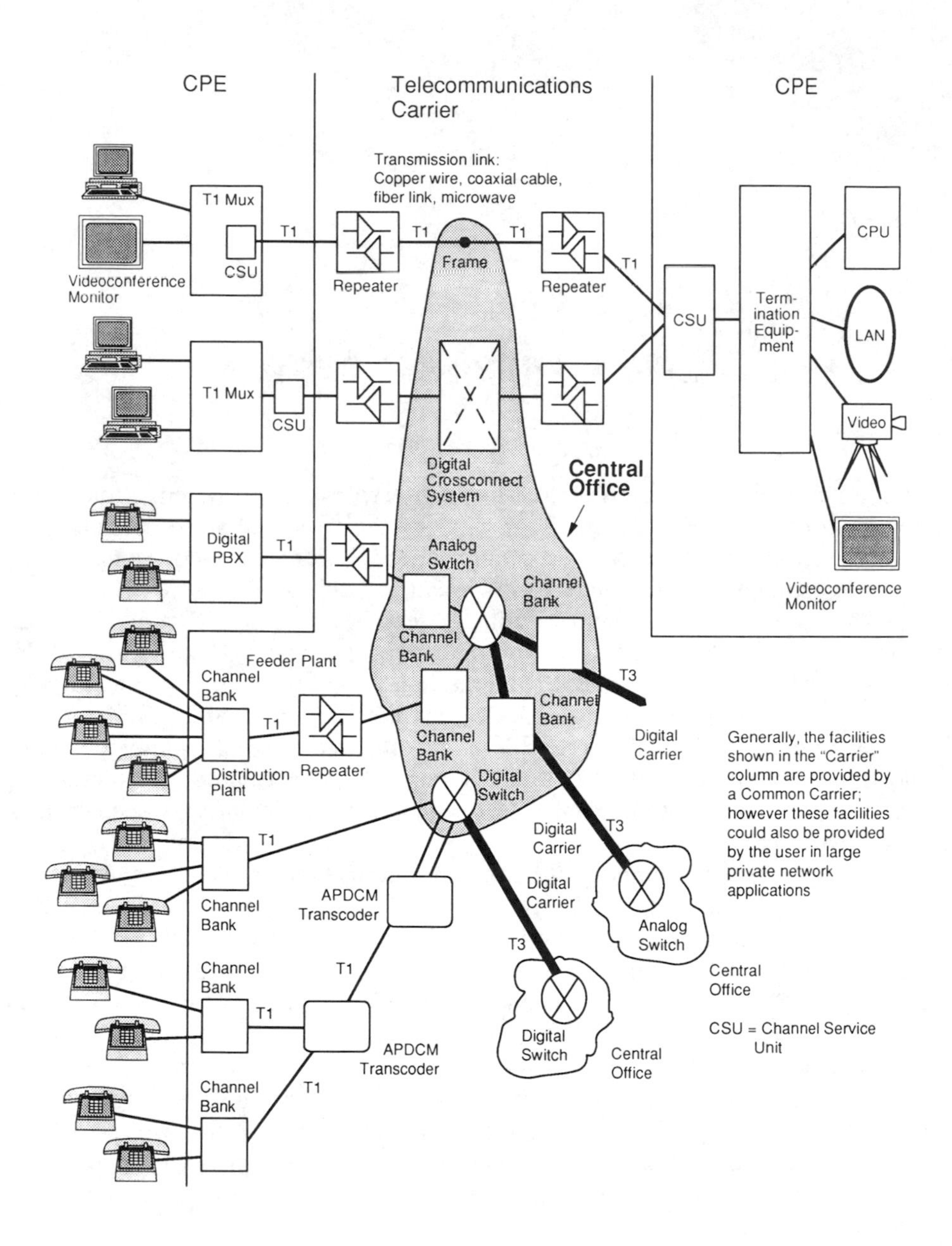

Figure 3.1 T-carrier applications.

3.1 CARRIER SYSTEMS

As we discussed in Chapter 1, in the early days of telecommunication, a medium such as copper wire carried a single information channel. Because of economic reasons, both in terms of construction costs and in terms of materials, new ways of packing multiple channels onto a single physical link were needed. The resulting system is referred to as a *carrier system*, or simply a carrier (the term *carrier* in this context should not be confused with the "telecommunication carriers" discussed at the end of Chapter 1). Carrier systems can be analog or digital. Analog systems are decreasing in importance, and are now a small fraction of the total number of carrier systems in the public switched network. The first digital system introduced into the public network was the so-called "T-carrier" system, which utilized twisted-pair wire. Digital signals are now transmitted from one location to another by facilities using a multitude of media, including paired cable, coaxial cable, radio, optical fibers, and satellite. Great progress has been made in all of these types of media as discussed in Chapter 1. A carrier system consists of a transmission component, a user interface component, and a user termination equipment component.

The *transmission component* is a transmission system carrying multiple channels; it in turn entails multiplexer equipment and a transmission link, as depicted in Figure 3.1. T-carriers (in the strict telephonic sense) are copper-based digital facilities that carry 24, 96, 672, or 4032 simultaneous PCM-coded voice (or voiceband data) channels, operating at 64 kb/s each (the T1, T2, T3, and T4 systems, respectively). In reality, many carrier systems today are based on fiber transmission. The correct nomenclature in describing generic digital carrier systems should be DS1, DS2, DS3, and DS4, where DS stands for *digital signal*. In particular, "DS" refers to the coding format used to transmit the information over the carrier system. Prior to 1977, carrier systems were employed exclusively by telephone companies. Beginning in 1977, but more widely after 1982, high-capacity facilities were tariffed by telecommunication carriers and made available to commercial customers. As of 1990, a number of BOCs also had a tariffed DS3 service, typically provided over fiber.

Carrier termination equipment includes telephone company equipment such as channel banks, transcoders, and digital cross-connect systems (described later); and CPE, such as T1 multiplexers and PBXs. Generally, CPE termination equipment must be interfaced to the carrier system via a *channel service unit* (CSU). Some of the functionality of these types of termination equipment as it relates to carrier technology, including the role of the CSU, will also be discussed in the chapter. Some issues pertaining to testing are also highlighted.

Below we discuss the traditional multiplex hierarchical scheme used in the U.S. telephone network (we use the word "traditional" because new broadband schemes are now emerging; these are covered toward the end of this chapter).

One of the achievements of the telephone engineers of the 1930s and 1940s was to define a hierarchical structure for "packages" of user channels, which could be treated as ensembles, and which in turn could be placed on any type of appropriately configured medium. This hierarchical concept was used initially for analog interoffice trunks.

A number of multiplexing schemes exist for placing multiple channels on one medium, as we alluded to in Chapter 1. The two that apply most directly to the telephone network are: *frequency division multiplexing* (FDM) and *time division multiplexing* (TDM). FDM is typical of an analog carrier system and is generally disappearing except in analog microwave transmission. TDM is typical of digital transmission; it lends itself well to computer interfaces. T-carrier systems use TDM.

3.1.1 Analog Signals: The FDM Hierarchy

While this chapter focuses on digital technology, a brief discussion of analog carrier systems is provided for completeness. Analog voice (or voiceband data) channels can be combined in the telephone network to provide a stack of signals that are suitable to be carried by a medium such as analog microwave and coaxial cable. These media operate at high frequencies, so that numerous voice channels can be carried by them; it would be wasteful to put a single channel on a coaxial cable. In addition, the basic frequency of the individual channels must be raised so that it is consistent with the medium at hand. Twelve steps make up the analog multiplex scheme of the U.S. telephone network. Some steps represent integral system levels; others are known as intermediate levels. Some of the principal levels are shown in Table 3.1. The analog hierarchy is now decreasing in importance because of the widespread emergence of digital technology.

3.1.2 Digital Signals: The TDM Hierarchy

The traditional TDM hierarchy is described as *DS level 0* through *DS level 4*. The 0 to 4 kHz nominal voiceband channels are first converted to a digital stream by PCM analog-to-digital techniques, and then stacked (multiplexed) onto higher bit streams. Each of the individual digitized 0.064 Mb/s channels is referred to as a DS0 level. In the United States, the traditional digital hierarchy uses 1.544 Mb/s for 24 channels, also called a *digroup* (for digital group), 3.152 Mb/s for 48 channels, 6.312 Mb/s for 96 channels, 44.736 Mb/s for 672 channels, and 274.176 Mb/s for 4032 channels [3.1] (see Table 3.2). As indicated, these are colloquially called T1, T2, T3, and T4, though this nomenclature is a misnomer: the correct identifiers are DS1, DS2, DS3, and DS4, respectively. These digital streams produced by channel banks and other multiplex equipment are (by design) independent of the target transmission media; in fact, in an end-to-end circuit, many different types

Table 3.1
FDM Multiplex Levels

12 voice-grade calls are multiplexed by a channel bank into a basic group.

5 basic groups are multiplexed by a group bank into a basic supergroup.

10 basic supergroups are multiplexed by supergroup bank into a basic mastergroup.

6 basic mastergroups are multiplexed by appropriate equipment into a basic jumbogroup.

2 intermediate levels of 3000 calls are multiplexed by appropriate equipment into an AR6A level.*

3 basic jumbogroups are multiplexed by appropriate equipment into an L5 level.

3 intermediate levels of 4200 calls are multiplexed by appropriate equipment into an L5E level.

Basic group = 12 message channels
Basic supergroup = 60 message channels
Basic mastergroup = 600 message channels
Basic jumbogroup = 3600 message channels
AR6A level* = 6000 message channels
L5 level = 10,800 message channels
L5E level = 13,200 message channels

*AR6A is a single-sideband (SSB) AM radio transmission system.

Table 3.2
TDM Multiplex Levels

24 voice-grade channels are multiplexed by a D-channel bank into a DS1 level.

48 voice-grade channels are multiplexed by a D-channel bank into a DS1C level.

96 voice-grade channels are multiplexed by a D-channel bank into a DS2 level.

2 DS1-streams are multiplexed by appropriate equipment into a DS1C level.

4 DS1 streams are multiplexed by appropriate equipment into a DS2 level.

28 DS1 streams are multiplexed by appropriate equipment into a DS3 level.

14 DS1C streams are multiplexed by appropriate equipment into a DS3 level.

7 DS2 streams are multiplexed by appropriate equipment into a DS3 level.

6 DS3 streams are multiplexed by appropriate equipment into a DS4 level.*

DS1 = 24 message channels
DS1C = 48 message channels
DS2 = 96 message channels
DS3 = 672 message channels
DS4 = 4032 message channels

*DS4 is not a level that has been standardized in ANSI or Bellcore documents.

of media may be encountered. The output of a T1 multiplexer or a channel bank at the DS1 level could be placed on copper facilities, which are designated as T1 systems; the output of a channel bank at the DS2 level could be placed on copper facilities, designated as T2; the output of a DS3 system could be placed on a fiber optic facility, designated FT3 (as discussed below). Thus, a T1 system must use a DS1 signal format; but a DS1 signal need not use a T1 facility. Another way of expressing this is that T1 technology is medium-specific. Figure 3.2 is a summary of the TDM hierarchy.

Three other traditional digital interfaces were used in early fiber optic systems: the FT2 rate at 12.624 Mb/s (twice a DS2); the FT3 rate at 44.736 Mb/s (same as DS3); and the FT3C rate at 90.524 Mb/s (this is slightly higher than twice a DS3 signal, because of overhead).

The rates corresponding to the TDM hierarchy were initially derived based on the information carrying capacity of copper wires and the spacing of manholes where repeaters could be located. The original intercity trunking application was motivated by the predicted exhaustion of the conduit infrastructure in case analog multiplexing schemes continued to be employed, and the resulting expenditure to expand the physical infrastructure.

As we will cover later, overhead or "stuffed" bits are associated with the aggregate multiplexed signal. Twenty-four 64,000 b/s channels result in 1.536 Mb/s; the DS1 signal at 1.544 Mb/s has 8 kb/s of overhead. Four 1.544 channels result in 6.176 Mb/s, compared to the DS2 signal at 6.312 Mb/s. Twenty-eight DS1 channels result in 43.232 Mb/s, compared with the DS3 signal at 44.736 Mb/s.

A similar TDM scheme is used in Japan. The European telephone system is based on 32-channel blocks (typically 30 voice channels and 2 signaling channels). The hierarchy is comprised of the following aggregate rates: 2.048, 8.448, 34.368, and 139.264 Mb/s [3.2]; see Section 3.2.3.

Digital level	Level's Bandwidth (Mb/s)	Transmission Facilities		
		Copper	Radio	Optical
DS4	274.176	T4M	DR18	
DS3	44.736		3 ARDS	FT3
DS2	6.312	T2		
DS1C	3.152	T1C,T1D		
DS1	1.544	T1,T1/OS	1 ARDS	
DS0	0.064			
Analog	4 KHz			

Figure 3.2 U.S. TDM digital hierarchy of interest to channel banks.

Standardized multiplexing equipment supports this digital hierarchy. For example, 4 DS1 are multiplexed by M12 equipment into a DS2; 7 DS2 are multiplexed by M23 equipment into a DS3; and 28 DS1 are multiplexed by M13 equipment into a DS3.

Standards

Prior to divestiture, AT&T set the various *de facto* standards and protocols for T-carrier transmission and termination equipment. Equipment manufacturers have had to meet or conform to these requirements. In the post-divestiture environment, ECSA in general, and Bellcore in particular, publicize industry requirements for manufacturers who wish to develop equipment that meets the standards of the U.S. (local exchange) telephone network. As discussed in Chapter 2, Bellcore holds regular TRIFs to publicize requirements, in addition to publishing technical advisories and technical requirement documents [3.3, 3.4].

3.1.3 T1 Copper-based Local Loop

This section describes the traditional physical configuration of a T1 copper-based line, as implemented in the feeder and distribution plants. Eventually, the service will prominently be provided over a fiber system and, therefore, some of the repeater discussion is no longer applicable. However, the interface part, discussed below, and the DS1 signal framing schemes, discussed in the following sections, continue to apply.

A T1 copper-based facility consists of office repeaters, exchange grade (22-gauge) cable pairs, and line repeaters in tandem to regenerate the signal after it has propagated along a section of the cable. Cable pairs must be conditioned, namely, bridge taps, build-out capacitors, load coils, and cable stubs must be removed from the cable. Four-wire facilities are used: two for the transmitting side and two for the receiving side). A typical T1 line is shown in Figure 3.3 [3.5].

The electrical specification for a DS1 signal at crossconnect points is known as the DSX-1 electrical interface. A description of the electrical characteristics of the DS1, DS1C, DS2, and DS3 signals can be found in ANSI specification T1.102-1987. The physical connection at the network interface is an 8-pin modular jack DA 15 S (ISO 4903). The bit rate is 1.544 Mb/s ±75 b/s, and the electrical encoding is "bipolar return to zero," and called "bipolar" henceforth.

At the CO end, the inputs and outputs of T1 lines appear at a DSX-1 patch and cross-connect bay that provides a signal access point. The DS1 signals at the DSX-1 are connected to office repeater bays that provide impedance matching, equalization, and signal regeneration (receiving side only), while applying power to the line. These repeaters are connected to line repeaters located in the feeder

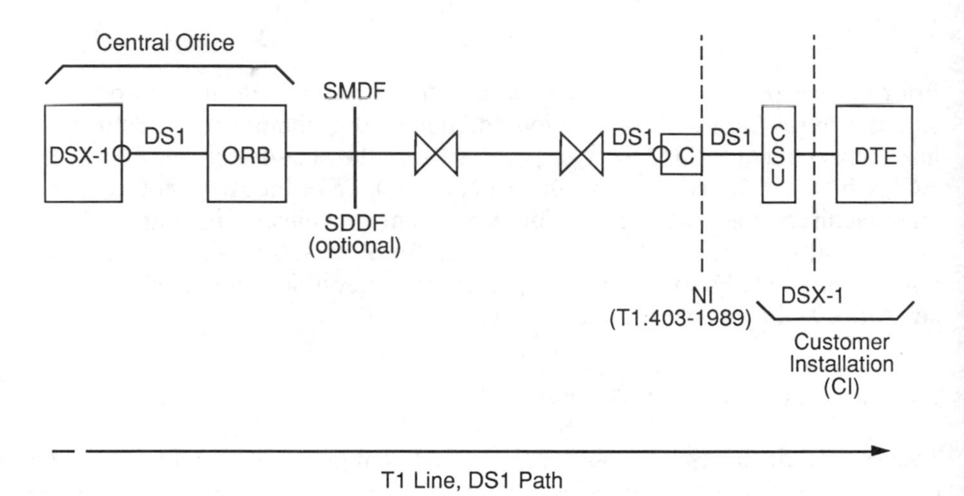

Figure 3.3 A DS1 line to the central office.

and distribution plant. The office repeater can be cabled to a distribution frame. Line repeaters provide timing and equalization to restore the pulse shape. The repeaters are powered from the office repeater bays. Repeaters are housed in apparatus cases located either in manholes or on poles. The first repeater section is nominally 3000 feet from the CO in order to control CO noise or cabling and equipment attenuation. The remaining line repeaters are installed at a nominal spacing of 3000 to 6000 feet.

Additional pairs are needed for maintenance purposes. A maintenance line is used to substitute for any in-service line by patching; a loaded order-wire pair permits communication between repeater locations or between repeater locations and the end of the span. A fault-locating cable pair for each set of repeater locations in tandem along the span is also required to sectionalize failures to a particular repeater span.

After being regenerated by the last repeater on the line, the DS1 signal crosses the *network interface* (NI), the demarcation point between the network and the *customer installation* (CI). The NI is now physically and electrically defined in ANSI T1.403-1989. The CI is equipment and wiring at the user's location, on the customer side of the NI. At the NI, DS1 signals pass through a DS1 connector (C); this is a device that the FCC permits a carrier to provide and install on the network side of the NI on the customer premises. The connector provides signal loopback to sectionalize a potential problem between the carrier's network and the CI. The connector will generally be located inside the user's building; it may also be mounted in an external location [3.5].

A CSU is typically the first component of CI equipment connected to the T1 line on the user's premises (originally, the CSU belonged to the network; later, the FCC ruled that all DTE, except multiplexers—such as M12—located on customer premises, must be customer provided). This device regenerates the DS1 signal received from the network, as well as the customer's signal to be given to the network, and presents the user with a DSX-1 interface. We will describe additional features of this device later.

3.1.4 DS1 Signal Format

Overview

This section describes how 24 voice channels are arranged so that they can be placed simultaneously multiplexed onto a single transmission facility. The following list identifies aspects associated with defining the DS1 format (details are provided in the sections that follow) [3.3, 3.4]:

1. *Channel numbering*. This refers to how the 24 channels are arranged as they appear in the combined bit stream. In the modern scheme, channels are numbered sequentially in one-to-one correspondence with the sequence of 8-bit time slots; older schemes employed different numbering methods. The 8 bits being discussed originate from the PCM sampling process for each voice channel.
2. *Information bits*. The information bits in each time slot are used in a variety of ways. In most instances, the 8 bits are used in a similar fashion in all 24 time slots, but exceptions occur where two or more uses are intermixed in

the 24 time slots. Three ways to utilize the information bits are voice, data, and transparency.

3. *Channel-associated signaling bits.* Channel-associated signaling bits are used for per-channel supervision and addressing. Three ways to signal are A,B-signaling; network control; and A,B,C,D- signaling.
4. *Code substitutions.* Under a number of conditions, a fixed code must be substituted for the contents of a particular time slot; three substitutions are zero-suppression codes, unassigned and idle channels codes, and alarm codes.
5. *Framing bits.* The framing bit (F-bit) in each frame allows the receiving equipment to synchronize (frame) on the incoming DS1 signal.
6. *Framing method.* Framing (also called synchronization) is needed for identification of digital information. It can be handled by a number of schemes, all involving, in one way or another, the F-bit.
7. *Yellow code.* The yellow code provides an alarm in the case of an outgoing signal failure. A number of ways to achieve this alerting exist.
8. *DS1 data links.* Within the DS1 signal format, a number of alternatives exist to provide a control-management data link of a few Kb/s.

3.1.4.1 Framing and Framing Bit

One of the challenges of any multiplexing technique is to establish the boundary of the octets within the overall combined stream. As indicated, a common approach is to have the 24 channels arranged in a round-robin fashion: each channel is transmitted in turn, in a linear sequence. Each channel is represented by 8 bits (octet or byte) in turn, corresponding to the PCM-sample word. For the receiving equipment to interpret the 24 octets properly, the receiving hardware needs a way to distinguish the beginning and end of each frame, after which the position and, hence, content of each octet can be determined. This is called *frame-level synchronization*. In modern parlance, this method of recovering information based on its location within a frame is called *synchronous transfer mode* (also known as TDM), compared to a nonpositional method called *asynchronous transfer mode*, discussed in Chapter 4.

A unique sequence of bits (known as a starting delimiter or flag) could be located at the beginning of each frame of 24 octets (which corresponds to 192 bits or 125 microseconds) to identify the beginning of the frame. For example, the sequence 100011011100 could designate the beginning of the frame. When the receiving hardware locates this string, the offsets for the individual octets can be easily computed [see Figure 3.4(a)]. In this fashion, the receiving equipment can make a firm decision at the end of each frame. With this scheme, a single flag will allow synchronization; put differently, a single frame suffices to recover the synchronization. (This is the method used in Europe.)

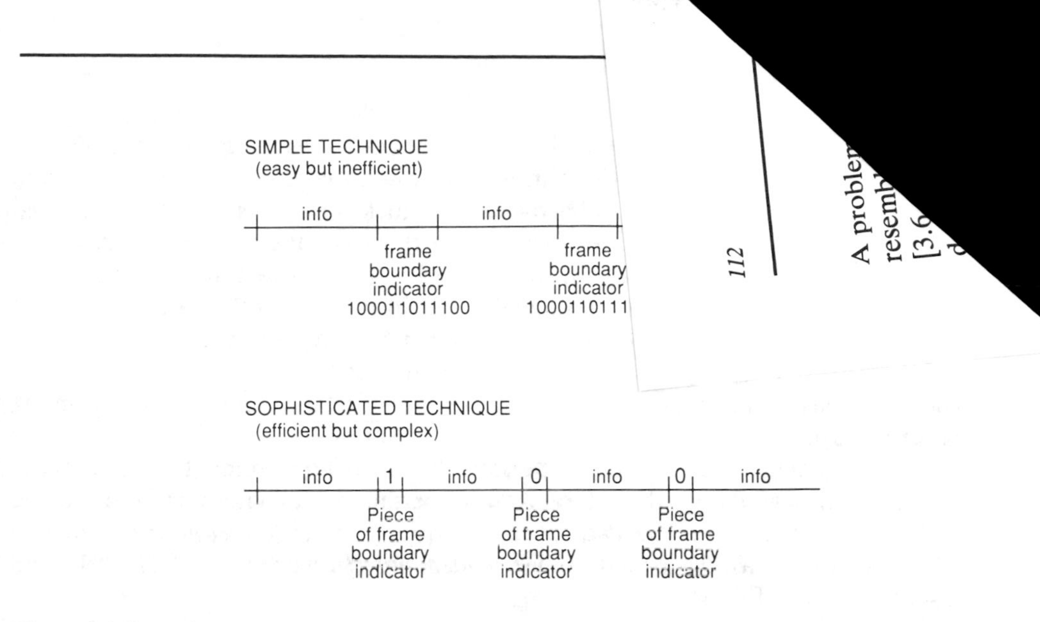

Figure 3.4 Frame boundary.

A 12-bit flag every 192 bits, however, represents an overhead of approximately 5%. The designers of the T1 carrier system in the early 1960s decided that this overhead was too high. Instead, they decided to sprinkle the unique word over 12 frames, thus only putting one of the framing bits (in turn) at the end of a frame [see Figure 3.4(b)]. By looking at one frame, we are not able to determine the position of the synchronization point because that bit will simply be a 0 or a 1, which alone does not impart much. Instead, the receiving equipment can synchronize by buffering the entire 12 frames (2316 bits). (Older equipment uses a more basic algorithm.) Then the equipment postulates the frame boundary at a bit x (assuming that bit was 1), then looks at 193 bits further to see if that bit was a 0, then looks 193 bits further to see if that bit was a 0, and so on until the entire 100011011100 flag is registered. If at any point an unexpected bit is found, this tells the equipment that the postulated boundary was in fact not correct. The hardware then postulates the boundary at bit $x + 1$ and repeats the process. If the flag is recovered, then the boundary is found; otherwise the process starts at $x + 2$, *et cetera*, until the boundary is located. Once the boundary is located, the decoding of all octets for the preceding 12 buffered frames can be accomplished.

The termination hardware is constantly looking for frame boundaries. A framing-pattern-sequence detector is employed. In its search for the true boundaries, the hardware postulates a certain boundary and then checks to see if the expected framing bits are found in the appropriate location in the stream (for example, if it has tracked the 10011 part of the word, 193 bits later it should be able to find a 0, *et cetera*). This type of rolling boundary (prior to lock-in) is part of the synchronization mechanism; the detector soon realizes that it has not yet found a frame, and it continues the search until the appropriate pattern is found.

n could arise if the signal in the channel provides a sequence that closely es the framing word (this can happen when the DS0 channels carry data , 3.7]). The synchronization hardware may lock on this "spurious" word and eclare synchronization at the wrong place (in this case, the link level protocol of the end-user system transmitting the data would have to correct the ensuing errors; for voice, such a mechanism is not available). Termination hardware is such that in the absence of channel noise errors, the maximum average reframing time will be less than 50 ms (maximum average reframe time is the average time to reframe when the maximum number of bit positions must be examined for the framing pattern [3.8]).

The framing bit in each frame permits the receiving terminal to frame on the incoming DS1 signal once these F-bits are accumulated over a horizon, as discussed in the previous section. The two formats currently in use are *superframe format* (SF), also known as D4 format, and *extended superframe format* (ESF). We can describe them as follows:

1. SF uses a 12-frame superframe horizon and divides the F-bits of each individual frame into alternating frame bits, F_t, and signaling bits F_s. F_t bits occur in the odd frames and alternate between "ONE" and "ZERO." The F_s bits occur in even frames and carry the pattern 000111 . . . , which identifies the signaling frames (frame 6 and 12). This same F-bit sequence is repeated in every 12-frame superframe. SF was the original T1 format.
2. ESF uses a 24-frame superframe horizon and subdivides the F-bits among a framing pattern sequence that identifies the location of frames 1 to 24, and other channels. ESF was developed in the early 1980s.

Superframe Format

Although we introduced the concepts above, we repeat them here for completeness. In SF one finds a framing bit every 193 bits; the 192 bits correspond to 24 conversations sampled with PCM-type methods generating 8-bit words (acoustical amplitude values being quantized at 256 discrete—but not equally spaced—intervals); the combined signal is "byte-interleaved," providing one frame. The superframe is a repeating sequence of 12 frames. The superframe, thus, contains 12 framing-signaling bits. One frame corresponds to 125 ms; one superframe, therefore, is 1.5 ms in duration. Each frame contains one synchronization bit to allow the receiving equipment to decode, demultiplex, and allocate the incoming bits to the appropriate voice channels. Each superframe contains a 12-bit flag, made up of individual bits coming from each of the 12 frames. The 12-bit flag is used for synchronization and for identifying frames number 6 and 12, which contain channel signaling bits [3.9] (see Figure 3.5). Carrier equipment (channel banks in particular) minimizes the bandwidth spent on VF signaling by putting information only every sixth frame.

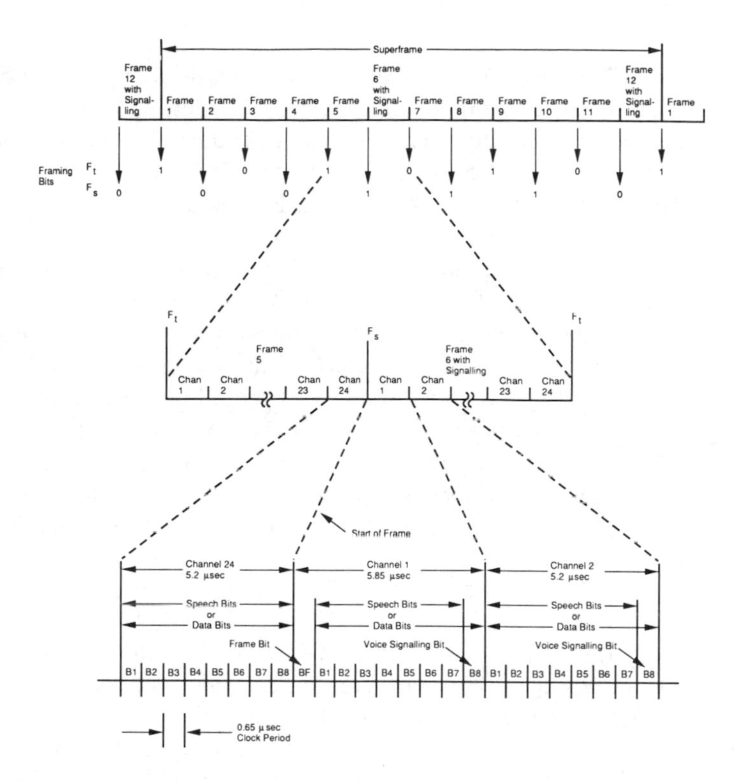

Figure 3.5 SF framing pattern.

Given the repeater technology of the 1960s, at least one bit in 15 bits of the combined stream (information plus signaling) had to be a "1," and at least three bits in 24 bits of the stream had to be a "1." This is required by the channel banks to keep track of the frame structure. The VF signaling method consists of time-sharing the least significant bit in the sixth and the twelfth frame, providing 7 and 5/6 bits for encoding the voice signal. The 12-frame SF has been described in Bell System PUB 43801, ANSI T1.107-1988 [3.10], and Bellcore TR-TSY-000510 [3.11] and TR-TSY-000499 [3.12].

Extended Superframe Format

The SF DS1 superframe format currently in existence grew out of the 1960s, based on the then-available technology. The basic challenge of DS1 transmission, as discussed earlier, is that of preserving bit and word synchronism. To do this, twenty-four 8-bit words (corresponding to samples for 24 conversations) are followed by one bit (the "193rd bit") for frame tracking and management; there are 8000 repetitions of this pattern in one second, which results into an additional 8 kb/s bandwidth "loss." The hardware of the time was relatively unsophisticated and required the stated number of bits to manage the channel. Modern hardware has progressed to the point at which only 2000 b/s are required; this implies that the other 6000 b/s become available for other service-related purposes. In the early 1980s, a new channel bank framing format, ESF, was announced. It used to be called F_e, or "F sub e," but the ESF nomenclature is now preferred. Most carriers have announced that, as CO equipment is replaced at the end of its useful life, new equipment supporting the ESF will be installed, and users will be offered the new format. ESF will allow use of the new bandwidth to provide more advanced services; among these services may be the user's ability to reconfigure his or her network in real time, using a data terminal. ESF benefits the end-user indirectly by providing a more reliable digital service. The present SF format will continue to be supported to protect the embedded base of equipment; the ESF is not compatible with the billions of dollars worth of installed channel banks and T1 multiplexers (some of the more sophisticated and newer channel banks-T1 multiplexers have a software programmable frame structure).

ESF was originally described in AT&T's Technical Publication 54016 [3.13]. That specification has been followed by ANSI T1.403-1989 [3.14], T1.107-1988 [3.10]; and Bellcore TA-TSY-000147, TR-TSY-000194, and TR-TSY-000499 [3.12]. The ESF has 24 frames in its definition of superframe, but only six bits in its framing pattern. This means that instead of being able to accumulate the synchronization flag every 1.5 ms as in the regular format, we can only obtain the resynchronization flag every 3 ms (the 193rd bits are now being looked at as an ensemble of 24-bit words). However, substantial progress has occurred in the VLSI area in the past decade, so that modern carrier equipment keeps timing more accurately; this implies that fewer bits are required for this housekeeping function. Of course, this means that the channel banks need to be upgraded or replaced to benefit from this functionality.

With ESF, 4000 b/s of the channel become available, without giving up any previous functionality or any additional bits. For what would this new available bandwidth be used? It will be used to facilitate the trend toward introduction of intelligent (automated) network surveillance functionality, to monitor the plant in a proactive mode in order to detect impaired conditions before they result in a total facility outage. Equipment providing this functionality will use this newly available bandwidth to communicate with the appropriate centralized computers, which can dispatch staff or take other corrective measures. The future is one of automation for the purpose of reduced manpower in the management of the telephone plant. The reason for this is not strictly an economic one: today's technology has progressed to the point where a fully automated plant is indeed possible and just around the corner. Enhanced forms of telemetry, of which the extended superframe is one example, will inevitably continue to emerge over the next few years, and network equipment will eventually incorporate the appropriate functionality.

Figure 3.6 depicts the ESF format. Only successive 193rd bits are shown explicitly (twenty-four 8-bit words appear between the signaling bits). The key difference is that the new superframe format is composed of a sequence of 24 frames, rather than 12. Six bits are in the frame synchronization word, rather than 12. The D-bit represents link-level data. The C-bit handles error checking and monitoring functions.

ESF properties

The properties of the framing format to be incorporated in new carrier equipment are as follows [3.15]:

- All previous functionality, including the VF-level signaling rules, remains available; no new bits from the 1.544 Mb/s signal are taken away from the user. Twenty-four frames have, in theory, to be examined by the termination equipment to establish synchronism and extract other channel information. This means that in the case of total loss of synchronization it takes longer under ESF to regain synchronization; however, with more sophisticated clocking mechanisms, this situation should occur rarely if a facility is operating correctly.
- A 4000 b/s data link (also called *embedded operations channel*) used for maintenance information, supervisory control, and other future needs, becomes available.
- A 6-bit CRC code is provided. The CRC is used for monitoring transmission quality of the DS1 facility; additional functions include false-framing protection and other performance functions. The CRC-6 is generated from the bits of the preceding frame (for the calculation, the framing bits of that frame are considered to be equal to "1"). This will allow the carrier equipment to inform the appropriate agency in the network that something appears to be degrading; the problem may thus be fixed before total failure occurs. The CRC-6 will detect 100% of all errors of 6 bits or less, and 98.4% of errors of more than 6 bits.

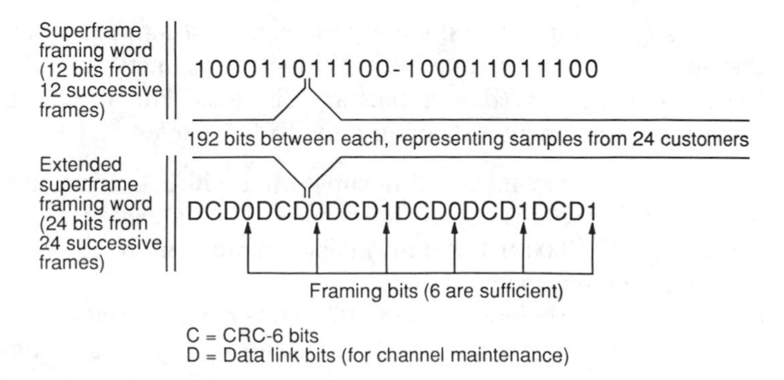

Figure 3.6 Extended superframe format.

The CRC-6 will also provide false-frame protection. This follows from the fact that if the wrong synchronization boundary is selected by the carrier equipment, the CRC will not calculate correctly; assuming the channel was not experiencing intrinsic problems, the hardware can assume that the wrong boundary was selected, and resume the search.

- In addition to the increased functionality just described, and implemented "outside" the user's signal, in the 193rd bit area, new VF-level signaling capabilities become available under ESF. Two additional signaling bits (robbed from the user signal) are made available to allow representation of up to 16 states. The bits are known as "C" and "D," which complement the "A" and "B" bits of the regular superframe format. Options are available (out of these 16 codes) for transparency (no robbed signaling bits), two-states signaling (A-bit only), four-state signaling (A and B bits), or sixteen-state signaling (A, B, C, and D bits are used).

  In the transparent signaling option, all 8 bits within a frame within the extended superframe are given to the user for his data. When the A-signaling mode is employed, the least significant bit of every sixth signaling frame is robbed to show the desired bit. When the A,B-signaling mode is employed, the first and third signaling frames are robbed to carry the A bits; the second and fourth signaling frames are robbed to carry the B bits. When the A,B,C,D-signaling mode is employed, the desired bits are shown by robbing the first, second, third, and fourth signaling frames, respectively.

  These channel-associated signaling bits are discussed more below. This 16-state voice signaling method is not consistent with the ISDN view of explicit message-oriented signaling, and its implementation in a widespread fashion is not clear.

  The 8000 b/s overhead bits are now reallocated as follows:

  1. 2000 b/s for framing (6 bits distributed over 24 framing bits; 333 such 6-bit words occur per second);
  2. 2000 b/s for CRC error and performance determination (6 bits distributed over 24 framing bits);
  3. 4000 b/s for telemetry and facility management and reconfiguration (12 bits distributed over 24 framing bits). With clear channel capability (CCC) using the ZBTSI method (both topics are discussed later), 2000 b/s are used to undertake this function, leaving only 2000 for telemetry.

Clearly, a period of transition will occur during which DS1 signals with the conventional SF and ESF coexist in the same CO. For example, if the embedded operations channel (4000 b/s channel) is not implemented, then the D bits are all set to 1 by the sending carrier system.

CSUs, discussed below, can actually help in the migration to an ESF-based plant because they mitigate the need to replace or upgrade the SF-type CPE already deployed. The CSU can, if appropriately engineered, provide conversion between SF and ESF signals, thus allowing end-users to retain the existing multiplexing equipment while taking advantage of the monitoring features available with ESF.

We note that some equipment, such as a CPE T1 multiplexer, needs its own internal network management channel. In theory the equipment could be 99.6% efficient on a DS1 link if the ESF format could be invoked (which depends on the telephone company facilities at hand), and network management information could be placed in the 4 kb/s channel discussed earlier. In practice, this data link must be left unused (for carrier purposes), so that the network management information must appear elsewhere in the DS1 channel, thereby "wasting" a small amount of bandwidth.

AT&T announced ESF in the early 1980s, but did not begin to deploy these facilities and achieve measurable penetration until the late 1980s. AT&T has publicly committed itself to implementation of ESF [3.16]. The ESF format has been widely adopted in the industry, involving changes to channel banks, digital switch interfaces and digital cross-connect systems [3.17, 3.18]. CSUs, which interface the transmission facility to the user's termination equipment, can store and display facility performance parameters, such as errored seconds, failed seconds, bit error rate, and others.

DS0 Synchronization

After the DS0 signals are demultiplexed and individually reassembled, they are just serial bit streams that need a reference point to give them meaning. The framing bit is identified by the receiving equipment (network equipment or CPE) to decode properly the DS1 signal. Clock signals allow the same function for a DS0 signal: a bit clock at 64,000 cycles per second tells when to examine the DS0 data stream for bit values. In Digital Data Service (DDS), a low-speed private line digital service introduced by AT&T in the mid 1970s, a byte clock at 8000 cycles per second tells when the 8-bit character from one channel of a DS1 signal frame in a DDS application is ready for decoding [3.19].

3.1.4.2 Information Bits

The 8 bits in each time slot within a frame (SF or ESF) can be used in a variety of ways. The three schemes for information bit use are as follows:

1. "Voice, Nearly Eight Bits." This scheme was shown in Figure 3.5 above. To accommodate signaling requirements while still maintaining 8 (or nearly 8) bits of information in each time slot, the least significant bit is robbed and used for channel-associated signaling in every sixth frame. This leaves only 7 bits for the coded voice circuit in that particular time slot. This scheme

results in five out of six frames containing eight bits, with the sixth frame containing only seven information bits. This is often referred to as "7-5/6 bit voice coding," or "nearly 8-bit voice coding." The framing bits in this superframe are used alternatively as terminal framing bits, F_t, and signaling framing bits, F_s. The coded voice channel normally contains 8 bits—except in those frames following a transition in the F_s bit. In the 24-frame ESF, the framing bits contain a framing pattern sequence that identifies the frames from 1 to 24; the coded voice channel contains 8 bits except in frames 6, 12, 18, and 24.

2. "Data, 23 Time Slots." This frame format is used exclusively for transmitting data by a telecommunication carrier (may not be the method used by a CPE multiplexer). Data transmission must be accommodated over this format, which evolved principally to carry voice streams. The scheme is depicted in Figure 3.7. It uses a 12-frame superframe and shares the F-bit between alternating F_s and F_t bits. In this format, the first 23 time slots contain data bytes of either 6 or 7 bits, and the 24th time slot contains a 6-bit fixed framing pattern, a yellow alarm bit, and a remote signaling bit. The 8 bits in each time slot are numbered sequentially. For 7-bit data bytes, bits b1 through b7 form a 56 kb/s channel. For 6-bit data bytes, bits b2 through b7 form a subrate data channel (less than 56 kb/s), with bit b1 used as a subrate synchronization bit. For both 6- and 7-bit data bytes, b8 in each slot is used for network control.
3. "Eight-Bit Transparency." This scheme allows switching of 8-bit data bytes through a digital switching machine. In this scheme, the signaling bits are not inserted into the DS1 signal, which permits full 8- bit data bytes. This is commonly called 64 kb/s CCC and is discussed later.

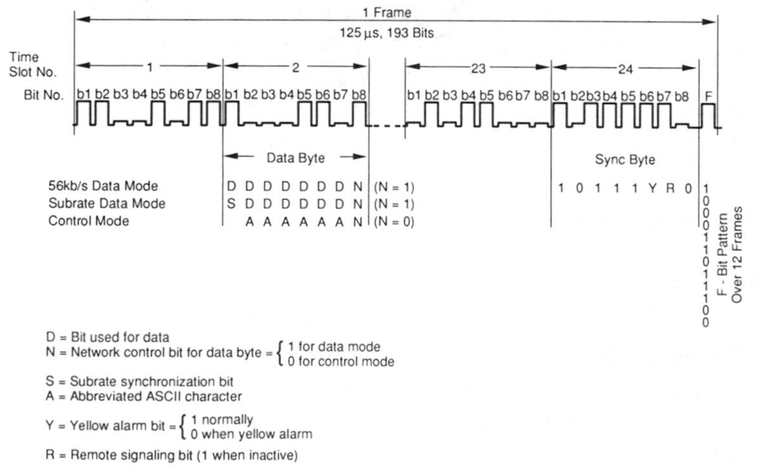

Figure 3.7 Data bytes within the DS1 signal.

Within the DS1 signal format, a number of alternatives exist to provide an administrative channel to transmit control and maintenance information between the ends of a DS1 line: (1) For data applications where the 24th time slot is used as a sync byte, 1 bit (b7) is reserved. This bit, often called the "remote signaling bit," "R," or "data channel" is used in DDS for the digital transmission surveillance system and provides the capability for an 8 kb/s channel. (2) Although not exactly a common data link in the sense of the preceding case, the data port method provides a data link between terminals. The data port does this by taking a full time slot of 8 bits in the DS1 signal, providing a link with a capacity of 64 kb/s. (3) ESF automatically includes the capability for a 4 kb/s facility data link in the F-bits.

3.1.4.3 Channel-associated Signaling Bits

The 193rd F-bit signaling described in some detail above deals with the management of the DS1 facility itself. Each individual voice channel requires its own signaling information (to indicate, for example, call setup or call completion). These signaling bits are contained in a small fraction of the time slots (in the b8 position) and are used for per channel supervision and addressing. Three schemes, already briefly discussed, are:

1. *A,B-Signaling*. This signaling scheme uses the 12- or 24-frame structure, and robs the b8 bits of each time slot in every sixth frame for signaling purposes, using them alternately to form two signaling channels. The two signaling channels in the SF format are designated A and B. These two signaling channels provide a four-state signaling capability for each voice channel, at a rate of 0.667 kb/s (8 kb/s, divided by 12) per signaling channel. The use of the signaling bits is equipment-dependent.
   In SF, the F_t bits alternate every frame, providing the sequence 101010. The receiving channel bank or PBX can identify this unique sequence in the incoming digital stream to maintain or reestablish frame boundaries. Frame 6 is identified by the fact that it occurs when the F_s is "1," preceded by three F_s that were "0"; frame 12 is identified by the fact that it occurs when the F_s is "0," preceded by three F_s which were "1;" the receiving channel bank can easily identify this sequence, in the absence of severe line errors or impairments, and thus identify the desired frames that contain VF signaling information. The signaling information must be transmitted along with the PCM samples; to achieve this, the channel bank will "rob" or "share" the least significant bit (b8) from the user data stream; consequently, this bit alternatively carries information or signaling data. For five frames, b8 will contain voice bits; on the sixth, it will contain a signaling bit. Thus, the sequence of bits for a given voice-data channel will be:

b8 = v; b8 = v; b8 = v; b8 = v; b8 = v; b8 = s; b8 = v; b8 = v; b8 = v; b8 = v; b8 = v; b8 = s;
(v = voice, s = signal). The sixth bit is also called the A-bit, and the twelfth the B-bit. These combination of bits will allow the end-user station equipment to carry out its signaling protocol, which involves indicating things as idle, busy, ringing, no-ringing, loop open, *et cetera*. For data applications, the A,B-signaling has no relevance; however, other types of in-band signaling may be required.

2. *Network Control Signaling*. This signaling scheme is used for data transmission in the DS1 format by a telecommunication carrier (may not be the method used by a CPE multiplexer). The last bit, b8, in each data time slot of every frame is used as a network control bit. This bit transmits channel status information; it is a "1" if data is being sent over the channel and a "0" in the control mode. In the control mode, an abbreviated ASCII character may be transmitted in the remaining bits.
3. *A,B,C,D-Signaling*. The four signaling channels in the ESF format are designated as A, B, C, and D. They may be optioned for only A, B, in which case they are the same as A and B for the SF format. When needed, the four signaling channels provide 16-state signaling capability for each voice channel at a rate of 0.333 kb/s (8 kb/s divided by 24) per signaling channel.

3.2 CHANNEL CODING METHODS AND ISSUES

A method is needed to apply the composite information signal (whether SF or ESF) onto the transmission channel. After considerable study in the early 1960s, the decision was made that the DS1 pulse train for direct application to a T1 copper facility, at the output of the multiplex-terminal or other termination equipment, should be encoded with a bipolar scheme. The choice of bipolar was intimately related to the characteristics of the copper medium. In bipolar coding, alternating positive and negative pulses represent one state (a binary "1"); absence of pulses represents the other state (a binary "0") (see Figure 3.8). For distances exceeding one mile on typical gauge copper facilities, regenerative repeaters are required, as discussed earlier.

Bipolar coding (also known as AMI, alternate mark inversion) has been selected for T1 links because of the following desirable properties, among others:

1. Clocking need not be absolutely perfect to be able to decode the bit; in Figure 3.8, the "sampling" need not be done exactly at the maximum signal value: any time energy is present, a "1" is recorded (even if that energy is not the maximum energy); if no energy is present, then a "0."
2. Any single bit error can be detected, and this is useful because copper loops are subject to a number of natural (e.g., lightning) and man-made (induction

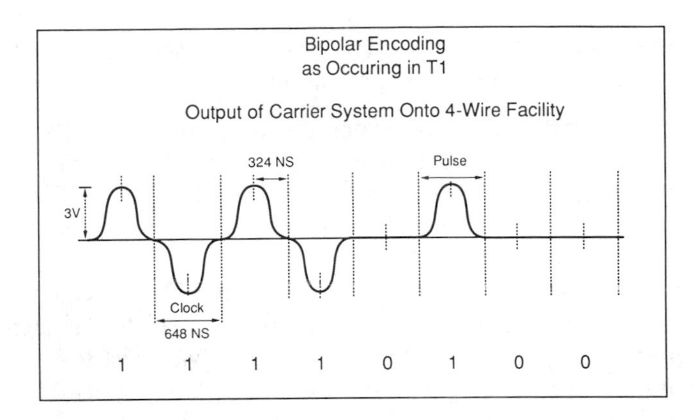

Figure 3.8 Bipolar coding performed by channel banks and digital repeaters.

motors, elevators, power lines) noises. If an error occurs in a 1 bit position, thereby converting it to 0, adjacent 1s will be of identical polarity, which is easily detectable because it violates the polarity rule. If an error occurs in a zero, converting it to a 1, two successive 1s of identical polarity will result, which also violates the polarity rule. These events are called *bipolar violations*.

3. The bipolar method does not build up a dc component on the metallic line.

The characteristics of the output signal of the termination device are:

- 1,544,000 clock intervals, each of 648 nanoseconds (the time between the vertical lines in Figure 3.8);
- "1" can either be a positive or negative pulse, lasting at least 50% of the clock interval (this is called the pulse duty cycle); the polarity of any pulse must be opposite of the polarity of the last pulse; in T1 the pulse is nominally 3 V in absolute amplitude;
- "0" is coded by no pulse; and
- At least 12.5% of the bits, on average, must be a "1" or repeater synchronism cannot be maintained (see below).

This coding method allows a BER in excess of 10^{-6} 95% of the times, with repeater spacing of approximately one mile. In fact, this scheme is fairly robust, and can be employed on transmission facilities that have poor S/N and, consequently, poor transmission characteristics. The 50-mile limitation on T1 facilities is due to the timing jitter accumulated in a long sequence of repeaters.

3.2.1 Ones-Density Issues

A repeater's job is to regenerate the signal; this is accomplished by sampling the bits at appropriate intervals. Repeaters derive the sampling timing from the user's bit stream. When too many zeros are received, no pulses feed the timing recovery circuitry and mistiming results. Three situations may then arise:

1. The repeater may reestablish synchronization when a sufficient number of one-bits are received; in the meantime, resulting link-level errors are presumably corrected by layer 2 protocols (HDLC, SLDC, Bisync, *et cetera*), which are associated with individual subrate channels, or even associated with the T1 multiplexer aggregate.
2. The repeater (particularly if of vintage technology) may generate a continuous alarm condition, notifying the CO that it has lost the signal.
3. The repeater may shut down completely, and will have to be reset from the CO, with a corresponding user outage. When a user puts a large amount of subrate channels on a single DS1 facility, an outage can be fairly severe, particularly if it is part of a backbone high-speed network.

Newer repeaters are more tolerant of long streams of zeros. However, a long span may consist of dozens of repeaters, of various generations of equipment; clearly the "weakest" repeater in a line will be the determining factor.

At this time the CPE T1 equipment is "encouraged" to incorporate built-in circuitry to ensure that the ones-density is met. CPE T1 equipment for connection to the public network must comply with Part 68 and Part 15 (Subpart J) of the FCC's Rules and Regulations. Part 15 deals with radiation or conduction onto power lines. Part 68 deals with effects of user equipment on the network [3.20]. The "harm to the network" issue is not an academic one: equipment improperly engineered may damage the network; cause problems to the user, or other network users; or result in erroneous billing. For example, when the CSUs or the user's T1 termination equipment fails to send keep-alive signals, some of the older metallic repeaters could introduce noise on other T1 circuits. Users of these circuits will be jeopardized until the offending circuit is found. Low ones-density will introduce jitter on the offending circuit, which translates into data errors and retransmission delays.

Until a 1987 FCC change, transmission equipment was legally required to maintain the ones-density ratio (more on the change is discussed in Section 3.3.2). The assumption on the DS1 stream was that at least one-eighth (12.5%) of the incoming bits were "1s"; if that ratio was not maintained, synchronization would be lost, and an error resulted (which in turn would have required a data link-level protocol-initiated retransmission to compensate, as discussed in Chapter 1). When the DS1 channel contains voice signals, "work-around" techniques such as changing the eighth zero to a one would not be a major problem because the distortion in the voice message is minimal and the ear is very forgiving (voice acoustical information is redundant). Hence, the approach of substituting a 1 in the 7th bit position of an all-0s word, would be acceptable for voice, because under both the A-law and the μ-law, the difference between a 00000000-coded sample and a 00000010-

coded sample is acoustically imperceptible to the ear (the reason for substituting the 7th and not the 8th bit is because the 7th bit is not affected by robbed-bit signaling). Even when the 00000000-coded sample represents a signal corresponding to the modulated output of a modem, the problem is bearable, as the receiving modem is not affected by this small change in the signal power spectrum. However, when the bits of a DS1 channel represent directly input digital data (such as DDS or ISDN), alteration of the stream is not acceptable, and a technical solution to be incorporated in the transmission hardware must be provided.

Code Substitutions

Under many conditions, a fixed code may be substituted by the network equipment for the contents of a particular time slot. Such substitutions occur in the form of: "0" code suppression, unassigned channel code–idle channel code, and certain alarm codes [3.3, 3.4].

1. *Zero Suppression Codes*. To eliminate the possibility of too many consecutive zeros occurring in the DS1 signal, various traditional substitutions are made for time slots containing all "0s," in both voice and data applications, or a "1" in the first bit position with the remainder containing all "0s," in data applications. The two methods are: (i) "b7 forced," and (ii) "b4 and b5 forced." In the first method, when the 8-bit voice code (b1 through b8) for any time slot consists of all "0s," the second from the last, b7, is forced to a "1," so that 00000000 becomes 00000010. In the second method, for either an all-data or combined data-voice DS1 format, the word 00000000 (b1 through b8 all "0s") is modified to 00011000, and the word 10000000 is modified to 10011000 for all data channels.
2. *Unassigned and Idle Channel Codes*. Three fixed codes are used to indicate unassigned or idle channels. (i) For channels that are not assigned, a fixed all "1s" code (b1 through b8) is inserted into the corresponding time slot (11111111). (ii) For either an all-data or combined data-voice DS1 format, the fixed code 00011000 (b1 through b8) is inserted into those DS1 time slots that represent unassigned channels. (iii) For 4ESS digital switches, the idle channel code 01111111 (b1 through b8) is inserted in the appropriate time slot. (Note that ESS is a registered trademark of AT&T.)
3. *Alarm Codes*. Two alarm codes are used to indicate an out-of-sync or trouble condition. The codes are 00011010 and 01111001, respectively.

Until now, a user of DS0 digital facilities in the United States has been restricted from accessing digitally the full 64 kb/s bandwidth of the channel [3.21]. In voice, the 8th bit has been used for in-band signaling as discussed earlier (though only the 6th and 12th occurrence is really affected). In speech, the ones-density requirement discussed earlier is met; this same condition cannot be guaranteed for data. Changing a customer data bit would not be acceptable, as that may imply a substantial difference in the received data (for example, 0001 and 1001 are quite different numbers). Therefore, in DDS applications, 1 bit out of 8 in the customer channel is reserved in each frame to inject a "1"; this bit is called the "network control bit," and is not available to the user. While this fixes the ones-density problem, it imposes a bandwidth restriction of 56 kb/s instead of a full 64 kb/s. Because the network control bit is "1" when customer data is being sent, the user data stream is not important; when a 7-bit control code needs to be sent for signaling purposes, the network control bit is "0"—the control code had at least one "1" by design; the 8th bit is used as a discriminant to pick out the control codes being sent. (Therefore, within a DDS channel the user can send as many zeros as he or she wants. Because every DDS channel in the DS1 stream has the network control bit, and every voice channel has no all-zeros sample, the density and consecutive-zeros restrictions are met by all DS0s, and so by the DS1.)

Clear Channel Capability (CCC)

The implication of the above factors has been the deployment of digital facilities providing user access to 56 kb/s, with an ensuing "loss" of 8 kb/s of usable bandwidth. At times, this has caused complications for the end-user, particularly when interfacing with an international facility (for example, a 64 kb/s satellite channel).

The ability to provide a CCC, also known as "Clear 64," is not a trivial one. Regenerators on T1 digital lines, and loop-timed DS1 terminations, require a specific density of pulses to retain clock synchronization. As discussed, the predominant method of ensuring adequate ones density at the DS1 output of digital channel terminating equipment today is to check each voice PCM byte for an all-zeros condition, and if it occurs, to force bit 7 of the PCM data byte to a "1." This technique, called *zero byte suppression*, clearly precludes a 64 kb/s CCC. To provide 64 kb/s CCC, a carrier system will need to employ a different method of ensuring that ones-density requirements are satisfied.

ISDN facilities will solve both problems of bandwidth and transparency. ECSA standards recommend *bipolar with 8th zero substitution* (B8ZS) capability for new carrier equipment. Some telecommunication carriers may also require a zero byte substitution method known as *zero byte time slot interchange* (ZBTSI), which only affects DS1 path termination equipment but introduces additional transmission delay. As a required option, a ZBTSI capability implemented or controlled via the data link of the ESF format must be available at both the remote terminal and the local terminal. Both B8ZS and ZBTSI introduce additional delay into the four-wire digital path.

The B8ZS Technique

The ones-density condition can be met by employing an algorithm that does not arbitrarily insert a one in the user's signal or create a bipolar violation. With CCC, the CSU will allow any input data stream; internal data coding via a B8ZS chip will ascertain that the synchronization bits will be present in the appropriate density. This advancement will benefit the end-user directly in additional bandwidth and fewer link-level errors. Tariffs for CCC were filed in 1989 by at least one BOC. We note, however, that general availability of Clear 64 should be expected no sooner than the early to mid-1990s. B8ZS appears to be a favorite of ISDN for provision of the Clear 64 service. The technique has been studied by the T1X1.4 subcommittee of ECSA.

With B8ZS, deliberate bipolar violations in the DS1 signal stream are substituted whenever eight consecutive zeros occur in the bit stream from which the DS1 signal is derived. This enhancement avoids the need to alter the PCM stream or the user-produced DS1 stream to control the "1s" density and distribution [3.19]. The 12.5% average "1s" density, and no 15-consecutive "0s" requirement within the inner workings of the DS1 facilities and the core equipment are dictated by those circuits that extract the clock from the received bipolar signal; even though the design of systems that can retain synchronism for longer periods of time is possible, the large embedded base of older equipment would still demand enforcement of the earlier restrictions. This is why the work-around must be applied at the periphery of the network using the B8ZS approach.

Figure 3.9 depicts graphically the principles behind this technique. The middle portion of the diagram shows the coded output of a carrier system without B8ZS; the lower portion shows the effect of B8ZS: two specific bipolar violations (two contiguous pulses of the same voltage polarity) are deliberately inserted, replacing the string of eight zeros. The first bipolar violation occurs on the fourth 0-bit (this

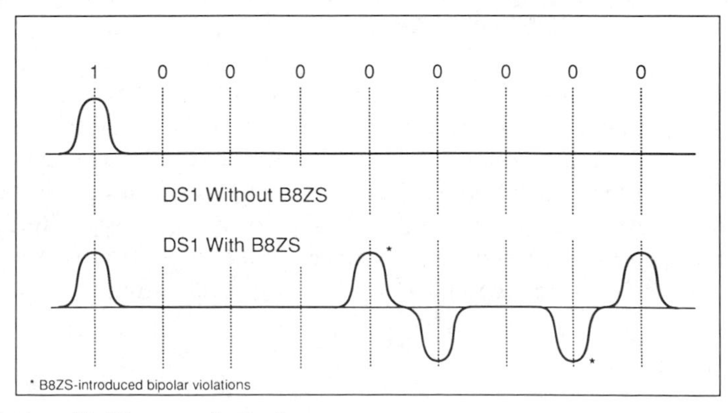

Figure 3.9 Bipolar with 8th zero substitution.

pulse is of the same polarity as the last pulse—this was a 1, which in this particular signal sequence had a positive polarity); the second bipolar violation occurs in the 7th 0-bit position. We note that the 8th 0-bit is also given a pulse, though normally a 0 would be coded without a pulse; this is done to keep the average dc power to zero because an equal number of positive and negative pulses is added.

As described earlier, bipolar violations are considered a sign of an impaired line and transmission degradation; how, then, can this pattern be accepted by the endpoint channel banks or other termination equipment? The answer is simply that the specific pattern just described at specific bit positions is recognized as a legal indication of eight successive zeros.

Since the early 1980s, new DS1 termination equipment manufactured for AT&T was required to include the CCC eliminating the ones-density and the consecutive-0s restrictions by employing the B8ZS method [3.15]. Equipment that accepts B8ZS can accept a standard non-B8ZS AMI stream, but not the other way. This means that a newer channel bank at one end of the link can be installed without having to replace both ends; eventually when both ends are replaced, then the link becomes a *bona fide* clear channel. The next couple of years should see a proliferation of this type of equipment, also with the direct incorporation of B8ZS hardware in the CTE or CSUs.

Some of the recent CPE, particularly in the T1 multiplexer arena, is engineered to be user-selectable in terms of line coding. SF or ESF can be selected in accordance with the properties of the available DS1 link. B8ZS can be selected in accordance with the properties of the available DS1 link. Users of carriers' DS1 facilities with B8ZS can select the B8ZS option on their equipment. This gives full use of the DS1, except possibly for the internal network management channel, which typically requires 16 kb/s.

We note that M13 telephone company multiplexers (combining 28 DS1 into DS3) attempt to fix bipolar violations they encounter. This means that the provision of an end-to-end channel supporting CCC necessitates the replacement or retrofit of these multiplexers. ISDN will utilize AMI, ESF, and B8ZS, as discussed in the next chapter.

The competitor to B8ZS, ZBTSI, seems to be less favored; this is due, in part, to the fact that it introduces a 1.5-ms delay in the final delivery of the bit stream [3.22]. ZBTSI is described in T1.107-1988, and is elaborated on in [3.12]. It meets the present DS1 pulse density restrictions by substituting a variable address word for bytes with all zeros, referred to as zero bytes. This technique requires an additional 2 kb/s of overhead information; this information is provided in the data link of the ESF, which is mandatory for this application.

3.2.2 DS2 Stream

Briefly revisiting the DS1 format, we can summarize the process by saying that 24 channels are byte-interleaved, plus the injection of one signaling bit (namely, the eight bits from a PCM sample from each call is treated as an ensemble). In contrast to this scheme, the DS2 signal is bit-interleaved from four incoming DS1 streams. The DS2 stream appears as follows:

bit 1 call 1 stream 1, bit 1 call 1 stream 2, bit 1 call 1 steam 3, bit 1 call 1 stream 4;
bit 2 call 1 stream 1, bit 2 call 1 stream 2, bit 2 call 1 steam 3, bit 2 call 1 stream 4;
. . .
bit 8 call 1 stream 1, bit 8 call 1 stream 2, bit 8 call 1 steam 3, bit 8 call 1 stream 4;
bit 1 call 2 stream 1, bit 1 call 2 stream 2, bit 1 call 2 steam 3, bit 1 call 2 stream 4;
. . . *et cetera*.

We note that after the sequence of eight segments for the first set of calls (first three lines just listed), the pattern moves on to a sequence of eight segments for the second set of calls (last line just listed), *et cetera*. Thus, a four-times bit interleave results. The resulting stream is a complex mix of bits coming from all 96 voice channels, and complex electronics are necessary the manage this ensemble of signals.

The DS2 format has been around for a number a number of years; however, in the future its importance will probably decrease; more transmission facilities now operate at the DS3 level. When channel banks output a signal at the DS2 rate, then it must be multiplexed one more time (seven-times bit interleave); thus a channel may first be multiplexed to the DS2 rate and then multiplexed again to the DS3 rate. This proves to be both inefficient and hard to manage because to recover a single DS0 channel, or even a DS1 stream, the entire DS3 stream must be demultiplexed. Proposals have been made for new DS3 formats, notably the SYNTRAN format, which would bypass the DS2 rate, and combine the channel bank outputs of 28 DS1s directly to the DS3 rate, as discussed later.

3.2.3 European Digital Systems

This section briefly describes some key features of the European digital system (countries in Africa and South America, including India and Mexico also employ this standard). These standards, based on CCITT's G.700 Series, are colloquially called E1 or CEPT-1.

The digital hierarchy is based on blocks of 32 PCM channels, which are aggregated to form a 2.048 Mb/s channel. Four of these aggregate signals are multiplexed to obtain a 8.448 Mb/s channel; four of these channels are multiplexed to obtain a 34.368 Mb/s; and four of these channels are multiplexed to form a 137.462 Mb/s channel.

The E1 system uses bipolar line coding, allowing clock information to be embedded in the data stream. However, different voltages, pulse shapes, and line impedances are specified in G.703, compared to the U.S. specification. It employs four copper wires, two to transmit and two to receive. While a μ-law is used in the United States for assigning the PCM encoding values, a different method, called the A-law, is used under the E1 method.

As indicated above, in the United States the user has traditionally not been "allowed" to send more than 15 consecutive zeros, and B8ZS is now being introduced. By contrast, E1 systems have always used a technique called *high-density Bipolar three zeros suppression* (HDB3) to guarantee the 1s density, regardless of the user's data [3.23].

Both framing and VF signaling are handled differently compared to the United States. The G.704 specification divides the 2.048 bearer into thirty-two 8-bit slots. The first slot, time slot #0, is used for synchronization and has the provision for maintenance channels and CRC in-service performance monitoring. Time slot #16 is used for voice channel signaling if required (user time slots or bits are not robbed to provide signaling) [3.23].

3.3 CHANNEL SERVICE UNITS

The CSU is a protective interface designed to connect customer-premises data equipment to a carrier's digital transmission line, either a DDS facility or a DS1/T1 facility. It is thus an interface between the public (switched) network and the user's digital telecommunication equipment (see Figure 3.10).

The basic function of a CSU is to ensure that the appropriate digital signal is maintained into and out of the network. Additional features of the CSU provide for test and monitoring access of the signals being sent and received by the DTE, in order to facilitate fault isolation. From a macroscopic point of view, the CSU equipment provides access to a digital line, as a modem provides access to an analog line. The functions of the analog and digital termination equipment are, however, widely different. In digital transmission, precise synchronization is essential and the CSU is a key element of this synchronization process. CSUs must comply with FCC Parts 15 and 68, and should comply with other ECSA and carrier-specific standards such as Bell Reference 62411. The purpose of FCC Part 68 is to ensure that individuals connecting to the public network do not harm the network, including creating an impact on other users, permanent damage to the network, billing protection, and affecting access for the hearing impaired. Circuit-derived power has been used to energize the critical circuitry in the CSU, which includes the continuity signal circuit, the pulse density, and the keep alive functions. However, circuit performance issues are not covered in Part 68, for example, BER levels.

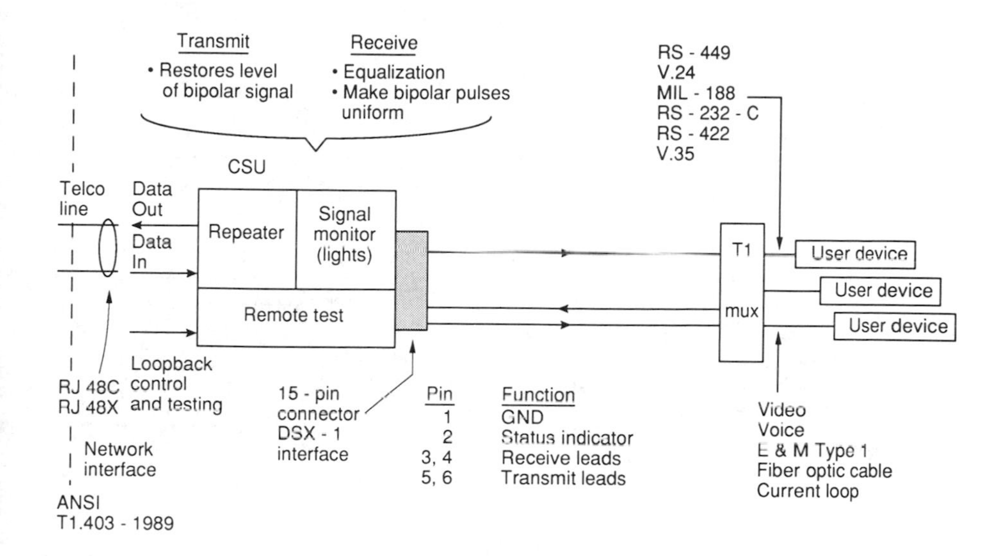

Figure 3.10 CSU internals.

The CSU has two sections: a transmitting side and a receiving side. The function of the transmitting side electronics is to regenerate the digital DS1 signal received from the user's equipment, to check for bit stream errors, and to apply the regenerated signal to a DS1 transmission facility (typically a copper T1 loop). The function of the receiving side electronics is to regenerate the DS1 signal received from the network, to check for remote loopback codes, and to apply the signal to the customer equipment. If the T1 network is private and uses local lines on noncarrier provided microwave, satellite, or fiber optics facilities, the CSU is not strictly necessary, although it may be useful for the purpose of testing.

Interconnection at the network interface is via one of the four Universal Service Ordering Code (USOC) connectors RJ48C, RJ48X, RJ48M, and RJ48H shown in Figure 3.10. The 8-pin connectors in parts (a) and (b) of Figure 3.11 have the same pin assignments, but the connector shown in (b) provides a physical loopback when unplugged. The 50-pin connectors in parts (c) and (d) are physically the same but have different pin assignments, as described in T1.403-1989 [3.14].

A number of newer, more sophisticated CSUs support both the standard bipolar coding (AMI) and the B8ZS line coding; in terms of the framing format, these CSUs support SF and ESF. They also allow code translation when the T1 facility and the user equipment use different line coding techniques. When this translation occurs, the CSU must continue to generate error-free transmission that meets the ones-density requirements, for performance reasons.

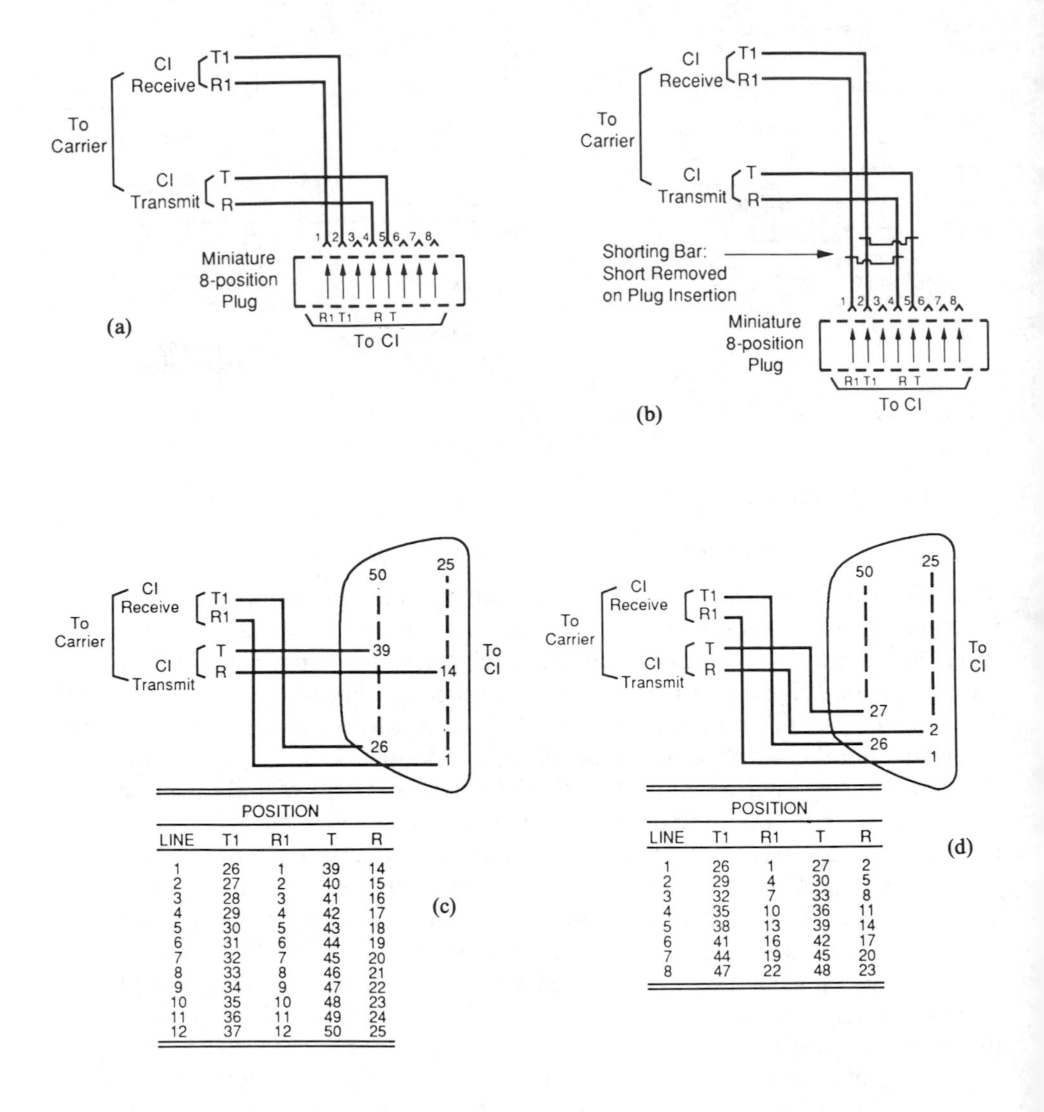

(c)

POSITION				
LINE	T1	R1	T	R
1	26	1	39	14
2	27	2	40	15
3	28	3	41	16
4	29	4	42	17
5	30	5	43	18
6	31	6	44	19
7	32	7	45	20
8	33	8	46	21
9	34	9	47	22
10	35	10	48	23
11	36	11	49	24
12	37	12	50	25

(d)

POSITION				
LINE	T1	R1	T	R
1	26	1	27	2
2	29	4	30	5
3	32	7	33	8
4	35	10	36	11
5	38	13	39	14
6	41	16	42	17
7	44	19	45	20
8	47	22	48	23

Figure 3.11 Connector pin assignments: (a) 8-position RJ48C; (b) 8-position RJ48X; (c) 50-position RJ48H; and (d) 50-position RJ48M.

Both the FCC and AT&T and BOCs have issued guidelines for manufacturing carrier terminal equipment. An area that has caused some ambiguity pertaining to CSUs has to do with keep-alive methods. These are patterns sent by the termination equipment (CSU in particular) to the CO, should the information signal disappear. Traditionally, three types of keep-alive signals have been possible:

Type 1: Transmission of unframed ones. This allows the carrier to detect a loss in framing, which in this method is interpreted as a link problem;
Type 2: Transmission of a framed signal. This prevents the carrier from detecting a false outage simple because a user disconnected a cable. A quasirandom signal source (QRSS, a standard 1,048,575-bit pattern) is used;
Type 3: Loopback transmission of received signal, without modifications.

Currently, different carriers employ different methods. Termination equipment may have to provide all three methods, with the appropriate one being invoked by the user, depending on the carrier's requirement. Standardization to a single method would reduce equipment costs.

3.3.1 Digital Link Testing Capabilities Provided by the CSU

Networks in general, and private networks in particular, work best when the facilities are able to rapidly detect and isolate line and equipment problems. Sophisticated CSUs can be a useful tool in the process by providing extensive onsite diagnostics in the form of jacks, LED displays, and loopback switches. Some T1 multiplexers have integral CSUs; the advantage of this approach is that the user can obtain the total system from one vendor, to minimize potential incompatibilities and facilitate responsibility assignment in case of a technical problem. CSUs facilitate testing both the span lines and the CPE. The CSU may employ in-band and out-of-band signals at the network interface to allow maintenance personnel to conduct span diagnostics locally or from a remote location. Front panel switches are often provided to allow users to generate line-loopback and test-loopback signals to diagnose local equipment problems. The typical signal access points on the CSU are:

- Line in, breaking toward the network interface receiver;
- Line out, breaking toward the network interface transmitter;
- Equipment in, breaking toward the CPE receiver;
- Equipment out, breaking toward the CPE transmitter.

These tests are service-affecting; namely, they disrupt normal communications. Other signal taps can be done while the facility is in normal operation.

In addition to the service-affecting loopback tests described, some CSUs (and many T1 multiplexers for that matter) can monitor the performance of the T1 link on a real-time basis while it is carrying normal traffic. The monitoring is continuous over time and thus is very effective. These CSUs can also generate alarms based on selectable thresholds (for example, loss of line signal for more than 200 ms), and report alarms to a central location. This saves time and money, as it avoids the need to connect external test equipment [3.16]. Additional discussion of testing is provided below.

3.3.2 Recent Changes to FCC Part 68

The FCC's Part 68 required until recently that CSUs must be powered from the network with a 60-mA current. Some carriers interpreted that requirement to be in effect only when the T1 is rendered on nonmultiplexed metallic facilities (i.e., true T1); for DS1 signals derived from carrier systems on fiber loops, the powering is intrinsically more complex to provide, and may have to be supplied locally. On October 19, 1987, the FCC issued a news bulletin stating they had "eliminated the requirement that carriers provide line power on 1.544 Mb/s service, as well as an associated requirement that terminal equipment connected to 1.544 Mb/s service contain a continuity of output capability as a registration condition under Part 68 of the rules." The paragraph they deleted, 68-318(b) of the FCC Rules and Regulations, required that signals that come into the DS1/T1 network from registered network channel termination equipment (NCTE) (1) provide one of three types of keep-alive signals, (2) provide pulse density maintenance, and (3) be powered by 60 mA from the network [3.24].

Prior to the rule specifying the specs for connection to the network (Docket 81-216), the FCC used an interim registration specification based on Bell System Specification 62411. This specification required such features as loopback, line powering, pulse density, and keep-alive signals. Docket 81-216 adopted most of these, with the exception of the loopback requirement (the issue was that the loopback has nothing to do with the question of harm to the network). Loopback was not required by law after 81-216 was incorporated into Part 68. The commission's original reasoning was that the market should dictate whether loopback would be required or not. Most CSUs provide loopback capabilities because loopback is an essential network management function.

In November 1985, a committee of ECSA requested that fiber optic circuits and circuits derived from pair gain systems should be exempt from the line powering requirement because of the difficulties of providing power over fiber cables. In January 1986 the committee recommended the drafting of a standard for dry circuits (i.e., non-CO power). In March 1986, five BOCs petitioned the FCC to modify the section on line powering. These changes were generally supported by the industry.

On October 29, 1986, the FCC issued a news release [Notice of Proposed Rule Making (NPRM)] stating that a change, identified as Docket 86-423, was proposed, deleting all of Section 68.318(b). In January 1987, several parties provided comments on the proposed change. Comments varied widely and did not support a common position. In October 1987, the FCC issued a report and order on the docket. In November 1987, the FCC issued an erratum identifying the specific implementation dates for Docket 86-423.

Although the BOCs only asked for the deletion of line power, in some arguments they stated that the keep-alive signals were not necessary. The FCC also proposed elimination of the pulse density requirements because some believed that this might no longer be an issue of harm to the network. The newer repeaters deployed since 1987 do not have synchronization problems, although systems already in the field may indeed continue to experience the problem.

Results

With the new docket, the power source has to be provided by the user if the service provider states that they will no longer supply the 60 mA after 1989. The carriers had to provide power until December 1989. Any new installation occurring after February 18, 1988, did not have to provide line power (see Table 3.3) [3.24].

Given the time lead, and the fact that many newer models of NCTE (including CSU) already have dual-powering capabilities, the powering issue proved not to be a critical problem to the user. The circuit would have to be taken down for a few minutes to implement the change.

The pulse density requirement was poorly defined in 68-318(b), and requests for clarifications had been filed. In announcing its decision, the FCC said "there should be a two-year transition period for carriers to adjust to the possibility of connection of equipment without continuity of output (pulse density maintenance)." Some manufacturers could interpret this to mean that they have no requirement to maintain pulse density. In the existing plant, low pulse density will lead to jitter, which, in turn, increases the BER. Specific carriers can still tariff a transmission offering that requires the density to be maintained (for example, as specified in PUB 62411). To achieve a viable BER, the customer equipment must continue to provide pulse density management.

In the future, the ANSI standards prepared by the EIA TR41 Committee and the ECSA T-1 Committee will be the basis for tariff requirements and equipment manufacturing. To ensure the quality of service, any equipment that is purchased should comply with these ANSI specifications, as well as the carrier's criterion. CSUs also provide a diagnostic function, as discussed earlier; in-service monitoring, statistic collection, and alarming are typically provided. Users should analyze their particular needs when selecting NCTE, from the basic Part 68 minimum requirement, to features such as pulse density maintenance, loopback, keep-alive signals, and diagnostic capabilities.

Table 3.3
FCC Part 68 Rule Change—Effect on NCTE

	BEFORE		AFTER		
Function	*Telco Equipment*	*User Equipment*	*Telco Equipment*	*User Equipment*	*Comments*
Power source	60 mA line	Fiber optic and MUX circuits[a]	None in 2 years	Local	[a]Some MUX and fiber optic circuits were user-powered by necessity.
Pulse density maintenance	n.a.	>12.5% <80 zeros	n.a.	>12.5%[b] <15 zeros	[b]Required for performance, not legal harm reasons.
Keep alive[c]	n.a.	All-ones or loopback	Use network AIS signal	n.a.	[c]If pulse density is maintained, this keep alive is not needed.
Loopback[d]	n.a.	Network or user generated	n.a.	Network or user generated	[d]Loopback is not required by law, but is required by some tariffs; some telcos are installing internal network loopbacks.

Note: n.a. = not applicable.

3.4 DIGITAL CARRIER EQUIPMENT

Several types of network equipment are used to provide digital connectivity. The equipment can be grouped into three general component categories: channel banks, multiplexers, and cross-connects (refer again to Figure 3.1).

Channels banks (also called terminals) take an analog input and transform it, through the use of sampling and encoding, into a digital stream. Eventually, with the ubiquitous availability of digital circuits right to the customer premises, and the general deployment of digital switches, this type of equipment will no longer be needed in the network.

Digital multiplexers provide interfaces between the different bit rates in the digital network. This requires that blocks of data streams be stacked before being applied to a high-capacity medium. Multiplexing equipment generally includes failure detection, alarm, and automatic protection-switching features, in addition to the intrinsic bit-interleaving and framing function. Digital multiplexers can be located in the network (telephone company-owned), or can be at the customer location as CPE, such as T1 multiplexers. Transcoders also belong to this class of equipment.

The *digital cross-connects* are the interconnection points for terminals, multiplexers, and transmission facilities. They are electronic equipment frames where cabling between the system components is cross-connected to provide flexibility for restoration, automated rearrangements, and circuit order work. These devices, initially used by telephone companies, are now being acquired by end-users, to facilitate (private) network rearrangement. Table 3.4 provides a listing of the key types of equipment.

3.4.1 Channel Banks

The channel bank is one of the more well-known types of termination-multiplexing equipment in the telephone plant. A digital channel bank digitizes analog voice (and radio program) signals, and multiplexes these channels into a higher hierarchical level. At the remote end, the opposite process is undertaken, demultiplexing the higher hierarchical level into individual channels [3.25, 3.26]. The distinction between channel banks and digital loop carrier systems is as follows: channel banks apply to the trunk side, while digital loop carriers applies to the loop side (to the feeder plant, in particular).

Channel banks and carrier systems can also be analog, supporting the older FDM hierarchy, as discussed earlier. Analog-type channel banks have reached a "mature" stage in their deployment, though a nontrivial number currently popu-

Table 3.4
Carrier Systems Termination Equipment

ADPCM Transcoder: Accepts two DS1-rate input PCM channels and code translates the two streams, using ADPCM, onto a single DS1 channel.

ADPCM Channel Bank: Equipment that directly accepts 48 analog voice calls, then digitizes and multiplexes them onto a single DS1; the digitization is done with ADPCM techniques.

PCM Channel Bank: Equipment that directly accepts 24 analog voice calls, then digitizes and multiplexes them onto a single DS1; the digitization is done with PCM techniques.

T1 Multiplexer: Equipment that accepts a large number of data stream on conventional computer interfaces (such as EIA RS-232-C), and multiplexes them in an efficient manner onto a DS1 transmission facility.

Data-Voice T1 Multiplexer: A T1 multiplexer that also accepts voice to be placed on the DS1 facility along with the data; while the digitization can be accommplished by PCM methods, it is normally done with ADPCM techniques.

lates the telephone plant, and may occasionally continue to be deployed where required. Users may have become familiar with the digital type—the well-known D3/D4 class; indeed the number of these devices has been increasing during the past 30 years.

Digital channel banks have two basic functions: They convert analog voice to digital code, and *vice versa*; and they combine or multiplex the resulting digital streams from several active sessions (voice or data) onto a single stream. TDM with byte interleaving is the norm. To achieve these tasks, digital channel banks have equipment to provide proper VF and signalization interfaces with the CO. In addition, they provide filters to limit the transmitter input frequency (300–3300 Hz), so that PCM techniques employing 8000 samples per second suffice for faithful signal reproduction, and provide means for controlling the timing and synchronization. The data communication practitioner may view channel banks as specialized T1 multiplexers. A variety of channel bank frame formats exist that correspond to generations of equipment (D1D, D2, D3/D4 Mode 3, and ESF); the D3/D4 is the most widely used format.

Channel banks generally consist of a common-equipment section (functions that are identical across all channels), and individual channel unit cards that interface to the specific type of voice (data) trunk in question (two-wire, four-wire, tie-line, *et cetera*). These channel units are characteristic to a particular channel, and their basic function is to provide the interface between the CO trunk apparatus and the user circuit terminating on the channel bank. The correct type of line card for the channel bank ensures proper transmission and network signaling functions for the particular circuit.

In digital loop carrier applications of channel banks, we generally find a remote module and a CO module.The remote module digitizes and multiplexes multiples of 24 voice channels. The CO module either reconverts the channels to analog (to interwork with an analog switch). A multiplexed link also may connect directly to a digital switch. We note that, except for private-line applications, the channel will be switched (fanned-out) at the CO according to the dialing instructions contained in the signaling bits. Eventually, the phone call will have to be converted by a far-end channel bank, back to an analog signal, until ISDN becomes available.

Channel Banks and T1 Multiplexer Comparison

Basic differences exist between T1 multiplexers and channel banks, though the similarities are substantial. The similarities include:

- T1 multiplexers drive DS1/T1 facilities. Channel banks also drive DS1 facilities. Bipolar coding must be provided by both pieces of equipment.
- Both T1 multiplexers and channel banks multiplex multiple lower speed channels into DS1-range channels.

- Both systems have to accommodate the 1s density rules; T1 multiplexers cannot interfere with the 193rd bit because it provides basic signaling functions and superframe synchronization (the frame synchronization requirement is more relaxed in the T1 multiplexer situation because it is of secondary importance compared to the establishment of the superframe boundary).

The differences include:

- Channel banks generally handle up to DS2 rates (ninety-six 64 kb/s streams) and operate point-to-point. Basic T1 multiplexers normally handle DS1 rates, and may be networked. High-end systems, called T1 resource managers (TRMs), can handle multiple T1 lines (a dozen or more) and provide major network management capabilities to the data communication manager.
- Channel banks are ideally suited for voice because their basic rate and bit pattern are matched to the management of voice signals, both for multiplexing at the DS1 rate, as well as in preparation for additional downstream multiplexing at DS2 or DS3 rates. T1 multiplexers are better suited to the management of subrate data streams, although voice applications are emerging. Most T1 multiplexers now also allow voice transmission using ADPCM, CVSD, or other techniques (PCM is also used, but less frequently). Also, T1 multiplexers are more flexible in managing the available bandwidth, often allowing software-based reconfiguration.
- A channel bank will only carry 24 channels on a DS1 facility; a T1 multiplexer will multiplex many more data channels onto a DS1 facility (typically ninety-six 9600 b/s circuits, or even upward of 200 low-speed data channels).
- The standard "input" rate of a channel bank is 64 kb/s (corresponding to PCM data rates); T1 multiplexers accept subrates (2.4, 4.8, and 9.6 kb/s), as well as superrates (generally up to 768 kb/s, for compressed video), and asynchronous data streams.
- Channel banks are designed to take analog input; T1 multiplexers are designed to take digital input (RS-232-C or similar).
- Channel banks provide A/D and D/A functions; T1 multiplexers simply provide bit-interleave functions, except for cases where they also accept voice input.
- Channel banks are designed for telephone company applications and require dc voltages to operate. T1 multiplexers are designed as CPE for end-user application and run on standard 110-V electrical systems.
- T1 multiplexers have more network management features, for port assignment, monitoring, accounting, and automatic rerouting. The TRMs, high-end T1 multiplexers, also include digital cross-connect functions (discussed later in this chapter), perform automatic rerouting, and terminate multiple DS1s or DS3s.

- T1 multiplexers can be networked in a mesh topology, while channel banks typically operate in a point-to-point arrangement.

3.4.2 Transcoders

If 32 kb/s ADPCM voice coding discussed in Chapter 1 were to emerge as a network replacement for the standard 64 kb/s PCM method, and if ISDN is able to accommodate it, then we would see the emergence of channel banks directly equipped to handle and code the 48 VF channels onto the DS1 signal. ADPCM channel unit cards for insertion on standard channel banks would evolve. This is the approach followed by a number of T1 multiplexer vendors with the inclusion of ADPCM channel unit cards; however, the problem is that the telephone network is not currently able to accept the signaling aspects of the new method. Therefore, for private off-net applications, we can apply this type of equipment; but to integrate the customer streams with the telephone network (in those cases where this is desirable or required), we need to employ a transcoder.

The function of a transcoder is to accept two DS1-rate input PCM channels and to code and translate the two streams, using ADPCM techniques, onto a single DS1 channel while still employing the SF/ESF framing formats of the DS1 standard. The exact name of the equipment is *ADPCM transcoder*. In reality, the equipment achieves a compression or concentration of the two streams onto one. Equipment that directly accepts 48 analog voice calls and digitizes and then packs them onto a single DS1 facility is also thought of as being of the transcoder family, though this is really a channel bank (the exact name of this equipment is *ADPCM channel bank*). The class of devices discussed here are variously called *T1 compressors*, *bit compression multiplexers*, *low-bit-rate voice (LBRV) terminal*, or *transcoders*. Transcoders should not be confused with *transmultiplexers*. The purpose of the latter is to provide a conversion between the digital TDM telephone hierarchy and the analog FDM hierarchy, as an intermediary evolutionary step toward an all-digital network; this type of equipment is intrinsic to a telephone plant, and is employed by the carriers.

The resulting grade of service of the voice channels on a transcoder will be somewhat diminished, compared to the standard service achieved by a traditional channel bank system. In particular, voice will experience some quality degradation, and the throughput of a channel employed for data applications will generally be limited to 4800 b/s.

In 1984, AT&T recommended an M44 transcoder format (based on the underlying D4 frame structure) that would allow interconnection of equipment from different vendors, given that these vendors adopted the proposed architecture. (AT&T's BCM 32000 has received the most attention in the trade press.) Transcoders started to appear in the 1982–84 time frame, though other forms were available earlier (to the carriers). After an initial surge of product announcements, the market slowed down. The transcoding method did not allow certain modems to operate at 9600 b/s over the compressed channel. Transcoders appear to have a limited future, and only a few thousand units have been deployed throughout the United States, according to experts [3.27]. Increased availability of bandwidth with wide deployment of fiber and ISDN (based on 64 kb/s channels) may neutralize this market in the early 1990s. ADPCM may reappear as a contender in the late 1990s, after the introduction and initial stage of ISDN, but not in the transcoder mode.

With the M44 coding format, up to 20 voice or non-DDS-type data channels feed into the user's transcoder. The transcoder encodes the 64 kb/s voice onto 32 kb/s streams; 44 voice or low-speed data channels (up to 4.8 kb/s) are aggregated onto the standard DS1 signal format. Four signaling channels at 32 kb/s each are reserved for necessary signaling functions. Because the output of the M44 scheme looks like a standard DS1 signal, it can go through existing CO equipment. The output of the M44 follows all the requirements of the DS1 bit signal. The multiplexer forms 11 voice channels and 1 signaling channel from each group of 6 standard PCM slots of the DS1. We note that because of the additional processing involved in compressing each PCM stream, the resulting 32 kb/s subchannels are not transparent channels (not to mention, "Clear 64"); as a result, this type of arrangement is limited in the type of traffic it can carry on a 2-for-1 basis [3.28]. In particular, modem-output type data is generally limited to 4.8 kb/s. Data channels at 9.6 kb/s as well as some facsimile, would require a full PCM slot (64 kb/s). For example, with a 9.6 kb/s data input to the M44, we would only be able to get 42 voice channels. The other two subchannels will be "consumed" by the one data channel.

The M44 is similar to a T1 multiplexer, and a poor one at that (because it only provides 44 channels and not 96 or more, which is typical of a regular T1 multiplexer). However, the advantage of this approach is, theoretically, that the compression scheme is a public, nonproprietary scheme. AT&T formally issued the interface specifications for the compression algorithm and format, so that independent vendors of multiplexing equipment could begin to design CPE M44 (AT&T's Technical Publications 54070 and 54015). The techniques used by the various T1 multiplexers are often proprietary and unique; this implies that one has to purchase both ends of the link from the same vendor [3.28].

One of the factors to take into account in transcoders is that of signaling. PCM channel banks have a well-defined way to transmit in-band signaling, with bit-robbing techniques in the least significant bit in the 6th and 12th frames. AT&T's approach involves placing the signaling data for 44 channels in four 32 kb/s slots dedicated to signaling-only functions. Many of the products now on the market for ADPCM use proprietary signaling schemes in an effort the maximize the bandwidth efficiency; this implies that consumers are forced to buy both ends of the link from the same vendor, which runs counter to the open channel bank philosophy embodied in the U.S. telephone plant. Some applications require no VF in-band signaling; in these cases, a transcoder can generally achieve a straight 2-to-1 compression.

Circuits that require signaling can be dealt with in two ways with a transcoder. One option is to let the transcoder "rob" bits to achieve the desired VF signaling. When the transcoder is alerted that a bit is stolen, it changes its voice coding algorithm to put out fewer voice bits. While this affects the quality of the voice less severely than simply robbing a bit altogether (as would be done in regular PCM), some degradation still occurs, particularly when multiple encodings are undertaken along the transmission path. The effective voice bit rate would be 30.67 kb/s (rather than the nominal 32 kb/s), and the bit stream is not compatible with digital cross-connect equipment. The second option (and the M44 recommended standard [3.29]) is to employ a bundling method. Rather than steal some bandwidth from each VF channel, a specific out-of-band channel is allocated to carry signaling information for a group of channels. AT&T's BCM 32000 uses a 384 kb/s bundle of eleven 32 kb/s channels and one 32 kb/s "delta" channel that carries signaling for the other 11 channels. The bundling function occurs after low bit rate encoding and before the compressed signal is placed on the line; this allows each of the 44 channels of the M44 to retain a full bandwidth of 32 kb/s. With divestiture, and the complexity of coordination, we do not know if this attempt of AT&T in the transcoder arena, namely, in the proliferation of the "public" M44 algorithm, will be successful.

3.4.3 Digital Cross-Connect Systems

A digital cross-connect system (DCS) is a computerized facility that allows DS1 lines to be remapped electronically at the DS0 level. It allows the assignment and redistribution of 64 kb/s channels among various T1/DS1 systems connected to the DCS, at the digital level, and can therefore be considered a DS1 switch. DCSs also provide per-channel DS0 test access in digital form. DCS is not a direct replacement for, nor an alternative to, any single previously existing network equipment [3.30]. It is, however, a versatile piece of equipment, which has several roles in a large telecommunication network. A DCS is sometimes called DXS or DACS. The first DCS was AT&T's Digital Access and Cross-connect System (DACS), and this is the reason why this type of equipment is occasionally known by the DACS acronym. By late 1980s, more than twenty manufacturers of DCSs existed [3.31]. The terms "slow switch," "nailed-up switch," and "channel switch" are also occasionally used, the last one especially in an ISDN context.

Telephone companies and commercial end-users with large private networks may benefit from a technology. DCS equipment has been used mostly by telephone companies (BOCs, independents, and IXCs) because it is ideally suited to large applications. A private network DCS performs all the functions performed by a DCS in a telephone company application. By and large, however, a sophisticated T1 multiplexer will provide similar functions, and may be better suited to a data communication network designer or administrator for building a private network, unless multimedia requirements are present. A user needs at least one T1/DS1 line before a private or telephone company-provided DCS even begins to apply. The need for a dozen T1/DS1 channels is the point at which the user should begin to consider this type of hardware. Several dozen DS1 lines would normally justify it; and several hundred lines will demand it. DCSs allow calls to be routed without having to be demultiplexed. The equipment allows rapid and inexpensive cross-connection of DS1 channels; the connections can easily be monitored, tested, disconnected, and reconfigured from local or remote terminals. A DCS serves the same function in the digital environment that a distributing frame served in an analog environment [3.32].

A three-step evolution with respect to this equipment began in the early 1980s. The first step involved the deployment of DCS equipment by the telephone companies, internal to the network and invisible to the end-user. The second step involves providing control access to the end-user, so that the end-user can reconfigure a network using the telephone company's DCS. The third step in the evolution involves the deployment of user-owned DCSs.

DCSs were employed at first by telephone companies as a way to administer "special" services, which includes voiceband data lines. Dedicated lines, which are the workhorse of data communications, do not generally run through a digital switch (or an analog switch, for that matter) because a switch must service every information bit of a connection through it and remap incoming PCM slots to a new set of outgoing PCM slots, which requires substantial switch resources. Voice calls are of relatively brief duration, so that internal resources are freed for other incoming calls. A dedicated line, on the other hand, requires a constant and continuous remapping of slots, effectively *ad infinitum*; this condition, described as a "nailed-up," can severely impact the ability of the switch to serve other customers, particularly if the number of nailed-up connections through the switch is high. The DCS allows these lines to be treated as any other telephone company bit-pipe, including the capability of rapidly reconfiguring and re-engineering the circuits, without having to connect them to the switch. If the dedicated lines were implemented simply as pairs of wires (or equivalent), modification of the circuit would be difficult, whether for maintenance purposes or redesign purposes; the DCS facilitates this process. DCSs are a step in the evolution toward the goal of totally automated special service provisioning.

A DCS serves as a hub in digital distribution networks. Bit-stream transparency for voice and data transmission is provided, except for signaling bit reinsertion on switched network voice channels. Several key applications exist for DCSs (whether carrier owned or end-user owned for a private network) including segregation, concentration, on-route circuit combination, elimination of back-to-back channel banks, improvement of plant utilization, elimination of nailed-up connections, testing, and transmission facilities (and other facilities) sparing [3.32].

Telephone companies have found these devices to be very helpful in improving existing services and even in providing new services, while at the same time reducing costs. End-users with large T1/DS1 networks can employ DCS-like services obtained either from the carrier (BOC or IXC) or from customer owned and maintained systems. End-users can derive the same benefits the carriers enjoy by employing DCS hardware.

3.5 TESTING DIGITAL CARRIER SYSTEMS

An increasing number of Fortune 1500 companies is utilizing T1/DS1 transmission facilities, either for point-to-point applications, or for more sophisticated backbone networks. The trend is toward the DS3 rate, and for the T1 multiplexer to become a corporate resource manager system, accepting input not only from computer data lines, but also PBX voice trunks, video-conferencing codecs, and other signal paths. As more and more of the corporation's communication traffic is put through a small set of discrete transmission facilities, the potential liability from either catastrophic failure of the communication vehicle (termination equipment or lines), or from a deterioration of the service due to a partial impairment of these same facilities, increases. This predicament raises the critical issue of testing and diagnostic procedures, and of equipment that must be put in place by a user of T1/DS1 facilities to prevent, monitor, or resolve problems that may arise.

Testing can be categorized as *in-service testing* or *service-disruptive testing*. In-service testing means that the test can be performed while the facility is carrying actual traffic data; for example, a drop-insert test on a DS0 slot and an accumulation count of synchronization slips are in-service operations. Service-disruptive testing means that the link must be taken out of service; for example, loop-around and direct BER testing. Key in-service and service-disruptive testing for T1 facilities and associated test equipment is discussed briefly below.

3.5.1 In-Service Testing

Access Testers. A basic type of testing equipment is the access tester. In addition to providing access to the DS0 slots and to the A,B-signaling bits, this equipment also provides monitoring for error performance, integrity checking, and bit pattern synthesis. Access testers are ideal for voice application, but can also be used for data applications in those cases in which the testers provide additional functions such as jitter measurements, BERT, *et cetera*. Add-drop type tests are greatly simplified by the use of access testers.

Bipolar Violation Testing. Noise and crosstalk can affect the bipolar signal to the point where a no-pulse condition is interpreted as a pulse condition (see Figure 3.9). The number of resulting bipolar violations can be counted. A high bipolar violation rate is generally indicative of some intrinsic carrier system problem. Hence, an instrument that measures this information can be useful in isolating the potential problem. Bipolar violation measurements are ideal for intra-LATA and inter-LATA T1 facilities (and less useful for non-telephone company facilities). Note that the B8ZS method employed to achieve clear-channel conditions uses deliberate violations; these intentional violations must not be counted as errors (test equipment for B8ZS lines must be appropriately configured).

Excess Zeros and Framing Errors Testing. Framing errors occur when the receiving equipment (or a repeater) is unable to recover the clock. This condition can be caused by a number of system impairments. Repeaters in DS1 lines require a specific density of 1-bits to maintain synchronism, as discussed earlier. While the CSU should guarantee that the appropriate rate is maintained, not all CSUs explicitly provide this function. Excess zeros can cause the repeaters to shut down and put the facility out of service. Hence, excess-zeros detection instrumentation can help identify potential problems.

Loss of synchronism leads to framing errors, in which an incorrect bit appears in the framing position (193rd bit); this precludes the correct decoding of the subchannels. One aspect of the framing error measurement is that it approximates the actual BER of the facility, even for high error rates. Thus, it can serve as an in-service approximation of the end-to-end performance of the facility without having to do the out-of-service testing.

Signal Levels. The DS1 signal must satisfy certain defined electrical levels at the repeater, CSU, and DCS points (as specified in ANSI T1.102-1987). Signals exceeding the voltage level specifications will generate crosstalk on other circuits (going through the same repeater); signals that are low will be masked by noise. We must be able to measure signal levels.

Jitter. Whether provided by the network or by the user equipment, timing must be within 50 parts per million (80 Hz at DS1 rates). Operation outside this range will result in jitter, which in turn implies timing slips and data errors. We must be able to measure this factor. Jitter is a displacement in time of the signal transitions compared to the ideal signal. The severity of the distortion depends on both the amplitude and the frequency of these displacements. The predominant source of jitter is multiplexers and regenerative repeaters. Network and user equipment will have a certain tolerance to jitter that will depend basically on the sophistication (hence, cost) of the timing recovery circuits. How well the jitter is controlled (and how much the various DS1 components can tolerate the jitter) is important in achieving a robust and reliable backbone T1 network. Jitter measurement is a sophisticated and fairly technical test; detailed knowledge of the various jitter specifications is required.

All-Ones Condition. This condition is also known as *alarm indication signal* (AIS). If a multiplexer detects too many errors in its framing pattern, it will transmit all ones in both the customer bits and the framing bits. Additionally, prior to the changes to FCC Part 68 shown in Table 3.3, an all-ones keep-alive signal pattern was required, even when no data are being transmitted, to keep the repeaters synchronized. Detection of this alarm condition of all-ones in the framing bit, when it should not exist, will indicate a failure of some network component.

CRC. The ESF format allows more sophisticated in-service tests to be conducted. ESF facilities are now being introduced by the carriers. Some of these newer tests are based on the CRC code embedded in the bit stream formed by the accumulation of the 193rd bit. The CRC provides a measurement of the quality of the line.

3.5.2 Service-Disruptive Testing

Many of the above in-service tests are blocked by any equipment that reinserts framing, such as a digital switch, a digital cross-connect system, or an echo canceler. At times, out-of-service testing is the only recourse available. These types of tests may involve two pieces of test equipment, one at each end, or a single instrument connected through a loop-around arrangement. An example of out-of service testing is BER testing.

3.5.3 ESF Testing Issues

As discussed above, ESF has a 4 kb/s facility data link (FDL), which allows access to performance data from remote devices. The communication capabilities of the FDL along with the error detection function of the CRC constitute a powerful mechanism for performance monitoring, maintenance, and network management. While AT&T's original ESF specification 54016 is followed closely in the ANSI specification T1.403-1989 accepted by the industry, differences exist in the way in which the performance data are accessed. Because of the market share of AT&T in the private line T1/DS1 segment, this issue is worth some discussion.

A 54016-configured CSU must monitor the CRC of the received data stream and store the performance information for the previous 24 hours. The local CSU collects performance information about the receiving side of the T1/DS1 link; the remote CSU collects performance information about the transmitting side of the link. The service-providing telecommunication carrier uses an intrusive line monitoring unit (LMU), which is interjected between the CSU and the circuit (i.e., it is connected in series). The LMU can send a request message to the CSU over the ESF FDL to retrieve the performance information stored by the CSU; this information is sent by the CSU using another FDL message.

A T1.403-configured CSU monitors the CRC of the received data stream and stores the performance information. In this case, however, the CSU automatically transmits a performance report message (PRM) over the FDL every second. The PRM broadcasts the performance information for the last four seconds; any device on the T1/DS1 link can access the PRM to observe the end-to-end service performance. Here the LMU is a low-cost nonintrusive device, connected in parallel. The CSU can now provide full performance monitoring to the user network manager [3.13]. This arrangement is fully reciprocal in the sense that it gives both the user and the service provider full visibility to the performance information for both directions.

At press time, the top three IXCs were proceeding with ESF implementations based on the 54016 specification; LECs were planning to use the T1.403 standard [3.13]. The IXCs have pledged to migrate to the ANSI standard in time. The existence of two approaches means that in the early 1990s situations will arise in which circuits need to support both methods, making it difficult for the end-user. Newer CSUs can handle both variations of ESF. As long as the network equipment will pass ESF framing, the user need only install ESF CSUs; this gives the user a way to accomplish full performance monitoring, but it does not give the user the benefit of rapid repair response by the carrier when ESF monitoring and maintenance is fully implemented.

3.6 DS3 SYSTEMS

Changes are now taking place in the transmission of digital signals in the telephone networks. The traditional hierarchy of digital signals discussed in the earlier part of the chapter, which has been with us for more than a quarter century, is changing due to the availability of fiber optic technology [3.17].

3.6.1 Activity at the DS3 Rate

Similar improved network monitoring and maintainability goals have been the rationale for defining an enhanced DS3 signal format. Three formats for DS3 were being discussed in the late 1980s [3.12]:

1. The traditional (asynchronous) M23 multiplex format, which is widely deployed;
2. A DS3 C-bit parity format (studied by the ECSA T1X1 Committee); and
3. The SYNTRAN format.

A fourth method for interworking with fiber optic systems is also emerging. This last method, described later in the chapter, is becoming the option of choice.

Driving factors for new schemes include demands for a higher grade of service and the need to reduce costs. Synchronous DS3 transmission allows: (1) bandwidth administration (provisioning), and (2) automated maintenance and testing.

Bandwidth Administration (Provisioning). The loop and interoffice networks need automated control over individual DS0s, as well as higher level channels such as DS1s. Synchronous transmission leads to a number of key new services needed by telephone companies such as circuit provisioning by time-of-day, circuit provisioning by destination, and customer-controlled reconfiguration. The ability to manage circuit provisioning in real time helps manage changing circuit requirements and growth rates.

Automated Maintenance and Testing. Synchronous transmission formats provide standardized maintenance channels embedded in the overhead of the transmission bit stream. These are called embedded operation channels (EOCs). EOCs help both the user and the telephone company automate the maintenance, control, and test functions that were previously performed locally and manually.

3.6.2 Approaches to Synchronous Transmission

Two methods to incorporate synchronous transmission into the network have evolved:

1. Define a synchronous format that will operate throughout the existing asynchronous transmission network and provide synchronous transmission benefits within the installed base. This is SYNTRAN (Synchronous Transmission). It operates at the standard DS3 rate of 44.736 Mb/s. SYNTRAN signals are compatible with any standard higher speed transmission facility (synchronous or asynchronous) [3.33].
2. Develop a transmission format that would be used in new or overlay systems. This is not intended to operate through existing asynchronous transmission networks. Such a system is SONET (Synchronous Optical Network). This system will accept DS1, DS1C, DS2, DS3, and CEPT-1 rate input tributary signals. (SONET is discussed is Section 3.8).

As of mid-1990, SYNTRAN did not appear to have a commercial future. This is due to the fact that (1) not many manufacturers developed equipment; (2) the cost of SYNTRAN equipment remained high, even compared with traditional M13/DCS alternatives; and (3) SONET is becoming available. This method is, nonetheless, discussed briefly below for completeness.

3.6.3 Existing Format: The Traditional M23 Multiplex DS3 Format

The existing DS3 signal format is a result of a multistep, partially synchronous, partially asynchronous multiplexing scheme (Figure 3.12). Three steps are involved:

1. The first step is the synchronous DS0 to DS1 multiplexing process in which 8-bit bytes from each of 24 DS0 channels are synchronously byte-interleave multiplexed to form a DS1 signal;
2. The second step is asynchronous, and interleaves four DS1 signals on a bit-by-bit basis to form a DS2 signal at 6.312 Mb/s. "Dummy bits" are pulse-stuffed in this step to allow for different timing references on the DS1 signals;
3. The third step, which is also asynchronous, multiplexes seven DS2 signals on a bit-by-bit basis along with pulse-stuffing to form a DS3 signal.

The M23 multiplex format provides for transmission of seven DS2 channels. Because each DS2 channel can contain four DS1 signals, a total of 28 DS1 signals (672 DS0 signals) is transported. The M23 multiplex format accounts for timing differences between the DS1, DS2, and DS3 signals with stuff bits inserted in the DS2 and DS3 bit streams. The presence or absence of stuffing bits in the DS3 bit stream is indicated by C-bits (21 per DS3 frame; this is equivalent to 197,358 b/s with a DS3 bit rate of 44.736 Mb/s). The M-bits and F-bits provide frame and subframe synchronization of the DS3 signal. The signal includes a parity check for simple performance monitoring (P-bits) and the capability of sending a yellow alarm (X-bits) (see Figure 3.13). The parity check provides end-to-end performance monitoring in situations wherever parity is not recalculated during signal transmission. Some equipment, however, recomputes the parity bit whenever the parity is found in error.

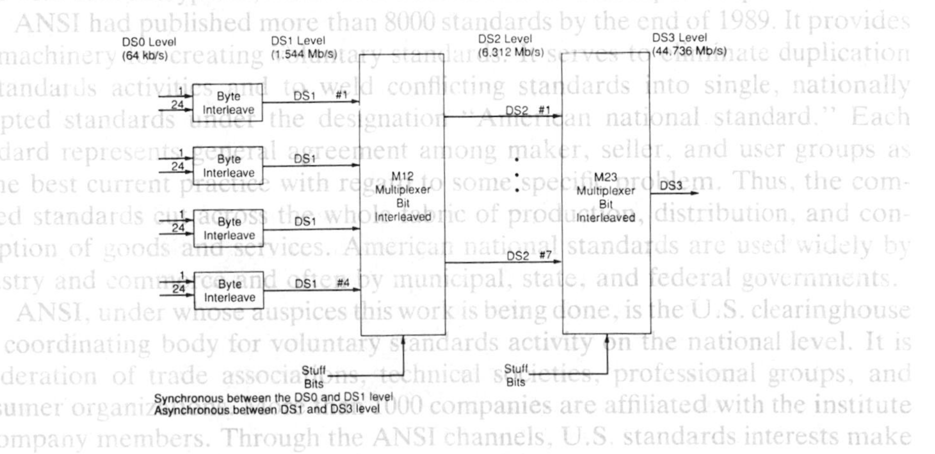

Figure 3.12 Digital multiplexing in the North American hierarchy.

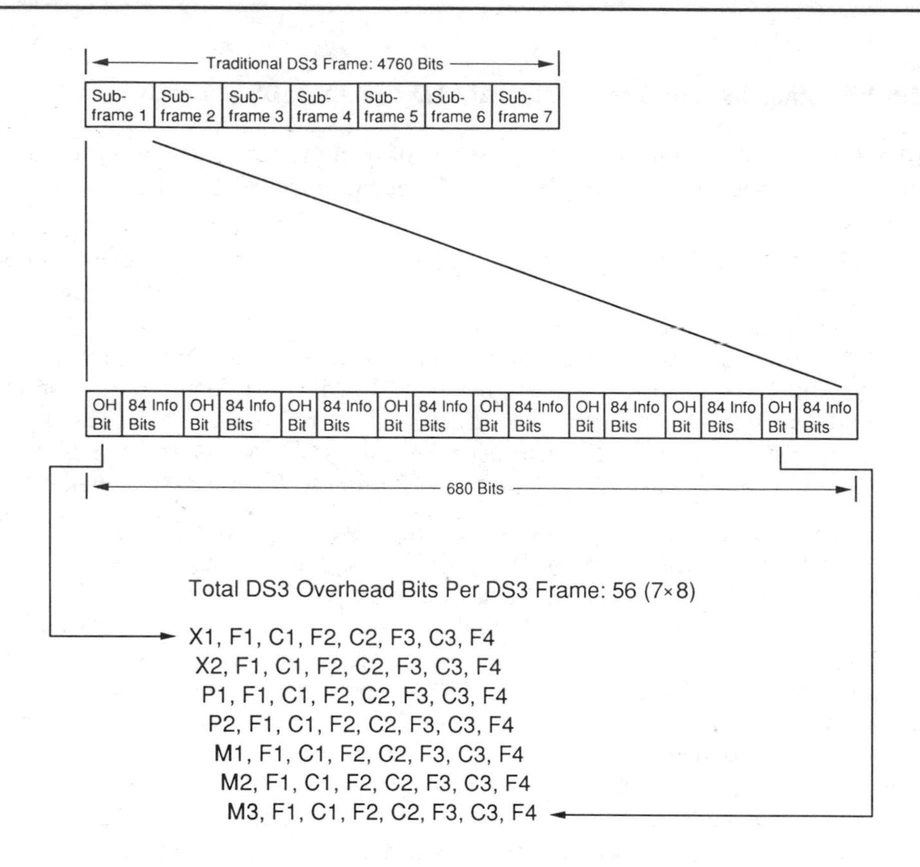

Figure 3.13 Traditional DS3 format.

3.6.4 C-Bit Parity Format

The DS3 C-bit parity format, studied in the late 1980s, changes the use of the C-bits to allow for an end-to-end performance measurement. This is accomplished by embedding a redundant copy of the originating parity bits in the C-bit channel. These bits are not modified by any equipment during transmission unless the information content of the DS3 signal is intentionally changed, for example, with a DS1 drop-insert. Therefore, the C-bit parity bits can be used to determine errors in the overall transmission path or any section of the path. The use of C-bits for functions other than the stuff indication is accomplished by deleting the need for stuff indication. The DS3 stuff bits are always used for stuffing in the C-bit parity format, and the DS2 signals are generated at exactly the bit rate needed to multiplex into the DS3 signal without compensation for timing differences. The DS2 bit rate used is 6.306272 Mb/s. All stuffing for timing differences is accomplished in the DS1 to DS2 multiplexing process [3.17]. Implementation of C-bit parity format involves changes to DS3 multiplexers, cross-connects, and switch interfaces. The advantage is improved network monitoring and maintainability.

In addition to the transmission of parity bits, the C-bit channel can be used for the following [3.17]:

- *Application Identification*. This allows for the identification of the actual format: M23, SYNTRAN, or C-bit.
- *Far-End Alarm Channel*. A 16-bit repetitive pattern can be transmitted to indicate the status of a terminal (loss of signal, out of frame, DS1 status, *et cetera*).
- *FEBE Function*. A terminal that is receiving framing or parity errors on the incoming DS3 signal will so indicate such error events by using the far-end-block-error (FEBE) bits in the C-bit channel of the outgoing DS3 signal.
- *Terminal-to-Terminal Data Link*. A 28.2 kb/s data link is available for maintenance information channel. A 16-bit repetitive pattern or a LAP-D (CCITT Q.931) protocol may be used. Line loopback commands may be transmitted on the data link as well as identification signals, such as DS3 path identification, terminal equipment type and location, *et cetera*.
- *Application Specific Data Link*. An 84.6 kb/s data link is available for carrier and end-user needs. The data link can be broken down into subchannels for this purpose, e.g., three 28.2 kb/s data links.

3.6.5 DS3 SYNTRAN Format

The M23 and C-bit parity formats are asynchronous formats. The DS1 frames slide within the DS3 frames because the DS3 frame period is not an even multiple of the DS1 frame period. ANSI Standard T1.103-1987 [3.34] defines a synchronous DS3 format, also known as SYNTRAN. SYNTRAN is a restructured DS3 signal format for synchronous transmission at the 44.736 Mb/s DS3 level of the North American hierarchy. The new format has been designed to be compatible with the existing base of DS3-oriented transmission facilities. This format combines 699 DS3 frames into a "synchronous superframe." The period of the synchronous superframe is exactly 595 DS1 frame periods. This allows DS1 frames to have fixed locations within the synchronous superframe. The main advantage of the SYNTRAN format is the simplification in equipment design and deployment that results from this synchronous access to the DS1 signals. For example, drop-insert of a DS1 signal can be performed directly on the particular DS1 signal. The asynchronous formats require demultiplexing and multiplexing of a DS2 channel for DS1 drop-insert. Because the need for DS2 demultiplexing is eliminated with SYN-

TRAN, the C-bits are again available for other purposes. The SYNTRAN format includes a CRC check and a 64 kb/s data link using the C-bits. The SYNTRAN format provides three key functions that are distinct from the older asynchronous DS3 format:

1. The SYNTRAN DS3 signal provides immediate identification and direct access to all DS0 and DS1 signals within the 44.736 Mb/s SYNTRAN signal. This capability facilitates economic add-drop and DS0/DS1 time slot interchange functionality.
2. The SYNTRAN format frees bandwidth used for the bit stuffing process required in the asynchronous DS3 signal. The freed EOC bandwidth, 1.2 Mb/s in capacity, is used to provide (i) high performance maintenance and test functionality to every SYNTRAN network element from a centralized remote location; (ii) remote provisioning of DS0 and DS1 signals throughout a SYNTRAN network; and (iii) improved DS3 level performance monitoring using a CRC-9 to improve the integrity and quality of the DS3 transmission signal throughout the network. This is similar to the CRC-6 option used in the DS1 ESF format. It provides fast and highly reliable bit error and burst error detection.
3. The format provides a one-step multiplexing scheme that eliminates the need for intermediate DS2 multiplexing stage. Thus, it permits simpler architectural designs for equipment.

Bellcore has published technical advisories for equipment designs required in the BOCs: (1) the features, functions, performance criteria, message formats, protocols, and commands required to ensure industry standard and multiuser compatibility, and (2) three network elements outlined in the advisories include an add-drop multiplexer, a digital cross-connect system, and a DDS interface.

The SYNTRAN format specifies two modes of operation: *bit synchronous* and *byte synchronous*. The bit synchronous mode provides for the transport of DS1 signals as a complete bundle. This mode is used when access to the individual DS0 channels is not required. The byte synchronous mode provides DS0-level access and facilitates digital signal processing at the DS0 level. Functions such as an integrated time slot interchange (digital cross-connect) can be directly implemented in this mode. The format was designed to be compatible with the existing asynchronous network so that synchronous transmission benefits could be provided through the existing network. A large and growing embedded base of digital microwave radio and high-speed fiber optic transmission facilities is in place, representing an investment of billions of dollars.

SYNTRAN Frame Structure

The asynchronous DS3 format used in the network employs a frame structure that includes seven rows of 8 × 85 bit blocks.The first bit in each 85-bit block is a special-purpose bit that precedes the 84 information bits. The special-purpose bits are termed X-bits, P-bits, M-bits, F-bits, and C-bits. The M-bits provide master frame identification (seven rows); the F-bits provide DS3 framing; the P-bits are parity bits that provide low-level BER detection; the X-bits provide a low-level communication channel within the DS3 stream; the C-bits are control bits and are used as pulse-stuffing indicators.

For the SYNTRAN DS3 signal to be compatible with existing DS3 transmission facilities that expect the asynchronous DS3 format, the M-, X-, P-, and F-framing bits in the SYNTRAN format are identical in meaning and in placement to the asynchronous DS3 format. The DS3 M frame is 106.4 μs long for both SYNTRAN and the asynchronous DS3 format. The SYNTRAN format creates a superframe that is 74.375 ms long to allow access to the 8-kHz-based 125-ms time base necessary for DS0 identification. A superframe contains 699 M-frames, which is equal to 595 × 125 ms.

Decoding a SYNTRAN signal involves first identifying the number of the M-frames (from 1 to 699), and referring to a lookup table to tell the time slot number at the beginning of the frame. In the SYNTRAN format, pulse stuffing is not needed, and the C-bits are used for the embedded spare capacity. Four C-bits are allocated for M-frame numbering. The other C- bits are used for the 64 kb/s embedded facility data link, the DS3 level CRC-9 performance monitoring function, and other EOCs.

DS3 Signals Summary

SYNTRAN, M23, and C-bit parity formatted signals are transport-compatible. They all use the 4760 bit frame, the M-bit and F-bit frame synchronization, and the P-bit parity check shown in Figure 3.13. The three formats, however, are not terminal-compatible. The new DS3 signal formats allow for improved monitoring and maintenance of a DS3 network including: end-to-end parity checks; sectionalized parity checks; loopback control and testing; alarm detection; idle path monitoring; path and equipment identification; and line and path performance monitoring.

A more reliable and maintainable network for transport of DS3 signals is being designed for the future. A variety of incompatible DS3 terminals will be found in the field for some time as new signal formats are implemented. However, SYNTRAN and C-bit parity formats appear to have limited, if any, commercial deployment potential.

3.7 END-USER APPLICATION OF DIGITAL TRANSMISSION TECHNOLOGY

As indicated earlier, major deployments of the digital technologies discussed thus far have occurred for private network applications. While at the beginning of the

1980s only a few Fortune 500 companies employed DS1 facilities, now approximately 80% of Fortune 1500 companies employ them. With the introduction of fractional T1, more medium-size companies will be employing digital technologies in their networks in the early 1990s, while waiting for the availability of ISDN facilities (channel switched service, in particular). Fractional T1 typically provides interoffice channels (IOCs) at 1/2, 1/4, and 1/8 of the bandwidth of a T1 channel.

1970s

DDS was introduced in the mid-1970s and represented the first digital service for private line (data) applications. Some observers now see the importance of DDS decreasing given the fractional T1 and ISDN alternatives that are becoming available, particularly from a cost perspective [3.35]. DDS offered rates from 2.4 to 56 kb/s, at typical data communication rates.

1980s

During the 1980s, users employed DS1 facilities in private networks to replace multiple voice-grade private lines (including DDS), which were used for data or voice applications. Digital backbone networks have now become the common solution to corporate communication problems. Sophisticated CPE multiplexers have become available to assist the end-user with management and reconfiguration of a network in near real time. Three applications of digital transmission technology have been: (1) private line consolidation using multiplexing (also known as *backbone networking*); (2) aggregate PBX access to CO (discussed in Chapter 10; this application is expected to migrate to an ISDN access in the near future); and (3) DS1-rate CPE device interconnection, for example, LAN-to-LAN gateways and computer channel extenders (discussed in Chapter 12) [3.36,3.37]. In the last case, no multiplexing occurs; instead, the devices use the entire DS1 bandwidth.

Initially DS1 was employed in large cross-section environments, possibly integrating voice and data trunks. Multiplexers were basic, being similar to channel bank technology. The channel bank channelization is not efficient in terms of bandwidth management in which a variety of data channels operating at different speeds exist. These corporate networks consisted of point-to-point links.

1990s

More recently, sophisticated multiplexers allowing networked systems, bit multiplexing (i.e., nonchannelized frames), ESF line coding, and even ISDN interfaces (for backup applications) have emerged.

Also recently, we have seen the commercial emergence of fractional T1 to serve smaller cross-section environments. In fractional T1, the user typically has a DS1 link to an IXC point-of-presence. The user only utilizes some predetermined number of slots in the DS1 signal, typically 1, 4, 6, or 12. At the IXC office, this user's stream is multiplexed over a digital facility (including possibly a DS1 channel) with the signal of another user going to the same remote city. On the long-haul circuit, the user is only charged an appropriate fraction of the cost of the entire DS1 link. The IXC will use DCSs at both ends to demultiplex the signal as needed. The user will need an appropriately configured multiplexer which can "drop" the signal in the appropriate slots so that the IXC can further process the stream. As of early 1990, several carriers and several equipment manufacturers were offering fractional T1 service. Initially, the multiplexers were using strict channelization, but we expect that in the following couple of years, nonchannelized fractional T1 equipment will become available [3.38].

DS3 service is also becoming more prevalent. Typically, in the late 1980s and early 1990s, DS3 lines were tariffed to cost from 4-to-9 T1 lines; thus users are finding the technology quite cost-effective. Applications include [3.39]:

- Replacement of multiple T1/DS1;
- Large section voice trunks for private voice networks;
- Backbone integrated voice and data networks;
- LAN interconnection (particularly for computer-aided design and computer-aided manufacture (CAD/CAM) applications).

Once a private DS3 line is installed, additional bandwidth can be used for other applications. DS3 facilities are increasingly available in many parts of the country, although they are still fairly expensive. With the extensive fiber-based plant now in place, DS3 capacity is readily available. These private DS3 systems now tariffed by many carriers employ the traditional DS3 format.

Commercial Trends

Major growth is predicted in all areas of private digital networks, as shown in Table 3.5. (See also Table 1.4.)

3.8 SONET: BEYOND DS3

3.8.1 Background

In 1988, four years of standards work culminated in the publication of a worldwide standard for optical communication. This standard is known as SONET in the United States, and *synchronous digital hierarchy* (SDH) elsewhere [3.40]. Optical

Table 3.5
Private Digital Network Market Growth
(1988–1994)

Year	*T1/DS1 Lines (IXC)*	*T1/DS1 Market Service and Equipment (billions $US)*[1]	*Fractional T1 Lines (IXC)*[2]	*T3/DS3 Market Service and Equipment (millions $US)*[3]
1988	5,000	6.6	700	40
1990	10,000	9.9	4,000	160
1992	15,000	14.1	15,000	650
1994	17,500	18.9	50,000	1,250

[1]From [3.46]
[2]Based partially on [3.47]
[3]Based on [3.48]; approximately 10% is for equipment and 90% is for service (for all years shown)

transmission systems are the economic choice for transporting large cross sections of traffic. A standardized optical interface is, however, required to facilitate deployment of this technology. SONET is a set of network interface standards aimed at enabling global network interconnection. SONET was originally conceived to provide a standard optical interface signal specification that would facilitate midspan meets. SONET now also defines a new multiplexing hierarchy that attempts to ensure equipment compatibility between offerings from different manufacturers. SONET is a new digital hierarchy ideally suited to handling fiber-based signals and at the same time allowing easy extraction of lower rate signals. SONET provides the carrier with a number of advantages. These include unified operations and maintenance, the capability for multivendor midspan meets, integral cross-connect functions within transport elements, and the flexibility to allow for future service offerings. In the early 1990s, telephone operating companies and IXCs will deploy network equipment meeting the new SONET standards [3.33].

The purpose of this section is to provide a summary treatment of SONET. More detailed information can be found in Ref. [3.11]. SONET was initially proposed in 1985 by Bellcore. It was first accepted as a national standard and then taken to CCITT. Parts of this standard, documented in ANSI T1.105-1988 (developed by the ECSA T1X1.5 Committee) and T1.106-1988, are now also defined in the following CCITT Blue Book Recommendations:

CCITT G.707, "Synchronous Digital Hierarchy Bit Rates";
CCITT G.708, "Network Node Interface for Synchronous Digital Hierarchy"; and
CCITT G.709, "Synchronous Multiplex Structure."

SONET standards are being defined in phases. Phase I, completed in 1987, included the rates and format definition (ANSI T1.105-1988) and the optical interfaces (ANSI T1.106-1988). Phase II, in ballot as of June 1990, included an electrical interface definition (reissue of T1.102-1987); an addendum to T1.105 defining protocols for data communication channel protocol suites; and a new standard for SONET OAM&P (Operations, Administration, Maintenance and Provisioning).

Phase III defines the message sets to be utilized over the data communication channels to carry out specific OAM&P functions (at press time a draft was considered possible for early 1991, with ballot in mid-1991).

SONET defines a hierarchy of rates and formats to be used by vendors, carriers, and end-users for optical transmission at and above the 51.840 Mb/s rate. The SONET hierarchy currently addresses transmission up to 2.5 Gb/s and can be extended, if necessary, to more than 13 Gb/s (for comparison note that commercially available fiber optic systems now carry up to 2.4 Gb/s).

The basic building block is an 810-byte frame transmitted every 125 μs to form a 51.840 Mb/s signal known as a *synchronous transport signal-level 1* (STS-1). At this rate, each of the constituent 8-bit bytes is equivalent to a 64 kb/s channel. The functionality of SONET is achieved by defining the basic STS-1 signal, and an associated byte-interleaved multiplex structure that creates a group of standard rates at N times the STS-1 rate, where N takes selected integer values from 1 to 255. Currently, the following values are defined: N = 1, 3, 9, 12, 18, 24, 36, and 48. For transmission over fiber facilities, an optical counterpart of the STS-1 signal, called the *optical carrier-level 1 signal* (OC-1), is defined. It is an electrical-to-optical mapping of the STS-1 signal. The OC-1 signal forms the basic SONET transmission building block from which higher level signals, such as OC-3 and OC-48, are derived. For example, OC-3 signal operates at 3 × 51.84 Mb/s, or 155.52 Mb/s. The line rates are defined in Table 3.6.

3.8.2 Overview of Key SONET Concepts

The STS-1 signal is divided into a portion assigned for transport overhead, and a portion that contains the synchronous payload. Synchronous payloads are payloads that can be derived from a network transmission signal by grouping integral numbers of bits in every frame (i.e., no variable bit stuffing rate adjustments are required to fit the payload into the transmission signal). The synchronous payload

Table 3.6
Basic SONET Rates

Optical Carrier Signal	*Data Rate*
OC-1	51.840 Mb/s
OC-3	155.250 Mb/s
OC-9	466.560 Mb/s
OC-12	622.080 Mb/s
OC-18	933.120 Mb/s
OC-24	1.244 Gb/s
OC-36	1.866 Gb/s
OC-48	2.488 Gb/s

carries the user's data stream. The *synchronous payload envelope* (SPE) is a 125-μs frame structure composed of STS path overhead and bandwidth for a payload. A *payload pointer* in the transport overhead indicates the location of the beginning of the STS SPE. The payload can be used to transport asynchronous DS3 signals, synchronous DS3 SYNTRAN signals, or a variety of sub-DS3 signals, such as DS1 signals (discussed further below).

Transport overhead is overhead added to the SPE for transport purposes. Transport overhead is, in turn, composed of line overhead and section overhead [3.11]. The transport overhead has been designed to accommodate several different functions, including maintenance, user channels, frequency justification, orderwire, channel identification, and growth channels (see Table 3.7).

In SONET, the overhead and transport functions are be grouped into layers. In order of increasing complexity from the viewpoint of hardware and the optical interface frame format, the layers are *photonic*, *section*, *line*, and *path*. The layers have a hierarchical relationship and can be considered either from the top down or the bottom up. The top-down approach is useful for providing a general introduction to the individual layers and their functionality. A layered approach for overhead has also been established whereby overhead bandwidth has been allocated to a layer based on the function addressed by that particular channel. This layered approach allows creation of equipment that need not access all layers of the overhead, thereby keeping costs down. Detailed descriptions of the function of all the overhead bytes can be found in Ref. [3.11].

A *path* at a given rate is a logical connection between the point at which a standard frame format for the signal at the given rate is assembled, and the point at which the standard frame format for the signal is disassembled (see Figure 3.14). For example, if 28 DS1s are multiplexed by a SONET multiplexer into an STS-1 signal, 28 paths will be end-to-end, while only one line will be associated with that system. The path overhead (POH) is overhead assigned to and transported with the payload, until the payload is demultiplexed. The STS POH consists of nine evenly distributed bytes per 125 μs starting at the first byte of the STS SPE. It is used for functions that are necessary to transport the payload. It provides for

Table 3.7
Transport Overhead Functions

Framing
Error detection
Orderwire
User channel for network provider
Data communication (for operations)
Pointer (for locating payload)
Automatic protection switching (APS) signaling
Maintenance signaling

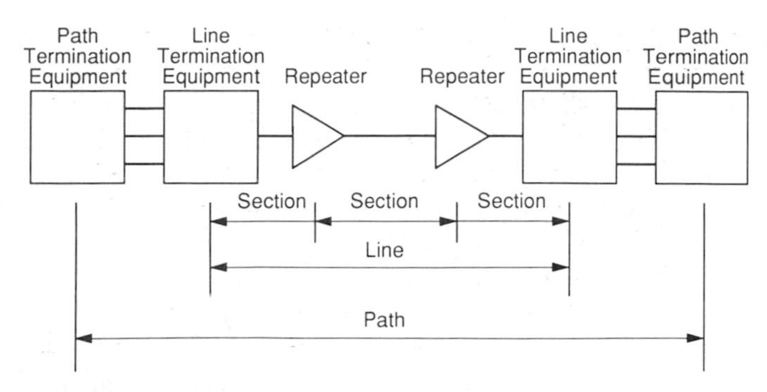

Figure 3.14 Paths, lines, and sections.

communication between the point of creation of an STS SPE and its point of disassembly. The POH is assigned to and remains with the payload until the payload is demultiplexed. The POH provides the following functions:

Path trace;
Error detection;
Payload composition (not yet well-defined as of 1990);
Maintenance signaling;
Far-end error information;
Path user;
Payload-specific "signaling."

A primary goal in defining these signals is to provide a synchronous hierarchy with sufficient flexibility to carry many different capacity signals. Because some services may need to be transported at a rate higher than the basic rate, a technique for linking several basic signals together to build a signal of varying capacity is needed. The STS SPE bandwidth can be combined from N STS-1s in order to carry an STS-Nc SPE. A *concatenated synchronous transport signal level N* (STS-Nc) is an STS-N line layer signal in which the STS envelope capacities from N STS-1s have been combined to carry an SPE, which must be transported not as several separate signals, but as a single entity. In the case of a superrate service, only one set of POHs is required, and is contained in the first STS-1 of the STS-Nc.

A *virtual tributary* (VT) is a structure designed for transport and switching of sub-STS-1 payloads; all services below the DS3 rate are transported within a VT structure. There are currently four sizes of VTs defined [3.30]. A key point of interest is the overlay of existing DS3 networks over the SONET network. SONET defines payload mappings for DS3, DS1, and DS0 transport, as well as other digital hierarchy rates; this is a critical compatibility issue with embedded base. SONET can be expanded to operate at higher rates and to support new payload types, if necessary. The following subsections provide additional details on some of these topics.

3.8.3 STS-1 Frame Format

The STS-1 frame consists of nine rows by 90 columns of 8-bit bytes, for a total of 810 bytes (6480 bits) [3.41]. With a frame duration of 125 μs, the STS-1 bit rate is 51.840 Mb/s (see Figure 3.15).

The first three columns are for transport overhead, and contain overhead bytes for section and line layers. Twenty-seven bytes are assigned for these functions, with 9 bytes for section overhead and 18 bytes for line overhead. The transport overhead of 3 × 27 octets, repeated every 125 μs (8000 times a second), is then 1.728 Mb/s.

The STS-1 SPE is nine rows by 87 columns consisting of 783 bytes. The STS POH consists of 9 bytes; the remainder, 774 bytes, is available for actual payload. In Figure 3.15 the STS PO is in column four, but it can float to other columns. STS POH is used to communicate functions from the point at which a service is mapped into the STS SPE to where it is delivered [3.11]. The payload pointer contained in the transport overhead designates the location of the byte where the STS-1 SPE begins. The SPE carries the actual "payload" or information being transported over the SONET facility. An STS-1 SPE (excluding transport overhead) has a capacity of 50.112 Mb/s and can be organized in a variety of ways. It is normally drawn as a rectangle 87 octets by 9 wide by 8 bits (1 octet) deep, repeated every 125 μs. Many of the mappings into this structure rely on VTs. The order of transmission of bytes is row by row, from left to right; in each byte, the most significant bit is transmitted first.

Overhead *data communication channels* (DCCs, also known as EOCs) are used to communicate alarm, maintenance, control, performance, and administrative data between SONET elements and to network management systems (also known as *operations systems*). See Figure 3.16. The DCCs are provided in the

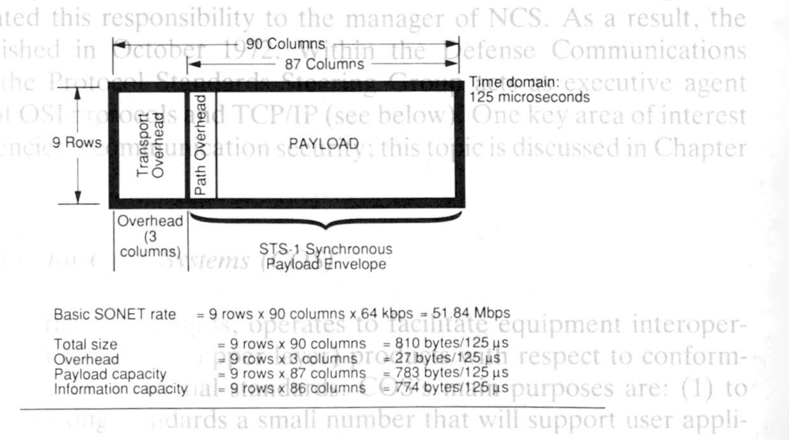

Figure 3.15 SONET STS-1 frame.

transport overhead capacity: 192 kb/s for communication between section layers of terminating equipment; 576 kb/s for line layer communication. The messages are now being defined, as part of the SONET Phase III activities [3.50]. While the lower layer protocols of the DCCs had been firmed up at the time of this writing, the message sets were not well advanced in their definition; the sets may become available in the 1991–92 time frame, with products by 1993–94.

3.8.4 Virtual Tributaries

As indicated earlier, the SPE may be further subdivided into smaller envelopes that correspond to lower-bit-rate signals within the SONET signal. Additional overhead is defined within the payload. The VT is a structure designed for transport and switching of sub-STS-1 payloads. The most common North American tributary (DS1) and international tributary (CEPT-1) have defined VT mappings. The less common tributaries DS1C and DS2 are defined as well (considering possible future needs, provision has been made for concatenated tributaries for services not yet defined). (Although not defined as a VT, there is a mapping to transport asynchronous DS3 signals within a SONET payload).

Table 3.8 summarizes the four VT mappings. The fundamental distinction between types of VTs is the bit rate. Of chief interest to North American users is the VT1.5 at 1.728 Mb/s. This is designed to transport a DS1 signal efficiently plus the required VT overhead. Figure 3.17 depicts the logical relationship between the various tributaries.

The VT mappings to the SONET frame are illustrated in Figure 3.18. In the 9-row structure of the STS-1 SPE, these VTs occupy 3, 4, 6, and 12 columns, respectively.

90 Bytes

				Path Layer Overhead (9 Rows)
Section Overhead	Framing A1	Framing A2	STS - 1 ID C1	Trace J1
	BIP - 8 B1	Orderwire E1	User F1	BIP - 8 B3
	Data Com D1	Data Com D2	Data Com D3	Signal Label C2
Line Overhead	Pointer H1	Pointer H2	Pointer Action H3	Path Status G1
	BIP - 8 B2	APS K1	APS K2	User Channel F2
	Data Com D4	Data Com D5	Data Com D6	Multiframe H4
	Data Com D7	Data Com D8	Data Com D9	Growth Z3
	Data Com D10	Data Com D11	Data Com D12	Growth Z4
	Growth Z1	Growth Z2	Orderwire E2	Growth Z5

Figure 3.16 SONET overhead.

Section Overhead		
Framing	A1, A2	F628 Hex (1111011000101000); to be provided in all STS-1 signals within an STS-N signal
STS-1 Identification	C1	Unique number assigned just prior to byte interleaving that stays with the STS-1 until de-interleaving (can be used in the framing and de-interleaving process to determine the position of other signals)
Section BIP-8	B1	Allocated in each STS-1 for a section error monitoring function (a bit interleaved parity 8 code using even parity)
Orderwire	E1	Used as a local orderwire channel, reserved for communication between regenerators, hubs, and remote terminal locations
Section User Channel	F1	This byte is set aside for the network provider's purpose; it is passed from one section level entity to another and is terminated at all section level equipment
Section Data Comm channel	D1, D2, D3	A 192 kbps channel used for alarms, maintenance, control, monitor, administration and other communication needs between section terminating equipment
Line Overhead		
Pointer	H1, H2	Indicates the offset in bytes between the pointer and the first byte in the STS SPE. It is used to align the STS-1 Transport Overhead in an STS-N signal (it is provided in all STS-1 signals within an STS-N signal)
Pointer Action Byte	H3	This byte is used for frequency justification purposes; it is used to adjust the fill of input buffers
Line BIP-8	B2	Allocated in each STS-1 for a line error monitoring function (a bit interleaved parity 8 code using even parity)
APS Channel	K1, K2	Allocated for Automatic Protection Switching signaling between two line level entities; also carries other management signals
Line Data Comm channel	D4-D12	Nine bytes (576 kbps) allocated for Line data communication for alarms, maintenance, control, monitor, administration and other communication needs; this is available for internally generated, externally generated, and manufacturer specific messages
Growth	Z1, Z2	For future use
Orderwire	E2	Allocated in this layer for an express orderwire between Line entities
Path Overhead		
STS Path Trace	J1	Used to transmit a 64 byte, fixed length string repetitively so that a Path receiving terminal can verify its continued connection to the intended transmitter
Path BIP-8	B3	Allocated for path error monitoring (a bit interleaved parity 8 code using even parity)
STS Path Signal Label	C2	Allocated to indicate the construction of the STS SPE (only two of the 256 codes are currently defined. 0: the Line Connection is complete, but there is no Path originating equipment; 1: used for payloads that need no further differentiation)
Path Status	G1	Allocated to convey back to an originating Path Terminating Equipment, the path terminating status and performance

Figure 3.16 continued.

Table 3.8
Virtual Tributary Rates

Tributary	*Carrying Capacity*(Mb/s)	*Carried Signal* (Mb/s)	*Service*
VT1.5	1.728	1.544	DS1
VT2	2.304	2.048	CEPT
VT3	3.456	3.088	DS1C
VT6	6.912	6.176	DS2

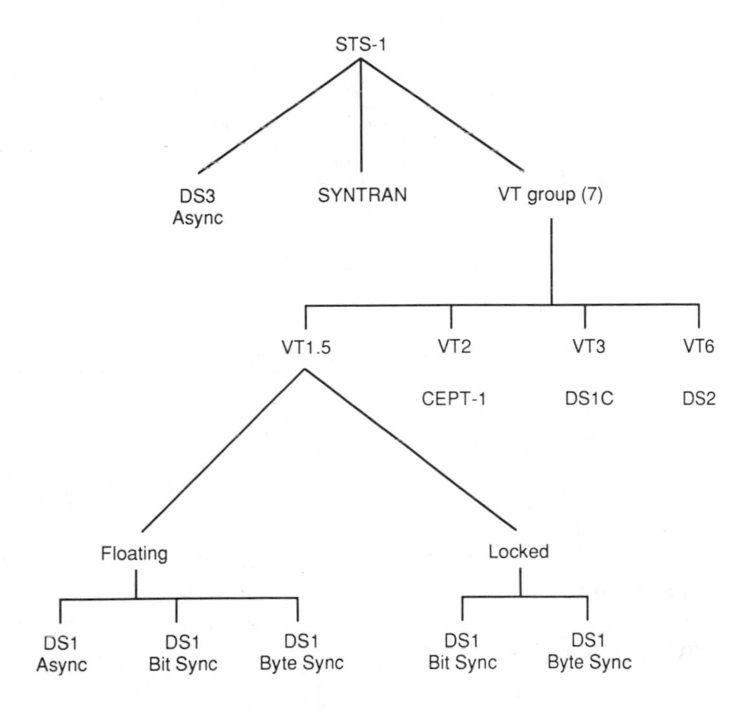

Figure 3.17 STS-1 SPE mappings.

To accommodate mixes of these VTs in an efficient manner, the VT-structured STS-1 SPE is divided into seven VT groups. Each VT group occupies 12 columns of the 9-row by 87-columns structure and may contain four VT1.5s, three VT2s, two VT3s, or one VT6. Reference [3.11] provides detailed information on these seven VT Groups. VT1.5 has two modes: locked and floating. Locked VTs require a fixed mapping of the VT payload into the STS1 SPE. Floating VTs allow dynamic alignment of a VT payload within the SPE.

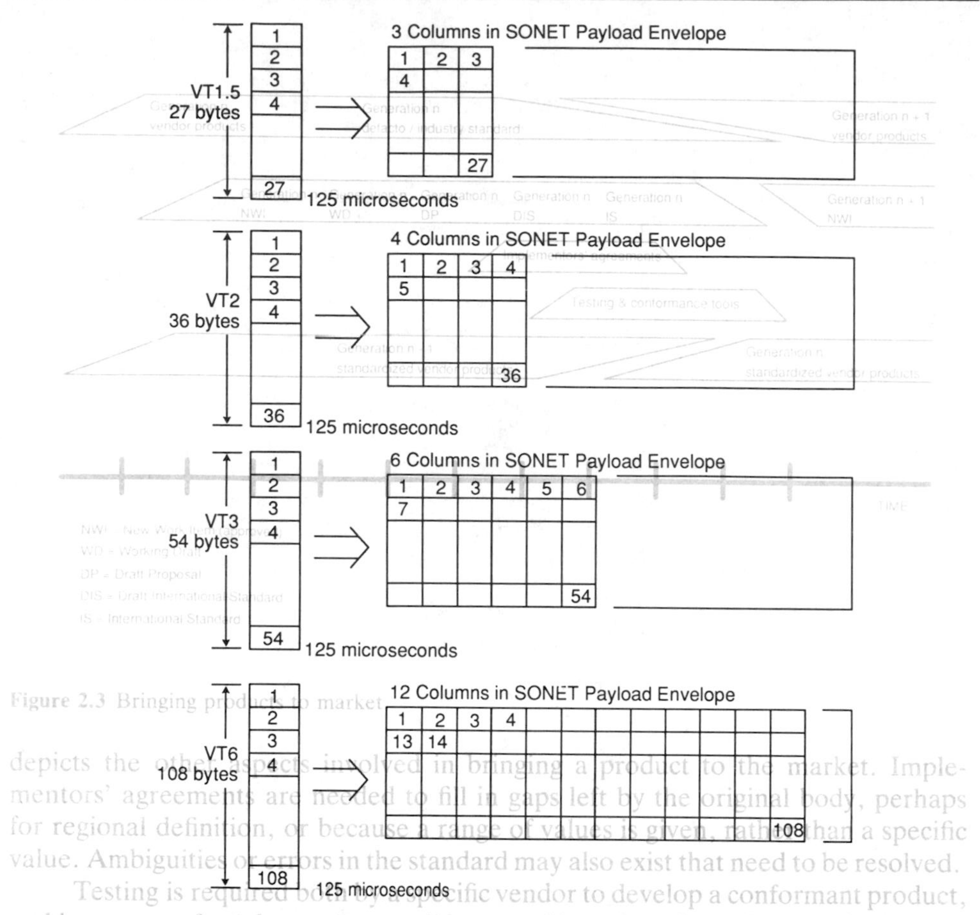

Figure 3.18 Virtual tributaries.

3.8.5 Other SONET Mappings

SONET allows the mapping of several network services into the SPE. These services include lower rate tributaries discussed in the previous section, as well as asynchronous DS3 and synchronous DS3 SYNTRAN signals, among others [3.41].

1. Asynchronous DS3 multiplexing.
   This option allows a mapping for the transport of asynchronous DS3 signals (M13 multiplexers can be used to multiplex 28 DS1s, 14 DS1C, or 7 DS2 signals into an asynchronous DS3, which is then transported through the SONET overlay facility as an STS-1 SPE). This option duplicates the functionality of transporting DS3 signals inherent in today's networks. A limitation of this option is the inability to directly access the DS1 and DS0 signal within the DS3 bit stream. Access to the DS1 and DS0 signals can be provided

with the addition of M13 demultiplexing equipment; rearrangement requires digital cross-connect equipment.

2. Synchronous DS3 (SYNTRAN) multiplexing. This option allows direct mapping of the synchronous SYNTRAN DS3 signal into the STS-1 payload. SYNTRAN provides direct user access to the DS0 and DS1 signals, as discussed earlier and also provides a means to interface directly to existing asynchronous DS3-oriented transmission facilities (DS0 is not observable "inside of SONET"; it is observable "inside SYNTRAN," which SONET can carry).

Table 3.9 provides a summary of the traditional and new digital signal hierarchies.

3.8.6 STS-N Frame Structure

As already indicated, the STS-N signal is formed by byte interleaving N STS-1 signals. The STS-N frame structure is depicted in Figure 3.19. The transport overhead channels of the individual STS-1 signals must be frame aligned before interleaving. The associated STS SPEs are not required to be aligned because each STS-1 will have a unique payload pointer to indicate the location of the SPE. The resulting STS-N is then passed through an electrical-to-optical converter.

3.8.7 Concatenated STS-1s

Superrate services are services that require multiples of the STS-1 rate, such as the broadband ISDN H4 channel. They are mapped into the STS-Nc SPE and

Table 3.9
Traditional and New Digital Signal Hierarchies

(24) DS0 → DS1
(2) DS1 → DS1C
(4) DS1 → DS2
(28) DS1 → DS3 asynchronous mode
(7) DS2 → DS3 asynchronous mode
(1) DS3 asynchronous mode demultiplexed, and → remultiplexed to DS3 synchronous mode
(3) DS3 asynchronous mode → DS4NA
(6) DS3 asynchronous mode → DS4
(6) DS3 asynchronous mode → DS4
(28) DS1 → OC-1
(14) DS1C → OC-1
(7) DS2 → OC-1
(1) DS3 asynchronous → OC-1
(1) DS3 synchronous → OC-1
(3) OC-1 → OC-3
Mixed combinations with VTs (for example, 24 DS1 and DS2 → OC-1)

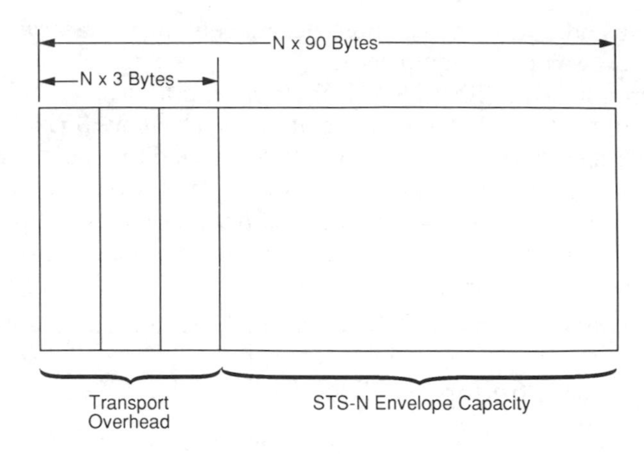

Figure 3.19 STS-N frame.

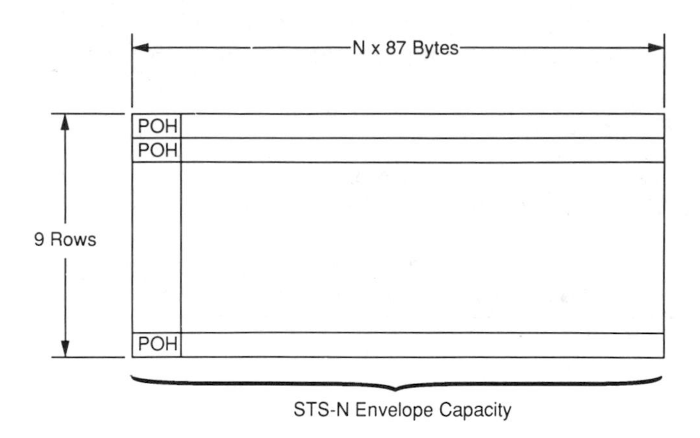

Figure 3.20 Concatenated STS-Nc payload.

transported as a concatenated STS-Nc whose constituent STS-1s are kept together. The STS-Nc is carried by an STS-M line signal where $M \geq N$ (typically M is used to depict a lower rate signal than N; however, this is the way the standard describes concatenation). The STS-Nc must be multiplexed, switched, and transported through the network as a single entity. A concatenation indication is contained in the STS-1 payload pointer. In a concatenated STS-Nc only one POH exists; in STS-N there are N lines and N paths. The STS-Nc SPE is depicted in Figure 3.20. It consists of N $\times$ 87 columns and 9 rows of bytes. The order of transmission is row by row, from left to right. As indicated, only one set of STS POH is required in the STS-Nc SPE. The STS-Nc SPE is carried within the STS-Nc such that the STS POH always appears in the first of the N STS-1s that make up the STS-Nc.

3.8.8 Network Equipment

SONET network elements (NEs) are pieces of equipment which must be embedded in the network to make the SONET concept a reality. SONET NE functions can be divided into two groups:

1. Multiplexing several STS-1 electrical signals into an STS-N signal, and performing the electrical-to-optical conversion;
2. Conversion of network signals into STS-1 payloads.

As of the end of 1989, Bellcore had issued five SONET equipment TA or TR documents as follows: ADM, DCS, switch interface, regenerator, and DLC with SONET interface; an ADM operating in the terminal mode becomes a SONET multiplexer. Equipment manufacturers are beginning to deliver SONET NEs; some of this equipment may provide additional functions in addition to those spelled out in the basic Bellcore TRs. In the discussion below an available product line is discussed to give the reader a sense of what the new digital equipment provides. These types of equipment are discussed for illustrative purposes only, and are covered here for pedagogical reasons.

Fiber Transmission System

These types of NEs multiplex N STS-1 signals to form an STS-N that is then converted to an OC-N optical signal. This NE can accept DS3 terminations in addition to the STS-1 signal interface.

Terminal Multiplexers

The terminal multiplexer (TM) NEs terminate several DS1 signals and assemble them into STS-1 payloads. One type of terminal may terminate 28 DS1 signals and functionally replace the M13 multiplexers in use today. These types of NEs are equipped with an integrated electrical-to-optical converter, providing an OC-1 optical signal. Another type may terminate 84 DS1 signals and include a byte interleave function that produces an OC-3 interface.

Add-Drop Multiplexer

Add-drop multiplexer (ADM) NEs are placed in series along a SONET route and allow DS1 signals to be added to or dropped from an STS-1 signal. The ADM has an east and a west direction, and DS1 terminations can be to or from either direction. One type of ADM provides add-drop access to one STS-1. It can be equipped with integrated optical converters for OC-1 compatibility. Another type of ADM may allow add-drop access to multiple STS-1 signals and be equipped with integrated multiplexers and optical converters. Add-drop multiplexers can also be equipped with time slot interchangers to allow digital cross-connect functions among the DS0 channels carried in the STS-1 payload.

Access Multiplexers

Access multiplexer (AM) NEs make use of the inherent byte visibility of the SONET format to terminate DS0-based services directly to an STS-1 payload. This is accomplished without the need to first multiplex them into DS1 signals and then into DS3 signals as is done today. AMs typically integrate the functions of a TM or ADM with the line circuit termination equipment of a digital loop carrier. One type of AM may terminate 672 DS0 channel equivalents on one STS-1 and may be equipped with integrated digital cross-connect functions.

Equipment Availability

As of mid-1990, approximately one dozen vendors had announced SONET products, including AT&T Technologies, Alcatel, and Fujitsu America [3.42], [3.43], [3.44], [3.45]. According to some projections, SONET-based facilities are expected to carry 40% of interoffice circuits by 1997 and 80% by the year 2000 [3.51].

REFERENCES

[3.1] ANSI T1.101-1987, Synchronization Interface Standards for Digital Networks, ANSI, New York, NY.

[3.2] G.H. Bennett, *Pulse Code Modulation and Digital Transmission*, Marconi Instruments Limited, White Crescent Press, Luton, 1978.

[3.3] Systems Interface, TR-TSY-000510, July 1987, Bellcore, Livingston, NJ.

[3.4] Miscellaneous LSSGR, TR-TSY-000530, July 1987, Bellcore, Livingston, NJ.

[3.5] B.A. Mordecai, personal communication, Bellcore, April 1990.

[3.6] W. Freyer, "Performing DS1 Signal Testing with a User-configured Test Set," *Communications News*, December 1984.

[3.7] *Bell System Technical Journal*, special issue on the D4 channel bank family, November 1982.

[3.8] Bell System Technical Reference, PUB 43801, November 1982.

[3.9] Coastcom, Inc., promotional literature.

[3.10] ANSI T1.107-1988, *Digital Hierarchy—Format Specifications*, ANSI, New York.

[3.11] SONET Transport Systems: Common Generic Criteria, TR-TSY-00253, Issue 1, September 1989, Bellcore, Livingston, NJ.

[3.12] Transport Systems Generic Requirements (TSGR): Common Requirements, TR-TSY-000499, Issue 3, December 1989, Bellcore, Livingston, NJ.

[3.13] P.T. Rux, "T1 Testing and the Standard Answer," *TE&M*, January 1, 1990, pp. 50 ff.

[3.14] ANSI T1.403-1989, *Carrier to Customer Installation—DS1 Metallic Interface*, ANSI, New York.

[3.15] W. Freyer, "Performing DS1 Signal Testing with a User-configured Test Set," *Communications News*, December 1984.

[3.16] M. Lefkowitz, "Intelligent CSUs for Performance Monitoring," *Telecommunications*, November 1986.

[3.17] K.H. Molow, "ABCs of Digital Formats," *TE&M*, June 15, 1989, pp. 70 ff.

[3.18] K. Stauffer and A. Brajkovic, "DS1 Extended Superframe Format and Related Performance Issues," *IEEE Communications Magazine*, April 1989, pp. 19–23.

[3.19] W. Freyer, "Doing DS1 Signal Tests with User-configured Sets," *Communications News*, November 1984.

[3.20] W. Buckley, "T1 Standards and Regulations: Conflict and Ambiguity," *Telecommunications*, March 1987.

[3.21] D. Minoli, "Phone Changes Benefit Users," *ComputerWorld*, May 12, 1986.

[3.22] J. Ingle, personal communication, Bellcore, April 1990.

[3.23] N. Richard, "Bridging Between E1 and T1," *Telecommunications*, January 1990, pp. 39 ff.

[3.24] W.J. Buckley, "T1 Standards Battles," *TE&M*, January 1988.

[3.25] M.H. Weik, *Communications Standard Dictionary*, Van Nostrand, New York, 1983.

[3.26] G. Langley, *Telephony's Dictionary*, Telephony Publishing Corp., Chicago (undated).

[3.27] R. Giddens, personal communication, Telco Systems, August 1986.

[3.28] "AT&T to Double its T1 Circuit Capacity," *Data Communications Newsfront*, April 1984.

[3.29] R. Adleman, and D. Sparell, "Taking a Bite out of Bits," *Telephone Engineer and Management*, October 15, 1984.

[3.30] R.F. Rey, ed., *Engineering and Operations in the Bell System*, 2nd Ed., AT&T, Murray Hill, NJ, 1984.

[3.31] K. Hohhof, "What is a Digital Crossconnect?," *CO Switching, Transmission and Network Services*, May 1986.

[3.32] R. Connolly, *et al.*, "Why use a Digital Access and Crossconnect?," *Telephone Engineer and Management*, March 1, 1986.

[3.33] J. Mille, "Beyond DS1," *TE&M*, January 1, 1988, pp. 63–65.

[3.34] ANSI T1.103-1987, *Digital hierarchy—Synchronous DS3 Format Specification*, ANSI, New York.

[3.35] M.F. Finneran, "Designing and Operating Digital Backbone Networks Using Fractional T1, T1, and T3," Communication Networks '90 Notes.

[3.36] E.E. Mier, "Adding to Your Net Worth with T-1-to-LAN Devices," *Data Communications*, September 1989, pp. 103 ff.

[3.37] S.L. Ledgerwood, "T-1 Outage Ahead: A Backup Strategy Could Save Your Job," *Data Communications*, July 1989, pp. 75 ff.

[3.38] E.E. Mier, "Fractional T1: Carriers Carve Out Bandwidth for Users," *Data Communications*, November 1989, pp. 84 ff.

[3.39] S. Fleming, "Get Ready for T-3 Networking," *Data Communications*, September 1989, pp. 82 ff.

[3.40] S. Fleming, "To Know SONET, Know Your VTs," *TE&M*, June 15, 1989, pp. 66–69.

[3.41] L. Campbell, and C. Engineer, "Standards for an Evolving Network," *TE&M*, July 1, 1988, pp. 50–58.

[3.42] "AT&T's SONET to NTT," *Lightwave*, August 1989, p. 3.

[3.43] "SONET sells," *Lightwave*, January 1990, p. 3.

[3.44] Alcatel Product literature, January 1990.

[3.45] T. Sweeney, "Focus on Broadband, SONET," *Communications Week*, April 16, 1990, p. 93.

[3.46] IRD Report #749, *T3, DS3, and Related Wideband Products and Services.*

[3.47] Vertical Systems Group information.

[3.48] Market Intelligence Research Company, *New Outlooks in the US T1 Equipment and Services Market,* Mountain View, CA, 1989.

[3.49] R.L. Gillan, "Advanced Network Architectures Exploiting the Synchronous Digital Hierarchy," *Telecommunications Journal of Australia,* Vol. 39, No. 3, 1989.

[3.50] B. Bowie, "How to Join the SONET Club," *Lightwave,* January 1990, p. 25 ff.

[3.51] H. Rausch, "SONET Expected to Carry 40% of Interoffice Circuits in 1997," *Lightwave,* January 1990, p. 23.

Note: Sections of this chapter are based on the following reports produced by the author:
"All About Channel Banks: Technology Briefing," DataPro Report CA80-010-902, November 1986.
"An Overview of ADPCM Transcoders," DataPro Report CA80-010-604, November 1986.
"T1 Test and Monitoring Equipment," DataPro Report CA30-020-701, July 1988.
"An Overview of Digital Crossconnect Systems," DataPro Report CA80-010-501, March 1987.
"An Overview of CSUs," DataPro Report CA80-010-401, July 1987.
"An Overview of T-Carrier Systems," DataPro Report CA30-010-304, March 1988.

Chapter 4
ISDN and BISDN

4.1 THE MOTIVATION FOR ISDN

During the 1970s, future telecommunication needs were recognized to be centered on the rapid growth and evolution of digital communication services, increased end-user control, and standardized network interfaces. User demands for voice, data, video, image, and text require the deployment of flexible public-switched transmission facilities to accommodate more easily the anticipated changing mix of services. Although voice services will continue to represent the dominant share of the telecommunication market, the greatest growth area in the next decade is expected to be in nonvoice services. Interest has increased in nonvoice transmission, including data, video, facsimile, image, and graphics information, which is not particularly suited to an analog network [4.1]. The concept of an integrated services digital network (ISDN) began to evolve in the early 1970s to support these market demands for universal service. ISDN represents the latest stage of evolution for telephone networks as discussed in Chapter 1 and provides end-to-end digital connectivity with access to voice and data services over the same digital transmission and switching facilities. ISDN provides a range of services using a limited set of connection types and multipurpose user-network interface arrangements. ISDN is now being positioned as the immediate technology on the horizon for the integration of various telecommunication technologies. Existing digital transmission and CO equipment can fairly easily be upgraded to provide ISDN. This, in turn, means that the service can be offered to the end-user in an economical way.

4.1.1 ISDN Goals

ISDN is directed primarily at the standardization of the interfaces and services of a public digital network. Currently, different types of interfaces to the telephone network exist for different services, including VF two-wire switched, VF two-wire dedicated, VF four-wire dedicated, DDS, T1 four-wire dedicated, *et cetera*. ISDN capabilities include the use of digital loops from COs to user premises, multichannel facilities operated using a standard demultiplexing scheme, and user-accessible control channels. The elimination of multiple analog-digital conversions increases the quality of transmission. In addition, a variety of implementation configurations is supported, including circuit switched, packet switched, and nonswitched connections. Many customer-premises systems can be linked through the ISDN interfaces.

In addition to providing an end-to-end digital path over a set of standardized user interfaces, ISDN provides the user with the capability to signal to the network using an out-of-band channel during all phases of a call. Standardized signaling channels allow control signals to be transmitted between the network and intelligent devices on the user premises. The call control employed for ISDN uses explicit messages, so that CPE devices can readily instruct and automatically control the connection with more flexibility. The direct out-of-band signaling conceivably makes possible a class of new services. For example, the use of out-of-band signaling between switching systems, as well as between the local switching system and the user terminal, makes it possible to provide network-wide versions of call management features, which are today confined to calls within an exchange [4.2].

Traditionally, when using the *public switched telephone network* (PSTN), a user's digital data are transmitted as analog signals, using a modem. The signals may then be digitized by a digital loop carrier system (as discussed in Chapter 3) at the 64 kb/s rate and transmitted to the destination office. The data are then converted back to analog form, for delivery to a user, whose modem reconverts the signals to digital data for computer and terminal use. The transmission process is naturally more efficient if the signals remain digital throughout the process. Also, a 64 kb/s digital interoffice channel is employed, even with modems operating at 300 b/s. With the ISDN connection, the user's CPE connects directly to a digital line, with the full 64 kb/s capacity of the carrier's interoffice channel now available to the user. Better error characteristics exist in this environment: Digital signals can be regenerated, while analog signals can only be amplified. In a noisy environment, both the analog signal and the noise are amplified. In addition, data multiplexed into the signaling channel can be directed to a packet switched network, enabling the user to utilize the data multiplexing advantages afforded by packet switching.

In 1984, the Plenary of the CCITT adopted the I Series recommendations dealing with ISDN. CCITT stated that "an ISDN is a network . . . that provides end-to-end digital connectivity to support a wide range of services, including voice and non-voice services, to which users have access by a limited set of standard multipurpose user-network interfaces." CCITT Study Group XVIII has already created a large number of recommendations covering the development and implementation of ISDNs, which can be found in the CCITT "Red Book" (1984–85) and in the "Blue Book" (1988–89). Many countries are now involved in the planning and deployment of ISDNs as the basis of their future telecommunication infrastructure. ISDN (or more precisely, islands of ISDNs) will evolve over the decade from the existing telephone network into a comprehensive and ubiquitous infrastructure by progressively incorporating additional functions to provide for both existing and new services [4.3].

A major effort is under way in Europe and Japan to bring ISDN to the market. In the United States, several trials have taken place in the 1986–89 time frame. The first commercial ISDN service of ISDN in the United States was provided by BellSouth in the spring of 1988; several BOCs and IXCs are now offering ISDN services (see a later section on projected penetration).

4.1.2 Thrust for ISDN

The current forces that are providing the thrust toward ISDN are common channel signaling, software-controlled products, and automated operations systems. Technical developments prompting the development of ISDN include:

- Very large scale integration (VLSI);
- High-capacity optical fiber transmission;
- High-speed switching; and
- Software, particularly SPC.

Networks already make extensive use of digital time-division techniques, rather than the earlier analog space-division switching. Digital switching systems gracefully interface with digital transmission systems (such as those discussed in Chapter 3), facilitating a digital end-to-end connection. Digital transmission technology in the form of trunk carrier systems and digital loop carrier systems is becoming prevalent. Common channel signaling over a separate packet switched network (but still part of the public switched network) is also becoming the norm, even for many local end offices. Explicit user access to functional signaling facilitates user-network control and user access to advanced intelligent services.

4.2 ISDN CONCEPTS

This section provides a description of key ISDN concepts. Two new ISDNs are evolving:

- Narrow band ISDN, providing DS0 and DS1 range digital bandwidth;
- Broadband ISDN (BISDN), providing DS3 and SONET range digital bandwidth.

Narrow band ISDN has been standardized beginning in 1984, and is further along in terms of deployment. BISDN had been (partially) standardized by 1988. The following sections deal with narrow band ISDN, while the latter part of the chapter deals with BISDN. Unless specifically noted to the contrary, the term "ISDN" refers here to narrow band ISDN.

4.2.1 ISDN Framework

The support of out-of-band signaling and the ability to activate services during a call imply a separation between the control information and the user information. The notion of plane—control plane (C-plane) and user plane (U-plane)—is introduced in ISDN to reflect this separation. The main rationale for protocols within the user plane is the transfer of information among user applications (digitized voice, data, and information transmitted between users). This information may be transmitted transparently through ISDN, or may be processed or manipulated (for example, conversion between μ-law coded PCM speech and A-law coded PCM speech). The main rationale for protocols within the control plane is the transfer of information for the control of user plane connections. Examples include controlling a network connection (establishment and clearing); controlling the use of an already established network connection; invoking a supplementary service (such as call forwarding). All control information that involves resource allocation or deallocation by ISDN pertains to the C-plane.

The ISDN framework advanced by CCITT aims at achieving [4.4, 4.5]:

- Full integration of C-plane procedures for all services (i.e., one set of protocols for call control); supplementary services and operational, administrative, and maintenance messages across all telecommunication services; and
- Decoupling of user information transfer requirements from C-plane transfer requirements. This allows for the possibility of defining telecommunication services having U-plane characteristics that are tailor-made only to the transfer needs of user information and not to those of C-plane information.

The bearer service capabilities offered by ISDN are: (1) speech; (2) 3.1-kHz audio; (3) 7-kHz audio; (4) high-speed end-to-end digital channels, basically at 64 kb/s but with (planned) subrates and superrates (e.g., 384 kb/s); and (5) packet-mode transmission. Telemetry-type services are also identified, but not emphasized in the current standards. These services may be provided as permanent, reserved, or switched-on-demand.

4.2.2 ISDN Reference Configuration

As indicated, a predicate of ISDN is that a small set of compatible user-network interfaces can economically support a wide range of user applications, equipment, and configurations. The number of user-network interfaces is kept small to maximize user flexibility through terminal compatibility (from one application to another, one location to another, and one service to another) and to reduce costs through economies in production of equipment and operations of both ISDN and other user equipment. However, different interfaces are required for applications with widely different information rates, complexity, or other characteristics, as well as for applications in an evolutionary stage. Another objective is to use the same interfaces even though different configurations exist (for example, single terminal versus multiple terminal connections, connections to a PBX *versus* connections into the network).

To support the goals just stated, three modeling concepts are defined in CCITT Recommendation I.411: *reference configurations, functional groups,* and *reference points.* Reference configurations are conceptual configurations useful in identifying various possible physical user access arrangements to ISDN. The next two concepts are used in defining reference configurations. Functional groups are sets of functions that may be needed in ISDN user access arrangements. In a particular access arrangement, specific functions in a functional group may or may not be present. Specific functions in a functional group may be performed in one or more pieces of equipment. Reference points are conceptual points dividing functional groups. In a specific access arrangement, a reference point may correspond to a physical interface between equipment, or a physical interface corresponding to the reference point may not exist. (Physical interfaces that do not correspond to a reference point are not the subject of ISDN user-network interface recommendations.)

Figure 4.1 shows the reference configuration with a topological overlay. Physical interfaces not included in CCITT's I.411 may appear at the R reference point (for example, CCITT X Series interface recommendations). In Recommendation I.411, no reference point is actually assigned to the transmission line because no ISDN user-network interface is required internationally at that location; for the United States, this interface is required. Figures 4.2 and 4.3 show examples of applications of ISDN reference configurations, based on I.411.

The ISDN user-network interface recommendations, defined in CCITT I.412 and discussed below in Section 4.2.2.3, apply to the physical interface at the reference points S and T.

4.2.2.1 Functional Groups

A description follows of the functions included in each functional group (see also Figure 4.1).

Network Termination 1 (NT1). This functional group includes functions broadly equivalent to layer 1 of OSIRM. These functions are associated with the proper

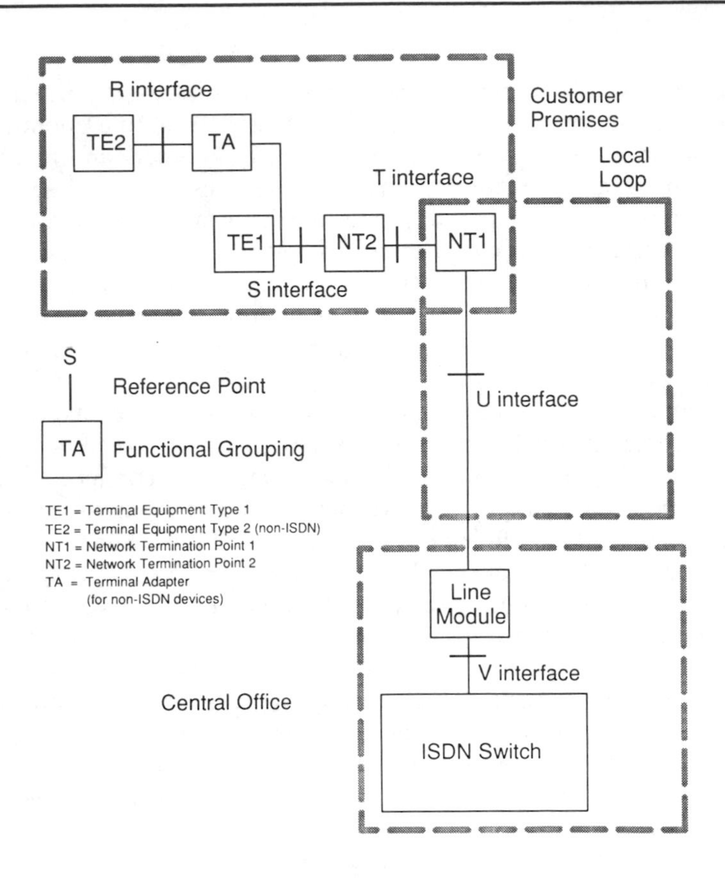

Figure 4.1 The ISDN reference configuration model (extended from CCITT I.441 for U.S. applications).

physical and electrical termination of the network, including line transmission termination; layer 1 line maintenance functions and performance monitoring; timing; power transfer; layer 1 multiplexing; and interface termination, including multidrop termination employing layer 1 contention resolution. (NT1 is customer equipment in the United States and network equipment elsewhere.)

Network Termination 2 (NT2). This functional group includes functions broadly equivalent to layers 1 and higher of OSIRM. Functions include layers 2 and 3 protocol handling; layers 2 and 3 multiplexing; switching; concentration; maintenance functions; interface termination; and other layer 1 functions. These functions are typically undertaken in such CPE as PBXs, LANS, terminal controllers, and multiplexers.

Terminal Equipment (TE). This functional group provides functions undertaken by such equipment as digital telephones, data terminal equipment, and integrated

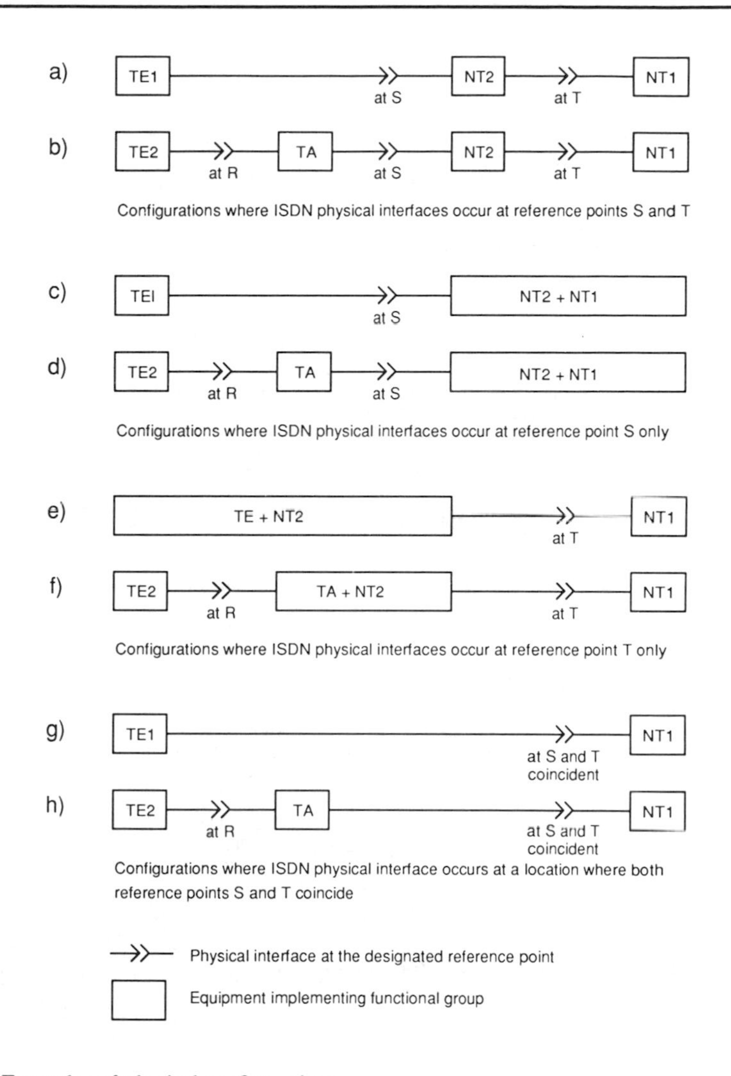

Figure 4.2 Examples of physical configurations.

voice and data workstations. Functions include protocol handling; maintenance functions; interface functions; and connection to other equipment.

Terminal Equipment Type 1 (TE1). This functional group includes functions belonging to the functional group TE, and has an interface that complies with the ISDN user-network interface standards.

Terminal Equipment Type 2 (TE2). This functional group includes functions belonging to the functional group TE but with a non-ISDN interface.

Terminal Adapter (TA). This functional group includes functions that allow a TE2 terminal to be served by an ISDN user-network interface.

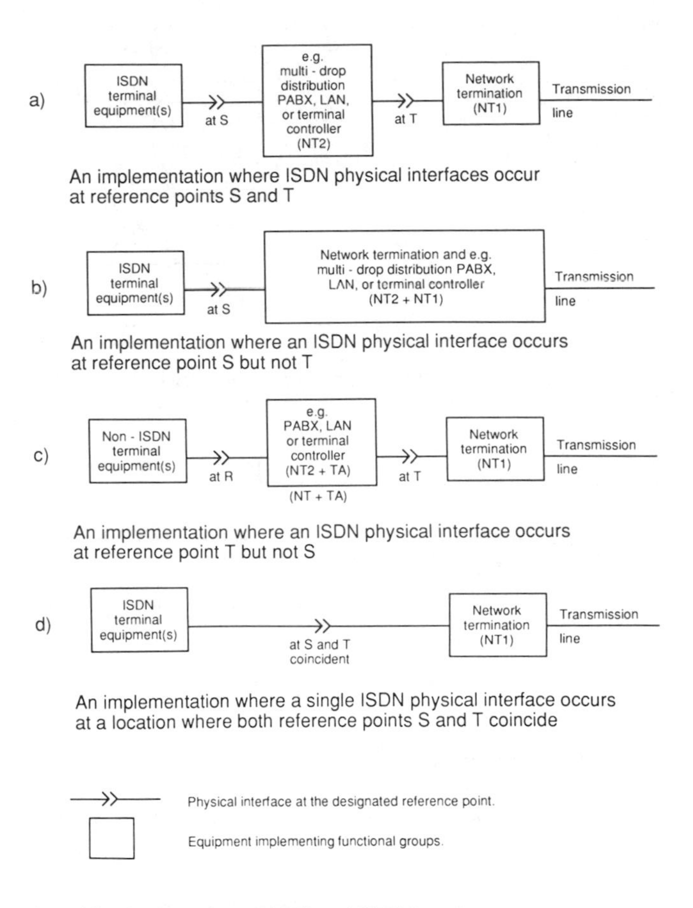

Figure 4.3 Examples of implementation of NT1 and NT2 functions.

4.2.2.2 Reference Points

A description of the reference points follows (see also Figure 4.1).

The *R reference point* is the functional interface between a non-ISDN terminal and the TA. This interface is likely to follow the RS-232-C or V.35 specification, depending on the bandwidth of the TE2 signal stream. This interface between the terminal and the TA is an interim solution. In the future, the TA may migrate into the terminal itself, giving rise to an "ISDN terminal."

The *T reference point* is the functional interface seen by the user's ISDN-configured equipment, as it connects to the NT2. The *S reference point* is the functional interface seen by the user's ISDN configured equipment as it connects the NT1. In practical terms, the S and T reference points can be thought of as identical. A large share of the ISDN standards concentrate on the activities associated with access to the S and T reference point. This reference point is also referred to, with a slight loosening of the language, as the S/T interface. Bit timing, octet timing, power feeding, activation and deactivation, request and permission to access the signaling channel for the purpose of transmitting data, and request and permission to busy one of the bearer channels will be performed across the S-T interface.

The *U reference point* (also called interface on the network side of the NT) is the functional interface of the access line with the NT, on the network side. Standardization of this interface is needed if the NT1s and the CO concentration module are to be manufactured by different vendors. (This reference point is not defined in CCITT standards.)

The *V reference point* is the functional interface between the line module (typically a CO terminal—carrier system) and the switch itself. (This reference point is not defined in CCITT standards.)

4.2.2.3 ISDN User-Network Interfaces

CCITT Recommendation I.421 defines three channel types available to the ISDN user: B channels, D channels, and H channels. The B channel is a 64 kb/s access channel that carries customer information, such as voice calls, circuit switched data, or packet switched data. It is a U-plane facility. The D channel is an access channel carrying control or signaling information and, optionally, packetized customer information. The D channel has a capacity of 16 or 64 kb/s, as discussed below. (The Red Book also suggested the use of the D channel for telemetry, but at present no special provisions have been made for it.) The D channel is principally a C-plane facility, although some U-plane activities can also be undertaken. The H channel is a 384 kb/s, 1.536 Mb/s, or 1.920 Mb/s channel that carries customer information, such as teleconferencing, high-speed data, or high-quality audio or sound program. The H channel is a U-plane facility.

ISDN defines *user-network physical interfaces* at reference points S and T that comply with the interface structures depicted in Table 4.1 (based on CCITT I.412 for narrow band ISDN). This interface is also called "user-network access arrangement" and "user-network interface structures". All the narrow band ISDN S-T interface structures listed in Table 4.1 can be provided on unloaded copper loops. $2B+D$ can be provided over two-wire loops; the others will require four-wire loops. BISDN utilizes fiber local loops.

The two most well-known interface structures, and the first to be commercially deployed in the United States, are as follows:

The *basic rate access* (also known as $2B+D$, with B = bearer, D = delta [4.6] or *basic rate interface*) is a standardized user interface to ISDN service providing access to two B channels and one D channel. This access provides three

Table 4.1
ISDN User-Network Interface Structures

Narrowband—Data Rates at the DS0 and DS1 Levels:	
2B + D	Two 64 kb/s channels, plus a 16 kb/s packet-signaling channel*
23B + D	Twenty-three 64 kb/s channels, plus a 64 kb/s packet-signaling channel†
30B + D	Thirty 64 kb/s channels, plus a 64 kb/s packet-signaling channel (Europe)‡
3H0 + D	Three 384 kb/s channels plus one 64 kb/s packet-signaling channel
4H0	Four 384 kb/s channels (signaling to be provided on another D channel interface)§
5H0 + D	Five 384 kb/s channels plus one 64 kb/s packet-signaling channel (Europe)
H11	Nonchannelized 1.536 Mb/s (signaling to be provided on another D channel interface)§
H12	Nonchannelized 1.920 Mb/s (signaling to be provided on another D channel interface)§
Broadband—High Data Rates Based on SONET as the Transport Technology:	
H21	32.768 Mb/s
H22	In the range of 43 to 45 Mb/s, integer multiple of 64 kb/s, and with 3 × H22 ≤ H4 (yet to be formally adopted)
H4	In the range of 132 to 138.240 Mb/s, integer multiple of 64 kb/s, H4 ≥ 4 × H21 (yet to be formally adopted)

* CCITT I.412 also lists a B + D and a D configuration as an allowed basic-rate access capability.

† CCITT I.412 also lists nB + D with $n < 23$, and 24B as allowed primary-rate access capabilities.

‡ CCITT I.412 also lists nB + D with $n < 30$, and 31B as allowed primary-rate access capabilities.

§ A primary rate interface may have a structure consisting of a single D channel (64 kb/s) and any mixture of B and H0 channels. In the case of a user-network access arrangement containing multiple interfaces, a D channel in one interface structure may also carry signaling for channels in other interface structures. When a D channel is not activated, its 64 kb/s capacity may be used, if desired, for the mixture of B and H0 channels, depending on the situation, e.g., 3H0 + 6B for a 1.544 Mb/s interface.

bidirectional, symmetric digital channels to the user's premises. Each of the B channels can be independently switched. Either or both B channels may be permanently connected in special service applications. Control over B channel connections for demand applications resides in the signaling messages passed via the D channel [4.2]. The customer can use all or parts of the two B channels and the D channel. For a speech call, the user equipment must digitize the user's speech and place the properly coded speech on the appropriate B channel. Analogously, the user speech equipment must decode B channel information and convert it to audible speech. Basic rate access terminals, such as PCs and integrated voice and data terminals, started to appear on the market in the late 1980s, but an embedded base of non-ISDN terminals may well exist for a long time to come.

The *primary rate access*—also known as $23B+D$ or *primary rate interface* (PRI)—is a standardized user interface to ISDN service providing access to 23 B channels and one D channel (for North America, Japan, and Korea; for Europe, a $30B+D$ structure is defined). The customer can use all or parts of the B channels and the D channel, which in this case has a bandwidth of 64 kb/s. Primary rate access is provided using time-division multiplexed signals over four-wire copper circuits (using standard regenerators as necessary), or on other media. As is the case with basic rate access, the D channel has a message-oriented protocol that supports call control signaling and packet data. Each B channel can be switched independently; some B channels may be permanently connected in special service applications. In addition to other applications, PRI ISDN is finding a niche in PBX to CO connections, as described in Chapter 10.

For single mountings, the NT will connect to the network through a miniature eight-position jack, identical to the RJ 45. The cord from the NT will terminate in a miniature eight-position plug. For multiple mountings, other connection arrangements may be appropriate. For basic rate access, specifications for the eight-position plug and jack are described in ISO 8877-1987, except for pin assignments. The jacks are equipped with the center two contacts (pins), which are used for the cable pair, commonly called tip (T) and ring (R).

Other Concepts

The term *bearer capabilities* describes the nature of the speech or data call. For the United States, six bearer capabilities will be developed [4.2]:

1. Circuit-mode unrestricted digital transmission, 8-kHz structured, demand, point-to-point, and bidirectional symmetric;
2. Circuit-mode unrestricted digital transmission, rate-adapted from 56 kb/s, 8-kHz structured, demand, point-to-point, and bidirectional symmetric;
3. Circuit-mode, 64 kb/s, 8-khz structured, usable for speech, demand, point-to-point, and bidirectional symmetric;
4. Circuit-mode, 64 kb/s, 8-khz structured, usable for 3.1-kHz audio information transfer, demand, point-to-point, and bidirectional symmetric;
5. Packet-mode, unrestricted, virtual call and permanent virtual service, point-to-point, and bidirectional symmetric; and
6. Circuit-mode, 64 kb/s, 8-kHz structured, usable for 7-kHz audio information transfer, demand, point-to-point, and bidirectional symmetric.

Echo cancellation is the technique used in North America for the ISDN digital subscriber line in which a record of the transmitted signal is used to remove echoes of this signal that may have mixed with and corrupted the received signal (see Figure 4.4).

4.3 ISDN STANDARDS

Currently, more than 70 ISDN standards are defined in CCITT I and Q Series, in addition to approximately 100 other secondary but related standards. These standards encompass thousands of pages of technical specifications. Also, some of the ISDN standards are specific to the United States. Table 4.2 provides a summary of the key applicable standards [4.7]. Table 4.3 provides a more complete listing, based on the Blue Book [4.8]. Note that some CCITT ISDN standards have both an "I" and a "Q" designation.

As implied earlier, the S and T interfaces are important because they represent the user's interface to ISDN. The standards described below apply to this interface (except where noted), and apply to the C-plane (except where noted).

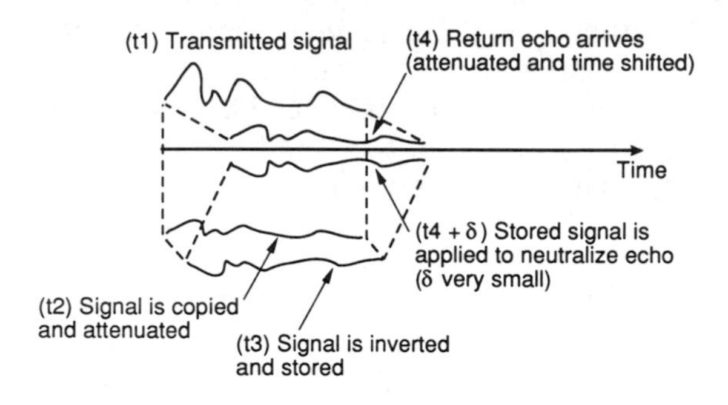

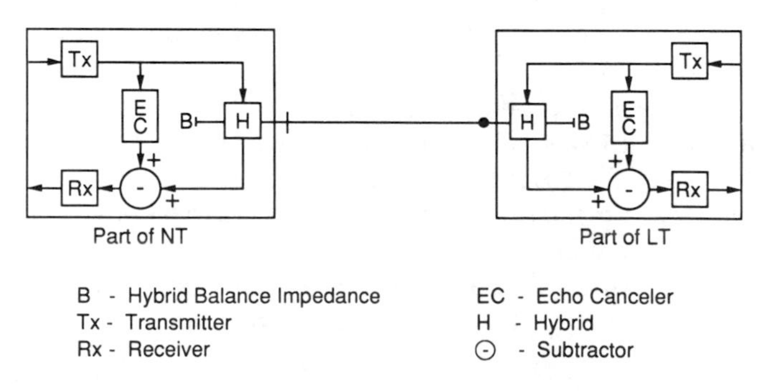

Figure 4.4 Echo cancellation.

Table 4.2
Key ISDN Standards

Basic Rate (Physical Layer)	CCITT I.420, I.430 (four-wire) ANSI T1.605-1989 (U.S. S/T), T1.601 (two-wire U.S. "U-interface") Bellcore TR-TSY-000397
Primary Rate (Physical Layer)	CCITT I.421, I.431 ANSI T1.403 and T1.408 (U.S. U-interface) (full number to be assigned mid-1990) Bellcore TR-TSY-000754
D Channel (Link Layer)	CCITT Q.920 (I.440) and Q.921 (I.441) ANSI T1.602-1989 (U.S.) Bellcore TR-TSY-000793
D Channel (Network Layer)	CCITT Q.930 (I.450), Q.931 (I.451), Q.932 (supplementary services) ANSI T1.607-1989 (U.S.), T1.608-1989 (U.S.), T1.610–1990 (supplementary services) Bellcore TR-TSY-000268
Rate Adaption	CCITT V.110 (Red Book), V.120 (Blue Book), ANSI T1.406 (U.S.)

The CCITT standards at the physical layer are I.420 (*basic rate access definition*), I.430 (*basic rate access layer 1*), I.421 (*primary rate access definition*), and I.431 (*primary rate access layer 1*). The narrow band ISDN transmission system (both basic rate access and the primary rate access) is designed to operate on twisted metallic cable pairs with mixed gauges. In the case of the basic rate access, the physical layer specified in I.430 manages a 192 kb/s full-duplex bit stream using time-division methods to recover the two B channels and one D channel (the remaining bits are used for framing and for physical layer control information). The *data link layer* is not defined for transparent B channels used for circuit switched voice or data; on the other hand, it is defined for the D channel. ISDN provides a specific protocol that the user can employ to signal the network. Currently, a three-layer protocol suite is defined. For narrow band ISDN at the least, the D channel employs a LAP-D link layer protocol; it is a version of the ISO HDLC Data Link protocol, and is specified in CCITT Recommendations Q.920 (I.440) and Q.921 (I.441). It provides for statistical multiplexing of three types of information: (1) signaling information for the management of the B channels; (2) X.25 packet switched service over the D1 channel; and (3) optional channels, used for telemetry or other applications (the protocols used here may not be Q.921, although they must share a common "core" with Q.921).

The *network layer protocol* for the signaling channel is specified in CCITT's Q.930 (I.450) and Q.931 (I.451) specifications; it provides the mechanism for establishing and terminating basic connections on the B channels and other network control functions. In the United States, T1.607-1989 and T1.608-1989 were close

Table 4.3
Listing of CCITT ISDN Standards

I.100 Series—General ISDN Concepts, Terminology, Methods:	
I.110	Preamble and general structure of the I Series Recommendations
I.111	Relationship with other recommendations relevant to ISDNs
I.112	Vocabulary of terms for ISDNs
I.113	Vocabulary of terms for broadband aspects of ISDNs
I.120	Integrated service digital networks (ISDNs)
I.121	Broadband aspects of ISDN
I.122	Framework for providing additional packet mode services
I.130	The method for the characterization of ISDN telecommunication services
I.140	Attribute technique for the characterization of ISDN telecommunication services
I.141	ISDN charging capability attributes
I.200 Series—Service Aspects:	
I.200	Guidance to the I.200 Series Recommendations
I.220	Common dynamic description of basic telecommunication services
I.221	Common specific characteristics of services
I.230	Definition of bearer service categories
I.231	Circuit mode bearer service categories
I.232	Packet mode bearer service categories
I.240	Definition of teleservices
I.241	Teleservices supported by an ISDN
I.250	Definition of supplementary services
I.251	Number identification supplementary services
I.252	Call offering supplementary services
I.253	Call completion supplementary services
I.254	Multiparty supplementary services
I.255	"Community of Interest" supplementary services
I.256	Charging supplementary services
I.257	Additional information transfer supplementary services
I.300 Series—Network Aspect:	
I.320	ISDN protocol reference model
I.324	ISDN network architecture
I.325	Reference configurations for ISDN connection types
I.326	Reference configurations for relative network resource requirements
I.32X	ISDN hypothetical reference connections
I.330	ISDN numbering and addressing principles
I.331	Numbering plan for the ISDN era (E.164)
I.332	Numbering principles for interwork between ISDNs and dedicated networks
I.333	Terminal selection in ISDN
I.334	Principles relating to ISDN numbers and subaddressing to the OSIRM addresses
I.335	ISDN routing principles
I.340	ISDN connection types
I.350	General aspects of quality of service
I.351	Recommendations in other series including network performance objectives
I.352	Network performance objectives for call processing delays

Table 4.3 continued
Listing of CCITT ISDN Standards

I.400 Series—User-Network Interface Aspects:	
I.410	General aspects and principles relating to ISDN user-network interfaces
I.411	ISDN user-network interface—Reference configurations
I.412	ISDN user-network interface—Interface structures and access capabilities
I.420	Basic user-network interface
I.421	Primary rate user-network interface
I.430	Basic user-network interface—Layer 1 specification
I.431	Primary rate user-network interface—Layer 1 specification
I.43x	Higher rate user-network interfaces
I.440	ISDN user-network data link layer general aspects (Q.920)
I.441	ISDN user-network data link layer specification (Q.921)
I.450	ISDN user-network interface layer 3—general aspects (Q.930)
I.451	ISDN user-network interface layer 3 specification for basic call control (Q.931)
I.452	ISDN user-network interface layer 3 specification for supplementary services (Q.932)
I.460	Multiplexing, rate adaption and support of existing interfaces
I.461	Support of X.21, X.21bis, and X.20bis based DTEs by an ISDN (X.30)
I.462	Support of packet mode terminal equipment by an ISDN (X.31)
I.463	Support of DTEs with V-Series type interfaces by an ISDN (V.110)
I.464	Multiplexing, rate adaption and support of existing interfaces for restricted 64 kbps
I.465	Support by an ISDN of DTEs with V-type interfaces with provision for statistical muxing
I.470	Relationship of terminal functions to ISDN
I.500 Series—Internetwork Interfaces:	
I.500	General structure of ISDN interworking recommendations
I.510	Definitions and general principles for ISDN interworking
I.511	ISDN-to-ISDN Layer 1—interwork interface
I.515	Parameter exchange for ISDN interworking
I.520	General arrangements for network interworking between ISDNs
I.530	Network interworking between an ISDN and a public switched telephone network
I.540	Interworking circuit switched public data networks and ISDN (X.321)
I.550	Interworking packet switched public data networks and ISDN (X.325)
I.560	Requirements to be met in providing telex service within the ISDN (V.202)
I.600 Series—Maintenance Aspects:	
I.601	General maintenance principles of ISDN subscriber access and subscriber installation
I.602	Application of maintenance principles to an ISDN subscriber installation
I.603	Application of maintenance principles to ISDN basic access
I.604	Application of maintenance principles to ISDN primary rate access
I.605	Application of maintenance principles to static multiplexed ISDN basic accesses

Note: Other CCITT standards applicable to ISDN are found in the E, F, G, H, Q, S, V, and X Series.

to final approval as of the end of 1989. These recommendations provide detailed specifications for national networks. For the packet switched service over the D channel, the network layer protocol is X.25. The layer 3 protocols for the optional channels are to be defined by CCITT in the future or are to be specified as national options.

CCITT Q.932 provides generic procedures for the control of supplementary services at the user-network interface (a supplementary service is a service beyond the basic connectivity; the latter, as indicated, is also known as a *bearer service*). In the Red Book (1984), these functions were included in Q.931, but in the Blue Book have been separated into a new standard. User-to-user signaling is also allowed (the applicable Bellcore document is TR-TSY-000845 [4.9]). This type of signaling could have applications in PBX-to-PBX signaling, as discussed in Chapter 10.

4.4 ISDN DATA COMMUNICATION ISSUES

Examples of CPE that can be connected to ISDN include [4.10]:

- ISDN terminals: integrated voice-data communication devices;
- Digital telephone and a discrete data device;
- Personal computers with internal ISDN adapter board. The ISDN PC card replaces a modem card; and
- Terminal adapters at the R reference point allow connection to non-ISDN data terminals to B1, B2, or D channels at the S-T interface point. The embedded base of existing non-ISDN terminals can be connected via a V.24/V.28 cable (alias RS-232-C). ISDN TA devices have been announced for synchronous terminals such as the X.21 or SDLC/SNA-compatible controller devices, and for asynchronous terminals.

For primary rate, PBX trunking arrangements seem to be the key application as of late 1990; use in host-to-host communication is also being explored. As of 1990, some T1 resource managers (high-end T1 multiplexers) had PRIs, to be used as a backup mechanism in case of T1 private line failure.

4.4.1 ISDN Rate Adaption

The types of devices described in the previous subsection may use rate adaptation (RA) techniques to connect non-64 kb/s streams to the B channel(s). Three CCITT RA techniques are recommended in I.460 [4.10]:

- X.30/V.110 bit repetition and positioning (similar to ECMA 102 RA). The RA is performed in three steps for asynchronous terminals at or below 19.2 kb/s, two steps for synchronous terminals at or below 19.2 kb/s, and one step for 48 and 56 kb/s synchronous terminals. For terminals with 64 kb/s synchronous transmission capabilities, an RA is not necessary. V.110 was specified in the Red Book (1984) and is stable, although there have been changes in the standard between the 1984 and 1988 versions.

- X.31 with X.25 Packet Layer Protocol (PLP) over LAP-B: for packet-mode use of a B channel.
- V.120 statistical multiplexing of several data connections in the circuit switched mode. V.120 is a packet-switching oriented multiplexing scheme that has been sanctioned in the CCITT Blue Book, and it applies to the R reference point. At the time of this writing, ANSI was in the process of accepting V.120 as the official U.S. standard for RA. V.120 offers the advantage of automatic speed matching and an error-checking option.

V.110 uses bit-oriented time-division multiplexing. V.120 is more like statistical multiplexing, but V.110 has been around much longer. V.120 will undoubtedly see far greater acceptance than V.110. As of end of 1989, V.120 was a stable specification.

4.4.2 Some Possible Limitations of ISDN for Data

Six years of research into cost-effective applications of ISDN to data communication to serve the embedded base of terminals and computers have been conducted; in spite of this, no clear-cut applications have surfaced as of 1990. Network delivery of the calling number of an incoming call, with possible interplay with a host database and display of a customer record simultaneously with the voice call, appeared to be the only "new" data service available at the time of this writing (at the technical level, automatic number identification could already be achieved in September 1958—when this author was entering the first grade in elementary school—and customers willing to pay for such service could probably have obtained it at that time, if they so desired [4.11]).

Currently, ISDN PC cards are expensive compared to (9.6 kb/s) modem cards: PC ISDN adapters implementing the basic rate interface cost around $1500 (*circa* 1990); by comparison, PC modem cards could be purchased for as little as $600 (note that a higher bandwidth is achieved with an ISDN card compared to that obtainable with a voiceband analog modem card; however, as discussed in Chapter 1, modems now can provide a high throughput of 38.4 kb/s). As of 1990, the effective speed with such ISDN PC cards was 19.2 kb/s asynchronous, given existing PC software [4.10]. V.120, however, is not limited to that speed: Vendors have reported sustained throughput of more than 60 kb/s using V.120. In fairness to TAs, 19.2 kb/s multicarrier modems may reduce data rates for line conditions or only give half-duplex [4.12]. The typical asynchronous line speed is what is already available and at a much lower cost with modems compared to an ISDN TA, which provides only asynchronous 19.2 kb/s or less.

Other applications (and appropriate interface hardware) may develop in the early 1990s to connect LAN gateways, T1 multiplexers, mainframes, and Group IV fax machines.

4.4.3 NIU Forum

The North American ISDN User Group Forum (NIU Forum) is one organization exploring applications of ISDN. The mission of the NIU Forum is "to create a strong user voice in the implementation of ISDN and ISDN applications; and, ensure the emerging ISDN is interoperable and meets user requirements" [4.13].

The NIU charter describes the mission as follows: "To broaden the utilization of ISDN-based technologies and to speed their development, NIST has established a Consortium to pool the resources of interested companies. The intent of this Consortium is to support the advancement of a technically efficient and compatible technology base for emerging ISDNs, on a national basis. The purposes of this Consortium are:

- To promote an ISDN forum committed to providing users the opportunity to influence the developing ISDN technologies to reflect their needs;
- To identify novel ISDN applications;
- To develop agreements and tests to accomplish effective, efficiency-enhancing, technical compatible ISDN technologies, on a nondiscriminatory voluntary, multi-vendor basis;
- To conduct appropriate research, including experimental prototype activities, and informational activities." [4.14].

NIU is chaired by NIST, and started operation in 1987. The group meets three or four times a year. NIU has three major working groups: a users group, an implementer's group, and an application analysis group. Each group has subgroups; for example, the users group is composed of small business, manufacturing, process, service, financial, government, and computer communication segments.

Users submit "business problems." All applications are accepted and prioritized by users. Technical discussions in various NIU committees follow, in an attempt to identify means by which ISDN could solve these needs. As of January 1990, 91 applications had been submitted; 41 were analyzed by the Applications Analysis Team; 30 progressed through the User-Implementer Team; and 5 reached the Application Profile Team. Special interest items include electronic data interchange and network management.

4.5 ISDN PHYSICAL TRANSMISSION METHOD

This section discusses the physical transmission method used at the *U reference point* for basic rate access in the United States. Information about this method is provided here to contrast it with the well-established AMI/SF method for DS1 links described in Chapter 3, as well as the AMI/ESF to be employed in the United States for the primary rate access.

In the United States, a publicly defined interface at the U reference point is specified. The FCC has ruled that the NT1 is CPE; hence, a single specification is highly desirable to facilitate commercial availability of the product. The U.S. standard that has evolved, ANSI T1.601-1988 [4.3], applies to a single *digital subscriber line* (DSL) consisting of a *line termination* (LT), a two-wire metallic cable pair, and an NT. The physical termination of the DSL at the network end is the LT; the physical termination at the user end is the network termination.

ANSI T1.601 applies to the basic rate access. This standard is based on the use of cables without loading coils. The ANSI interface standard for implementation of the physical layer U interface was written to provide a minimum set of requirements to provide for satisfactory transmission between the network and the NT, while conforming, wherever possible, with the I Series of CCITT Recommendations. It presents the electrical characteristics of the ISDN basic access signal appearing at the network side of the NT. It also describes the physical interface between the network and the NT. T1.601 was complete at the time of this writing, although there was ongoing work on addenda (mostly dealing with management).

The scope of the ANSI T1.601 standard for the United States is as follows:

1. It describes the transmission technique used to support full-duplex service on a single twisted-wire pair;
2. It specifies both the input signal with which the NT must operate and the output signals that the NT must produce;
3. It defines the line code to be used, and the spectral composition of the transmitted signal;
4. It describes the electrical and mechanical specifications of the network interface;
5. It describes the organization of transmitted data into frames and superframes; and
6. It defines the functions of the operations channel.

The transmission system uses the *echo canceler with hybrid* (ECH) principle to provide full-duplex operation over a two-wire subscriber loop. With the ECH method, as illustrated in Figure 4.4, the *echo canceler* (EC) produces a replica of the echo of the near-end transmission, which is then subtracted from the total received signal. The system is intended for service on twisted-pair cables, for operation to 18-kilofoot (5.5 km). The ECH method is contrasted below with *time compression multiplexing* (TCM), another method that has been around for a few years, particularly in a PBX environment. In TCM (also called *the ping-pong method*), data are transmitted in one direction at a time, with transmission alternating between the two directions. To achieve the stated data rate, the subscriber's bit stream is divided into equal segments, compressed in time to a higher transmission rate, and is transmitted in bursts that are expanded at the other end of the link to the original data rate (the actual data rate on the line must be more than twice the data rate required by the user). A short guard band period is used between bursts going in opposite directions to allow the line to settle down. The use of this technique in ISDN, which has a total user data rate of 144 kb/s, would require more than 288 kb/s to be transmitted over the copper loop; this is not always possible, particularly for loops longer than three miles.

With echo cancellation, used in the United States for ISDN basic rate access, digital transmission occurs in both directions within the same bandwidth simultaneously. Both transmitter and receiver are connected through a device (known as a hybrid) that allows signals to pass in both directions at the same time. The problem is that echo is generated by the reflection of the transmitted signal back to the user. Both a near-end echo and a far-end echo occur. The near-end echo arises between the sender's hybrid and the cable; the far-end echo is from the receiver's hybrid device. The magnitude of the echo signal is such that it cannot be ignored; EC is used to overcome this problem. With this technique, an estimate of the composite echo signals is calculated at the transmitter end and a signal of that value is subtracted from the incoming signal, effectively canceling the echo.

4.5.1 The 2B1Q Line Coding

To provide basic rate service economically, the digital loop must be implemented without conditioning the plant, special engineering, or special operations. A line code is the electrical representation of digital signals. The previous chapter discussed the classical AMI/SF line coding for T1 links. A newer method is used in basic rate access ISDN lines. The selection of a line code is critical to the performance of the loop; digital loop carrier systems for ISDN must have a BER of better than 10^{-7}. The line code determines both the transmission characteristics of the received signal and the added near-end crosstalk noise levels on other pairs in a multipair cable.

The ECSA assessed a number of codes as candidates for the ISDN U interface; see Table 4.4 [4.15]. The ISDN basic rate line code now specified by the ECSA for North America is called *2B1Q* (2 binary, 1 quaternary). This is a four-level pulse amplitude modulation (PAM) code without redundancy. The ISDN user-data bit stream, comprised of two 64 kb/s B channels and a 16 kb/s D channel, entering the NT from the S-T interface (i.e., entering the S-T interface toward the NT) and the bit stream on the network side are grouped into pairs of bits for conversion to quaternary symbols. In the sequel, quaternary symbols are also called *quats*. In each pair of bits thus formed, the first bit is called the sign bit and the second is called the magnitude bit. Figure 4.5 shows the relationship of the bits in the B and D channels to quats. The B and D channel bits are scrambled before

Table 4.4
Partial List of Possible ISDN Line Codes

Code	*Description*
2B1Q	The code converts blocks of two consecutive signal bits into a single four-level pulse for transmission. The information rate is double the baud rate. All four possible combinations of two information bits map into a quaternary symbol; this block code is said to be saturated. The 2B1Q code is specified for North America; other countries have adopted different line coding techniques.
3B2T	The three binary, two ternary code uses a three-level transmission instead of a four-level; this results in a lower baud compression than 2B1Q. The code maps the eight possible combinations of three bits into the nine possible combinations of two ternary symbols. This block code is not saturated (there is a small amount of redundancy).
4B3T	The four binary, three ternary code maps the sixteen possibilities of four information bits into twenty-seven combinations of three ternary symbols. This block code is not saturated (there is a fair amount of redundancy, which can be favorably exploited).
AMI	Code reverses the polarity of the mark (1) from that of the previously transmitted mark. It is, in effect, a ternary code, but performs the coding on a bit-by-bit basis, providing a bit rate that equals the baud rate; this is inefficient.
Dicode	The incoming bit stream is passed through a digital filter prior to modulation for transmission.
MBD	In the modified duobinary code, each information bit leads to a transmitted pulse in a time slot, followed by an identical pulse of opposite polarity in a time slot one time interval removed. Bits arriving for transmission, one time slot removed from each other, result in overlapping transmitted symbols. It is, in effect, a ternary code that uses bit-by-bit encoding; it is redundant without baud reduction.
Biphase	The transmitted signal has equal positive and negative pulses. It has half the pulsewidth of dicode. This is a binary code that is saturated and offers no baud reduction.

coding. See Figure 4.6 for a functional description of the coding, framing, and scrambling operations of the transceiver.

Politics has played a bigger role than technical issues in delaying the 2B1Q line code standard. Two "camps" debated on the interface: one camp with European roots was pushing the 4B3T, a U interface transceiver approved for use in Europe. Another camp led by U.S. manufacturers was pushing the AMI line code. Neither side was able to convince the other to agree. In late 1986, British Telecom made a compromise proposal. The proposal was accepted because it met the

Time →

Data	B_1				B_2				D
Bit Pairs	b_{11} b_{12}	b_{13} b_{14}	b_{15} b_{16}	b_{17} b_{18}	b_{21} b_{22}	b_{23} b_{24}	b_{25} b_{26}	b_{27} b_{28}	d_1 d_2
Quat # (relative)	q_1	q_2	q_3	q_4	q_5	q_6	q_7	q_8	q_9
# Bits	8				8				2
# Quats	4				4				1

Where:
b_{11} = first bit of B_1 octet as received at the S/T interface
b_{18} = last bit of B_1 octet as received at the S/T interface
b_{21} = first bit of B_2 octet as received at the S/T interface
b_{28} = last bit of B_2 octet as received at the S/T interface
$d_1 d_2$ = consecutive D-channel bits
(d_1 is first bit of pair as received at the S/T interface)
q_i = ith quat relative to start of given 18-bit 2B+D data field

NOTE: There are twelve 2B+D 18-bit fields per 1.5 ms basic frame.

Figure 4.5 2B1Q encoding of $2B + D$ bit fields.

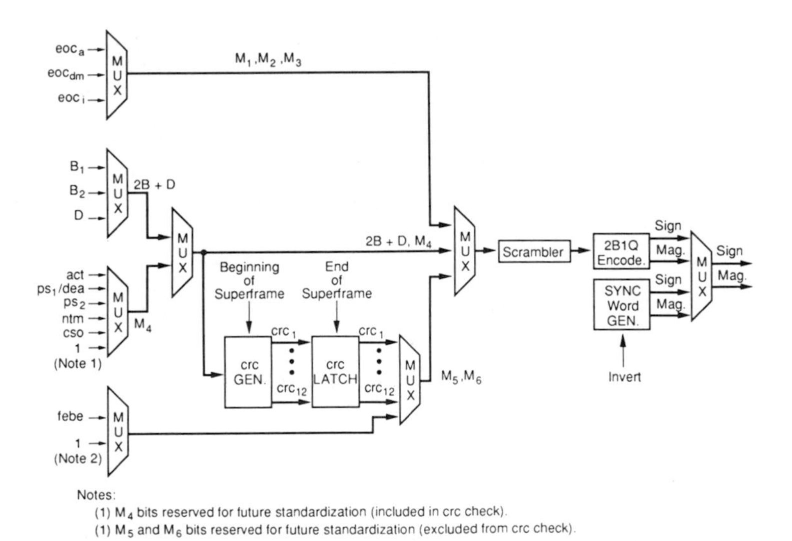

Figure 4.6 DS1 framer functional description.

technical requirements, and from a political and commercial perspective it was neutral [4.16].

The line coding scheme adds adaptive digital signal processing to smooth out interference on a line, such as those described in Chapter 1. This allows the use of existing twisted pairs for a distance of about three miles. The 2B1Q technique was selected at a T1D1.3 meeting in August 1986; however, chips to implement the line coding scheme only started to appear in 1989. Until these chips are incorporated into actual products, the existing AMI method will be used by the switch vendors (such as AT&T). To some potential users, such differences will be insignificant: the TAs or PC cards can plug into the S-T side of the NT1-NT2 for AMI or 2B1Q circuits with no discernible differences. However, users getting both AMI and 2B1Q circuits (provided over, say, two different switches or two transitional implementations) would have to keep track of the differences in order to ascertain that connectivity is achievable [4.17].

The 18-kilofeet issue may be more pressing for rural environments: As indicated in Chapter 3, a fair number of loops, particularly those close to new developments, suburbia, and commercial environments, are already on digital loop carrier feeder systems. The 18 kilofeet would then be measured from the digital loop carrier and not the CO. The principal problems impacting a digital loop are [4.15]:

- Impulse noise;
- Intersymbol interference;
- Echo noise;
- Quantizing noise; and
- Near-end crosstalk.

The intersymbol interference and the near-end crosstalk problems were the most important factors used by ECSA in selecting the 2B1Q code.

In 2B1Q, each successive pair of scrambled bits in the binary data stream is converted to a quaternary symbol to be output from the transmitter at the interface, as specified in the Table 4.5. The four values listed under "Quaternary Symbol" in the table are to be understood as symbol names, not numerical values. At the receiver, each quaternary symbol is converted to a pair of bits by reversing the table, descrambled, and finally formed into a bit stream or bit streams representing B and D channels, and the M channel bits for maintenance and other purposes.

Figure 4.7 is an example of 2B1Q pulses over time. Square pulses are used only for convenience of display and do not in any way represent the specified shape

Table 4.5
2B1Q code

First Bit (Sign)	*Second Bit (Magnitude)*	*Quaternary Symbol (Quat)*
1	0	+3
1	1	+1
0	1	−1
0	0	−3

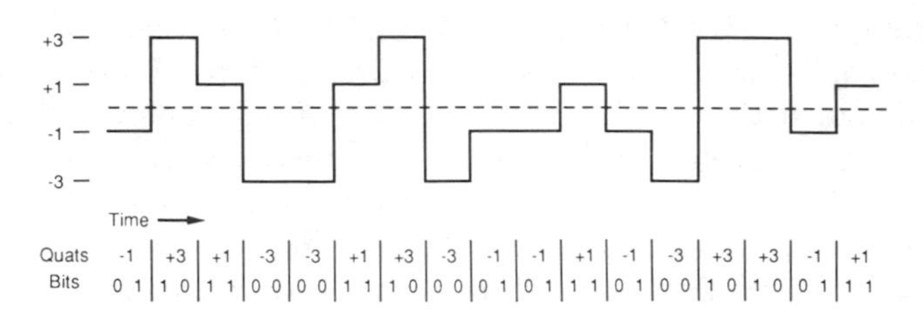

Figure 4.7 Example of 2B1Q Quaternary Symbols.

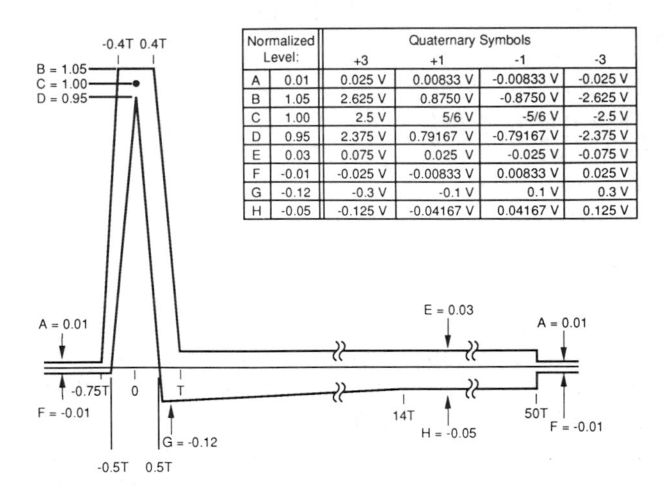

Normalized Level:		Quaternary Symbols +3	+1	-1	-3
A	0.01	0.025 V	0.00833 V	-0.00833 V	-0.025 V
B	1.05	2.625 V	0.8750 V	-0.8750 V	-2.625 V
C	1.00	2.5 V	5/6 V	-5/6 V	-2.5 V
D	0.95	2.375 V	0.79167 V	-0.79167 V	-2.375 V
E	0.03	0.075 V	0.025 V	-0.025 V	-0.075 V
F	-0.01	-0.025 V	-0.00833 V	0.00833 V	0.025 V
G	-0.12	-0.3 V	-0.1 V	0.1 V	0.3 V
H	-0.05	-0.125 V	-0.04167 V	0.04167 V	0.125 V

Figure 4.8 Normalized pulse from NT appearing at Interface.

of real 2B1Q pulses. The transmitted pulse must have the shape specified in Figure 4.8. The pulse mask for the four quaternary symbols is obtained by multiplying the normalized pulse mask shown in Figure 4.8 by 2.5, 5/6, −5/6, or −2.5 V.

4.5.2 Functional Characteristics

The information flow across the interface point utilizes frames and superframes as shown in Figure 4.9. As shown in the figure, a frame is composed of 120 quaternary symbols. The nominal time for the frame is 1.5 ms. The first nine symbols of the

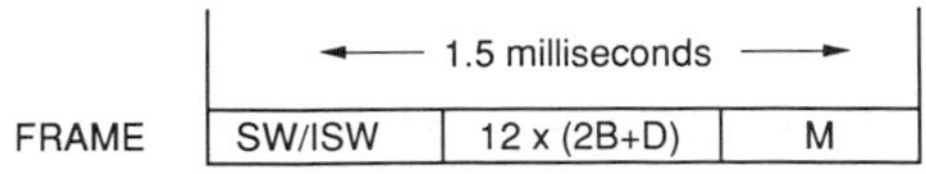

	← 1.5 milliseconds →		
FRAME	SW/ISW	12 x (2B+D)	M

Function	Sync Word	2B+D	Overhead
# Quats	9	106	3
Quat Positions	1-9	10-117	118-120
#Bits	18	216	6
Bit Positions	1-18	19-234	235-240

Frames in the NT-to-Network direction are offset from frames in the Network-to-NT direction by 60±2 quats.

Symbols & Abbreviations:

quat	= quaternary symbol = 1 baud
-3, -1, +1, +3	= symbol names
2B+D	= Customer data channels B_1, B_2 and D
SW	= Synchronization Word (9-Symbol Code)
	= +3 +3 -3 -3 -3 +3 -3 +3 +3
ISW	= Inverted (or complementary) Sync Word
	= -3 -3 +3 +3 +3 -3 +3 -3 -3
M	= M-Channel Bits, M_1 -M_6

Figure 4.9 ISDN basic acess 2B1Q DS1 1.5-ms basic frame.

frame are a synchronization word (SW), with the quaternary symbols in the following sequence:

$$SW = +3 +3 -3 -3 -3 +3 -3 +3 +3$$

Frames are organized into superframes: eight frames (12 ms) constitute a superframe. The first frame in the superframe is to be identified by inverting the polarity of the SW in this frame. The inverted synchronization word is:

$$ISW = -3 -3 +3 +3 +3 -3 +3 -3 -3$$

The first frame in the superframe of the signal transmitted from the NT is the next frame following the first frame in the superframe of the signal received from the network. The 2B1Q code offers the greatest baud reduction of the codes considered, and has an intersymbol interference-near-end crosstalk performance 2 to 3 dB better than the other codes.

At least five manufacturers expected to bring U interface chips, with the 2B1Q line coding scheme, into the U.S. market by the end of 1990. AT&T, which was the first with a two-chip 2B1Q line code device in early 1989, was planning to offer it in production quantities in 1990. Some areas of 2B1Q line code devices may at first be done in firmware; these include how to handle information from the maintenance channels and the embedded operations channel. Other manufacturers include AMD-Siemens, Motorola-Northern Telecom, NEC Electronics, and National Semiconductor.

4.5.3 Primary Rate Line Coding

The North American primary rate was still evolving at the time of this writing. This standard is being worked on by ECSA's T1E1.2, with a vote expected in 1992 [4.17]. This specification is critical for ISDN-compatible PBXs: if a manufacturer follows the CCITT ISDN specifications, but does not include the North American agreements, lack of interoperability can easily result.

The U.S. primary rate line coding will be consistent with the U.S. standard for DS1 interfaces (AMI/ESF, B8ZS, and DSX-1), i.e., with ANSI T1.403-1989.

4.6 C-PLANE (SIGNALING) PROTOCOLS

Section 4.3 briefly described the C-plane protocol needed by a user. This section provides additional details.

As described in Chapter 1, and as applied to ISDN, the OSIRM physical layer (layer 1) consists of requirements on the physical connection, power transfer, line transmission sending and receiving signals, timing, framing, multiplexing, maintenance, and performance monitoring. The data link layer (layer 2) exists between a point on the user's premise and a point at the local switching system. It provides the function of message delimitation into frames, error detection, and correction through retransmission of errored frames, and subdivision of the D channel into a multiplicity of logical channels. A set of coded messages is available at layer 3 to be exchanged by user equipment and the local switching system for call establishment and disconnection control (these messages take precedence over user-data packets on the D channel). Pertinent ISDN issues related to this C-plane protocol suite are discussed in this section.

4.6.1 D Channel Data Link Layer Operation

Layer 2 offers two capabilities in support of ISDN. The first capability provides user device addressing; the other capability provides for the transport of (signaling) messages between a particular device and the switch. These two capabilities are discussed in turn below.

4.6.1.1 Data Link Connection Identification

Every user device must have an address not shared by other devices. The layer 2 address consists of two values: (1) a *SAP identifier* (SAPI), which points to the particular type of processing required, and (2) a *terminal endpoint identifier* (TEI), which is a user device identifier [4.6]. The entire address is referred to as the *data link connection identifier* (DLCI), with DLCI = (SAPI, TEI). Each different DLCI defines a different link. The DLCI is a pure data link concept; it is used internally by the data link layer entity and is not known by a layer 3 entity.

A data link layer SAP is the point at which the data link layer provides services to layer 3 (as discussed in Chapter 1, each layer in OSIRM communicates with the layers above and below it across an interface; the interface is realized at one or more service access points [4.18]). Associated with each data link layer SAP is one or more data link connection endpoint(s), as depicted in Figure 4.10.

Cooperation between data link layer entities is governed by a peer-to-peer protocol specific to the layer. For information to be exchanged between two or more layer 3 entities, an association must be established between these layers via the data link layer using a data link layer protocol. This association is called a data link connection (DLC). DLCs are provided by the data link layer between two or more SAPs. Layer 3 requests services from the data link layer via service primitives. The same is true, in turn, for the interaction between the data link layer and the physical layer.

Typically, each user terminal is given a distinguishing TEI; the TEI is assigned by the network or is entered into the user equipment (for example, by the manufacturer) [4.18]. The SAPI is used to identify the service access point on the network side and the user side of the user-network interface. The SAPI identifies the traffic type. For example, the SAPI value of 0 directs the frames to layer 3 entities for call control procedures; a SAPI value of 16 indicates a packet communication procedure.

At layer 3, a data link connection endpoint (depicted in Figure 4.10) is identified by a data link connection endpoint identifier as seen from layer 3, and by a connection identifier seen from the data link layer itself. This gives rise to the concept of a *connection endpoint identifier* (CEI). The CEI is used to identify message units passed between the data link layer and layer 3. The CEI consists

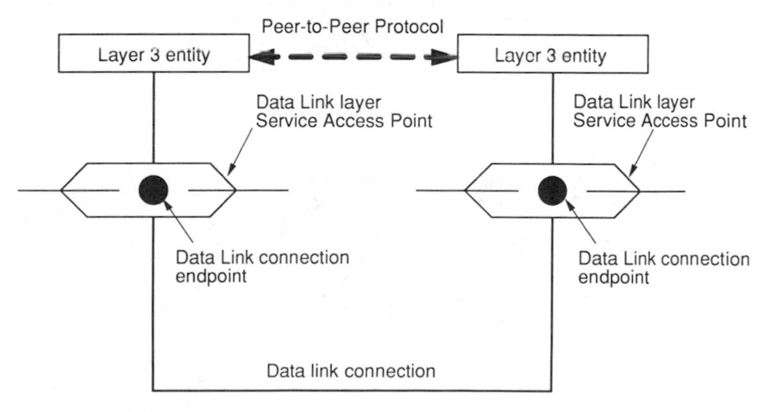

Figure 4.10 Peer-to-peer relationships.

of the SAPI and the *connection endpoint suffix* (CES), as shown in Figure 4.11 [4.7]. More general addressing issues are described in Chapter 13.

4.6.1.2 Transmission of Signaling Messages

The second layer 2 capability provides bidirectional transfer of network layer call control and X.25 PLP messages between the switch and a user device. None of these messages is interpreted by layer 2 during the provision of the data link layer service.

Data link layer functions provide the means for information transfer between multiple combinations of data link connection endpoints. The information transfer

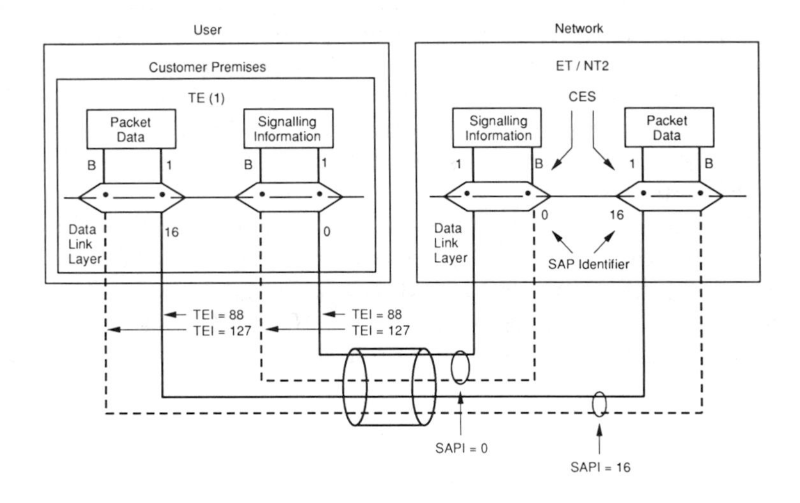

Key:
----- Broadcast data link connection (B)
——— Point-to-point data link connection
DLCI = SAPI + TEI
CEI = SAPI + CES

Note 1: The management entity is not shown in this figure.
Note 2: The selection of the SAPI and TEI values is based on Recommendation Q.921.

Figure 4.11 Relationship between SAPI, TEI, and DLCI.

may be via point-to-point data link connections or via broadcast data link connections. In the case of point-to-point information transfer, a frame is directed to a single endpoint; in the case of broadcast information transfer, a frame is directed to one or more endpoints. Two types of operation of the data link layer are defined for layer 3 information transfer, which can coexist on a single D channel: unacknowledged and acknowledged.

Link Access Procedure D

LAP-D is the data link standard developed for ISDN; it specified the link access protocol to be used over the D channel. LAP-D (CCITT Q.921, ECSA T1.602.1989) is based on LAP-B, which, in turn, is based on HDLC. LAP-D has to deal with two levels of multiplexing. First, at a subscriber location, multiple user devices may share the same physical interface. Second, within each user device, multiple types of traffic may exist, including packet switched data and signaling. To accomplish this type of multiplexing, as discussed above, LAP-D employs a two-part address, consisting of the TEI and the SAPI.

LAP-D includes functions for [4.7]:

- The provision of one or more data link connections on a D channel. Discrimination between the data link connections is by means of the DLCI contained in each frame;
- Frame delimiting, alignment, and transparency, allowing recognition of a sequence of bits transmitted over a D channel as a frame;
- Sequence control, to maintain the sequential order of frames across a data link connection;
- Detection of transmission, format, and operational errors on a data link connection;
- Recovery from detected transmission, format, and operational errors;
- Notification of the management entity of unrecoverable errors; and
- Flow control.

For the case of unacknowledged operation, layer 3 information is transmitted in unnumbered information frames. At the data link layer, these frames are not recovered if transmission or format errors are detected; also, flow control mechanisms are not defined. The frames can be sent to a specific endpoint or broadcast to multiple endpoints associated with a specific SAPI.

For acknowledged operation, layer 3 information is transmitted in frames that are acknowledged at the data link layer. Error procedures based on retransmission of unacknowledged frames are provided. In the case of errors that cannot be corrected by the data link layer, a report to the management entity is made. Flow control procedures are also defined.

LAP-D Frame Structure

The basis for all layer 2 D channel (C-plane) communication with a user ISDN device is the exchange of information contained in frames. The frame structure includes the partitioning of different frame types into fields and, for complex fields, the partitioning of the fields into subfields. It also includes the range of values permitted in each field or subfield [4.6].

Layer 2 supports three frame types to undertake the two types of operations discussed above: (1) I (information) frames, (2) S (supervisory) frames, and (3) U (unnumbered) frames. The frame format type identification field consists of bits 1 and 2 in octet 4 for the S and U frames, and of bit 1 in octet 4 for the I frames. The values of these bits uniquely identify the frame type. When a frame is received at either end of the ISDN user-to-CO link, the value of these bits is used to define one of the three different sets of default field boundaries required to interpret the whole frame.

Figure 4.12 depicts the three types of frames [4.6]. Type I is distinguished by the presence of a 0 at bit position 1 in octet 4 (types S and U have the value 1 at this bit position). The switch uses this format for the transfer of network layer information, which is contained in the portion of the frame labeled "Information" in Figure 4.12. The other fields and subfields of this type are used for addressing to indicate polling, to indicate a command or response frame, and to convey the number of the frame assigned by the transmitter and the number of the frame next expected to be received.

The type U frame is distinguished by the presence of the value 1 at bit positions 1 and 2 in octet 4. A subtype of U frames (known as UI frames) is used by the switch for the transmission of network layer information using DLCI = 127; one example is the SETUP message described later.

The type S frame is distinguished by the presence of the value 0 at bit position 2 and the value 1 at bit position 1 in octet 4. These frames are used to signal receipt of an out-of-sequence frame, to provide acknowledgment of received I frames, to poll the other side, and to respond to polls.

Reference [4.6] provides a detailed description of the operation of layer 2 for ISDN deployment in North America. The specification describes the frame structure, including all the fields, commands and responses, exchange of information, DLCI management, error detection and recovery, the layer 3 interface, and services provided by the switch.

4.6.2 D Channel Network Layer Operation

4.6.2.1 Network Layer Functions

CCITT's Q.930 (I.450), Q.931 (I.451), and Q.932 (I.452) provide details on the ISDN user-network functional interface at layer 3. The layer 3 protocol provides

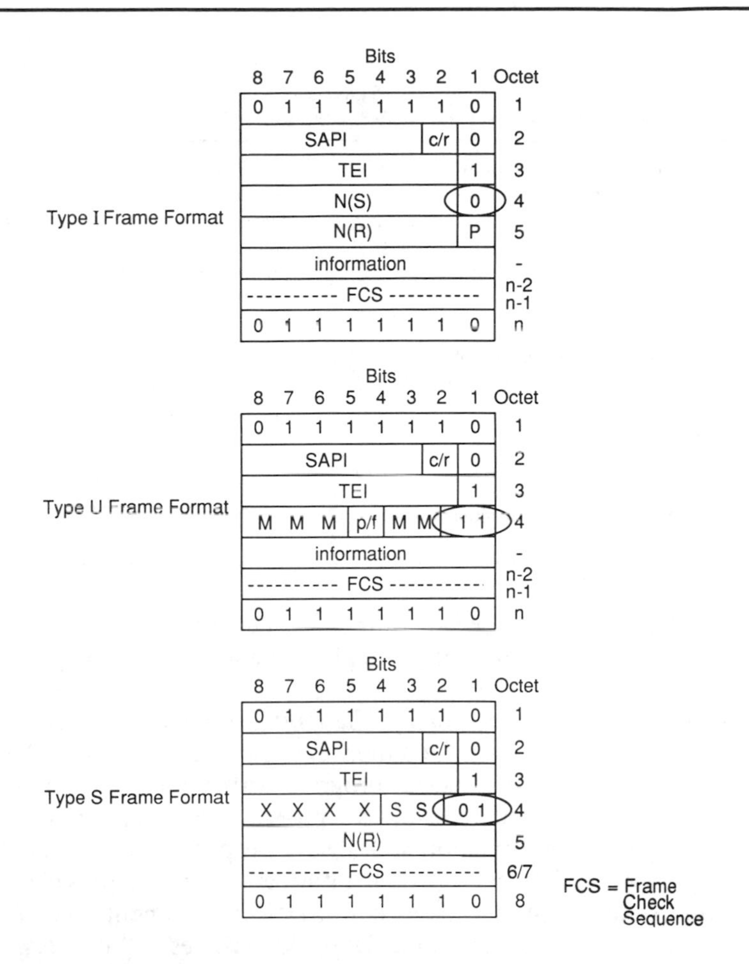

Figure 4.12 Three types of Layer 2 frames.

the means to establish, maintain, and terminate network connections across an ISDN between communicating applications entities. In addition, it provides generic procedures that may be used for the invocation and operation of supplementary services. Q.931 specifies the procedures for establishing, maintaining, and clearing network connections at the ISDN user-network interface. These procedures are defined in terms of messages exchanged over the D channel of both basic and primary rate access.

Connection control by the user of ISDN requires the application of layer 3 protocols for control of circuit switched or packet switched connections [4.19]. Layer 3 provides the user with the functions associated with the establishment and operation of a network connection. Table 4.6 describes some of the key layer 3 functions in general, and ISDN (Q.931) in particular. The list of Table 4.6 is not exhaustive, and it is not intended to imply that all functions are provided on both

Table 4.6
Key Network Layer Functions for the ISDN User-Network Interface

Processing of primitives for communicating with the data link layer
Generation and interpretation of layer 3 messages for peer-level communication
Administration of timers and logical entities (e.g., call-references) used in the call control procedures
Administration of access resources including B channels and packet layer logical channels
Checking to ensure that services provided are consistent with user requirements
Routing and relaying
Network connection control
Conveying user-to-network and network-to-user information
Network connection multiplexing
Segmenting and reassembly
Error detection
Error recovery
Set message sequence
Congestion control and user data flow control
Restart

the terminal and the network side of the user-network interface. The Q.931 specification is close to 400 pages of detailed information; only the general outline of the standard is covered here.

CCITT ISDN recommendations support a functional approach toward signaling, with some stimulus information elements. Originating ISDN equipment will receive information about call progress within the ISDN message-oriented protocol, as well as some in-band call progress information. (For voice calls, certain call progress information will always be provided in-band, for example, "audible ringing"; other call progress information can be provided in-band as a subscription option.) [4.2]. In the non-ISDN environment, the originator of a speech call and certain circuit-mode data calls receives call progress signals (e.g., dial tone, busy tone, announcements) in-band to advise the user of events regarding the call.

4.6.2.2 Q.931's Network Layer Messages

The ISDN call control messages defined by Q.931 are identified in alphabetical order in Table 4.7 (for packet-mode connections refer directly to Reference [4.19]; channel-switched ("semipermanent") connections are for "further study"). A brief description of these call-control messages follows. One key message, SETUP, is discussed in more detail. Call control is described in Q.931 for circuit-mode, packet-mode, and user-to-user signaling.

For the United States, Bellcore's TR-TSY-000268 [4.2] identifies 15 layer 3 messages, as marked in Table 4.7.

Table 4.7
ISDN Call Control Messages

ALERTING: This message is sent by the called user to the network and by the network to the calling user to indicate that called user alerting has been initiated.

CALL PROCEEDING: This message is sent by the called user to the network or by the network to the calling user. In the network-to-user direction, it indicates that the requested call establishment has been initiated and no more establishment information will be accepted.

CONGESTION CONTROL: This message is sent by the user or network to indicate the establishment or termination of flow control on the transmission of USER INFORMATION messages.

CONNECT: This message is sent by the called user to the network and by the network to the calling user to indicate call acceptance by the called user.

CONNECT ACKNOWLEDGE: This message is sent by the network to the called user to indicate the user has been awarded the call. It may also be sent by the calling user to the network to allow symmetrical call control procedures.

DISCONNECT: This message is sent by the user to request the network to clear an end-to-end connection, or is sent by the network to indicate that the end-to-end connection is cleared.

FACILITY: This message is now described in Q.932.

INFORMATION: This message is sent by the user or the network to provide additional information. It may be used to provide information for call establishment or miscellaneous call-related information.

NOTIFY: This message is sent by the user or network to indicate information pertaining to a call, such as user suspended.

PROGRESS: This message is sent by the user or the network to indicate the progress of a call in the event of interworking or in relation to the provision of in-band information and patterns.

RELEASE: This message is sent by the user or the network to indicate that the equipment sending the message has disconnected the channel (if any) and intends to release the channel and the call reference, and that the receiving equipment should release the channel and prepare to release the call reference after sending a RELEASE COMPLETE message. The meaning may be modified in the context of supplementary services (e.g., conferencing).

RELEASE COMPLETE: This message is sent by the user or network to indicate that the equipment sending the message has released the channel (if any) and call reference, the channel is available for reuse, and the receiving equipment will release the call reference.

RESUME: This message is sent by the user to request the network to resume a suspended call.

RESUME ACKNOWLEDGE: This message is sent by the network to the user to indicate completion of a request to resume a suspended call.

RESUME REJECT: This message is sent by the network to the user to indicate failure of a request to resume a suspended call.

SETUP: This key message is sent by the calling user to the network and by the network to the called user to initiate call establishment. This message is discussed more below. Figure 4.13 depicts the general format of the SETUP message.

SETUP ACKNOWLEDGE: This message is sent by the network to the calling user (or by the called user to the network, but not in the United States) to indicate that call establishment has been initiated, but additional information may be required.

STATUS: This message is sent by the user or the network in response to a STATUS ENQUIRY message or at any time during a call to report certain error conditions.

STATUS ENQUIRY: This message is sent by the user or the network at any time to solicit a STATUS message from the peer layer 3 entity (sending a STATUS message in response to a STATUS ENQUIRY message is mandatory). (Bellcore modifies this for the United States.)

SUSPEND: This message is sent by the user to request the network to suspend a call.

SUSPEND ACKNOWLEDGE : This message is sent by the network to the user to indicate completion of a request to suspend a call.

SUSPEND REJECT: This message is sent by the network to the user to indicate failure of a request to suspend a call.

USER INFORMATION: This message is sent by the user to the network to transfer information to the remote user. This message is also sent by the network to the user to deliver information from another user. This message is used if the user-to-user transfer is part of an allowed information transfer.

Message type: SETUP
Significance: global
Direction: both

Information Element	Direction	Type	Length (Octets)
Protocol discriminator	both	M	1
Call reference	both	M	2 - *
Message type	both	M	1
Sending complete	both	O (Note 1)	1
Repeat indicator	both	O (Note 2)	1
Bearer capability	both	M (Note 3)	4 - 13
Channel identification	both	O (Note 4)	2 - *
Facility	both	O (Note 5)	2 - *
Progress indicator	both	O (Note 6)	2 - 4
Network specific facilities	both	O (Note 7)	2 - *
Display	n→u	O (Note 8)	Note 9
Keypad facility	u→n	O (Note 10, 12)	2 - 34
Signal	n→u	O (Note 11)	2 - 3
Switchhook	u→n	O (Note 12)	2 - 3
Feature activation	u→n	O (Note 12)	2 - 4
Feature indication	n→u	O (Note 12)	2 - 5
Calling party number	both	O (Note 13)	2 - *
Calling party subaddress	both	O (Note 14)	2 - 23
Called party number	both	O (Note 15)	2 - *
Called party subaddress	both	O (Note 16)	2 - 23
Transit network selection	u→n	O (Note 17)	2 - *
Low layer compatibility	both	O (Note 18)	2 - 16
High layer compatibility	both	O (Note 19)	2 - 4
User-user	both	O (Note 20)	Note 21

M = Mandatory
O = Optional

Figure 4.13 SETUP message content.

4.6.2.3 Q.931's Network Layer Message Format

The layer 3 format for the messages just described is shown in Figure 4.14 [4.19]. Every message contains:

- The protocol discriminator. This is used to distinguish messages for user-to-network call control from other messages. See Figure 4.15. (No other message is carried in the case of SAPI 0.)
- The call reference. This is used to identify the call or facility registration-cancellation request at the local user-network interface to which the particular

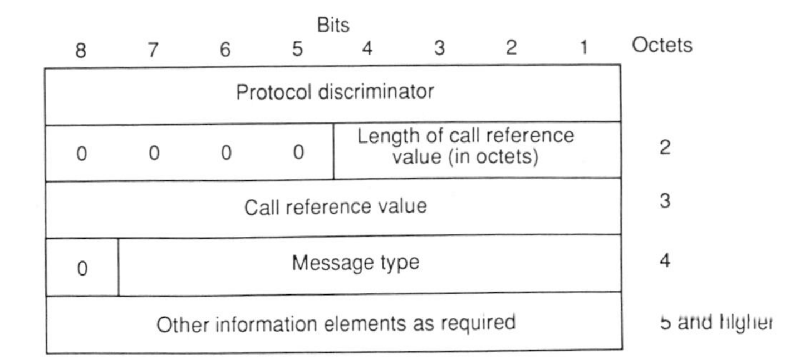

Figure 4.14 General message orginination example.

	Bits 8765 4321	
through	0000 0000 / 0000 0111	not available for use in the message protocol discriminator
	0000 1000	Q.931/(I.1451) user-network call control messages
through	0001 0000 / 0011 1111	reserved for other Network layer or layer 3 protocols, including Recommendation X.25 (Note)
through	0100 0000 / 0100 1111	national use
through	0101 0000 / 1111 1110	reserved for other Network layer or layer 3 protocols, including Recommendation X.25 (Note)
		All other values are reserved.

Note - These values are reserved to discriminate these protocol discriminators from the first octet of a Rec. X.25 packet including general format identifier.

Figure 4.15 Protocol discriminator.

message applies (the call reference does not have end-to-end significance across ISDN).

- The message type. The purpose of this field is to identify the function of the message being sent (see Figure 4.16).
- Other information elements, as required. A great multitude of detailed call information is contained in this field. See Figure 4.17 for an example.

The first three information elements are common to all the messages and are always present, while the last item is specific to each message type.

4.6.2.4 Call Handling Message Sequence

In addition to specifying the messages and the message format, we must specify the message sequence between the network and the user. Figure 4.18 depicts, for

Bits 87654321	
00000000	escape to nationally specific message type; see Note
000-----	*Call establishment message:*
00001	- ALERTING
00010	- CALL PROCEEDING
00111	- CONNECT
01111	- CONNECT ACKNOWLEDGE
00011	- PROGRESS
00101	- SETUP
01101	- SETUP ACKNOWLEDGE
001-----	*Call information phase message:*
00110	- RESUME
01110	- RESUME ACKNOWLEDGE
00010	- RESUME REJECT
00101	- SUSPEND
01101	- SUSPEND ACKNOWLEDGE
00001	- SUSPEND REJECT
00000	- USER INFORMATION
010-----	*Call clearing messages:*
00101	- DISCONNECT
01101	- RELEASE
11010	- RELEASE COMPLETE
00110	- RESTART
01110	- RESTART ACKNOWLEDGE
011-----	*Miscellaneous messages:*
00000	- SEGMENT
11001	- CONGESTION CONTROL
11011	- INFORMATION
00010	- FACILITY
01110	- NOTIFY
11101	- STATUS
10101	- STATUS ENQUIRY

Note - When used, the message type is defined in the following octet(s), according to the national specification.

Figure 4.16 Message types.

example, the specification of the only valid message sequence for an ISDN call origination [4.2]. The ISDN originating user sends a SETUP message to the switch, providing all required information in the SETUP message. The next potentially valid message for a successful call associated with the originating user is the CALL PROCEEDING message. Figure 4.18 also depicts the only valid terminating message sequence for a successful call. With respect to the terminating ISDN user, an incoming call will cause a SETUP message to be sent to the called user equipment. If the called user equipment does not respond within a certain period of time (specified by appropriate timers), a second SETUP message will be sent. The called user equipment may optionally respond with a CALL PROCEEDING or ALERTING message before responding with a CONNECT message. The only valid response sequences to a SETUP message for a successful call are:

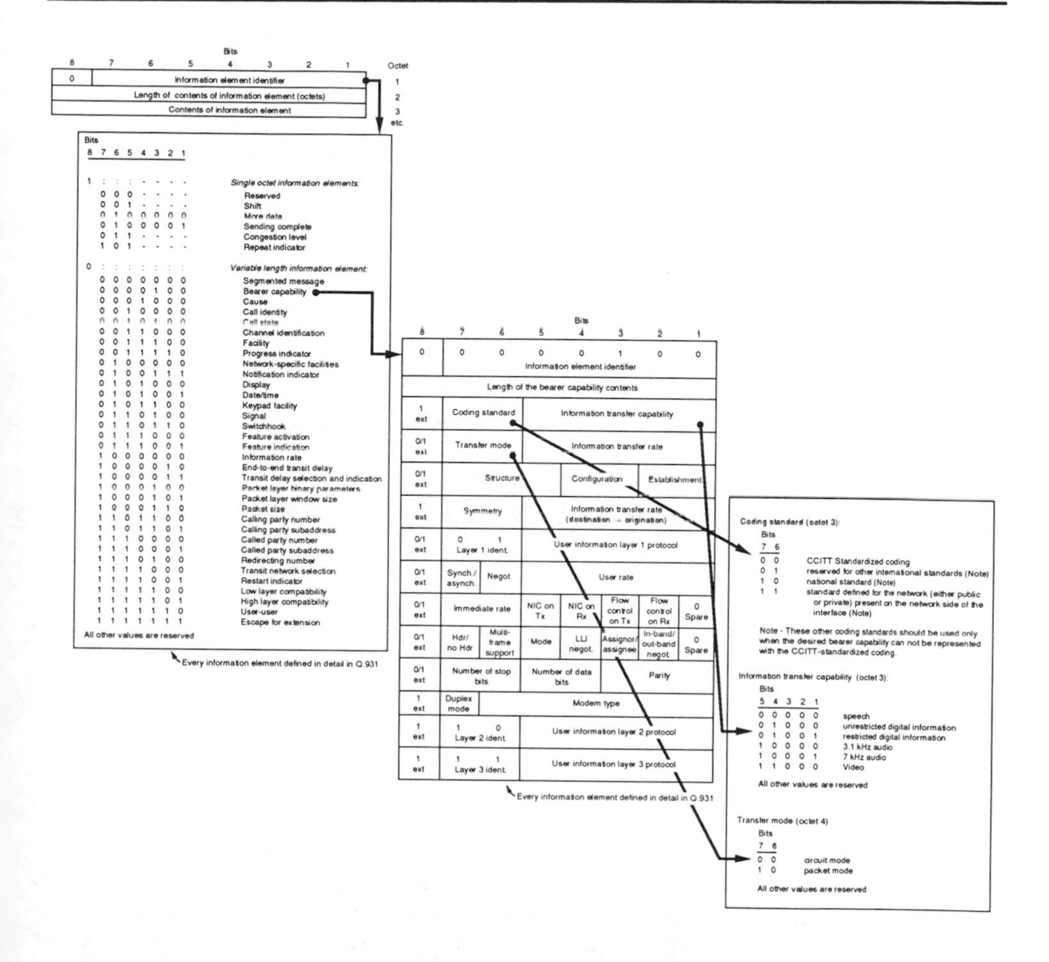

Figure 4.17 Format and example of variable length Q.931 information element message

1. CALL PROCEEDING, ALERTING, CONNECT;
2. CALL PROCEEDING, PROGRESS, CONNECT;
3. CALL PROCEEDING, CONNECT;
4. ALERTING, CONNECT; and
5. CONNECT.

At any time during an in-progress call, either of the two users of the network may choose to end the call by beginning a clearing message sequence. When a user requests release from a call, a request is made of the network, which then completes a message sequence with the requesting party and initiates a second release sequence to the other party (if there is one on the call). The network may initiate clearing of the call, either of a single user during an origination sequence or of both users if the call proceeded to the point where two interfaces are involved.

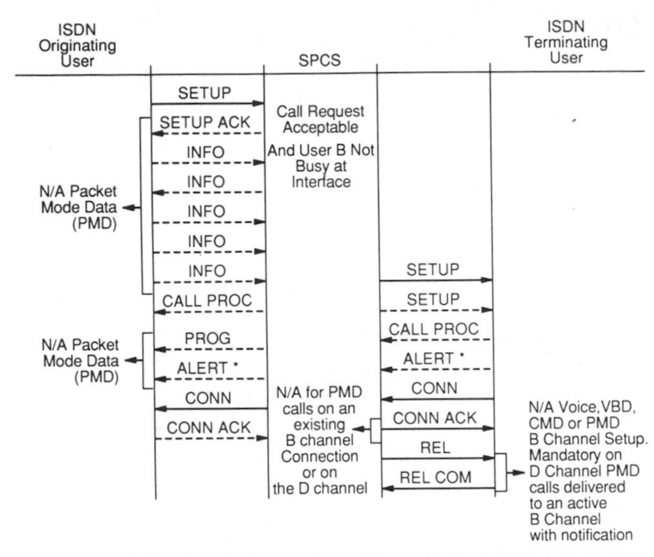

Figure 4.18 Valid message sequences.

Figure 4.19 depicts an example of an originating SETUP message; this provides a more intuitive view of one network layer message described in the previous section. Reference [4.2] provides a 300-page discussion of ISDN network layer C-plane requirements for design of ISDN-compatible network and termination equipment for the United States.

4.6.2.5 Q.932 Messages

This section briefly identifies the messages employed in CCITT Q.932 to control supplementary services at ISDN user-network interfaces (the protocol operates at layer 3 of the C-plane at the S-T reference point, and assumes that the lower layers employ I.430, I.431, and Q.921).

FACILITY: This message may be sent to request or acknowledge a supplementary service.

HOLD: This message is sent by the network or the user to request the hold function for an existing call.

HOLD ACKNOWLEDGE: This message is sent by the network or the user to indicate that the hold function has been successfully performed.

87654321	Meaning
00001000	Protocol discriminator
00000001	Call reference (value is 1 octet)
00101011	allocated by user side, value is 43
00000101	message type = SETUP
00000100	Bearer capability
00000011	length of 3 octets
10000000	speech
10010000	64-kbps circuit-mode
10100010	μ-law
01101100	Calling party number
00001001	length of 9 octets
01000001	local ISDN number
10100000	presentation prohibited
00110111	digit 7
00110101	digit 5
00111000	digit 8
00110101	digit 5
00110000	digit 0
00110111	digit 7
00110000	digit 0
01110000	Called party number
00001000	length of 8 octets
01000001	local ISDN number
00110111	digit 7
00110101	digit 5
00111000	digit 8
00110101	digit 5
00110000	digit 0
00110010	digit 2
00110000	digit 0

Figure 4.19 Originating SETUP message coding example.

HOLD REJECT: This message is sent by the network or the user to indicate that the hold function has been denied.
REGISTER: This message is sent by the network or the user to assign a new call reference for noncall-associated transactions.
RETRIEVE: This message is sent by the network or the user to request the retrieval of a held call.
RETRIEVE ACKNOWLEDGE: This message is sent by the network or the user to indicate that the retrieval function has been successfully performed.
RETRIEVE REJECT: This message is sent by the network or the user to indicate the inability to perform the requested retrieval function.

Supplementary services identified in CCITT I.250 through I.257 include the following:

Number identification
Calling line identification presentation
Malicious call identification
Call offering
Call transfer
Call forwarding on busy
Call completion
Call waiting
Call hold
Multiparty
Conference calling
Three-party service
Community of interest
Closed user group
Private numbering plan
Charging
Credit card calling
Calling collect
Additional information transfer
User-to-user signaling.

4.7 ISDN PACKET SERVICES

Packet-mode bearer services supported by ISDN are described in Recommendations CCITT I.232 and I.462 (X.31); the latter specifies the procedures for virtual call and permanent virtual circuit bearer services.

In addition to these traditional X.25 packet modes, new packet-switching technologies are under consideration in ISDN. Four potential services proposed for standardization by the CCITT in 1988 in Recommendation I.122 ("framework for providing additional packet-mode bearer services") are:

1. Frame relaying 1[1] (FR-1—no functions above core data link functions are terminated by the network; if needed, such functions are terminated only by end-to-end peers);
2. Frame relaying 2 (FR-2—no functions above the core data link functions are terminated by the network; I.441* upper functions are terminated only at the endpoints). This implies that user equipment must support a specific data link protocol. I.441* is I.441/Q.921 extended to cover I.122 requirements. More recently, the FR-1 and FR-2 distinction has been eliminated; this dis-

1. The term "frame relay" is now also used.

cussion follows the original standard. The deployment of frame relays has also been divorced from ISDN;

3. Frame switching (the full Recommendation I.441* protocol is terminated by the network); and
4. X.25-based additional packet mode (the full Recommendation I.441 protocol and the *data transfer part* (DTP) of Recommendation X.25 PLP are terminated by the network). Here, the U-plane is similar to X.25 PLP, and the C-plane is based on I.451.

FR-1 can be provided over permanent virtual circuits (PVCs) or switched virtual circuits. With PVCs, no call set-up establishment is needed on a per-packet or -session basis because the address fields are agreed on when the user subscribes to the service. In FR-1, the network has no knowledge of the end-to-end protocol.

Due to the novelty of frame relaying methods, this section concentrates on it rather than on the traditional X.25 methods. I.122, however, is a framework document; additional work is needed before frame relaying service is fully standardized.

The data link layer core functions are:

- Frame delimiting, alignment, and transparency;
- Frame multiplexing-demultiplexing using the address field;
- Inspection of the frame to ensure that it consists of an integer number of octets prior to zero bit insertion or following zero bit extraction;
- Inspection of the frame to ensure that it is neither too long nor too short; and
- Detection of transmission errors.

The principle of separation of the user and control planes for all telecommunication services has been established as a fundamental goal of the ISDN protocol reference model, as described in CCITT I.320. Under ISDN, the SVC packet-mode bearer services described above have the following characteristics:

1. All C-plane procedures, if needed, are performed in a logically separate manner using protocol procedures that are integrated across all telecommunication services. Namely, Q.931 will be used to set up and tear down the service; the C-plane is used to establish DLCI-to-global address mapping.
2. The U-plane procedures share the same layer 1 functions based on Recommendations I.430 and I.431. Moreover, they share the same core procedures.

On the user side, Recommendations I.430 or I.431 provide the layer 1 protocol for the U- and C-planes. The C-plane uses the D channel with Recommendations I.441 and I.451 extended as the layer 2 and 3 protocols, respectively. In the case of PVCs, no real-time call establishment is necessary and any parameters are agreed on at subscription time. The U-plane may use any channel on which the user implements at least the lower part (the core functions) of Recommendation I.441 with appropriate extensions (I.441*). Since 1991 the core functions of I.441 have been known as Q.922, LAP-F, or T1.618.

4.7.1 Frame Relaying Service

The term "relay" implies that the layer 2 data frame is not terminated or processed at the endpoints of each link in the network, but is relayed to the destination, as is the case in a LAN. In contrast to CCITT's X.25-based packet switching, in *frame relaying service* (FRS), the physical line between nodes consists of multiple data links, each identifiable by the address in the data link frame. FRS's major characteristics are out-of-band call control and link layer multiplexing. Unlike the (X.25-based) X.31 packet-mode services, FRS integrates more completely with ISDN circuit-mode services because of the out-of-band procedures for connection control described in the previous two paragraphs [4.20]. At the time of this writing, FRS was expected to become an ANSI standard, but much additional supporting standardization is necessary before the service can be offered in a carrier-vendor independent fashion.

FRS is based on the frame structure employed by the ISDN D channel LAP-D to provide statistical multiplexing of different user data streams within the data link layer (layer 2). In contrast, X.25 multiplexing takes place at the packet layer (layer 3). In other words, a feature of FRS is to have the virtual circuit identifier, currently implemented in the network layer (layer 3) of X.25, at the link layer (layer 2) so that fast-packet switching can be accomplished more easily. In the X.25 environment, when a data call is established, the virtual circuit indicator is negotiated and used for the duration of the call to route packets through the network. The indicator is enveloped within the layer 2 header-trailers (as discussed in Chapter 1), which must be processed, before it is exposed (this processing involves more than just stripping the header-trailer; for example, it involves error detection and correction). In the OSIRM environment, layer $n+1$ protocol information is enveloped inside layer n information. In LANs, the routing of the actual packets is accomplished directly at layer 2: The data packets are supplied with a 48-bit destination address, which is readily available, and which is used to route the data physically to the intended destination. Also, error recovery is not necessary in a LAN as a packet flows by a station on its way along the bus or ring. In FRS, only the lower sublayer of layer 2, consisting of such core functions as frame delimiting, multiplexing, and error detection, are terminated by a network at the user-network interface. The upper procedural sublayer of layer 2, with functions such as error recovery and flow control, operates between users on an end-to-end basis. In this sense, a user's data transfer protocol is transparent to a network.

Thus, FRS proposes to implement only the core functions of LAP-F on a link-by-link basis; the other functions, particularly error recovery, are done on an end-to-end basis. Indeed, the capabilities provided by the transport layer protocol (CCITT X.214 and X.224, ISO 8072 and 8073) accommodate this transfer of responsibilities to the boundaries of the network. FRS is a connection-oriented service because routing is based on establishing virtual circuit indicators.

Figure 4.20 [4.4] shows the partition of the data link layer. The figure is a direct application of the ISDN protocol reference model to the packet-mode communication discussed above, and shows the user-network interface protocol architecture. As shown in Figure 4.20 for both FR-1 and FR-2, the network terminates only the "core" aspects of the data link protocol (I.441* core). The terminals in FR-2 terminate the full data link protocol. The terminals in FR-1 terminate the core aspects of the data link protocol (I-441*). What they terminate above core aspects is a user's option. See the table associated with Figure 4.20.

Frame relay uses routing based on the address carried in the frame; the frame itself is defined by CCITT Q.922. Figure 4.21 depicts the lower core LAP-F frame [4.21]. The "remainder" of the data link layer functions, above the core functions, need to be defined into a peer-to-peer protocol, at least for FR-1. Work is currently under way in the standards bodies (CCITT SG XI and ECSA T1S1) to complete

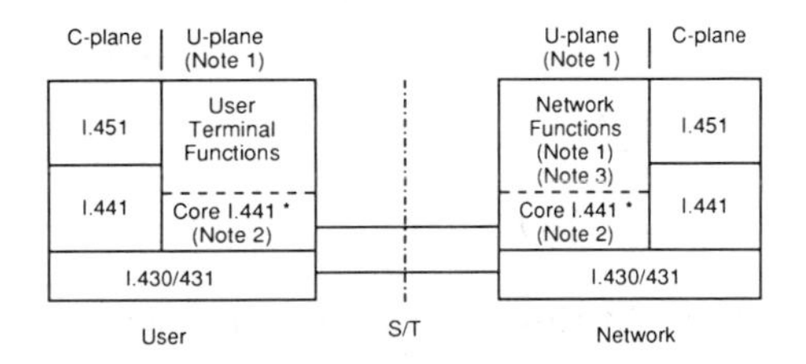

Note 1 - The U-plane functions applicable to each bearer service are given in Table below.
Note 2 - The core functions of Recommendation I.441 are described in text.
Note 3 - The U-plane functions provided by the network at the S/T reference point are determined by the network after negotiation with the user, based on the requested bearer service and associated parameters. These functions are user-selectable for each call. A network may choose not to implement the full set of options. These functions may not be available one by one. So far only three groupings have been identified:

a) the null set,
b) the upper part of Rec. I.441, and
c) the upper part of Rec. I.441 and the data transfer of X.25 PLP.

U-plane functions applicable to each bearer service

Bearer Service	User Terminal (Note 1')	Network
Frame Relaying 1	I.441* Core (Note 2')	I.441* Core
Frame Relaying 2	I.441*	I.441* Core
Frame Switching	I.441*	I.441*
X.25-based Additional Packet Mode	I.441* X.25 DTP	I.441* X.25 DTP

Note 1' - Additional user-selectable functions may be implemented.
Note 2' - I.441* is I.441 with appropriate extensions. The use of the extensions may depend on each bearer service and is for further study according to the Blue Book.

Figure 4.20 User-network interface protocol architecture (original CCITT standard).

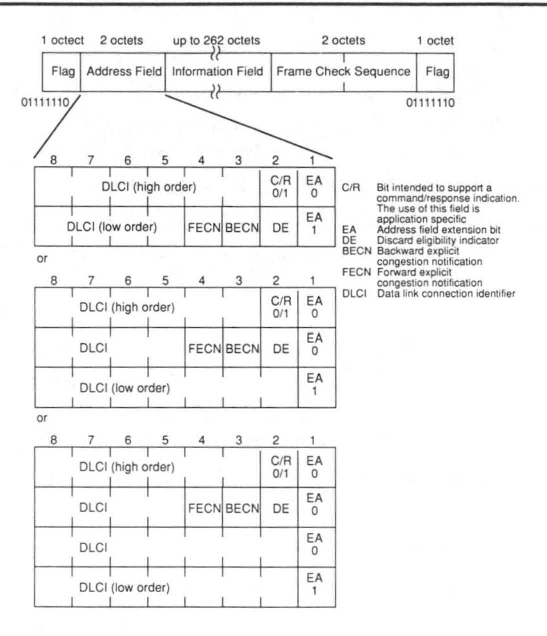

Figure 4.21 LAP-F frame format (T1-618).

the standardization effort. At this writing, for FR the core appeared to be going to be somewhat enriched to support congestion control [4.12]. Frame relay is now receiving considerable user, vendor, and carrier attention in the United States, particularly for LAN interconnection. However, only a permanent virtual circuit (pre-ISDN) version will see early deployment.

As of 1991 several documents have been issued by ANSI in reference to frame relay usage in the United States. These are:

1. T1.606-1991: Frame Relaying Bearer Service—Architectural Framework and service description;
2. Addendum to T1.606: Frame Relaying Bearer Service—Architectural Framework and service description;
3. T1.618-1991: Core Aspects of Frame Protocol for use with Frame Relay Bearer Service;
4. T1.617-1991: Access Signaling and PVC Management.

The data transfer phase of the frame relay bearer service is defined in T1.606. The protocol needed to support frame relay is defined in T1.618; the protocol operates at the lowest sublayer of the data link layer, providing "core aspects." The frame relay data transfer protocol defined in T1.618 is intended to support multiple simultaneous end-user protocols within a single physical channel. This protocol provides transparent transfer of user data and does not restrict the contents, format, or coding of the information, or interpret the structure. The addendum to T1.606

defines congestion management strategies; it covers both network and end user mechanisms and responsibilities to avoid or recover from periods of congestion.

Since the beginning of 1991, a number of vendors (now over 50) have joined the Frame Relay Forum. The goal of the Forum is to ensure some degree of interoperability of new products. The need to offer interoperable products is critical, and vendors realize that users may not be willing to deploy technologies that locks them in with systems which could become obsolete in a year or two; cisco Systems, Digital Equipment Corp., Northern Telecom, and StrataCom jointly developed a frame relay specification on which product development can be based until complete international standards become available. The interim specification, announced in September 1990, was based on the emerging ECSA standard, but had additional management features and broadcasting. By the end of 1991, a final set of standards became available. The ANSI/CCITT Standards include the extensions developed by these vendors and by the Forum. Approximately three dozen vendors have announced or delivered frame relay products; about half a dozen carriers are establishing PVC-based service. SVC service may become available in 1994–95.

4.7.2 Frame Relaying 1 Service Description

The basic bearer service provided is the unacknowledged transfer of frames from S/T to the S/T reference point. More specifically, in the U-plane:

- It preserves their order as given at one S/T reference point if and when they are delivered at the other end. (Because the network does not terminate the upper part of I.441*, sequence numbers are not kept by the network; networks should be implemented in a way that, in principle, frame order is preserved.)
- It detects transmission, format, and operational errors.
- Frames are transported transparently (in the network), only the address and FCS field may be modified (some bits being defined in the address field for congestion control may also be modified); and
- It does not acknowledge frames (within the network).

4.7.3 Frame Relaying 2 Service Description

The basic bearer service provided is an unacknowledged transfer of frames from S/T to the S/T reference point. More specifically, in the U-plane:

- It preserves their order as given at one S-T reference point if and when they are delivered at the other end. (Because the network does not terminate the upper part of I.441*, sequence numbers are not kept by the network. Networks should be implemented such that, in principle, frame order is preserved.)
- It detects transmission, format, and operational errors.

- Frames are transported transparently in the network. Only the address and FCS field may be modified.
- It does not acknowledge frames (within the network).
- Normally the only frames received by a user are those sent by the distant user.

4.7.4 Frame Switching Service Description

In the U-plane, frame switching:

- Provides for the acknowledged transport of frames.
- Detects and recovers from transmission, format, and operational errors.
- Detects and recovers from lost or duplicated frames.
- Provides flow control.

Note: These distinctions were not carried forward in the latest standards. Evolving services resemble FR-1 without the control plane.

4.7.5 Frame Relay Network Interworking

In the future, FRS may be provided by different carriers, network providers, or telephone operating companies. These FR networks will need to be interconnected among themselves as well as with existing *packet switched public data networks* (PSPDNs). Any interconnection should be such that it allows users on different networks to communicate with each other in a uniform way, as though the users were all on a single network. To achieve these goals, standardized procedures must be established for the interworking between (1) FR networks (FR-FR) and (2) FR networks and X.25-based networks (FR-X.25). A FR connection may pass through multiple networks, including both private and public networks. For public networks, both local exchange and interexchange carriers may be involved.

In a multivendor environment, the network of a given carrier or provider may consist of equipment from different vendors or different types of equipment from the same vendor. Two approaches for interworking can be used: network-to-network (NW-NW) interworking and node-to-node (ND-ND) interworking. In the NW-NW method, the signaling and transport procedures at an internetwork interface are standardized. Within a given network, internodal signaling is an internal matter. Traditionally, telephone exchanges in a public switched telephone network are interconnected using the ND-ND strategy. This is the philosophy behind SS7 (CCITT Recommendation Q.701-Q.795), discussed in Chapter 9. In contrast, the NW-NW approach has been used for the interworking of data communication networks. As specified in Recommendation X.300, the protocol between individual data switching exchanges (DSEs) within a given network is an internal matter. For interworking purposes, those DSEs that are in the same network are considered to be only one intermediate system involved in a multi-network data call [4.20].

In a PSTN, and also in a *circuit switched public data network* (CSPDN), the data transfer protocol of a user connection is transparent to the network. Processing by the network is required primarily during the call setup and teardown phases. In a PSPDN, connection management involves not only the establishment and release of a user connection, but also the integrity of individual packets in the data transfer phase. This means that internodal signaling is much more complex in a PSPDN than in a CSPDN. In the interconnection of PSPDNs based on Recommendation X.75, the heterogeneity of the internal protocols of different networks is preserved. In a FR network, in addition to call processing, per-frame handling is also needed. By exploiting the advances in transmission technology and end-system intelligence, the network does not need to guarantee delivery of frames (the end-system will sort out possible problems). Consequently, the amount of per-frame processing required is significantly less than that for the packets in a PSPDN.

The service of a FR network is closer to that of a CSPDN in the sense that, except for the multiplexing and error detecting functions, a user's data transfer protocol is transparent to a FR network. With relatively little per-frame processing, the internodal transport signaling required in a FR network, for data transfer purposes after call establishment, should be relatively simple.

4.8 ISDN EVOLUTION

ISDN has the potential to offer users the same or better services, features, and cost advantages through the public networks than previously have been available with private networks. Also, ISDN may be able to offer to small business and residential users certain economically priced services and features through the public networks that traditionally have been targeted for the business community. However, industry has expressed some concerns that it has never been demonstrated adequately that ISDN will promote user options and flexibility or replace existing networks with an equally or more cost-effective alternative [4.1]. For example, ISDN-ready voice stations are currently an order of magnitude more expensive than regular phone sets ($675 to $990 compared to $40). Four vendors had sets available in 1990: AT&T, Fujitsu, NEC, and Northern Telecom.

A number of factors are necessary for ISDN to have a favorable and growing market in the United States. Implementation of standards, availability of feature-rich services, and an innovative regulatory environment are some of the key factors that ISDN will need to achieve market penetration in the United States. Other factors include:

- Wide-scale development of ISDN VLSI chip technology;
- Finalization of an ISDN numbering (while E.164 needs no further work, issues concerning how the numbers will be allocated need to be resolved); and
- ISDN pricing.

Although deployment of ISDN was moving closer to reality at the end of 1990, vendors of compatible chip sets continue to expand their offerings at a rather slow pace. Reportedly, they are waiting to see how the standards-setting process will go. Standards development is slowed not only by technical issues but also by the politics of the process. ISDN functions that have not been fully defined in the standards are usually implemented in software or firmware; until these are firm, semiconductor companies are reluctant to put a lot of functionality into silicon. As requirements for functionality change, either because of standards or widespread user demand, firmware can be modified more easily than the printed circuit board.

Currently, approximately 120,000 ISDN circuits are installed in the United States. More than 50 trials were either under way or completed. Virtually all BOCs already sell ISDN as a special assembly, and general tariffs are being filed [4.17]. Inter-LATA ISDN service is also beginning to appear; in addition to AT&T, both MCI and US Sprint have operating CCSS7 networks, which are essential to ISDN.

Fewer than 2% of the lines served by the BOCs will be "fully equipped" by 1994 to deliver ISDN; these deployment projections are contained in filings accompanying the BOC's comments to the FCC on interstate rates of return (FCC Docket 89-624)[4.22]. (The meaning of the phrase "fully equipped" is not totally clear.) By contrast, 75% of the 120 million access lines will have access to SS7,which is a key component for providing intelligent services (see Chapter 9). See Table 4.8, which is compiled from information provided in References [4.22] and [4.23].

Table 4.9 provides the status of ISDN deployment plans as of early 1990 [4.24].

In 1990, one view of the role of ISDN for data communication was the following [4.25]:

> The battle for the minds of ISDN users in the business environment will be won or lost based largely upon vendors' ability to convince data users of the technology's value . . . Vendors must find ways to make a strong case for ISDN deployment to the real end users who are also being courted by T1 multiplexer and LAN vendors . . . The use of ISDN technology for voice applications will not produce significant revenues to justify investment . . . the significant revenues are going to come from the data communication customers; the ISDN trial applications also support this conclusion . . . Vendors could clearly state ISDN's benefits to their own operations and specific business goals, but the researchers found that they were much less able to

Table 4.8
Projected ISDN Penetration

Regional BOC	*ISDN Lines in 1990**	*Number of Users in 1990†*	*Percentage of Lines in Column 1 at Customer Sites*	*Projected 1994 ISDN Lines*	*Percentage of Total Access Lines**
Ameritech	15,677	50	30	n.a	0.7
Bell Atlantic	57,290‡	28	92	n.a	1.2
BellSouth	2,275	>10	32	1.6 M	7.9
NYNEX	17,785	12	68	n.a	0.3
Pacific Telesis	1,718	6	52	42.2k	0.26
Southwestern Bell	22,049	15	89	n.a.	0.64
U S WEST	6,515	25	81	n.a	0.45
TOTAL	123,309	>146	78	2.4M	2.00

Notes: n.a.—not available.
*Numbers are approximate and subject to debate.
† Based on [4.24].
‡ 5000 according to [4.24].

> articulate the real-life benefits to users . . . it may take a different breed of spokesperson to persuasively explain the potential benefits of ISDN to the real end users.

Creative data communication experts and practitioners must be empowered by these vendors to identify and implement network-based data applications over the ISDN platform. Table 4.10 provides a view of the current penetration of data applications of ISDN (based partially on [4.25]).

4.9 BROADBAND ISDN

The need for services employing bit rates greater than 2 Mb/s was already clear when the I Series recommendations were written, especially because LANs already provided 10 Mb/s at the time (in fact, ISDN still has a void in its accommodation of LAN speeds, and some observers claim that this will hinder ISDN and BISDN in the data arena). In the 1984 version, other H channels and interface types above 2 Mb/s were slated for further study. Therefore, CCITT Study Group XVIII installed the "Task Group on Broadband Aspects of ISDN." In the current study period (1989–92), Working Party 8 of the CCITT SG XVIII has the charter to develop standards for general aspects of BISDN.

In comparison to several dedicated networks, ISDN services and network integration have major advantages in terms of economic planning, development, implementation, operation, and maintenance [4.26]. While dedicated networks

Table 4.9
LEC ISDN Deployment Status or Plans (as of 1990)

BOC	*Tariff Status*	*ISDN Deployment Plans*	*Signaling System 7 Plans*	*Switches*
Ameritech	Tariffed in Illinois and Ohio; other states to follow in 1990–91	To be available from 80% of COs in January 1991	Being introduced in metropolitan areas	AT&T, NTI, Siemens
Bell Atlantic	Planning to file 3Q1990	Marketing in 1991; major delivery in 1992	80% of COs to be equipped by 1991	AT&T, NTI, Siemens
BellSouth	Planning to file 4Q1990	General availability starting January 1991	Available in all major metropolitan areas by end of 1990	AT&T, NTI; tests on Siemens
NYNEX	Filed November 1990	General availability starting in 1991	Available in Boston in 1991, and in New York by 1993	AT&T, NTI
Pacific Telesis	California tariffed; Nevada has no announced plans at time of publication	PRI in Pacific Bell as of Jan 1991; BRI service in Reno-Carson area in 1991	Nevada Bell to introduce starting mid-1991; Pacific Bell in 1Q1992	AT&T, NTI
Southwestern Bell	Preliminary tariffs filed in Texas; Oklahoma pending in 1990; other states to follow in 1991	To be determined by market	Will be available from 300 COs to 70% of customers by 1994	AT&T, NTI
U S WEST	Planning to file in 1990	Available in 12 major cities; rest to be determined by market	Trials in 1991; available in 1992	AT&T, NTI

require several distinct subscriber access lines, the BISDN access can be based on a single optical fiber. Large-scale production of highly integrated VLSI system components will lead to cost-effective BISDN solutions. BISDN is aimed at both business applications and residential subscribers.

To meet the requirements of possible future broadband services, BISDN is designed to be flexible. A variety of potential interactive and distribution broadband services is contemplated for BISDN [4.26]:

- Broadband video telephony and videoconferencing;
- Video surveillance;
- High-speed unrestricted digital information transmission;
- High-speed file transfer;
- High-speed high-resolution facsimile;
- Video and document retrieval service; and
- TV distribution (with existing TV or HDTV).

Table 4.10
Current Penetration of Data Applications of ISDN (based partially on [4.24])

	1989	*1991*	*Cumulative 1994*
Number of ISDN Data Terminals	50 k	185 k	1.5 M
Terminal Price	2.3 k	2.0 k	2.3 k
ISDN Terminal Market	115 M	390 M	3,500 M
Number of ISDN TAs	50 k	95 k	700 k
Cost of TAs	1.5 k	1.3 k	800
ISDN TA Market	65 M	104 M	560 M
Total ISDN Data Equipment Terminal Market ($US)	180 M	494 M	4,160 M

This list represents "bearer services" (OSIRM layers 1 to 3) and "teleservices" (full OSIRM specifications layers 1 to 7).

The goal of BISDN is to define the user-network interface to the network, which meets varied requirements, particularly when the traffic mix is highly dynamic. The reference configuration of Figure 4.3 is sufficiently general to be applicable not only for basic access and primary rate access, but also for broadband access; both reference points S and T are valid for broadband access [4.8]. To emphasize the broadband aspects, the notations for reference points and for functional groupings with broadband capabilities are appended with the letter B (e.g., TB).

4.9.1 BISDN Protocol Model

The BISDN protocol model is shown in Figure 4.22. The *adaptation* layer shown in the model is service dependent. It supports the higher layer functions of the user and control planes, as well as the connections between BISDN and non-BISDN interfaces. Information is mapped by the adaptation layer into lower layer cells. At the transmitting end, information units are segmented to be inserted into these cells. At the receiving end, the information units are reassembled from the underlying cells. Any adaptation-layer specific information (e.g., data field length, time stamps, sequence number, *et cetera*) that must be passed between peer adaptation layers is contained in the information field of the lower layer.

4.9.2 Asynchronous Transfer Mode

BISDN work started in earnest in 1985. Worldwide activities to evolve ISDN into an optical-fiber-based broadband network have resulted in the first baseline documents, in particular, CCITT Recommendation I.121. A method known as the

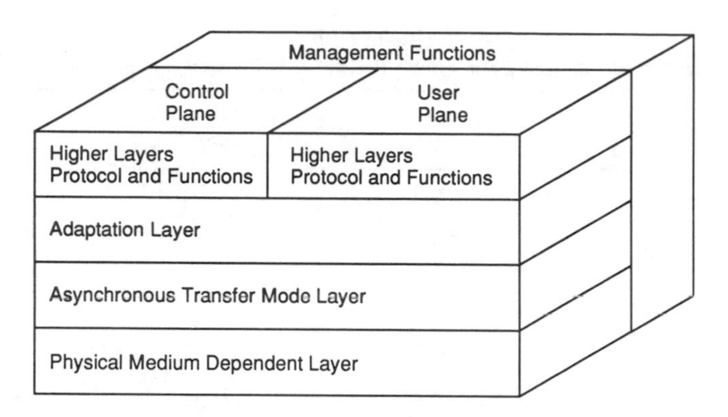

Figure 4.22 BISDN protocol model.

asynchronous transfer mode (ATM), discussed below, has been strongly promoted as the transport structure for future broadband telecommunication networks, within standards bodies CCITT Study Group XVIII (internationally) and T1S1 (North America). The T1S1.5 subcommittee of the ECSA addresses all aspects of BISDN and provides guidelines to other T1 subcommittees on related topics. Recommendation I.121 targets ATM as the desirable technology for BISDN. Two BISDN access rates will be standardized: one at about 150 Mb/s; another at about 600 Mb/s.

ATM is a high-bandwidth, low-delay packet-based switching and multiplexing technology; it is a connection-oriented mechanism, although it is designed as a basis for supporting both connectionless and connection-oriented services. For ATM, a number of functions of the layer 2 protocol are removed to the edge of the network. Bare bones layer 2 capabilities are supported, in addition to layer 1 functions (clocking, bit encoding, physical medium connection).

ATM is meant to be supported by any digital transmission hierarchy or system, including existing DS3 systems (for some but not all of the BISDN services). ATM is not an asynchronous transmission technique. *Transfer mode* refers to the switching and multiplexing process; transfer mode logically resides on top of the transmission layer (this transmission layer corresponds to the physical medium dependent sublayer of the OSI physical layer). Connections through the BISDN networks can be conventional circuit-switching paths or may be based on the ATM high-speed packet transfer techniques (provided that the appropriate switching technologies will be available).

In ATM, the information is packed into fixed-size cells of relatively short length. Agreement on 48 octets was reached in mid-1989 (see Section 4.9.4). Accelerated procedures are being used for BISDN, with standards scheduled for 1990; more complete standards will be finalized by 1992. These cells are identified and switched throughout the network by means of a label in the header—hence, the term *label multiplexing* is also used. In OSIRM terminology, ATM provides core layer 2 functions.

Initially, other proposals were advanced for the transport mode, including *synchronous transfer mode* (STM) and a hybrid of ATM and STM. STM basically implies an underlying circuit switched transport structure based on TDM. STM is, to a certain extent, a natural extension of the current public network and of the narrow band ISDN interface standard. However, after rounds of discussions, international agreement has been reached on ATM [4.27].

In ATM, information to be transferred is packetized into fixed-size slots called *cells*. Cells are identified and switched by means of a label in the header (see Figures 4.23 and 4.24). The term "asynchronous" refers to the fact that cells allocated to the same connection may exhibit an irregular recurrence pattern, as cells are filled according to the actual demand (see Figure 4.23). ATM allows for bit rate allocation on demand, so the bit rate per connection can be chosen flexibly.

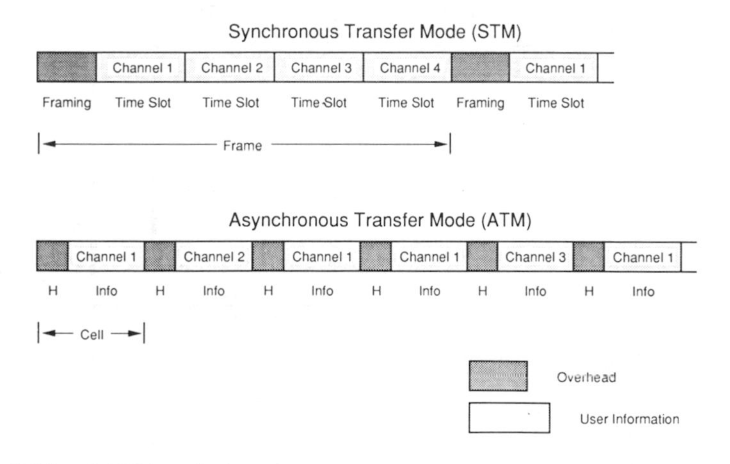

Figure 4.23 STM and ATM methods.

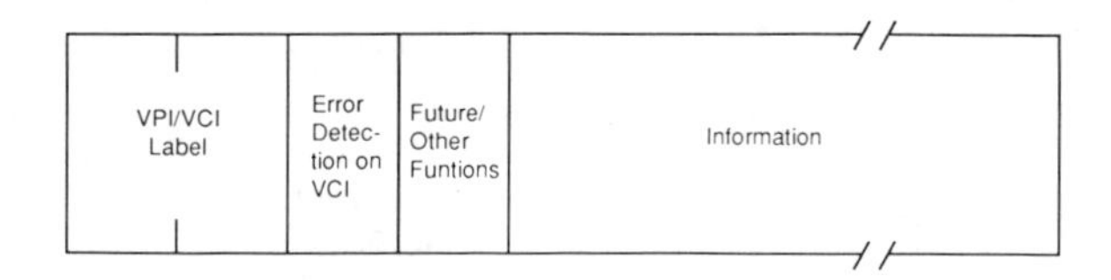

Figure 4.24 Generic structure of an ATM cell.

In addition, the actual "channel mix" at the broadband interface can dynamically change. ATM allows for labeled channels with any fixed rate in the range of some kb/s up to the total payload capacity of the interface.

The ATM layer is common to all services. Its functions are represented by the ATM cell header functions. Thus, the boundary between the cell header and the cell information field corresponds to the boundary between the ATM layer and the higher layers. The ATM header contains the label, which contains a *virtual path identifier* (VPI) and a *virtual channel identifier* (VCI), and an error detection field (see Figure 4.25). Error detection on the ATM level is confined to the header. The label that is used for channel identification, in place of the positional methodology for assignment of octets, is inherent in traditional synchronous digital systems (as discussed in Chapter 3). A special label value will be used to indicate unassigned cells. ATM technology can flexibly support a wide variety of services with different information transfer rates: It is suitable for both *constant bit rate*

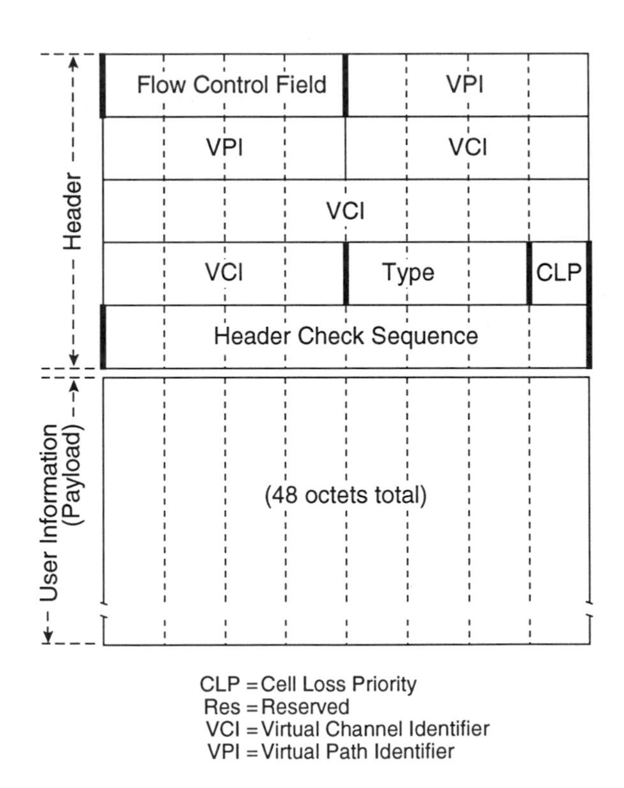

Figure 4.25 ATM cell format.

(CBR) and *variable bit rate services* (these are also sometimes called isochronous and bursty streams, respectively).

ATM can be viewed as a limiting case of packet-switching technology: ATM deals with procedures for allocating bandwidth at the user interface and allocating the bandwidth to various user services. ATM is similar to packet switching but with the following differences:

1. No error control and flow control on the links—responsibility is moved to the edge of the network.
2. Cells (packets) have a fixed and small length. No variable cells are allowed. This framework allows very high speed switching nodes because the logical decisions are straightforward. Cell length has now been agreed to be 53 bytes (48 for user information and 5 for overhead).
3. Unlike "packet switching" in LANs, the communication in ATM is connection-oriented—all information is transferred via a VCI, which is assigned for the duration of the session.
4. The header provides only limited layer 2 functionality.

The VPI-VCI is locally significant to the user interface, but may undergo translation as it is being transported to another interface. The VPI-VCI constitutes a label used to allocate transmission resources; the label, rather than the position of a frame, does the allocation. At call setup, the VCI value can be associated with a particular quality of service. Usable capacity can be dynamically assigned; the network can take advantage of statistical fluctuations while at the same time providing an established grade of service. Multiple rate-dependent overlay networks are obviated, facilitating integration.

The introduction of ATM networks will be a complex task that will take considerable time to be completed. In addition to economical reasons, technical obstacles may arise to hinder immediate overall ATM deployment:

1. ATM parameters (header definition and length among others). Some believe that service considerations (interworking with embedded base) should be used to optimize the parameters; others wish to stress new requirements (HDTV, CAD-CAM graphics, *et cetera*) and maximize flexibility.
2. Harmonization between SMDS, IEEE 802.6, and ATM. SMDS is a Bellcore-proposed DS1-DS3 switched transmission capability for LAN interconnection (see Chapter 15).
3. Interworking with existing (e.g., DS3) and new (e.g., SONET) digital transmission hierarchies. The substantial investment in the existing transmission plant is a major transition consideration in deploying ATM. The evolution must consider transmission, switching, signaling, and service modules, among others.
4. The interface connecting a customer to a BISDN network, the user-network interface (UNI). Both transmission and ATM parameters affect the usable capacity of an interface, limiting services that can be provided.

Other issues concern congestion control and cell sequencing.

A point-to-point connection-oriented service rides directly over the transfer layer mode: ATM performs the relaying functions required to transport the user information through the network [4.26]. ATM (or other transmission-switching), however, is somewhat hidden from the data communication user, who only sees the interface.

Synchronous Transfer Mode

Originally, STM, based on time-division switching and multiplexing (which depend on the temporal position of data within a frame superstructure), was considered for BISDN. STM can be simpler to administer, but is not as efficient as ATM; the relationship is similar to TDM *versus* STDM. STM-based systems are optimally suited to fixed-rate services; under a dynamically changing composition of services, ATM is better suited. With ATM, usable capacity is assigned on demand. However, the ATM approach requires that many new problems be solved. For example, the impact of possible cell loss, cell delay, and cell jitter on the quality of all broadband and other services needs to be determined. STM would have facilitated early deployment of BISDN, because of the compatibility with the embedded base of plant technology [4.26]. I.121 states that in the early stages of the evolution of BISDN, some interim combinations of STM and ATM may need to be adopted in certain countries to facilitate early penetration of digital services.

4.9.3 User-Network Interface and Access

The general reference configuration for the broadband UNI is the same as that of narrow band ISDN. In Figure 4.26, several terminals are connected (via appropriate terminal interfaces at the SB reference points) with the subscriber-premises network, which accesses the local network itself via a standardized interface. The

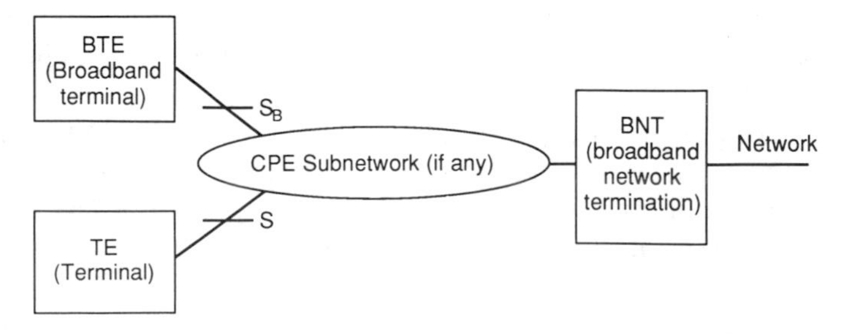

Figure 4.26 The BISDN interface model.

broadband network termination (BNT) is the boundary between the local access network and the subscriber-premises network.

Due to uncertainties about service bit rates, simultaneous service mix, future improvement of video-coding, HDTV requirements, *et cetera*, designers have decided to consider two interface candidates: (1) around 150 Mb/s or (2) around 600 Mb/s. The 150 Mb/s interface (with symmetrical information transfer capability) is conceived for ATM use and effectively supports the flexible breakdown of the interface payload into several (labeled) channels on demand. This interface type is useful for business customers who want to access a variety of interactive services. The 600 Mb/s interface may be the required solution for certain needs, for example, simultaneously conveying several HDTV programs to a residential customer (it can be asymmetrical, with 150 Mb/s being a sufficient transfer capability in the direction toward the network). The feasibility of a 600 Mb/s interface is somewhat unclear for the immediate future, thus, the hybrid solution (ATM-STM usage) with subdivision of the 600 Mb/s interface into smaller ATM modules (e.g., 150 Mb/s) is also under consideration.

4.9.4 BISDN Status

CCITT Study Group XVIII is working toward a detailed recommendation via the accelerated procedure. The status of agreements in the U.S. industry and the understanding of issues related to BISDN at the time of this writing were contained in References [4.28] and [4.29]. Thirteen BISDN standards were approved in November 1990. (Nine of these were reissued in 1992, as were five new ones.)

While the ATM avenue to BISDN is now certain, several issues related to the parameters and the underlaying transmission system for the UNI for connecting a user to the BISDN network remain to be resolved as of 1990. A single user interface, independent of the underlying technology, is desirable; this will facilitate the development of inexpensive network termination equipment, which in the end will determine the commercial viability of BISDN. This interface cannot be fully defined until the ATM parameters (layer 2) and the transmission parameters (layer 1) are settled because they both affect the usable capacity of the interfaces and, hence, dictate the service capabilities that can be provided.

At the time of this writing, a number of agreements on some of the ATM parameters had been reached, as follows [4.30]:

- ATM will consist of a 5-octet header and 48-octet information field. This is a compromise between the positions held previously by the two major technical groups working on BISDN, ECSA's T1S1 and ETSI. The former had argued for a header of 5 octets and for an information field of 64 octets; the latter argued for a header of 4 octets and for an information field of 32 octets (48 is a mathematically perfect compromise).

- The UNI header will include (see Figure 4.25): (1) a 4-bit generic flow control field, which may be used to achieve some of the functionality of an access control field for multiaccess interfaces; (2) a 24-bit label of which 8 to 12 bits can be used for the VPI, and the remaining 16 to 12 bits are for the VCI; (3) a 3-bit payload type field; (4) a 1-bit cell loss priority field; and (5) an 8-bit header error check field.
- At the physical layer, BISDN will use SONET. A variation, asynchronous time division (ATD) was favored at the time of this writing by some factions in the ETSI. The UNI speed has been set at 155.520 Mb/s for both approaches, with an interface transfer capacity for carrying cells containing user information set at 149.760 Mb/s. Thus, the information payload capacity in 125 ms is 2346 octets, and the total information payload provided to the user is 135.979 Mb/s.

After the ATM parameters are defined, and the transmission technology is selected, emphasis will shift in the direction of upper layer capabilities, including adaptation of higher layer functions to ATM, new service-specific functions including multimedia, and signaling [4.30]. The goal is to complete the necessary standards by early 1993.

ATM experiments are planned in Belgium, with participation from various carriers and manufacturers. The experiment will last for five years, and the goal is to deploy a complete system including terminals and transmission equipment to assess the viability of the technology [4.31]. ATM multiplexers and switches are less dependent on considerations of the bandwidths involved in particular services. In theory, the label is the only field that is essential in the ATM header to perform switching and multiplexing functions. As of mid-1992, many switch vendors had announced products to support high-speed cell relay (i.e., for transmission and switching of ATM cells across a private or public network).

Comparison of this layout with the layout of the proposed MAN standard (Chapter 15) indicates that although strong similarities exist, there are also non-trivial differences, making interworking a challenging task.

4.9.5 Relationship with SONET

SONET, described in Chapter 3, is a physical layer (layer 1) standard in its frame definition, just as the superframe and the extended superframe of DS1 is a physical specification. SONET is oblivious to what goes on beyond a single physical link, so it does not take an end-to-end routing view (although it allows straight-through connections to compose an end-to-end physical link). SONET is also oblivious to the content of the information bit stream.

ATM, as discussed, has a provision for virtual circuit indication, allowing end-to-end routing. In X.25 packet switching, this routing function is a layer 3 capability. In general, data link layer routing is also allowed by layer 2 protocols, although this is not a commonly implemented option in WANs. (It is, however, common in LANs, and is the basis for FRS.) ATM does aim at establishing an end-to-end path over which the bits can travel from origination to destination, which is intrinsic with the connection-oriented nature of virtual circuits and traditional telecommunication carrier philosophy. The routing capability of ATM can be interpreted as a network layer capability or a link layer capability. Some people do think of ATM as another version of fast-packet switching, however, the community is interested in trying to keep ATM as physical and low level as possible, suggesting a layer 2 interpretation of the ATM functions [4.32]. This interpretation implies that ATM can use SONET, at the lower layer, as a physical conduit. See Reference [4.30] for a more detailed discussion. SONET is a synchronous standard in the sense that the position of the information within the frame determines who owns that information. ATM is asynchronous in the sense that the position of the information does not establish ownership; a header field is added to each block of information to identify who owns the data in the block. A more familiar analogy is with a TDM and a statistical multiplexer.

REFERENCES

[4.1] "The Growing Importance of ISDN," TE&M, March 15, 1989, pp. 53–57. (Based on "Growing Importance of ISDN," a market research study conducted by Market Intelligence Research Co., Mountain View, CA).

[4.2] "ISDN Access Call Control Switching and Signaling Requirements," TR-TSY-000268, Issue 3, May 1989.

[4.3] "American National Standard for Telecommunications—Integrated Services Digital Network (ISDN)—Basic Access Interface for Use on Metallic Loops for Application on the Network Side of the NT (Layer 1 Specification), ANSI, New York.

[4.4] CCITT Recommendation I.122, Blue Book, 1988–89.

[4.5] CCITT Recommendation I.320, Blue Book, 1988–89.

[4.6] "ISDN D-Channel Exchange Access Signaling and Switching Requirements (Layer 2)," TR-TSY-000793, Bellcore, October 1988.

[4.7] ANSI T1.602-1989, "ISDN Data-Link Layer Signaling Specification for Application at the User-Network Interface, ANSI, New York, 1989.

[4.8] Integrated Services Digital Network, CCITT Blue Book, Fascicles III.7, III.8, and III.9, 1989.

[4.9] "User-to-User Signaling with Call Control," TR-TSY-000845, Bellcore, December 1988.

[4.10] Jaap Van Till, "The A-ISDN Proposal to Bridge 'Personal Computers' and 'ISDN'," *Computer Networks and ISDN*, 1989, Vol. 17, 149–152.

[4.11] D.H. Pennoyer, "Automatic Number Identification," *Bell System Technical Journal*, September 1958, pp. 1295 ff.

[4.12] R. Ephraim, personal communication, Bellcore, April 1990.

[4.13] D. Norem, "NIU Forum—An Overview," COMNET 90, Washington, D.C., February 6–8, 1990.

[4.14] Memorandum for NIU Forum members, cooperative R&D agreement, April 27, 1990.

[4.15] J.W. Lechleider, "Line Codes for Digital Subscriber Lines," *IEEE Communications Magazine*, September 1989, pp. 25 ff.

[4.16] R. Winton, "The Chip Manufacturer's View of ISDN," *Business Communications Review*, July 1989.

[4.17] P. R. Strauss, "An Update on U.S. Services and Standards," *Data Communications*, May 1989, pp. 129 ff.

[4.18] W. Stallings, *Handbook of Computer-Communications Standards*, Macmillan Publishing Company, New York, 1987.

[4.19] CCITT Recommendation Q.931, Blue Book, 1988–89.

[4.20] Wai Sum Lai, "Network and Nodal Architecture for the Interworking Between Frame Relaying Services," *Computer Communication Review*, January 1989.

[4.21] G. Tom, personal communications, Bellcore, March 1990.

[4.22] *Intelligent Networks News*, March 1990, Telecom Publishing Group, Alexandria, VA.

[4.23] *The Added Dimension*, Vol. 1, No. 5, April 1990, MAK Associates, Ridgefield, CT.

[4.24] D. Bushaus, P. Travis, "Users Find ISDN Pays Off," Telephony's ISDN Supplement, February 1990.

[4.25] "ISDN Vendors Must Court Data Users," *Telephone News*, March 19, 1990 (based on a Phillips Telecommunications report).

[4.26] R. Handel, "Evolution of ISDN Towards Broadband ISDN," *IEEE Network*, January 1989.

[4.27] J. Gecher and P. O'Reilly, "Conceptual Issues for ATM," *IEEE Network*, January 1989.

[4.28] Lin, Spears, Yin, "Fiber-based Local Access Networks," *IEEE Communications Magazine*, October 1989.

[4.29] T1S1 Broadband Baseline Document, Ed. R. Sinha, Bellcore, November 1989.

[4.30] S.E. Minzer, "Broadband ISDN and Asynchronous Transfer Mode," *IEEE Communications Magazine*, September 1989, pp. 17–24.

[4.31] M.D. Prycker, "Evolution from ISDN to BISDN: A Logical Step Toward ATM," *Computer Communications*, June 1989, Vol. 12, No. 3.

[4.32] G. Geer, personal communication, Bell Communications Research, September 1989.

Chapter 5

Transmission Media and Radio Services

5.1 INTRODUCTION

The demand for mobile voice and data communication has risen dramatically in the past decade. A large consumer market is emerging not only because of the demand for mobile phones for cars and boats, but also because of the demand for portable devices that can be carried in a briefcase or purse. More than 8% of the residents of the top 25 U.S. cities had a cellular phone in 1990, and in Chicago, Detroit, Washington, D.C., and Boston, that figure is 11% [5.1]. In 1989, more than 570 cellular telephone systems were operating in 340 markets; the capital investment in cellular systems in the United States at that time was $4.5 billion. In 1990, three million cellular subscribers existed [5.2, 5.3, 5.4]. This technology has a large market potential if the monthly charges can be lowered to a reasonable level. Estimates allow that if the charges decrease from the present $120–$150 to $30–$50 per month, 10% of all U.S. households would subscribe to the service. Industry analysts forecast that by the year 2000, 20% of the telephones in the United States will be cellular or wireless [5.5].

Passengers on domestic flights and trains can now call the office or home; estimates show that 15,000 aircraft are equipped with airborne telephones [5.6]. Satellites provide communication to ships and offshore oil platforms in the form of telexes, voice, compressed video, and data. In addition to this international mobile service, efforts are under way to deploy domestic satellite systems to send and receive signals from trucks, trains, and boats [5.7]. Two-way radio systems for aviation, marine, public safety, industrial, transportation, and amateur applications grew from 1.5 million units in 1979 to 2.5 million in 1984, and more since.

The United States currently has more than 107 million workers. While 50% of them work in offices, according to the U.S. Census Bureau, the other 50% work outside an office. This group of "office-less" workers may benefit greatly from the introduction of portable communication, leading to improvements in productivity. For example, a rented car can now be returned without having to go to the company's desk; the company's clerk transmits the pertinent information to the mainframe computer over a radio link. More than 500,000 laptop computers were in use in 1988, with a projected growth to 15 million units by 1992. Office workers can also benefit from portable communication.

This chapter assesses some of the available radio technologies, in particular as applied to data communication; a practical outlook is followed in this chapter. Figure 5.1 depicts the areas of applicability of the technology in a network context. The systems discussed support relatively low data rates and can be utilized for inquiry and response applications. For large throughput applications (graphics, bulk-file transfers, *et cetera*), terrestrial-based systems, fiber in particular, may be preferable.

5.1.1 The Electromagnetic Field

The electromagnetic environment that surrounds us, in free space or within a physical medium, is the mechanism through which communication takes place. Figure 5.2 depicts the electromagnetic spectrum, parameterized on the underlying primitive of frequency. The radio section covers from around 0.1 MHz to about 100 GHz (the correct pronunciation is as in "gigantic," not "guilty"). This section includes long-wave radio, commercial radio, TV, radar, and microwave and satellites.

Electromagnetic signals may be propagated in three ways. First, they may be passed through an electrical conductor. This mode of transmission, however, becomes impractical at higher frequencies because of the tendency of the signal to migrate to the outer surface of the conductor, a condition known as "skin effect." The second transmission mode is via a channel called a *waveguide*, which allows the signal to be radiated into a confined space, then guided by the walls of the physical medium. Both of these approaches are examples of *guided transmission*. The third transmission mode is to radiate the signal into the atmosphere or into space. This is called *unguided transmission*. Much greater control can be exerted over guided systems compared to unguided "over-the-air" radio systems. Transmission over copper twisted-pair is representative of the first mode; fiber and coaxial cable are representative of the second mode; microwave and free-space infrared transmission are representative of the third mode. Radio-based systems do not require physical rights-of-way. Such systems are especially effective for communicating over barriers such as bodies of water or mountainous terrain, and for communicating when the endpoint or points are mobile. The radio spectrum is highly regulated, not only by the FCC domestically, but also by international organizations, including the International Frequency Registration Board (IFRB) and CCIR, discussed in chapter 2.

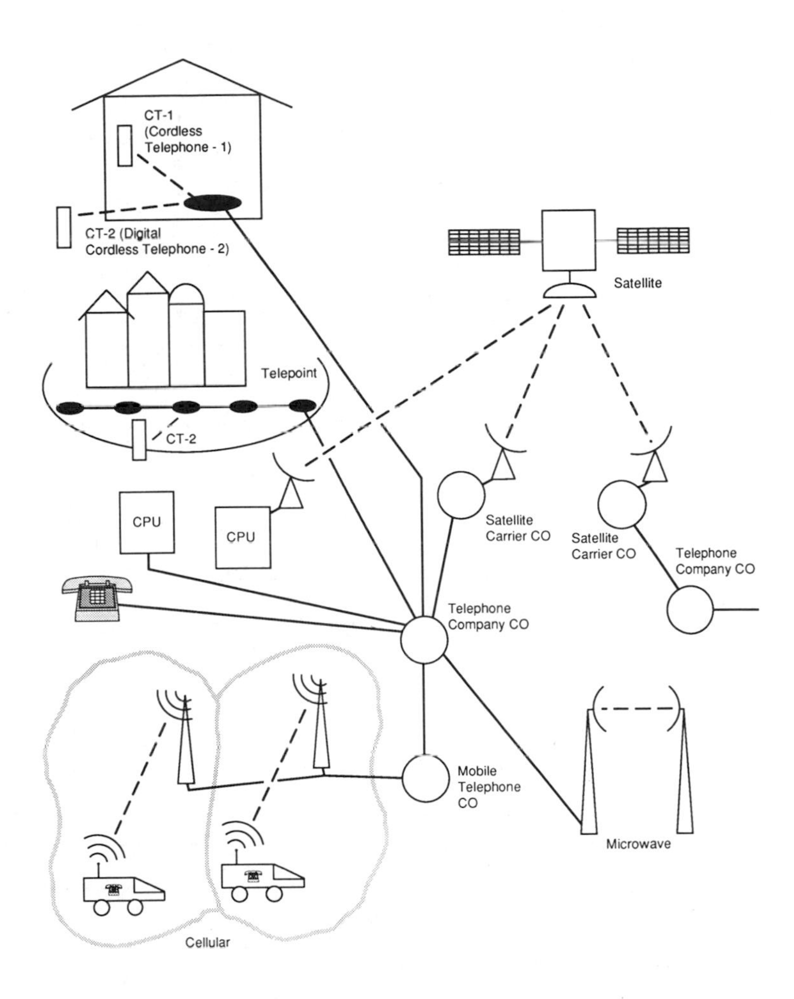

Figure 5.1 Some examples of radio-based telecommunication.

5.1.2 Examples of Wireless Transmission Systems

Wireless communication can take place in (1) the radio band and in (2) the infrared-optical band. Infrared uses the same type of light sources and detectors as fiber optic components. Similar to microwave technology, an infrared transmitter-receiver pair is mounted in a stable fashion to maintain alignment. By far, radio is the most versatile technology of the two, particularly because at low frequency we do not need line of sight (LOS); we can communicate across the world by bouncing

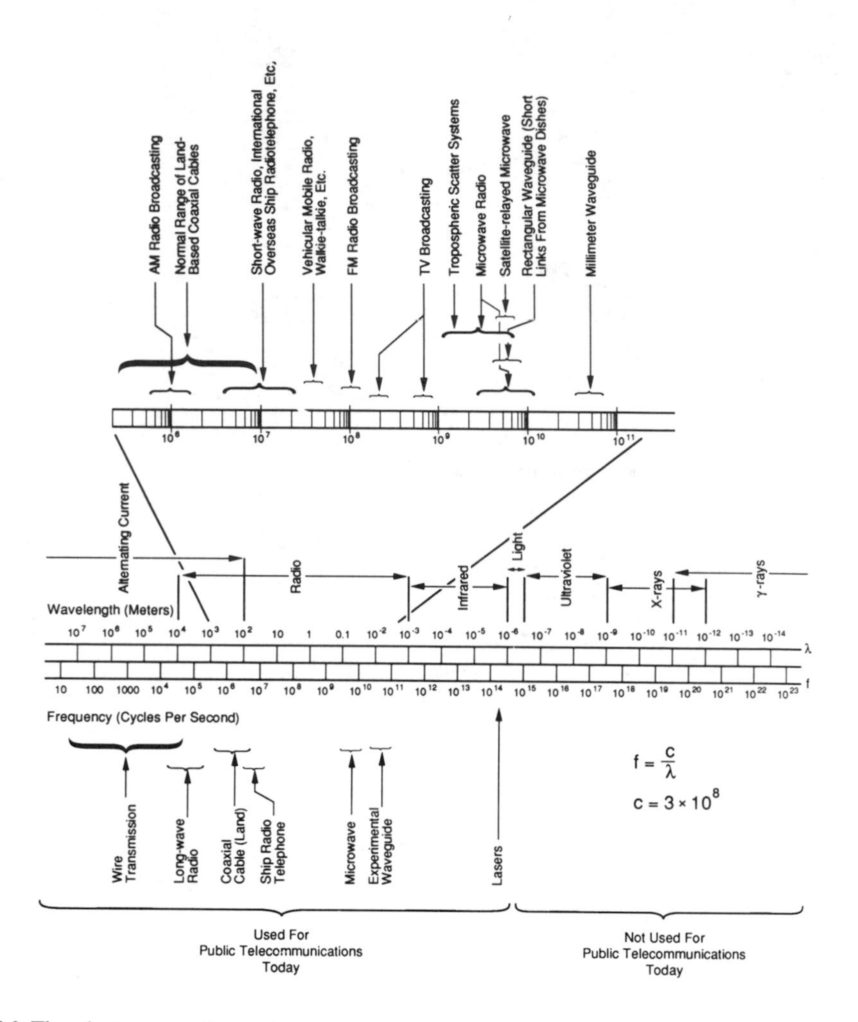

Figure 5.2 The electromagnetic spectrum.

off the radio waves on the ionosphere and troposphere. At higher frequencies (FM, TV, and microwave), LOS is required. In addition to the LOS constraint, higher frequencies are more affected by atmospheric conditions, such as rain or snow.

The objective of high-frequency wireless communication systems is to impress information on a signal transmitter at a given location, to send that signal through the air to a receiving location, and to detect and isolate the original information, so that it can be passed to a receiving user. The transmitter must generate the microwave or infrared-optical signal and direct it to an antenna system or optical source that concentrates the power in the direction of the user; this concentration reduces the vulnerability of the system to interference or obstruction. A similar signal concentration at the receiver side, through a directional antenna or an optical collector with a small field of view (FOV), helps mask out competing signals and other forms of interference.

5.1.3 Typical Throughput

To broadcast information, it must be superimposed on a carrier signal, microwave, or lightwave. This process is called "modulation," as discussed in Chapter 1. The total range of frequencies used for transmission is called the information channel's *bandwidth*. The bandwidth of a channel determines its information-carrying capacity, as described by Shannon's equation of Chapter 1. The practical bandwidth also depends on the frequency of the carrier. A bandwidth of 100 MHz on a carrier frequency of 1000 MHz (1 GHz) would be a 10% variation on the carrier—a wide range for receiver technology to handle. The same 100 MHz would be only 1% at 10 GHz (the Ku band of satellite communication), and a minuscule 0.00003% in the infrared range (about 330,000 GHz). Clearly, providing high bandwidth at high carrier frequencies is easier. Light has a higher frequency than microwave, giving it the ability to provide large bandwidths.

Digital bandwidths for atmospheric infrared systems can be as high as 45 Mb/s; bandwidth on microwave systems can be as high as 45 Mb/s for digital systems and 25 MHz for analog systems, though the combined bandwidth of the basic band can be higher (for example, 500 MHz for satellite links, as discussed in the next chapter). Figure 5.3 depicts the bandwidth characteristics of several transmission media [5.8].

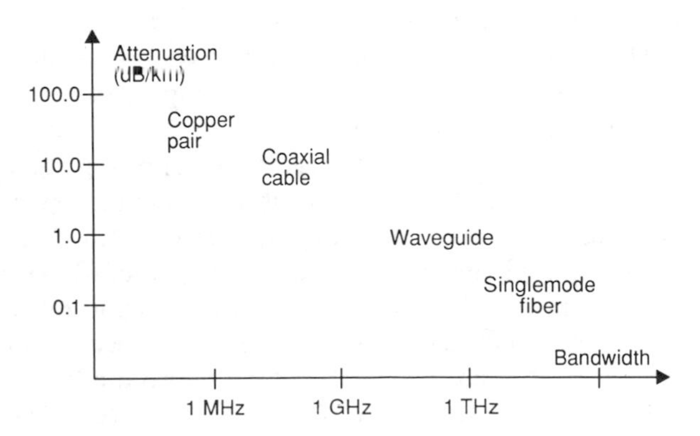

Figure 5.3 Bandwidth characteristics of some common media.

5.1.4 Transmission Technologies

Today, fiber is the workhorse of domestic telecommunication, followed in importance by microwave; satellite is a remote third place, and radio and atmospheric infrared are "niche" technologies. The following three chapters discuss in some detail the following transmission systems: satellite and microwave (Chapter 6), fiber (Chapter 7), and atmospheric infrared (Chapter 8). This chapter covers radio bands, followed by a survey of a number of radio technologies available for use, followed by a more detailed discussion of three specific radio technologies: cellular, packet, and FM subcarrier.

5.2 RADIO TRANSMISSION CHANNEL

The received signal power of unguided radio-based transmissions is a function of transmitter power, antenna patterns, path length, physical obstructions, and several other factors. The received signal is also affected by diverse factors such as meteorological and even extraterrestrial phenomena. Use of theoretical results must be supplemented by observations; this is in direct contrast to cable-based systems where the theory is much more useful in predicting the transmission characteristics exactly, and the outside effects are generally less important.

The following two subsections describe the characteristics of the frequency bands below 3 GHz and those above 3 GHz. See Table 5.1 for a brief summary of these radio bands. Figure 5.4 depicts the modes of wave propagation; these modes are referred to in the discussion below and in Table 5.1.

5.2.1 Frequency Bands Below 3 GHz

At the low end of the spectrum, below 3 kHz, is the ELF (extremely low frequencies) band. This band is used by the military to communicate with submarines underwater. ELF, which the Navy started to develop in the 1960s, relies on the principle that low frequency waves can penetrate much further in seawater than high frequency waves. Conventional naval radio, transmitting at a few tens of kHz, reaches a depth of 30 feet. ELF ground-to-sub transmission operates between 72 and 80 Hz and can reach depths of several hundred feet. An ELF antenna, which transmits signals that travel first through the air and then down into the ocean, relies on rocks: to produce a detectable signal, the antenna must form a loop in the vertical plane several miles across. This requires an electrical current than runs along cables near the earth's surface and dives deep into the bedrock to create the rest of the loop. The bedrock needs high electrical resistance to achieve the strongest signal possible; these types of rock are found mostly in Wisconsin and Michigan. The antennas are inherently inefficient: 1.4 MW of input power produces ELF

Table 5.1
Summary of Internationally Based Radio-Frequency Bands

Frequency Band	Frequency Range	Propagation Modes	Systems/Uses/ Characteristics
ELF (Extremely Low Frequency)	Less than 3 KHz	Surface wave	Worldwide, military and submarine communication
VLF (Very Low Frequency)	3-30 kHz	Earth-ionosphere guided	Worldwide, military and navigation
LF (Low Frequency)	30-300 kHz	Surface wave	Stable signal, distances up to 1500 km
MF (Medium Frequency)	300 kHz - 3 MHz	Surface/sky wave for short/long distances, respectively	Radio broadcasting. Long distance sky-wave signals are subject to fading
HF (High Frequency)	3-30 MHz	Sky wave, but very limited, short-distance ground wave also	3-6 MHz: Continental; 6-30 MHz: Intercontinental. Land and ship-to-shore communications
VHF (Very High Frequency)	30-300 MHz	Space wave	Close to line-of-sight over short distances. Broadcasting and land mobile
	30-60 MHz	Scatter wave	Ionospheric scatter over 900-2000 km distances
UHF (Ultra High Frequency)	300 MHz-3 GHz	Space wave	Essentially line-of-sight over short distances. Broadcasting and land mobile
	Above 300 MHz	Scatter wave	Tropospheric scatter over 150-800 km distances
SHF (Super High Frequency)	3-30 GHz	Space wave	The "workhorse" microwave band. Line-of-sight. Terrestrial and satellite relay links
EHF (Extremely High Frequency)	30-300 GHz	Space wave	Line-of-sight millimeter waves. Space-to-space links, military uses, and possible future use

radio waves with just 2 W of power. This, in turn, limits the transmission bandwidth to a few bits per second. For this reason, highly coded messages that are predefined are sent. For example, "RTO" might mean "Read *Telecommunication Technology Handbook* overnight." A new system for communicating with submarines is now being studied. This system is based on laser signals from airplanes or satellites [5.9].

In the VLF (very low frequency) band from 3 to 30 kHz, signals are propagated in a guided mode between the earth and the ionosphere. This mode of propagation produces stable signals and good coverage. However, only a limited and specialized use is made of this band because of the high cost of the large transmitting facilities required, the very limited bandwidths available, and the susceptibility of the signals to atmospheric noise disturbances. Consequently, the VLF band is used mainly for global broadcast and for the transmission of time signals.

The LF (low frequency) band, from 30 to 300 kHz, is the next highest frequency band and has many of the same characteristics as the VLF band. It is used primarily for the transmission of both medium-range navigation and time signals.

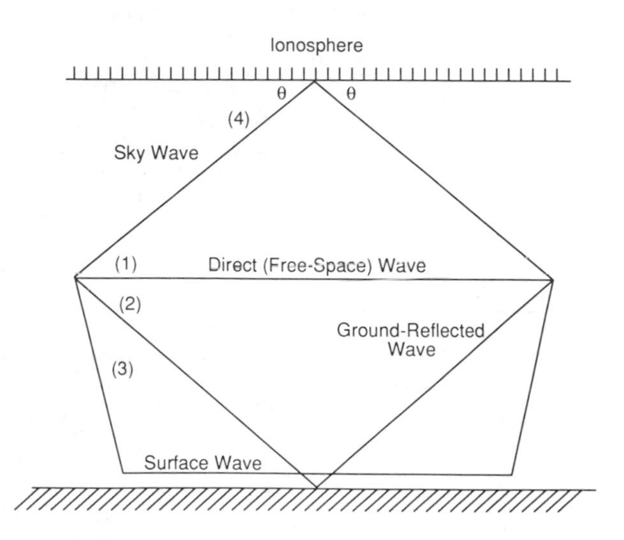

Figure 5.4 Transmission paths between two antennas.

The next band, the MF (medium frequency) band from 300 kHz to 3 MHz, contains the familiar AM commercial radio broadcast band. These signals are propagated via surface waves for relatively short distances, or by sky waves, reflected by the ionosphere, for much longer distances. In addition to AM broadcasting, this band has been used for intercontinental and ship-to-shore communication.

The HF (high frequency) band covers the spectrum from 3 to 30 MHz and is used extensively for shortwave broadcasting and other long-distance point-to-point intercontinental communication. The use of surface-wave transmissions is limited to relatively short distances, and nearly all communication is made via the sky-wave mode. Signals in this band are heavily dependent on ionospheric conditions. Low-data-rate teleprinter operations in both government and commercial applications are still found in this band. However, with the exception of international shortwave broadcasting, many of these uses have been shifted to satellite and other more reliable and wider bandwidth systems.

The VHF (very high frequency) band, ranging from 30 to 300 MHz, is used heavily for television broadcasting, FM broadcasting, cellular telephone transmission, aeronautical, and maritime mobile applications. Propagation in this electromagnetic region is confined to almost LOS distances ("direct wave"). Historically, data communication within this band has been very limited and it was only in 1983 that the FCC removed data from its secondary status. Increased use of data transmissions, at rates up to about 10 kb/s, is expected in land mobile and related applications.

The UHF (ultra high frequency) band ranges from 300 MHz to 3 GHz. Television broadcasting has a substantial allocation in the lower portion of this band, followed by land mobile applications. Most of the land mobile frequency allocations are within this band and below 1 GHz. The land mobile communication range varies substantially with frequency, power, antenna height, intervening terrain, buildings, foliage, *et cetera*, so that the useful range is generally under approximately 40 miles. The band above 1 GHz is used for a number of purposes, including radar.

5.2.2 Frequency Bands above 3 GHz

The SHF (super high frequency) band between 3 and 30 GHz is the real "workhorse" for radio systems in terms of traffic capacity. In the United States, microwave radio bands in the 4-, 6-,10-, 12-, 13-, 18-, 23-, and 28-GHz regions of the SHF range are available for point-to-point use by private or common-carrier systems. Generally speaking, the total frequency allocation in each of these bands expands with increasing frequency and goes as high as 2000 MHz in the 28-GHz band. These frequency bands are, in turn, subdivided into individually assigned channels whose bandwidths range from as low as 10 MHz to as high as 220 MHz.

Clear-path point-to-point communication is the norm in the SHF band. Frequencies in this range are generally unsuitable for terrestrial broadcasting or land mobile purposes because coverage to those points with obstructed paths is difficult to achieve. Received signal levels vary considerably due to atmospheric conditions, especially on terrestrial paths. Generally speaking, the severity of signal fading increases with increasing frequency and path length. Fading is caused by multipath and inverse path bending, which is produced by changes in the refractive index of the atmosphere. Multipath fading results from interference between a direct wave and another wave, usually a reflected wave from the ground or from atmospheric layers. Such fading can be in excess of 30 dB for periods that can be seconds or even minutes long. In addition, for frequencies above approximately 10 GHz, fading due to rain attenuation must be taken into account. Repeater spacings for microwave transmission systems vary from about 1 to 30 miles in order to deal with some of these problems. The shorter paths are generally required at the higher frequencies where multipath effects and rain attenuation are the greatest.

In the case of satellite-to-earth paths, high elevation angles achievable with most satellites and at most latitudes minimize path obstruction possibilities and atmospheric effects, including rain attenuation. Rain effects are less severe due to the shorter paths through the atmosphere. Rain attenuation can add noise temperature—up to the rain's temperature, say 290 K—to the receiving system noise temperature. Rain attenuation varies very little with frequency, so that in-band frequency diversity is not effective. Geographically separated ground stations can provide good space diversity improvements on earth-to-satellite paths because the heaviest rains tend to occur in relatively small geographic areas where simultaneous heavy attenuation at both locations is unlikely.

5.3 RADIO-BASED DATA COMMUNICATION SERVICES

A survey of commercially available radio telecommunication systems follows, starting with the most common applications.

5.3.1 Microwave Transmission

Microwave transmission takes place in the 3- to 20-GHz range. It has been used for decades to carry telephone and data traffic across the United States and the rest of the world. Microwave transmission is a LOS technology. As indicated, repeater spacings (towers) are generally 1 to 30 miles apart (and often 20 to 30 miles), depending on topography, frequency, *et cetera*. The 30 miles is dictated by the curvature of the earth and by the attenuation effects of the atmosphere. Coaxial-based systems require repeaters every 1 to 4 miles; in a 2000-mile trunk, this would require 500 to 1000 signal manipulations, which introduce distortions. Optimally, microwave only requires only 60 regenerations.

Microwave antennas are directional. The received signal at one tower is reamplified and retransmitted on a new frequency to the next tower in the chain. In addition to being affected by atmospheric conditions, the spectrum is very congested, particularly in metropolitan areas. Microwave is the most popular type of "bypass." An FCC licence is required to operate standard microwave transmission equipment. Microwave technology is discussed further in Chapter 6.

5.3.2 Satellite Transmission

Satellite transmission is a type of microwave communication. Here the "first tower" is the sending earth station; the "intermediary tower" is the satellite amplifier; the "last tower" is the receiving earth station. The satellite can be 22,300 miles in space (where it remains fixed relative to a point on the ground, even though the satellite is moving through space at 7500 miles per hour) without signal attenuation because the radio ray only has to traverse the atmosphere perpendicularly for 4 to 6 miles. The notorious 250-ms delay arises because of the propagation time of the energy through space (186,000 miles per second, or 44,640 miles in 0.238 seconds).

Modern commercial satellite communication is confined to two major bands: C and Ku. The C band is at 3.7 GHz for the downlink and 5.9 GHz for the uplink (as noted above, the output frequency from any "tower" is not the same as the input frequency). The military band is at 7.2 GHz for the downlink and 7.9 GHz for the uplink. These frequencies, however, are congested because they are also being used by the terrestrial microwave; therefore, two new bands are becoming prevalent (as technology permits): the Ku and Ka bands. Ku is around 12 to 14 GHz, and Ka is around 18 to 30 GHz. Most early satellite transmission used the C band, a frequency at which human-made technological and astronomical noise sources were less damaging, and the technology for communication was available. However, two major problems with C-band transmission exist: (1) microwave services cause interference with the satellite link and (2) the antenna size required to provide narrow uplink beams to support closely spaced satellites is large.

Other bands include the L band (1.53 to 2.7 GHz), the S band (2.5 to 2.7 GHz), and the X band (7.25 to 7.75 and 7.9 to 8.4 GHz) [5.10].

In general, higher frequencies are more directional in propagation, but are subject to significant transmission loss, and may be reflected from nearby objects. At the Ku band, atmospheric conditions exhibit a great effect on the error performance. Additionally, degradation at the Ku band occurs from water vapor in the atmosphere. This problem can be overcome by higher power transmission, by diversity techniques via alternative routes or frequencies, and by forward error correction (FER) coding techniques combined with diversity. For users of small antenna systems, the higher bands have some advantages. At C band, small terminal use is generally limited to receive-only applications; these are 2 to 4 feet in diameter for data and 10 to 15 feet in diameter for video. At Ku band, an interactive terminal becomes feasible for 4- to 6-ft antennas. In the long term, satellite communication is likely to move increasingly to higher frequencies because existing terrestrial applications limit the range of C-band transmissions that can be used for satellite communication. The C and Ku bands are allocated a total of 500 MHz of bandwidth shared by 10 to 16 transponders. For each individual transponder, 36 MHz is typical for C-band operation. At Ku band, 85 MHz of bandwidth per transponder is common.

Satellites have been used to transmit both voice and data. Channels have traditionally been provided on a point-to-point private line basis, but new methodologies using random access are beginning to be implemented. In particular, satellites that can broadcast directly to 2-ft dishes (one-way service), and 4-ft dishes (two-way service) are emerging. These are the so-called *very small aperture terminals* (VSATs).

Satellite technology is discussed further in Chapter 6.

5.3.3 Cellular Radio Transmission

Cellular radio is a technique that provides mobile telephony by using the allotted radio band in an efficient manner. Cellular telephone service involves the subdivision of a service area into a number of smaller cells to facilitate frequency-reuse and the ensuing increase of the number of potential users. This is an example of SDM, described in Chapter 1. The difference compared to the older "radiotelephone" is that microcomputer-based switches can "follow" a moving object by way of power measurements from the numerous antennas that constitute the cellular system and change the frequency according to the object's cell. Each cell requires an antenna operating in the UHF band (the same one used for television for the higher channels), at a power output just sufficient to cover the area of the cell. The antennas are connected by landlines to a mobile telephone central office (MTCO), which, in addition to the switching function, has the job of "tracking" the users. When the user (car or personal portable phone) moves from one cell to another, the MTCO must hand off a call in progress from the original antenna to another appropriate antenna. This handoff can also occur with some switches even if the user is actually stationary; this is done to achieve load sharing among the various towers. This system handoff is caused by peaks in traffic: As the switch approaches the maximum threshold, it will hand off a number of cellular conversations to an adjacent cell, regardless of the user's status. Cellular telephones are progressing toward a second generation in terms of size and features [5.11].

Currently, cellular telephony is based on analog transmission, but second-generation digital systems will become available in the United States beginning in 1992. Digital cellular allows the best quality with the least amount of bandwidth, while improving frequency reuse. A digital system in a city that currently has 200,000 to 300,000 analog subscribers will eventually be able to support two to three million digital subscribers [5.12]. Digital cellular phones were being tried out in 1990 and the trials were expected to continue into 1991. Commercial service should become available toward the end of 1991 or the beginning of 1992 [5.12]. According to one approach, the bandwidth offered to the customer would be reduced from 30 kHz to the equivalent of 10 kHz in the early digital systems (early 1990s), and then to the equivalent of a 5-kHz channel (late 1990s). This will be done initially through a three-time-slot TDMA approach similar to the one used in Japan (i.e., three pairs of people will be able to use a 30-kHz radio channel simultaneously). The second phase will provide an additional doubling of the number of users [5.12]. While the majority of the industry has adopted in principle the TDMA standard for digital cellular, another approach, code-division multiple access (CDMA), may emerge as a competitor [5.13]. This more sophisticated method, based on spread spectrum techniques (discussed in Chapter 6), can supposedly provide a three- to fourfold improvement in capacity over TDMA. The FCC is looking favorably toward CDMA. Two wireline carriers (NYNEX and Pacific Telesis) were trying CDMA systems in 1990.

In addition to the digital transmission standards, other standards such as subscriber unit performance standards and base station performance standards were being developed at the time of this writing. The move to digital will be a slow and carefully planned effort. Analog channels may be converted to digital one at a time to minimize the disruption to the thousands of users with embedded analog equipment. Both analog and digital use will grow simultaneously because digital systems will at first be confined to major metropolitan areas where the need for spectrum relief is the greatest. The same cell structure will be kept, and the digital equipment will be installed next to the analog. When a digital customer presents a request for service, the system will recognize that the user has a digital mobile terminal and will assign a digital channel. If a digital channel is not available, then an analog channel will be assigned (the new mobile telephones will be hybrids, able to handle both signals).

Currently, cellular telephones can only be used when they are sufficiently close to cellular antennas; they cannot be used in many rural areas in the United States or outside major cities in the rest of the world. In 1990 Motorola, Inc., announced that it was planning (subject to funding availability and regulatory approval) to build by 1994 a global mobile telephone system, based on low altitude, low weight satellites. The project, named "Iridium" after the 77-electron element iridium, requires at least 77 satellites to ensure that one satellite will always be within LOS. Each satellite will cover 37 cells on the ground; each cell will have a radius of 400 miles and will be able to support more than 330 simultaneous conversations with existing frequency allocations. The project was estimated to cost $2.5 billion. The telephone sets were planned to cost $3500 and the service to range from $1 to $3 per minute of usage. Up to five million users could subscribe to such service by the year 2000.

Cellular radio has been used mostly for voice, but it is also being used to carry data from car-based terminals. The issue is one of cost: The average per minute charge for a channel is around $0.50, implying $30 an hour or $250 a day. Some attempts are under way to allocate one channel that can be accessed for data use in a statistical or multiplexed mode, rather than a circuit switched mode. Cellular technology is discussed in more detail later in this chapter.

5.3.4 Packet Radio Transmission

Packet radio networks represent an extension of packet switching technology into the environment of mobile radio. Packet radio is a powerful technique developed for the military in the late 1960s and 1970s. It used hand-held devices to allow secure and reliable communication under unpredictable and hostile environments [5.13 through 5.24]. These networks are intended to provide data communication to users located over a broad geographic region, where wire connection between the source and the destination is not practical or cost-effective [5.25].

Packet-switching networks, first developed in the 1960s, have become widespread in commercial, military, and academic environments. To make the advantages of packet switching available to mobile users (particularly important to the military, where mobility is key), the Defense Advanced Research Projects Agency (DARPA) initiated in 1972 a research effort to develop a radio-based system [5.25, 5.26]. The effort continued into the early 1980s. This system was to provide an efficient means of sharing the broadcast radio channel while at the same time allowing for the incomplete and changing connectivity that is typical of a mobile environment. Development of this technology involved a large number of technical issues, including integration of high-rate modems and powerful microprocessors into a small low-cost unit, channel sharing protocols, and algorithms for routing packets in a highly dynamic network. The packet radio concept as we know it today was invented by R. E. Kahn, when he came to DARPA in 1972. The earlier packet radio system developed by Abramson at the University of Hawaii was a single-channel system (for the inbound channel); the outbound channel had a single user, the mainframe. In the system designed by Kahn, the single shared channel was used for all traffic.

As the interest in packet radio networks grew, research and development programs were initiated to apply the technology to a number of environments. Packet radio networks now make use of ground and airborne mobile radios operating from 16 to 400 kb/s, amateur-radio packet radio networks, Navy, and now commercial satellite communication. A packet radio network consists of a number of packet radios that use radio to integrate the units into a network.

A packet radio consists of a radio, an antenna, and a digital controller. The radio provides connectivity to a number of neighboring radios, but typically is not in direct connectivity with all radios in the network. Therefore, the controller needs to provide for store-and-forward operation, relaying packets in order to accomplish connectivity between the originating and destination users. The simpler satellite networks involving the use of a single satellite do not include store-and-forward operation [5.25]. Packet radio is intended to support mobile users, although some of the amateur, commercial, and satellite applications deemphasize this capability.

The concepts developed for packet radio eventually were applied to LANs. This technology is discussed in more detail later in this chapter.

5.3.5 FM Carrier Transmission

In the 40- to 60-MHz range, FM techniques can be used to transmit wireless digital data for commercial, industrial, medical, and military applications. The technique used is FM carrier modulation, resulting in up to 19.2 kb/s to 0.5 miles (6-ft antennas are required). Typical applications include LANs, robotics, telemetry (particularly from hostile environments), emergency, and transient applications (drilling, construction, *et cetera*). Because of the power limitations imposed by the FCC, the actual applications and market deployment are small.

5.3.6 FM Subcarrier Transmission

FM subcarrier is a relatively new technique for data transmission that allows commercial FM stations in the 88- to 108-MHz region to use a portion of the allocated but unused bandwidth for the distribution of data. The mode of operation is simplex (one direction only), but the appearance of duplex transmission is achieved by a microprocessor built into the user terminal, which selects incoming data specified by the user at the keyboard. Delivery of stock quotations and other financial market data is the leading application of subcarrier data technology and will continue to be so for some time. FM subcarrier is labeled the low-cost solution to the provision of near-real-time information. This technology is discussed in more detail later in this chapter.

5.3.7 Teletext Transmission

Teletext techniques employ an unused portion of the TV bandwidth (the blanking interval) to broadcast data. Some applications are the transmission of captions for the hearing impaired and the transmission of programming schedules between the TV affiliates. A typical application of this blanking-interval bandwidth has been the transmission of graphical information directly to the TV set.

5.3.8 TV-Embedded FM Subcarrier Transmission

This is a technique that differs somewhat from both the FM subcarrier and the teletext technologies, and yet it is related to them. It involves the use of a spectral portion of a commercial TV channel to carry a secondary signal (this could be a stereo or other type of signal). Typical equipment allows audio, data (up to 19.2 kb/s), or voice to be carried on the same microwave baseband as a video signal, particularly in a satellite environment. When used for TV-associated audio, this subcarrier technique eliminates delay distortion and lip-sync problems that occur when video and audio are transmitted separately, as is the case in some satellite systems. When used for nonvideo, auxiliary services can be transmitted over the same route and facilities employed for the video portion; for example, this technique can be added to CATV earth stations to receive data through the existing satellite transponder.

5.3.9 Pagers

One-way paging techniques provide low-cost alerting to a mobile population. Local radio paging has been offered by radio common carriers and telephone carriers (known in this context as *wireline common carriers*) since about 1950. Local and regional paging systems operate at 150 and 450 MHz. (More than 10,000 paging frequencies are in use worldwide; some pagers will scan this plethora of frequencies and retrieve information addressed to it.) The four types of pagers are tone only, tone and voice, numeric display, and alphanumeric display (available since the mid-1980s). Paging is the precursor of FM subcarrier and may, in fact, use it as the means of distribution. In the late 1980s, 5.5 million pagers were in operation on private and common carrier systems. By the end of 1990, 10 million units were projected to be in use.

The early tone-only pagers transmitted only a beep. This requires a telephone-based callback for more information because no provision exists for delivering messages. The tone and voice pager can receive a 10- to 20-s voice message over the pager's speaker. Callers speak the message into their telephone sets; the message is then sent through terrestrial links and eventually radio links to the paged terminal. Modern pagers also receive a small message (20 to 40 characters) and store up to 10,000 characters [5.27]; these pagers are a relatively simple method of sending E-mail to people on the move. Hence, pagers can function as a pocket mailbox, capable of receiving messages and obviating the need to call the bureau, facilitating faster decision-making. A sophisticated pager provides tone, voice, numeric, or alphanumeric display paging, or any combination. It also offers hard-copy messages (printer is optional). Alphanumeric pagers are more expensive than tone-only or tone-and-voice pagers. Some tone-only pagers can be bought for under $100, while alphanumeric pagers cost more than $300.

Of the two nationwide paging systems, one operates in a 900-MHz land-mobile channel; the other utilizes FM subcarriers at 88 to 108 MHz. The 900-MHz system is the newer of the two. Ten thousand nationwide users of pagers existed in 1988 [5.28]. International, and even global, paging is under discussion. INMARSAT technology is contemplated (one million subscribers are projected for the year 2000) [5.29].

5.3.10 Radiodetermination

Radiodetermination is the positioning of an object (a boat, a car, *et cetera*) from digital pulses sent from a satellite. The maritime and military sectors have had this service for a number of years; continental U.S. (CONUS) service is becoming available [5.30, 5.31]. Accuracy at better than 50 m for commercial applications is achievable.

5.3.11 Digital Cordless Telephones

Digital cordless telephones are now becoming available. These second-generation cordless phones, known as *CT-2*, may find applications in the office environment, as a wireless PBX, and then advance to the residential phone, as costs decrease. Compared to cellular technology, CT-2 is cheaper in infrastructure, equipment, and air time costs. CT-2 can support 5000 users per square kilometer compared to 25 users for cellular technology. Using 100 MHz of bandwidth in a FDMA arrangement, each telepoint site (i.e., public space outside the office or home) can serve up to 100,000 units. The technology is being introduced in Europe. The average handset cost is about $200 compared to $500 for the least expensive portable cellular telephone; the air time is less than one-third the charges for cellular service (in the United Kingdom). Issues pertaining to standards and regulation may preclude early deployment in the United States. In addition, spectrum allocation is required. Some proponents have advocated using bandwidth allocated for pagers or for the experimental air-to-ground telephone system available on commercial aircraft.

CT-2 technology works as a one-way telephone system, somewhat similar to cellular in its use of radio frequencies, but it does not incorporate a cell site structure to allow for roaming. Users carry personal handsets and employ radio base stations for a range of 150 to 600 feet. Calls are routed through the local exchange networks, and billing is charged according to air time usage, as for cellular [5.32]. In the United Kingdom, the Ministry of Trade and Industry granted licenses in 1989 to four consortia for the establishment of CT-2 systems; one of the systems was slated for operation as of January 1990.

The handsets can be activated when they enter the telepoint site, scattered with base stations. After entering their personal identification number, the caller can use the phone within the range of the base station. Users can also purchase base stations for the home; in 1990, however, the cost of these base stations was higher than that of first-generation cordless phones introduced in the early 1980s.

A limitation of the system is lack of incoming call capability [5.4]. In addition, one cannot use the system while driving a car. Because the public phone system is not as reliable and ubiquitous in Europe as it is in the United States, and because cellular hardware is twice as expensive, CT-2 will probably become popular there. As of 1990, uncertainty existed as to whether the U.S. market would support the introduction of this technology. The introduction of TDMA digital cellular in 1992 should bring the cost of cellular down in the United States, further limiting the market pull for this technology.

5.4 DATA COMMUNICATION OVER RADIO CHANNELS

This section analyzes three radio data services in some additional detail.

5.4.1 Motivation

The market for mobile data communication has grown substantially in the past five years, tracking the trend in mobile services. The increased mobility of the work force, along with the need to remain in touch with a centralized computer, has created new opportunities for wireless data communication. Transmission of data over FM radio channels, which have been used for voice applications for quite some time, has become more common with the growth of paging, cellular, and packet technologies. Radios are now being used in a number of situations as alternatives to copper and coaxial cable to connect computers for data transmission purposes, particularly in mobile environments. The advantages of this technology include mobility, lower installation cost, and transmission over a large area.

Mobile data systems are being used by field personnel in many U.S. companies, allowing them to complete work orders in the field as soon as the job is finished, and to receive their next assignment without having to return to the office or find a public phone [5.33]. On the other hand, radio systems have some limitations that must be addressed and handled. Some of these limitations include the following: susceptibility of radio systems to noise from machines and machinery (computers, automobiles, elevators, power lines, *et cetera*), and from atmospheric conditions; bounded distances over which the signal can travel, making the systems applicable to MANs, but not directly WANs (unless repeaters are used); and finally the issue of security. Well-designed radio systems manage all of these problems in a satisfactory way.

A number of large corporations have built private nationwide mobile radio data systems. Some of these applications use data over cellular systems, others employ private systems. The decision to do one or the other is based on size: A small locality requiring 10 to 20 terminals may find a cellular system to be the most cost-effective solution; a locality requiring hundreds of terminals may require a separate system. Small users not wanting or able to afford a private system may use other techniques such as data over cellular or public packet radio. The major types of mobile data systems are cellular telephone, direct data on dispatch radio, packet radio, and FM subcarrier. Each of these is examined below.

5.4.2 Data Transmission over Cellular Links

The transmission of data over cellular networks is now becoming common, particularly as cellular becomes more widespread; see Table 5.2 [5.34]. To date, most uses are still for special applications, rather than generic solutions [5.35]. Users include real estate agents, insurance salespeople, salespeople in general, journalists, physicians, and attorneys. The applications involve the use of laptop computers connected to cellular phones that allow the user to enter or retrieve data automatically from a central computer without having to find a stationary telephone, which is often inconvenient and at times impossible.

Transmission speeds were around 2400 b/s in the mid-1980s. Some high-end cellular modems utilizing the data compression and error-correction techniques

Table 5.2
Cellular Telephones in Use

Year	*Units (thousands)*
1983	5
1984	125
1985	330
1986	650
1987	1115
1988	1624
1989	2800
1990	3800*

Portable phones were 20% of the equipment base in 1990.†
1989 Revenue: $3.2 billion.†
1989 Investment: $4.5 billion.†
Average monthly bill: 250 minutes of air time per user, at $125 (same usage in 1984 cost $150).†

*Projected.
† From Reference [5.4].

discussed in Chapter 1 can now transmit up to 16.8 kb/s (example CellBlazer Telebit). These data rates are reasonable for applications involving small- to medium-size file transmission and are excellent for inquiry-response and facsimile applications. A number of manufacturers offer portable computers with a built-in modem; most computers have internal power supplies for remote operation. These computers can be used over the cellular data systems. Several cellular providers now offer data service over cellular by the provision of MTCO equipment (i.e., a modem pool). By placing cellular modems at the switch, customers need only one cellular modem per car telephone; without MTCO equipment, the user needs two modems, one for the mobile unit and the other for the landline telephone set at the stationary end. Some carriers are also offering encrypted data service [5.2, 5.36]. Use of proprietary multicarrier modulation with an error-correction algorithm has produced modern cellular modems that in 1989 could transmit up to 30,000 b/s [5.37].

5.4.2.1 Applications

Many police departments have installed cellular phones in the patrol cars to obtain better voice communication and to obtain data capabilities. Voice communication over traditional radio dispatch involves verbal protocol codes, which can hinder dialogue. Conversation over cellular does not require these codes and is more direct; in addition, the police are able to call specific individuals, victims, *et cetera*. The data communication system allows the generation of electronic offense and arrest reports by field officers using portable computers; communication of the electronic reports to a central computer system using cellular telephones; and the retrieval of information in the field. Not only has the report writing and generation process been streamlined, but patrol officers have become more efficient in their duties [5.38].

These data systems will ease congestion on existing voice radio networks, thus eliminating the need for expansion of the existing voice systems. Traditional dispatch requires a human operator at the base; this not only requires staffing, but also represents a constrictive bottleneck. For example, checking a license plate during rush hour is difficult because the hundreds of officers cannot all call up the dispatcher simultaneously. Only one word can be transmitted, orally, in a half second, while 170 words can be sent over a 16,800 b/s channel in the same amount of time. Installed systems have already provided substantial productivity improvements. Reference [5.39] reports "two police officers with a terminal spotted more stolen cars that an entire division previously could." Similar productivity gains are likely to be realized by commercial application of wireless technologies.

Mobile data networks for the public safety sector are being introduced around the country. Fire fighters can use the system to determine the location of water hydrants, building access, and building content while rushing to a fire; building inspectors can obtain information on building codes and construction plants; water department personnel can input and access data from meters; the police can retrieve building histories and locations of exits and entrances, and they can identify automobile license plates. The city of Dallas has built a system that provides its police, fire fighters, and city officials with mobile data communication [5.39, 5.40]. Documented examples of cellular data usage include the following [5.2, 5.36, 5.41]:

- Reporters for several newspapers are using briefcase units that contain a cellular phone, a laptop computer, and a modem.
- Some ambulances transmit EKG data from a patient in the ambulance to the hospital to which the vehicle is directed. The operators find the system to be reliable and cost-effective because it relieves the hospitals from having to own a radio tower and a base station for dedicated radio channels.
- Physicians and emergency medics can access patients' files or transmit telemetric data.
- Sales and service people use cellular data transmissions; the system is used to report when a job has been completed, or to order parts immediately and make quotations in the field.
- Realtors can access multiple listings on their portable computers; this allows realtors to access directly from the field information about updated multiple listings on the data base at the office through a cellular phone [5.42].
- Fax machines are being used to send documents between mobile locations.
- Cellular emergency call boxes provide their own automatic status report over telemetry data channels with cellular access.

- Remote image transmission allows black-and-white photos to be sent over cellular telephone networks. Major users include federal law enforcement officials. This technology has been used primarily for surveillance and for classified operations.

5.4.2.2 Cellular Radio Data Transmission Equipment

Because of ambient electromagnetic noise, the motion of the vehicle, and the cell handoff process intrinsic in cellular telephony, data errors are likely to occur. The more sophisticated equipment incorporates advanced data communication protocols, in addition to FER software. The handoff procedure is generally between 0.5 and 1 s. For voice, this momentary pause does not represent a problem. For a data channel operating at 16,800 b/s, a 1-s interruption equates to the loss of approximately 1700 characters, or 20 lines of text in an 80-column document. The FEC algorithms ensure data integrity. However, the issue is not so much one of data loss, but one of the fact that most modems would interpret the interruption as a loss of carrier indicating a trouble on the line. In other words, the modem would break the session with the host, so that the transaction is held in a suspended mode. The user-to-host session must then be reestablished, which is time-consuming. Even under the best condition, this leads to data link level errors, which result in automatic retransmissions, again affecting the response time. One of the issues associated with the protocol required to maintain error-free transmission over cellular is that of standardization [5.43].

When a cellular call is put through, the cellular modem acts as an encoder, sending the data over cellular through the area's MTCO and then via a landline to the user's computer. The stationary device, interfaced with the landline modem, decodes the cellular data for the landline modem to accept. The equipment may employ automatic repeat request retransmission schemes, in addition to FEC methods.

The two modes of installing equipment are shown in Figures 5.5 and 5.6. In Figure 5.5, the user needs to purchase and install the unit as a front-end to the target computer; the unit will manage the protocol for noise and handoff protection. In Figure 5.6, the unit can be provided by the cellular carrier, as part of a modem pool. While this approach is advantageous to the user because it does not impose maintenance responsibilities and capital outlays, the service will cost more on a per-minute basis because the carrier must recuperate the investment [5.44].

5.4.3 Direct Data on Two-way Radio Channels

Private radio systems for data transmission can employ two basic techniques: traditional two-way radio, as routinely used for dispatch applications, and packet radio.

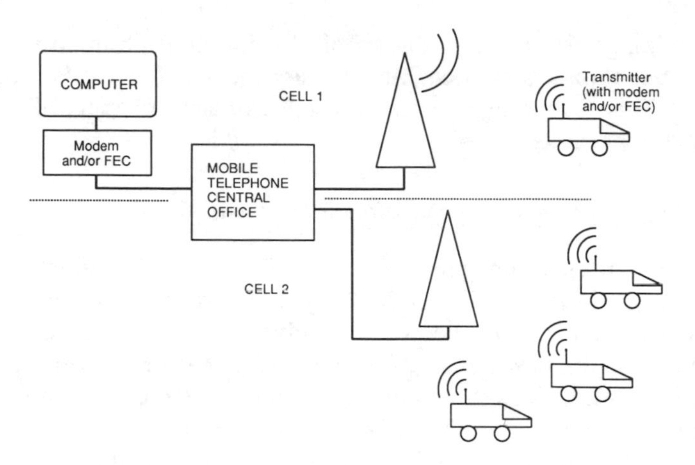

Figure 5.5 FEC protocol units at the computer site.

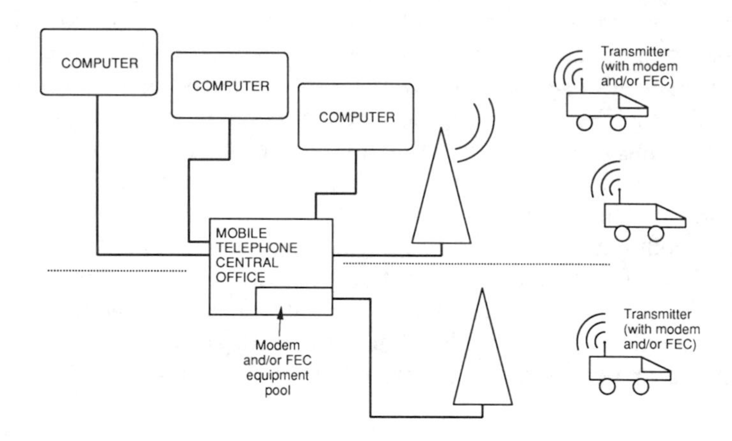

Figure 5.6 FEC protocol units at the CO.

Voice radio dispatch services (conventional two-way mobile radio) are simple two-way broadcast systems that operate on a single channel, usually in a half-duplex mode. Only one voice path is available for communication; therefore, transmission and reception is not possible at the same time. Dispatch is appropriate for situations in which people communicate among a group equipped with the same radio. Users of radio dispatch services are usually in vehicles on the road, sea, or air; they may be on an assignment for the delivery of some good or service, or on patrol. Radio dispatch services are a vital element for taxis, ambulances, police cars, fire trucks, and fleets of delivery vehicles. The dispatcher or control operator is located at the headquarters office where calls for service are monitored and retransmitted over the single channel. Mobile units and control stations form a network, where everyone hears what is being said or sees the data being transmitted. All users receive the message, which may be addressed to a single party or to several parties if the emergency requires such dissemination, as in the case of police and fire vehicles or ambulances. Privacy is not required, and indeed, may not be desirable. Conventional radio dispatch systems are rather expensive, primarily because relatively few vehicles can be serviced by voice within the bandwidth allocated.

Because all of the users share the same channel, conversations must be brief and the efficiency of the system depends on the dispatcher, who can control the use of the single channel. Addressing schemes are used in more sophisticated systems; for these systems, each vehicle can be assigned a digital code. Prior to sending a message, the dispatcher "dials"the code and only the vehicle with that particular code will pick up the message; the other receivers will be turned off by internal hardware [5.45]. Because of these limitations of voice dispatch systems, data systems are finding acceptance in the area of dispatch. Radio has been used for a number of years by the petroleum industry, gas and electric utilities, and other users who needed to transmit data from remote points where no telephone lines exist or where it would be cost-prohibitive to run such lines. The approach is to connect a telephone-type modem to a land mobile radio. This combination has worked well, provided the system was properly engineered to handle the expected data volume, the electromagnetic noise, and the reception patterns [5.46]. Data rates are in the 300 to 1200 b/s range, and errors are controlled via retransmission procedures. The data channels would have to be dedicated to this single function—if not permanently, then at least for the duration of the data transmission.

Dispatch radio is a two-way half-duplex system operating like a party line: data can be transmitted, but the access of the end-users wanting to "grab" the channel must be controlled to avoid breaking into the data stream. One solution is to dedicate a small radio subchannel to each user, for his or her sole data use. This FDM approach may at times be inefficient and costly. A good access mechanism is needed that automatically controls who gets the channel and when. Such techniques exist and are called *random access*, discussed in the next section.

Straight tapping of a mobile radio system, without some protocol for error correction, may present problems with noise, distortion, interference, and fading factors that are typical of radio channels. Two types of protection were discussed above for the cellular technology (and are also applicable to packet radio technology); other systems can be devised for mobile radio applications, loosely based on level 2 data link control protocols.

New FCC frequency allocations at 800 MHz and the use of multichannel trunking techniques allow a user to set up one efficient trunking system for all mobile communication in a metropolitan area. Trunking reduces congestion because the channels are shared rather than being dedicated to a single application or part of a company or department.

A natural gas utility is reported to be making effective use of mobile data technology to cut down their CGI-rate ("can't get in"). They are deploying a computer-aided dispatch system that, among other features, has a customer-call-ahead feature to ascertain that the customer is home before a visit is made. The field service person transmits the next intended destination to the system operator via a mobile data terminal; the system initiates a computer-generated call to the intended customer by accessing the customer file and retrieving the necessary information; the computer-generated voice informs the customer that a visit is pending over the next 30 minutes to one hour. If no answer is received, the field person will choose another customer. The system consists of about 250 mobile terminals [5.38]. The data system relieves congestion, which occurs when this operation is attempted over voice radio channels, particularly during the busy hour; long queues normally develop when using voice. Even when the field person can get through, the voice approach is slow and error prone. The data system allows the repair technician to have access to the customer file and trouble report on a mainframe computer, which aids efficiency. Real-time tracking of service person location and status from both the base and from the supervisor vehicle becomes possible, which in turns facilitates objective measurements of craft productivity. The mobile system is handled by a front-end adjunct, which controls operation and funnels the data to the mainframe.

5.4.4 Packet Radio Systems

Packet radio data transmission is considered by some as a breakthrough that is inexpensive, reliable, flexible, and allowed on a secondary basis in almost any part of the spectrum [5.47]. Packet is not a new technology; it is only new to commercial land mobile applications. Data packet radio techniques for nongovernment applications were first applied on a large scale by amateur-radio people, as a way to reach remote counterparts all over the globe using text, rather than the potentially noisy voice method. Business commercial systems are more complex, and more robust equipment is required. During the past five years, extensive research has been conducted in the development of a new generation of packet radio devices and system technology. As a result, higher throughput, range, accuracy, and security have been achieved [5.46].

Packet radio systems bundle a screen of data into a packet, and then transmit it at high speed to a similarly equipped station. Packets carry addressing information so that they can be selectively addressed to the desired destination station; they can also be repeatered along a physical route equipped with repeaters to achieve geographic distance. In addition, packets contain error-checking information, to provide error correction even on a very noisy channel [5.46]. Figure 5.7 depicts a packet radio environment.

Packet radio systems can accommodate both asynchronous or synchronous data transmission. The application of packet radio to synchronous data is technically simpler, assuming fixed repeaters and users. Packet radio wireless modems replace the regular modem and the terrestrial line for such synchronous protocols as BSC and SDLC. Packet radio can be achieved over standard 25-kHz radio channels; special modems employing bipolar modulation techniques achieve 9600 b/s transmission over 16-kHz bandwidths available on land mobile channels. FM signals in the VHF and UHF band have a maximum range, under the best set of conditions, of 60 miles (this applies whether the system is dispatch-based or packet-based).

Packet technology is spectrum efficient, sending bursts of data over very small temporal intervals, and then relinquishing the channel for other usage, such as voice applications or a multiuser shared facility. Applications in which the flow is point-to-multipoint, the traffic is bursty, and the volume is moderate are ideally suited to this technology. Thin route configurations include credit card verification, automated teller machines, security, and intrusion detection.

Packet radio links are managed using *carrier sense multiple access* (CSMA) techniques. This is a decentralized method (namely, without a polling master station) in which each station needing to send a message senses the channel until it is found to be free. This method reduces the collision rate to a practical minimum. Those collisions that do occur (because of the uncertainties associated with the propagation time), are resolved by an end-to-end acknowledgment protocol. When

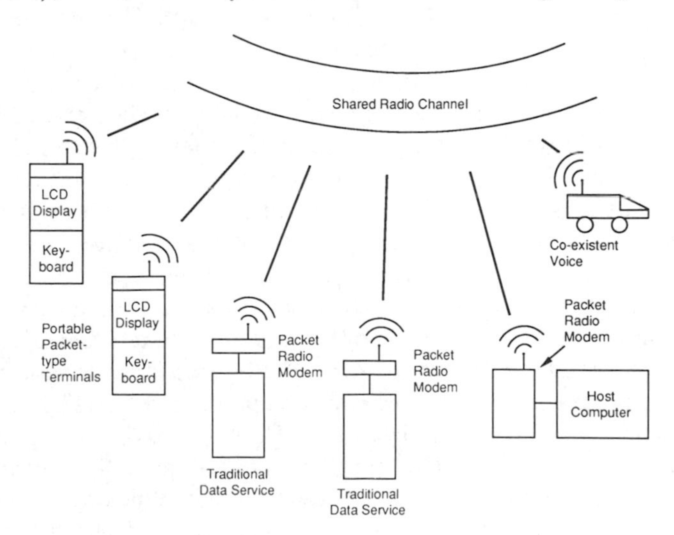

Figure 5.7 A packet radio configuration.

the amount of data that a terminal needs to transmit is small, the CSMA algorithm is a good one to use: the collision rate is low, the channel throughput is high, and data can coexist on a voice system. This is the same packet technology being used by many VSAT systems now being introduced, except that the repeater is a satellite in the sky rather than at some terrestrial tower (the technology is used in the uplink of the system).

Packet radio modems can also provide buffering and flow control procedures, in addition to the traditional virtual circuit management, in which several terminals can share the same host port. The virtual circuit features allow thousands of terminals to be served with a few central computer ports. Packet radio systems reduce frequency congestion because many independent users can share a single radio channel simultaneously, similar to a LAN environment. To extend the range, existing voice repeaters are employed or inexpensive digital repeaters are set up using a simplex radio and a packet controller. Typical payback periods for packet radio technology vary depending on the application; two to four years is not uncommon.

5.4.4.1 An Example

Large nationwide private radio networks have been deployed by corporations such as IBM and Federal Express. IBM is reported to have the largest private packet radio network; it has been in operation since 1984 and serves 20,000 portable terminals in more than 300 metropolitan areas. The portable terminal used has a full keyboard and two lines of LCD display, each with 54 characters; it weighs 30 ounces [5.35]. All members of IBM's service force carry the terminal wherever they go. The technician can interrogate IBM's inventory data base to determine availability of a required part and its cost; if available, they can order it and give the customer a delivery date. The host computer then automatically produces the customer's invoice. The terminal is also employed by the technician to obtain information on the next dispatch; the technician can obtain a clear readout of the address (rather than the old over-the-phone verbal method), pertinent contact point information, and a precise description of the problem.

The network is comprised of 1000 base stations and antenna sites (which include transmitters, receivers, and controllers) distributed throughout the United States. Each antenna covers from 20- to 30-square miles, depending on the territory. The data are routed to a central computer in Kentucky, where they are sent over landlines to the appropriate host CPU requested by the user. The devices can even be used to send E-mail between two users of the mobile terminal anywhere in the United States (the central computer will route the message to the appropriate destination). Incoming messages for the technicians are automatically queued in the portable device until they can be retrieved. The system operates on 800-MHz

FM channels, in the frequency group allocated for other mobile services. The stations broadcast at 45 W and the portable terminals at 4 W. The terminals output more power than a cellular phone because they are intended to operate within a building, where all the steel beams can absorb a lot of power; signal loss within a building can range from 0 to 40 dB. The additional terminal power, however, does not create much interference (which would be the case in cellular) because the data are transmitted in spurts and are highly multiplexed (in fact, proponents have called for piggy-backing packet radio data on all mobile communication channels, in a background mode, similar to spread spectrum).

5.4.4.2 Typical Packet Equipment

Packet radio systems can operate over existing two-way radio by adding a packet radio controller. By connecting a controller to a two-way radio at one end, and to any computer with an RS-232-C interface at the other end, the user is provided with a packet radio link. Data radio modems replace both the modem and the point-to-point dedicated line. Packet radio controller-modem combinations exist for operation at 2400, 4800, and 9600 b/s. The systems are meant to be transparent to the host computer, so that no modification to the host software is required; the controller appears to be a modem to the device cabled to it. Carrier sensing protocols that detect the presence of on-channel data or voice signals are employed. This allows the data transmission to coexist with voice traffic on the same channel; voice traffic preempts data transmission, which is bufferable at the packet controller. Packet radio technology costs around $4000 per end.

Packet radio technology can also be applied to a LAN environment. In fact, the first such system developed at the University of Hawaii in the late 1960s was a LAN using random access for a shared inbound channel. In this application, the terminals are scattered within a building or several buildings in the same general proximity. A typical modem with built-in radio transceiver is able to support up to 256 devices operating at a channel speed of 9600 b/s. The cost of each modem for the LAN application is in the $800 to $1000 range, depending on the antenna. Note that by comparison traditional LANs can cost from $250 to $500 per port. In some cases, radio will be cost-effective; in other cases it will not.

A number of antennas may be used for the packet radio: a 9-in. whip antenna that mounts on the back of the modem and provides a LOS range of about 0.5 miles; a magnetic-mount antenna (to go on top of a metal file cabinet or a car) can provide a 3-mile range; a building-mount antenna provides a LOS range of 30 miles. If obstructions are in the way, the range will be less than the LOS maximum. The actual range depends on the local terrain, the city landscape, and the antenna type [5.48].

5.4.5 Data over FM Subcarrier Channels

Data over FM subcarrier channels is an example of simplex data transmission. The commercial FM broadcast band spans the 88- to 108-MHz region of the radio band. This area is divided into 100 channels, each 200-kHz wide. The technique of subdivision is FDM, which implies that each user (or operator) is granted full-time private usage of a given band (in contrast to TDM, which is common in data communication in which different users share the same channel for a portion of the time). The operator must still obey certain regulations within the allotted band [5.49]:

1. The user must stay away from the boundary of the given channel slot, so as to provide a guardband between this user and the adjacent user (this is analogous to many construction laws that require a certain distance from the road, or from the edges of the lot).
2. Power constraints exist. FM modulation produces spectral lines (power spikes at integer harmonic multiples of the modulating signal) that, in principle, spread out infinitely both left and right of the carrier or central frequency. If the modulating power is within given limits, and appropriate filtering equipment is used, the importance (intrusion) of these spectral lines to the adjacent users is small; if the power exceeds the allowed limits, these harmonics start to interfere and disrupt the neighboring users.

Given an arbitrary 200-kHz channel, say, the 97.0- to 97.2-MHz channel, the center frequency (in this case, 97.1 MHz) is called the carrier. It identifies the particular station, and acts fundamentally in the modulation process. The 100 kHz to the left (half the bandwidth) is kept clear for the guardband (see Figure 5.8). For the monophonic portion of the transmitted program (the so-called "left + right

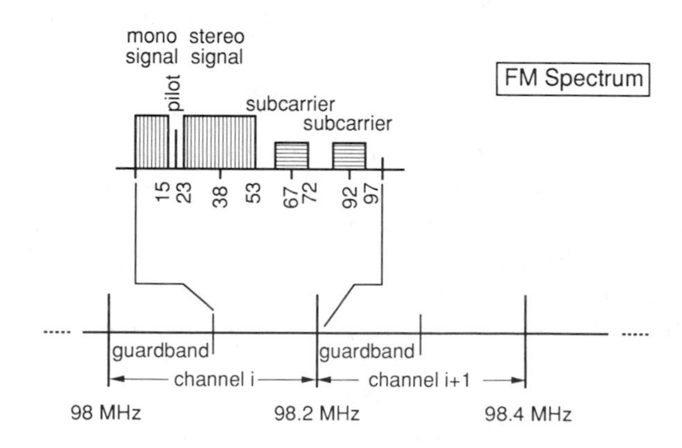

Figure 5.8 FM basic bandwidth allocation.

signal), 15 kHz (0 to 15 kHz) is used. Fifteen kilohertz is used to provide good fidelity for the transmitted acoustical signal; hearing can in theory take place in the 20- to 20,000-Hz range, but so few people can hear beyond 15 kHz that the decision was made to cut off the original acoustical signal at that value. In fact, until the introduction of the digital compact disk, it was not easy to record and press, or transfer to tape, signals with bandwidth exceeding 15 kHz. By comparison, AM provides only 5 kHz.

At 19 kHz, we find the *pilot*, which is a signal that informs the receiving circuitry that a stereophonic signal is present and that it should activate the appropriate circuitry. From 23 to 53 kHz, we find the stereo portion of the signal, namely the left-right component; 38 kHz is called the *stereo subcarrier* because it is the center frequency for that component of the signal. This means that from the 53- to the 100-kHz mark, the FM operator has unused bandwidth that can, under suitable engineering considerations (related again to modulation), be used for other purposes. The conventional allocation of that bandwidth is to create two additional subcarriers, one centered at 67 and 10 kHz wide; the other at 92 kHz, again 10 kHz wide. These subcarriers can each carry 5 kHz of additional audio programming, for example, a language translation, or "elevator" music; or 9.6 kb/s of data. Theoretically, these subcarriers could carry more data: in fact, on a standard voice-grade line with only 3 kHz of bandwidth, 14.4 kb/s, or 19.2 kb/s if the channel is clean, can be sent.

These two allocations of the carrier are not unique. Another management of that bandwidth would, for example, provide a combined 19.2 kb/s data stream. Another way would be to FDM the bandwidth from 53 to 100 kHz into five subbands, realizing 4.8 kb/s from each sub-band. Another proposal would employ the entire space above 53 kHz to achieve a data channel operating at 38.4 kb/s. Of course, some standard must emerge for the service providers, the equipment manufacturers, and the end-users to be able to operate and interact effectively.

To date, traditional subcarrier positioning has been used. This technology is known under the acronym "SCA," which stands for Subsidiary Communication Authorization (this was the FCC seal of approval to exploit the available bandwidth). Licensed owners of FM stations have tried for some time to utilize the unused capacity for applications other than the distribution of the basic audio programming. Initially, this capacity was resold for the distribution of commercial background music, such as Muzak. This use has been exploited since 1955 when the FCC allowed FM stations to use their subcarriers. Under the 1955 provision, a broadcast station could employ multiplexing techniques to divide its signal into a main channel and several subchannels. One subchannel carries the stereo signal, as noted; the other could be used for background music and for reading services for the visually impaired [5.50, 5.51].

In 1983, the FCC deregulated the use of the subcarrier. The FCC allowed more direct modulation techniques (digital rather than analog-acoustical); increased the portion of the unused spectrum that could be used for subcarrier services (consistent with Figure 5.8); and reduced the voluminous paper records that had been required to that time on subcarrier programming. Prior to deregulation, only analog broadcasts (such as background music) were permitted. After deregulation, the subcarrier could be used for any legitimate communication purpose. These changes not only allowed a more efficient use of the scarce FM band, but also made the subcarrier technology much more attractive to the data users.

Before 1983, only 5% of the FM broadcasters used the subcarrier for data purposes, and FCC authorization was on a case-by-case basis [5.52]. As of 1985, about 150 FM stations used subcarriers to transmit data. Approximately 50% of these stations are in the top 30 cities. Aggressive forecasts that by 1987 more than 500 stations would be employing the subcarrier for data applications did not prove to be true [5.52].

Thousands of broadcasters in the United States cover virtually every city and municipality in the United States. The following list provides a summary of the available stations as of 1988:

AM Radio	4742
FM Radio	3681
FM Educational Radio	1168
Total Radio	9591
UHF TV	359
VHF TV	536
UHF Educational TV	173
VHF Educational TV	114
UHF Low Power TV	107
VHF Low Power TV	203
Total TV	1492

5.4.5.1 FM Subcarrier Transmission Methods

As with commercial radio programming, on which this technology is based, the derivable data service is one way and suffers from a number of problems (reception and bandwidth); yet, it has a number of useful applications, particularly as a technique to reach many users in an economical way. The owner of the FM station would contract with some third party to supply some type of data base information; this could be stock quotations, theater events, flights schedules, paging, *et cetera*. A data line runs from the data base supplier to the FM station. The FM owner would purchase some fairly inexpensive equipment ($5000 to $10,000) to modulate—and possibly compact and encrypt—the data onto the subcarrier. The signal is then broadcast along with the main programming from the main antenna, covering a 20- to 40-mile radius. For the owner of a normal radio set, the subcarrier information is totally ignored. A subscriber to the service is provided with a special receiver, tuned to the subcarrier, that can pick up and decode the information. The receiver can be fairly inexpensive ($150 to $300). Figure 5.9(a) depicts this basic configuration.

The second variant occurs when the FM or TV station receives the signal from the data base supplier from a satellite link, as shown in Figure 5.9(b). The remote terminals can have an auxiliary link into a packet switched network, as shown in Figure 5.9(c). This provides duplex communication, but generally in a non-real-time mode. Of course, technically, data can be received on an FM subcarrier inbound link. Also, with a hard-wired terminal connected to the public (or other network), communication can be in real time back to the data base supplier.

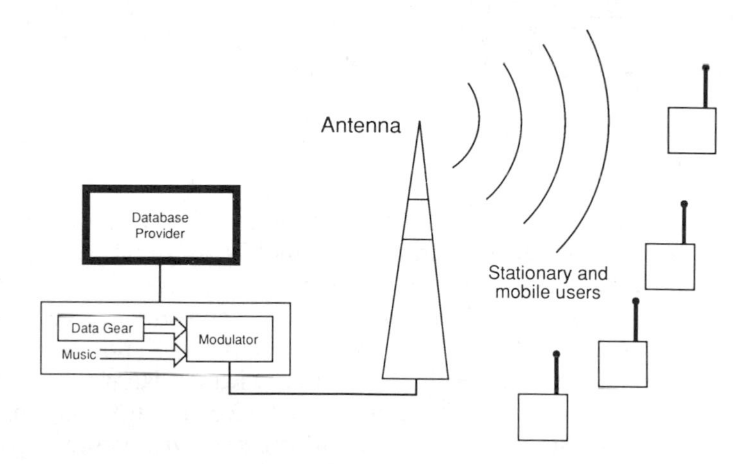

Figure 5.9 Subcarrier systems: (a) landline to data base provider.

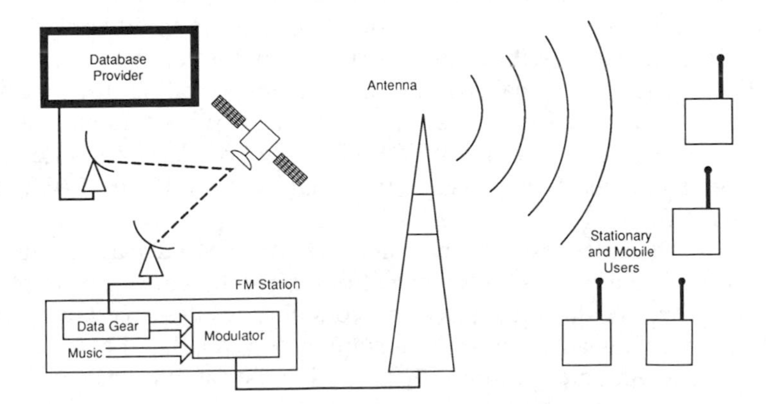

Figure 5.9 Subcarrier systems: (b) satellite link to data base provider.

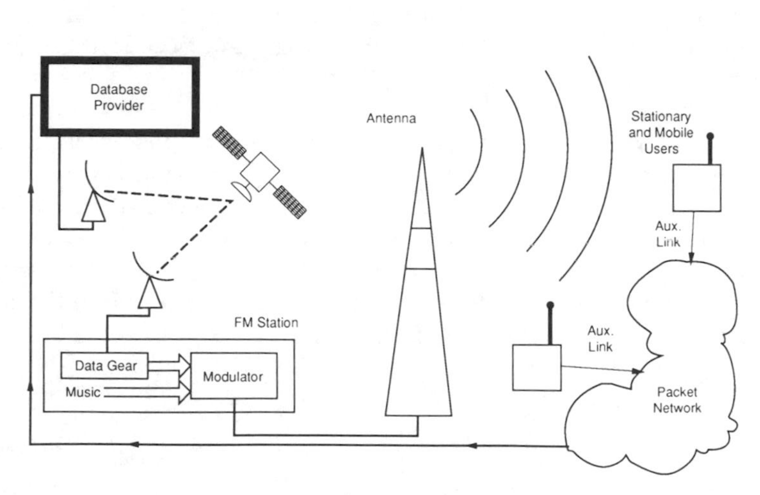

Figure 5.9 Subcarrier systems: (c) users have an auxiliary link into a packet network for two-way (non-real-time) communication.

The more common application is when the mobile terminal is not connected permanently to a land-based network; rather, the mobile terminal may receive notification that some large E-mail message has been received for that user, and provide a brief summary. The user would then, at his or her convenience go to a phone, and via an RJ-11 jack or other linkage, dial the data base and receive the long message over the land-based network. Another application is for the user to enter data into the portable terminal (for example, service orders), and then, when convenient, connect to the phone network and transmit (if this user had a burst-terminal rather than an FM subcarrier terminal, the data could be sent without having to connect to a land-based network).

Figure 5.9(d) shows how a group of FM operators could agree to "interconnect" into a nationwide system and provide national coverage. An example would be a national paging or E-mail system, or a system providing stock market quotes in major cities across the United States. Another example is National Public Radio's information network, which uses WESTAR IV to broadcast news items, business information, and emergency messages to affiliate stations [5.50]; PBS is also active in this area [5.51].

Figure 5.9(e) depicts a system in which the FM subcarrier information is distributed by satellite. This finds applications in rural areas; for example, midwest farmers may require the latest prices on commodities items (grains, pork bellies, *et cetera*). (In some cases where the terrain is very regular, FM stations can also cover a wide serving area, without the need for the satellite dish.) This approach is also useful for nationwide paging. The satellite antenna is a 2-ft dish, and in

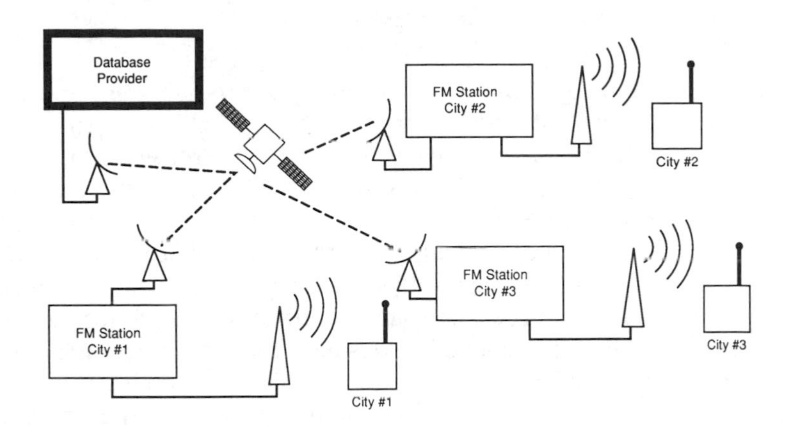

Figure 5.9 Subcarrier systems: (d) nationwide coverage.

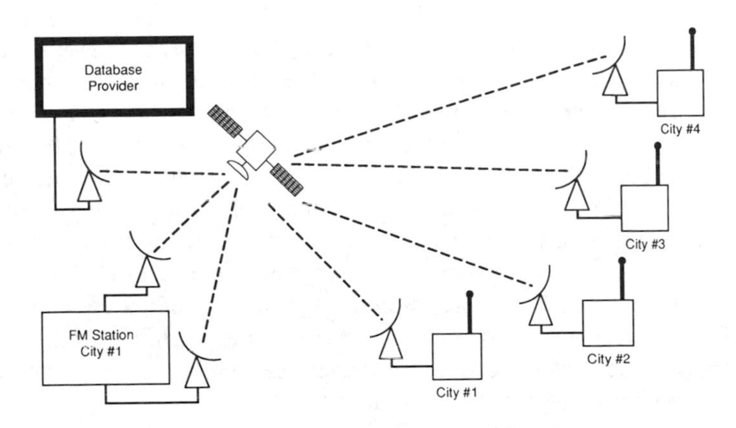

Figure 5.9 Subcarrier systems: (e) nationwide or regional coverage directly to satellite-based user terminals.

some buildings (i.e., older buildings in which window glass does not contain lead) can be located within an office and pick up the signal through the window. This is the same technology being employed by direct broadcast satellites (DBS) and VSATs discussed in Chapter 6. Most of the national radio networks now distribute their programming between affiliates by way of satellite links. The existence of this equipment, already in place at the FM stations, means that a national FM subcarrier system can be achieved easily with minimal incremental expenditures.

5.4.5.2 *Modulation and Error Correction*

As indicated, the type of modulation common prior to deregulation was based on an analog modulation of the subcarrier. Because the FCC only permitted indirect FM subcarriers, the modulating signal had to be converted to audio-frequency signals. This resulted in 2.4 kb/s data streams, similar to those obtained with ordinary low-speed modems; the subcarrier was treated as a leased channel, with the audio output of the modem being "summed" to the normal program signal for ordinary modulation. At the receiving end, the process would be reversed. This approach is both expensive (because each user needs a receiver for the entire signal and a modem) and inefficient (low 2.4 kb/s bandwidth). Direct modulation by the digital signal is more cost-effective. Digital modulation eliminates the modem to convert the data into audio pulses. Thus, rather than imparting (modulating) the carrier with audio tones, the carrier is imparted directly with the digital string, as discussed in Chapter 1. Any of the digital techniques are available to subcarriers. FSK is common; PSK and QAM are also being considered because of their higher efficiencies (packing more data onto the subcarrier) [5.52].

In a wireless voice system, one bit error in 10^{-4} bits can be tolerated. This level of error rate is unacceptable, however, for data transmission. The communication system then requires an error control mechanism. In a metropolitan environment with considerable human-made electromagnetic noise and multipath problems, the need for error correction is imperative, especially for financial data carried on the FM subcarriers. No rigorous measurements have yet been made on the uncorrected BER of FM subcarrier channels; indeed, this is dependent on complex topographical issues. Some suppliers claim uncorrected error rates in the 10^{-6} to 10^{-7}; by way of comparison, uncorrected error rates on the dialup network are generally in the 10^{-5} range, while conditioned dedicated lines have a 10^{-6} figure of merit. The error-correction scheme would build on that basic error rate and improve it by several orders of magnitude. However, the true uncorrected error rate is generally felt to be around 10^{-4}.

5.4.5.3 *Addressing and Security Issues*

Five families of sender-receiver mapping schemes are available:

1. *Deterministic*. This can be frequency-division or time-division multiple access, in which each user knows to look into a particular time or frequency "slot" to obtain data.
2. *Random Access*. No user has a dedicated "slot." Each user requiring bandwidth from a common channel attempts to "grab" it as needed. This is typical of the burst packet radio technology discussed earlier.

3. *Reservation.* A user requiring access to the channel posts a reservation (possibly via a random access channel). The recipient is informed as to the "slot" to which he or she should listen.
4. *Broadcasting, Round-Robin, and Address or Token Passing.* These are three flavors of the same basic technique.
5. A combination of the above.

FM subcarrier technology uses a combination of FDMA and broadcast-address parsing: A given user or receiver or service must first tune to a particular FM station offering the service (this is the FDMA part). It then uses address passing to select the message of interest. This approach to the problem of switching a message to the proper end terminal makes use of an address added to the message that is then broadcast to all the terminals on the system [5.53]. The terminals, each of which recognizes (owns) a unique address, have hardware circuitry, firmware, or software that decodes the address of all messages broadcast and accepts only those having the proper address. This technique has actually been used for more than 20 years in all sorts of systems, beginning with the IBM BSC protocol and including CATV systems, paging services, and LANs.

The address can be (see Figure 5.10) device specific, page specific, or record specific. In the "device-specific" mode, the address of the intended user is embedded in the stream, preceding and delineating the data destined for a unique (or "closed group" of) user. In fact, this technique can also be used to enforce security, such as session management and system access. The terminal firmware (machine instructions not accessible by the user) can be instructed by way of this unique

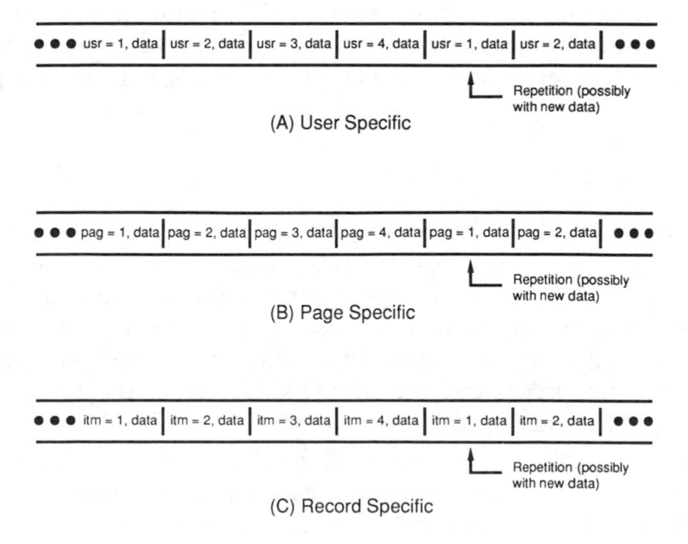

Figure 5.10 Addressing schemes for subcarriers.

address *not* to turn itself on; this is a means of excluding users from the systems who have not paid their bills, or have discontinued service.

In the "page-specific" mode (typical of teletext), each broadcast page has an address or identifier. Every user receives every page and can conceivably decode each page. However, page 1 may have an identifier or address of "Theater"; page 2, "Sports"; page 3, "Weather"; page 4, "News"; *et cetera*. The user, by selecting a specific page, is telling his terminal to ignore (at this particular time) all pages except the one of interest; all pages are read, but only the one of interest is posted to the user. The pages are continuously repeated in a round-robin fashion, say, every 2 minutes; allowing, for example, a total of 100 possible pages to flash by and be grabbed by the user before being updated 2 minutes later.

In the "record-specific"mode, every item of data is preceded logically by an address. Every user can conceivably read every record, but only the items of interest to the user (which have been programmed into the device) are trapped and posted. This is typical of the market data applications.

In financial transactions, a stock symbol (3 characters or 24 bits) is issued plus a stock quotation (say, five characters such as 132 5/8; this is 40 bits). Generally, other overhead and other information are also being transmitted (for example, number of shares involved in the transaction). Using 25 characters per transaction (200 bits), 45 events per second can be posted into a 9600 b/s channel. In 60 seconds, information on 2700 stocks can be posted in a round-robin fashion; after 60 seconds, new information on the same 2700 stocks is posted, and the user terminal is "refreshed." Note that the data will be (in this example) up to 1 minute late. If a single stock symbol is programmed into the terminal by way of the real-time display, then only that symbol is tracked; if several symbols are programmed (to track a portfolio), then as the data "swing by" on the round-robin loop, they are read off by the terminal device and placed into RAM for later access by the terminal.

5.4.5.4 Applications

The new market for FM subcarrier may benefit radio common carriers (RCC) and FM broadcasters. Numerous opportunities exist for both the lessors and the lessee. FM subcarriers can generate revenues for the owners with minimal hardware investment: The costly equipment is already in place for an FM station (transmitters, antenna, real estate, *et cetera*). In addition, the owners are already familiar with communication issues and interconnects, facilitating their transition into data distribution [5.54]. Some applications follow.

Paging Services. With numerous FM stations in a given metropolitan area, service can be extended to a larger market (i.e., larger than the traditional professionals market). In addition, a nationwide service is possible, as delineated in Figure 5.9(d). Digitally encoded "short messages" can be broadcast easily.

Utility Load Management. FM receivers mounted on advertisement boards and nonessential electrical devices can, in case of peak load, be turned off or controlled from a centralized location. They can also control traffic lights to manage the flow at different times during the day.

Text Transmission. The most common utilization of the subcarrier technology in the data area has been for the transmission of text material (financial information, bulletins, schedules, *et cetera*) to a specific group of individuals, or to a given individual.

Facsimile. Broadcast of maps, drawings, graphics, prints, *et cetera*, to portable facsimile devices.

Electronic Mail Delivery and Electronic Publishing. Printed material can be transmitted from a single location to any number of metropolitan locations (stationary or mobile). Typical applications include retail chains, banks, and supermarkets. Newsletters or bulletins that have very high perishability can be distributed easily using a subcarrier.

Market Data (stock quotes and commodities prices). This has been a very successful and early application of the technology. The terminal devices can be hand-held devices the size of a pocket calculator that carry a monthly service maintenance fee for stock exchange charges.

5.4.5.5 Limitations and Problems of Subcarrier Technology

FM subcarrier technology has a number of limitations, as discussed below.

One Way. FM subcarrier is really a one-way service. A semblance of interactive access is achieved by provision of a buffer in the terminal device, which can be queried from the keyboard. Nonetheless, the service remains one way, unless an auxiliary return path via a land-based network is provided. By way of contrast, packet radio is two ways.

Data Delays and Discrete Pagination. Data are posted in a round-robin fashion, meaning that delays in data refreshment occur. This delay is generally in the 1- to 2-minute range. For some applications this delay can be severe (example: financial data); in other cases, it will not matter (example: E-mail distribution). The permissible delay in the cycle and the channel bandwidth determine the maximum number of users in the "user-specific" addressing mode, and the number of pages in the "page-specific" addressing mode.

Bandwidth. With 9.6 kb/s (or even 38.4 kb/s) channels, bandwidth limitations exist for some applications. For example, if an application involves broadcasting a 20-page individualized newsletter to 15 users, with the letter not being available before 12:00 and with the requirement that the letters be distributed by no later than

12:15 , we quickly realize that a 9.6 kb/s channel would not suit this application. Two solutions involve (1) data compression and (2) more sophisticated modulation techniques to achieve more throughput (bits per hertz).

Reception and Multipath. Reception can be a problem, particularly in the presence of skyscrapers. The signal strength issue can be alleviated with externally mounted quality FM antennas, which cost around $200. At least one company claims to have also developed a microprocessor-based indoor antenna that can address the problem. Of course, this assumes that the receiving terminal is stationary. No solutions exist for portable devices, except for better receivers with signal processing chips. Multipath results from secondary reflections of the broadcast signal from buildings, hills, and other objects. While this is already an issue in audio and video reception, the situation is aggravated in the data arena. Additional signal processing electronics at the receiver are required to deal with the problem. Other techniques employ a method called "spread spectrum."

Shadowing. FM subcarrier is a LOS technology; obstructions between the transmitting tower and the receiver (such as buildings) will reduce the signal strength.

Single Tower. The subcarrier system employs a single FM tower within a given city. This may imply reception problems, particularly in downtown business districts. Small noise on the audio portion of the FM band is generally tolerated (example: noise from a downtown area due to multipath); for data, this noise cannot be tolerated. Adequate range and penetration has been a problem with subcarrier. Unlike RCCs, radio stations cannot build additional antennas to offer better range and penetration [5.54]. More sensitive receivers are being built; the more sophisticated ones scan for the strongest FM subchannel and lock into it.

Crosstalk (Stereo to SCA). The modulation process creates harmonics (spectral lines that extend beyond the nominal bandwidth of the stereo subcarrier 1). This signal leakage can compromise some data systems, particularly if the transmitting equipment does not have stringent parameters. The desirable crosstalk attenuation is usually −60 dB.

Security. All wireless technologies are affected by security problems. Very few systems, if any, employ encryption, though the discipline has progressed to the point where encryption can be added to a system for $50.

Some of the benefits of subcarrier include broad geographic coverage, nationwide scope, small startup costs, and reliability (FM transmitters and resulting service have extremely high availability—99.99%).

REFERENCES

[5.1] *Mobile Phone News*, January 18, 1990, p. 10.

[5.2] D. Purcell, "Listening to Non-Voice Cellular," *Communication*, April 1987.
[5.3] W.P. Finnegan, "Cellular's Demise is Premature Thinking," *Telephone Engineer & Management*, August 1, 1986.
[5.4] *Mobile Phone News*, March 1, 1990, pp. 1 and 4.
[5.5] J.G. Lucas, Telestrategies, promotional material.
[5.6] "Air-to-Ground Service is Ready to Take Off," *Telephone Engineer & Management*, May 1, 1986.
[5.7] J. Walker, *Mobile Information Systems*, Artech House, Norwood, MA, 1990.
[5.8] P.S. Henry, "High-capacity Lightwave Local Area Networks," *IEEE Communications Magazine*, October 1989, pp. 20 ff.
[5.9] T.A. Heppenheimer, "Staying in Touch with Subs," *High Technology*, March 1987, pp. 66–67.
[5.10] *Electronics*, September 11, 1980.
[5.11] P. Petersen, "Wireless Helps Out in the Field," *Telephone Engineer and Management*, January 1, 1986.
[5.12] *Mobile Phone News*, April 12, 1990, p. 5.
[5.13] D. Minoli, "Packet Length Considerations in Carrier Sense Multiple Access Packet Radio Systems," INTELCOM 80 conference record. See also F. Baumgartner, "Code Division Multiple Access," *Communications*, February 1990, pp. 26 ff.
[5.14] D. Minoli, "Analytical Models for Initialization of Single Hop Packet Radio Networks," *IEEE Transactions on Communication*, special issue on Digital Radio, December 1979, Vol. COMM-27, pp. 1959–1967.
[5.15] D. Minoli, "A Taxonomy and Comparison of Random Access Protocols for Computer Networks," in *Data Communication and Computer Networks*, ed. S. Ramani, pp. 187–206.
[5.16] D. Minoli, "Digital Voice Communication over Digital Radio Links," *SIGCOMM Computer Communication Review*, October 1979, Vol 9, No. 4, pp. 6–22.
[5.17] D. Minoli, "Packet Radio Monitoring via Repeater-on-Packets," *IEEE Transactions on Aerospace and Electronics Systems*, July 1979, Vol. 15, No. 4, pp. 466–473.
[5.18] D. Minoli, "Initialization Time for Packet Radio Networks with a Small Number of Buffers," *ALTA Frequenza*, October 1979, Vol. XLVIII, No. 10, pp. 653–628.
[5.19] D. Minoli, "A Closed Form Expression for Initialization Time of Packet Radio Networks," *Frequenz*, May 1979, Vol. 33, No. 5, pp. 126–133.
[5.20] D. Minoli, "Monitoring Mobile Packet Radio Devices," *IEEE Transactions on Communication*, February 1979, Vol.COMM-27, No. 2, pp. 509–517.
[5.21] D. Minoli, "Combinatorial Issues in Mobile Packet Radio," *IEEE Transactions on Communication, December 1978, Vol. COMM-26, pp. 1821–1826.*
[5.22] D. Minoli, "An Approximate Analytical Model for Initialization of Single Hop Packet Radio Networks," *1978 IEEE Canadian Conference on Communication and Power*, pp. 107–110.
[5.23] D. Minoli, "Analytical Models in Monitoring Mobile Packet Radio Devices," 28th IEEE Vehicular Technology Conference Record, March 1978, pp. 110–118.
[5.24] D. Minoli, "On Connectivity in Mobile Packet Radio Networks," 28th IEEE Vehicular Technology Conference Record, March 1978, pp. 105–109.
[5.25] *Proceedings of IEEE*, special issue on Packet Radio, January 1987.
[5.26] R.E. Kahn et al., "Advances in Packet Radio Technology," *Proceedings of the IEEE*, November 1978, Vol. 66, No. 11, pp. 1468 ff.
[5.27] D. Cervenka, "Don't Call It a Pager," *Communications*, October 1989, pp. 24 ff.
[5.28] *McGraw-Hill Yearbook of Science and Technology—1988*, McGraw-Hill, New York, 1988.
[5.29] T. Kerver, "Global Paging by Satellite," *Satellite Communications*, March 1990, pp. 17 ff.
[5.30] J.C. Bell, "Mobile Satellite Brought Down to Earth," *Communications*, November 1989, pp. 67 ff.

[5.31] M.A. Rothblatt, *Radiodetermination Satellite Services and Standards*, Artech House, Norwood, MA, 1987.
[5.32] K. Van Lewen, "Cellular's CT-2 Competition," *Communications*, September 1989, pp. 61 ff.
[5.33] J. Avery, "Radio Networks: Data Transfer's New Wave," *PC Week*, January 28, 1986.
[5.34] H. Shosteck, "Why Cellular Is Taking Off," *TE&M*, August 1, 1988, pp. 29–31.
[5.35] S. Tisch, "Going Mobile: Radio Data Keeps Users on the Move," *CommunicationsWeek*, March 23, 1987.
[5.36] S. Tisch, "Combined with Cellular, Data Saves Money," *CommunicationsWeek*, March 30, 1987.
[5.37] D. Gibson, "Advanced Connectivity Products for Dial-up data Communications," COMNET 1990, Washington, D.C., February 1990.
[5.38] *Communication News*, special issue on Two-way Radio Communication, April 1987.
[5.39] S. Tisch, "Dallas Uses New Mobile Data Net in Unique ways for City Agencies," *CommunicationsWeek*, October 27, 1986.
[5.40] MDI Corporation trade press advertisements.
[5.41] D. Minoli, "An Overview of Radio Technologies for Data Communication," DataPro Report CA70-010-501, October 1987.
[5.42] K. Poli, "The Big Switch to Data over Cellular," *Communications*, April 1987.
[5.43] A. Morant, "What's Ahead in Cellular?," *Telephone Engineer & Management*, February 15, 1986.
[5.44] J. Bush, "Firms Finding Solution to Problems of Data via Cellular Radio," *Data Communication*, October 1986.
[5.45] H.S. Dordick, *Understanding Modern Telecommunication*, McGraw-Hill, New York, 1986.
[5.46] S. Beeferman, "Packet Radio Can Be a Cost-Effective Alternative to Dedicated Land Lines for Datacomm Networks," *Communications News*, August 1986.
[5.47] G. Dennis, "Packet's Progress," *Communications*, January 1987.
[5.48] M.D. Stone, "Modems Take to the Airwaves," *PC Magazine*, January 14, 1986.
[5.49] D. Minoli, "All about FM Subcarrier Communication," DataPro Report CA70-010-301, May 1986.
[5.50] A. Reiter, "Who's Who in SCA?," *Personal Communications Magazine*, July–August 1983, pp. 10 ff.
[5.51] SCA: Radio Subcarrier Report, January 1986.
[5.52] D. Waters, "Surfacing: FM Radio Provides a New Alternative for Carrying Data," *Data Communications*, July 1985, pp. 173–181.
[5.53] L. Lewin, ed., *Telecommunications: An Interdisciplinary Text*, 1984.
[5.54] S. Goldman, "Two Dozen Ways You Can Profit from Subcarriers," *Personal Communications Magazine*, July–August 1985, pp. 10–12.

Chapter 6
Satellite and Microwave Transmission Systems

6.1 INTRODUCTION

Commercial satellite communication began in 1965. Many advances in the technology and in its economics have occurred since then. Satellites now carry voice, data, and video traffic. Efforts are under way to provide digital services and HDTV over satellite links. Recently, a resurgence of interest in satellite-based data communication has occurred due to the potential of VSATs.

Satellite communication offers certain advantages, such as broadcast capabilities and mobility. The technology, however, is under pressure from other high-quality, high-capacity media, particularly fiber optics. Difficulties in the late 1980s in delivering payloads into space have affected the availability of spare capacity. Additionally, satellite orbits are close to being fully utilized. Other limitations include a lack of intrinsic security and propagation delays. Nonetheless, many communication applications have emerged.

This chapter covers, among other topics, types of satellites; applications of satellites; technical aspects of the transmission channel; access schemes; transmission impairments; orbital and earth coverage issues; transponders; earth stations; and VSAT technology. A short treatment of microwave technology is found at the end of the chapter.

6.1.1 Types of Satellites

A number of different satellites and satellite services exist, as shown in Table 6.1 [6.1]. The most important are discussed below [6.1 through 6.8].

Communication. These satellites act as relay stations in space for radio, telephone, and broadcast communication. Figure 6.1 depicts the physical structure of two

Table 6.1
Satellite Service Designations

Abbreviation	*Service*	*Abbreviation*	*Service*
A	Terrestrial amateur	P	Passive
AMSS	Aeronautical mobile satellite service	RA	Radio astronomy
ARN	Terrestrial radionavigation	RDSS	Radiodetermination Satellite Service
AS	Amateur satellite	RL	Terrestrial radiolocation
BC	Terrestrial broadcasting	RLSS	Radiolocation satellite service
BSS	Broadcasting satellite service	RN	Terrestrial radionavigation
EEX	Earth exploration satellite	RNSS	Radionavigation satellite service
F	Terrestrial fixed	S	Secondary service
FSS	Fixed-satellite service	SFT	Standard frequency & time satellite
ISL	Intersatellite link	SO	Space operation
ISM	Industrial, scientific & medical	(SOS)	Distress & safety operations
LMSS	Land mobile satellite service	SR	Space research
M	Terrestrial mobile	STL	Studio-transmitter link
MMSS	Maritime mobile satellite service	TV	Television
MSS	Mobile satellite service	WA	Terrestrial meteorological aids
MSXA	Mobile satellite (except Aeronautical mobile satellite)	WS	Meteorological satellite
MXA	Terrestrial mobile (except aeronautical mobile)		

types of communication satellites. Commercial communication satellites lie in the geostationary (geosynchronous) orbit at approximately 36,000 km (22,320 miles) over the equator, as shown in Figure 6.2. (To be more precise, *geosynchronous* refers to the orbit in which the speed of a satellite's orbit is synchronized with the speed of the earth's rotation; *geostationary* refers to a geosynchronous satellite with a zero angle of inclination, so that the satellite appears to hover over a spot on the earth's equator.) Small variations in the mass and shape of the earth affect the orbit, requiring that the satellite be "repositioned" under spacecraft power to regain the proper position. More than 120 communication satellites occupy the 165,000-mile circumference of the geosynchronous orbit.

The characteristics of geosynchronous satellites include [6.10]:

- Broad earth coverage (corresponding to approximately one-third of the earth);
- Orbital period of 24 hours; and
- Round-trip delay of approximately 500 ms for half-duplex communication.

The location of the satellite is nominally defined by the longitude (degrees west) of the point on the earth's equator over which the geostationary satellite appears to be positioned.

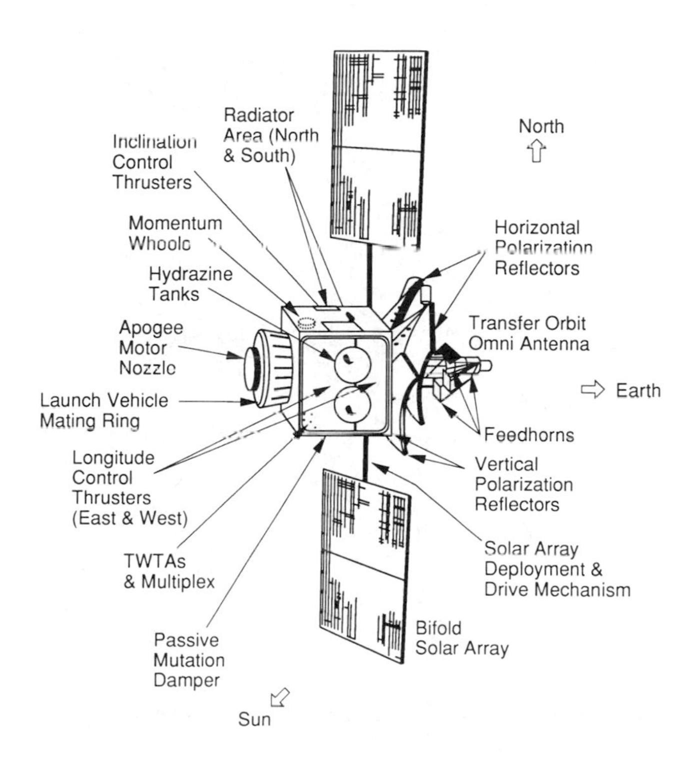

Figure 6.1 The physical structure of a satellite: (a) three-axis stabilized satellite.

Most satellite antennas must point to a predetermined area of coverage throughout the entire life of the satellite. The method used to control the antenna depends on whether the satellite is spin-stabilized or three-axis stabilized. The former satellites are stabilized by the induced rotation of the satellite according to the principle of the gyroscopic effect. The antenna is mounted on a counter-rotated platform so that it appears to be stationary in reference to the earth's surface. In the three-axis stabilized satellite, the antenna is mounted on a limited-motion gimbal, giving it flexibility in pointing; this antenna system, however, requires a more complex control system compared to the spin-stabilized system [6.11]. Three-axis stabilized satellites have larger solar cell arrays compared to the surface of a spin-stabilized satellite, which allows them to have more operational power.

Satellites that are not in the geosynchronous orbit require earth stations with movable antennas to track them; this is not the case with geosynchronous satellites. The United States is the largest user of satellite resources. In the United States, several organizations provide a spectrum of communication services over geostationary satellites operating around 4 and 6 GHz and around 12 and 14 GHz.

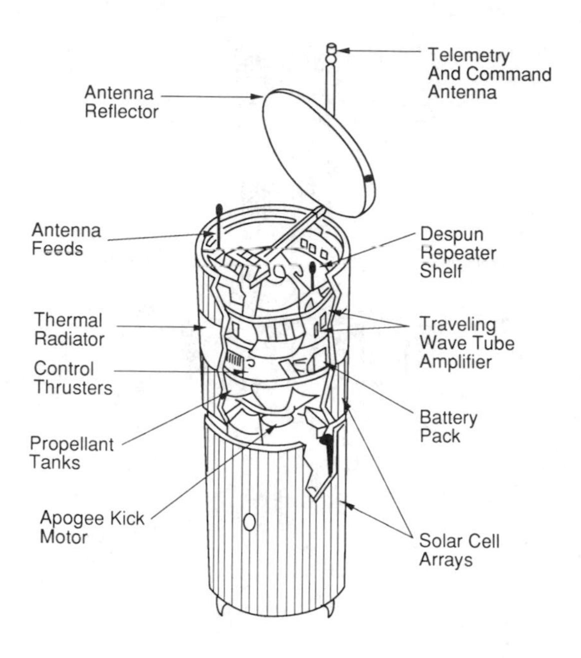

Figure 6.1 The physical structure of a satellite: (b) spin-stabilized satellite.

Additional satellites in this frequency band are already precluded by the lack of suitable orbit slots. Consideration is being given to higher frequency operation (around 20 to 30 GHz).

Some of the factors affecting the applications of the technology to business communication include orbital spacing (only about 60 degrees of total equatorial orbit space is suitable for domestic satellite use), earth station cost, antenna size, transponder power, and security. These issues will be discussed in the following sections.

Navigation. These satellites provide a reference so that aircraft and ships can establish their position [6.12]. Another application is a positioning service for ground vehicles anywhere in the United States. The service is known as radio-determination satellite service (RDSS). A hub computer spotting signal is transmitted via satellite to a receiver-transmitter unit in a moving vehicle, which will then respond when its individual number identification is called out. This provides the central station with the data it needs to calculate the vehicle's precise location.

Worldwide revenues for the provision of RDSS are forecast at $200 million for 1992 and $1 billion for 1996. In the United States alone, the forecasts are for

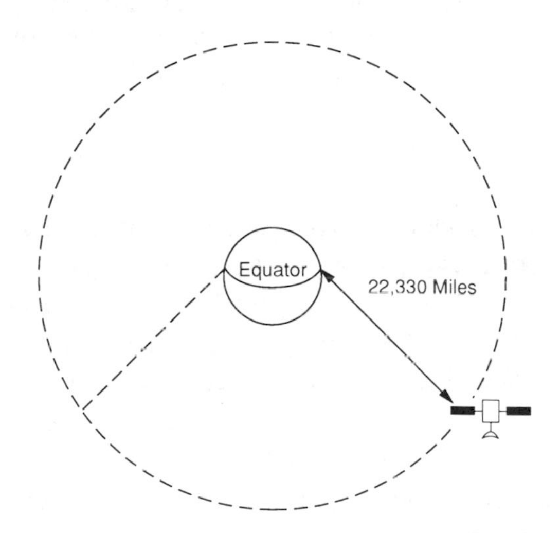

Figure 6.2 Geosynchronous orbit.

130,000 units by the end of 1990 (each unit selling at $1000 to $2000); by 1995, the price per unit is expected to drop to less than $500 [6.13].

Weather. These satellites monitor the atmosphere and surface conditions of the earth as an aid to weather reporting and forecasting.

Surveillance. These satellites observe and record details about and movement of features and objects on the earth's surface.

Electronic Surveillance. These satellites relay or record signals that are broadcast on radio frequencies from objects on or near the earth's surface.

Research. These satellites provide platforms for scientists to observe phenomena or to conduct experiments.

Earth Sensor. These satellites record and relay emissions and reflections of energy from the earth's surface [6.14]. Some of the tasks that can be performed via remote sensing include:

- Meteorology and Climatology: weather and atmospheric observation and prediction;
- Navigation: position locating for mobile vehicles in the air and on the ground; and
- Military and Defense: reconnaissance.

Reference [6.15] provides a listing of value-added service providers.

Photo Reconnaissance. These satellites are military (and now also commercial) spacecraft equipped to record and transmit high-resolution images of surface features and objects. This service is also being considered for the news media [6.16].

Space Stations. These satellites are spacecraft equipped to accommodate crews.

6.1.2 Altitude Bands

Several orbital bands in addition to the geosynchronous band are used by satellites.

21,750 to 22,370 Miles. The most distant band is the location of the geosynchronous satellites, most of which are used for broadcast, communication and data relay, surveillance, and weather observations.

13,700 to 21,750 Miles. This band is not used extensively.

6200 to 13,700 Miles. This is the ideal altitude for the precision orbits of navigation satellites, which are the exclusive occupants of this band.

3100 to 6200 Miles. A sparsely populated band, used for scientific satellites.

1250 and 3100 Miles. This altitude band is not used very much.

630 to 1250 Miles. A wide cross section of types occupies this altitude band, including surveillance, electronic monitoring, and communication.

300 to 630 Miles. Most earth observation and roving weather sensor satellites orbit at this altitude, which is typified by satellites orbiting at high inclinations in order to survey wide areas of the globe. The band is also populated with navigation and surveillance satellites.

90 to 300 Miles. This low altitude band is ideal for imaging reconnaissance, scientific satellites, and manned space stations.

According to the National Aeronautics and Space Administration (NASA), 6772 major human-made objects were circling the globe at the beginning of 1990. Of these, only 1870 are actual payloads; the rest is debris such as boosters, shrouds, mounting brackets, *et cetera* (bolts and other small metal fragments are not included in the count). Most of the actual payloads are still working. The Soviet Union claims 1136 of these satellites; the U.S. claims 556; Japan, 37; INTELSAT, 37; other countries and consortia, 104.

6.1.3 Apogee and Perigee

All orbital paths are eccentric to varying degrees. Satellite operators define the most distant point of an orbit as its *apogee* and the nearest point as its *perigee*.

The difference between these two distances describes the degree of eccentricity of the orbit.

6.1.4 Inclination

The angle at which a satellite's orbit is tilted in relation to the earth's equator is called its *inclination*. A satellite in an equatorial orbit has an inclination of zero degrees, whereas one whose orbit crosses the earth's poles has an inclination of 90 degrees. A satellite can have an inclination of more than 90 degrees, and some sun-synchronous satellites that maintain the same ground track throughout the year have inclinations of as much as 98 degrees.

Satellite operators try to select an inclination best suited to the job the spacecraft is expected to perform. Photographic reconnaissance satellites offer the best example of how orbital inclination affects a spacecraft's performance. Satellites usually are inclined 96 degrees, permitting them to view territory from the Arctic to the Caspian Sea several times a day. Soviet reconnaissance satellites do not need to look so far north and south, so most of them orbit at inclinations of 67 degrees or less.

Satellite orbits obey Keplerian laws and are elliptical by definition (the circular orbit of a geosynchronous satellite is a special case of the elliptical orbit). The equation of the period of revolution in a circular orbit, in hours T, as a function of the earth's radius R and the satellite's altitude h is:

$$\begin{aligned} T &= (1/3600)(2\pi)[(R + h)^3/(3.99 \times 10^{14})]^{1/2} \\ &= 0.872 \times 10^{-10}[(R + h)^3]^{1/2}. \end{aligned}$$

For a geosynchronous satellite, $h = 35{,}810{,}000$ m and $R = 6{,}378{,}000$ m (R is clearly fixed), giving $T = 24$ h. This equation assumes a perfectly spherical earth, without any external perturbations to the orbit. In reality, the earth is not a perfect sphere, and the force of the moon's and the sun's gravity has a substantial impact on orbital dynamics; actual orbital equations must take these factors into account [6.11]. The circular velocity of a satellite in circular orbit is:

$$V = [(3.99 \times 10^{14})/(R + h)]^{1/2}.$$

For a geosynchronous satellite, this is 10,800 km/h.

6.1.5 Advantages of Satellite Communication

The advantages of satellites include:

1. The superiority of satellites for point-to-multipoint transmission makes them ideal for geographically dispersed broadcast TV (which requires transmission from a single location to many affiliate stations or to satellite dishes on roofs) and for private networks (e.g., teleconferencing or communication between corporate headquarters and branch locations).
2. Satellite antennas are relatively easy to install (or can even be mobile, on a small truck) and enable any ground station to become a network node. This makes satellites ideal for reaching remote or thinly populated areas where fiber optic cables would not be economically feasible. Fiber optic systems also require the establishment of a right-of-way for the cable to be laid, which can be difficult and costly. Currently, Ku-band satellite ground facilities do not require FCC clearance (C band does require a permit).
3. Satellites can be easily reconfigured while in orbit to cover different geographical areas.
4. Total network failures are unlikely with satellite systems, except for catastrophic failure of the satellite itself. Storm damage to individual antennas will not affect the rest of the network.

Recent satellite advances affecting the comparison are:

1. An increased number of 14/12 gigahertz-band satellites have been placed in orbit. These have greater amplifier power than earlier satellites.
2. Improved solid-state amplifiers, which are replacing traveling-wave tubes, are more reliable, compact, and cheaper than the older devices.
3. On-board facilities for switching and reusing frequencies have been included, which will increase satellite capacity.
4. Mobile satellite communication services are being planned [6.17 through 6.19].
5. VSAT systems are increasingly being introduced to handle point-to-multipoint data needs.

Judging by current and planned expansion of both satellite and fiber optic technologies, they are both likely to continue to play significant roles in telecommunication [6.20, 6.21].

6.1.6 Satellite Regulation and Management

International Telecommunication Union

The ITU, headquartered in Geneva, Switzerland, has the responsibility for regulating international satellite communication, primarily through the CCIR and the IFRB, as discussed in Chapter 2. The ITU periodically convenes a World Administrative Radio Conference (WARC) as well as a Regional Administrative Radio

Conference (RARC). These groups have defined the frequencies to be used for satellite transmission. The first WARC took place in 1971; at that meeting,the IFRB of the ITU was charged with coordinating and authorizing orbital locations, compiling technical data on each satellite system, arbitrating disputes among member governments, and resolving potential mutual-interference problems [6.11]. The IFRB carries on the day-to-day activities of the ITU related to registration of frequencies and radio interference problems. The detailed duties of IFRB include [6.22]:

1. Recording the frequency assignments made by various countries (including date, purpose, and technical characteristics of each assignment) for the purpose of ensuring formal international recognition of such assignments;
2. Recording the position assigned by various countries to geostationary satellites;
3. Advising member countries of potential interference problems for the purpose of "equitable, effective and economic use of the geostationary orbit."
4. Performance of any additional duties related to the assignment and utilization of frequencies or geostationary orbit as provided for in the Radio Regulations of the ITU, or as directed by a competent entity of the ITU.

The ITU Radio Regulations (discussed in Chapter 2) govern numerous aspects of the provision of satellite radio services. Among the most important functions of the ITU relating to satellite services is its authority to define new radio services and allocate frequencies for such services. Frequencies allocated to specific satellite services and their regulatory status are listed in the International Table of Frequency Allocations, Radio Regulations, Section IV, Article 8.

The international Radio Regulations include allocations for the following satellite services (also see Table 6.2): aeronautical mobile satellite, amateur satellite, fixed satellite, intersatellite, land mobile satellite, maritime mobile satellite, meteorological satellite, mobile satellite, radio navigation satellite, space operation, space research, and standard frequency and time signal satellite. The following services are defined in the Radio Regulations but do not have specific allocations within the Table of Allocation: aeronautical radio navigation satellite, maritime radio navigation satellite, and radiodetermination satellite.

Services that are considered "primary" or "permitted" have equal rights in operating in the designed frequency band. Services that are considered "secondary" are permitted to operate in the designed frequency band only if they do not cause harmful interference to stations or the primary services to which frequencies are already assigned, or may be assigned in the future. Secondary services cannot claim protection from harmful interference from stations of a primary service to which frequencies have already been assigned or may later be assigned. They can, however, claim protection from interference from stations of the same or other secondary services to which frequencies may be assigned at a later date [6.22].

Table 6.2
ITU Recognized Satellite Radio Services, as of May 1985

Service	*Description*
Aeronautical Mobile Satellite Service	A mobile satellite service in which mobile earth stations are located on-board aircraft. Survival and emergency positioning stations also are considered part of this service.
Aeronautical Radionavigation Satellite Service	A radionavigation service in which earth stations are located on-board aircraft.
Amateur Satellite Service	A radiocommunication service using satellites for the same purpose as amateur radio service.
Broadcasting Satellite Service	A radiocommunication service in which signals transmitted or retransmitted by space stations are intended for direct reception by the general public (individual reception as well as community reception; also known as *direct broadcast satellite*, DBS).
Earth Exploration Satellite Service	A radiocommunication service between earth stations and one or more satellites, which may include links between satellites, whereby (1) information relating to the characteristics of the earth and its natural phenomena is obtained from active or passive sensors on earth satellites; (2) similiar information is collected from airborne or earth-based platforms; (3) such information may be distributed to earth stations within the system concerned; (4) platform interrogation may be included. This service may also include feeder links necessary for its operation.
Fixed-Satellite Service	A radiocommunication service between earth stations at a specified fixed point when one or more satellites are used; in some cases, this service includes satellite-to-satellite links, which may also be affected in the intersatellite service. The fixed-satellite service may also include feeder links for other space radiocommunication services.
Intersatellite Service	A radiocommunication service providing links between artifical earth satellites.

International Telecommunication Satellite Organization (INTELSAT)

Another important source of international law governing the use of communication satellites is the Agreement Relating to the International Telecommunications Satellite Organization (INTELSAT). The INTELSAT system, which provides international telecommunication services throughout the world, was brought into existence by this treaty. In 1964, ten countries agreed to join with COMSAT Corporation (discussed below) to form a single global system. INTELSAT soon introduced a plan for a worldwide system of geosynchronous satellites located over the Atlantic, Pacific, and Indian oceans. In June 1965, the INTELSAT I satellite,

Table 6.2
ITU Recognized Satellite Radio Services, as of May 1985 *continued*

Service	*Description*
Land Mobile Satellite Service	A mobile satellite service in which transportable earth stations are located on land.
Maritime Mobile Service	A mobile satellite service in which transportable earth stations are located on-board ships; survival craft stations and emergency position-indicating radio beacon stations may also participate in this service.
Maritime Radionavigation Service	A radionavigation satellite service in which earth stations are located on-board ships.
Meteorological Satellite Service	An earth exploration satellite service for meteorological purposes.
Mobile Satellite Services	A radiocommunication service: (1) between mobile earth stations and one or more space stations, or between space stations used by this service; (2) between mobile earth stations by means of one or more space stations. This service may also include feeder links necessary for its operation.
Radiodetermination Satellite Service	A radiocommunication service for the purpose of radiodetermination (location and positioning) involving the use of one or more space stations.
Radionavigation Satellite Service	A radiodetermination satellite service used for the purpose of radionavigation. This service may also include feeder links necessary for its operation.
Space Operation Service	A radiocommunication service concerned exclusively with the operation of spacecraft, in particular, space tracking, space telemetry, and space telecommand.
Space Research Service	A radiocommunication service in which spacecraft for other objects in space are used for specific or technological research.
Standard Frequency and Time Signal Satellite Service	A radiocommunication service using space stations on earth satellites for the same purpose as those of terrestrial standard frequency and time signal service. This service may also include feeder links necessary for its operation.

better known as "Early Bird," was placed in service, establishing the first satellite pathway between the United States and Europe and, for the first time, making live transoceanic television programming possible. INTELSAT was formally established in 1973, when the following two agreements entered into force: (1) the Agreement Relating to the International Telecommunication Satellite Organization, describing the organization's purposes, functions, and structure; and (2) the Operating Agreement Relating to INTELSAT, which encompasses detailed technical, operational, and financial aspects of the organization.

The guiding principles of the INTELSAT organization, as set forth in the preamble to the agreement, are identified as:

- [to facilitate that] communication by means of satellite should be available to the nations of the world as soon as practicable on a global and nondiscriminatory basis;
- to continue the development of telecommunication satellite systems with the aim of achieving a single global commercial telecommunication network that will provide expanded telecommunication services to all areas of the world and that will contribute to the world peace and understanding;
- to provide, for the benefit of all mankind, through the most advanced technology available, the most efficient and economic facilities possible consistent with the best and most equitable use of the radio-frequency spectrum and of orbital space;
- to permit all peoples to have access to the global satellite system and those state members of the ITU so wishing to invest in the system with consequent participation in design, development, construction, including the provision of equipment, establishment, operation, maintenance, and ownership of the system.

Under the agreements, INTELSAT's primary objective is to provide, on a commercial basis, the space segment required for "international public telecommunications services" of high quality and reliability, on a nondiscriminatory basis. Services may also include domestic public telecommunication services between areas separated by areas not under the jurisdiction of the state concerned, and between areas not linked by any terrestrial wideband facilities and separated by natural barriers (the domestic role of INTELSAT is of secondary importance, and does not apply to the United States).

INTELSAT is now owned and operated jointly by more than 118 nations. INTELSAT operated 14 satellites at the end of 1989, which carry the bulk of the international telephone, television, and data communication among some 180 nations and territories. There has been a hundredfold growth in international traffic in the past two decades [6.23]. The INTELSAT satellite system is the dominant carrier of all international communication, handling about two-thirds of all transoceanic traffic. The system consists of high-capacity communication satellites stationed in geosynchronous orbit. The satellites in the global system have 348 international earth station antennas operating with them. Earth stations are owned and operated by entities in the countries where they are located. A growing number of countries are also building earth stations within their boundaries and are using INTELSAT satellites to improve their domestic communication systems.

Satellite services provided by INTELSAT include voice, data, facsimile, live television, and International Business Satellite Services (IBSS) digital paths. The designated U.S. participant to INTELSAT is COMSAT Corporation, which holds the largest investment share (24.7% as of January 1, 1987).

International Maritime Satellite Organization (INMARSAT)

In 1979, the INMARSAT system was deployed as a global maritime satellite communication system. INMARSAT has three satellites positioned over the Atlantic, Indian, and Pacific oceans, with large terrestrial footprints, and is able to provide maritime and land mobile satellite communication services throughout the world. The international organization is composed of 57 member states and parties. Worldwide mobile packet switched data transmission, for two-way messaging and other applications, using low cost mobile earth stations suitable for installation and use on any type and size of mobile platform was demonstrated in 1989. Low profile stubby antennas are used by the mobile stations. INMARSAT is closely modeled after the INTELSAT multilateral agreement. The two pertinent documents are the Convention of the International Maritime Satellite Organization and the Operating Agreement of the International Maritime Satellite Organization. These were signed in 1976 and entered into force in 1979. Similar in structure to the INTELSAT treaty, the INMARSAT Convention set forth the basic policies, functions, and structure of the organization; the Operating Agreement describes the financial, technical, and operational aspects. The preamble to the Convention notes that a very high proportion of world trade is dependent on ships, that satellites could provide a significant improvement in communication services to these ships, and that an international organization dedicated to making this service available on a nondiscriminatory basis through the employment of advanced technology was necessary to make these services a reality. INMARSAT is dedicated to the provision of satellite capacity (space segment) to meet the needs of improved maritime communication. The technology was based on the Marisat system, initially developed by COMSAT to provide services to the U.S. Navy, and later evolved into a commercial service to domestic and foreign interests [6.22].

Communication Satellite Corporation (COMSAT)

In the United States, in accordance with the Communication Satellite Act, COMSAT was formed as a private company in February 1963 to pioneer the development of a commercial international communication satellite system. Under the 1962 Satellite Act, COMSAT is authorized to [6.22]:

- Plan, initiate, construct, own, manage, and operate itself, or in conjunction with foreign governments or business entities, a commercial communication satellite system;
- Furnish, for hire, channels of communication to U.S. communication common carriers and to other authorized entities, foreign and domestic; and
- Own and operate satellite terminal stations, when licensed by the commission.

These activities are known collectively as COMSAT's "statutory mission." In support of these activities, COMSAT is authorized to engage in the following:

- Conduct or contract for research and development related to its mission;
- Acquire the physical facilities, equipment, and devices necessary to its operations, including communication satellites and associated equipment and facilities, whether by construction, purchase, or gift;
- Purchase satellite launching and related services from the U.S. government;
- Contract with authorized users, including the U.S. government, for the services of the communication satellite systems; and
- Develop plans for the technical specifications of all elements of the communication satellite system.

COMSAT provides international communication services to and from the United States through its INTELSAT Satellite Services (ISS). ISS provides space segment services only, which are accessed by customers' earth stations. These stations are located in Maine, Pennsylvania, West Virginia, Washington, California, Hawaii, Guam, American Samoa, the Northern Mariana Islands, the Federated States of Micronesia, the Marshall Islands, and the Republic of Palau. Users of COMSAT Corporation's global communication services gain access through any of three types of U.S. earth stations: (1) large, multipurpose national gateway stations; (2) mid-sized, urban gateway stations; or (3) small earth stations located on or near customer premises.

U.S. customers make arrangements with overseas INTELSAT signatories, their local telecommunication organizations [Post Telephone and Telegraph companies (PTTs)], and user administrations to provide for overseas space segment services matching those provided by ISS.

Federal Communications Commission

The FCC authorized operation of domestic satellite systems by nongovernmental entities, starting in 1970, with Docket 16495. All legally, technically, and financially qualified entities were allowed at that time to enter the field. The commission imposed some restrictions on AT&T and COMSAT for three years. AT&T, for example, was initially limited to provide only certain services; COMSAT could lease capacity to AT&T under the same tariff terms applicable to other carriers. These restrictions sought to discourage monopolistic access to satellite facilities.

In the domestic satellite field the FCC originally adopted, and continues to follow, a flexible leasing policy, giving considerable freedom to applicants to design satellite systems as they saw fit, within regulatory constraints of the ITU. To let the market evolve to meet customer needs, the FCC chose not to prescribe any particular form of ownership for satellite ventures, or any type of satellite system arrangement, except for the applicants to follow prudent business judgment.

As indicated above, the radio spectrum available for satellite operation is allocated to separate radio services that are defined in the international Radio

Regulations. After an international allocation has been made, the FCC must subsequently implement the international allocation domestically by incorporating it in the FCC's Rules and Regulations. However, a substantial amount of spectrum dedicated internationally in the Table of Allocations is not available for public and commercial operation in the United States because it is allocated exclusively to government satellite operations.

WARC-92

The agenda for WARC-92, the first general spectrum allocation conference to be held since 1979, was tentatively set at ITU meetings held in 1989. Topics include HF band allocation for broadcasting service; land mobile service and radiodetermination; and recommendations on the use of geostationary orbit and the planning of services utilizing it. In December 1989, the FCC issued a Notice of Inquiry that contained a summary of the most significant outstanding issues for U.S. concerns. Additional allocations in the 1- to 3-GHz range for mobile communication will be examined.

A U.S. request for allocation of additional spectrum for mobile satellite services, or for a reallocation of existing mobile satellite service spectrum between services, will also be examined. The FCC, following a U.S. reservation to the land mobile satellite service allocation, declined to adopt the ITU allocation of the 1545- to 1559-MHz and 1646.5- to 1660.5-MHz bands to the aeronautical mobile satellite service. Instead, the FCC reaffirmed its decision to allocate these frequencies domestically to the generic mobile satellite services. Furthermore, the FCC recently proposed an amendment to the U.S. Table of Frequency Allocations, by reallocating an additional 33 MHz of L-band spectrum in the 1530- to 1544-MHz and 1626.5- to 1645.5-MHz bands for maritime mobile satellite service to generic mobile satellite services. The U.S. request for additional mobile satellite services spectrum is expected to be controversial at WARC-92 [6.24].

6.1.7 Components of a Satellite Communication System

Bouncing a radio signal off a satellite for reception at one or more stations is one basic way to achieve communication; in fact, the early satellites were of this passive type. However, the loss of signal power associated with such a signal reflection is very large. Unlike radar, communication requires that the signal carry intelligence. The use of very high-powered microwave transmitters in short radar-like bursts would make the transmitter stations prohibitively expensive. The problem of transmitter power and received signal quality was solved in the early 1960s through the use of the relay concept already used in terrestrial microwave applications.

A combination receiver-transmitter, called a *transponder*, is installed in the satellite. The transponder receives the signal from the transmitting earth station, reprocesses it, amplifies it, and retransmits it to the receiving station or stations. This active system allows even small satellite transmitters to provide high-quality transmission to a wide area. The satellite transponder receives information from an earth station (called the uplink) and rebroadcasts it at a different frequency to the receiving earth station (the downlink) to avoid interference with itself. The single antenna handles both input and output signals: a frequency-selective waveguide separates the frequencies of the receiving band from the frequencies of the transmitting band and channels the transmitted signals to their destinations [6.11] (see Figure 6.3).

After amplifying the signal, the satellite antenna typically radiates a broadly spread beam of energy in an earthbound direction. A satellite is placed in geosynchronous orbit in a location that is far enough from other satellites to prevent interference. Thus, a satellite path requires a properly positioned transponder, a set of compatible earth stations, and a transmission technology. During the late 1970s and early 1980s, most major operational satellites (and terrestrial LOS microwave systems) used analog FM. However, the current trend in satellite systems is to employ digital methods.

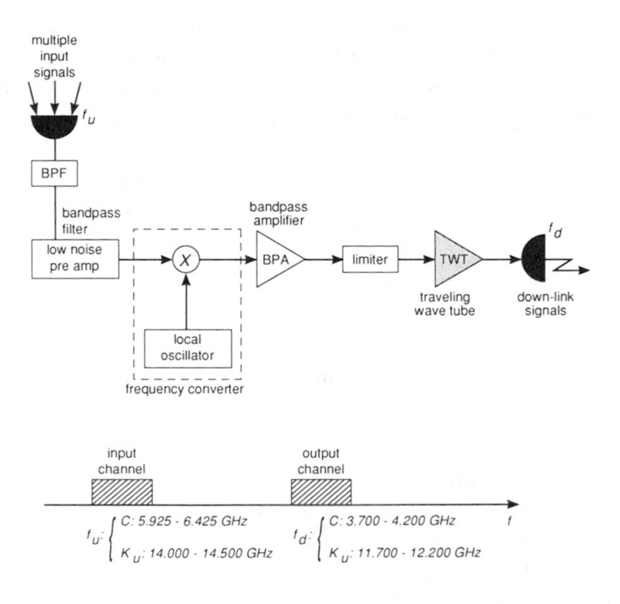

Figure 6.3 Block diagram of a typical commercial satellite.

6.2 APPLICATIONS OF SATELLITE TECHNOLOGY

6.2.1 Common Business Applications

While applications of satellite communication other than video broadcast have not met the optimistic projections of domestic vendors, international long-distance projected communication traffic volume is increasing. Generally, the more a satellite application resembles simple substitution for a high-capacity terrestrial service, the more the application can be expected to be economically threatened in the future. Mobile satellite applications for both voice and data offer a point of differentiation. Impromptu transportable earth stations to support temporary communication centers are one example. Marine applications are less affected by the antenna issues because shipboard space for antenna systems is more readily available. From the industry's point of view, the "door-to-door" promise of the fixed, small earth stations represents the key to the future of satellite data communication. The technological support for small earth station applications is relatively new, and the applications are only now being developed. The purpose of this section is to identify some of the applications of satellite technology. Discussions about some of the most important follow.

Video Teleconferencing (VTC)

Sixty-five percent of all international teleconferencing is relayed via satellite. More than 100 orbiting or scheduled satellites have as their primary application some form of VTC.

Satellite News Gathering (SNG)

This service permits live reporting through truck-mounted, transportable uplinks communicating back to stations equipped with fixed receive-only terminals. An outgrowth of this trend in news gathering is the development of national and regional cooperatives that deploy a sharing arrangement using one central SNG system.

Point-to-Point Satellite Data Transmission

In traditional satellite applications, the satellite provides a channel to support what appears to the user to be a standard switched or leased communication path. Where very large and powerful earth station facilities are used to support high-capacity transponders, the satellite service is normally sold through carriers.

As the satellite's design goals move from mass-capacity support to the support of larger numbers of earth stations, the environment becomes one in which users provide earth stations for their facilities and no carrier tail circuits are required. Carrier-oriented point-to-point satellite access will become less attractive in the future, accelerating the trend away from satellite use as a carrier service.

The use of satellites in point-to-point systems with customer-owned earth stations, however, raises a number of technical issues. The high cost of transponder space mandates efficient earth station management of segments or time slots. Most demand-access techniques require that a terrestrial circuit link each earth station with the master station to support requests for channel access. The cost of the arbitration technology sets a lower boundary on the capacity of a small earth station and results in the terrestrial control link having greater capacity than the satellite path. Earth stations with a capacity equivalent to a few T1 links are economical in terms of capacity for investment, and earth stations with a capacity of less than two 56 kb/s links are probably not reasonable. The exact capacity "break-even" point will be determined by the cost trends in earth stations, by LEC rates, and by long-distance rates. Note that T1 costs have been going down substantially since 1987, particularly in the midwest where there has been a glut of inter-LATA fiber capacity. International services, either voice or data, can be expected to remain economical via satellite through the early 1990s, but transoceanic fiber cables are already being deployed.

Multipoint Satellite Data Transmission

In multipoint satellite applications, a single master earth station supports multiple slave stations, similar to a situation in which a host supports multiple terminals. This, in effect, makes the satellite and earth stations a large multidrop line (see Figure 6.4). A terrestrial multipoint application is characterized by a central host facility that serves as the source of most outbound traffic and the destination of most traffic generated elsewhere. Such applications tend to produce much more outbound flow than inbound flow; terminal screens are larger than operator inquiries. The master earth station acts not only as the coordinator of remote channel access but as the "host" facility. The traffic from the central point, called "forward channel traffic," is sent to the satellite and may be addressed to any or all remote locations. "Back-channel" traffic from the remotes may be of very low volume, permitting contention-controlled access. A master station accepts requests from a large number of remotes for information. These requests are queued to a computer that retrieves the data and sends it to the uplink of the master station. Each remote station receives all blocks of data, but skips those that the local user has not requested [6.25].

Multipoint systems are practical as long as the back-channel usage is not sufficiently high to warrant sophisticated access control techniques. Once traffic

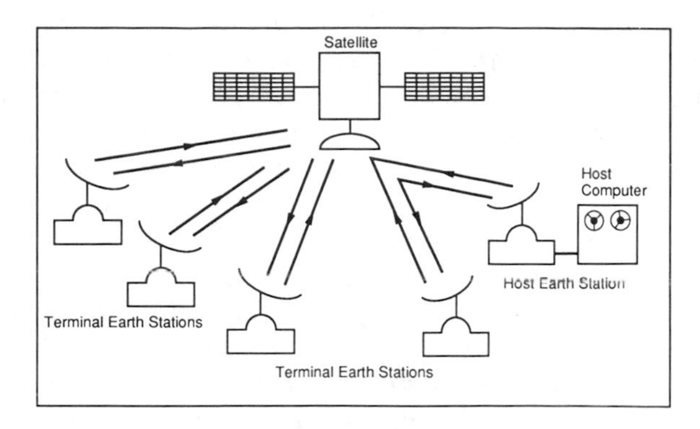

Figure 6.4 A multipoint satellite network.

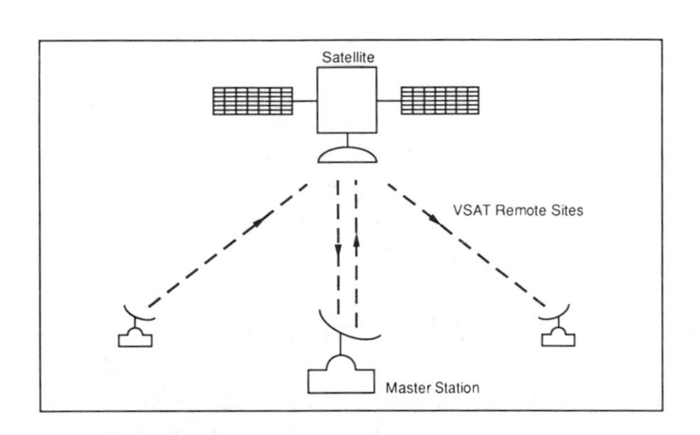

Figure 6.5 Dual hop for end-to-end communication.

volume becomes high enough to affect the efficiency of contention access systems, more sophisticated demand multiplex methods are needed. In many cases, providing the remote stations with back-channel communication via some other technology, such as terrestrial leased line, dialup lines, or even public packet network, is much easier than attempting to support it through the satellite.

Transmissions between end stations are handled by allowing the master station to act as a relay, as shown in Figure 6.5. This involves an extra satellite hop for such transmissions because the message must go from source to satellite to master, then return.

One of the principal limitations of satellite multipoint technology is the fact that it requires a different type of user-system interaction than that of conventional terrestrial communication environments. Most users would expect to be able to use current host and terminal devices for economic reasons. Protocols, such as in IBM's SNA protocol suite, are rich in control transmission exchanges between terminal and host. This type of short, frequent message creates congestion on a contention-oriented satellite environment. One way to eliminate the problem is to incorporate a form of protocol converter in each terminal-supporting earth station. This protocol would carry on the interactive control dialogue with the local device and would present only data on the satellite link.

Even packet-switching protocols such as those used by LANs must employ special techniques to adjust to the satellite environment. The classical CSMA/CD techniques of listening for direct interference will not work in a satellite environment because the earth stations cannot directly receive uplink signals from others because the power is too low. Instead, each station monitors the downlink and "listens" for its transmitted traffic. If it does not appear within a specified delay interval, the station assumes that the traffic was garbled through a collision.

6.2.2 Direct Broadcast Satellite (DBS)

A DBS environment is essentially a small earth station multipoint system with a zero volume of back-channel traffic. In this mode, the earth station antenna system needs only be large enough to acquire a reasonable signal from the satellite. Using an antenna that can be as small as 3 feet in diameter (and providing the satellite is operating in the high frequency band), earth station costs can be less than $1000 in 1990 and $300 in 1994. The front end of the customer's receiver will be mounted on the antenna and will down-convert the high frequency signal to 1 to 2 GHz, a frequency that can be carried over relatively inexpensive cable to the TV receiver for further processing [6.26]. DBS problems include technical issues on standards and encryption to prevent unauthorized interception of signals.

A DBS operates in the 12.2- to 12.7-GHz range for the downlink, and the 17.3- to 17.8-GHz range for the uplink feeders. Interference from other applications is likely to be a problem with small, inexpensive earth stations. A very high effective isotropically radiated power (EIRP) is also needed to support these earth stations, typical in the 52-dBW EIRP range. This would probably require a 200-W satellite transmitter system, and 6 to 8 transponders (compared to 12 to 16 transponders for a normal satellite).

In the early 1980s, the FCC believed that space broadcast of TV signals directly to users would serve public communication needs and, consequently, authorized construction of domestic DBS-TV systems. The FCC then began accepting applications for the building of these systems. DBS systems are not to be confused with traditional C-band and Ku-band reception, which is fairly common today. The FCC also proposed a high power satellite zone within the geosynchronous orbit. Originally, this zone was to be located on the arc between 97 and 93 degrees west longitude, at 1.5 degree spacing. Later the decision was made to have an eastern portion from 75 to 79 degrees west longitude, and a western portion from 132 to 136 degrees west longitude; the spacing was set at 2 degrees [6.26].

DBS deployment has experienced problems and delays. DBS transmission may be costly, being estimated at $20 million per year per TV channel (transponder). In addition, it may face competition from terrestrial coaxial or fiber CATV. In the United States, approximately 50% of the 90 million television households are served by cable; by comparison, only 1% of the British homes have access to cable. DBS service in the United States was scheduled to begin in 1986; however, due to delays, the service had not yet started at the end of 1989. DBS applicants have asked for extensions and revisions to their original FCC applications. Approximately one dozen companies have applications on file with the FCC to build and launch DBS systems [6.26]. A 108-channel DBS providing digital TV and audio in the United States is planned to be launched in 1993. Dubbed "Sky Cable," the $1 billion system will be operated by four media companies, including NBC. Sky Cable aims at delivery of digital audio and HDTV to 12- to 18-in. receivers costing only $300 [6.27].

As of early 1990, the Europeans had two functioning DBSs, Astra-1 and TDF-1 (the programming for TDF-1, however, was still being decided). The Japanese are currently broadcasting HDTV and other programming on their DBSs (BS-2a and BS-2b; BS-3 was scheduled for initiation in 1990, bringing the households with access to DBS to 3 million). In spite of this modest success, some observers question its future commercial viability [6.28].

6.2.3 Satellite Application Trends

New communication applications for satellites are appearing. However, in the foreseeable future, the bulk of the international traffic will remain the conventional speech telephone signal.

Video

Video will experience a leveling off of growth, but will remain the major application for satellites. Proponents of video claim it will represent 50 to 65% of all satellite usage in the future. Many video signals are now scrambled and transmitted using a technique called B-MAC. MAC (multiplexed analog component) signals are time-multiplexed with a digital burst containing digitized sound, video synchronization, and other signals.

Data

Industry analysts predict data to be the number one growth area for satellite applications, with little chance of saturation because it is "transponder efficient," requiring relatively small amounts of bandwidth. VSAT technology may become a common method of transporting data.

Voice

Voice will remain in "third place" through the digitization techniques that could effectively reduce 64 kb/s voice to 16 kb/s. The echo and delay conditions inherent with satellites will continue to discourage use in this area.

6.3 TRANSMISSION CHANNEL

This section describes the basic radio bands used for satellites.

6.3.1 Satellite Radio Transmission Bands

Of the two major frequency bands currently used for domestic fixed-satellite commercial earth stations, the lower frequency one is the C band. The C band employs the 3.7- to 4.2-GHz region in the satellite-to-ground (downlink) direction and 5.925 to 6.425 GHz in the ground-to-satellite (uplink) direction. The C band is used by approximately 20 domestic satellites in geosynchronous orbit. The second band is the Ku band. The Ku band uses the 11.7- to 12.2-GHz region downlink and the 14.0- to 14.5-GHz region uplink, and also has approximately 20 domestic satellites in service.

Other important bands are as follows: The Ka band, which may find future use for fixed-satellite services, has a frequency allocation of 17.7 to 21.2 GHz (downlink) and 27.5 to 31.0 GHz (uplink). Broadcasting satellites (DBS) are allocated the 12.2- to 12.7-GHz band (downlink) and 17.3- to 17.8-GHz band (uplink). Mobile satellite has a primary allocation in the band from 20.2 to 21.2 GHz (downlink) and 30.0 to 31.0 GHz (uplink), and as a secondary allocation in the bands 19.7 to 20.2 GHz (downlink) and 29.5 to 30.0 GHz (uplink) [6.22].

Tables 6.3, 6.4, 6.5, and 6.6 [6.1, 6.11] provide a summary of key bands, and Table 6.7 provides a more comprehensive listing of allocations.

The rest of this chapter deals with transmission at the C and Ku bands. Substantial differences in the operational characteristics for the two bands include the following.

Environmental. Rain, snow, and water vapor attenuate radio signals differently, with the Ku band being about five times more susceptible to these problems than the C band. Most Ku-band designers resort to extra power margins (5 or 6 dB typically), dynamic power sharing, varying data rates, or variable-rate FEC coding to minimize the impact of environmental losses.

Table 6.3
Satellite Frequency Bands (Major Bands Only)

Band	Uplink (Bandwidth)	Downlink (Bandwidth)	Use
C	5.925–6.425 GHz (500 MHz)	3.700–4.200 GHz (500 MHz)	Traditional geosynchronous satellite communications, used by Canada, France, Indonesia, and U.S. INTELSAT
	5.725–6.275 GHz (550 MHz)	3.400–3.900 GHz (500 MHz)	INTERSPUTNIK, USSR
X	7.925–8.425 GHz	7.250–7.750 GHz	Government and military telecommunication satellites
Ku	14.000–14.500 GHz (500 MHz)	10.700–11.200 GHz (500 MHz)	INTELSAT, EUTELSAT, and USSR satellites
K	14.000–14.500 GHz (500 MHz) 14.000–14.250 GHz (250 MHz)	11.700–12.200 GHz (500 MHz) 12.500–12.750 GHz (250 MHz)	Telecommunication satellites, used by Canada, France, and United States
Ka	27.500–31.00 GHz	17.700–21.200 GHz	Technology research satellites, used by Europe, Japan, and United States

Terrestrial Interference. Terrestrial sources can cause mutual interference that necessitates prior coordination between users in the same vicinity (within 100 km of each other). Three factors mitigate this problem:

1. Automated computer processing has reduced the time and cost of the coordination exercise.
2. Spread spectrum modulation (discussed in Chapter 5 and further below) reduces susceptibility to the interference typically encountered by 12 to 30 dB.
3. The small size of the antenna allows it to be readily placed in naturally shielded locations (such as below parapets, behind air conditioners, on the ground or pole-mounted while protected by building walls).

Adjacent-Satellite Interference. FCC rulings in the mid-1980s required satellites to be spaced 2 degrees of longitude apart over the equator for U.S. service, compared to the previous 4 degrees. At this close spacing, small earth antennas may transmit and receive measurable signals to and from the satellites nearest to the target satellite. The higher Ku-band frequency results in a narrower antenna beam width, thereby deteriorating signal discrimination. This difference is mitigated at the C band through the use of spectrum spreading.

Cost. Amplifiers used in the micro earth stations are less expensive at C band. Recent networks have been averaging around $10,000 per site for the licensing,

Table 6.4
Mobile Satellite Allocations

Frequency Range (GHz)	*Use Allowed*	*Status*	*Notes*
1.530–1.533	Land mobile satellite	P	—
1.530–1.535	Maritime mobile satellite	P	—
1.530–1.535	Space operations	P	—
1.533–1.544	Land mobile satellite	S	Low data rates
1.535–1.544	Maritime mobile satellite	P	—
1.544–1.545	Mobile satellite	P	—
1.545–1.555	Aeronautical mobile satellite	S	Use by the public
1.545–1.559	Aeronautical mobile satellite	P	—
1.555–1.559	Land mobile satellite	P	—
1.558–1.610	Radionavigation satellite	P	—
1.610–1.6265	Aeronautical mobile satellite	footnote	RDSS uplink
1.6265–1.6315	Land mobile satellite	S	Low data rates
1.6265–1.6455	Maritime mobile satellite	P	—
1.6315–1.6345	Land mobile satellite	P	—
1.6345–1.6455	Land mobile satellite	S	Low data rates
1.6455–1.6465	Mobile satellite	P	—
1.6465–1.6600	Aeronautical mobile satellite	P	—
1.6465–1.6565	Aeronautical mobile satellite	S	Use by the public
1.6565–1.6605	Land mobile satellite	P	—
1.6600–1.6605	Aeronautical mobile satellite	P	—
1.6605–1.6684	Space research	P	—

Key: P = primary; S = secondary; footnote indicates status in Table of Frequency Allocations.

equipment, and installation. Comparable licensing and installations at Ku band cost approximately $15,000 per site. The Ku-band satellite also is more expensive, with an in-orbit cost of $130 million compared to a typical $85 million for a C-band satellite.

Licensing. Every transmitting earth station requires a license granted by the FCC before it can begin operation. For the Ku band, the licensing process does not have a mandatory cochannel terrestrial (shared-frequency) coordination but is otherwise similar. Both C-band and Ku-band sites must perform adjacent-satellite coordination (the C band must do terrestrial coordination as well). The "prior coordination notice" is required at C band and is recommended at Ku band. This is because, in urban areas, the 11-GHz terrestrial microwave band may cause adjacent-band desensitization of a Ku-band satellite receiver. For C-band spread spectrum applications, the FCC process has been substantially streamlined to allow rapid processing of hundreds of applications each week.

Table 6.5
Allocations above 31 GHz

Service	*Additional Assignments (in GHz) (in the range of 31 to 275 GHz)*
Amateur satellite service	47–47.2, 75.5–76, 76–81 (S), 142–144, 144–149 (S), 241–248 (S), 248–250
Broadcasting satellite service	40.5–42.5
Earth exploration satellite service*	65–66
Fixed satellite service	37.5–40.5, 42.5–43.5, 47.2–50.2, 50.4–51.4, 71–75.5, 81–84, 92–95, 102–105, 149–164, 202–217, 231–241, 265–275
Intersatellite service	32–33, 54.25–58.2, 59–64, 116–134, 170–182, 185–190
Mobile satellite service	39.5–40.5, 43.5–47, 50.4–51.4, 66–74, 81–84, 95–100, 134–142, 190–200, 252–265
Radionavigation satellite service	43.5–47, 66–71, 95–100, 134–142, 190–200, 252–265
Space research*	31–32.3, 34.2–35.2, 50.2–50.4, 51.4–59, 64–66, 86–92, 100–102, 105–126, 150–151, 164–168, 174.5–176.5, 182–185, 200–202, 217–231, 235–238, 250–252
Standard time & frequency satellite	31–31.3 (S)

*In addition there are many frequency allocations for passive sensors. See key in Table 6.4.

6.4 TRANSMISSION IMPAIRMENTS

In addition to the terrestrial and adjacent-satellite interference, the uplink and downlink are subject to several other forms of degradation, including attenuation of the signal, due to the distances involved and to atmospheric absorption; obstruction of the signal by terrestrial objects; interference by satellite transits close to the solar disk; and propagation delay.

6.4.1 Attenuation

The EIRP is the ratio of the transmitted power of the satellite as it would appear at the earth station receiver to that which would be produced by a totally nondirectional antenna with the same transmitter, usually measured in dBW. In essence, EIRP measures the gain of the antenna system; because satellite power is limited, a high EIRP is very desirable (typically, 30 to 40 dB).

Attenuation has the effect of lowering the strength of the signal in relation to the level of noise present on the channel, as discussed in Chapter 1. The most practical way to compensate for these losses is to increase the EIRP of the transmitter and the gain of the receiver. This is most easily done through the use of a

Table 6.6
Allocations below 2.5 GHz

ADDITIONAL ASSIGNMENTS (in MHz)
(in the range of 100 MHz to 2.5 GHz)

Aeronautical mobile satellite service
1545–1559, 1610–1626.5 (footnote),
1646.5–1660.5

Amateur satellite service
21–21.45, 24.89–24.99, 30.05–30.01,
144–146, 1260–1300, 2300–2450

Broadcasting satellite service
614–806 (FM-TV)

Earth exploration satellite service
401–403 (S), 1530–1535 (S)

Land mobile satellite service*
1530–1533, 1533–1544 (S, low data rates),
1555–1559, 1626.5–1631.5 (S, low data rates)

Land mobile satellite service* (cont'd)
1631.5–1634.5,
1634.5–1645.5 (S, low data rates),
1656.5–1660.5, 1660–1661

Maritime mobile satellite service
1530–1544, 1626.5–1645.5

Meteorological satellite service
136–138, 400–403 (S), 460–470, 1670–1710

Mobile satellite service*
235–322 (footnote)
335.4–399.9 (footnote),
406–406.1, 608–614 (S),
806–896 (footnotes), 1544–1545,
1645.5–1646.5

Radionavigation satellite service
149.9–150.5, 399.9–400.15, 1215–1260,
1558–1610

Space operations
30.05–30.1, 137–138, 148–149.9, 400–402,
1427–1429, 1525–1535, 2025–2160,
2200–2290

Space research
136–138, 138–143.6 (S), 400–401,
1400–1427, 1660.5–1668.4, 2025–2120,
2220–2290, 2290–2300

Standard time & frequency satellite service
400.05–400.15

*Some allocations took place after January 1, 1990. See key in Table 6.4.

higher gain antenna system. At high frequencies, an almost direct relationship exists between the antenna size and the effective gain in both the transmitting and receiving directions.

Spread spectrum technology can also increase the effective gain by making the entire system less vulnerable to low-level noise. Attenuation is more pronounced at higher frequencies (Ka and Ku bands experience more loss than C band). However, the higher antenna gains possible at the higher frequencies largely compensate for this effect.

6.4.2 Obstruction Interference

Obstruction interference has proven to be an increasing problem because earth station locations are more likely to be integrated with normal business environments. Satellite transmission is LOS; in northern latitudes (in the northern hemisphere), where the angle of the antenna to the horizon is small, buildings, signs, or other obstructions can block the path. Few areas have building codes to protect satellite access [6.38].

The Ku band is subject to obstruction scatter interference from rainfall because even liquids become radio-reflective at high frequencies. This scatter effect can be a major problem in areas where annual rainfall is high.

6.4.3 Interference

Interference to microwave transmissions may come from random radio noise generated in space, from terrestrial sources, or from competing users of microwave frequencies. C-band transmission is most subject to competing station interference, but is less subject to other forms of noise. In general, galactic and solar interference is greater at the Ku band, but that band has little competing microwave use.

Most interference problems are unintentional and are caused by lax uplinking procedures. Everyone wants to use smaller antennas, but the smaller the antenna, the wider the beam width. The 2-degree spacing could become a problem if interference cannot be controlled. In response to this concern, the FCC formed the Reduced Orbital Spacing Advisory Committee in the spring of 1985. The committee was asked to advise the FCC on setting new rules, some of which include: (1) the establishment of new design standards for antennas to operate without interference in a 2-degree environment; (2) a suggested program of formal training and certification of earth station operators, not currently licensed; and (3) coding of all video transmissions so that signal "leakage" can quickly be traced to its source.

6.4.4 Propagation Delay

As indicated, a signal takes 240 ms to travel to a satellite in geosynchronous orbit and back. In the case of interactive transmission of data, that propagation delay causes the greatest problems. The sending station transmits a block of the message and awaits the positive reception indication (ACK) or retransmit request (NAK) from the receiver. Each block is delayed by 0.24 s in transit. The effect is to add 0.48 s of delay between message blocks. Depending on the message and block size, this can reduce the link's effective transmission rate drastically, down to around 10%.

Modern "sliding window" protocols such as SDLC and HDLC permit multiple data blocks to be outstanding and unacknowledged without stopping the sender from transmitting. Most forms of sliding window protocol use a window size of 8, and may also go as high as 128. If the time required to send this many blocks is less than twice the satellite delay (or 0.48 s), the sender will be flow

Table 6.7
Space Frequency Allocations for North and South America (Region 2) above 2.5 GHz and below 31 GHz

Satellite Service (Designation)	*Frequency (GHz)*	*Wavelength (mm)*	*Bandwidth (MHz)*	*Transmission Direction*	*Sharing Services*
Fixed (FSS)	2.5–2.69	112–120	155	2-way	BSS
	3.4–4.2	71–88	800	S-E	SR
	4.5–4.8	63–67	300	S-E	EEX
	5.85–7.075	42–51	1230	E-S	SFT (at 6.427 GHz)
	7.25–7.75	39–41	500	S-E	MSS
	7.9–8.4	36–38	500	E-S	WS
	10.7–11.7	26–28	1000	S-E	SR
	11.7–12.2	25–26	500	S-E	IMSS
	12.2–12.7	24–25	500	S-E	SR (deep space)
	12.7–13.25	23–24	550	S-E	
	14.0–14.5	21	500	E-S	BSS (with coordination)
	18.1–21.2	14–17	3100	S-E	
	27.0–31.0	10–11	2500	E-S	
Broadcasting (BSS)	2.5–2.69	112–120	155	S-E	FSS
	11.7–14	21–26	500	S-E	SR
	14–14.8	20–21	500	E-S	EEX
	17.3–18.1	17	800	2-way	IMSS
	22.5–22.55	14	50	S-E	RL
	22.55–23	13	450	S-E	F M IMSS (with coordination)
Space Research (SR)	3.1–3.3	91–97	200	S-E	RLSS
	4.204–4.4	68	4	S-E	EEX
	5.25–5.35	56–57	100	S-E	FSS
	5.65–5.725	52–56	75	S-E	BSS
	7.125–7.135	42	75	S-E	IMSS
	8.4–8.5	36	100	S-E	EEX (deep space)
	8.55–8.65	35	100	S-E	
	9.5–9.8	31–32	300	S-E	EEX (active)
	10.6–10.7	28	100	S-E	
	12.75–13.25	23–24	500	S-E	
	13.25–13.4	22–23	150	S-E	
	13.4–14.3	21–22	600	E-S	
	14.5–15.35	20	550	E-S	
	15.35–15.4	20	550	E-S	
	16.6–17.1	18	500	E-S	
	17.2–17.3	18	100	E-S	
	18.6–18.8	16	200	S-E	
	21.2–21.4	14	200	S-E	
	22.21–22.5	14	290	S-E	
	23.6–24.0	13	400	S-E	
Earth Exploration (EEX)	2.655–2.7	112	35	S-E	FSS
	3.1–3.3	91–97	200	S-E	BSS
	4.204–4.4	71–88	800	S-E	SR
	5.25–5.35	56–57	100	S-E	RLSS
	15.35–15.4	20	50	E-S	SFT
	17.2–17.3	18	100	E-S	MSS

Table 6.7 continued
Space Frequency Allocations for North and South America (Region 2) above 2.5 GHz and below 31 GHz

Satellite Service (Designation)	*Frequency (GHz)*	*Wavelength (mm)*	*Bandwidth (MHz)*	*Transmission Direction*	*Sharing Services*
Earth Exploration (EEX)	18.6–18.8	16	200	S-E	SR (active)
	21.2–21.4	14	200	S-E	
	22.21–22.5	14	290	S-E	
	23.6–24.0	13	400	S-E	
	24.05–24.25	13	200	E-S	
	25.25–27.5	11–12	2250	E-S	
Radiolocation (RLSS)	3.1–3.3	91–97	200	S-E	EEX
	5.25–5.35	56–57	100	S-E	SR
	8.55–8.65	35	100	S-E	
	9.5–9.8	31–32	300	S-E	
Standard Frequency & Time (SFT)	4.2–4.204	71	4	S-E	RLSS
	13.4–14.0	21–22	600	E-S	EEX
	20.2–21.2	14–15	1000	S-E	SR
	25.25–27.0	11–12	1750	E-S	MSS
	30.0–31.0	10	1000	S-E	FSS EEX (space-to-space)
Aeronautical Mobile (AMSS)	4.5–4.8	57–60	250	S-E	
	15.4–15.7	19	300	E-S	
Amateur Satellite (AS)	5.725–5.85	51–52	50	S-E	
	10.45–10.5	30	50	S-E	
	24.0–24.05	13	50	S-E	
Space Operation (SO)	7.125	42	4	E-S	
Mobile (MSS)	7.25–7.375	41	500	S-E	FSS
	7.9–8.025	38	500	E-S	SFT
	19.7–20.2	15	500	S-E	EEX
	20.2–21.2	14	1000	S-E	
	29.5–31.0	10	500	E-S	
Meteorological (WS)	7.45–7.55	40	500	S-E	FSS
	8.175–8.215	37	500	E-S	
	9.975–10.025	30	50	S-E	
	18.1–18.6	17	500	2-way	
Land Mobile (LMSS)	14.0–14.5	21	600	E-S	FSS BSS SR

Key: AMSS = aeronautical mobile satellite service; AS = amateur satellite; BSS = broadcasting satellite service; EEX = earth exploration satellite; FSS = fixed satellite service; LMSS = land mobile satellite service; MMSS = maritime mobile satellite service; MSS = mobile satellite service; RLSS = radiolocation satellite service; SFT = standard frequency and time satellite service; SO = space operation; SR = space research; WS = meteorological satellite.
Note: Primary, secondary, and permitted services are not indicated.
Source: Compiled and edited by Dennis N. Ricci, Artech House, Inc., based on chart prepared for *Satellite Communications* by Walter L. Morgan, Communications Center, Clarksburg, MD.

controlled, and the effective data rate of the path will be reduced. This means that delay will not be a factor if the sending time for a block is less than one-seventh of that delay, or about 70 ms. Communication systems may have flow control and error control procedures at the transport layer. SNA, for example, defines a pacing limit as one of the generation parameters for the network. Most SNA devices, including the 3274 controller, accept only small pacing limits to minimize transmission buffering. This may result in flow control at the higher level even when link-level sequence numbering has been increased to modulo-128 to minimize delay effects.

In TDMA systems, users may experience additional delays due to frame buffering in the interburst intervals, a delay that rapidly increases with the number of earth stations, and can actually reach several seconds in some designs. Delays of this magnitude make the channel useful only for broadcast applications because neither conversation nor data exchange is possible under extreme delay conditions.

6.4.5 Partial Solutions to Satellite Problems

Coding digital data using FEC techniques resolves most noise-associated errors (FEC was discussed in Chapter 1). This technique appends some number of additional bits on each digital byte, generally 4 per 8-bit byte. The value of the appended bits is based on processing of the prior 8 bits, so that even if an error in the composite 12 bits occurs, sufficient information redundancy permits the original 8-bit data portion to be recovered. FEC can simplify the protocols used and maintain a higher rate of transmission.

The problems of propagation delay can be minimized by proper protocol design, but equipment using those protocols is already in service. Some environments affected by satellite propagation delay can be handled through the use of a satellite delay compensator. This device operates at each end of the satellite path and provides a "local acknowledgment" of information sent before that information ever reaches the satellite path. The compensator essentially accepts and acknowledges each message on behalf of the other station, taking responsibility for its error-free delivery to the opposite side of the link. Compensators with FEC to prevent undetected errors can increase the efficiency of older protocols in satellite environments.

6.5 SATELLITE ACCESS SCHEMES

The primary factor that affects the capacity of a satellite is the limit in the number of transponders that may be active. A single satellite could contain transponders for the C and Ku bands, providing it with considerably greater channel capacity than any single band could provide. However, the multiband satellites are inherently more expensive because of the multiple antenna systems involved. One approach to obtain greater effective bandwidth is to employ frequency reuse. The reuse of a frequency, accomplished through spatial reuse or polarization reuse, multiplies the effective number of channels without increasing the required bandwidth. Spatial reuse of satellite frequencies takes advantage of the directional characteristics of large antenna systems. The coverage zone of that antenna will be small. Multiple antennas with coverage zones that do not overlap can use the same frequency, thus providing more effective bandwidth (see Figure 6.6). Polarization reuse takes advantage of the fact that microwave signals, which are oriented in space in a different way than the receiving antenna, will not produce a significant received signal. An earth station with a vertically polarized antenna will receive microwave information from a satellite antenna of the same polarization. A competing use of the frequency using horizontal polarization would not contribute significantly to the interference level. Spatial and polarization reuse are normally used jointly, with cross-polarization zones placed adjacent to each other so that the boundary areas will not experience interference.

After applying polarization, the problem of dealing with the competing use of the same transponder remains. A satellite transponder must be made available to multiple users to make communication cost-effective. At any point in time, multiple earth stations must be able to transmit information from several users on the uplink, and receive on the downlink in a way that does not cause mutual interference. Satellite access schemes are used to achieve this objective.

Generally, one transponder should not be dedicated to one user. Subdividing a transponder to make it available to several channels and to several earth stations must involve a set of rules to prevent collision of the users. The management of the transponder is called multiple access techniques (MATs). The two basic tech-

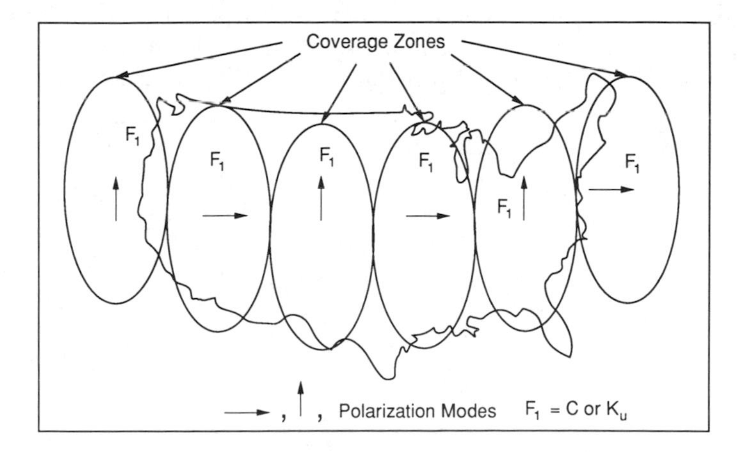

Figure 6.6 Frequency reuse.

niques for the allocation of transponder resources are fixed access and demand access. Each of these can be further segmented as FDMA and TDMA.

6.5.1 FDMA

FDMA assigns each user a specific uplink and downlink carrier frequency within the bandwidth of the transponder, as shown in Figure 6.7. The potential for interference from adjacent channels and the intermodulation distortion of multiple signals passing through the satellite simultaneously are the major shortcomings of this method. FDMA requires a reservation mechanism so that requests for channel capacity can be made through dynamic assignment of transmitting frequency. The satellite channel remains unused during the tens of milliseconds required to switch frequencies. As a result, this multiple-access scheme is optimized for high duty cycle (80 to 100%) traffic, with few channel assignment changes.

Until the 1980s, satellites offered dedicated channels via FDM, and most users allocated these channels on a point-to-point basis. This resulted in inefficient utilization of valuable bandwidth. The traveling-wave tube amplifier (TWTA) receivers in satellites generate a significant amount of intermodulation distortion when operated at full power and with multiple users. To prevent this, the transponder is "backed off" by 3 dB, reducing the distortion but also reducing the EIRP of the satellite. This causes an inefficient use of the satellite transponder, a serious drawback in an environment where transponder space and satellite orbital slots are at a premium.

6.5.2 TDMA

The TDMA system is more complex than FDMA. In TDMA allocation, earth stations share a complete transponder rather than a segment of a transponder. However, they do so for a very short time interval, as shown in Figure 6.8. Each

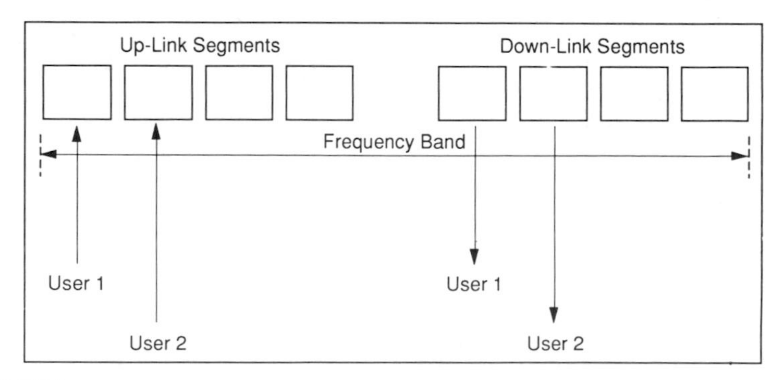

Figure 6.7 FDMA concepts.

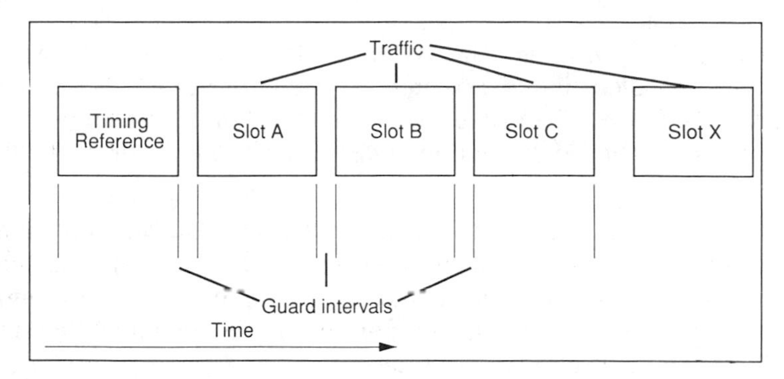

Figure 6.8 TDMA concepts.

earth station is assigned a time slot in a frame. A frame header containing clock timing and supervisory information is supplied by the satellite. Each station then provides a burst of data in the predefined time slot. A guard interval separates the slots to ensure that small differences in propagation delay or station timing precision do not cause collision. Burst lengths may be regulated to provide stations with a variable portion of transponder bandwidth proportional to their traffic requirements.

TDMA provides each station with full use of the transponder for an interval, so no intermodulation distortion results and no backoff is required. The timing of stations, however, must be precise to prevent collision, and information must be buffered. TDMA tends to result in higher earth station costs than FDMA because of the sophistication of the transmission timing. The trend toward TDMA as the *de facto* standard for satellite allocation control is clear. It represents the best way to utilize critical satellite resources for high-capacity applications such as carrier trunk service.

6.5.3 Demand Access

Demand access is an improvement to the basic fixed-assignment logic of either FDMA or TDMA. Demand allocation in TDMA is simpler to implement, requiring only a method of assigning time slots to active channels. The four principal variations in demand allocation follow.

Polled Allocation Systems. These involve a master earth station issuing a poll to all other stations, inviting them to use the transponder if they have traffic. Systems with traffic may seize the channel in response to a poll, or respond with a negative indication if they have none. The system is applicable to certain specialized multipoint environments, but is not generally suitable for high-capacity service with large numbers of earth stations.

Central Allocation on Demand. This system identifies a master station as the controller of time slot allocation and permits other stations to contact that station to reserve a slot for traffic. The connection for reservation may be a terrestrial path to eliminate propagation delay and expedite slot allocation, or a satellite supervisory channel.

Distributed Allocation. This is a system of contention that seizes a channel, and resorts to arbitration in the case of collision. The distributed system is more robust in terms of immunity from failure if the master station is lost, and more easily applied to multiple countries or to situations where designating a "master" might cause political or organizational difficulties.

Contention. In this system, the satellite transponder space is treated similar to a CSMA/CD local network with each earth station transmitting when the link appears to be free. The system was popularized in the University of Hawaii ALOHA packet radio network, as discussed in Chapter 5. Contention systems are not useful for generalized carrier service applications, but may be helpful in specialized multipoint satellite environments.

6.5.4 Spread Spectrum

Spread spectrum increases the channel bandwidth of the signal to make it less vulnerable to interference. It refers to the transmission of a signal using a much wider bandwidth than would normally be required or to the use of narrow signals that are frequency-hopped through the various frequency segments available to the transponder. This approach is called spread spectrum multiple access (SSMA) or code-division multiple access (CDMA). Spread spectrum technology was first adopted for military communication to prevent deliberate jamming or interference resulting from battle conditions [6.29, 6.30]. CDMA operates in three modes: direct sequence (DS), frequency hopping (FH), and time hopping (TH). In the sequel we focus on DS techniques.

Most interference, deliberate or accidental, affects communication because the information transmitted is condensed to a relatively small range of frequencies. An interference source active at the same frequency would produce signals that would mix with the actual communication channel to create errors. If, however, the information bearing signal is dispersed over a wider range of frequencies, the noise impulses and interference affect only a portion of the total information channel and can be filtered.

For commercial applications, spread spectrum technology permits the use of small antennas (1.2 to 1.8 m in diameter). FCC regulations governing interference are related to power per unit bandwidth (power density). Either increasing the antenna size or the signal bandwidth would reduce the power density to acceptable levels. Rather than use a large antenna (which increases costs by a factor proportional to at least the square of the diameter), signal bandwidth is increased to reduce power density and, thus, interference. Spreading mitigates the effects of interference from adjacent satellites that otherwise cannot be avoided at C-band satellite frequencies with small antennas. In point-to-multipoint C-band satellite networks, spread spectrum modulation keeps transmissions originating at adjacent satellites from interfering with the signal coming from the primary satellite to the micro earth station. The inbound signals (from the micro earth stations to the master earth station via the satellite) rely on spread spectrum to permit transmitting with very small antennas, which could otherwise cause interference with adjacent satellites. This antenna use is accomplished by dispersing what could be troublesome signal-power concentrations in narrow bandwidths, and spreading the energy to make it appear to adjacent satellites as a tolerable addition to background noise. See Figure 6.9 for a geometric view of this issue.

The benefits achieved by spread spectrum communication relate to the set of trade-offs that governs earth station cost, antenna size, and satellite spacing. Small, inexpensive earth stations using conventional technology would have difficulty communicating in a C-band satellite environment. The small antenna would have a broad pattern of reception that would make reception of the correct signal very difficult. The same small antenna would have a low EIRP, so it is unlikely that an inexpensive transmitter would generate a signal that the satellite could receive. See Figure 6.10 for a geometric view of this issue. Spread spectrum communication allows the small antenna of an inexpensive earth station to extract the correct codings from a received signal that contains information from several satellites. The uplink signal from that same earth station, similarly coded, can be received easily in spite of its low power.

In CDMA, the information spectrum is spread into a bandwidth many times wider than the bandwidth of data alone using a pseudorandom noise sequence of "microbits." In this technique, each data bit is encoded with a binary sequence of up to several thousand microbits. Any receiver can recover the original information by using the same sequence to decode the encoded data bits. Any other well-selected pseudorandom sequence (selected mathematically) increases the noise

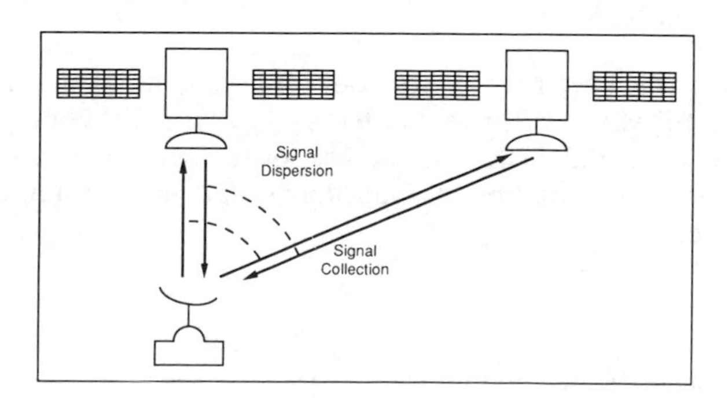

Figure 6.9 The use of small antennas increases the signal dispersion and collection.

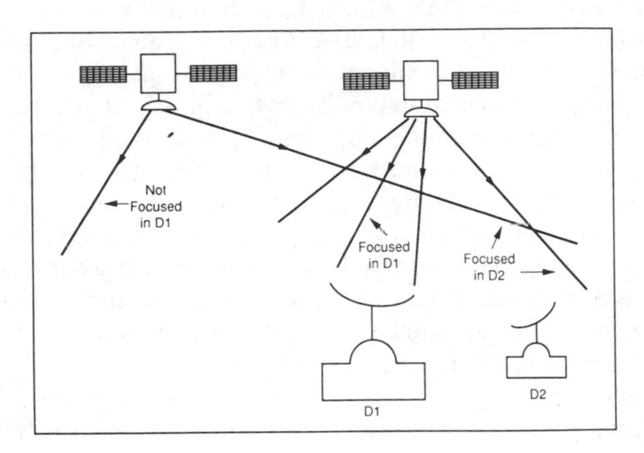

Figure 6.10 The use of increased signal bandwidth with spread spectrum decreases focusing problems.

slightly in the receiver and, therefore, degrades the performance of the receiver by a negligible amount. The aggregate of code-noise increments that increase the receiver noise to the point that reduces the signal-to-noise ratio of the receiver to its threshold of usable operation (usually a rate of 1 bit in 10^7) limits the number of stations simultaneously transmitting in a channel. In a typical network, about 120 stations may be active at one time in a shared channel. The BER increases about 1 order of magnitude for each 25% increase in the number of active stations above the 120 nominal.

The CDMA process is cost-efficient for low duty cycle, low average data rate earth stations. The signals occupy the same frequency channel (hence, require no frequency agility), are time independent (hence, require no network-wide synchronization), and suffer much less from collision-caused retransmission delays.

6.5.5 Transmission Summary

If a case requires high-capacity, continuous transmission of voice and video information, analog FDMA is still a viable technology. In most commercial point-to-point applications, digital TDMA is the best of the current technologies; for low power earth station and multipoint applications, spread spectrum communication offers advantages.

6.6 TRENDS

Many present-day satellite communication systems provide a repeater channel with fixed characteristics and play a passive role in the communication link. However, advanced techniques can be used to provide improved performance in this link and to allocate the satellite's resources dynamically to optimize system efficiency.

Recent satellite systems can include a number of new or relatively new technologies: frequency reuse; onboard processing and switching; use of spot beams; cellular coverage; rapidly scanning antenna beams; and increased transponder power.

6.6.1 Frequency Reuse

Increases in satellite capacity can be achieved by reuse of the available frequency spectrum, as discussed earlier. By reducing the antenna beam's size, which increases the satellite antenna gain, different beams can be directed to illuminate different areas. Orthogonal polarization in a polarization diversity system can also increase the system capacity. Antennas can be made to respond to only one polarization; thus, the same frequency band can be used twice within the same coverage area. Digital modulation techniques (which have a higher spectral efficiency), multiplexing methods, and advanced signal processing techniques such as ADPCM and vocoding (Chapter 1) can further improve the bandwidth utilization of satellite systems.

6.6.2 Onboard Processing and Switching

In a large-scale multiuser distributed system with multiple beams on the satellite, the ability to connect all the users in an effective fashion becomes a difficult task. Satellites can improve the traffic carrying capacity by incorporating satellite switched TDMA (SS-TDMA). With an onboard processor, rapidly changing interconnections can be achieved. This can be accomplished by changing the transmission paths or by dynamically rerouting information. A major cause of the reduction in satellite communication capacity is the inability, in a fixed coverage system, to divert unused capacity in one region to alleviate demand in another region. Onboard processing can be used to reallocate satellite resources to accommodate needs. This provides a method for coupling multiple arrow beams. Depending on the trade-off emphasis, simpler, lower cost earth stations can be realized or, alternatively, reliable, long-lived, and lower cost spacecraft can be used with existing earth stations [6.31, 6.32].

SS-TDMA will eventually supplant conventional "transparent" repeaters with combined onboard amplification and switching, making possible "electronic switching" in space. Because the switching may be controlled from the ground, changing connectivity will be slow. In the future, the switching will increasingly be handled by the satellite on the basis of information contained in the transmitted data bursts, and switching rates will increase.

Frequency reuse programs create a type of communication environment in which multiple conversations may be carried out on different satellite transponders. When these programs are coupled with demand allocation of channel capacity, situations arise in which an earth station in one zone or on one transponder wants to converse with an earth station served by another. Satellite switching of transmission channels is a technology that can deal with the complex transmission environments created by demand access, multizone operation, and frequency reuse programs. In a satellite switching environment, the input of several transponders is passed through a switching matrix that can rearrange the channel allocations among the transponders. The process is similar to the switching that takes place in a telephone switching office. The phrase "switch in the sky" has been used to refer to a satellite that can switch spot beams on and off and switch channels between the beams.

6.6.3 Spot Beams

Highly directional antennas are being used increasingly. Relatively narrow beams are called *spot beams*. They receive over a similarly narrow angle. The smaller the coverage of the antenna, the greater the power received from it in the area it covers. A single large antenna dish can produce many spot beams. Many small feed horns are positioned so that their signals are reflected in narrow beams by the dish, just as a huge concave mirror could be made to reflect narrow light beams from several bulbs [6.32]. This technology was first introduced in the late 1970s.

Different antenna feeds can be switched on and off, thereby selecting the spot beam to be used. Figure 6.11 shows a single-beam satellite [6.33]; Figure 6.12 shows an example of spot beams. Not only do satellites with multiple narrow-beam antennas give a higher EIRP, but the same frequency can also be reused several times for different portions of the earth. For example, if a coverage area, such as CONUS, is covered by 10 spot beams rather than 1 beam, without changing the RF power per transmission channel, the flux density at the ground terminal will be increased by a factor of 10 [6.34]. Also, the satellite can contain more transponders for a given fixed bandwidth. If a satellite has many spot beams, onboard equipment that can switch signals from one antenna to another is desirable. "Space division" implies multiple spot beams from the satellite and some capability to switch between the beams [6.35].

6.6.4 Cellular Coverage

The terrestrial mobile telephone business became a large-scale success after cellular frequency reuse was introduced to increase effective system capacity. In the same fashion, increasing the capacity of a satellite communication system for point-to-

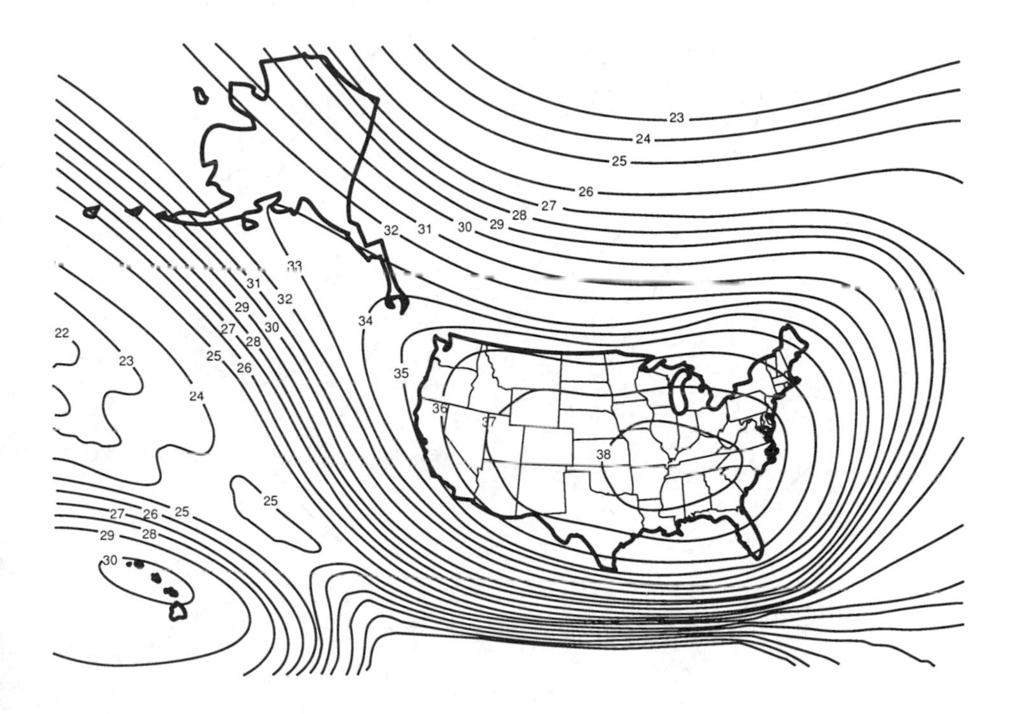

Figure 6.11 WESTAR 4 EIRP contours (direct-polarized transponders).

point communication is possible by utilizing cellular frequency reuse. This method is being advocated by some [6.34].

A satellite antenna beam diameter that permits 13 cells over CONUS would be practical. This system would permit four frequency uses by means of geographical separation in addition to the use of both orthogonal polarizations in each beam. By doing this, the 500-MHz Ku-band allocation will be used eight times, resulting in a total effective bandwidth of 4000 MHz; this is equivalent to 96 transponders with 36-MHz capacity. If this concept is ever realized, it will probably be by the year 2000.

6.6.5 Scanning Antenna Beams

Traditionally, satellite antenna beams have always been fixed. Except for minor movement to redirect the beams as needed to change network reconfigurations, beams remain pointed toward given coverage areas. A technique that will allow optimum use of satellite capacity is a scanning spot beam system. Essentially, SS-TDMA techniques are combined with fast-moving electronically steered antenna beams. This approach would use fixed spot beams for coverage of major cities, with a phased-array antenna scanning the remainder of the country. The sweep

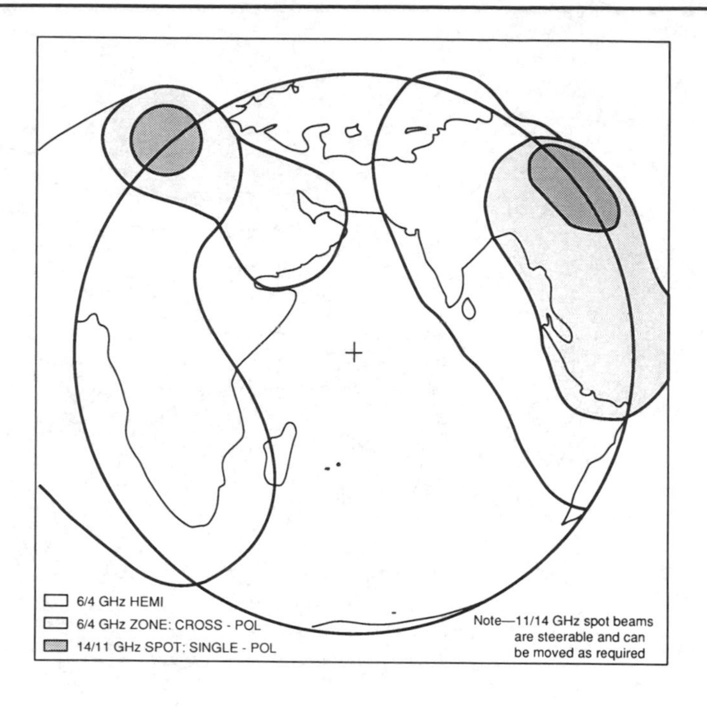

Figure 6.12 Indian Ocean region transmitting beam coverage.

rate for the system may be 0.01 s, and stations are polled by a TDMA system. The system will allow a more efficient use of the radio spectrum because the beam is limited to 1% of the nation at any instant as opposed to a beam covering the entire country. This allows frequency reuse techniques that increase satellite capacity by up to three times while actually lowering the required EIRP by as much as 20 dB. This concept has been used in radar systems for a long time, but its application to satellite is new. Because the hardware technology to implement these satellite communication systems exists, we anticipate their feasibility for operation or launch to be sometime in the 1990s [6.32].

6.6.6 Increased Transponder Power

Due to the critical nature of boosting power and alignment control, there is a limit on weight during a launch, at time of separation, and within the parking orbit. The dilemma of satellite design is that enhanced signal strength and more powerful performance essentially must be sacrificed for the factors that contribute to the life span of the vehicle, which is generally projected to be 7 to 10 years. This has initiated the trend toward more powerful satellites. The targeted range for the satellite's increased strength lies between 40 and 50 W, with maximum projections aimed at a super-powered 200-W transponder. In the late 1980s, the typical transponder power (for commercial satellites) was 5 to 7 W at C-band frequency and up to 20 W (though generally lower) at Ku band. Predictions are that the advent of the higher watt compatibles should foster much greater use in the Ku-band spectrum for applications such as video teleconferencing, business television, and direct broadcast services.

The issue of reliability is a major area of concern regarding more powerful transponders. The life span of the satellite that carries more power might become more difficult to ensure if the technological advancements result in more complexity or a greater aggregate weight. The dangers could outweigh the benefits because misalignment of the satellite's antenna and possibly its solar cells would result in premature consumption of onboard power and even loss of the vehicle.

6.6.7 Satellite's Long-Term Future

This section enumerates some possible future developments in the satellite field.

Antenna "space farms" are being considered that would provide a variety of coverage: beams transmitting to entire hemispheres, focused regional beams, and very narrow spot beams for high-speed data broadcasting. Computer control could include analysis of antenna performance to indicate the most efficient operating parameters.

Tightening the satellite's parking orbit could reduce the differences in propagation delay, thereby eliminating a portion of the synchronization overhead and improving throughput. Using the inherent propagation as a "delay line" and recirculating data between the satellite and its earth station, or between satellites, could prove effective for data storage on a temporary basis [6.38].

Mobile communication requires a sophisticated or large satellite antenna, and the amplifiers must be powerful to focus the power in narrow spot beams and boost the overall capacity. Therefore, research is being done on electronically steerable phased-array antennas for maximizing power.

By the mid-1990s it will be possible to purchase a 3-ft terminal for a single two-way voice channel for about $1000. Slightly larger terminals equipped for four to eight channels may cost $3000; and terminals with 12 to 16 voice channels may be in the $5000 range. These terminals will cost as much as one pays today for office equipment such as a PC, a facsimile machine, or a laser printer [6.34].

Hybrid satellites are being researched. INTELSAT first designed satellites capable of C- and Ku-band operation. In the past few years, hybrid dual-band satellites capable of receiving an uplink in the C band and retransmitting a Ku-band downlink or vice versa have emerged. The C-band 11-W TWT power will remain, thus requiring dishes the same size as those currently being used. Significant improvements over the previous generation of this satellite will include extended

Ku coverage. Fifty-watt TWTs will enable extension of the Ku band. Its design life will be 12 years, with a minimum service life of 10 years.

Superconductivity might be used for future satellites (because it is cold in space) for power storage or for less power-consuming electronics.

The interworking of satellite technology and ISDN was scrutinized in 1990. INTELSAT has voiced concerns that a number of CCITT ISDN Recommendations have not been aligned to the operation of ISDN over satellite links. In some observers' view, three major areas of concern are:

- Communication protocols;
- Connection-processing delay and timers;
- Quality of service and routing.

In terms of protocols, parameters considered optional in CCITT Recommendations may be essential for satellites. The satellite transmission delays might exceed ISDN timers. Some quality of service requirements in ISDN would favor the choice of terrestrial links over satellite links. These issues are expected to be addressed by CCITT, CCIR, and T1S1 during 1991 and 1992.

6.7 ORBITS AND COVERAGE CAPABILITIES

Satellite positioning considerations are a critical element of satellite communication [6.36]. A satellite must be above the horizon to both earth stations to relay information. This restriction limits the satellites used for U.S. domestic service to a band from approximately 80 to 140 degrees west longitude, a range of 60 degrees. Spacing of 2 to 4 degrees is required to eliminate interference between adjacent satellites. The total span for domestic satellite positioning is, therefore, limited to about 30 satellites at 2 degrees as shown in Table 6.8. Table 6.9 provides a more complete listing.

The required spatial separation depends on:

- Earth station and satellite antenna beam widths and sidelobe levels;
- Carrier frequency of transmission;
- Modulation technique used; and
- Permissible performance degradation due to interference.

Population density along the geosynchronous satellite strip has been growing in line with demand for space-based communication services. Satellites are beginning to be spaced 2 degrees apart. A side effect could be an increase of signal interference.

6.8 TRANSPONDER TECHNOLOGY

A terrestrial microwave relay station is a pair of antennas with a connecting signal switch and repeater. A satellite transponder is a similar repeater with the addition

Table 6.8
Satellite Spacing Plan
(Implementation through 1990)

Orbital Slot	*C Band*	*C and Ku Bands*	*Ku Band*
146	Aurora 2 (Alascom)		
144	Westar 7		
142	Aurora 1 (Alascom)		
140	Galaxy 4 (Hughes)		
138	Satcom 1R		
136		Spacenet 4	GSTAR 3
134	New C-band		Comsat B
132	Galaxy 1		Westar B
130	Satcom 3R		Galaxy B
128		ASC-1	
126	Telstar 303		MART-MARI B
124	Westar 5		Fed-Ex B
122	New C-band		S8S-5
120		Spacenet 1	
106–119	Orbits reserved for Canadian and Mexican use.		
105			GSTAR 2
103			GSTAR 1
101		Ford 1	
99	Westar 4		S8S-1
97	Telstar 301		S8S-2
95	Galaxy 13		S8S-3
93		Ford 2	
91	Westar 3		S8S-4
89	New C-band		New Ku-band
87		Spacenet 3	
85	Telstar 302		RCA KU-2
83		ASC-2	
81	Satcom 4		RCA-KU-1
79	Westar 2		MART-MARI A
77			Fed-Ex A
76	Comstar D-4		
75			Comsat A
74	Galaxy 2		
73			Westar A
72	Satcom 2R		
71			Galaxy A
69		Spacenet 2	
67	Satcom 6		RCA KU-3
64		ASC-3/4	
62	Satcom 7		S8S-6

Table 6.9
Satellite Performance Chart (Courtesy of Communications Center, Clarksburg, MD)
(a) NORTH AMERICAN DOMESTIC SATELLITES

Name	*Operator*	*Launch Dates*	*Transponders per Satellite*	*Bandwidth (MHz)*	*Uplink (GHz)*	*G/T (dBi/K) at edge*	*Downlink (GHz)*	*EIRP (dBW) at edge*	*Lifetime (years)*	*Orbit Location (longitude)*
Anik C1 to C2	Telesat Canada	85,83,82	16	864	14.0–14.5	+3	11.7–12.2	47	10	107.5,110,117.5W
Anik D1 to D2	Telesat Canada	82,84	24	864	5.925–6.425	0	3.7–4.2	38	8–10	104.5,111.5W
Anik E1 to E2	Telesat Canada	90	24	864	5.925–6.425	0	3.7–4.2	40	10	107.5,110.5W
			16	864	14.0–14.5	+3	11.7–12.2	47		
Aurora I	Alascom, Inc.	82	24	864	5.925–6.425	−4	3.7–4.2	32–38	10	143W
Aurora II	Alascom, Inc.	91	24	864	5.925–6.425	−4	3.7–4.2	34–40	12	137W
ASC 1 & 2	Contel-ASC	85,90	18	864	5.925–6.425	−4	3.7–4.2	34 & 36	8–10	129,83W
			6	432	14.0–14.5	−1	11.7–12.2	40		
Contelsat-1,-2,-3	Contel-ASC	93,93,Spare	24	864	5.925–6.425	−4	3.7–4.2	35	10	101,129W
			16	864	14.0–14.5	+2	11.7–12.2	46–49		
Galaxy I-IV	Hughes Comm. Galaxy, Inc.	83,83,84,91	24	864	5.925–6.425	−2	3.7–4.2	34.5	9–12	133,74,95,141W
Galaxy V-VI	Hughes Comm. Galaxy, Inc.	93,91	24	864	5.925–6.425	−2	3.7–4.2	34.5	9–12	64,91W
Galaxy IR,IIR,IIIR	Hughes Comm. Galaxy, Inc.	93,94,94	24	864	5.925–6.425	−2	3.7–4.2	TBD	10	133,74,95W
Galaxy A & B	Hughes Comm. Galaxy, Inc.	93,94	24	864	14.0–14.5	+2	11.7–12.2	47–49.5	12	99,131W
Gstar I-IV	GTE Spacenet Corp.	85,86,88,90	16	864	14.0–14.5	−2	11.7–12.2	38–48	10[a]	121,105,125,64W
Gstar IR	GTE Spacenet Corp.	94	24	864	14.0–14.5	1	11.7–12.2	43.8	10	121W
Morelos A & B	Mexico	85,85	18	864	5.925–6.425	+2	3.7–4.2	36–39	10	113.5,116.5W
			4	432	14.0–14.5	+0.1	11.7–12.2	44		
Satcom IR,IIR	GE Americom	83,83	24	864	5.925–6.425	−6	3.7–4.2	34	10	139,72W
Satcom IIR & IV	GE Americom	81,82	24	864	5.925–6.425	−4	3.7–4.2	34 & 32	11	131,81W
Satcom C-1[b]	GE Americom	93	24	864	5.925–6.425	−4	3.7–4.2	34	TBD	138W

Table 6.9 (continued)
Satellite Performance Chart (Courtesy of Communications Center, Clarksburg, MD)
(a) NORTH AMERICAN DOMESTIC SATELLITES

Name	*Operator*	*Launch Dates*	*Transponders per Satellite*	*Bandwidth (MHz)*	*Uplink (GHz)*	*G/T (dBi/K) at edge*	*Downlink (GHz)*	*EIRP (dBW) at edge*	*Lifetime (years)*	*Orbit Location (longitude)*
Satcom C-3,C-4[c]	GE Americom	92,93	24	864	5.925–6.425	−6	3.7–4.2	34 & 32	12	130,81W
Satcom K1-K2	GE Americom	86,85	16	864	14.0–14.5	0	11.7–12.2	43–49	10	67,81W
Satcom K3	Crimson Associates	89	16	864	14.0–14.5	+2	11.7–12.2	45–49	10	85W
Satcom H-1	GE Americom	93	24	864	5.925–6.425	−2	3.7–4.2	34	10	79W
			16	864	14.0–14.5	+2	11.7–12.2	42		
SBS 1-2	Comsat General	80,81	10	430	14.0–14.5	−2	11.7–12.2	41	7[d]	74W
SBS 3	MCI	82	10	430	14.0–14.5	−2	11.7–12.2	41–43	8–9[d]	95W
SBS 4	IBM/Hughes	84	10	430	14.0–14.5	−2	11.7–12.2	40	9	91W
SBS 5	IBM/Hughes	88	14	870	14.0–14.5	−5	11.7–12.2	40	10	123W
SBS 6	IBM/Hughes	90	19	817	14.0–14.5	0	11.7–12.2	42	10	72W
Spacenet I-IV	GTE Spacenet Corp.	84,84,88,90	18	864	5.925–6.425	−4	3.7–4.2	34 & 36	10	120,69,87,141W
			6	432	14.0–14.5	−2	11.7–12.2	39		
Spacenet IR,IIR,	GTE Spacenet Corp.	93,93,	24	864	5.925–6.425	−2	3.7–4.2	35	10	103,69W
IIIR		Spare	24	864	14.0–14.5	−1	11.7–12.2	40.8		
Spotnet 1,2,3	National Exchange	93,93	24	864	5.925–6.425	−5.9	3.7–4.2	31	10	135,76W
	Satellite, Inc.	Spare	18	3753	14.0–14.5	+5	11.7–12.2	56–59		
Telstar 301–303	AT&T Communications	83,84,85	24	864	5.925–6.425	−5	3.7–4.2	32–36	10	97,85,123W
Telstar 401,402,	AT&T Communications	92,93	24	864	5.925–6.425	−2	3.7–4.2	31–38	10	97,89W
Telstar 403		Spare	24	864	14.0–14.5	−2	11.7–12.2	40–47		
Westar III	Hughes Comm. Galaxy, Inc.	79	12	432	5.925–6.425	−7.4	3.7–4.2	33	8	91W
Westar IV & V	Hughes Comm. Galaxy, Inc.	82,82	24	864	5.925–6.425	0	3.7–4.2	35–37	10	99,122.5W

[a] Except Gstar III, which is 3 years.
[b] Will replace Satcom IR in 1993.
[c] Will replace Satcom IIIR and IV in 1992 and 1993, respectively.
[d] Lifetime may be extended by using the COMSAT maneuver.

Table 6.9 (a) broadside

Table 6.9 (continued)
Satellite Performance Chart (Courtesy of Communications Center, Clarksburg, MD)
(b) EXISTING AND NEAR-FUTURE FOREIGN SATELLITES

Name	*Operator*	*Launch Dates*	*Transponders per Satellite*	*Bandwidth (MHz)*	*Uplink (GHz)*	*G/T (dBi/K) at edge*	*Downlink (GHz)*	*EIRP (dBW) at edge*	*Lifetime (years)*	*Orbit Location (longitude)*
					ASIA					
Asiasat 1 (Alias Westar VI)	Asia Satellite Telecommunications	94	24	864	5.925–6.425	−1	3.7–4.2	34	12	TBD
BS 2b (Yuri 2b)	MOPT, Japan	86	2	140	14.0–14.5	+2.4	11.7–12.2	46–55	3	110E
China Sat	P.R. China (PRC)	86	1	36	6	TBD	4	50	3	103E
CS-2a & 2b (Sakura 2)	NASDA-NTT (Japan)	83,83	2	360	5.925–6.425	−8	3.7–4.2	29	5	127 & 136E
			6	780	27.55–30.05	−5	17.75–20.25	37		
CS-3a & 3b (Sakura 3)	NASDA-NTT (Japan)	92,94	2	360	5.925–6.425	TBD	3.7–4.2	TBD	10	TBD
			10	1300	27.55–30.05	TBD	17.75–20.25			
Insat IB & IC	Indian Space Res Org	83,88	2	72	5.855–5.935	−6	2.555–2.635	42	7	74,94E
			12	432	5.935–6.425	−6	3.710–4.200	32		
Insat II	Indian Space Res Org	93+	24	432	5.850–6.425	−6	3.7–4.2	32	7	TBD
			2	432	5.9	−6	2.5	42		
JC Sat-1,-2	Japan Comm Satellite	89,90	32	864	14.0–14.5	+6	12.25–12.75	48	10–12	150,154E
Palapa B1, B2-P, B2-R	Perumtel (Indonesia)	83,87,90	24	864	5.925–6.425	−5	3.7–4.2	34	8	108,113,118E
Superbird	Space Comm Corporation (Japan)	89	12	432	14.0–14.5	TBD	12.5–12.75	TBD	10	124,128E
			23	TBD	27.55–30.05	TBD	17.75–20.25	TBD		
					AUSTRALIA					
Ausdsat 1 to 3	Aussat Pty Ltd	85,85,87	15	675	14.0–14.5	−3	12.25–12.75	36–47	8	156,160,164E
Aussat B1 to B3	Aussat Pty Ltd	92,93,97	15	810	14.0–14.5	TBD	12.25–12.75	45	10	156,160,164E
			1	14	1.6	TBD	1.5	TBD		
					EUROPE					
Astra 1A	Soc. Europeene des Satellites (SES)	88	16	400	14.25–14.5	+3	11.2–11.45	50	10	19.2E
Astra 1B	Soc. Europeene des Satellites (SES)	93	16	400	14.25–14.5	+3	11.2–11.45	50	10	TBD
Atlantic Sat	Hughes/Ireland	90	24	1296	12.75–13.25	+9	10.7–11.2	48	10	31W
			5	135	17	+9	11.746–12.054	61.5		
BSB	British Sat. Broad.	89+	5	135	17	+7	11.785–12.015	62	10	31W
DFS (Kopernikus)	Deutsche Bundespost	90	3	270	14.0–14.25	9.6	12.5–12.75	49	10	23.5E
			7	308	30	7.7	20	48		
			1	90	14.25–14.50	8.9	11.450–11.700	49		

Table 6.9 (continued)
Satellite Performance Chart (Courtesy of Communications Center, Clarksburg, MD)
(b) EXISTING AND NEAR-FUTURE FOREIGN SATELLITES

Name	*Operator*	*Launch Dates*	*Transponders per Satellite*	*Bandwidth (MHz)*	*Uplink (GHz)*	*G/T (dBi/K) at edge*	*Downlink (GHz)*	*EIRP (dBW) at edge*	*Lifetime (years)*	*Orbit Location (longitude)*
Ekran-M (Statsionar-T)	USSR	87+	1	25	6.2	−19	0.714–0.754	50	1–2	99E
Eutelsat I (ESC-I)	Eutelsat	83,84,87	12–14	800	14.0–14.5	−5.3	10.95–11.70	35–41	10	13, 7 & 16E
ECS 1, 2, & 4					14.0–14.5	−3	12.5–12.75	41		
Eutelsat II (ESC-II)	Eutelsat	90+	14	1250	14.0–14.5	−5.3	10.95–11.70	45–51	10	36, 3 & 19E
						−3	12.5–12.75	45		
Italsat 1, 2	Italy	90,92	9	828	30	6–16	20	42–45	10	13E
Olympus	ESA	89	2	54	17.7	+6	12.5	63	5	19W
			3	780	30	TBD	20	52		
			4	144	13 & 14	1.1	12.507–12.598	44		
TDF-1, 2	France	88+	5	135	17.3	+5	12.1	60	9	19W
TV-SAT-1, 2	Germany	87+	5	135	17.7	+7	12.0	60	7	19W
Sarit 1, 2	Italy	91,92	5	135	17.7	TBD	12.0	57	10	19W
Telecom I	Telecom France	83+	2	80	5.925–6.420	−16	3.7–4.195	28–35	7	8,5,3W
			2	240	5.925–6.420	−15	3.7–4.195	26		
			2	72	7.98–8.095	−13	7.255–7.37	33		
			6	216	14.0–14.25	+8	12.5–12.75	44		
Telecom II	Telecom France	91,92,94	10	TBD	5.925–6.425	−16	3.7–4.2	TBD	10	3,5,8W
			5	360	8	−13	7	26		
			11	396	14.0–14.25	+8	12.5–12.75	46		
Tele-X	Notelsat	89	2	126	14.0–14.5	TBD	12.5–12.75	60	5	5E
			2	54	17.7–18.1	TBD	11.7–12.5	TBD		
					MIDDLE EAST					
AMS-1,-2 (AMOS)	Israel	92,93	17	612	5.85–6.55	+7	3.5–4.2	41	TBD	15E
			12	432	12.75–13.25	+11	11.2–11.7	38–45		
Arabsat IA, IB	Arab. Sat. Comm Consortium	85+	25	825	5.925–6.435	−6	3.7–4.2	31	7	19.26E
			1	33	5.925–6.435	−6	2.54–2.65	41		
					SOUTH AMERICA					
Brasilsat	Embratel*	85,86	24	864	5.925–6.425	+2 to −4	3.7–4.2	33	8	65, 70W

*Empresa Brasilerira de Telecommunicacoes.

Table 6.9 (continued)
Satellite Performance Chart (Courtesy of Communications Center, Clarksburg, MD)
(c) INTERNATIONAL SATELLITES IN THE FIXED SATELLITE SERVICE (FSS)

Name	*Operator*	*Launch Dates*	*Transponders per Satellite*	*Bandwidth (MHz)*	*Uplink (GHz)*	*G/T (dBi/K) at edge*	*Downlink (GHz)*	*EIRP (dBW) at edge*	*Lifetime (years)*	*Orbit Location (longitude)*
Intelsat V (F-1 to F-8)	INTELSAT	80+	4	136	5.925–6.425	−14.6	3.7–4.2	23–9	7	63,66,174,180E 1,18.5,21.5,34.5 50,53W
			5	180	5.925–6.425	−14.6& −21.6 −21.6	3.7–4.2	20–26		
			1	41	5.925–6.425	−21.6	3.7–4.2	20.5		
			10	720	5.925–6.425	−11.6& −14.6	3.7–4.2	23–26		
			4	308	5.925–6.425	−11.6& −14.6	3.7–4.2	23–26		
			2	144	14.169–14.241	−3 & 0	10.95–11.7	38–41		
			2	154	14.004–14.081	−3 & 0	10.95–11.7	38–41		
			2	482	14.259–14.498	−3 & 0	10.95–11.7	38–41		
Intelsat VA (F-10 to F-12)	INTELSAT	85+	10	360	5.925–6.425	−14.6	3.7–4.2	20–26	7	60,174,177E 1,24.5,27.5,34.5W
			2	82	5.925–6.425	−21.6	3.7–4.2	20–26		
			13	936	5.925–6.425	−11.6& −14.6	3.7–4.2	20–26		
			4	308	5.925–6.425	−11.6& −14.6	3.7–4.2	23–26		
			2	144	14.169–14.241	−3 & 0	10.95–11.7	38–41		
			2	154	14.004–14.081	−3 & 0	10.95–11.7	38–41		
			2	482	14.259–14.498	−3 & 0	10.95–11.7	38–41		
Intelsat VA (IBS)	INTELSAT	88+	10	360	5.925–6.425	−11.6& −14.6	3.7–4.2	20–26	7	18.5, 53W
			2	82	5.925–6.425	−11.6& −14.6	3.7–4.2	20–26		
			13	936	5.925–6.425	−11.6& −14.6	3.7–4.2	20–26		
			4	308	5.925–6.425	−11.6& −14.6	3.7–4.2	23–26		
			4	288	14.0–14.5	−3 & −8	10.95–11.7	38–41		
			2	154	14.0–14.5	−3 & −8	10.95–11.7	38–41		
			2	482	14.0–14.5	−3 & −8	10.95–11.7	38–41		
			4	288	14.0–14.5	−3	11.7–11.95	TBD		
			2	154	14.0–14.5	−3	11.7–11.95	TBD		
			4	288	14.0–14.5	−8	12.5–12.75	TBD		
			2	154	14.0–14.5	−8	12.5–12.75	TBD		
Intelsat VI (F-1 to F-5)	INTELSAT	89+	12	432	5.85–6.425	−21.6&−11.6	3.625–4.2	20–26	10	60, 63E 24.5,27.5,34.5E
			2	82	5.85–6.425	−21.6&−11.6	3.625–4.2	20–26		
			26	1872	5.85–6.425	−21.6&−11.6	3.625–4.2	23–26		
			4	288	14.0–14.5	TBD	10.95–11.7	TBD		
			2	154	14.0–14.5	TBD	10.95–11.7	TBD		
			2	300	14.0–14.5	TBD	10.95–11.7	TBD		
Intelsat VII (F-1 to F-5)	INTELSAT	92–94	26	1512	5.85–6.425	−21.6& −11.6	3.625–4.2	26–36	TBD	174,177,307,325.5, 342,359E
					5.85–6.425	−21.6& −11.6	3.625–4.2			
			10	880	14.0–14.5	TBD	10.95–12.75	41–46		

TBD: To be determined.

Table 6.9 (continued)
Satellite Performance Chart (Courtesy of Communications Center, Clarksburg, MD)
(d) OTHER INTERNATIONAL SATELLITES

Name	*Operator*	*Launch Dates*	*Transponders per Satellite*	*Bandwidth (MHz)*	*Uplink (GHz)*	*G/T (dBi/K) at edge*	*Downlink (GHz)*	*EIRP (dBW) at edge*	*Lifetime (years)*	*Orbit Location (longitude)*
Celestar I & II	McCaw Communications	94,95	102	3672	14.0–14.5	+12	11.7–12.2	46–48	11	170,70E
Finansat 1 & 2	Financial Satellite	93,94	24	864	5.925–6.425	0	3.7–4.2	37.6	10	47, 178W
ISI	International Satellite, Inc.	TBD	34	1836	14.0–14.5	+2 to +5	10.7–10.9	42–44	10	56, 58W
					14.0–14.5		11.7–12.2			
Orion	Orion Satellite Corporation	TBD	44	432	14.0–14.25	TBD	11.45–F11.7	TBD	TBD	37.5,40,47W
				864	14.0–14.50	TBD	11.7–12.2	TBD		
				432	14.25–14.5	TBD	12.5–12.75			
Pacstar-A &-B	Pacific Satellite, Inc.	90	17	TBD	6.495–7.075	+5 to +8	3.61–4.19	34–38	10	167E, 175W
			4	TBD	14.0–14.5	+5	11.7–12.2	43.5		
			4	TBD	14.0–14.5	+5	12.5–12.75	43.5		
PAS-1	PanAmSat	88	18	864	5.925–6.425	−11	3.7–4.2	31	8.5	45W
			6	432	14.0–14.5	−4	11.48–11.92	43		

Table 6.9 (continued)
Satellite Performance Chart (Courtesy of Communications Center, Clarksburg, MD)
(e) INTERNATIONAL SATELLITES IN THE FIXED SATELLITE SERVICE IN GOVERNMENT USE

Name	*Operator*	*Launch Dates*	*Transponders per Satellite*	*Bandwidth (MHz)*	*Uplink (GHz)*	*G/T (dBi/K) at edge*	*Downlink (GHz)*	*EIRP (dBW) at edge*	*Lifetime (years)*	*Orbit Location (longitude)*
DSCS-III	US Air Force	82+	1	395	7.9–8.4	−15	7.25–7.75	25	10	60 & 175E
			4	TOTAL	7.9–8.4	−16, −1	7.25–7.75	23–44	10	13 & 135W
			1		7.9–8.4	−15	7.25–7.75	23–37.5		
			1		0.3–0.4	−17	0.225–0.26	26		
Fitsatcom	US Navy	78+	12	0.7	0.29–0.32	−16.7	0.24–0.27	16.5	5	23,93 & 100W
			1	TOTAL	0.29–0.32	−16.7	0.24–0.27	26–28		72.5, 172E
			1		0.29–0.32	−16	0.24–0.27	27		
			9		0.29–0.32	−16	0.24–0.27	26, 28		
Leasat	Hughes Comm	84+	5	0.7	0.29–0.32	−8	0.24–0.27	6.5	10	105, 15, 23W
	Serv. Inc.		6	TOTAL	0.29–0.32	−18	0.24–0.27	26		
	(for US Navy)		1		0.29–0.32	−18	0.24–0.27	28		
			1*		8	−20	0.24–0.27	26		
Milstar	US Dept of Defense	89+	TBD	TBD	44	TBD	20	TBD	10	TBD
NATO III	NATO	76+	1	152	7.95–8.162	−14	7.25–7.437	31 & 37	7	30,18,13.5,52W
			1	TOTAL	7.95–8.162	−14	7.25–7.437	31 & 36		
			1		7.95–8.162	−14	7.25–7.437	31 & 36		
TDRS	Contel	83+	1(MA)	331	2.2875	−13.7	2.10	26.3	TBD	41, 171W
	(for NASA)		2 (SSA)	TOTAL	2.2–2.3	+8.9	2.0–2.1	44.0–47.7		
			2(KSA)		15	+24.6	13.8	47.4–52.2		
UHF Follow-On	US Navy	92+	TBD	TBD	0.29–0.32	TBD	0.24–0.27	TBD	14	TBD

* Fleet broadcast

Table 6.9 (continued)
Satellite Performance Chart (Courtesy of Communications Center, Clarksburg, MD)
(f) COMMERCIAL MOBILE SATELLITES

Name	*Operator*	*Launch Dates*	*Transponders per Satellite*	*Bandwidth (MHz)*	*Uplink (GHz)*	*G/T (dBi/K) at edge*	*Downlink (GHz)*	*EIRP (dBW) at edge*	*Lifetime (years)*	*Orbit Location (longitude)*
Geostar (on GTE)	Geostar Corporation	87	1	TBD	1.61–1.627	TBD	12	TBD	See GTE, Gstar and Spacenet	125, 87W
Geostar RDSS	Geostar Corporation	91+	4	16.5	1.61–1.627	TBD	2.483–2.500	TBD	10	100,70,130W
			1	16.5	6.525–6.542	TBD	5.142–5.225			
Inmarsat 2	INMARSAT	91+	1	TBD	1.6	−15.0	−1.5	TBD	10	15,26W,64.5E
Maritime Comm. Subsystem (MCS)	INTELSAT and INMARSAT	82+	1	7.5	1.6	−15.0	1.5	27	7	60,63,66,179E 18.5,21.5E (on Intelsat-Vs)
Marecs A and B	ESA & INMARSAT	81+	1 ea.	5.5	1.6	−13.3	1.5	34	7	27.9W,177.8,64.5E
Marisat	Comsat General and INMARSAT	76+	1 ea.	4	1.6	−17	1.5	30–33	TBD	73,176.5E,15W
AMSC and M-SAT	(US and Canada)	93+	TBD	TBD	1.6	+3	1.5	56	7–10	106.5,101,62,137W
			TBD	200	13.0–13.2	−4	10.7–10.9	36		

of specialized components to increase reliability, improve performance, and control any special transponder features. The trend has been toward satellite designs that provide the largest number of transponders and the largest number of channels possible. Transponder receiver sensitivity is not a major factor to most satellite applications because the earth station transmitter can be very powerful. However, the transmitter portion of the transponder has been subject to considerable design optimization. An almost linear relationship exists between the power of the satellite transmitter and the total mass of the satellite. Because larger satellites are more expensive to construct and to orbit, attempts have been made to increase the effective power of the satellite transmitter through the use of highly directional antenna systems. These have the positive effect of increasing effective radiated power in the coverage area, but the negative effect of further limiting the size of that area.

Early transponders were heavily dependent on traveling-wave tube amplifier technology, but solid-state components based on GaAs technology are becoming available, replacing TWTA systems. These solid-state systems are more reliable, have lower power requirements, and generate less cross-channel distortion. Advances in solar cell technology and battery systems are also helping to improve the power available to transponders, making satellites with high EIRPs more practical.

Earth station design and transponder design are interrelated in that trade-offs in one area are often made to provide additional latitude in the other. In general, the power and sensitivity of the earth stations are increased so that the mass and complexity of the satellite can be reduced.

A modern satellite, such as Hughes Communications' JCSAT2, launched December 31, 1989, carries 32 Ku-band transponders [6.27].

6.8.1 Transponder Internals

Figure 6.13 shows the schematic structure of a typical transponder. The uplink beam from the earth station is received by the satellite antenna and conveyed to the microwave receiver tuned to the uplink frequency. The receiver feeds a channel processor component, which provides the kind of channel amplification, enhancement, and switching that is required by the access control technique used (TDMA, FDMA). The result is passed to the downlink transmitter, which feeds the antenna system.

FDMA satellites receive a range of uplink segments in a cluster around the basic uplink frequency. These signals are applied from the microwave receiver to a mixer stage, which mixes the uplink signal with a local oscillator signal to produce the downlink frequency. Successful relaying of the uplink information requires that the individual uplink segments be matched in power and relative frequency to minimize cross-distortion and interference in the transponder. When the mixing

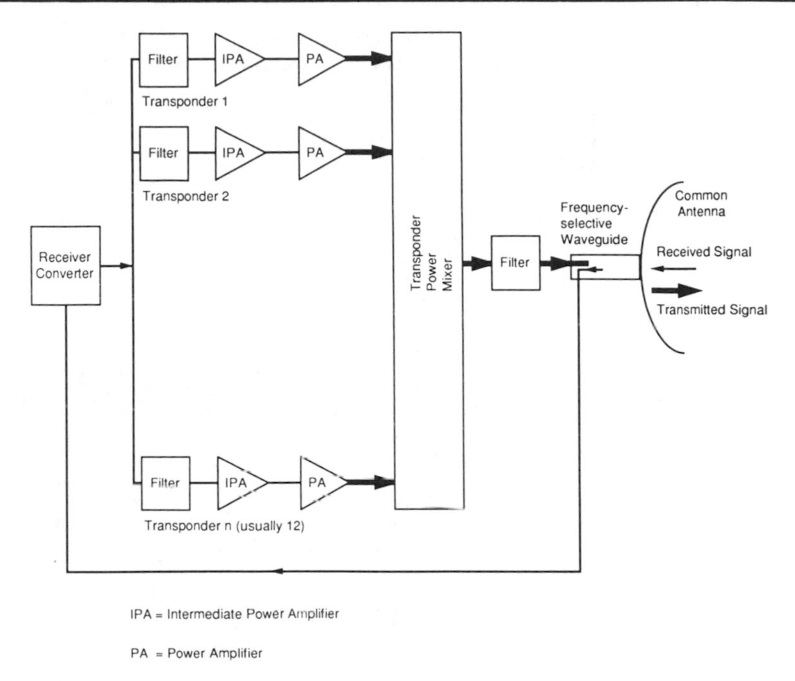

Figure 6.13 Block diagram of a typical commercial satellite.

process is nonintelligent, the uplink frequencies and downlink frequencies must have the same relationship to one another.

In TDMA operations, the output of the microwave receiver feeds an input multiplexer that separates the multiple channels that make up the burst frame. Each of these channels can be separately amplified, and each may be cross-connected in the satellite to create a downlink frame with a structure that is different from that of the uplink.

At C band, the transponder spacing is at 40-MHz intervals, which allows a 2-MHz guardband between each side of the transponder, or a total guard of 4 MHz. An example is the frequency of the transponders on Galaxy 1: transponder 1 is 3720 MHz and transponder 2 is 3760 MHz. Of this 40-MHz bandwidth, 36 MHz is used by the video and audio requirements, plus 2 MHz on either side, which makes up the full 40-MHz frequency spread. See Figure 6.14 for the North American frequency plan for transponders on a C-band satellite.

Most transponders have an effective bandwidth of less than a 10-MHz usable frequency spectrum. The open area, after the required 6-MHz color bandwidth for the TV segment, is where many secondary satellite signals are located, as shown in Figure 6.15.

6.8.2 Coverage Zones and Transponder Design

Simple satellites have a single operating "zone" in which both transmitting and receiving earth stations are located. The satellite antenna is designed to cover this

North American C-Band Satellite
Freqency/Transponder Conversion Table

Channel or Dial Number	Uplink Frequency in MHz	Downlink Frequency in MHz	Satcom 1R, 2R, 3R 4 & 5	Telstar 301,301&303 Comstar 3 & 4	Anik D	Galaxy 1, 2 & 3	Westar 4 & 5	Spacenet 1 & Westar 2 & 3 Anik B
1	5945	3720	1 (V)	1V (V)	1A (H)	1 (H)	1D (H)	1 (H)
2	5965	3740	2 (H)	1H (H)	1B (V)	2 (V)	1X (V)	
3	5985	3760	3 (V)	2V (V)	2A (H)	3 (H)	2D (H)	2 (H)
4	6005	3780	4 (H)	2H (H)	2B (V)	4 (V)	2X (V)	
5	6025	3800	5 (V)	3V (V)	3A (H)	5 (H)	3D (H)	3 (H)
6	6045	3820	6 (H)	3H (H)	3B (V)	6 (V)	3X (V)	
7	6065	3840	7 (V)	4V (V)	4A (H)	7 (H)	4D (H)	4 (H)
8	6085	3860	8 (H)	4H (H)	4B (V)	8 (V)	4X (V)	
9	6105	3880	9 (V)	5V (V)	5A (H)	9 (H)	5D (H)	5 (H)
10	6125	3900	10 (H)	5H (H)	5B (V)	10 (V)	5X (V)	
11	6145	3920	11 (V)	6V (V)	6A (H)	11 (H)	6D (H)	6 (H)
12	6165	3940	12 (H)	6H (H)	6B (V)	12 (V)	6X (V)	
13	6185	3960	13 (V)	7V (V)	7A (H)	13 (H)	7D (H)	7 (H)
14	6205	3980	14 (H)	7H (H)	7B (V)	14 (V)	7X (V)	
15	6225	4000	15 (V)	8V (V)	8A (H)	15 (H)	8D (H)	8 (H)
16	6245	4020	16 (H)	8H (H)	8B (V)	16 (V)	8X (V)	
17	6265	4040	17 (V)	9V (V)	9A (H)	17 (H)	9D (H)	9 (H)
18	6285	4060	18 (H)	9H (H)	9B (V)	18 (V)	9X (V)	
19	6305	4080	19 (V)	10V (V)	10A (H)	19 (H)	10D (H)	10 (H)
20	6325	4100	20 (H)	10H (H)	10B (V)	20 (V)	10X (V)	
21	6345	4120	21 (V)	11V (V)	11A (H)	21 (H)	11D (H)	11 (H)
22	6365	4140	22 (H)	11H (H)	11B (V)	22 (V)	11X (V)	
23	6385	4160	23 (V)	12V (V)	12A (H)	23 (H)	12D (H)	12 (H)
24	6405	4180	24 (H)	12H (H)	12B (V)	24 (V)	12X (V)	

Polarization for each transponder denoted in parenthesis.

Figure 6.14 North American C-band satellite frequency and transponder conversion table.

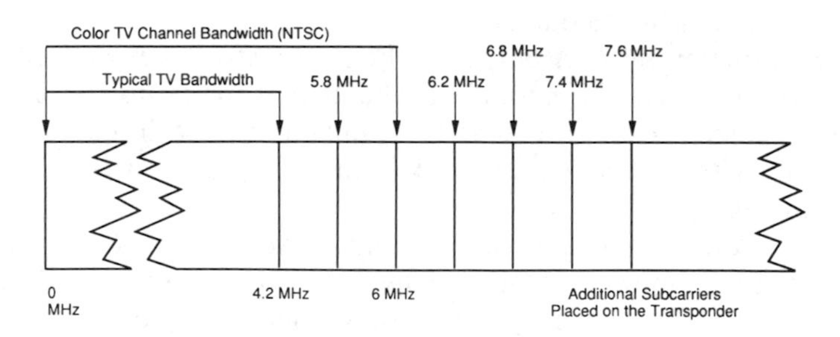

Figure 6.15 Standard color TV carrier with audio subcarriers. Program audio is on the 6.2- or 6.8-MHz subcarrier.

zone. This type of use of satellite radiated power may be wasteful, however, because many applications would not involve earth stations evenly distributed in such a large area. Transoceanic satellites, for example, could normally be expected to serve earth stations clustered on either side of the ocean, but would waste considerable energy by broadcasting to the area of the ocean itself.

A satellite with nonuniform coverage has three major options: (1) The antenna radiation pattern can be distorted so that the radiated energy can be directed only to the shape of the earth station distribution. (2) The satellite could use multiple transponders and antenna systems, one for each zone. This would permit the satellite to service multiple fixed zones, but would require some form of internal switching. (3) The satellite transponder coverage zones could be made variable through the steering of the satellite or positioning of the antenna system. If the satellite has a single steerable zone, the coverage can be changed to accommodate differences in usage that may result from time or other changes. Multizone, steerable satellites can provide focused coverage to a number of areas and can vary these areas at will. Steering systems are clearly not applicable to situations in which many locations must be covered alternately at close intervals. They may, however, be useful in broadcasting applications for which a scheduled, short-interval transmission may be needed.

6.9 EARTH STATION TECHNOLOGY

6.9.1 The Earth Station

Earth station cost is the most important single factor affecting the future of satellite communication. The economic advantage of satellite communication is based on its ability to span large distances and serve discontinuous user bases without intermediate repeater expense. Thus, technical advancements that can reduce the per-earth-station cost and encourage the deployment of earth stations will add to the popularity of satellite transmission. High earth station costs will tend to limit the number of earth stations currently available, making satellite transmission vulnerable to terrestrial alternatives such as fiber optics [6.37].

Three major classes of earth stations are currently available:

1. The mass-capacity station is designed for carrier or large user applications. This earth station typically services a user community large enough to require feeder line access to the earth station. Earth stations in this class cost in the millions of dollars.
2. The middle-range earth stations are designed for direct corporate applications. Earth stations in this class cost roughly $100,000.
3. The low end includes 2- to 3-m antennas used for VSAT applications. The cost for VSATs ranges from $10,000 to $20,000, depending on features and applications.

6.9.2 Antenna Size

Aperture refers to the cross-sectional area of the antenna exposed to the satellite signal. Antennas used in satellite communication are parabolic in shape to focus the energy to a spot at which it can be collected and processed.

Antenna size is directly related to antenna gain—a key system parameter. As antenna size increases, antenna gain increases. Antenna gain is an important consideration because system performance is dependent on the received signal strength relative to noise. A common expression in assessing an earth station's receiving performance is the "figure of merit," or G/T. This defines the sensitivity of an earth station. It is expressed as the ratio of antenna receiving gain by system noise temperature (system noise temperature is expressed in kelvins).

A high antenna receiving gain and minimal system noise is desirable. A large antenna will provide a greater G/T than a smaller antenna, but the antenna size may be limited depending on systems implementation. For example, a small 1.8-m antenna with a receiving antenna gain of 44 dB and a system noise temperature of 300 K (25 dB/K) would have a G/T equal to 19 dB/K; whereas a large 18-m antenna having a receiving antenna gain of 56 dB and a systems noise temperature of 300 K (25 dB/K) would have a G/T equal to 31 dB/K.

6.9.3 Amplifier Equipment

The amplifiers used in an earth station are an important system consideration [6.38].

Receiving-Side Amplifiers

On the earth station receiving side, a low noise amplifier (LNA) is used to amplify the received signal while introducing as little noise as possible. LNAs have a noise level associated with them that is used in determining the receiving system noise temperature. Again, the receiving system noise temperature is a parameter used in determining the earth station's receiving figure of merit, or G/T. Typical noise temperatures for a C-band LNA are 165 to 300 K, and for a Ku-band LNA, 70 to 90 K. The lower the LNA noise temperature, the better the receiving system performance.

Transmitting-Side Amplifiers

On the transmitting side of an earth station, a power amplifier is used to amplify the transmitted signal. The power amplifier used in conjunction with an antenna needs to produce a strong enough signal for the receiving earth station's G/T to produce a high-quality signal. The combination of transmitting power from the power amplifier and the gain from the antenna produces an EIRP in the direction of the satellite equal to the sum of the two.

The EIRP of an earth station is a key parameter in determining the station's capabilities and system performance. The EIRP that an earth station must produce depends on the communication traffic the earth station needs to support. An earth station's transmitting EIRP and receiving G/T will determine the data rate it is capable of supporting: a small earth station will not be able to support as high a data rate as a larger earth station with a greater EIRP and G/T. A trade-off occurs in the communication link between the transmitting EIRP and receiving G/T required to support a given data rate. Within a network, a number of small earth stations may be intermixed with large earth stations. In this case, the small earth stations may not be able to communicate directly with each other because the small earth stations cannot produce the required transmitting EIRP or receiving G/T. Often, then, the small earth stations may have to communicate through a large earth station, which would act as a relay between the small earth stations. Table 6.10 illustrates the various parameters of an earth station with different antenna sizes and power amplifiers. A small C-band earth station with a 1.8-m antenna has a G/T of 14.4 dB, while the same earth station with a 9.0-m antenna has a G/T of 28.9 dB. A small Ku-band earth station with a 1.8-m antenna and a 2-W power amplifier produces an EIRP of 49 dBW; whereas the same earth station with 9.0-m antenna and a 200-W power amplifier produces an EIRP of 82.4 dBW.

Approximately 30 vendors manufacture satellite earth station equipment. Three of the major vendors by market share are Harris Corporation, GTE Corporation, and M/A-Com.

6.10 VSAT TECHNOLOGIES

One of the technological advances on the space side of satellite communication has been increased transponder power, resulting in improved signal strength. This implies an acceptable signal quality even when smaller and less expensive earth

Table 6.10
Earth Station System Performance Parameters

	Antenna Diameter (m)			
Parameter	*1.8*	*6.1*	*9.0*	Units
Receiving antenna gain (C Band)	39.4	50.3	53.9	dB
Receiving antenna gain (Ku Band)	44.5	55.4	59.1	dB
G/T (C Band)	14.4	25.3	28.9	dB/K
G/T (Ku Band)	19.5	30.4	34.1	dB/K
Transmitting antenna gain (C Band)	38.2	49.1	53.0	dB
Transmitting antenna gain (Ku Band)	46.0	55.9	59.4	dB
EIRP (10.0-W amplifier) (C Band)	48.2	59.1	63.0	dBW
EIRP (125.0-W amplifier) (C Band)	59.2	70.1	74.0	dBW
EIRP (2.0-W amplifier) (Ku Band)	49.0	58.9	62.4	dBW
EIRP (200.0-W amplifier) (Ku Band)	69.0	78.9	82.4	dBW

stations are used. Another improvement discussed earlier is the capability to transmit the signal in spot beams that offer more accurate coverage to specific sites. The term *VSAT* refers to the size of the antenna dish, which is usually 1.2 or 1.8 m in diameter. The network topology is a star.

Equipment prices have dropped to the point at which a 1.2-m transmitting-receiving antenna now costs around $10,000 [6.39]. Users of VSAT networks can take advantage of the inherently broadcast nature of satellite communication. They can add or move sites without concern for signal loss or increased monthly charges because the signal is being beamed continuously across an entire area. (However, moving does incur installation charges.) The savings possible with a VSAT network have been reported to be as high as 50% over the cost of leased lines under special conditions [6.40 through 6.48], but may, in fact, break even under average network topologies (more on this below). VSATs were first introduced in 1980, but in the early years, the VSAT industry encountered technical problems [6.39].

Observers now claim a third generation of VSATs. The first generation of technology covered the late 1970s and early 1980s. In that phase, the basic viability of the technology was demonstrated. Initially, the systems could receive only; later, with the development of random access schemes, two-way transmission became possible. The second generation, covering the mid-1980s, saw the application of VSATs to data communication, the beginning of the network management tools, and the general switch from C band to Ku band. The third generation of VSAT technology appeared in the late 1980s. The major features of the third generation are: standardization (at least up to the network layer, via X.25 protocols); improved network management; and software definition, allowing the ability to add new features or to upgrade networks quickly and less expensively, compared to the earlier hardware-based approach [6.49]. Disagreement exists over the types of networks the VSAT solution may suit. Most analysts agree that these networks should have a star configuration, in which a host computer site communicates with a number of geographically dispersed remote sites. As a rule, the outbound volume should be greater than the inbound volume. However, other configurations can be accommodated as well. In designing VSAT systems, careful consideration should be given to any traffic dependencies in the arrival stream. Because the inbound channel is typically based on random access schemes, which are subject to instability under certain traffic conditions, these considerations are important [6.50]. VSAT applications are primarily for data, and include point-of-sale, credit authorization, inventory control, and remote processing. Video applications account for about 20% of the total VSAT traffic, and voice represents about 5% [6.39]. VSATs are generally not used as an alternative to either 56 kb/s DDS or FT1 and T1/DS1 backbone networks; instead, they can collect and distribute data in point-to-multipoint applications and possibly feed this traffic into a terrestrial backbone. Typical VSAT networks are optimized for an outbound-inbound traffic ratio of 4:1 [6.39].

The FCC allocated the frequencies from 11.7 to 12.2 GHz and from 14.0 to 14.5 GHz—portions of what is commonly referred to as the Ku band—for primary use by fixed-satellite communication, as discussed in previous sections. With Ku band, some adjustment for rain attenuation can be made by using a stronger signal. However, in areas subject to very heavy rainstorms, sometimes nothing works except to wait for the signal quality to recover sufficiently and begin transmission anew.

At least one vendor has offered a VSAT network at the C band. This vendor has claimed that the lower cost of C-band receiving-site equipment, together with its tolerance of heavy rain, makes it the frequency band of choice. The C-band earth stations initially supported data rates up to 1200 b/s inbound. The chief disadvantage of C-band transmission is that it shares the frequency band with terrestrial microwave systems. The frequency congestion can also make the process of locating an interference-free zone for installation of the VSAT quite a challenge. At C band, the spread spectrum signal processing technique that enables the VSATs to ignore interfering signals must be used.

The FCC had formerly required licensing procedures for satellite earth stations that became demanding and cumbersome for networks of the size made possible by VSAT technology. In the past few years, the FCC has streamlined the process for both the C and Ku bands. Planners of new Ku-band networks can make a single application for all earth stations of the same type on the network.

VSAT networks include capabilities that make it possible to monitor, on a real-time basis, major aspects of the network, from performance parameters to traffic distribution. Along with the network control that such advanced network management techniques afford comes the capability to print reports on network statistics.

6.10.1 VSAT Network Components

A VSAT network consists of three major elements (see Figure 6.16): (1) the master earth station (MES), (2) a number of remote VSAT earth stations, and (3) a host computer site.

The MES, an intelligent node, is the communication hub for the rest of the network. A key to the successful use of small earth stations is the star network topology, which enables the powerful MES to compensate for the relative weakness of the VSAT end of the channel. The large MES transmits a strong signal to the satellite, so that the receiving VSATs can capture a high-quality signal. The transmitting signal of the VSATs is relatively low powered, and the MES is needed to receive it, especially in a point-to-point network in which the VSATs are communicating with each other. The MES antenna is generally from 5 to about 9 m in diameter. The MES must be designed specifically to support the type of equipment in use at the customer's host site. Most of the vendors' systems are designed to accommodate IBM SNA-based equipment and are plug compatible. The MES performs a variety of essential functions, including transponder monitoring and

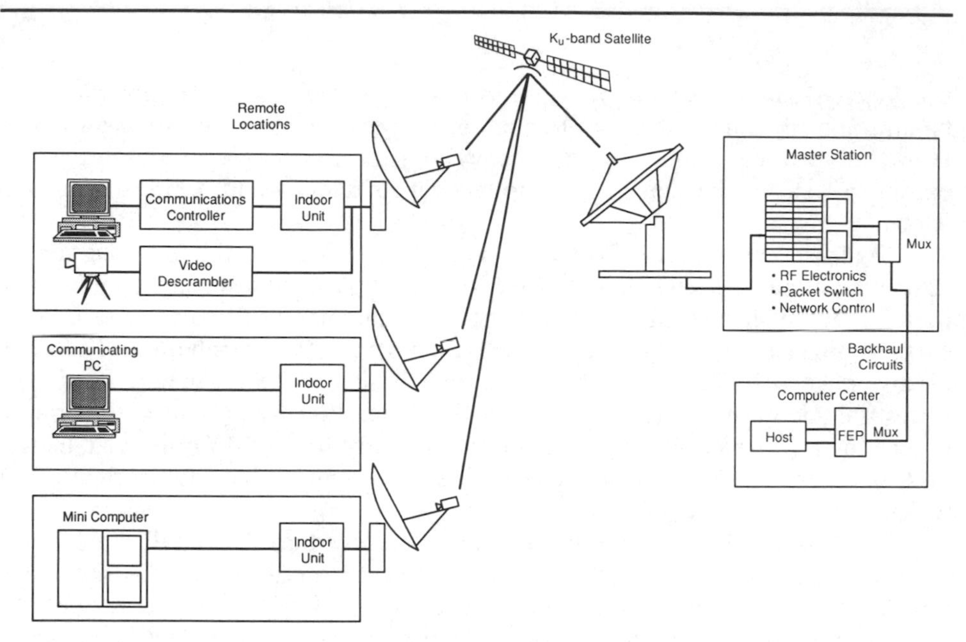

Figure 6.16 Star topology of a typical two-way VSAT network.

host interfacing. In addition to its antenna dish, the hub earth station consists of RF-IF electronics, the network switching system, and a network management computer. A key question is that of whether the MES for a network should be shared among several organizations or dedicated to one organization. The average network payback period for a network with a dedicated MES is estimated to be about five years, and it may be as short as one year in some cases. The cost of the hub—up to $1 million—places a fully private VSAT network out of the reach of many organizations. However, potential users will have access to full-service shared-hub network offerings from satellite carriers and others.

The use of microprocessors for the implementation of many of the micro earth station's functions is the most significant factor in lowering costs. The earth station is comprised of outdoor equipment and indoor equipment. The outdoor unit is the antenna dish, RF electronics package, and mounting frame. The indoor unit is a desktop box, sometimes called a controller, which typically incorporates the modulator-demodulator, a microprocessor for data communication functions, and a microprocessor to provide protocol handling for interface to terminal equipment. The transmitting-receiving dish with its mounting frame typically weighs from 100 to 250 pounds.

The number of sites required to justify the choice of VSAT is variable. The vendors offering shared-hub network services claim there is no minimum number of sites (the emergence of shared-hub services offered by carriers is an important element in the size of the market). Whether or not the cost of a VSAT solution can be justified for a given organization is a complex decision and must finally be made on the basis of a detailed analysis of traffic patterns. For an organization needing low-volume, medium-speed, interactive data communication to a number of remote sites, VSAT may prove to be cost-effective; vendor-provided calculations should always be reviewed analytically. Figure 6.17 provides a point of departure for the comparison, based on calculations published by a VSAT vendor (see [6.39] for specific assumptions).

Some estimate that for 100 VSATs in the system the amortized monthly cost will be $625, while for 2000 stations the cost will be $380 per month [6.51] (for a rough comparison, note that for distances around 500 miles, the cost of a dedicated voice-grade line is approximately $1 per mile per month). These VSAT cost calculations, however, do not take into account that the space segment will have to be increased substantially if the traffic increases proportionally with the number of stations. Namely, if 100 VSATs generate ϵ erlangs of traffic (or bits per second), then 2000 stations may well generate 20ϵ erlangs; this would imply that, unless the system is greatly underutilized with 100 stations, the number of transponders or the power will have to be increased appropriately. Power is not free on a satellite, and the statement that satellite communication is independent of the (terrestrial) distance is a misconception. In addition, these calculations use a rather long amortization period of seven years or more. The technology is moving so rapidly that this would imply that a user would soon be locked into an outdated system: If the

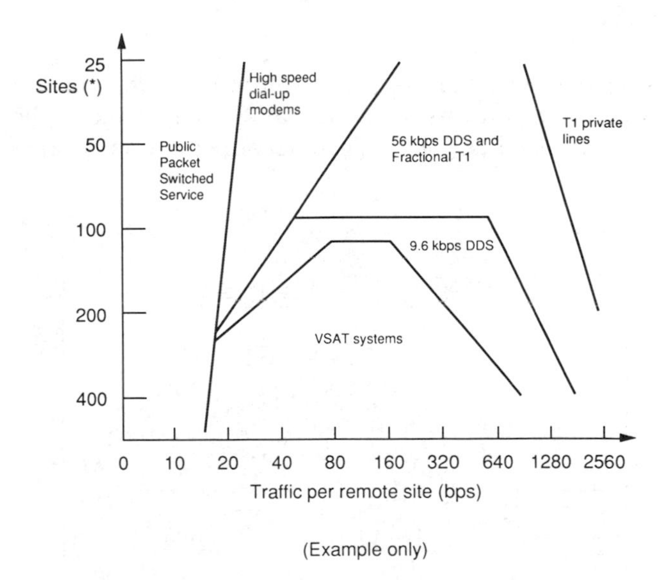

Figure 6.17 Appropriateness of various transmission technologies at the retail level.

payback is achieved only after seven years, the project is probably not worth pursuing because at the end of the seven years the user will have to replace the system (seven years of wind, acid rain, and frost would wear out the outdoor antenna) without being able to utilize the system for a period extending beyond the minimum payback period. Finally, the engineering and installation costs may be nontrivial.

Network Configuration

Four possible configurations of a VSAT network are shown in Figures 6.18 through 6.21. They are:

- A one-way shared-hub network (Figure 6.18);
- An interactive shared-hub network, with a central host site that communicates with the VSAT sites through the MES (Figure 6.19);
- A point-to-point shared-hub network, in which the VSAT sites communicate with one another through the MES (Figure 6.20); and
- A fully private interactive network with a dedicated MES at the host computer site (Figure 6.21).

6.10.2 VSAT Market

The VSAT market did not begin to develop until late 1984, when one vendor, Equatorial, entered the market with its one-way C-band service. The technology and pricing for the Ku band and interactive networks did not mature until 1985.

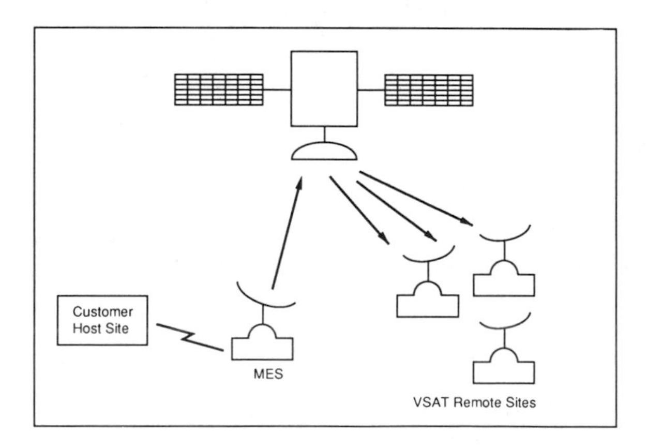

Figure 6.18 A one-way VSAT network configuration.

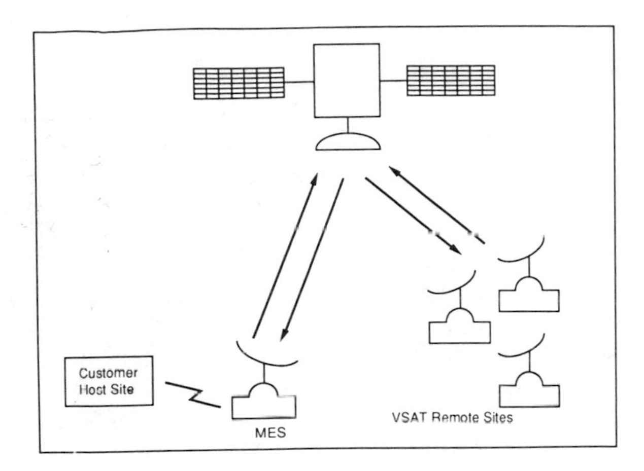

Figure 6.19 A two-way VSAT network configuration.

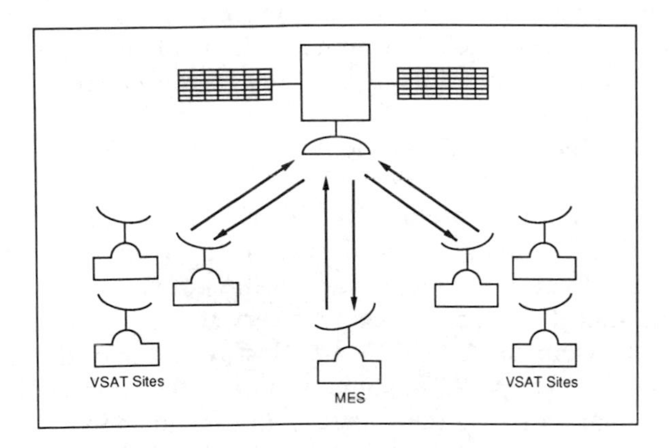

Figure 6.20 A point-to-point VSAT network configuration.

This market became highly visible with the early announcement of Federal Express's intention to construct its ZapMail network, with the number of sites projected at 25,000. The announcement in October 1986 that Federal Express was abandoning the ambitious project was a major setback for the industry. Particularly affected was equipment vendor M/A-Com, whose Telecommunication Division had won the equipment contract.

From 1984 to the present, the number of vendors providing end-to-end VSAT network services increased from 1 or 2 to more than 18. VSAT vendors include

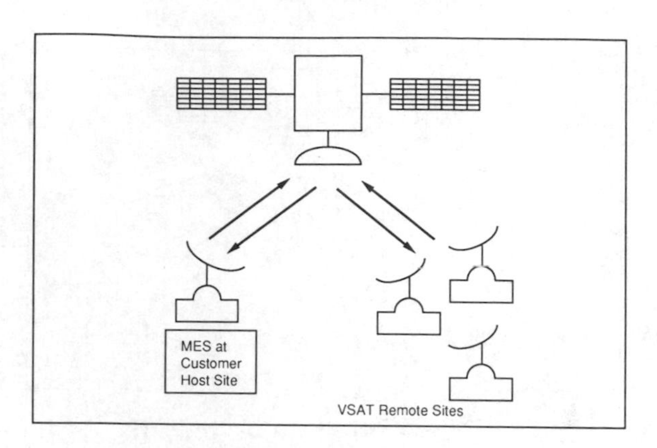

Figure 6.21 A private interactive VSAT network configuration.

Comsat Technology Products, Contel ASC (Equatorial), Fairchild, GTE Spacenet, Harris, M/A-Com-Hughes, Microtel, NEC America, Satellite Technology Management, Satellite Transmission Systems, Scientific-Atlanta, Sky Switch Satellite Communication, Tridom-AT&T, Tymenet-McDonnell Douglas, Cylix Communication Corporation, and Vitalink. According to the Gartner Group, the 1990 market share for full-service VSAT vendors was as follows: Hughes Network Systems, 41%; GTE Spacenet, 17%; Scientific Atlanta, 16%; Tridom-AT&T, 14%; Contel ASC, 6%; and all others, 6% [6.52].

International Data Corporation (IDC), of Framingham, Massachusetts, estimated the number of potential VSAT sites in the United States at 400,000 to 600,000. According to IDC's analysis, 25% of the potential VSAT sites are in the financial services industry segment. Another very active segment is the retail business, including, for example, 7-11, K-Mart, Best, Child World, Levi Strauss, and others. Their use would be VSAT communication between their headquarters and their stores. The market size in the United States is shown in Table 6.11. Initially, the VSAT market was estimated at $1 billion by 1990; but that figure was then revised downward to $1 billion by 1995; that market figure may not be achieved [6.39, 6.53, 6.54].

Thirty thousand "documented" VSAT locations exist at about 30 major companies (see Table 6.12 [6.51]). Trade press figures may, however, be inflated. This author has been involved in installing a system for a company with 325 nationwide sites; however, only about one-third to one-half of the sites could be fitted with VSATs; yet, the trade press lists this company with 325 systems. Typical reasons for the inability to install a VSAT at a location (particularly for banks and brokerage firms with downtown offices) include no southern exposure or elevation to the satellite, obstruction from nearby buildings, strict building or zoning codes, an

ties,
been u
chnology i

arted in 1922 wh
nd *r*anging (radar)
orldwide. Multiplexing
ower frequency voice cha
ater become a key aspect of
neers succeeded in commu-
Channel. In 1947, America's
ration. This was a New
parabolic dish antennas car-
voice conversations between
San Francisco using approxi-
ntinental microwave systems
ls on hops averaging 25 miles

usiness today fall in the range
have been allocated by inter-
es; other portions have been
such as long-haul microwave
nd, starting in the 1960s, for
yday of microwave: telephone
cross the United States without
e early 1980s led to the devel-
s microwave transmitters that
ting the need for the separate
crowave" transmitters. Starting
reasingly being carried by fiber
ajor east-west and north-south
KCs.
n digital microwave. Advanced
e by VLSI, will increase digital
frequency microwave equipment
ectrum above 23 GHz to use.

rs were based on vacuum tube
emain in service. Klystron micro-

y

wagen

rgill, Rose Stores

ton, Farm Credit

.
st, is 50 locations.

ne building, a
at face value,
00 plus 30 ×

ucts until the
, particularly
tworks takes
tion and so-
ners. For the
n nationwide
ts.

fiber fa
ly digital) h
escription of the

microwave transmitter
prototype *ra*dio *d*etection
, radio research accelerated
equency carrier signals with man
ed in 1927. This development woul
crowave systems. In 1933, European e
reliably via microwave across the English
commercial microwave network began c
ork–to–Boston system consisting of 10 towers ar
rying New York television to Boston, and multiplex
the two cities. In 1948, New York was linked wit
mately 100 microwave towers. By the 1950s, trans
were routinely handling more than 2000 voice chan
and, in some cases, extending as far as 140 miles.

The microwave frequency bands of interest to
of 2 to 23 GHz. Portions of the microwave band
national agreements to special interests and purp
used for years by existing communication service
systems carrying telephone and television traffic
satellites, as discussed above. The 1970s were the
calls, television shows, and data messages could no
utilizing a microwave radio link at some point. T
opment of digital microwave systems; this invol
can be directly modulated by data signals, elimi
and expensive modem required by older "analog m
in the mid-1980s, telecommunication traffic was i
optic facilities, especially in the proximity of the
corridors of fiber-based systems installed by the

The 1990s should see evolutionary advances
bandwidth compression techniques, made possib
microwave's bits-per-hertz figure of merit. Higher
should also become feasible, opening the radio s

6.11.1 Microwave Transmitters

Until 15 years ago, most microwave transmitte
oscillators called *klystrons*, many of which still r

wave transmitters generally operate in the lower frequency portions of the m[illegible] wave bands (2 to 6 GHz). Between 1960 and 1975, the klystron tube became workhorse of microwave technology. It developed over the years into a high[illegible] reliable device.

As the 4-GHz band became congested in recent years, higher frequency microwave generators were needed. The klystron was incapable of producing reliable microwave energy at frequencies much beyond the 4-GHz range. Therefore, a strong demand grew throughout the 1970s for equipment capable of operating at or above 12 GHz. One solution to the higher frequency problem was to reintroduce the TWT, which is capable of generating the higher microwave frequencies with the advantage of being able to accept a broad modulating bandwidth. However, because of expense and fragility, TWTs today are utilized primarily in satellite applications.

Most modern generators of microwave energy today consist of some form of Gunn diode, oscillating in a metallic structure called a *resonant cavity*. The physical and electrical dimensions of the resonant cavity are related to the wavelength of the energy to be produced. The operation of Gunn diodes (invented in 1963) depends on the application of a direct current voltage (called a *bias*) across a wafer made of gallium (Ga) and arsenide (As). The wafer is engineered such that it becomes, under the bias voltage stimulation, a type of semiconductor device known as a *field effect transistor* (FET). Using tunneling, FETs are capable of generating reliable but low power microwave energy as high as 50 GHz. The GaAs FET solid-state oscillator is likely to remain the method of choice for several years.

One disadvantage of the Gunn diode is that it produces only a single fixed frequency. By feeding the Gunn diode's output into a resonant cavity that can be mechanically or electrically tuned, however, microwave equipment manufacturers have been able to produce "frequency agile" Gunn diodes across the entire 2-to 23-GHz band. Another disadvantage of the Gunn diode is that, in its present form, it produces just enough power for use in a practical business system. A moderate rainfall between transmitter and receiver on a 3-mile point-to-point system will cause serious path degradation. Increased power at the transmitter can reduce rainfall problems significantly. Very heavy rainfall will cause a 23-GHz system very serious problems on a half-mile link, such as unacceptable high BERs, fading, and static.

6.11.2 Modulation Methods

Most microwave radio systems traditionally use FM, briefly described in Chapter 1. The modulating signal can be analog or digital. *Analog microwave* refers to a microwave system with an analog incoming modulating signal; *digital microwave* refers to a microwave system with a digital incoming modulating signal.

FM analog modulation is still being used because it is reliable, it is a mature technology, and it is inexpensive. Analog FM is about four times more efficient than digital modulation in terms of spectrum conservation: an analog FM transmitter can deliver more information than a digital transmitter, all else being equal. However, all types of analog modulation, including FM, are subject to background noise. Each time the signal passes through a repeater, the problem is aggravated because the noise is cumulative, degrading the signal-to-noise ratio. This limits the number of repeaters that can be accommodated on a given link.

Many of the newer 18- and 23-GHz systems utilize digital modulation. Given fading conditions of gradually increasing severity between transmitter and receiver, an analog microwave system user would immediately notice a degradation in quality, while the user employing digital microwave would not notice anything until the fade reached a certain level of intensity (the "threshold"), beyond which the digital microwave signal would quickly disappear. Aside from this fading advantage, digital modulation schemes have two other major strengths: (1) They are impervious to noise, again up to a threshold: No matter how many repeater stations are added to the network, the signal is delivered clean at the remote end. (2) They are easily interfaced to digital feeds such as DS1 copper and fiber systems without going through a modem.

The current limitation of digital modulated microwave radios is that, compared to older, more conventional AM, SSB, and even FM systems, digital modulation achieves less throughput. Voice encoding schemes, which use the non-PCM methods discussed in Chapter 1, could in the future mitigate this limitation. Advances in technology have steadily increased the number of voice circuits transmitted per channel using FDM/FM. Today, for example, 1800 and 2400 voice circuits are routinely carried on 20- and 30-MHz channels in the popular 4- and 6-GHz common-carrier bands, respectively. More recently, single-sideband radios have been introduced that permit the transmission of 6000 voice circuits in the 30-MHz channels of the 6-GHz band. The use of digital radios for microwave transmission has increased. This growth has been driven by the increased use of digital switching in both public and private networks, by lower multiplexing and demultiplexing costs, as well as by other advantages of digital transmission (this avoids the need for analog-to-digital and digital-to-analog conversion).

The maximum data rate that can be transmitted over a channel depends on the channel bandwidth and the spectral density or transmission spectrum efficiency achieved. Because of spectrum scarcity and regulatory requirements, digital radios must achieve a high level of spectral density. The most commonly used digital radios today achieve a spectral density of 3 bits/Hz of bandwidth using an 8-level PSK modulation technique. This allows the transmission of a 90 Mb/s signal, for example, in the 30-MHz channels of the 6-GHz band (90 Mb/s is approximately equal to 1200 voice channels). QAM densities of 6 bits/Hz are possible. Such systems compare favorably in terms of spectrum efficiency with analog FDM/FM

radios, but they fall far short of the new analog single-sideband radios. Di
radios with capacities of up to 274 Mb/s per radio carrier are available for use
the 18-GHz band where wider bandwidth channels are available. For the 18-GH
band, the required tower spacings are in the 2- to 3-mile range.

6.11.3 Microwave Antennas

Microwave antennas radiate only a few watts of power, under FCC rules. Thus, the sending or receiving signal needs to be concentrated with a concave dish antenna. As with satellite transmission, the size of the antenna becomes smaller as frequency increases. At 1 GHz (for military applications), the parabolic antennas can be 50 to 100 ft. At 2 GHz, we find antennas ranging from 6 to 9 ft. At 12 GHz, the typical antenna size is 4 ft.

Two antennas must have a clear LOS to operate properly. Midway between the transmitter and receiver, the beam can spread a few dozen feet (at 18 and 23 GHz); this area must be completely clear of obstructions, including trees, buildings, billboards, *et cetera* (this is called the *Fresnel zone*). In constructing a microwave system, note that the microwave beam can bend away from the theoretical LOS because of atmospheric conditions such as air temperature, humidity, and pressure. In addition, rain will cause fading even over short hops (2 miles at 23 GHz) and may cause a total outage in cases of heavy activity. At higher frequencies, rain absorption becomes a problem: Rain drops are heated as they absorb and consume microwave power. The signal-to-noise ratio will degrade under these conditions.

6.11.4 Connection to an Outdoor Antenna

The antenna of a microwave system is connected to its indoor electronics using waveguides or coaxial cable.

Waveguides are rigid hollow pipes with circular, oval, square, or rectangular cross sections. The diameter depends on the frequency on the signal being transported: Lower frequencies require larger cross sections than higher frequencies. In the gigahertz area of microwave, the waveguides typically range from 0.5 to 3 in. While waveguides can transport microwave for some distance, the attenuation is fairly sharp as a function of distance, particularly at the higher frequencies (18 and 23 GHz) and are thus kept to the minimum distance possible, typically in the hundreds of feet.

Coaxial cable is even worse in terms of attenuation compared to waveguide; therefore, it is used only for distances of less than 100 ft, particularly at the higher frequencies. The advantage of coaxial cable is that it is flexible and, hence, easier to install (however, bends also affect the signal quality).

.1.5 Microwave Repeaters

If either a LOS condition is not available or the distance is too long (from a topographic or rain-precipitation perspective), then repeaters must be used. Repeaters receive the sender's signal, reamplify it, and in the case of digital modulation, regenerate it and then retransmit the power-booted signal to the next repeater or to the receiver (see Figure 6.22).

6.11.6 Microwave Receiver

A receiver is functionally the inverse of a transmitter. The receiving antenna captures several square feet of the incoming wavefront. The signal is focused by the concave dish onto a feedhorn. The horn transforms the free-space signal into an electrical signal. The signal is run through a down converter, to lower its frequency, and subsequently demodulated. Again the distance of the waveguide from the antenna to the electronics must be as small as possible, due to the large attenuation and signal degradation introduced at this stage.

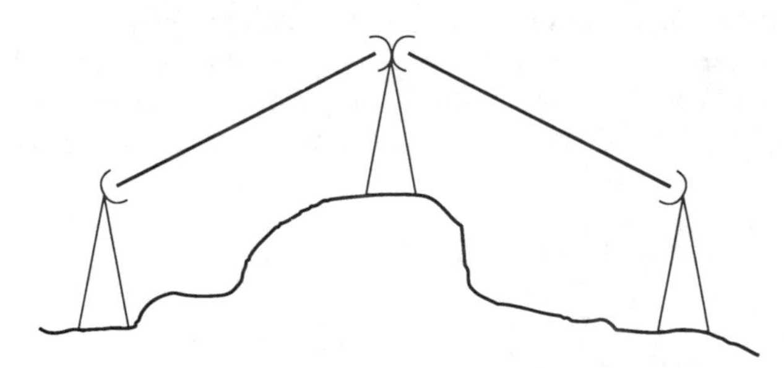

Figure 6.22 Repeaters to eliminate LOS problems.

REFERENCES

[6.1] W.L. Morgan, "Satellite Communication Frequencies," *Satellite Communications*, May 1990, special insert.

[6.2] K. Feher, ed., *Digital Communication: Satellite/Earth Station Engineering*, 1986.

[6.3] E. Fthenakis, *Manual of Satellite Communication*, McGraw-Hill, New York, 1984.

[6.4] T.P. Harrington and B. Cooper, Jr., *The Hidden Signals on Satellite TV*, 2nd ed., SAMS, Columbus, OH, 1986.

[6.5] H.E. Hudson, ed., *New Directions in Satellite Communication: Challenges for North and South*, Artech House, Norwood, MA, 1985.

[6.6] M.H. Kaplan, *Modern Spacecraft Dynamics and Control*, John Wiley & Sons, New York.

[6.7] L. Martinez, *Communication Satellites: Power Politics in Space*, Artech House, Nor MA, 1985.

[6.8] W.L. Pritchard and J.A. Sciulli, *Satellite Communication Systems Engineering*, Prentice-H West Nyack, NY 1986.

[6.9] R.D. Rosner, ed., *Satellites, Packets, and Distributed Telecommunication*, Lifetime Learning Publications, Belmont, CA, 1984.

[6.10] B.R. Elbert, *Introduction to Satellite Communication*, Artech House, Norwood, MA, 1987.

[6.11] R.Z. Zaputoycz, "Satellite Communication: The Fundamentals," *Journal of Data and Computer Communications*, Winter 1990, Vol. 2, No. 3, pp. 26 ff.

[6.12] P.K. Enge et al., "Differential Operation of the Global Positioning System," *IEEE Communications Magazine*, July 1988, pp. 48–60.

[6.13] "Industry Watch," *Satellite Communications*, August 1988, p. 14.

[6.14] R.J. Aamoth, "Eyes in the Sky," *High Technology Business*, July 1988, p. 16.

[6.15] "Remote Sensing Service Providers," *Satellite Communications*, February 1989, pp. 38 ff.

[6.16] E. Marshall, "A Spy Satellite for the Press?," *Science*, December 4, 1987, pp. 1346 ff.

[6.17] G.K. Noreen, "Dawn of a New Era in Mobile Satellite Communication," *Telecommunications*, June 1986, pp. 61–64.

[6.18] J.G. Gardiner, "Satellite Services for Mobile Communication," *Telecommunications*, August 1986, pp. 34–41.

[6.19] "Tests Prove Mobile Satellites Work," *Newsletter Digest*, August 1988, pp. 46–47.

[6.20] "Satellites Have Revolutionized Communications," *Communications News*, March 1988, pp. 24 ff.

[6.21] T. Shimabukuro, "Satellites Are Creating the Corporate Village," *Communications News*, March 1988, pp. 26 ff.

[6.22] C.A. Meagher, *Satellite Regulatory Compendium*, Phillips Publishing, Potomac, MA, 1985.

[6.23] T. Chidambaram, P. Nadkarmi, "Quantitative Methods in the Evaluation of Coverage and Connectivity Features in a multi-beam Communications Satellite," 1st ORSA Telecommunications SIG Conference, March 12–14, 1990, Boca Raton, FL.

[6.24] R.R. Rodriguez, S. Baruch, "WARC-92: An Exercise in Spectrum Allocation," *Satellite Communications*, May 1990, pp. 22 ff.

[6.25] E.B. Parker, J. Rinde, "Transaction Network Applications with User Premises Stations," *IEEE Communications Magazine*, September 1988, pp. 23 ff.

[6.26] "The Changing Face of DBS," *Satellite Communications*, May 1989, pp. 39 ff.

[6.27] Uplink (Hughes Communication), The Aegis Group, Warren, MI, Spring 1990.

[6.28] T. Kerver, "DBS: An Uncertain Future," *Satellite Communications*, May 1989, pp. 46 ff.

[6.29] J.E. McDermott, "Spread Spectrum Communication," *Radio-Electronics*, April 1987, pp. 55–58.

[6.30] D. Minoli, K. Schneider, "An Optimal Receiver for Code Division Multiplexed Signals," *Alta Frequenza*, July 1978, Vol XLVII, pp. 587–591.

[6.31] D. Minoli, "Satellite Onboard Processing of Packetized Voice," ICC 1979 Conference Record, pp. 58.4.1–58.4.5.

[6.32] F.M. Naderi, W.W. Wu, "Advanced Satellite Concepts for Future Generation VSAT Networks," *IEEE Communications Magazine*, September 1988, pp. 13–22.

[6.33] *The 1989 World Satellite Directory*, 11th Annual Ed., Phillips Publishing, Potomac, MD.

[6.34] R. Stamminger, S. Simha, "Next Generation Communications Satellites," *Satellite Communications*, November 1989, pp. 69 ff.

[6.35] J. Martin, *Communication and Satellite Systems*, Prentice-Hall, Englewood Cliffs, NJ, 1978.

[6.36] D.M. Jansky, M.C. Jeruchim, *Communication Satellites in the Geostationary Orbit*, 2nd Ed., Artech House, Norwood, MA, 1987.

[6.37] "An Overview of Satellite Earth Station Equipment," DataPro Report CA50-010-801.

] "Satellite Communication: Technology Briefing," DataPro Report CA50-010-301, September 1989.

.39] T. Khan, "Comparing VSATs, DDS and Public Packet Service," *Business Communications Review*, January 1990, pp. 43 ff.

[6.40] D. Chakraborty, "VSAT Communication Networks, An Overview," *IEEE Communications Magazine*, May 1988, pp. 10–24.

[6.41] X.T. Vuong et al., "Performance Analysis of Ku-band VSAT Networks," *IEEE Communications Magazine*, May 1988, pp. 25 ff.

[6.42] D. Raychaudhuri, K. Joseph, "Channel Access Protocols for Ku-band VSAT Networks: A Comparative evaluation," *IEEE Communications Magazine*, May 1988, pp. 34 ff.

[6.43] W.H. Highsmith, "VSAT Technology—Exploiting the RF Link for User-Driven Innovation," *Telecommunications*, October 1988, pp. 51 ff.

[6.44] D.M. Chitre, J.S. McCoskey, "VSAT Networks: Architectures, Protocols, and Management," *IEEE Communications Magazine*, July 1988, pp. 28 ff.

[6.45] J. Stratigos, R. Mahindru, "Packet Switch Architectures and User Protocol Interfaces for VSAT Networks," *IEEE Communications Magazine*, July 1988, pp. 39 ff.

[6.46] A.L. Mcbride, C.V. Cook, "VSAT Maintenance and Installation," *IEEE Communications Magazine*, September 1988, pp. 36 ff.

[6.47] W. Garner, "VSAT Technology for Today and for the Future," *Communications News*, March 1988, pp. 36 ff.

[6.48] T. Bennett, "The Test at Prudential-Bache," *Satellite Communications*, October 1987, pp. 33 ff.

[6.49] S. Nowick, "VSATs' Evolving Third Generation," *Satellite Communications*, February 1990, VSAT supplement.

[6.50] D. Minoli, "Aloha Channels Throughput Degradation," 1986 Computer Networking Symposium Conference Record, pp. 151—59.

[6.51] S. Borthick, "VSAT Market Gathers Momentum," *Business Communications Review*, January 1989, pp. 33 ff.

[6.52] *Communications Week*, February 26, 1990, p. 32.

[6.53] J. Foley, "Recent Contracts Demonstrate User Commitment to VSAT Technology as a Private Network Tool," *Communications Week*, November 7, 1988, pp. 28–29.

[6.54] *Fiber Optics News*, February 8, 1988, p. 7.

[6.55] D.Kenney, "Microwave Communications: Technology Briefing," DataPro Report CA60-010-302, 1990.

Chapter 7
Fiber Optic Technology and Transmission Systems

7.1 INTRODUCTION

Fiber optic technology is revolutionizing communication [7.1 through 7.3]. Optical fiber is now employed for a large number of applications including long-distance networks, interoffice trunks, intraoffice wiring (distribution frames, remote switching modules, *et cetera*), feeder plants, private user networks, teleport access, building wiring (LANs and risers), and underwater cables. Other applications include railroad networks, power company networks, oil and gas pipelines, and CATV. When first introduced in the early 1980s, optical fiber systems operating at 45 Mb/s were considered high speed; currently, 565 Mb/s systems are common in the public switched network, having been introduced in the mid-1980s. Off-the-shelf equipment provided by a number of vendors operates at 1.12 Gb/s [7.4]. A number of manufacturers have introduced systems operating at 1.7 Gb/s, and some have conducted field trials at 2.4 Gb/s under typical outside plant conditions. Most of the currently installed single-mode optical fiber can be upgraded to this transmission bandwidth by replacing the termination electro-optical equipment. Experimental systems carrying 20 Gb/s have been demonstrated [7.5]. Some feel that these capacities may be high for today's environment, and that a real market will only develop later [7.6].

The terms "fiber" and "optical fiber" are used interchangeably. A system based on optical fiber techniques is also called a "fiber optic system." The field is referred to as "fiber optics."

This chapter provides a survey of the fiber optic field. Light sources, light transmission, optical fiber types, and detection systems are covered. Issues affecting the selection of cable for network design are covered, and technology trends are discussed. Proposed techniques for deployment of optical fiber in the local loop are also surveyed.

Brief History of Fiber Optics

The laser was invented in the 1960s and the concept was refined in the next few years. Transmission through optical fiber was proposed in the literature in 1966 [7.7]. The first low-loss optical fiber was announced in 1970; it was silica-based and had an attenuation of 20 dB/km. By 1986, the attenuation had been reduced to 0.154 dB/km (for dispersion-shifted optical fiber discussed later); this is close to the theoretical minimum for silica, which is 0.13 dB/km at 1.55 mm [7.8] (see Figure 7.1).

As of the late 1980s, the nation's long-haul carriers had nearly reached completion of their route upgrade. The BOCs have also reached substantial penetration in the intracity and intercity trunk routes (long-haul circuit mileage is only around 10% of the total U.S. telephone plant circuit mileage; 90% is the local loop). The next likely area to experience widespread deployment of optical fiber is in the feeder plant as discussed in Chapter 1. Fiber-based systems have experienced price drops in the 20% per year range for the past several years (combined cable and electronics equipment). The price of single-mode fiber strand is now around 10 to 20 cents a foot, down 50% from 1980. Multimode fiber strand is around 10 to 30 cents a foot, depending on the application. Finished cable costs depend on the type of sheath, application, and number of fiber strands in the cable.

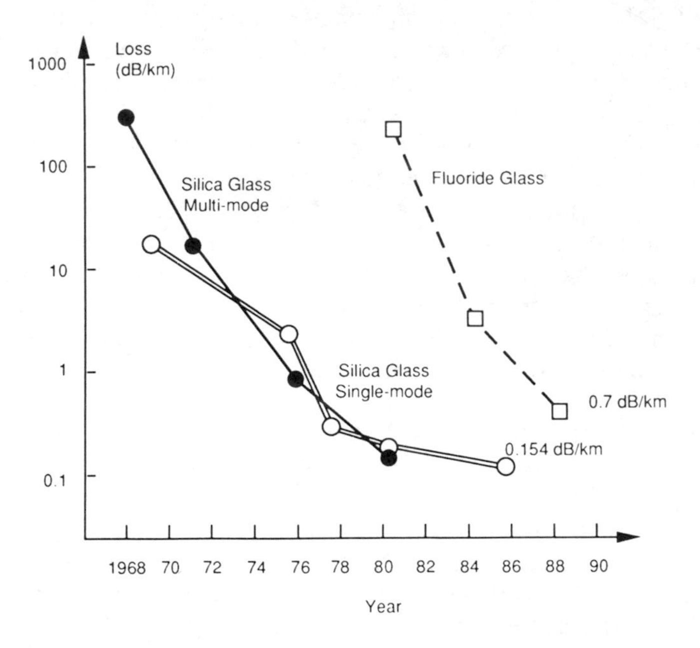

Figure 7.1 Progress of optical fiber in terms of loss over the years.

Table 7.1 lists some desirable intrinsic features of optical fiber, modifie Reference [7.8].

7.2 FIBER OPTIC COMMUNICATION COMPONENTS

A formal prototype for a communication system will assist the discussion by identifying the areas in a system in which fiber optic components come into play. As shown in Figure 7.2, a basic communication system consists of a transmitter, a receiver, and a channel. Fiber-laser technology affects all three areas. Our description of these elements emphasizes those aspects suitable for optical fiber usage and replacement; however, the diagram and the discussion are applicable to other types of communication systems.

7.2.1 Message Origin

The message origination process may take several physical forms. Often it involves a transducer, which converts a nonelectrical message into an electrical signal. Common examples include microphones for converting sound waves into currents, TV video cameras for converting images into currents, and a teleprinter for converting keystrokes into a coded electrical pattern. When data are transferred between computers or parts of a computer, the message is already in electrical form; this situation also arises when an optical fiber link comprises a portion of some larger system (for example, fibers used in the ground portion of a satellite communication earth station, or fibers used in relaying CATV signals).

7.2.2 Modulator

The modulator carries out two functions. First, it converts the electrical message into the proper format. Second, it impresses this signal onto the wave generated by the carrier source. To impress a digital signal onto a light carrier, the modulator need only turn the source on or off at the appropriate times. The ease of constructing digital modulators makes this format attractive for optical fiber systems. Contrary to popular belief, optical fiber is not intrinsically a digital medium; only the manner in which it is employed—the modulator and demodulator technology—makes it "virtually" digital. Analog optical fiber links have been used for transmitting multichannel TV. [Analog fiber optic solutions in this area offer considerable advantages over equivalent coaxial systems in terms of repeater spacing and cable size. Analog transmission also allows simpler implementations compared to digital alternatives with respect to cost, power consumption, and complexity (uncompressed digital TV would require from 100 Mb/s per channel to 1 Gb/s per channel for HDTV).]

Table 7.1
Intrinsic Advantages of Optical Fiber

...all size	Outer diameter of strand approximately 1/10 mm. Fiber cable (including sheaf): 2 to 5 mm. Copper cable: 3 to 6 mm; coaxial cable 10 to 12 mm.
Light weight	The specific gravity of silica is 25% that of copper (the latter has a specific gravity of 8.9); in addition, much less material is used. The weight of the finished cabled fiber is 10 to 30% that of a copper cable.
Physical flexibility	Optical cable can easily be bent, allowing it to be installed in the conduit already in place for other types of cable.
Free from oxidation (rust)	Glass materials are stable chemically, so that they do not rust, as is the case with metals. As a consequence, optical fibers can endure adverse environments (such as at the bottom of an ocean) better than metal cables.
Low loss	The loss of an optical fiber is now below 0.5 dB/km, allowing unrepeatered links of 60 km for 30-dB loss. By comparison, the loss of coaxial cable is around 20 dB/km (requiring repeaters every 1 to 2 km).
Large bandwidth	This quantity is measured as the product of the bandwidth in hertz and the unrepeatered distance. A commercial graded-index fiber provides from 1 to 10 GHz × km; single-mode fiber provides several tenfold GHz × km.
High density bandwidth	This refers to the bandwidth as a function of the cross-sectional area of the cable. The transmission capacity of fiber per cross-sectional area is about 100 times that of paired cable, and 10 times that of coaxial cable.
Bandwidth upgradability	Using WDM techniques (described later) the transmission rate can be upgraded up to one order of magnitude, while utilizing the existing optical fiber.
Electromagnetically robust	Optical fiber is free from electromagnetic induction (glass is a good dielectric and is immune to electromagnetic induction).
Very low crosstalk	Very little light escapes from the fiber or is absorbed through the cladding, implying good crosstalk characteristics.
Resistance to high temperature	The melting point of silica is about 1900°C, far above that of copper. Optical fiber cables are, therefore, more resistant to high temperatures.
Photonic, not electronic	Fibers do not generate sparks, so they can be used in flammable or explosive environments.
Difficult to tap	Fibers cannot be tapped too easily; a conversion to the electrical domain is typically required to branch a tributary (this can also be viewed as a disadvantage).
Availability of material	Copper is limited and must be mined. Silica is abundant.

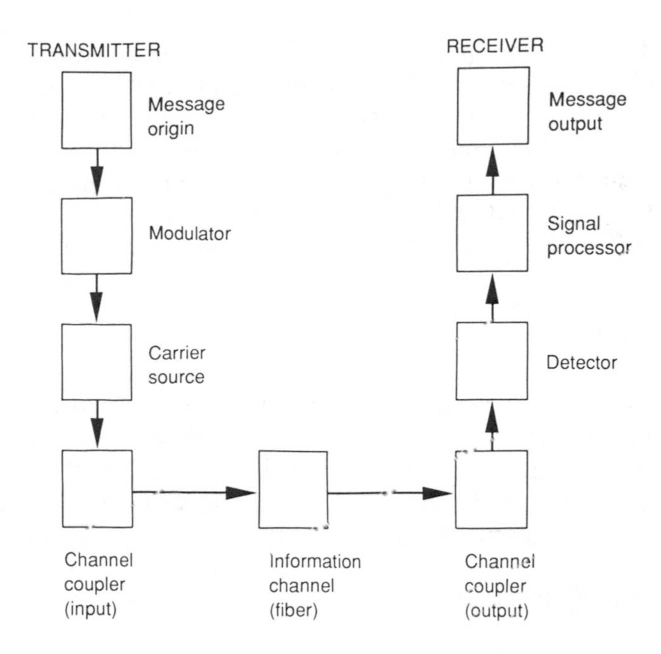

Figure 7.2 A model of a communication system.

7.2.3 Carrier Source

The carrier source generates the wave on which information is carried. This wave is therefore called a *carrier*. Three optical source technologies exist to carry information over a fiber: laser diodes (LDs), surface-emitting light-emitting diodes (SE-LEDs), and edge-emitting light-emitting diodes (EE-LEDs). The surface emitter is the most common type of LED. Ideally, the light source should provide stable-frequency waves with sufficient power for long-distance propagation. Actual LDs and LEDs differ somewhat from this ideal; they emit a range of frequencies and generally radiate only a few milliwatts of average power. Transmission losses continually decrease the power level along the optical fiber, so the lack of sufficient source power limits the length of the communication link. Repeater spacing parameters and goals are intimately related to this fiber loss–source power relationship; research will continue to attack both of these areas by improving the performance of optical fibers and of the carrier source technology.

LEDs and LDs are small, light, and consume moderate amounts of power. They are relatively easy to modulate, that is, to impress electronic information on the input side of the transducer for the purpose of converting from electric power to optical power on the outgoing side. The optical output power takes the shape of the input current coming from the modulator; information being transmitted is contained in the variation of the optical power. This process is called *intensity* ulation (IM). The viability of optical fiber systems is essentially due to the fact t the performance of three basic materials, silica (the fiber), gallium arsenide ne emitter), and silicon (the detector), is compatible within the same range of he light spectrum. LDs and LEDs have been constructed in the past decade that radiate at frequencies where glass fibers are efficient transmitters of light, that is, where fibers have low attenuation. Without this match between source frequency and fiber low-loss region, fiber optic communication would not exist.

LED and LD Issues

While the performance characteristics of LEDs do not compare with those of LDs, LED light sources are less expensive. To date, LDs have been used almost exclusively whenever a long-haul or feeder system has been installed; in terms of price, LEDs appear more suited for distribution plant applications. The power transferred to the optical fiber is one of the key issues related to LED systems; hence, the choice of LDs for long-haul systems. Several suppliers of electronic components are now qualifying or offering single-mode transmission equipment that operates with LEDs.

Greater use of LEDs in the next one to three years will make optical fiber systems cheaper and easier to install. LEDs are not as susceptible to environmental factors as LDs; LDs operate within narrow ranges of temperature and humidity. The telephone companies have had to put necessary optoelectronics in specially built "huts" or "controlled environment vaults," which can cost from $100,000 to $500,000. LEDs will help reduce these environmental limitations. Experimental systems transmit at 565 Mb/s over 2 miles [7.9]. At lower data rates, the mileage is higher (20 miles at 3 Mb/s; 10 miles at 12 Mb/s; 8 miles at 45 Mb/s; 5 miles at 140 Mb/s [7.10]).

A suggestion was made in the beginning that the distribution plant to residential homes be based on LEDs, particularly if this network upgrade was to take place in the early to mid-1990s. Ninety percent of the U.S. installed feeder routes are within the LEDs' current range [7.11]. Industry experts are now saying that utilization of LDs instead may be a better option. Conceivably, the LED range can be extended to 15 miles in the near future for fairly high bandwidths. This will also allow the LED-based systems to be employed for interoffice applications; a large number of COs are within five to eight miles [7.4]. However, single-mode fibers equipped with the more costly LD optoelectronics are now only 5 to 10% more expensive than multimode LED systems. Single-mode fibers have growth advantages and are the preferred choice for long-haul systems [7.6, 7.9]. Laser and low cost receivers for loop applications priced around $200 by 1992 were being discussed as nearly possible [7.10, 7.12]. The target cost for a transceiver for loop applications is $100.

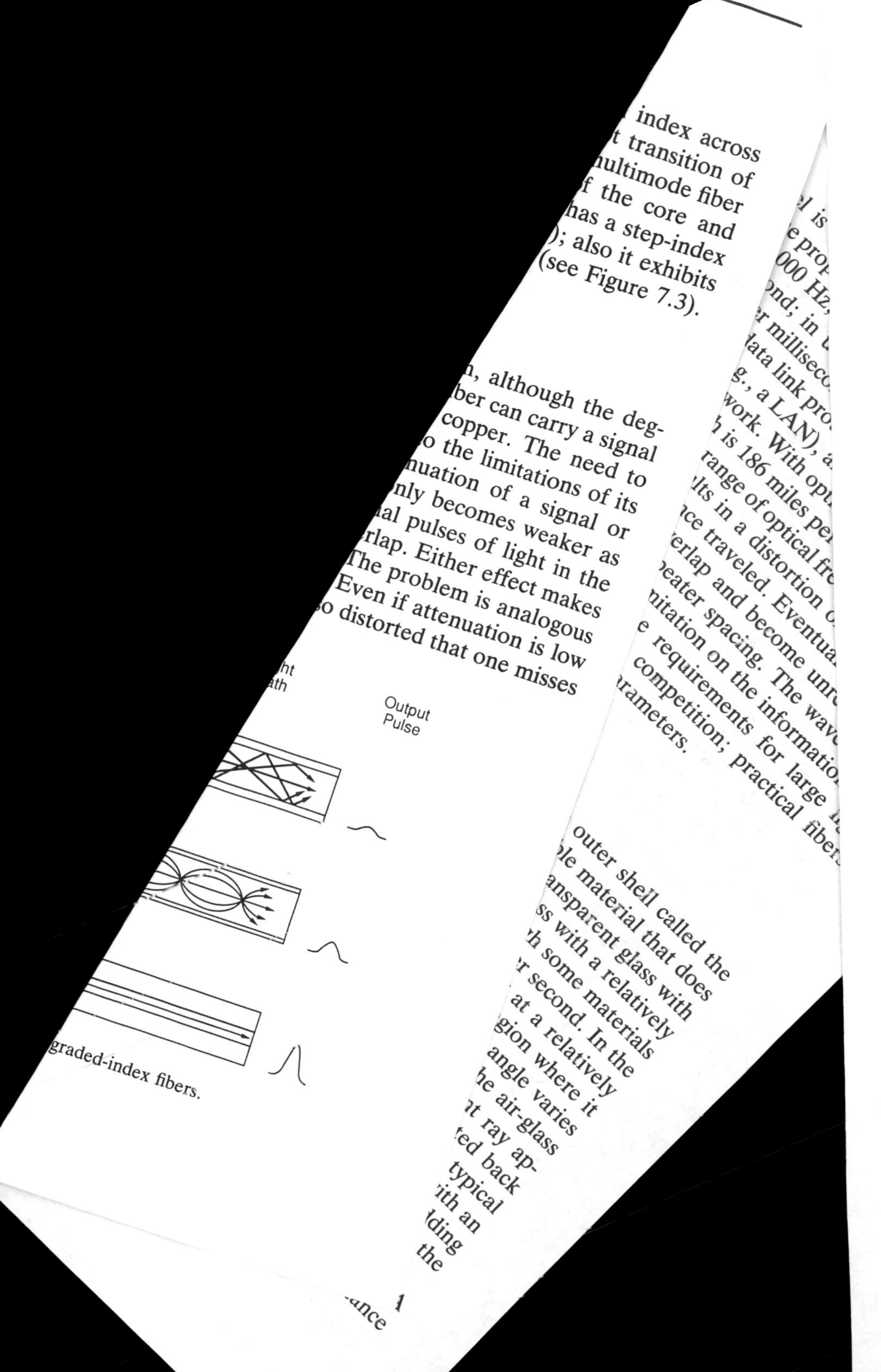

Some early optical fiber technologies such as multimode fiber and higher lo splices are now considered to be liabilities by some carriers. In retrospect, som carriers think they should have waited a little to be able to employ the more advanced systems. Because LEDs in general only transfer a small amount of power to the fiber, LEDs must be used for modest bit rates (say, between DS3 and DS4) or short distances. For baseband LAN applications and even many broadband LANs, both of these conditions are met (data rates are generally no more than a DS3 and distances of 500 to 1000 ft). The cheapest hardware solution for LANs may then be multimode fiber with LEDs. Future broadband LANs for residential applications may require 155 Mb/s (HDTV plus some data and voice). At these rates and typical CO distances of 2 to 5 miles, LED solutions reach their limits.

LD reliability concerns remain, particularly if the LDs are considered for the local loop (distribution plant). The measure of interest is the lifetime hours without serious degradation at temperatures reaching 70°C. Lasers for transoceanic cables (such as TAT-8) are rated at 10 million hours for an area where the temperature is considerably cooler. Current-technology LDs operate at 70°C for a few tens of thousands of hours (1 year equals 8784 hours). Laser lifetimes are becoming longer with improved technology, but remain difficult to compare, particularly for long-wavelength lasers. Not only do we need to look at optoelectronic endurance, but also mechanical and environmental factors. For systems being designed now for installation a few years down the line (such as residential loops), use of LDs over LEDs may be preferred, considering the price trend. The question is one of whether the improvements in LD cost and reliability will occur soon enough.

LEDs will appear in backbone LAN FDDI products because of the technical specifications of that standard. As of 1990, FDDI standards were almost complete. FDDI employs a token passing network using two pairs of fibers operating at 100 Mb/s. The standards encompass the first two layers of OSIRM. This technology provides a fiber backbone for interconnecting multiple LANs independent of the protocols used by the various LANs in the system. The FDDI standard has developed to the point at which products based on it could start to reach the market in the early 1990s. FDDI is discussed in more detail in Chapter 15.

New Light Sources

Production and development of electro-optic components today are concentrating on 1300-nm sources and detectors using InGaAs materials. These devices are driven by requirements of the long-haul telecommunication industry. For data communication applications with distances of approximately 1 mile and data rates of 100 to 400 Mb/s, the laser sources are not mandatory; in fact, they are still too expensive. The sources applicable to this environment are LEDs (both at 800 and 1300 nm) and a new proposal, the short wavelength laser [7.13].

The 800-nm LED is a candidate only for low performance link
suffers from low coupled power into the optical fiber, high fiber tra
and dispersion. An LED normally couples only 3% of the li
compared to 50% or more for lasers. At 1300 nm, the LED is
the fiber has lower attenuation and dispersion in this regio

The short wavelength laser (780 to 830 nm) has hi
optical fiber, but must face high fiber transmission lo
this type of laser is well suited for data communic
width capacities in excess of 1 GHz, requires re
order of 5 mA) due to its high internal ef
photodetector usable for automatic powe
perature and lifetime variations. The sh
receiving major attention as a result
disk systems in the early 1980s. To
at quantities in excess of two million d
$4 and $10. Reliability, which was a pr
A number of prototype systems using short
and the laser may become popular as a poten
links [7.13].

7.2.4 Channel Coupler (Input)

The coupler (or interface) feeds power into the information ch
radio or television broadcasting system, this element is an ant
in Chapters 5 and 6. In an optical fiber system, the coupler must ef
the modulated light beam from the sources (LDs or LEDs) to the
This transfer is not a simple accomplishment without relatively large re
power (attenuation) or somewhat complicated optical geometry and coupler
One difficulty arises because of the small size of the individual fiber strands, w
have diameters on the order of 50 μm (millionths of a meter). Even if the fiber
big enough to intercept all of the light rays emitted by the source, the light will not be entirely collected because of the difference between the radiation and acceptance optic-geometry angles.

7.2.5 Transmission Channel

The transmission channel is the path or physical medium between the transmitter and receiver. In fiber optic communication, a glass (or plastic) fiber is the channel. Desirable characteristics of the transmission channel include low attenuation and a large light acceptance angle. Low attenuation and efficient light collection are particularly necessary for transmission over long paths. Another important prop-

erty of the information channel is the propagation time of the waves traveling along it. In traditional systems, the propagation time is a function of the frequencies being carried; for example, at 10,000 Hz, the signal velocity for a loaded copper line is about 15 miles per millisecond; in unloaded copper (digital carrier), the velocity is approximately 100 miles per millisecond. The higher velocity is desirable because it helps data transfer under a data link protocol utilizing ARQ, it improves performance in a CSMA/CD system (e.g., a LAN), and minimizes psychoauditory issues pertaining to delays for a voice network. With optical fibers, the propagation speed approaches the speed of light, which is 186 miles per millisecond. However, a signal propagating along a fiber contains a range of optical frequencies and divides its power along several ray paths. This results in a distortion of the propagating signal. The spreading increases with the distance traveled. Eventually, the spreading is so great that adjacent pulses begin to overlap and become unrecognizable. This physical issue will dictate the allowable repeater spacing. The wave velocity dependence on frequency and path results in a limitation on the information rate, whether the modulation is digital or analog. The requirements for large light acceptance angle and low signal distortion are in competition; practical fibers represent a design compromise between these two parameters.

How Light Travels in a Fiber

An optical fiber consists of a cylindrical inner core and an outer shell called the *cladding*. The cladding is enveloped by a protective outer cable material that does not play a role in the light propagation. The core is made of transparent glass with a relatively high index of refraction; the cladding is made of glass with a relatively lower index. Refraction occurs because light moves faster through some materials than others. The maximum speed is in vacuum at 186,280 miles per second. In the case of a glass-air interface, a light ray approaching the boundary at a relatively steep angle has its direction suddenly changed as it encounters a region where it can travel at a faster speed. The amount of change in the refracted angle varies as a function of the incident angle up to a certain critical angle. For the air-glass interface, this is about 42.5 degrees. If the angle at which the incident ray approaches the boundary is less than this critical angle, the light is reflected back into the glass at an angle of reflection equal to the angle of incidence. A typical core-cladding critical angle is 8.5 degrees. When light enters from the end with an encoded information signal, multiple internal reflections occur at the core-cladding interfaces, allowing the light to be guided into the fiber core and emerge from the other end.

Two major subdivisions of optical fibers are multimode and single mode. A *mode* is a group of rays bouncing through the fiber at a given incidence-reflectance angle.

Step-index and Graded-index Fiber

Multimode fibers are differentiated by the profile of the refraction index across the fiber's diameter. The step-index multimode fiber has an abrupt transition of refraction index at the core-cladding boundary. The graded-index multimode fiber has an index of refraction that reaches a peak in the center of the core and monotonically tapers off to a lower value. The single-mode fiber has a step-index profile, but has a much smaller diameter (an order of magnitude); also it exhibits a smaller difference in index between the core and the cladding (see Figure 7.3).

Signal Attenuation

Fiber degrades signal quality, as is true of a copper medium, although the degradation is much less than that of copper; this is why optical fiber can carry a signal over a longer distance or at much higher bandwidth than copper. The need to economize will always drive a system's performance close to the limitations of its components. Two general kinds of degradation are attenuation of a signal or waveform and the distortion of its shape. The light not only becomes weaker as it goes through a long segment of fiber, but the individual pulses of light in the signal may become broadened or blurred so that they overlap. Either effect makes the signal more difficult to decipher at the remote end. The problem is analogous to conversing with a friend at the end of a long tunnel: Even if attenuation is low (the voice can be heard), the sound waveform may be so distorted that one misses some of the words.

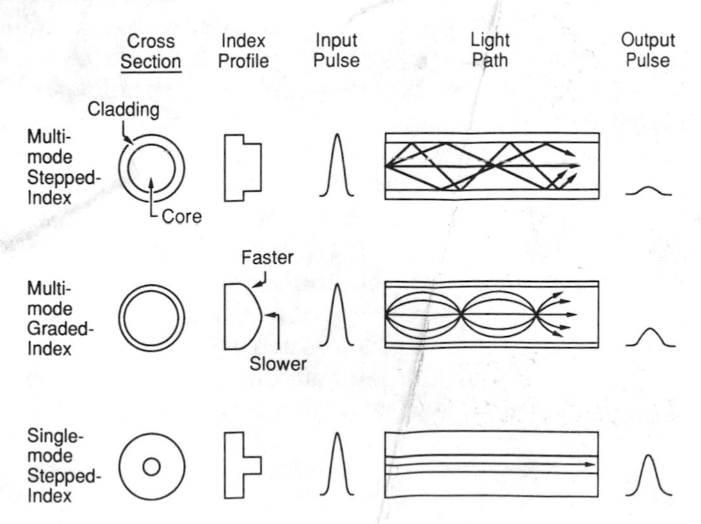

Figure 7.3 Single-mode and multimode stepped- and graded-index fibers.

Dispersion is the name of the principal cause of waveform distortion. The two principal types of dispersion are *modal dispersion* (also known as intermodal dispersion) and *chromatic dispersion* (also known as intramodal dispersion).

Modal Dispersion

When a very short pulse of light is put into an optical fiber, all of the optical energy does not reach the far end at the same time. This causes the exit pulse to be broadened or dispersed. In a simple step-index multimode fiber (which has a large, uniform core), rays of light propagate through the core at different angles. Rays that go straight through have a shorter path than those at a large angle that bounce off the inside wall of the core; hence, the straight rays traverse the fiber faster. This is called modal dispersion. In fiber, the problem can be reduced by using a graded-index fiber. High angle rays, with a longer physical distance to travel, spend more time in regions of the glass with a lower index of refraction where the speed of the light is faster. This compensating effect gives all rays almost the same transit time; the result is that graded-index fiber has a much higher information capacity than step-index fiber. The compensation is never perfect, and a practical graded-index multimode optical fiber still cannot carry large data rates more than a few kilometers before distortion makes the data unintelligible.

The optical fiber is a waveguide that oscillates in a number of different "modes," like a guitar string. These modes of oscillation carry energy down the fiber, but they do it at different speeds. Hence, the name modal dispersion. A typical fiber with a 50-μm core may have several hundred modes. If the core is made smaller, fewer modes will carry energy. If it is made small enough, there is only one mode. Modal dispersion thus disappears. With a single-mode fiber, the information capacity is much higher, however, because the core is so small, the light sources are more expensive, and connectors are required.

The step-index multimode fiber enables transmission of any light ray that enters the optical fiber within an angle of acceptance. The numerical aperture is a measure of the fiber's ability to accept light. High order modes travel over a longer distance than low order modes; therefore, they arrive out of phase at any given point along the fiber. This causes intermodal delay distortion, and thus limits the bandwidth capacity of the fiber. In a graded-index multimode fiber, the rays travel through curved paths. The more sharply curved paths of the higher order modes are primarily in the low refraction areas near the cladding, so that they propagate faster and arrive at the same time as low order modes. Light propagation in the step-index multimode fiber is depicted in Figure 7.4.

Only rays of light that enter the optical fiber passing through the longitudinal axis within a certain angle, called the *angle of acceptance*, are able to travel within the core. Three modes are shown in the figure; in reality, hundreds of discrete modes typically propagate in a single fiber. Figure 7.4 also illustrates the major

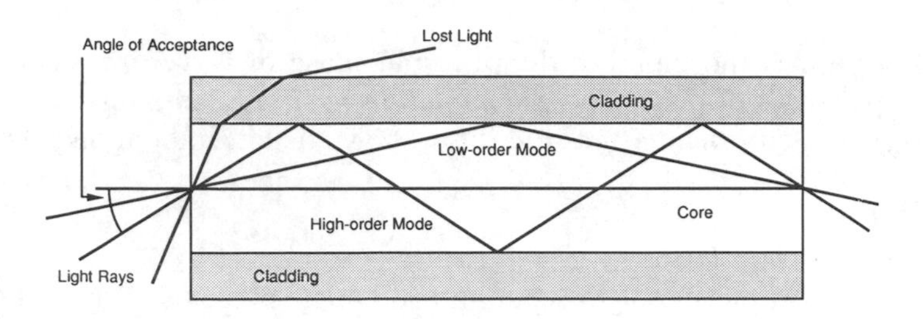

Figure 7.4 Step-index multimode fiber.

limitation of step-index fibers. The low order modes, those launched into the fiber at small angles with respect to the axis, travel a shorter ray distance to reach a given point along the fiber than do the high order modes, which are launched at large angles. Waves that start at the same instant arrive out of phase at the remote end, causing distortion. Optical fibers are assigned dispersion ratings expressed in either (1) bandwidth, in megahertz per kilometer, or (2) in time spread, in nanoseconds per kilometer.

Graded-index fibers were developed to overcome these limitations. The refraction index changes gradually across the core's diameter to guide the various modes along curved paths. Because the index of refraction is essentially an expression of the speed at which light propagates in a material, the gradations of index reduce the delay suffered by the high order modes: Although they zigzag nearer the cladding, they travel faster while in that region and thus do not fall behind the low order modes, which deviate less but travel mainly in the slow area in the middle of the core.

Another method of avoiding intermodal distortion is to select a refraction index and a small enough core radius to obtain a waveguide that cuts off all angular incident modes and enables propagation only straight down the center axis of the optical fiber, that is, to use single-mode fiber. With the major source of the problem eliminated, these single-mode fibers offer a much greater usable bandwidth. Due to the reduced core dimensions, however, single-mode fibers require more care to install. Additionally, they must be driven by laser transmitters, making their interface circuitry more complex and expensive than that needed by multimode fibers, which can be driven by LEDs.

Chromatic Dispersion

Even in a single-mode fiber, all the energy in a pulse does not reach the remote end at the same time. Although only one mode is used, the pulse is usually composed of a small spectrum of wavelengths, and different wavelengths travel at different speeds in the fiber. This is known as *chromatic dispersion*. Chromatic dispersion is defined as the slope of the transit time versus wavelength curve for a given fiber substance. Single-mode fibers are not directly immune to intramodal delay distortion, but one property of silica glass provides a way to avoid the problem. In the visible spectrum of light, material dispersion causes longer wavelengths to travel faster than shorter ones. But in the near-infrared region around wavelengths of 1.1 to 1.3 μm, the opposite begins to happen: Longer wavelengths begin to travel slower. Two general ways to reduce this effect follow.

Reducing the effect of chromatic dispersion is analogous to eliminating modal dispersion by using single-mode fiber: Use a light source with only one wavelength. This is not altogether possible, though, because every source has some spectral spread. The simplest lasers to manufacture are multilongitudinal-mode lasers, which produce light at a number of different wavelengths spread over 5 to 10 nm or more. More sophisticated designs (distributed feedback, external cavity, *et cetera*) reduce the spectrum to a single line, narrow enough such that chromatic dispersion can be ignored in most applications. These devices, however, are expensive. This approach is being applied in many long-distance systems; cost and reliability have even driven designers to shorter distance systems.

The second and more common way to reduce the effect of chromatic dispersion is to choose the transmitter center wavelength and the fiber design so that the effect is minimized. Silica-based optical fiber has low loss at 1300 nm, and even lower loss at 1550 nm, making these good operating wavelength regions. Fortunately, silica also has very low chromatic dispersion in the vicinity of 1300 nm. Therefore, two signals at 10 nm (a typical laser line width) apart in wavelength will have little difference in velocity if they are close to 1300 nm. In contrast, a laser pulse with a 1-nm spectral width centered at 1550 nm will be broadened by 16 ps after traveling 1 km. If the spectral width is 10 nm and the fiber length is 10 km, the pulse will be broadened by 1600 ps (1.6 ns) [7.14].

Zero-Dispersion Point

Negative values of chromatic dispersion have the same effect as positive values (the sign indicates whether the longer or shorter wavelengths travel faster). The optimal center wavelength for the transmitter is the point where the chromatic dispersion is zero; this value varies from optical fiber to optical fiber and is an important parameter for fiber manufacturers to control. Another parameter of interest is the slope of the chromatic dispersion near zero called the zero-dispersion slope, S_0. This is a measure of how rapidly the chromatic dispersion increases as one moves away from the zero-dispersion point (ZDP).

The ZDP occurs naturally in pure silica glass at 1.27 μm. Because, in principle, single-mode fibers work with a single coherent wavelength, one way to exploit the ZDP is to find a laser that emits light at 1.27 μm. However, the search of greatest overall efficiency triggers other factors. Usually, glass waveguides suffer from losses due to Rayleigh scattering, which occurs because of density and compositional variation within the glass. Rayleigh scattering lessens as wavelengths grow longer. To take advantage of this, dopant materials can be added to the glass until its ZDP is shifted into the range between 1.3 and 1.6 μm. Many formulations of glass reach their lowest possible absorption in this range. (Below this, molecular thermal scattering becomes a problem.) Some care in manufacture must be taken to avoid impurities in the glass that increase absorption. A typical light-absorption curve for a high silica fiber is shown in Figure 7.5.

Dispersion-shifted Fiber

For long-distance applications, operation at 1550 nm is desirable because of the lower attenuation compared to 1300 nm. However, simple step-index single-mode fiber shows a large chromatic dispersion at 1550 nm. More sophisticated index profiles can be designed so that the point at which chromatic dispersion is zero falls near 1550 nm instead of 1300 nm. This optical fiber is called a *dispersion-shifted fiber* [7.14]. Segmented-core fiber is an experimental new fiber that is designed to operate as a low-loss single-mode fiber at 1300 and 1550 nm while maintaining the zero-dispersion wavelength in the 1550-nm window. In addition, the fiber performs as a high bandwidth "few-mode" fiber near 800 nm where inexpensive light sources are available. The segmented-core fiber allows mode equalization over a large range of wavelengths. The same fiber can be used in high performance long-haul or in subscriber-loop applications at the longer wavelengths.

Physical Characteristics of Commercial Fibers

Multimode fiber was the first type available commercially; early long-haul telecommunication applications employed multimode. In the mid-1980s, a major shift

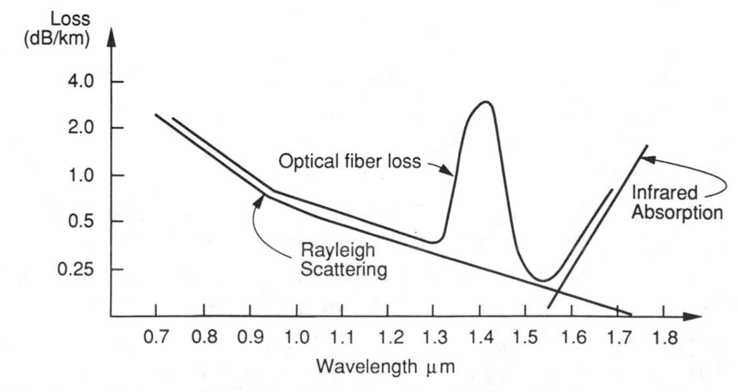

Figure 7.5 Light attenuation as a function of wavelength, and the main sources of loss.

away from multimode to single-mode fiber occurred. Many observers called for the imminent death of multimode. Single-mode fiber is, in fact, technically superior, and is thus the choice for long-haul applications. Multimode fiber operates at wavelengths of 850 and 1300 nm; single-mode operates at 1300 and 1550 nm. Multimode fibers are 50 to 100 μm in glass diameter, have attenuation losses of 1.2 to 2.2 dB per mile, and achieve a maximum range of 12 miles. Single-mode fibers are 10 μm in diameter, have attenuation losses of 0.5 to 1.0 dB per mile, and thus have an unrepeatered range of 35 miles or more. Experimental fibers have even lower attenuation (see below). Most vendors indicate that they now only manufacture 5 to 10% of their entire stock in the multimode grade (mostly for data communication applications).

Multimode fiber is available in a number of strand configurations: 50/125, 62.5/125, 100/140 μm (a 85/100 configuration is also available). The smaller figure represents the nominal core diameter; the larger figure, the nominal cladding diameter. Once coated, all three forms have a diameter of 250 μm (0.25 mm). The *attenuation* of this type of fiber is a function of the core diameter and the operating wavelength: At 850 nm, the attenuation in dB/mile is 4.5, 7.5, and 7.5, respectively, for the fibers listed. At 1300 nm, the attenuation in db/mile is 1.5 and 4.5, respectively, for the first two types. (By comparison, single-mode has an attenuation of generally 0.5 dB/mile at 1550 nm and 1.0 dB/mile at 1300 nm.) The *bandwidth* of multimode fiber is also a function of the core diameter: The minimum bandwidth in MHz-km is 400, 150, and 100, respectively, for the three types.

The FDDI standard could give a boost to multimode fiber for data communication applications. Multimode fibers are employed (at least initially) at distances up to 1.3 miles (2 km). Multimode 62.5/125 μm (core diameter/cladding diameter) fiber is specified; however, a single-mode specification has been recently added (as discussed in Chapter 15).

While multimode was introduced earlier, no one now installs it for long-haul applications, which require the low attenuation losses afforded by single-mode fiber. However, the smaller core of the single-mode fiber requires more care to align during a splice, which results in more time (a splice typically takes 15 to 30 minutes per strand). Aligning the larger core of the multimode fiber is simpler and requires less time and instrumentation. Figure 7.6 provides some idea of the achievable bandwidths of the two technologies.

Most experts think that single-mode fiber will be better in the end even for local loop applications. When we consider the potential services to the home, which may include three or four video channels, about half a gigabit per second is required. In Figure 7.6, note that this would limit the applicability of multimode to less than 2 miles. A cable installation normally has a 30-yr life; multimode is now already close in bandwidth to its theoretical maximum, and as a mature technology may not be ideal for deployment. Single-mode fiber will probably be used to deliver BISDN to users. Single-mode fiber is also better in terms of mi-

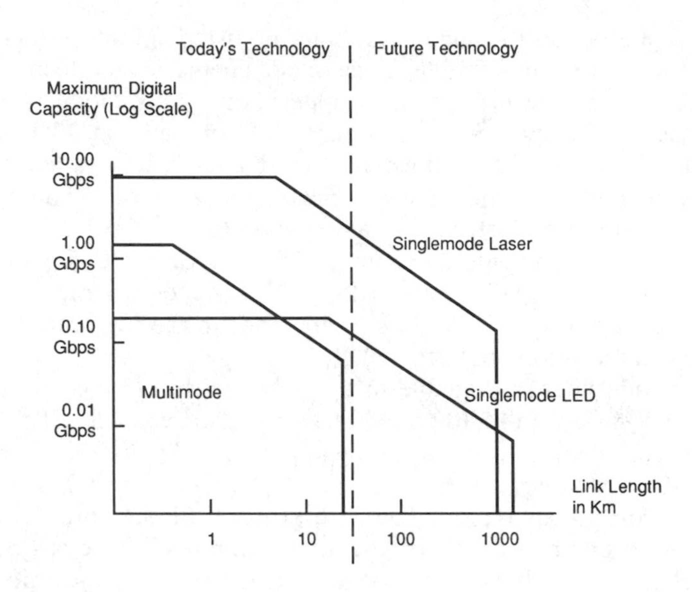

Figure 7.6 Comparison of theoretical information carrying capacity.

crobending tolerances, allowing cable manufacturers to design more types of cables. For home applications, smaller equipment and enclosures give rise to sharper bends, and single-mode fairs better under bending. Single-mode fiber is approaching multimode fiber in cost, as already indicated [7.15].

Multimode fiber deployment in the United States stood at about 100,000 km in 1989, a figure that has not changed much since 1985. By contrast, single-mode fiber stood at 2.2 million kilometers in 1989, and was expected to surge about 25% in 1990 [7.16].

Pricing

Pricing trends are shown in Figure 7.7. While the figure depicts prices of the fiber strand, cable costs have followed the same trend.

Plastic Fibers

Plastic optical fibers were introduced in the early 1980s with mixed success, due to mechanical, optical, and heat sensitivity problems. Improved products are now being introduced that find applications in short data links, illumination, and sensing. The new durability and flexibility makes them applicable to factory applications

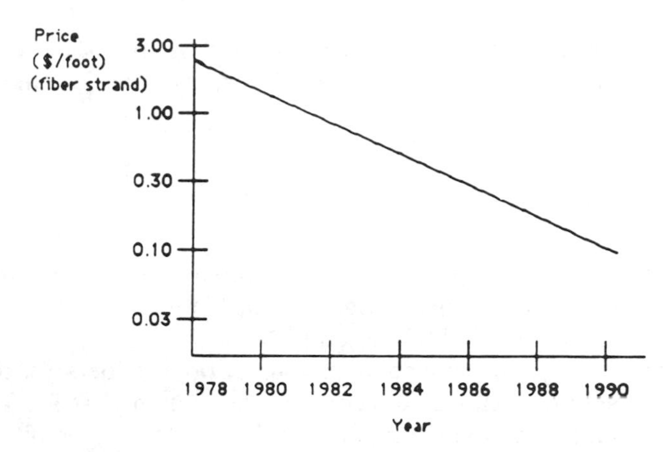

Figure 7.7 Fiber pricing trends.

(robotics) and automobiles; frequent flexing and vibrations would make glass fiber unsuitable for these applications. Another advantage is ease of termination of the fiber and splicing, reducing the system cost. The diameters of these plastic core fibers are generally 100 to 1000 μm (1 mm), in steps of 100 μm; cladding adds 50 to 100 μm. Because the core is large, connectorization can be very rapid (typically requiring only 1 min).

For short runs in which a high loss can be tolerated (for example, short data links for terminals and process control devices), these fibers may be cost effective. For example, a cable of plastic fiber 150 ft in length costs around $25, compared to a single-mode glass fiber cable at $200 [7.17]. These fibers, however, have severe attenuation as indicated by the following list and, therefore, must be appropriately matched to the right application; also they operate at a shorter wavelength than glass fibers.

Wavelength	Attenuation (dB/km)
400	400 to 600
500	250 to 325
600	200 to 300
700	550 to 600
800	1500 to 1600.

In comparing systems, note that the installation cost tends to dominate the equation, so that the ultimate bottom line may be skewed. For example, if the installation cost is $10 a foot (which is not atypical), then 150 ft of installed plastic fiber will run $1525 compared to $1700 for glass fiber, which is a mere 13% difference.

Some typical systems can carry 80 kb/s at a distance of 250 ft, 1 Mb/s at 90 ft, and 5 Mb/s at 30 ft. IBM 3270 type terminals are generally within 200 to 300 ft from the controller; however, these operate in the MHz range. Thus, if the attenuation of plastic fibers could be somewhat reduced compared to the above list, the market for this type of fiber cable could expand. Work is under way in this area.

Manufacturing Techniques

Fiber strand is made from a preform of several components. The two major ways of manufacturing optical fibers are the double-crucible method and the modified chemical vapor deposition (MCVD) method [7.18].

The original technique developed for step-index optical fibers was the *double-crucible method*. One platinum crucible contains the molten glass for forming the fiber core. This crucible is nested inside another that contains the molten glass for the cladding. Both crucibles have drawing nozzles, arranged so that the clad fiber can be drawn from the intersection of the melts. Graded-index fibers are made by depositing materials by means of vapor deposition on or within a starting tube of glass for creating a preform rod, in which the varying chemical content at different depths controls a varying index of refraction. The preform rod is then drawn long and thin to achieve the final fiber.

The second major technique is the MCVD. The process starts with a glass tube that is rotated over a flame of controlled temperature while a chemical vapor is introduced at one end. The vapor carries chemicals that are deposited on the interior of the glass, a process known as *deposition*. The deposited chemicals form a tube composed of many layers of glass inside the original tube. When the process is complete, the tube is collapsed under heat into a solid glass rod, now called the *preform*. The preform is placed on a drawing tower where the fiber is heated to the melting point and drawn into fiber strand. Figure 7.8 depicts the two-step process [7.8].

Other fabrication techniques have evolved in recent years. Advances have been made in all of the major processing techniques during the late 1980s. Progress in fabrication rates and materials has made possible improved fiber designs and enhanced optical properties. Vapor phase processing consists of four major techniques: plasma chemical vapor deposition (PCVD), MCVD (already discussed), vapor phase axial deposition (VAD), and outside vapor phase deposition (OVD).

In the PCVD process, which is primarily practiced by Philips, advances have been made in the late 1980s in terms of rate, size, and water content. Single-mode preforms yielding 800 km of optical fiber have been projected, but 80-km preforms are the largest possible now. For the MCVD process, practiced by AT&T and others, little information has been reported in the late 1980s with respect to rates and preform sizes. However, some process enhancements have been documented in the areas of collapse time and the incorporation of rare earths. The most important advance in the VAD process has been the development of an all-synthetic

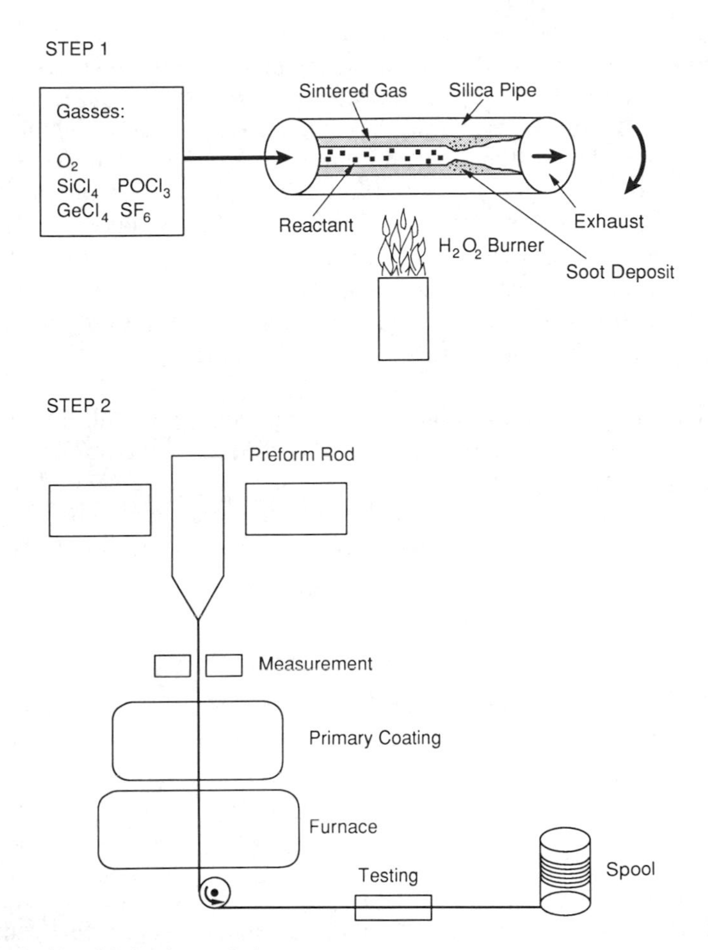

Figure 7.8 Basic fiber fabrication technique.

single-mode process. Higher rates have been achieved through process advances such as multiflame torches and multiple cladding torches. Eccentricity and strength have also been improved. For the OVD process, practiced primarily by Corning Glass Works, recent advances in deposition rate and preform size have been reported; rates in excess of 10 g/s and preforms yielding more than 160 km of fiber have been demonstrated [7.19].

A new process called intrinsic microwave-heated chemical vapor deposition (IMCVD) can triple the amount of optical fiber that can be drawn from a preform. Microwave speeds the deposition of the soot on the glass. The microwave source replaces the oxyhydrogen torch as the heat source, giving a more homogeneous heating to the whole preform body, in contrast to the former method which is capable of heating only the surface [7.20].

The rate at which fiber can be drawn from the preform clearly controls the productivity and cost of the finished product. Improved automatic manufacturing systems are now becoming available that can draw and coat at 600 m/min; this is roughly eight times faster than conventional methods. In the future, such systems may become prevalent [7.21]. Fabrication techniques based on the reaction of vapor phase starting materials to form ultrapure glasses have proven to be the most effective methods for the production of optical fibers; this is especially true for the silica-based glasses, and is being investigated for fluoride-based glasses [7.19]. The advantages of vapor phase processing include not only high purity, but also flexibility, allowing the fabrication of index-complex profiles to match the application.

Future advances in vapor phase processing techniques will probably concentrate on two objectives: reduction of overall cost and special applications. Higher rates, larger blanks, and possibly lower cost dopants will be needed to bring fiber to the home economically. Work will also continue on special fibers such as those optimized for use in long links, fibers for sensors, and radiation-resistant fibers for military applications.

7.2.6 Channel Coupler (Output)

In the optical fiber system, the output coupler directs the light emerging from the fiber onto the light detector. This light is radiated in a pattern in accordance with the fiber's specifications.

7.2.7 Detector

The information received must be taken off the carrier wave. In an electronic system, this process is called "demodulating the signal" and is performed by electronic circuitry. In the optical fiber system, the optical signal received is converted into an electric current by a photodetector. Semiconductor photodiodes of various designs are used. The current developed by these detectors is proportional to the power in the incident optical signal. Because the information is contained in the optical power variation, the detector output current contains the desired information. Key properties of photodetectors include small size, economy, long life, low power consumption, high sensitivity to optical signals, and fast response to quick variations in the optical power. Light detectors having these characteristics are readily available.

Termination Equipment Trends

Optoelectronic equipment is required at both ends of a fiber transmission link to interface it to the rest of the communication system. This equipment incorporates the devices discussed above (detectors, receivers, couplers, *et cetera*). However, this equipment is increasingly reaching a limit in terms of its ability to meet demands for higher rate multiplexing. The major limitation in achieving increased bandwidth has been the upper limit set by chromatic dispersion; the limit is induced by a combination of the source spectral width and the fiber dispersion. To overcome this limitation, research is being conducted to develop reliable single-frequency lasers and dispersion-free fibers. Researchers are also looking at advanced optical technology. Over the years, advances were made in both electronic TDM and in optical wavelength division multiplexing [7.22].

Experiments at 10 Gb/s over 50 miles have been documented in the United States and Japan. Nonetheless, this experiment's bit rate represented slower improvements than in the past, when bit rates approximately doubled each year. In another experiment, AT&T Bell Labs used optical time-domain multiplexing to send 16 Gb/s signals over 8 km of single-mode fiber at a low error rate. The idea behind optical time-domain multiplexing is to interleave bits optically from multiple sources, a process that requires very short laser pulses and extremely precise timing on both ends [7.23].

For impressing the electrical signal onto a single-mode lightwave carrier, broadband gallium-arsenide polarization modulators achieve coupling efficiencies to 70%. On the receiver end, photodiodes with bandwidths in excess of 100 GHz have been demonstrated. As researchers push the limits of direct modulation, interest is also turning to external modulation. Coherent transmission, one of the most efficient optical techniques under consideration, requires a polarization-insensitive receiver so that the received signal and the local oscillator can be precisely matched [7.24].

Optoelectronic Integrated Circuits

An optoelectronic integrated circuit (OEIC) is an optical and electrical circuit composed of active and passive components, including coupling between optoelectronic devices for the provision of signal processing functions. Advances in fabrication techniques make it possible to combine the two types of devices (e.g., transistors and photodetectors) into one circuit to create a synergistically integrated optoelectronic complex. This marriage of lasers, detectors, and transistors could result in devices that are smaller and more reliable than discrete electrical and optical components. Instead of several circuits and components, we have a single circuit. In the future, OEICs may include amplifiers, equalizers, pulse reshapers, and photodetectors.

This packaging process, with its intrinsic advantages in terms of size, power consumption, and cost, follows the trend on the electronic side of the business. Electronics has moved from components circuitry (late 1950s), to integrated circuits (early 1960s) with several transistors in one enclosure, to large-scale integration (early 1970s) with 4-bit type micros, to very large scale integration (late 1970s) with 16- and 32-bit type micros employing more than 200,000 transistors on a chip. Efforts are currently under way to move on to a new phase. The benefits of this progression are obvious. The integration efforts in optoelectronics started about ten years ago. These efforts have not yet reached the level of maturity of their electronic counterpart, thus providing fertile ground for advancement.

Fibers can transmit more information in terms of intrinsic bandwidth than conventional optical sources or detectors can transmit or receive, leading to an electronic bottleneck. Chips with integrated optoelectronic transmitters (laser and electronic housekeeping circuitry for laser control, monitoring, stabilizing, and signal processing) that alleviate this bottleneck problem are starting to become commercially available.

7.2.8 Signal Processor

This stage includes amplification and filtering of the electrical signal. Any undesired frequencies are blocked from further travel. An ideal filter passes all frequencies contained in the transmitted information and rejects all others. Proper filtering maximizes the ratio of signal power to unwanted power, the signal-to-noise ratio. Because of unavoidable noise, some probability of error will always exist in this process; the residual errors will manifest themselves as the BER.

For a reliable communication system, the BER should be very small. In fact, fiber optic systems provide very low BER; namely, equal or less than 10^{-9}. This is sometimes partially achieved by FEC encoding techniques. By way of comparison, the public switched network generally provides a BER of 10^{-5}; conditioned private lines have a 10^{-6} rating.

Additional digital circuitry will take care of other housekeeping functions, including bit synchronization.

7.2.9 Message Output

Action at this stage depends on the intended purpose of the system. If the signal is to be delivered for human consumption, the electrical signal is transformed into a sound wave or a visual image. If the signal is to proceed into another communication stage (for example, a multiplexer or a channel bank), the electrical form of the message emerging from the signal processor is directly usable for appropriate reinjection. This process arises when the fiber system is only a part of a larger network, as in a fiber link between telephone exchanges or a fiber trunk line carrying a number of television programs. In these last two systems, the processing includes distribution of the electrical signals to the proper destination. The message output device is an electrical connector from the signal processor to the upstream system.

7.3 FIBER OPTIC CABLE

Many types of optical fiber cables exist. In fact, more than 25 different types of fiber strand are currently commercially available, with more being developed; cable varieties exceed this number. Table 7.2 is a listing of some common fiber types. In conjunction with the cabling options available, we can envision why the number of cable configurations is large [7.25].

The typical situation is for some large fiber manufacturer (such as Corning Glass) to produce fiber strand, and a cable manufacturer to package the fiber into a suitably designed cable. To appreciate the scope of the market, note that more than 100 manufacturers are providing cable products in the United States [7.26]. Cables come in different forms depending on the ultimate application, installation method, and desired (bandwidth) cross section. Applications include:

Table 7.2
Fiber Types

Fiber Types
1.3 μm single-mode fiber
1.5 μm single-mode fiber
Dispersed-shifted fiber
Hard-clad silica fiber, step index
Large core fiber, step index
Multi-mode fiber, graded index, 50/125
Multi-mode fiber, graded index, 62.5/125
Multi-mode fiber, graded index, 85/125
Multi-mode fiber, graded index, other
Multi-mode fiber, step index, 100/140
Multi-mode fiber, step index, 200/230
Multi-mode fiber, step index, 400/430
Multi-mode fiber, step index, other
Non-silica glass fiber
Plastic fiber
Plastic-clad silica fiber
Mid-infrared fiber, chalcogenide glass
Mid-infrared fiber, fluoride glass
Polarization retaining fiber
Radiation-hardened fiber
Ultraviolet fiber (less than 0.4 μm)
Visible light fiber
YAG fiber
Zinc selenide fiber

- Inside (data communication):
  - LANs
  - Mainframe channel-to-channel extensions
  - Coaxial multiplexer for replacement of multiple coaxial cables;
- Outside (telecommunication):
  - Loop feeder
  - Long-haul trunking
  - Underwater links
  - Umbilicals between switches and remote switching modules;
- Military; and
- Aircraft, ship, and automotive usage.

Inside cables have less stringent environmental requirements compared to outdoor applications, but the fire codes are tighter. Also, the bandwidth requirement is generally lower (a current maximum of 100 Mb/s compared to 560 Mb/s typical of long-haul applications). Military cables have to be hardened to a variety of physical and radiation threats. Cables for aircraft must be light. For installation, the following may be employed: (1) duct or conduit; (2) direct buried (regular or steel armor sheath); or (3) aerial.

In terms of bandwidth, as few as 2 or 4 and as many as 144 fiber strands are in one cable. Each combined application and installation method results in a cable configuration optimally suited to that combination. The cable assembly is more a function of the application than the type of fiber. The inner member can be made of steel or dielectric. The nylon buffer tubes are generally gel-filled to keep out any moisture or potentially corrosive liquids that might find their way into the cable. The polyethylene jacket may be reinforced with Kevlar material to provide mechanical protection for the fibers. Pulling tensions of 2000 to 3000 N are not uncommon (multiply newtons by 0.225 to obtain pounds); cable for aircraft applications requires a tensile strength around 1000 N. If the maximum pulling tension is exceeded at installation time, the fiber may become damaged, which would void the manufacturer's warranty. To prevent potentially damaging tensile force, the weight and length of the cable being pulled through a conduit must be limited. This means shorter cable spans and more vaults where the segments are spliced together. In turn, splices degrade the optical signal, imposing the requirement for more repeaters, and consequently raising the cost.

Considerations in Selecting Cable

One of the considerations in the purchase of cable is the standard cable length: Longer cable lengths minimize the splice points, thereby reducing installation and maintenance costs while improving signal quality. Multimode fiber is generally available in 2-km rolls; single-mode is easily available up to 5 km, and some vendors go up to 12 km. The length of the cable roll should not be confused with the unrepeatered length: The latter specifies at what distance a new repeater is required—the cable between two repeaters may have been spliced several times.

In addition to the maximum allowable pulling tension that can be applied (around 2500 N), the minimum bending radius is also of concern when purchasing cable. Fiber strands subjected to severe bending will become damaged, and even if not damaged, they will introduce severe attenuation to the signal. Packaged cable cannot be bent into a small loop; this must be taken into account when trying to install a cable in a small conduit (particularly for high-rise applications where 2-in. pipes and sharp bends may be common). Minimum bending radii vary between 10 and 20 times the cable diameter.

To appreciate the physical aspect of fiber cable, note that a 4-strand nonmetallic cable will typically weigh 120 kg/km (approximately 400 lb/mile). A 12-strand nonmetallic cable will typically weigh 270 kg/km (approximately 900 lb/mile). Steel-armored cable will generally weigh twice as much as nonmetallic cable. The diameter of 4-strand nonmetallic cable is around 1 cm (about 3/8 of an inch); a 12-strand cable is around 1.5 cm (about 5/8 in.). Armored cable is 1.5 in. in diameter.

The advantage of placing several strands in a single buffer tube is that the diameter and weight are less sensitive to the strand count than the numbers just given. For example, up to 36 strands can be packed in a cable 1 cm in diameter (1.6 cm armored) and weigh only 120 kg/km.

Cables for Aerial Suspension

Suspending optical fiber cable aerially typically provides a substantial reduction in installation cost, as compared to buried cable. However, the tension on suspended cable is much higher, both because of gravity as well as wind and icing. Vibrations and "dancing" are common. For this reason, cables for suspended applications must be specifically designed. A typical approach is to provide a double-reinforced jacket of reinforced polyethylene, concentric with the first. The pulling tension is now in the 30,000-N range (6000 lb).

Water Protection

The pneumatic resistance of air core optical fiber cable is so great that protecting this cable against water entry by means of pressurized dry air is usually impractical. Should a major leak occur, not enough air could get through the cable to prevent water from entering the leak. Consequently, most telephone companies install filled fiber optic cables in the buried underground and aerial environments. Some telephone companies protect the integrity of their optical fiber splice closures by using re-entrable encapsulating compounds and double-closure systems. The encapsulating compounds can be difficult to use and remove when reentering the splice; double-closure systems can be large and expensive. Some other telephone companies install remote moisture sensors in the splice closures and use a monitoring system to check and report on their status [7.27].

Nontelecommunication Cables

Cables for nontelecommunication applications have different characteristics than cables for long-haul or outdoor use. These cables may be used as a means of internal distribution either for LANS, point-to-point data links using fiber optic modems, factory-floor instrumentation interconnection, or patching purposes. The environmental requirements are relaxed compared to outdoor cables (temperature, humidity, physical protection, *et cetera*), but other factors such as flexibility, weight, and fire resistance come into play.

These cables also need to be compatible with commercially available connectors. They have to be flexible and lightweight for installation in trays, ducts, or ceiling spaces; sharper bends are typical for these applications. Achievable cable weights are 10 to 20 kg/km (30 to 60 lb/mile), with a diameter of about 0.5 cm. Typical installations may be in a building conduit or riser. These cables generally have a tight buffer construction to maximize flexibility and handling.

Because of the difficulties of tapping optical fiber (see below), LANs may employ the passive or active star technology, in which several fibers radiate from a central point to interconnect the desired number of devices, much like a traditional IBM cluster arrangement of controller and terminals. Some proponents claim that passive stars are inherently more reliable. Passive stars generally connect up to 64 devices; larger systems have to use active, or at least partially active (hybrid), systems.

Splicing Methods and Implications

Splicing and connectors are receiving attention because their costs play an important practical role in fiber deployment, particularly for LAN applications. A number of optical fiber splicing methods exist. Debate arises as to which method is superior in terms of splicing time, attenuation introduced by the splice, durability of the splice, and cost of the equipment to perform the splice. No single method is better all the time, and the prospective user should consider the issue carefully. Splicing requires some expensive equipment: Single-mode fusion equipment costs in the $30,000 to $40,000 range; rotary splicing equipment costs in the $20,000 range. Optical time-domain reflectometers (OTDRs) that perform appropriate tests cost around $20,000; enclosed environmentally controlled vans and other support hardware are also required. The final decision must be based on productivity considerations; namely, the cost per slice and the splice durability. The decision as to which technology to employ is not clear *a priori*.

The basic steps in the splicing process are fiber-end preparation, alignment, and retention. Methods for preparing fiber ends include controlled breaking, sawing, and polishing. Several techniques for alignment exist, including v-grooved substrates, snug-fitting circular tubes, loose-fitting square tubes, precision rods for splicing individual fibers, and multigrooved substrates for group-splicing of multifiber ribbons [7.28].

Fusion splicing joins two fiber ends by butting them, forming an interface between them, and then removing the common surfaces so that no interface remains between them when lightwaves are propagated from one fiber to the other (no refraction or reflection at the former interface will occur). The protective coating of the fiber is removed, the fiber ends are prepared by scoring and breaking, and heat is applied until the fibers melt and fuse together. Electric arc and gas flame are the two predominantly used heat sources. The resulting slice will be strong, especially with flame fusion, and losses can be low with either arc or flame fusion. Loss due to the splice is in the 0.30-dB range. Splicing takes 20 to 30 min. Fusion splicing capabilities have improved in the past couple of years.

In the rotary splice technique, the two ends of the two single-mode fibers are put in glass sleeves; the fiber cores are initially off axis. The ends of the fibers and sleeves are polished; at this point, the polished ends are put into a triangular alignment sleeve and rotated until the cores, in perfect alignment, allow maximum light throughput; light intensity is measured with appropriate instrumentation. The fiber ends are not fused or cemented because alignment retention is mechanical. This technique can be generalized to allow joining ribbons of 12 fibers; this group method is about 15 times faster than individually splicing the 12 fibers. Loss due to the splice is in the 0.20-dB range (lower than that achievable with typical fusion splicing methods). Splicing takes 15 to 20 min.

Tapping Methods

While a large percentage of the light transmitted over an optical fiber is confined to the core, a small part of the radiated field actually travels in the cladding region adjacent to the core. Most optical tapping methods rely on this principle to gain access to the signal. Access is accomplished by causing some physical or mechanical deformation to the cladding. The fiber may be notched with a laser beam or fused to another fiber. The tapping operation is a delicate procedure requiring extreme precision. Tapping must generally be performed in the field after the installation of the fiber under less-than-ideal conditions. A problem intrinsic with many tapping methods, limiting such applications as LANs, is that these mechanical solutions produce one-way taps: light can be removed from the fiber, but not injected. Mechanical taps also weaken the fiber. A number of proposed solutions that would eliminate these limitations are now under study [7.29]. Until these techniques are perfected, passive or active star networks are the only real alternatives for LAN applications.

7.4 NETWORK DEPLOYMENT

7.4.1 Long-Haul

If we examined a U.S. map of installed optical fiber (such as the one of Reference [7.30]), we would see the following seven major deployment strategies to date:

- A major East Coast corridor from Boston to Miami, connecting every major city en route, involving a number of carriers (AT&T, MCI, US Sprint, Lightnet, and SouthernNet).
- A major Midwest to South corridor from Chicago–Toledo–Detroit to Atlanta and New Orleans, connecting every major city en route (AT&T, US Sprint, and NTN carriers).
- A major Midwest to Texas corridor from Milwaukee to Dallas and Houston, connecting every major city en route (AT&T, US Sprint, and LDX carriers).
- A north East-West corridor from Boston to Seattle, via Chicago (US Sprint and NTN in the Midwest segment).
- A central East-West corridor from Philadelphia–Washington to San Francisco–Los Angeles via Chicago and Salt Lake City (AT&T, US Sprint, and Wiltel).
- A south East-West corridor from Atlanta to Los Angeles via Dallas (AT&T, US Sprint, and MCI to Dallas).
- A West Coast corridor from San Diego to Seattle (AT&T, US Sprint, and MCI to San Francisco).

By early 1989, AT&T had approximately 23,000 miles of fiber; MCI had approximately 14,000; and US Sprint had approximately 23,000.

7.4.2 Cost Considerations for Long-Haul

In the long-haul segment, the parameters that have affected the calculations of what technology to employ have been route mileage and cross section of the circuit. Some of the figures for the late 1980s follow to give a sense of the decision-making process [7.31].

At a cross section of about 6000 voice circuits, microwave costs $4 per circuit mile for routes of 250 miles, and $3.50 for routes of 2000 miles. Fiber costs $6.50

and $6.25, respectively. Satellite costs $16 and $2, respectively. Everything else being equal, microwave would be used for the short route and satellite for the long route.

At a cross section of about 20,000 voice circuits, digital microwave costs $3 per circuit mile for routes of 250 miles, and $2.50 for routes of 2000 miles. Fiber costs $2.75 and $2.50, respectively (for a 135 Mb/s system). Satellite costs $13 and $1.6, respectively. Again, everything else being equal, fiber would be used for the short route and satellite for the long route.

At a cross section of about 40,000 voice circuits, analog microwave costs $3.75 per circuit mile for routes of 250 miles, and $1.75 for routes of 2000 miles. Fiber costs $1.40 and 1.20, respectively (for a 565 Mb/s system). Satellite costs $12 and $1.50, respectively. Everything else being equal, fiber would be used for both routes. At higher cross sections, fiber is universally better.

7.4.3 BOCs

The deployment of fiber by the BOCs in the late 1980s (end of 1988 data) was as follows, based on Reference [7.32] (these data have not been independently verified with the BOCs):

Ameritech	180,000
Bell Atlantic	240,000
BellSouth	340,000
NYNEX	300,000
Pacific Telesis	190,000
Southwestern Bell	210,000
U S WEST	125,000

(See also Reference [7.44]).

7.4.4 Cost Considerations for the Feeder Plant

Complex economic calculations for the feeder plant continue to be required before the decision to utilize fiber rather than copper is made. The cost is not yet at a stage where the choice is automatically and universally fiber; however, more and more calculations lead to that solution. Two key topological factors influence the choice, given today's fiber and termination costs: span length (distance from CO) and required capacity. (A secondary factor is conduit space; when very little space is left, fiber must be used regardless of cost.) If we need to take 500 Mb/s to a certain location with no existing facilities because, for instance, a large industrial park is being built, then fiber would probably be employed because of the distance from the CO. If we need to provision a feeder route that is 15 miles long and carries several DS1s, then again fiber is the choice because copper would require several repeaters.

The break-even point in the early 1990s between copper feeder systems and fiber, with everything else being equal, was around 9000 ft (1.5 miles): If less than 9000 ft, use copper; if more than 9000, use fiber. This mileage has been decreasing for some time because of the downward trend in the cost of fiber cables, eventually making fiber ubiquitous (in the late 1980s, the break-even point was at 2.5 miles [7.9]). The prove-in distance also varies between different telephone companies. In the late 1980s, Pacific Bell was reported to employ fiber only beyond 18,000 ft; BellSouth employed 12,000 ft at the time; other users have employed 4000 ft as the cutoff point [7.34].

A copper-wire local loop cost $547 in 1989; it is expected to remain at that level in inflated dollars through 1999. A fiber-based loop (double star, using subscriber carrier in the feeder route) cost $2779 in 1989; it was expected to cost $1357 in 1995. A fiber-based loop with an active pedestal cost $828 in 1989; it was expected to cost $565 in 1995 [7.35]. See Figure 7.9. Some observers conclude that fiber will not replace copper pair at the same cost within the decade of the 1990s, although in the next five to seven years the LECs will probably deploy fiber to the curb ("to the pedestal") in a shared architectural configuration to serve upscale residences and small businesses [7.35]. Also see Section 7.8.

7.4.5 Bandwidth Augmentation on Existing Fibers

Once the optical fiber is in the network, the carrying capacity of the system can be upgraded easily. Currently, at least two ways are used to achieve this upgrade (with coherent transmission technology being a future approach): new, higher rate

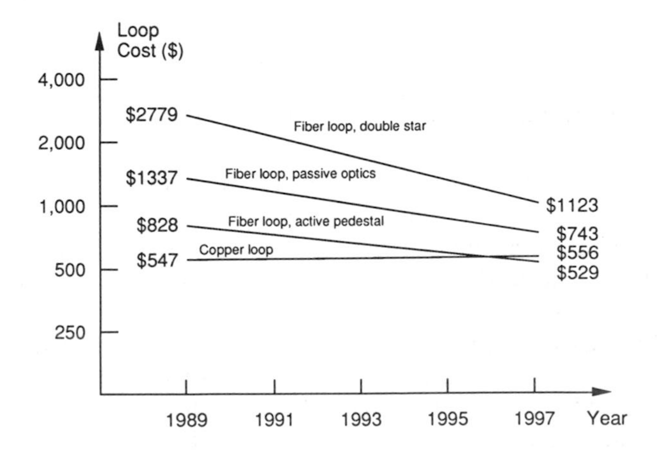

Figure 7.9 Cost of fiber local loop over the years.

regenerators and wavelength-division multiplexers (WDMs). The preferred approach until now has been to use regenerators that double the bit (clock) TDM rate. However, to retrofit for an existing system, all of the terminal equipment has to be replaced, which can be expensive (of course, when building a new system from scratch, the faster technology would be built right into the termination equipment). However, once a system built on older technology has been in the field for a number of years, the WDM approach may be cheaper. For a 20% increase in system price, the WDM approach can double the capacity of a link [7.21].

One of the major advantages of fiber installations is that the bandwidth of an existing link can be upgraded easily by replacing the termination optoelectronic equipment. Costly replacement or reinforcement of fiber cable is not required. In fact, some observers believe that leaps in the speed of fiber optic transmission are outstripping the demand for these high capacities. A system operating at 560 Mb/s can be upgraded with new equipment to 1.12 Gb/s. Other systems operating at the 810 Mb/s can be upgraded to 1.62 Gb/s. Some Japanese manufacturers have laboratory prototypes that triple the 1.12 Gb/s to 3.6 Gb/s. Developments in gallium-arsenide laser technologies for the termination equipment indicate that speeds around 10 Gb/s on single-mode fiber-cables may be possible in five to ten years.

Another way to increase bandwidth on an existing cable is to use WDM (instead of investing in more costly high-speed systems discussed above). WDMs based on diffraction grating or fused couplers can increase the transmission capacity by two, four, or six. Diffraction grating for 16-channel WDM systems has already been demonstrated in the laboratory [7.36]. WDMs can, for example, double the capacity of an existing cable route by allowing one signal on a 417 Mb/s system to travel at 1310 nm and a new signal at 1550 nm. WDM with a WDM filter involves two lasers operating at two different wavelengths and a receiver that is able to receive two wavelengths; thus, we have two complete systems, each operating at the base capacity (say, 417 Mb/s). A normal installation involves two fibers: one for each direction. Some WDM applications obtain a bidirectional signal on a single fiber by sending a west-east signal at 1300 nm, and an east-west signal at 1550 nm.

7.5 FIBER OPTIC STANDARDS

According to one source, more than 400 proposed or actual standards exist in the United States for optical fiber components and systems; the same source estimates more than 1000 standards worldwide [7.25]. Some of the U.S. organizations issuing standards pertaining to fiber-cable are the U.S. Department of Defense, Electronic Industries Association (EIA), Insulated Cables Engineers Association, Institute of Electrical and Electronics Engineers, National Fire Protection Association,

Underwriters Laboratories, and ANSI Accredited Standards Committee X3. Some of these were described in Chapter 2. Some of the typical cable standards are UL910, IEEE 383 (flammability), and MIL-SPEC (military applications).

While the capacity of an installed system can be incrementally increased as the demand and the technology allow by replacing the terminating optoelectronics, some potential problems exist in this upgrade. One approach to growth (followed for example by Fujitsu) goes from an existing 405 Mb/s to an 810 Mb/s system, and then to a 1.7 Gb/s system. Another approach (followed, for example, by Ericsson) goes from 565 Mb/s to 1.2 Gb/s and then to 2.4 Gb/s. These data rates are inconsistent and interconnection among systems of different rates is difficult. A standard, particularly for interface purposes between different systems, is required [7.4].

Transmission Standards

Efforts are under way to define a digital hierarchy that is ideally suited to handle fiber-based signals and at the same time allow easy extraction of subsignals. The approach has been called SONET or SDH, as discussed in Chapter 3.

Standards for Dispersion-Shifted Fiber

Dispersion-shifted fiber, optimized for use at the high end of 1550 nm compared to the normal 1300 nm, offers a number of advantages, including low dispersion, low intrinsic loss, and superior resistance to bend-induced attenuation. These advantages make this type of fiber ideal for undersea installations because of the increased repeater spacing and reduced costs. The fiber was first offered commercially at the beginning of 1985 by Corning.

In spite of its advantages, however, widespread use of dispersion-shifted fiber has been slow to develop [7.37]. While the relative unavailability of 1550-nm hardware in support of the fiber has played a role in the deployment, an equally significant deterrent has been the general lack of standards necessary to support measurement and performance criteria. Until recently, characteristics of a dispersion curve that would uniquely identify dispersion-shifted fiber have never been formally documented. Activities by the EIA and CCITT are providing much-needed support in this area. To provide a commercial basis for agreement, the EIA has adopted a dispersion-shifted definition consistent with existing products. A second proposal submitted to the CCITT provides a means of calculating dispersion for dispersion-shifted fibers across several fiber designs. Although work remains, significant progress is being made.

7.6 TECHNOLOGY TRENDS

Progress continues to be made in the components area; for example, transmitter-receivers complying with the FDDI standard for LAN backbones. Integrated circuits continue to show price improvements. Recent emphasis has been on packaging (for example, color-coded connectors). New lasers and LEDs have been introduced recently (for example, EE-LEDs for 1300- and 1550-nm transmission in the local loop and LAN environment). The price of the 1300-nm version with one pigtail and three outlets is around $200; prices for the 1550-nm version are around $600. New test equipment and instrumentation have also reached the market [7.38]. Integration is a key technology for achieving the next generation of broadband optical communication systems such as ISDN subscriber loops, high-speed links, LANs, cable television systems, and optical interconnections; the industry is pursuing this aspect with renewed emphasis [7.39].

7.6.1 Near-Term Trends

In this section, some near-term trends (one to four years) that are expected to affect the outlook of fiber and communication are discussed.

Cost Reduction to End-Users

The continued installation of fiber in the network will reduce the cost (not the price) of long-haul communication to an estimated 2.6×10^{-8} cents/bit over the next few years [7.40]; these cost benefits will not necessarily be significant to the end-user because the major cost component is in the administration (billing, forecasting, and maintaining) of the plant, and in the endogenously more costly local access.

MANs

The BOCs will begin to provide broadband services (45 to 155 Mb/s) on a tariffed basis in the early to mid-1990s under the MAN architecture or SMDS, which is a precursor of BISDN. Extensive optical fiber facilities are already available in most business centers in the United States (see Chapter 15).

Competition from Satellites

Much has been said in the trade press about VSAT systems. In truth, however, not only will satellite technology and services have no effect on the fiber market in the future in terms of market share, but, in fact, fiber may affect the future of satellites, where the two compete. Many companies (including the IXCs) want to stop using satellites for voice and data applications because of security, bandwidth, and delays issues. Satellites do have an important role for mobile communication, maritime communication, and one-way TV, audio, or data broadcasting to hundred or thousands of sinks. But for point-to-point applications, fiber is superior. VSAT technology may become more expensive in the future, due to increased payload delivery costs, and it may then be used only for those applications for which no alternatives exist; where broadband fiber facilities exist, they may often be the preferred choice.

Transoceanic Cables

Five cables are planned for the North Atlantic route between 1988 and 1992, and three are planned across the Pacific by 1995. Major business opportunities exist for suppliers of underwater-grade fiber cables. These projects will require 40,000 miles of cable, 200,000 miles of fiber, and 10,000 regenerators. At least one company plans to provide transatlantic video transmission over the fiber cable more economically than over satellite facilities [7.21]. A number of fiber optic cables are in the process of being installed, or have just entered service: TAT-8 across the Atlantic, HAW-4 to Hawaii, and TPC-3 across the Pacific.

TAT-8 is the first long-haul undersea system to use digital transmission, optical technology, integrated circuits in the repeaters, an undersea branching capability, and a supervisory system capable of in-service monitoring and control for the undersea repeaters [7.41]. It connects the United States to the United Kingdom and France, and became operational in 1988.

The 8270-mile Pacific link, connecting the U.S. mainland, Hawaii, Guam, and Japan, was put into service in April 1989. The cable is owned by 30 countries (AT&T owns the largest share at 35%). The cable consists of three pairs of single-mode fibers, providing 280 Mb/s (equivalent to 4000 DS0 voice circuits) on each circuit [7.42].

TAT-9, a fiber optic submarine cable connecting the United States to the United Kingdom, Spain, and France is planned to become operational in 1991. "Private" submarine cables are also being installed. PTAT-1 connecting the United States and Bermuda to the United Kingdom and Ireland became operational in 1989. A second cable, PTAT-2, connecting the United States to the United Kingdom and Italy, is planned for 1991.

Short-wave LDs and Long-wave LEDs

Long wavelength LDs have dominated single-mode fiber applications for long-haul systems, and short wavelength LEDs have been common in emerging short-haul applications (multimode systems for data communication). Other sources can also be short-wave LDs and long-wave LEDs.

Short-wave LDs are used in compact disk (CD) players. Japanese companies are now mass producing CD-player lasers (at the rate of half-a-million a year) to the point at which the cost has gone down to $6 per device. Fiber producers and carriers may find this technology hard to ignore. A number of technical issues have to be resolved before the LDs can be exported from the CD players to telephone transmission.

Long-wave LEDs operate in the 1300- to 1600-nm range (near infrared). While these devices have some advantages (easy to manufacture, more reliable, less expensive, and less sensitive to environmental factors), there are also disadvantages (dimmer and diffuse output, slower modulation, and broader spectra). This technology has the same general bandwidth limitations of normal LEDs. Their future, however, is in military applications because of the increased reliability. Another possibility is with the FDDI standard for data applications because of the technical requirements of the latter.

Polarization-Preserving Fibers

Polarization-preserving optical fiber is a specialty fiber in which stress is added to the core to preserve separate modes of polarized light. Major applications include the sensor-gyro area (of interest to avionics), oil well drilling, and the military. Substantial market growth is expected in this area over the next few years.

Plastic Fibers

These optical fibers find applications in automotive, aircraft, and other related illumination and sensing tasks, as described earlier. The fiber can even be tied in knots and still work. It has a temperature range of between −10 and 120°F. At present, this market is dominated by the Japanese, but U.S. firms are getting into the business, and production should increase over the next few years. The market is currently estimated to be between five and ten million dollars. The car of the 1990s will contain approximately 30 ft of fiber; 80% will be silica-based, 10% plastic-clad silica, and 10% plastic. The auto industry will spend $400 million in 1990 in fiber, and perhaps $1 billion in 1995 [7.36]. Munich's Productronica '87 was the site for the first successful implementation of a plastic optical fiber cable linking multiple computers [7.43]. Computers communicated at 10 Mb/s on a star Ethernet LAN using polymer fiber developed jointly by Codenoll and Hoechst of Germany. These polymer fibers are very transparent over a wide range of wavelengths; they have high strength and are resistant to heat, humidity, and a number of chemicals. They are of the large-core multimode type.

Some proponents say that "what is glass today will be plastic tomorrow." Polymer fibers are less expensive to manufacture; it is expected that they will be half the price of glass fiber and less than the cost of twisted-pair. Because of the larger core, these fibers are easier to handle and to connect; connections can be made in minutes using a razor blade. However, plastic fibers will be limited to LAN applications because of their transmission limitations: 150 to 4500 ft and hundreds of Mb/s, compared to 30- to 40-mile unrepeatered distances at bandwidths of 1 to 10 Gb/s for standard single-mode fibers [7.44].

Coherent Methods

With coherent transmission methods, many more channels can be multiplexed onto a fiber, compared to IM: Bandwidth can be increased by two or more orders of magnitude. Coherent modulation techniques are similar to FM. At present, a bulky external laser is required; efforts are under way to reduce the size and increase the reliability. A trial was conducted by AT&T in 1989–90 using a coherent system carrying 1.7 Gb/s over 22 miles; 30 Gb/s systems are already being discussed [7.45].

7.6.2 Long-Term Trends

Optical Switching

At present, a growing interest in photonic switching and signal processing has become apparent. Much of the stimulus comes from the worldwide installation of optical fiber communication systems and the rapidly increasing demand for broadband services. In the late 1990s, photonic switching may become an attractive alternative to electronic switching. A photonic switch has the advantage of transmission channel data rate independence and can handle throughputs far in excess of a conventional electronic switch. Photonic switching would facilitate the introduction of BISDN, discussed in Chapter 4.

A 32 × 32 photonic switch was reported by Suzuki in 1987 [7.46]. In 1990, AT&T announced that it was planning to bring photonic switching technology, now in the prototype stage, to the market within five years. The technology may at first be applied to SONET-based digital cross-connect systems. The five-year plan is considered to be aggressive even by the vendor [7.47].

More sophisticated multi-terabit photonic switching will be achieved in the middle to late 1990s.

Light-Metal Fluoride Fiber

Research now under way may provide in the next decade unrepeatered links in the 3000- to 10,000-km range. Silica fiber is approaching its theoretical limit in terms of attenuation (but not bandwidth). Theoretical projections for light-metal fluorides show possible losses below 0.01 dB/km. In recent years, losses in nonsilica-based fibers have been reduced from 1000 dB/km to around 1 dB/km. More work needs to be done to lower this number to its theoretical limit [7.28].

Photonic Computing

As a result of a U.S. Air Force assessment of its technical needs, photonics has been earmarked as one of the most important future technologies for research. The Photonics Center at the Rome Air Development Center was reported to be embarking on a project having the long-range goal of developing an all-optical computer [7.48]. A laboratory-grade photonic computer was announced by AT&T in March 1990. Free-space lightwave photonics was used as an optical interconnect method that can replace copper, or even fiber, interconnects in supercomputers [7.45].

7.7 FIBER-BASED CATV

CATV operators see the need to deliver as many as 60 video channels over fiber to subscribers in the near future. On the other hand, at the beginning of 1990, the CATV industry still saw the need for technical advancements before they can commit to the technology. To install fiber they claim that the following must be achieved [7.49]:

- At least 40 channels initially with more later;
- Ranges up to 15 miles;
- Carrier-to-noise ratio of 55 dB; and
- Carrier-to-intermodulation ratio greater than 65 dB.

Currently, with specially selected optical devices, all four of these requirements can be combined in a single fiber system.

Fiber optics is being used for the transmission of broadband signals using both analog and digital techniques. In the short term, analog optical channels have excellent transmission properties for broadband analog signals and are more suited to television because the TV set can be interfaced directly to the transmission medium. Digital signals require expensive demultiplexers and digital-to-analog converters to drive a standard television terminal. The digital implementation is more applicable to data transmission (which by definition is already in digital format), though digital video has its own role, particularly for high-quality editing processes and special effects.

Some progress was achieved in the late 1980s in attempts to have the CATV industry use fiber for programming distribution. A number of companies have shown fiber-based product prototypes that promised audio and frequency modulation transmission alternatives that meet quality and cost requirements. For example, a system to send 60 analog video channels in FM or a mixture of FM video and digital voice and data over one fiber for at least 12 miles without a repeater has been announced [7.50]. In 1989, a vendor announced an 80-channel AM fiber optic trunking system that can integrate with existing coaxial systems [7.51]. A fiber can carry an AM signal where the intensity of the light carrier is varied in accordance with the incoming signal (in contrast to the on-off method used in digital modulation; however, note that, as discussed in Chapter 1, the ASK method intrinsic in today's IM technique used in fiber transmission is ultimately a form of amplitude modulation). For a discussion of AM, FM, and digital video transmission products on fiber optic systems, refer to Reference [7.52].

Some proponents claim that the age of fiber optics in CATV has arrived. The technology is being used with greater frequency in supertrunking applications, showing that the technology can withstand the rigors of the CATV operation environment. Increasing digitization of TV sets and VCRs coming from Japan is another force making fiber an attractive system for transmission of video.

Fiber systems appear to be a strong near-term candidate for wide-band CATV distribution. Optical fiber communication systems have several useful technical properties that make them suited for urban distribution. The most likely scenario for the immediate future, however, is the one in which fiber optics will not replace coaxial cable entirely, but will be used in conjunction with coaxial when cost-effective. The decision is generally an economic one. Coaxial cable will remain the dominant transmission medium for the U.S. CATV industry over the next few years, but fiber optic systems will increasingly penetrate the market. The U.S. CATV cable market is growing at a rapid rate—from $250 million in 1981 to nearly $700 million in the early 1990s. Plant mileage will grow from 234,000 km installed in 1982 to 360,000 km installed by the late 1980s.

Several designs for optical fiber distribution systems and a number of experimental field trials providing video services directly to subscriber domiciles have taken place in Japan, Canada, France, the United Kingdom, Germany, Denmark, and the United States in the past eight years. None of the trials reached significantly large numbers of subscribers, although they provided valuable experience for the design of future larger systems. The challenges facing widespread introduction of fiber optics are:

- Large existing investment in coaxial cable systems in major urban areas representing top markets;
- Coaxial cable systems that are underutilized and are capable of carrying much more information than they currently do; and
- Current systems that use well-developed, relatively inexpensive technology and are directly compatible with consumer television electronics.

Optical fiber networks normally tend to be implemented with some sort of star topology, rather than tree-and-branch, which is typical of CATV (telecommunication applications employ a point-to-point topology; LANs are generally ring-oriented). Optical fiber does not lend itself too easily to the tree-and-branch configuration because of the difficulty of splitting and tapping signals. Thus, the video distribution networks of the early 1990s will be using both optical fiber and coaxial cable with elements of both tree-and-branch and switched-star topologies to provide a network that is matched to the requirements of the service to be provided.

Currently, the optimal topology for distributing TV signals to subscribers, in terms of cost, is a tree-and-branch coaxial cable network; however, the fiber alternative is considered reasonable by industry experts in view of the flexibility of the resulting plant and the technical improvements it affords. In the near term, implementation decisions between coaxial cable and fiber optics for new CATV networks will be made on economical as well as technical grounds. The economics favor the retention of coaxial-cable systems for the reasons listed above. Coaxial-cable technology is well developed and is compatible with broadcast television technology. On the other hand, fiber-based AM-CATV systems are being developed for CATV trunking systems to replace coaxial cable trunks. In addition, they are being considered for distribution of CATV to the "curb" (see below) for systems that take fiber to the curb.

7.8 FIBER LOOP ARCHITECTURES

Entering 1990, about a dozen and a half architectures for bringing optical fiber to the customer have evolved. Of these, the switched single-star and the double-star most closely follow the traditional loop-CO switch architecture of the existing telephone plant. Other architectures include the Raynet passive bus and the QPSX MAN topology (Chapter 15). These topologies provide decentralized switching. British Telecom is considering a passive star. Three BOCs were performing trials with the Raynet bus in 1990 [7.53].

Some of the newly discussed architectures provide fiber to the curb, rather than all the way into the home. In this configuration, the fiber part of the network terminates near the premises and the last drop is copper-based. Other "fiber-thin" topologies are also being discussed.

All architectures have advantages and disadvantages. Key features of any such architecture must be remote troubleshooting, remote grooming, high bandwidth, and provision of value-added services. We do not yet know which architecture will eventually be implemented, or whether several architectures will evolve, possibly for different applications. Some are in favor of the double-star connection-oriented BISDN because this architecture is the one closest to today's plant. Others prefer ring-based connectionless systems (defined in Chapter 1), which afford distributed switching. These architectures may be preferred by the data community because of their preference for a connectionless LAN interconnection system. Such a connectionless approach to broadband already has three manifestations: FDDI, IEEE 802.6 MAN, and SMDS (discussed in Chapter 15).

7.8.1 Major Proposed Architectures

Figure 7.10 depicts four major classes of architectures being considered:

1. Traditional switched star;
2. Active double star;
3. Passive double star; and
4. Hybrid star-bus.

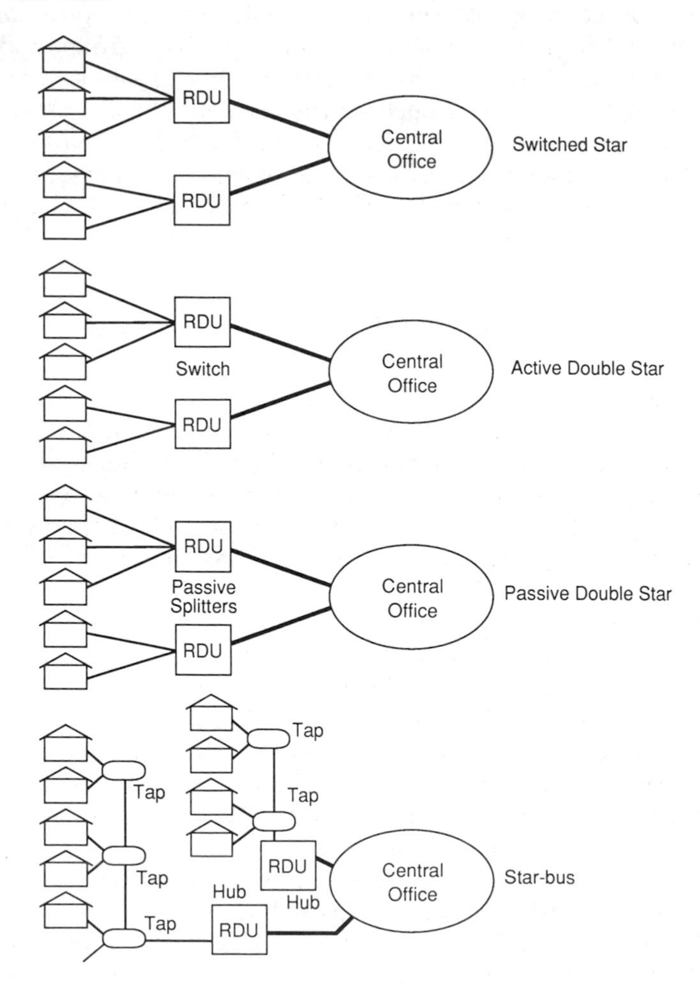

Figure 7.10 Four major classes of proposed fiber architectures.

Switched-star systems provide individual paths from the CO directly to each home; each path carries signals destined to a specific subscriber. The CO provides the switching function. This is the traditional method for delivering telephone signals to customers. Interoffice traffic is switched at the CO, sent over a feeder plant to a residential distribution unit, and then delivered over the distribution plant to the subscriber, as discussed in Chapter 1. When this architecture is extended to a fiber loop, a high-capacity fiber feeder system brings a signal close to the customer residence. Demultiplexer equipment then splits out the signals destined to different users. Signals are distributed over a fiber-based distribution plant in such a fashion that each "last mile" facility only carries the signal intended for that subscriber. The limitation of this system is that the premises-based optoelectronics costs are high, and are not shared between users.

The active double star allocates some active signal switching to residential distribution units, reducing the amount of fiber required. Note, however, that the cable itself is only about 10% of the fiber loop cost (about 45% is in the optoelectronics and 45% in the electronics [7.54]). The broadband switch at the CO sends a multiplexed broadband signal to the residential distribution unit (typically at 2.488 Gb/s). At the distribution unit, the signal is demultiplexed; each premise may be provided with four video signals and telephone signals over a single fiber. This approach still needs premises-based optoelectronics. In addition, another conversion is needed before a user can connect to an existing TV set: The digital TV signal must be converted to analog. One way to eliminate this conversion is to send the video signal in analog FM format.

A passive double star uses a splitter to route signals for a specific home onto the appropriate fiber. It shares electronics and the physical medium by multiplexing at the CO signals intended for several users in a proximity; the passive splitter at the distribution unit is used to recover the intended signal and loads it onto the appropriate fiber connecting the premise to the distribution unit. The passive star and star-bus architectures may be more appropriate if one-way CATV is the major service to be provided over fiber; switched star or active double stars are more appropriate if interactive video on demand is the major service.

The star-bus architecture differs from the others by using coaxial cable for the last 100 ft. This approach allows sharing of the optoelectronics and requires less total fiber. By using the Raynet architecture, signals are delivered to as many as 50 homes from each of several distribution lines fed by a hub serving several thousand homes. Signals are converted from the optical to the electrical domain at tap sites; these tap sites allow the sharing of the optoelectronics costs between several domiciles. Each hub is fed by a trunk fiber from a CO or a CATV headend. In the initial configuration, only standard TV service would be provided; in more advanced schemes, a mix of broadcast and on-demand services would be provided.

The cost of providing full broadband capabilities (including HDTV on demand) with today's technology may be around $5000 to $10,000; this cost may go down to $2000 over the next few years [7.35, 7.55]. In the meantime, architectures providing fiber to the curb and taking fiber to pedestals, taps, or remote terminals, while delivering the signal to the actual home over coaxial or twisted-pair cables, are being studied. These architectures aim at reducing costs by varying the placement of switching and multiplexing functionality [7.54]. Economies are obtained by serving a few customers from a single residential distribution unit. Other approaches (favored by the CATV industry) involve avoiding switching and using the fiber with analog modulation of several video signals (this, however, is not consistent with the BISDN model).

The final disposition as to which architecture is the best compromise between cost, features, and industry preference will probably not be reached until 1996. The active star configuration is the basis of today's telephone company networks, using copper at least part of the way; simply replacing the medium with fiber would provide a viable architecture. However, other possible methods may be more cost-effective in some situations.

Issues concerning the architecture selection are complex and may be partially based on LEC's corporate business strategies, particularly with regard to TV distribution (and also, whether it should be dedicated, as with CATV systems, or switched, as with BISDN).

7.8.2 Major Fiber Loop Trials, circa 1990

Table 7.3 provides a listing of the known fiber loop trials entering 1990. Most trials employ single-mode fiber.

New England Telephone began a trial of the Raynet bus in 1989, providing fiber-to-the-curb services to 100 homes in Lynnfield, Massachusetts [7.53]. A multimode 85/125 μm fiber providing 2 Mb/s was used in the trial. The goal is to provide connectivity for $1300 or less. The bus architecture allows cost sharing of the electronics, located every three or four houses. The electrical signal is delivered to each customer's home via the existing telephone copper dropwire (recall that twisted-pair can deliver up to 4 Mb/s over short distances without expensive signal processing hardware). Shared optoelectronics brings the cost down, compared to other systems, which currently range from $2000 to $5000 [7.56]. At present, the trial does not include TV delivery, which would require more bandwidth than achievable with the copper dropwire. At the very least, these fiber architectures can be considered as transitional mechanisms to BISDN.

British Telecom is planning a nationwide telephone over fiber local loops by 1993 [7.57]. For the United Kingdom, a passive optical network architecture was being tested in 1989. Fiber is used all the way to the home. The evolutionary architecture provides 2.048 Mb/s initially, and can upgrade to BISDN rates in the future, using WDM techniques and single-mode fiber. Two megabits per second

Table 7.3

Holding company	*Operating company*	*Developer*	*Cable television company*	*Timeframe*	*Location*	*Number of residences*	*Switch*	*Electronics supplier*	*Transmission system/equipment*	*Transmission mode*	*Cable and fiber supplier/type*	*Notes/Services*
BellSouth	Southern Bell	Heathrow Development Corp.	Heathrow Telecommunications	9/89 Integrated voice & video	North Orlando, Fla.	55 now, 256 targeted	Northern Telecom DMS-100	Northern Telecom	LEDs and laser diodes	Singlemode digital	Northern Telecom for central office to residences. Optical Cable Corp. within homes	
BellSouth	Southern Bell	Genstar Southern Development Inc.	Hunter's Creek Cablevision, Genstar and Scientific-Atlanta	1986—CATV began, upgrades continue	South Orlando, Fla.	250		Scientific-Atlanta	LEDs	Single-and multimode	AT&T SM 48-fiber cable from headend to selector node; 5 multimode 144-fiber cables	
BellSouth	South Centeral Bell	Boyle Investment of Memphis	POTS only	11/88	Riveredge, Tenn.	54 so far, 99 targeted	AT&T 1A-ESS	AT&T System SCC-5 system and terminal	FT-series 1.7-gigabit/sec transmission system	Singlemode digital, analog for POTS	1 singlemode fiber	Same technology as is installed in Memphis telephone system
BellSouth	Southern Bell		POTS only	Began 8/89	Governor's Island, Lake Norman, N.C.	42 targeted	AT&T 5-ESS	AT&T	AT&T Series 5 SLC	Singlemode digital	1 singlemode fiber	
BellSouth	Southern Bell		POTS only	4th quarter 1989	Lakeview Terrace, Charleston, S.C.	100	AT&T 5-ESS	AT&T	AT&T Series 5 SLC	Singlemode digital	1 singlemode fiber	
BellSouth	Southern Bell		POTS only	Began 8/89	The Landings, Skidaway Island, Savannah, GA.	192	AT&T 5-ESS	AT&T	AT&T Series 5 SLC	Singlemode digital	1 singlemode fiber	
BellSouth	Southern Bell		POTS only	Began 8/89	Hunter's Creek II, Orlando, Fla.	117	AT&T 5-ESS	AT&T	AT&T Series 5 SLC	Singlemode digital	1 singlemode fiber	
BellSouth	Southern Bell		POTS only	3/89	Coco Plum, Miami, Fla.	45 initially 300 eventually	AT&T 5-ESS	AT&T	AT&T Series 5 SLC	Singlemode digital	1 singlemode fiber	
BellSouth	Southern Bell		POTS only	1st quarter 1990	The Summit, Columbia, S.C.	285	AT&T 5-ESS	AT&T	AT&T Series 5 SLC	Singlemode digital	1 singlemode fiber	
BellSouth	Southern Bell		POTS only	1st quarter 1990	Morrowcroft, Charlotte, N.C.	50 now, 90 eventually	AT&T 5-ESS	AT&T	AT&T Series 5 digital	Singlemode digital	1 singlemode fiber	
GTE Corp.	General Telephone of California	Existing community		2nd quarter 1990	Cerritos, Calif.	705 targeted; 600 for POTS only, 100 for Stell TV, and 5 for "jukebox video"	Amer. Light. Sys.		Analog video digital POTS	36 video channels and POTS over fiber optic supertrunk; hybrid video/voice/data over singlemode fiber		
				2nd quarter 1990	Cerritos, Calif.			GTE Labs		Digital	1 singlemode fiber	Switched video/video on demand
				2nd quarter 1990	Cerritos, Calif.			AT&T Net. Sys.		Digital	1 singlemode fiber	Voice, no TV
GTE Corp.	General Telephone of California	Existing community	POTS	1989	Cerritos, Calif.	Existing network	GTE 2EAX			Analog	Twisted pair	
GTE Corp.	General Telephone of California	Existing community	Apollo Cablevision is leasing bandwidth	Main Street and pay-per-view 2nd quarter 1990	Cerritos, Calif.	250–300 targeted now, potentially 16,000		GTE Services Corp.	Digital	Coaxial cable		

Table 7.3 (continued)

Holding company	*Operating company*	*Developer*	*Cable television company*	*Timeframe*	*Location*	*Number of residences*	*Switch*	*Electronics supplier*	*Transmission system/equipment*	*Transmission mode*	*Cable and fiber supplier/type*	*Notes/Services*
Contel	Contel Service Corp.		POTS only	4th quarter 1988	Ridgecrest, Calif.	100 targeted	AT&T 5-ESS	AT&T Phoenix, Ariz.	AT&T Network Systems	Singlemode digital	Singlemode fiber	House wiring as it normally is
Ameritech	Illinois Bell		POTS only	Late 1989	Chicago's northwest suburbs	300				Digital	2 singlemode fibers	Fiber optics set up in "active pedestal" format to get cost down to POTS
Bell Atlantic	New Jersey Bell	Rieder and Sons	POTS only	8/88	"Princeton Gate," South Brunswick N.J.	50 so far, 104 targeted	AT&T 5-ESS	AT&T Network Systems	Laser diodes	Digital	Singlemode fiber	4 dual lines per residence
Bell Atlantic	Bell of Pennsylvania	Rehab	Helicon	1st quarter 1989	Perryopolis, Pa.	80–100	Alcatel analog	Alcatel N.A.		Analog video; digital voice over multimode	62.5/125-micron multimode fiber	
Southwestern Bell	Southwestern Bell Telephone	Cedar Creek Properties Inc.		10/89	Cedar Creek, Olathe, Kan.	260			AT&T Series 5 SLC	Digital	Singlemode fiber	Fiber-to-the-pedestal
Southwestern Bell	Southwestern Bell Telephone		POTS only	1989	Leawood, Kan.	50–100	AT&T AESS	AT&T DDM-1000 time-division multiplexer	Laser diodes	AT&T digital subscriber loop carrier, digital/analog	AT&T singlemode fiber/twisted-pair copper "mixed" cable	
Southwestern Bell	Southwestern Bell Telephone		Sammon CATV	4th quarter 1989	Mira Vista, Ft. Worth, Tex.	80		Amer. Light. Sys.	Amer. Light. Sys.	ALS FM analog video and POTs	2 singlemode fibers	Fiber-to-home switched video
U S WEST	Northwestern Bell		POTS only	2nd quarter 1989	Mendota Heights, Minn.	97 targeted	AT&T 5-ESS	AT&T Network Systems			Singlemode fiber	
U S WEST	Mountain Bell		POTS only	1990	Desert Hills, Scottsdale, Ariz.	102 targeted	AT&T 5-ESS	AT&T	AT&T Series 5 SLC	Digital	Singlemode fiber	
British Telecom	British Telecom		Single-line telephony passive optical network (TPON); voice only	Start 9/90	Bishops Stortford, U.K.	128 customers	BT	BT	BT passive WDM (TPON)	Singlemode digital	1 singlemode fiber	
British Telecom	Britsh Telecom			Start 9/90	Bishops Stortford, U.K.	125 business customers	BT	BT	BT passive WDM (TPON)	Singlemode digital	1 singlemode fiber	Business TPON; voice only
British Telecom	British Telecom			Start 9/90	Bishops Stortford, U.K.	128 customers	BT	BT	BT passive WDM (TPON)	Singlemode digital/copper	1 singlemode fiber	Single-line street TPON; voice only
British Telecom	British Telecom		Broadband distributed star; voice, TV and other services	Start 3/90	Bishops Stortford, U.K.	125 residential customers	BT switched star	BT	BT passive WDM (TPON)	Singlemode digital	1 singlemode fiber	Similar to Westchester cable TV network in U.K.
NYNEX	New England Telephone		POTS only	1989	Lynnfield, Mass.	100		Raynet			85/125-micron multimode fiber	Fiber-to-the-curb
Ameritech	Ohio Bell			4Q90	Columbus	300		Raynet	<OC-1 <OC-2			POTS FTTC
Bell Atlantic	C&P of Virginia			5/90	Cascades	77-FTTC						POTS
				Video: 4Q90	Louden County VA.	49-FTTH		BBT		switched digital video		video
Alltel					Piper Glen Charlotte N.C.	50		RTEC	OCM-6	850nm LED		FTTC POTS video
Contel				in service	Sydney, NY			AT&T				POTS
Contel				in service	Wyoming, Minn.			AT&T				POTS
Contel	Contel of Calif.		ContelVison	1991	Rancho Las Flores Calif.	350 init. 15,000 ult.		AT&T RTEC Raynet		AM-video		FTTC POTS video

with 850-nm LEDs on single-mode fiber up to 5 km (3 miles) is within the envelope of feasibility and technical practicality. Remote concentrators are used to achieve one or more of the following six implementation strategies:

1. TDM multiplexing of several users over the backbone to the CO;
2. WDM multiplexing of several users over the backbone to the CO;
3. TDM as above, but fiber is terminated a few hundred feet from the home, and coaxial cable is used for the drop line;
4. Same as 3, but with WDM in the backbone;
5. WDM multiplexing and coherent optical technology in the backbone; and
6. Low cost superluminescent LEDs in place of the standard LEDs.

As of 1990, the BOCs continue to be barred from offering any service other than telephony and from owning CATV systems in their own territory. This may be a retardant factor in establishing an all-fiber plant because the BOCs must prove-in the technology only on POTS and data revenues.

In conclusion, four strategies for fiber to the home are possible:

1. If the emphasis is on voice service and DS1-rate data, then the passive bus network is the most cost-effective as of the beginning of 1990.
2. If the emphasis is on dedicated video broadcast (CATV) technology, then the double star fiber network with analog TV transport is the most cost-effective as of the beginning of 1990.
3. If the emphasis is on BISDN services, then the double star with digital transport is the most cost-effective as of the beginning of 1990.
4. If connectionless data are the emphasis (LAN interconnection), then MAN-like rings may be the best choice as of the beginning of 1990.

REFERENCES

[7.1] D. Minoli, "Fiber Optics Communication: Issues and Trends," DataPro Report CA40-010-101, 1989.

[7.2] D. Minoli, "Fiber Optics Communication: Markets and leaders," DataPro Report CA40-010-201, 1989.

[7.3] D. Minoli, "Fiber Optics Communication: Technology Briefing," DataPro Report CA40-010-301, 1989.

[7.4] "Fiber Optics Continuing Growth," *Communications Week*, April 7, 1986.

[7.5] G. Friesen, "Optical Fiber for Subscribers," *CO Magazine*, March 1986.

[7.6] "Fiber Optics after Long Distance," The Yankee Group, Boston, July 1986.

[7.7] C.H. Kao, G.A. Hockham, "Dielectric-Fiber Surface Wave-Guides for Optical Frequencies," *Proceedings of IEE*, July 1966, Vol. 133, pp. 1151–1158.

[7.8] H. Murata, *Handbook of Optical Fibers and Cables*, Marcel Dekker, Inc., New York, 1988.

[7.9] S. Esty, "Single-Mode Fiber Applications," *CO Magazine*, March 1986.

[7.10] *Lightwave*, November 1986.

[7.11] "Local Loop Requires New Hardware," *BOC Week*, September 22, 1986.

[7.12] C. Lin, personal communication, Bellcore, May 1990.

[7.13] R.L. Soderstrom et al., "A Miniaturized Fiber Optic Laser Receptacle Using a Compact Disc Laser as a Low Cost, Reliable Source," FOC/LAN '87 Conference Record.
[7.14] P.S. Lovely, "The Principles of Chromatic Dispersion," *TE&M*, February 15, 1989.
[7.15] S. Scully, "The Loop Spurs New Interest in Multi-Mode," *Lightwave*, February 1987.
[7.16] *Lightwave*, March 1990, p. 9.
[7.17] P. Susca, "Plastic Fiber's Dramatic Resurgence," *Lightwave*, February 1987.
[7.18] *Byte*, December 1984.
[7.19] A. Morrow, "Advances in Optical Fiber Fabrication Using Vapor Phase Processing Techniques," *Optics News*, January 1988.
[7.20] J. Kreidl, "Ericsson Method Speeds Manufacturing of Fiber," *Lightwave*, August 1989, p. 10.
[7.21] *Lightwave*, December 1986.
[7.22] H. Rausch, "Optical TDM Overtakes Electrical," *Lightwave*, March 1988.
[7.23] J. Hecht, "Pushing the Limits of Fiberoptic Technology," *Lasers and Optronics*, January 1988.
[7.24] H. Rausch, "Setting the Stage for 1988," *Lightwave*, January 1988.
[7.25] *Fiberoptic Product News 1986–87 Buying Guide*, Torrance, CA.
[7.26] *Fiber Optics Magazine 1987 Handbook and Buyers Guide*, Boston, MA.
[7.27] A.J. Ross, K. Sontag, "Manitoba Telephone Keeps FO Cables Dry with Special Monitoring System," *Telephony*, April 1987.
[7.28] D.J. Reinder, "Photonics—Fast Track for Tomorrow's Communications," *AT&T Technology*, Vol. 2, No. 1, 1987.
[7.29] J. Swartz, "Tapping Optical Fibers," *Lightwave*, January 1987.
[7.30] KMI's fiber optic map of long-haul systems, Newport, 1986.
[7.31] *Fiber Optics News*, March 31, 1986.
[7.32] *Fiberoptic Marketing Intelligence*, Kessler Publisher, Newport, RI, June 1990.
[7.33] *Uplink (Hughes Communications)*, Warren, MI, Spring 1990.
[7.34] *Fiberoptic Marketing Intelligence*, Kessler Publisher, Newport, RI, December 1986.
[7.35] *Telephone News*, February 19, 1990, p. 4 (discussion on the report "Optical Fiber in the Local Loop," by Dittberner Associates, Bethesda, MD).
[7.36] *Lightwave*, October 1986.
[7.37] "Fiber Optics Trends," *Photonics Spectra*, January 1988.
[7.38] H. Rausch, "Optical TDM Overtakes Electrical," *Lightwave*, March 1988.
[7.39] S. Scully, "Fiber LANs: More Vigor than Sales Figures Show," *Lightwave*, March 1988.
[7.40] W. A. Morgan, "Spotlight on Fiber Optics," *Business Communications Review*, June 8, 1986, and September 10, 1986.
[7.41] H. Rausch, "Setting the Stage for 1988," *Lightwave*, January 1988.
[7.42] G. Kotelly, "Undersea Lightwave Cable Leapfrogs Pacific Ocean," *Lightwave*, June 1989, p. 11.
[7.43] *Fiber Optics News*, December 7, 1987.
[7.44] *Fiber Optics News*, October 12, 1987.
[7.45] *Lightwave*, March 1990, p. 1.
[7.46] *Fiber Optics News*, May 11, 1987.
[7.47] "News of the Week," *Telephony*, April 23, 1990, p. 8.
[7.48] *Fiber Optics News*, January 4, 1988.
[7.49] *Fiber Optics News*, February 8, 1988.
[7.50] R. Olshansky, "Fiber Technology Steals Cable-TV Show Spotlight," *Lightwave*, January 1988.
[7.51] "80-channel Trunk," *Lightwave*, August 1989, p. 3.

[7.52] S.A. Etsy, "Video Transmission on Fiber-Optic Systems," *Lightwave*, May 1990, p. 30.
[7.53] S. Scully, "Fiber-to-the-Curb Gains Status, Might Preempt Home Connection," *Lightwave*, June 1989, p. 1.
[7.54] G. Kim, "Subscriber Loop Architectures Considered," *Lightwave*, October 1989, pp. 24 ff.
[7.55] S. Salamone, "Phone Companies and Suppliers Looking for Marketable Fiber Loop Strategies," *Lightwave*, October 1989, p. 30 ff.
[7.56] G. Kotelly, "Fiber-to-the-Curb Cuts Delivery Cost," *Lightwave*, June 1989, pp. 5 ff.
[7.57] J. Kreidl, "British Fiber Loop Trial to Deliver Telephone Service," *Lightwave*, June 1989, p. 10.

Chapter 8
Free-Space Infrared Optical Communication Systems

Free-space infrared optical communication systems, employing air as the transmission medium, are a practical and cost-effective solution in a number of niche communication situations. Infrared systems eliminate the need for and expense of cabling and the installation challenges that may be associated with hard-wired systems. No FCC licensing is required for infrared transmission, and no rights-of-way are needed. On the other hand, infrared communication has a number of limitations, notably LOS requirements, short range, and severe susceptibility to weather conditions such as fog [8.1].

Microwave is a more "understood" technology, has a longer range, and is less affected by some weather conditions. However, the high frequency microwave bands suitable for metropolitan communication are congested; in addition, microwave links are more easily intercepted, affecting security. Granted the trade-offs of these conflicting considerations, end-users are now beginning to explore the opportunities afforded by infrared systems. In addition to outside applications, infrared is also being used inside as a way to connect PCs to LANs with wireless links, and in factories to link robots. While infrared component technology continues to grow rapidly for use in optoelectronic equipment for fiber-based systems, its applications in atmospheric communication have developed slowly.

This chapter presents an overview of wireless infrared communication systems. Typical applications are provided. The chapter also discusses the advantages and limitations of the technology.

8.1 APPLICATIONS OF FREE-SPACE INFRARED TECHNOLOGY

Infrared communication systems fill a niche for short-range, cable-free, easily deployed, and cost-effective communication for voice, data, and video. The technology has been available commercially since the late 1960s. Infrared is just another type of electromagnetic emission. Figure 5.2 in Chapter 5 depicted the electromagnetic spectrum; the figure shows the frequency range of infrared communication compared to the more familiar microwave communication. As can be seen in this figure, infrared transmission occurs at a considerably higher frequency than conventional microwave. Due to the higher frequency, infrared signal propagation is much more directional than microwave; this results in a number of advantages to the user, but it also has some disadvantages. Infrared signals are loosely referred to as "light signals" because they are very close to the visible range of the spectrum, as seen in Figure 5.2. On the other hand, infrared, typically at 8300 Å (0.83 μm or 830 nm; 1 Å = 10^{-10} m) is not visible to the human eye because the eye's response is between 6900 Å (red) and 4300 Å (violet); traditional infrared systems operate at 8300 Å, just below the visible area [8.2].

Infrared is a short-range alternative to microwave, radio, cable, or fiber optic facilities, especially in harsh or difficult environments. Atmospheric infrared links are also ideal for electrically noisy environments such as power stations, reactors, and automated factories. Because the signal is transmitted LOS through the air, no permits, cables, wires, fibers, ducts, or conduits are required. Thus, crossing streets, alleys, railroad rights-of-way, or building-to-building can be done with ease. At the application level, infrared communication systems can be considered complementary to microwave. The current commercial infrared systems are capable of delivering DS3 data rates (44.736 Mb/s) over a full-duplex optical path to a range of a half mile, under nearly any weather conditions. This makes the technology ideal for local building clusters, communication with a temporary site local to the main building, and other applications.

A number of users have employed infrared as a quick solution for installation of temporary T1 links between adjacent buildings. A T1 line for buildings that are as little as 500 yards apart may cost $12,000 a year [8.3]. An infrared system will generally have a one-year payback period. The suppliers of infrared systems have developed successful applications in factory and refinery communication (where the sealed nature of the units makes them ideal for use in atmospheres that might be explosive), in the area of video distribution, in manufacturing control as a means of transmitting commands to heavy equipment, and in traditional communication applications where cable access is impractical, such as freight yards, harbors, and airports. Industrial controls for automated warehouses are another example of a low data rate application. One use of the technology is for controlling large cranes; because cranes tend to move back and forth quickly and constantly, the control cables wear out and may require expensive replacements; light beams are an ideal substitute. The military has also been interested in this technology. The concept of free-space atmospheric optical communication initially came out of the military need for secure, short-range voice communication on the battlefield; research and development of this technology began some 30 years ago [8.4].

8.2 ADVANTAGES AND DISADVANTAGES

The high frequencies of infrared-optical transmission have beneficial effects. As the carrier frequency of a transmitter increases, its waves are more directional. Light, far above microwave frequencies, is the most directional of all. Microwave transmission from 10 to 23 GHz results in a coverage area, outward from the transmission antenna, consisting of signal cones of 10 to 15 degrees. At a distance of 5000 ft, this would create a reception zone almost 900 ft in diameter.

Lasers can produce a coherent beam of light that is very narrow. An infrared laser lens-based or mirror-based optical collector can have extremely small angular coverage of less than one degree. An infrared laser signal at 5000 ft would produce a beam width of only 3 to 5 ft.

High directionality makes light ideal for concentrating information into a beam that can be sent accurately between users who are very closely placed, without resulting in mutual interference. Typical free-space infrared systems have a beam width of only a meter at normal operating distances. The high atmospheric absorption of infrared, coupled with its highly directional tight beam and receiver FOV, makes it unlikely that stations within even 10 ft of one another will interfere. On the other hand, point-to-multipoint transmission with a single transmitter becomes impractical because spreading the beam to cover a large area in which multiple receivers may be located would reduce the number of photons captured by a receiver below the minimum needed for detection.

The combination of the tight transmission beam and the low receiver FOV make the infrared carrier difficult to intercept. Because interception of a transmission requires that an intruding antenna be placed in the receiver zone, the relative difficulty of infrared interception can be easily appreciated. Also microwave radiates to space where it can be detected by spy satellites; infrared does not because the power of the terrestrial systems is too low to travel several hundred miles to space.

While directionality is a benefit, the high frequency vulnerability to atmospheric attenuation is not. The transparency of any object to electromagnetic radiation decreases sharply as the frequency increases. Nearly any form of microscopic particle is opaque to light and will cause scattering and loss of energy from the beam between the receiver and the transmitter. Fog, rain, and all the phenomena that obstruct normal vision also obstruct infrared communication paths. Figure 8.1, from Reference [8.5], depicts the effects of weather on the received signal, in terms of attenuation. High winds (more than 60 miles/h) can also be a problem because they raise dust particles and sand, which affect visibility. Under good conditions, a BER of 10^{-8} to 10^{-9} can be achieved; in bad weather, the error rate increases and total signal dropouts (fading) can occur. In contrast, microwave transmission is affected by other types of interference that do not affect

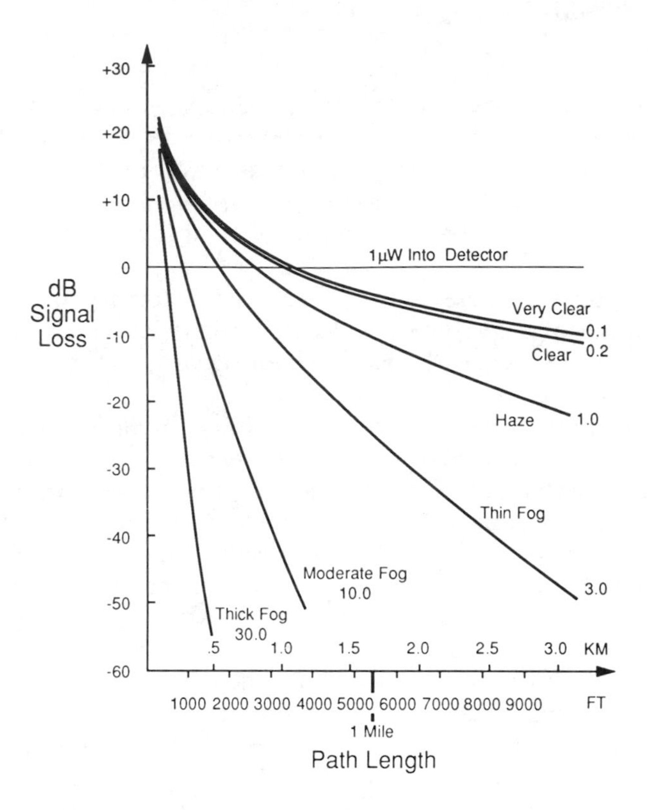

Figure 8.1 Effect of weather on infrared signals.

infrared; for instance, cofrequency signals from other microwave systems (terrestrial and satellite), human-made noise, and cosmic noise.

Three major effects of atmospheric influence on electromagnetic communication are *absorption*, *scattering*, and *shimmer*. Absorption is not important for infrared [8.6]. The scattering of light by the atmosphere is a function of the number and size of the scattering particles suspended in the air, as well as the wavelength of the light being transmitted. Fog, dust (sand), and smog all affect visibility and, hence, the attenuation parameters of an infrared beam. However, because the particle density of fog is greater than that of smog, fog is usually the limiting factor of an infrared communication system. Variations in the air temperature and air motion create differences in the air density (and, hence, the index of refraction of the air) along the path of the beam. This gives rise to the phenomenon known as "shimmer" or "scintillation"; the phenomenon is most often recognized as the "heat waves" that rise from a heated surface, for example, a highway. The effect of shimmer on the infrared beam is to direct some of it out of its intended path, with the air acting like a lens. A bright sunny day with high temperatures and relatively low humidity causes the worst shimmer; a relatively high humidity reduces the shimmer. Having the transmitted beam installed several yards above any heated surface along its path (including a tar roof at the installation site) helps considerably. Good microwave transmission is hard to achieve if the signal runs right over a river for several miles (i.e., parallel to the river and over it) because of the substantial exposure of the beam to water vapor. Similarly, good infrared transmission is also difficult to receive if the signal runs right over a highway for several miles (i.e., parallel to the road and over it) because the black surface intensifies the absorption of the sun's energy and its release into the air as heat, which causes shimmer. Such problems as rain, snow, and sleet, which in the past have been considered as difficulties, are actually not as severe for infrared as generally thought. Conversely, rain is a much more serious problem for the 18- and 23-GHz radios.

In some geographies, where rain and fog are limited, reliable communication at distances of two to three miles may be possible. For paths under two miles, the operation range will be approximately twice the minimum visibility, while for greater ranges the distances converge: 1.2 times maximum visibility at five miles and one-to-one beyond seven miles.

Another consideration has to do with temporary obstructions. For example, if the system is mounted in a roof a few feet from the ledge and birds rest on the ledge, the signal may be broken; the same occurs for birds in flight cutting through the beam, airplanes, people on adjacent roofs, *et cetera*. Infrared is also subject to other types of transmission interference. The sun directly illuminating an infrared detector can "swamp" the detector and impair communication. This fact may make certain link paths impractical, for example, paths leading directly into a rising or setting sun. Infrared is also subject to reflection, creating multipath propagation interference.

The transmission and attenuation of optical beams between their transmitters and receivers is quite a complex subject and is beyond the scope of this text. Reference [8.6] provides additional detailed information for the interested reader. However, note that the uses of infrared communication are primarily directed toward the closer ranges of one-half to one mile, while microwave can be used at distances up to 30 unrepeatered miles. Hence, trade-offs need to be considered between the two technologies when considering the installation of a private communication system. The absence of FCC licensing requirements, the ability to operate multiple systems in close proximity, and the inherent invulnerability to other types of radiation, including their own types of radiation, are major advantages for optical communication in urban environments. Infrared systems can be used to transmit indoors or out of doors, and can transmit through clear unrippled window glass. The infrared light beam is invisible to the naked eye, making the system difficult to tap as there is nothing to indicate that transmission is taking place.

Most commercial free-space infrared systems are point-to-point in nature. Multipoint infrared data systems, in which a central node broadcasts to multiple slaves on one channel while the slaves compete on a TDMA basis for return access, has been tried experimentally but not exploited commercially. The power of an infrared emitter is low compared to that of a microwave transmitter, and the only reason reception is practical at all is that the beam is concentrated. Because the distances involved in optical communication are so small, no propagation delay occurs, but some links are subject to dropouts due to the fading effects of fog, rain, or obstructions. Data communication links that employ infrared optical paths can easily deal with the low BERs they generally provide, but may be affected by the relatively long fade intervals. Users may need to tune the communication parameters of the protocols used to ensure that a link that has temporarily "faded" will not be declared down.

8.3 HARDWARE COMPONENTS

The issue of modulation is the point at which the parallels between microwave and optical communication begin to diverge. In microwave, modulation is generally applied through electronic means; frequency or phase shifts in the carrier caused by the information are generated by changing electrical characteristics in the device that generates the carrier. For infrared systems, VLSI devices with direct optical coupling are just now being developed, but the available technology cannot process light (photons) in the same way that it processes microwave signals carried as electrons. The light output of an optical laser, however, can be varied by changing the current through the emitter diode; this is also true of LEDs. Optical modulation can also be provided by placing a modulation component in the light beam's path. Polarization and other optical effects may be caused by electrical effects in some materials, making these materials a basis for optical modulation. These materials are generally not yet used for normal communication at this time.

Infrared in the range of 8000 to 10,000 Å is simply another form of light and is processed in the same way. Emitter technology may be based on coherent light generated by laser diodes, or on noncoherent LED-generated light. Receivers are generally photodiodes sensitive to infrared frequencies. A simplex system employs a single emitter-collector pair, while a full-duplex system has two pairs of elements, one for each channel.

Laser systems emit a very tight beam of light, which is polarized and limited sharply in its bandwidth. The light intensity of a laser can be managed by controlling the power applied. The ease of modulation translates into wide variations in the signal state representing the 0s and 1s of data, making it easy to detect and demodulate the signal in a robust fashion at the opposite station. Laser optical systems have theoretical ranges of up to ten miles under good conditions.

The light levels generated by LEDs are small when compared to that generated by lasers, and the losses associated with modulating LEDs are higher. A general relationship exists between the current applied to an LED and the intensity of its light, but the curve is not as linear as that of a laser. In addition, LEDs have a relatively slow switching time, making it difficult to pulse them; typical LEDs with a rise time of about 1 ms can support data rates of the order of 1 Mb/s. Outdoor applications generally employ LD technology; indoor systems use LEDs.

Neither the output of a LD nor that of a LED presents any physical risk: The lasers used in communication systems radiate power at low levels and the reduction in average power is often accomplished through "pulsing" the laser. This results in high power levels during the pulse, which facilitates detection of the signal, but the low pulse duration eliminates exposure risks. Terrestrial atmospheric laser communication systems on the market today use laser diodes operating in the milliwatt range (from 100 to 1000 mW), a power level too low to cause damage to tissue except in extreme cases of exposure. (Infrared transmission systems using optical fiber employ lasers operating at relatively higher power because the fiber confines the beam and eliminates risk of burns or eye damage.) Nonetheless, an LD emitter should not be looked at directly at a distance of less than a few feet. All LD systems are required to have warning decals, which caution users not to look into the aperture at close distance because the eye is the only part of the body in proximity to the light source that could be affected by such a beam. Most laser systems are rated as Class IIIb by the National Center for Devices and Radiological Health, meaning that they are not dangerous unless viewed directly or via reflection (Document 21CFR 1040 entitled "Performance Standards for Laser Products," amended August 1985). Typically, an atmospheric system puts out several hundred times less light energy than a regular flashlight. LED systems present no harmful effects at all: Neither the FCC nor the Bureau of Radiological Health places any restrictions on LED products, and these are as safe as any infrared remote control device for home TV.

Optical and infrared receivers can generally be classified as direct photon detectors or optical interference detectors. In the former technology, a photodiode in the receiver traps photons from the transmitter beam and converts them to electrons, which can then be amplified and processed to extract the signal; this technology is common today. Interference systems use the same principle as radio and television reception. The incoming light beam is optically mixed with a reference light signal; the interference field is then detected directly as the information channel. This technology may become more prominent in five to ten years.

Direct detection is an easy and inexpensive method, but it has some limitations. The photon detectors will operate well with LED or other noncoherent sources, but are more sensitive to background noise. The sensitivity of a photodiode photon detector peaks in the desired infrared range, but in this region other light sources may interfere with reception, for example, incandescent light for indoor applications, or daylight for external applications (the signal emission from fluorescent lights does not interfere because its output at 9000 Å is minimal). Placing an infrared color filter in the collector for the photodiode can help reduce the noise from other light sources, but it will also attenuate the real signal.

8.4 PRAGMATIC ISSUES AFFECTING THE CHOICE OF TECHNOLOGY

Infrared communication links will provide services in data rates from low-speed data at 9600 b/s to 45 Mb/s at distances ranging from a few yards to up to ten miles. The information transferred may be in video, voice, or data, and no particular restrictions exist for the protocols or formats of data exchanged. Data rates in the DS1 (1.544 Mb/s), DS1C (3.186 Mb/s), DS2 (6.312 Mb/s), DS3 (44.736 Mb/s), and Ethernet 802.3 range are available. Installation of infrared systems is simple and expeditious; without trenching, licensing, or permits, the system (which consists of two transmitters and two receivers) can be set up and aligned in a matter of hours, providing quick communication for permanent or temporary sites. For example, a company may have to move a computer center to some nearby building. The user may find that an infrared link is ideal while waiting for the permanent copper or fiber facility. The cost of infrared is relatively small so that even a system that is up for only six months could still be cost-effective—even after deployment of the permanent facilities the infrared link could be used as a backup link if the main link were to fail. Infrared systems for permanent outdoor applications are normally rugged and weatherproof and operate from 0 to 150°F. The mean time before failure (MTBF) of some good systems is 25,000 h, and the mean time to repair (MTTR) is 30 min (with spares available). Modern infrared systems can be interfaced directly with T1 CPE systems such as T1 multiplexers, channel banks, and LAN transceivers.

Unrepeatered infrared communication is suitable for distances up to as far as ten miles under ideal conditions, but most applications are for two miles or less, to avoid the atmospheric-induced outages discussed earlier. Most users report that performance beyond two miles requires unusually stable atmospheric conditions. If longer distances are required, two infrared hops using a total of four nodes (one at the origination site, two at an intermediary point, and one at the termination site) could be employed. The repeater mode also allows for interconnection of sites that do not have direct LOS because of some obstruction that could be avoided by a polygonal line (see Figure 8.2).

The characteristics of infrared transmission have caused it to be considered in two communication environments: (1) as a microwave substitute and (2) as the basis for LANs. If issues of range are discounted, 18-GHz microwave can do very little that infrared channels cannot do at least as well [8.4]. In crowded metropolitan

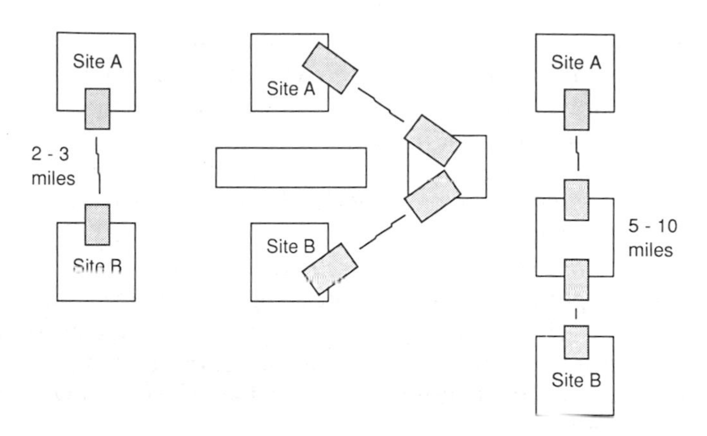

Figure 8.2 Repeater-based infrared systems.

environments, distance issues are minimal and the crowding of microwave space may make infrared a preferred choice. Considerable time is required to obtain a microwave license, due to frequency congestion; the typical interval is 90 to 180 days [8.7].

Large office complexes with open space can use infrared links to link computers and peripherals into a network, for example, PCs to a mainframe or cluster controller. Indoor applications generally employ LED technology. The data rates available from LED systems are somewhat lower than those of traditional cable LANs already in use, but this disadvantage is more than offset by the elimination of wiring and the ease of moving equipment from one place to another within the office. Shop floor applications of infrared communication already exist, applications in which the communication points are fixed in location or move in a track so that the angular positioning of the transmitter and receiver does not change with the motion.

The infrared transmission of information is immune to electromagnetic interference (EMI) from other microwave sources; however, office communication poses some significant challenges to infrared systems. As already indicated, fluorescent lighting is not a problem because its energy drops to almost zero in the infrared range, but incandescent lights can cause severe interference because they peak in energy at the infrared frequency range. This problem is complicated by the difficulty of achieving precise beams and small FOVs with noncoherent LED systems, and with accurate and consistent aiming of the components of the system. In spite of these difficulties, many of the infrared vendors are experimenting in the LAN market.

8.5 COST OF THE TECHNOLOGY

An entry-level infrared system (two transmitters and two receivers) can be purchased for as little as $5000, and systems capable of megabit-range communication can be installed for $20,000 or less. This is less than the price of most microwave systems of equal capacity. Models that handle voice and data tend to cost somewhat more than systems intended for video only. Some manufacturers also offer the equipment on short- or long-term rentals.

Infrared communication is not a household word. The technology has not been easy to sell, a problem that many feel has been exacerbated by the prominent warnings that government regulations require for laser LD systems.

Estimates vary, but infrared vendors feel that about half of the point-to-point private microwave applications now active or being considered could also be served with infrared devices. This would suggest that a market as large as $40 million per year could be developed from this source alone. Current estimates would place the actual market at less than half this value, but the market is now growing more rapidly than in the past.

REFERENCES

[8.1] D. Minoli, "An Overview of Infrared Communications," DataPro Report CA70-010-201, January 1988.

[8.2] D. Halliday, R. Resnick, *Physics*, Part 2, John Wiley and Sons, New York, 1962.

[8.3] "Effects of the Atmosphere on Infrared Communications," Isher, Inc., technical papers.

[8.4] "Lasers for Land Mobile?" *Communications*, April 1986.

[8.5] Literature from American Lase Systems, Goleta, CA.

[8.6] J.C. Thebault, "Optical Atmospheric Communications," *Computing Canada*, August 7, 1986.

[8.7] J. Evans, "Corporate Growth: A Telecom Time Bomb," *NetworkWorld*, August 11, 1986.

Chapter 9
Switching and Signaling Systems

9.1. OVERVIEW

Switches are a fundamental component of the telephone network. They allow a subscriber to connect with any remote subscriber on the network by specifying the address of the remote destination. At a high level, a modern switch consists of (1) the switching matrix, (2) the "common control," (3) subscriber line cards, (4) trunk line cards, and (5) maintenance channels. Switching matrix technology can be analog, including crossbar and reed relay systems, and digital, based on TDM techniques. SPC allows the use of computer software to control the switching matrix, rather than depending on hardware to perform this function, as was the case until the early 1960s. SPC offers flexibility, particularly in the provision of advanced voice or data services. In 1989, approximately 50% of the BOC switches and 95% of the major IXC switches were digital systems [9.1]. Figure 9.1 depicts the Bell System traditional switching hierarchy, prior to divestiture on January 1, 1984. The hierarchy of Figure 9.1 is not necessarily the hierarchy employed today by AT&T or other major IXCs. The pressures of competition—and the ensuing risk to billions of dollars of revenues—made a number of the network design models developed by transmission and switching engineers of the past decades inadequate for the business needs of the 1990s, particularly in terms of short call setup times. The hierarchies of the IXCs are now guarded and are not disclosed to public scrutiny.

As discussed in Chapter 1, two LEC end offices can be connected in a number of ways:

1. Via direct trunks between local switches (which are within a few miles or have high cross-traffic requirements);
2. Via interoffice trunks through a local tandem switch; and
3. Through a toll center tandem with trunks provided by one of several possible IXCs.

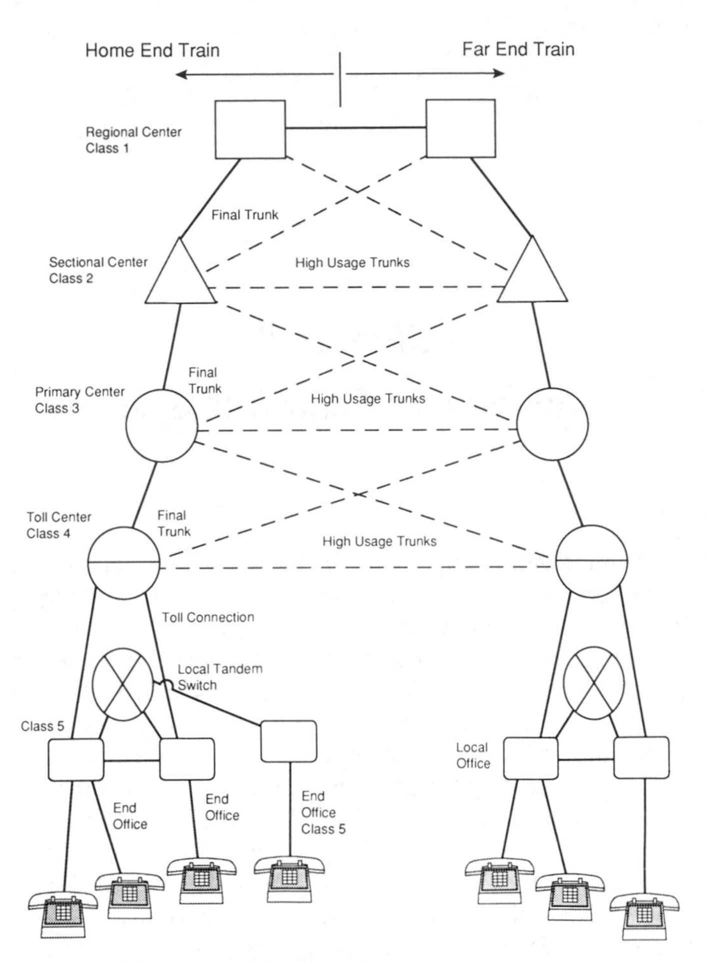

Figure 9.1 Traditional Bell System switching hierarchy.

Signals are messages pertaining to call management generated by the user or by network entities. *Signaling* is the act of transferring this information among remote entities. Early switching systems employed signaling that consisted of dc levels. Later, between the 1930s and 1960s, single- and multiple-frequency tones were used, both in an in-band fashion (within the spectrum of the voice), as well as out-of-band (beyond the spectrum of the voice, but below 4000 Hz). Since the mid-1970s, signaling networks for interoffice applications have begun to employ a separate dedicated data communication network for areas in which this could be proved-in economically. Here the signaling is both out-of-band and over a common channel. Common channel signaling is becoming critical for the provision of advanced services.

This chapter covers a number of issues pertaining to switching and signaling and a discussion of intelligent networks that will be deployed by carriers in the 1990s.

9.2 SWITCHING SYSTEMS

The public switched telephone network is designed so that any user can be directly connected to any other user on the network. To make these connections economical and practical, switching systems are used. Switching systems concentrate many users onto relatively few distribution paths and provide a connection over the distribution path to the called party. The terms *concentration*, *distribution*, and *expansion* relate to functions that must be performed in order to connect any inlet over a path to any outlet.

A circuit switched call is comprised of four formal phases. These phases are carried out through the services of the switch. The phases occur in the following order: call request, call confirmation, information transfer, and call clearing [9.2]. The call clearing phase may be initiated during any of the three other phases. These call phases are defined as follows:

Call Request Phase. A call with specific parameters is requested by the calling party. This call request is processed by, and routed through, the network. If it cannot be accepted by the network, the request is then delivered to and processed by the called party.

Call Confirmation Phase. Acceptance of the call is reported by the called party, unless this party does not accept the call. Final arrangements are made through the network for that call. The call confirmation is returned to the calling party.

Information Transfer Phase. Information can be exchanged between calling and called parties in accordance with the characteristics of the applicable call type.

Call Clearing Phase. Any network or party involved in a call has the ability to clear that call in any phase of the call. At the time a call is cleared, any network involved in the call would immediately abort the current phase and report the call clearing to the adjacent network or party, unless they were already informed of that clearing. Once call clearing is locally completed, any resource used for that call can be reused by the network for other calls.

The phases described here were implicitly utilized in Chapter 4, while describing the Q.931 ISDN standard, although the phases also apply to non-ISDN networks.

9.2.1 Common Control

Telephone switches now employ common control. A common control system can determine the existence of a path through the switching matrix, for connecting available inlets and outlets, before actually trying to commit resources for the path between them. In this environment, the system control can, if desired, look ahead to avoid tying up machine resources and paths for a destination that may already be busy. (While this capability exists in networks using common channel signaling, it is not generally used because the expense related to deploying and using this feature outweighs the savings in reduced trunk holding time.)

The common control works in conjunction with a connection switching matrix (analog or digital). The control can be assigned to an incoming call as required. It receives the dialed digits and then sets up the path through the switching matrix according to hard-wired or stored program rules. These rules provide for variation in the handling of local and long-distance calls, for choosing an alternative route for an interoffice call in case the first route chosen is busy, and for trying the call again automatically in cases of blocking or faults in the switching path for both interoffice and intraoffice calls.

Figure 9.2 depicts some internals of a common control SPC switch. It is representative of a switch such as AT&T's No. 2 ESS. More modern switches (such as AT&T's No. 5 ESS) have a distributed architecture that puts some of the functions (if needed) in remote switching modules, which are satellite switches located outside the CO.

9.2.2 Space-Division Switching

Space-division switching was originally developed for the analog environment and has been carried over into digital technology by a number of manufacturers because the fundamental switching principles are the same whether the switch is used to

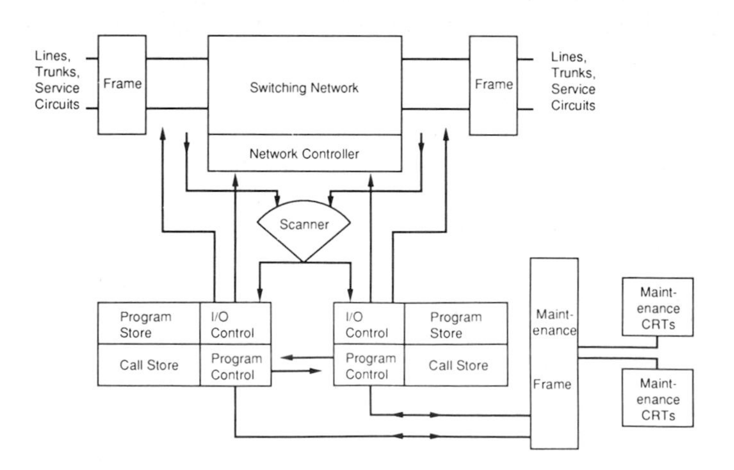

Figure 9.2 Common control SPC switch.

carry analog or digital signals. A space-division switch is one in which the signal paths are physically separate from one another ("divided," i.e., separated in space). Each connection requires the establishment of a physical path through the switch that is dedicated to the transfer of signals between the two endpoints. The basic module of this type of switch is a metallic crosspoint or semiconductor gate that can be enabled and disabled by a control unit. To establish a path in a single-stage network, only the gate needs to be enabled.

This space switching methodology has a number of limitations:

1. The number of crosspoints grows with the square of the number of input ports N; this can be costly for large N and results in high capacitive loading on any path.
2. The crosspoints are inefficiently utilized (at most, N out of N^2).
3. The loss of a crosspoint prevents connection between the two devices involved.

(The failure described in point 3 above would generally not be seen by an end-user, unless the user were actually engaged in a conversation at the time of the failure.)

To ameliorate these limitations, multiple-stage switches are employed. N input lines are broken up into N/m groups of m lines. Each group of lines goes into a first-stage matrix. The output of the first-stage matrices becomes input to a group of second-stage matrices, *et cetera*. The last stage has N outputs; thus, each device attaches its input line to the first stage and its output line to the last stage.

One has j second-stage matrices, each with N/m inputs and N/m outputs. The number of second-stage matrices is a design decision. Each first-stage matrix has j outlets so that it connects to all second-stage matrices. Each second-stage matrix has N/m outputs so that it connects to all third-stage matrices (see Figure 9.3). In a multistage network, a free path through all the stages must be determined, and the appropriate gates enabled.

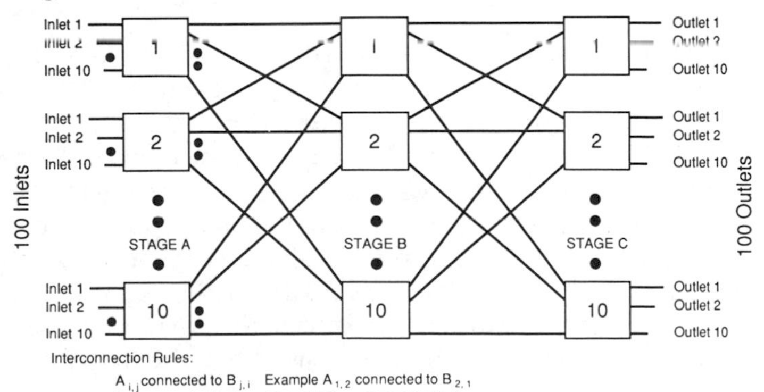

Figure 9.3 An example of a multiple-stage switch.

As indicated in point 1 above, a direct crosspoint matrix to interconnect 100 inlets to 100 outlets would require 10,000 crosspoints. Figure 9.3 shows that 3000 crosspoints are sufficient in a multistage arrangement (each of the ten "building blocks" in each of the three stages has $10 \times 10 = 100$ crosspoints, for a total of 3000).

While a multistage network requires a more complex control scheme, this arrangement has the following advantages over a single-stage matrix: (1) The number of crosspoints is reduced, increasing crossbar utilization; and (2) more than one path exists in the network to connect two endpoints, increasing reliability.

9.2.2.1 Crossbar Switch

Crossbar switches are electromechanical space-division switching devices that make connections in a telephone system by moving parts. The basic patent on crossbar switching was issued in 1915. The first major crossbar switching system (No. 1 Crossbar) was introduced in 1938 (also see Table 9.1 in Section 9.2.4). Some electromechanical crossbar switches are still in use in the United States, and they are quite common in other parts of the world.

As the name suggests, a lattice of crossed bars is involved. Each vertical bar carries a set of contacts that is connected to those carried on a horizontal bar when magnets are activated to move the two bars. Contact is made where the two bars cross—hence, the name "crossbar." Its contacts will latch; that is, they will stay attached after the magnets are deactivated and contacts can be made elsewhere by the other bars. Of course, when the communication is over, the contacts can be disconnected.

9.2.2.2 Reed Relay Switch

The reed relay is a small, glass-encapsulated, electromechanical space-division switching device. These devices are actuated by a common control that selects the relays to be closed in response to the number dialed and sends pulses through coils wound around the relay capsules. The pulses change the polarity of plates of magnetic material alongside the glass capsules. The contacts open or close in response to the direction of magnetization of the plates, which is controlled by the polarity of the pulse sent through the windings. Because the contacts latch, no holding current is required for this type of crosspoint, but separate action is required by the common control to release the connection when one party or the other hangs up. Reed relays were an important part of the SPC switching systems of the 1960s and early 1970s.

9.2.3 Time-Division Switching

With the advent of digitized voice and synchronous TDM techniques, both voice and data can be transmitted using digital signals. This has led to a major change in the design and technology of switching systems. Modern digital systems rely on computerized control of space- and time-division elements. Virtually all circuit switches manufactured today use digital time-division techniques for establishing and maintaining communication paths. Digital switching techniques employed in today's telephone switches are based on the use of synchronous TDM. Time-division switching involves the partitioning of a lower speed bit stream into pieces that share a higher speed stream with other bit streams. The slots are manipulated by control logic to route data from input to output. Synchronous TDM permits multiple low-speed bit streams to share a high-speed line. A set of inputs is sampled in turn. The samples are organized serially into slots to form a recurring frame of N slots. A slot may be a bit, a byte, or a larger block. In synchronous TDM, the source and destination of the data in each time slot are known. Therefore, address bits are not necessary in each slot.

Three methodologies are employed in time-division switching [9.3]:

1. *TDM Bus Switching*. In this arrangement, all lines are connected to a bus. Time on the bus is divided into slots. A path is created between two lines by assigning repetitive time slots. This is the simplest form of time-division switching. The size of switch is limited by the data rate on the bus. This architecture is common for small- and medium-sized data switches and PBXs. PBXs are CPE-based switches for intrapremises voice—and incidental data—communication; they also provide user access to the public network. PBXs are treated in more detail in the next chapter. Each input line deposits data in a buffer; the multiplexer scans these buffers sequentially, taking fixed size groups of data from each buffer and sending them out on the line. One complete scan produces one frame of data. For output to the lines, the reverse operation is performed, with the multiplexer filling the output line buffers one by one.
2. *Time-Slot Interchange* (TSI). In this arrangement, all lines are connected to a synchronous TDM multiplexer and a synchronous TDM demultiplexer. A path is created by the interchange of time slots within a time-division multiplexed frame. The size of switch is limited by the speed of the control memory. TSI can be used as a building block in multistage switches. A TSI unit operates on a synchronous TDM stream of time slots, or channels, by interchanging pairs of slots to achieve full-duplex operation (see Figure 9.4). The input lines of N devices are passed through a synchronous time-division multiplexer to produce a TDM stream with N slots. To achieve the interconnection of two devices, the slots corresponding to the two inputs are

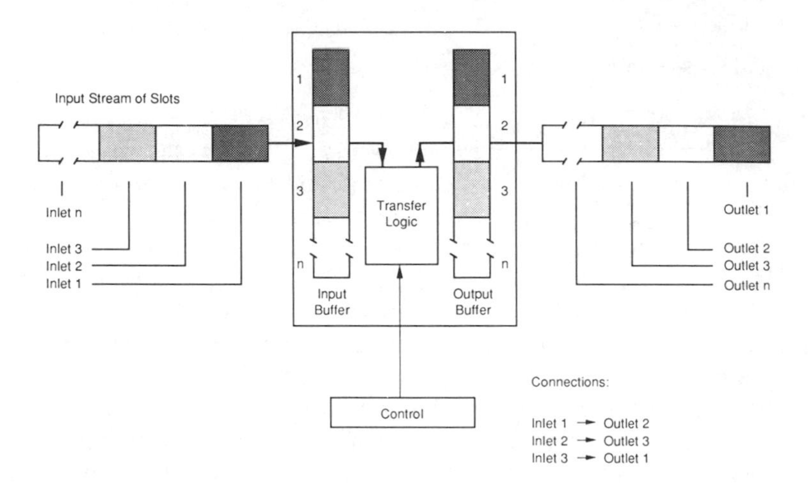

Figure 9.4 Time-slot interchange.

interchanged; the resulting stream is demultiplexed to the outputs of the N devices. This results in a full-duplex connection between pairs of lines. A real-time data store, with a width equal to one time slot of data and a length equal to the number of slots in a frame, is used. An incoming TDM frame is written sequentially, slot by slot, into the data store. An outgoing TDM frame is created by reading slots from memory in an order dictated by an address store that reflects the existing connections.

TSI is a simple way of switching TDM data. However, the size of such a switch, in terms of number of connections, is limited by the memory access speed. To keep pace with the input, data must be read into and out of memory as fast as it arrives. So, for example, if 12 sources are operating at 64 kb/s each and the slot size is 8 bits, we would have an arrival rate of 96,000 slots per second. For each time slot, both a read and a write are required. In this example, memory access time would need to be $1/(96{,}000 \times 2)$ seconds, or about 5.2 ms. A TSI unit can support only a limited number of connections. Also, as the size of the unit increases, for a fixed access speed, the delay at the TSI unit increases.

3. *Time-Multiplex Switching* (TMS). This arrangement entails a form of space-division switching in which each input line is a TDM stream. The switching configuration may change for each time slot. It is used in conjunction with TSI units to form multistage switches. Delays increase slightly by the use of multiple stages, but this allows much larger capacity.

   To work around the bandwidth and delay problems of TSI, multiple TSI units must be used. To connect a channel on one TDM stream (going into one TSI) to a channel on another TDM stream (going into another TSI), some form of interconnection of the TSI units is needed. The purpose of the interconnection is to allow a slot in one TDM stream to be interchanged with a slot in another TDM stream. Multistage networks can be built up by concatenating TMS and TSI stages. TMS stages, which move slots from one stream to another, are referred to as S (space), and TSI stages are referred to as T (time) (see Figure 9.5).

9.2.4 Deployment of Switching Technology and Switches

Figure 9.6, based on Reference [9.4], provides a relative measure of the introduction of switch technology over the years for a typical industrialized country. Note how technology is accelerating its pace: The span of utility, measured in years, from introduction to abandonment is decreasing with each new generation of equipment. Figure 9.7 depicts the exact percentage of three switching technol-

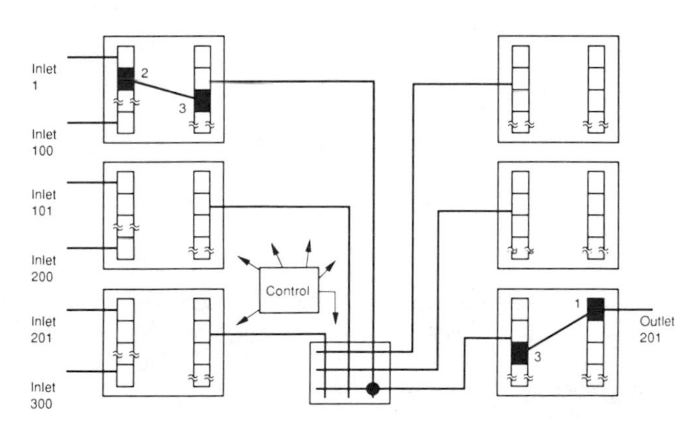

Figure 9.5 Time-space-time switching.

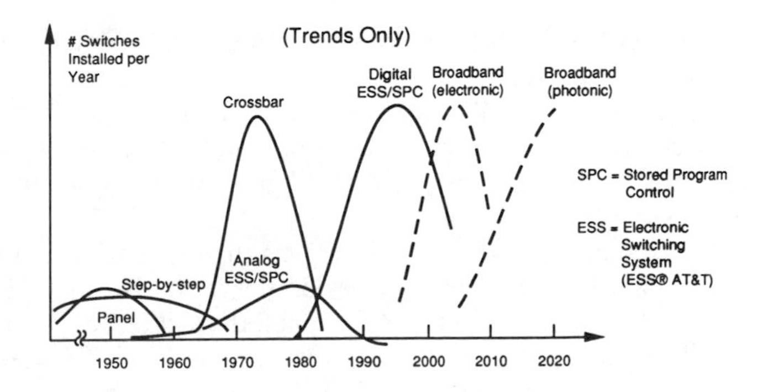

Figure 9.6 Switch technology over the years for a typical industrialized country.

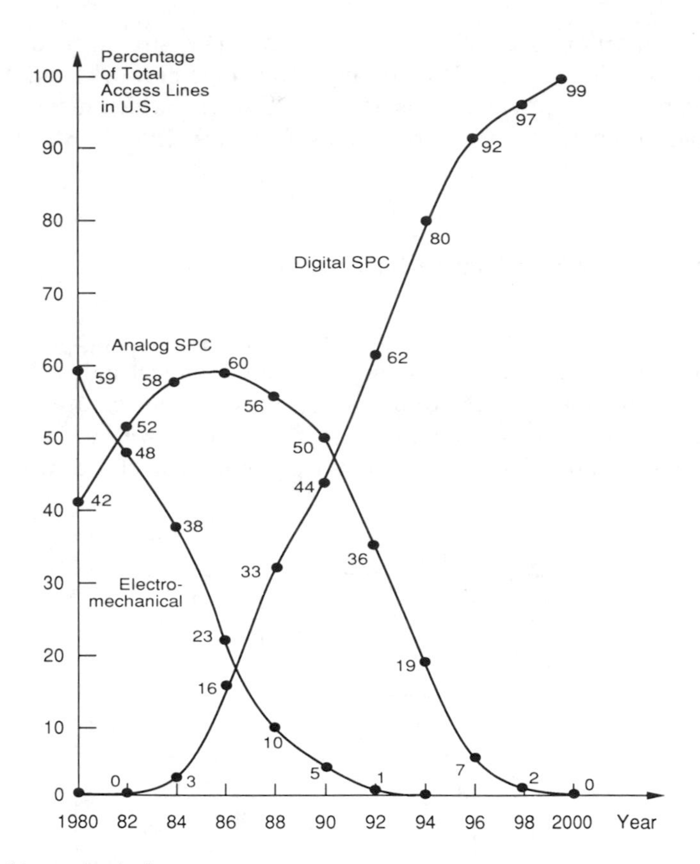

Figure 9.7 Switching technologies.

ogies in the United States, from both an historical perspective as well as a forecast perspective [9.5].

Table 9.1 depicts the major switching systems in the United States up until the AT&T divestiture, and related features.

NTI's DMS-10 was the Bell System's first digital local CO. Stromberg-Carlson's digital office is one of the most widely deployed switch by independents for small offices. Until 1985, the most actively deployed analog central office was AT&T's 1/1A ESS [9.6].

Table 9.2, based on Reference [9.1], depicts the population of BOC switches (these data have not been independently verified with the BOCs). The percent of digital switches has increased in the late 1980s, and the switch base will be primarily digital by 1993, as indicated in the table. Approximately 18,000 class 5 switches exist in the United States. In 1986, the BOCs had approximately 9400 switches and served 85% of the U.S. telephones; other LECs had 9000 switches and served

Table 9.1
Major Switching Systems in the U.S., up to Divestiture and Related Features

Electromechanical Switching Systems			
Step-by-step	Invented 1919		Deployed after 1926. Still around, rural areas, unsophisticated.
Panel	Invented 1921	Local and tandem use	Deployed starting in the 1920s. Last retired in 1982.
#1 Crossbar	Introduced 1938	For large offices	Common Control.* Major deployment starting in the 1940s.
#5 Crossbar	Introduced 1948	Last installed 1977	Served 40% of Bell lines in 1970. Originally local, but also used for toll.
#4 Crossbar	Introduced 1943	Toll	
#4A Crossbar	Introduced 1953	Toll	Largest system, until 4ESS. Upgraded in the 1960s and early 1970s.
Crossbar Tandem	Introduced 1955	Toll	Modified in 1958. Facilitated introduction of DDD.
Electronic Switching Systems			
1ESS**	Introduced 1965	SPC, analog space division	10K–65K lines, local
2ESS	Introduced 1970	SPC, analog space division	2K–10K lines, local
3ESS	Introduced 1976	SPC, analog space division	5K lines, local
1AESS	Introduced 1976	SPC, analog space division	Local and toll
Digital Switching Systems			
DMS 10	Introduced 1981	SPC, digital	200–6K lines, local
4ESS	1976	SPC, digital	Large capacity, toll
5ESS	Introduced 1982	SPC, digital	1K–100K lines, local

*All switching systems except Step-by-step and panel have common control.

**ESS is an AT&T trademark

15% of the telephones [9.7]. Rural digital COs were among the first to prove economical, even for the predivestiture Bell System.

Table 9.3 provides some information on the size of the BOCs networks [9.8] (these data have not been independently verified with the BOCs).

Table 9.2
Switching Technology to 1993

Region	Techn.	Percentage of Lines by Technology 1989	1990	1991	1992	1993	Percentage of Switches by Technology 1989	1990	1991	1992	1993
Ameritech	DIG	35	38	40	43	44	53	61	67	69	72
	1A/1E	57	56	55	53	52	26	26	23	23	20
	OTH	5	5	4	4	4	11	10	8	7	7
	EM	3	1	1	0	0	10	3	2	1	1
Bell Atlantic	DIG	44	48	51	58	60	67	71	77	79	82
	1A/1E	51	48	46	40	38	23	22	21	19	17
	OTH	4	4	3	2	2	8	6	2	2	1
	EM	1	0	0	0	0	2	1	0	0	0
BellSouth	DIG	42	46	53	58	64	65	75	84	86	88
	1A/1E	46	44	41	38	34	17	16	10	8	7
	OTH	9	8	6	4	2	12	9	6	6	5
	EM	3	2	0	0	0	6	0	0	0	0
NYNEX	DIG	50	60	64	68	74	42	58	70	78	84
	1A/1E	34	31	29	27	24	11	11	11	10	8
	OTH	2	2	1	2	2	14	13	12	7	7
	EM	14	7	6	3	0	34	17	8	4	1
Pacific Telesis	DIG	30	34	38	41	44	38	47	55	66	74
	1A/1E	59	57	55	53	51	30	30	30	28	23
	OTH	3	3	2	1	1	4	1	0	0	0
	EM	8	6	5	5	4	28	22	15	6	3
Southwestern	DIG	23	26	29	31	35	26	28	31	35	38
	1A/1E	62	60	59	58	57	17	17	16	15	15
	OTH	5	4	4	4	3	13	13	13	12	12
	EM	10	10	8	7	5	44	42	40	38	35
U S WEST	DIG	28	34	41	50	57	32	40	48	55	63
	1A/1E	57	52	49	45	42	15	15	15	15	15
	OTH	3	3	1	1	0	6	5	4	4	4
	EM	12	11	9	4	1	47	40	33	26	18
Total Regions	DIG	37	42	46	51	55	46	54	63	67	72
	1A/1E	52	49	47	44	42	19	19	17	16	14
	OTH	5	4	3	3	2	10	8	7	6	5
	EM	6	5	4	2	1	25	19	13	11	9

Key: DIG = digital; OTH = other; EM = electromagnetic.

9.2.5 Equipment

The principal manufacturers of high-capacity switches for the U.S. market are: AT&T Technologies, Northern Telecom, GTE (now AGCS), Siemens, and Er-

Table 9.3
BOCs' network statistics

BOC	*Total access lines 1989* ($M)*	*New Construction expenditures in 1989* ($B)*	*Total investment in plant in 1989* ($B)*	*Number of employees in 1989**
Ameritech	15.8	1.9	24	72.5 K
Bell Atlantic	16.8	2.5	18	71.6 K
BellSouth	17	3	35	87.5 K
NYNEX	16	2.3	29	97.4 K
Pacific Telesis	13.3	1.3	16	66.4 K
Southwestern Bell	12	1.3	23.4	57.7 K
U S WEST	12.1	1.8	N/A	58 K

*All numbers are estimates from reference [9.8].

icsson, in that order. New entrants include NEC and Stromberg-Carlson. ITT (now Alcatel) is not currently active in the U.S. market.

AT&T's 5ESS now supports more than 20 million lines in the United States. It can handle from 1000 to 100,000 lines, and 650,000 peak-hour call attempts. NTI's DMS 100 now supports more than 20 million lines in the United States. It can handle from 2000 lines to 100,000 lines, and 700,000 peak-hour call attempts. NTI's DMS 10 currently supports more than 3 million lines in the United States. It can handle from 300 lines to 50,000 lines, and 350,000 peak-hour call attempts. NEC's NEAX 61E now supports approximately 1 million lines in the United States. It can handle from 100 lines to 80,000 lines, and 600,000 peak-hour call attempts. Ericsson's AXE supports more than 2 million lines in the United States. It can handle from 100 lines to 200,000 lines, and 800,000 peak-hour call attempts. GTE's GTD-5 EAX supports more than 7 million lines in the United States. It can handle from 500 lines to 150,000 lines. Siemens' EWSD now supports approximately 3 million lines in the United States. [9.9]. EWSD can handle from 200 lines to 100,000 lines, and 1 million peak-hour call attempts. Stromberg-Carlson's Century DCO currently supports an estimated 1 million lines in the United States. It can handle from 100 lines to 70,000 lines ([9.10] with update; now merged with Siemens).

Many if not all of these switches can now support ISDN interfaces, digital CENTREX, and remote satellites. These large switches can cost between $5 million and $10 million, depending on the configuration.

9.2.6 The Future of Switches

Today's switches operate with throughputs in the gigabit range ($64{,}000 \times 100{,}000 = 6.4$ Gb/s). To provide BISDN, the switches handle from 1000 to 3000 times the throughput. New switching architectures are required to handle these traffic streams [9.11–9.17].

Traditionally, switching has been done at a central location by a centralized device. In contrast, LANs achieve switching (which is, effectively, the ability of any customer in the network to reach any other customer) in a distributed fashion. The same technology is also employed in MANs. The day may come, within 25 years, when centralized switches may disappear (this is a highly speculative view). In that case, a CO capability may be utilized for servicing the physical star loop plant (logically such plant may be a ring), and for providing servers for value-added services (for example, directory, E-mail, *et cetera*) [9.18].

9.3 SIGNALING

9.3.1 Need for Signaling

A network, whether public or composed of a group of privately interconnected nodes, would be of limited use unless users are able to communicate their needs for service. In addition to this user-network signaling, signaling capabilities between various components of the network are needed. Signaling systems and related equipment are used by all public and private telecommunication networks, with the exception of some types of data communication networks, which use their own mechanisms.

For traditional telecommunication services over the public switched network or over a private voice network, signaling refers to the mechanism necessary to establish a connection, to monitor and to supervise its status, and to terminate it, through the transmission and switching fabric of the underlying network. Formally, signals are messages generated by the user or some internal network processor, pertaining to call management. Signaling is the act of transferring this information among remote entities, including the communication handshake protocols and the "semantics" conventions. The signaling network is the collection of physical transport facilities that carry the signals.The signaling equipment performs the functions of alerting, addressing, supervising, and providing status in both private and common carrier networks [9.19].

Even early networks had some kind of signaling. For example, in the early days of telephony (the 1880s to the 1900s), a ringer with a crank attached to the telephone set was used to inform a central human processor—the operator—that the user wanted some service; the operator would obtain instructions orally from the user, typically a destination name, and then proceed to render manually the required service by electrically connecting the two parties. Eventually, this human switch was replaced by an automatic switch. In this environment, the user would inform the system that a connection to a local party was required by lifting the handset; the destination address was coded by way of the dialed number. Long-distance service, however, continued to be manual with respect to addressing and connection until the introduction of *direct distance dialing* (DDD) in the early 1950s. Traditionally, POTS (*plain old telephone service*) consisted of establishing an end-to-end transparent path between the user and some remote party (circuit switching). Recently, more sophisticated services have started to emerge (800 service, credit card calling, remote call forwarding, to list a few public services). In the future, a whole host of new services will begin to emerge and be available in the public and private networks. Today, a network totally dedicated to signaling is being deployed. ISDN will utilize this separate signaling network to provide sophisticated supplementary service to the user. The ability to signal the network is critical and the users needs must be described unambiguously and efficiently.

9.3.2 Signaling Functions

Signals can be classified by three methods: topologically, in the sense of where (or to what types of facilities) they apply to; functionally, in the sense of what they are intended to do; and physically, in the sense of an actual network implementation [9.20, 9.21].

Topologically, we have (see Figure 9.8):

- *Inter-CPE Signaling*: Advanced signaling between sophisticated CPE equipment, such as T1 multiplexers, which employ their own supervisory link embedded digitally in the data stream; or systems employing a dedicated data link not directly associated with a specific set of transmission facilities (for example, a centralized telemetry data collection system monitoring the network).
- *Intra-CPE Signaling*: Call handling interactions between CPE components, typically telephone sets and the PBX.
- *Customer Line Signaling*: Interaction between the customer and the switching system serving that customer.
- *Interoffice Trunk Signaling*: Exchange of call-handling information between switching processors in the network.
- *Special Services Signaling*: Certain systems employing telecommunication "special services facilities" (tie lines, FX lines, *et cetera*) have requirements that span across both the last two categories. This is also the case when alerting stations on a nonswitched private line arrangement.

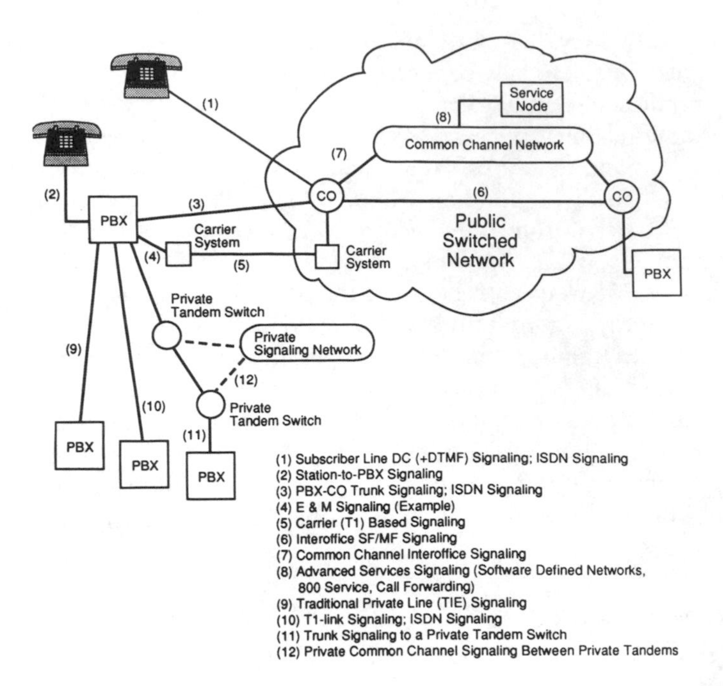

Figure 9.8 Various signaling environments.

- *Advanced-Services Signaling*: Signaling between switching nodes in the network and support processors at some other location in the network, for the provision of an advanced service such as 800 service, credit card calling, and others. This type of signaling required to complete a call is in addition to the customer line signaling and the interoffice trunk signaling listed above.

Customer line signaling and interoffice trunk signaling are the more well-known instances of signaling in a traditional public switched network. Signaling in support of advanced services started to become important in the 1980s, and will increase in importance in the 1990s.

Functionally, we have:

- *Supervisory*: Monitoring the status of a line or circuit to determine its state (busy, idle, *et cetera*). In a metallic loop, supervisory signals are indicated by the hook status (on or off) of the loop. These supervisory signals allow designated equipment at the CO or at the CPE switch to recognize origination and termination of a call.
- *Addressing*: Transmitting routing and destination signals over the network. These signals are generally dial pulses and tone pulses over local loops and trunks. More recently, with common channel interoffice signaling, these signals can also be data streams over the associated signaling links.

- *Alerting*: Advising the addressee of the arrival of an incoming call. Ringer currents can be used to drive audible bells, tones, or lights.
- *Billing*: Generation and collection of billing information pertaining to a specific call.

Physically, we have six generic signaling methods: direct current, in-band single-frequency tone, in-band multifrequency tone, out-of-band tone, digital code, and common channel; detailed information on these methods is provided in the sections that follow.

Table 9.4 provides a summary of signaling environments in the public switched network [9.21].

9.3.3 Functional *versus* Stimulus Signaling

This signaling method implied by the audible tones is geared to communication with humans. When computer equipment started to be connected to the plant in the mid-1960s, it had to "learn" to "hear" ring tones, busy tones, *et cetera*. The reception of tones implies "stimulus signaling"—actions have to be deduced from certain stimuli, such as a current flowing, a tone present for a length of time, *et cetera*. Functional signaling allows a data packet to be passed back to the computer with a coded message that explains what happened or is happening to a call, rather

Table 9.4
PSN Signaling Environments

Functional perspective:
- Supervision
- Addressing

Topological perspective:
- Customer loop
- Interoffice trunks
- Special services (including PBX trunks and tie lines)

Signaling systems:
- Facility-dependent:
  - dc
  - Out-of-band (N1 carrier)
  - Digital (T1 carrier)
- Facility-independent:
  - In-band tones
    - SF (supervision, addressing, or both)
    - MF (addressing over interoffice trunks)
    - DTMF (addressing in POTS customer loop)
  - Functional signaling (ISDN customer loop)
  - Common channel signaling

than having to rely on tones. Only with functional signaling of the type specified in Q.931 [9.22] (complemented with appropriate application-level messages) can advanced new data communication and telecommunication services be realized.

In "functional signaling," a device is instructed what to do by a data packet that actually contains the instructions in message form. Common channel signaling discussed below provides an example of functional signaling. In functional signaling, appropriate interconnection rules and procedures that allow two entities to exchange information must be defined by a formal protocol, rather than electrical or acoustical levels. Obviously, a voice system ultimately requires that a signaling protocol with the user exist, using tones. However, under functional signaling, the entire system can operate with data messages, which are then converted to a tone at the point closest to the user, if this is human. This can be done as far away from the network as the phone set: The set may receive a data packet that instructs it to apply a locally generated busy tone or dial tone, *et cetera*.

Functional signaling has existed for years in packet data networks; now it is being incorporated in the telecommunication environment. Consider, for example, call setup over an X.25 packet interface. An X.25 configured terminal initiates the procedure by presenting a *set asynchronous balanced mode* (SABM) packet-message to the network via the local packet switch; the network responds with an unnumbered acknowledgment (UA) packet-message. This provides for the logical link setup for LAP-B. The X.25 procedure for logical channel establishment, data transfer, and disconnect follows below. An X.25 call request packet-message containing appropriate information (parameters, address, *et cetera*) is presented to the network by the terminal; the network will forward the request to the appropriate target switch (via internal protocols, routing, flow control, *et cetera*). The remote switch, in turn, sends a SABM packet-message to the remote terminal; when the terminal returns the UA, the switch proceeds with an incoming call packet-message. If the remote terminal is able and willing to accept incoming traffic, a call accept packet-message, with appropriate embedded parameters, will be returned to the remote switch, which returns it to the local switch over the network. The local switch will then send a call complete packet-message to the originating station, informing it that data can now be transferred. Disconnect is initiated by sending a clear request packet-message to the local switch, which propagates the request to the remote switch, while at the same time sending a clear confirmation packet-message to the local terminal. The local terminal responds with a disconnect packet-message, which is acknowledged by the local switch with a UA packet-message. On reception of the clear request packet-message, the remote switch will send a clear indication packet-message to the remote terminal. This will issue a clear confirmation and then a disconnect packet-message; the remote switch completes the disconnect procedure by issuing a UA packet-message to the remote terminal.

As noted, all signaling activities are carried out by providing messages that contain descriptive and operative parameters among the various network entities, including the end-user.

9.3.4 In-Band and Out-of-Band Signaling

Traditional in-band signaling places supervisory and address instructions in the same stream as the actual user information. The signaling is not independent of the voice channel itself; not only does the possibility of unwanted interaction between the two (called a "talk off," where the voice signal becomes confused with signaling tones and can cause a disconnect) exist, but also the possibility of fraud [9.23]. (When the user's voice contains enough vocal energy at 2600 Hz to actuate the tone-detecting circuits in the signaling set, interference will occur—people with high-pitched voices may encounter this problem.) For data applications, in-band signaling has been a nuisance: Equipment (modems in particular) had to be engineered to avoid certain portions of the spectrum where various signaling tones are placed. The one-to-one relationship between the information channel and the supervisory channel meant the provision of a signaling mechanism for every voice channel; because of the required number, such a mechanism has to be simple and inexpensive. In turn, this leads to an inflexible system implementation, low signaling capacity in terms of supervisory interexchanges, and great expense if it is to evolve in parallel with network technology and architecture.

9.3.5 Problems with In-Band Signaling

In-band signals are vulnerable to fraud and crosstalk, and are relatively slow. Another problem is known as "glare," which occurs when both ends of a circuit are seized, due to the propagation and processing delays. The way around this problem is the use of one-way signaling on trunks; however, on small trunk groups, this solution is not economical. When glare occurs, the equipment is unable to complete the connection; the circuit must take a time out (which is quite inefficient) and a reorder (fast busy tone to indicate equipment or trunk unavailability) is issued to the user [9.20]. Glare can affect some CPE networks, notably, privately networked PBXs. (Glare is also possible with out-of-band or common channel systems, but is more prevalent with in-band signaling.)

9.3.6 Traditional Signaling Mechanisms

9.3.6.1 Trunk Side

Trunk signaling (i.e., signaling between switches) for the public telephone network occurs per-trunk in four ways (in contrast to CCS discussed below; see Figure 9.9):

1. Direct current signaling;
2. In-band pulse signaling (also known as ac signaling);
3. Out-of-band pulse signaling; and
4. Internal digital signaling, such as the T1 signaling mechanism.

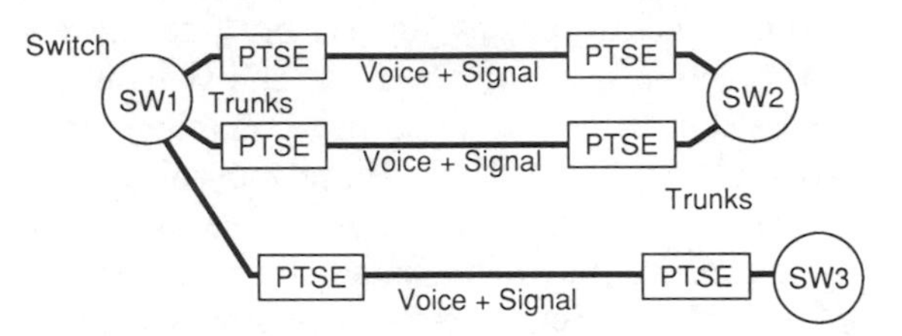

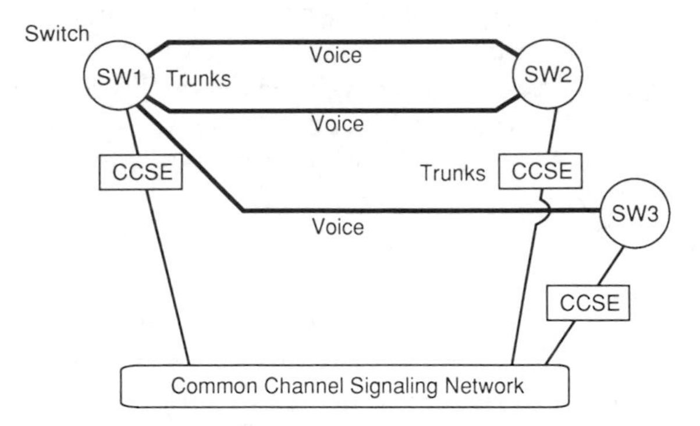

Figure 9.9 Per-trunk signaling *versus* CCS.

dc Signaling

Direct current (dc) signaling is accomplished through the use of two electrical states called "on-hook" and "off-hook." A third state may sometimes exist as a transition state. In a dc signaling arrangement, the supervisory or address signaling is done by superimposing one or more dc states on the same conductors that are used for voice transmission.

Currently, two types of per-circuit in-band trunk supervisory signaling methods are available: loop reverse battery (based on subscriber line signaling methods) and E&M signaling.

The first method is applicable to trunks that require call origination and seizure at only one end (one-way trunks, for example, direct outward dialing trunks). The method employs open and closure signals from the originating end, and reversal of battery-ground from the terminating end. At the originating end, on-hook is indicated by an open circuit; off-hook by a bridged circuit. At the terminating end, on-hook is indicated by a ground on the tip lead and −48 V on the ring lead of the circuit, and off-hook is indicated by −48 V on the tip lead and ground on the ring lead [9.24].

On two-way trunks (in which call origination is possible at both ends), E&M signaling is required. The E&M interface is one of the most common interfaces for signaling between a switching system and the transmission facility that connects it to another switching system. A number of variations of E&M signaling exist. The BOCs have standardized four E&M lead signaling interface arrangements and have tentatively standardized a fifth. The three types used in the United States are known as Type I, II, and III, respectively. The basic method (Type I) provides for signaling from the switch toward the transmission facility over the M-lead by presenting a ground (0 V) for on-hook and −48 V for off-hook. The signaling from the transmission facility to the switch occurs over the E-lead, with an open circuit for on-hook and ground for off-hook [9.24].

Direct current signaling for interoffice applications has been reduced in an evolution involving first ac signaling and then CCS. In interoffice applications, upgrading the signaling is a relatively easy job: The old signaling equipment is merely replaced at two ends with new signaling equipment (two pieces of equipment).

ac Signaling

Carrier systems started to appear in the early 1920s, typically in the trunk side of the telephone system. "Carrier" refers to the multiplexing of several channels over a single transmission system, as discussed in Chapter 3. A 1000-Hz tone interrupted 20 times a second was initially used to convey signaling information [9.25]. The use of this single-frequency tone for signaling is called single-frequency (SF) signaling. Automatic switching machines were widely deployed in the 1920s and 1930s, putting an end to operator-established local calling. Automatic local dial had spread fairly rapidly by the early 1940s. However, all toll calls were placed by operators until the 1950s. Originally, the time necessary to establish a call was 1 to 7 min; a need for faster processing was the driving factor for the introduction of DDD. DDD service began in 1951 and spread quickly; by 1960, more than 50% of the users had this capability. DDD was generally available by the late 1960s.

SF eventually utilized a 2600-Hz tone. SF can provide both supervisory and dial-pulse addressing. Because a single frequency is available (on-hook being presence of tone and off-hook absence of tone), addressing must be exchanged by pulse-type interruption of the signal (one interruption represents the digit 1, two interruptions represent the digit 2, *et cetera*). This interruption process is slow, with ten interruptions a second being the maximum speed. Also, time delays must be allocated for tone recognition, to allow for distortion caused by facility noise or transmission degradation, and to prevent false signaling possibly introduced by talk-off (interference with speech) [9.21].

N-carrier analog systems for short-haul applications were introduced in the late 1940s and 1950s; some continued to be deployed in the early 1960s. In the initial design, a tone at 3700 Hz was used for signaling; this out-of-band tone provided complete protection against talk-off. Later (in some N-carrier systems), in addition to single-frequency supervision, multifrequency (MF) address signals were sent by a selection of two frequencies in the range from 700 to 1700 Hz [9.25].

MF pulses convert addressing digits (initially input by the subscriber in most applications) by coding them with combinations of two out of six fixed frequencies between 700 and 1700 Hz. The six frequencies are 700, 900, 1100, 1300, 1500, and 1700 Hz. Fifteen combinations are possible when picking two items out of six: ten combinations are for the digits and five combinations are used for auxiliary status codes. For example, the digit 1 is coded by a 700 combined with a 900-Hz tone; the digit 2 is coded by a 700 combined with an 1100-Hz tone; *et cetera*. MF is used only for interoffice signaling for called number address signaling, calling number identification, ring-back, and coin control [9.24]. Digits are sent on interoffice trunks at the rate of seven digits per second, compared to dual-tone multifrequency (DTMF, used for addressing in POTS customer loops, as discussed in the next section) and SF signaling, both of which are transmitted at a lower speed (DTMF signaling over the local loop takes place at a lower speed due to the potentially noisy condition of the loop). Therefore, MF signaling used for addressing in the process of setting up a call requires substantially less time than would be required with SF signaling. Note that MF tones do not encounter the talk-off problem (hence, special selection of the frequencies was not critical) because voice transmission from the calling party is inhibited during MF signaling.

Signaling on Digital Systems

Digital carrier systems (notably T1s) have used a different type of signaling, involving framing and VF-robbed bits to achieve the desired signaling, as discussed in Chapter 3. Appropriately configured line cards on the carrier equipment have provided the necessary interface between the digital signaling method internal to the carrier equipment, with other types of signaling external to the carrier system. These line cards have also provided the special signaling arrangements required for special services applications.

Coexistence of Signaling Systems

The public switched network is a harmonious union of many different but coexistent technologies. Hence, a mixture of digital signaling is typically found along with

SF signaling and CCS. The goal is to migrate the plant to a CCS technology, though digital signaling will survive for a long time. Because of this mix of signaling methodologies, translations are required. Translations are actually data bases stored in the switch memory that contain information about the type of signaling and other characteristics of the trunk (translations also apply to the subscriber loop, particularly for class-of-service description and for call routing).

9.3.6.2 Loop Side

Subscriber Line Signaling over Metallic Facilities

At present, a majority of the local loops still involve a metallic segment: The feeder portion has employed carrier for a number of years and is now being complemented with fiber, but the distribution plant (the segment of the plant closest to the user) has continued to use copper because of the requirement of the installed phone sets to employ dc signaling (the only exception has been for large commercial customers, where fiber local loops have been installed to provide the appropriate bandwidth).

Direct current signaling is mandated not by the fact that the transmission facility is metallic, but by the fact that the receiving station can only operate with dc signaling. To upgrade signaling on the local loop would require replacing equipment for a large number of customers (approximately 100 million in the United States). This is fairly expensive and may occur only gradually under ISDN if the customer can be induced to buy a new phone set in exchange for the ability to obtain new services.

With dc signaling, taking the phone off the hook closes the electrical path between the tip and ring leads, allowing current to flow. This arrangement is also called "loop start," and is indeed used on all subscriber loops that terminate in actual phone sets. Metallic loop signaling provides for continuous application of dc voltage, nominally at 48 V, from a power supply (referred to as a "battery"), together with a current-sensing mechanism at the CO (or carrier system). When the loop is implemented on a carrier feeder system, the user's station equipment is connected by a pair of wires to the carrier's remote terminal (near the user), rather than to the CO; the carrier system is responsible for providing the metallic loop signaling protocol. Note that in these digital loop carrier applications, the switch may still employ metallic loop signaling to the CO-based carrier terminal (if any); the terminal will convert the signaling into internal digital signaling, and reconvert it to dc signaling in the proximity of the user [9.21].

The address information, when coded with dial pulses, consists of short on-hook pulses occurring as interruptions in the normal off-hook loop supervision current at a rate of 10 pulses per second. The number of dial pulses in a stream equals the value of the intended addressing digits (10 pulses, with exceptions, signal the number 0). This loop signaling technique is clearly slow.

DTMF

The addressing portion of the signaling handshake between the user station and the switch uses either dial pulses or DTMF (the supervisory portion of the loop signaling over the distribution plant will continue to use dc until the deployment of ISDN). In DTMF, translation equipment is required at the CO switch to convert the tones to interoffice addressing signals. All modern switches are capable of receiving DTMF (addressing signals are, in turn, transmitted over interoffice trunks as SF dial pulses, MF codes, or CCS).

Push-button dialing was introduced to accelerate call setup; it became available in 1963. This made voice frequency in-band signaling a feature of the local loop plant, as it had been for some time for the interoffice trunk plant. As mentioned above, DTMF is the coding of digits with tones. It is used for part of the subscriber loop signaling (the addressing portion). The differences between DTMF and MF are:

- DTMF is used only on customer loops or PBX trunks to the CO; MF is used for interoffice;
- The set of frequencies is different; and
- The speed at which the pulses are sent is different.

With push-button DTMF, the dialing digits are encoded as a combination of two sinusoidal tones selected out of eight tones. The tones are 697, 770, 852, 941, 1209, 1336, 1477, and 1633 Hz. The first four are considered to be a frequency group, and the last four is another frequency group; all digits are represented with a frequency from the first group and a frequency from the second group. For example, the digit 1 is coded by a 697 combined with a 1209-Hz tone; the digit 2 is coded by a 697 combined with a 1336-Hz tone. Special selection of these base frequencies was undertaken with the aid of harmonic analysis to minimize the interference with voice and the talk-off problem and to differentiate from MF.

DTMF allows sixteen combinations, as seen in Figure 9.10 (the states A, B, C, and D are reserved for "future use" [9.24]). The frequencies on the top row in the figure are high group frequencies; the frequencies in the columns are low group frequencies.

All switching systems that interface with end-users require 20-Hz ring generators supplying 90 V to the loop. Other audible tones are also required. Some of the common audible tones used to communicate with the end-user are shown in Table 9.5 [9.26].

Note that in digital offices, the tones are generated by digital synthesis, under the control of software modules. More details on these tones can be found in CCITT's Specification E.180, or in Reference [9.27].

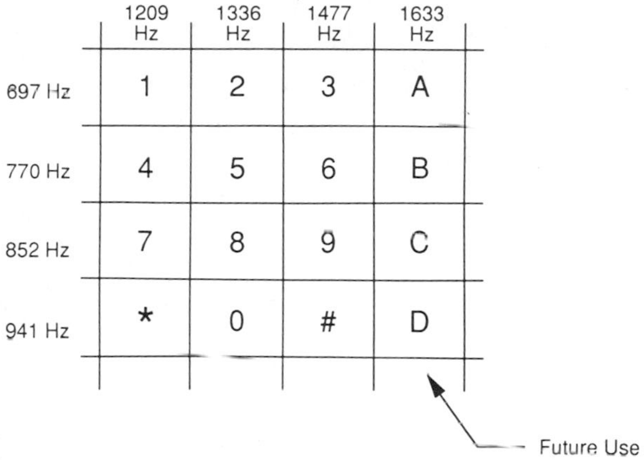

	1209 Hz	1336 Hz	1477 Hz	1633 Hz
697 Hz	1	2	3	A
770 Hz	4	5	6	B
852 Hz	7	8	9	C
941 Hz	*	0	#	D

Future Use

Figure 9.10 DTMF plan.

Table 9.5
Signaling Tones

Dial tone	350 Hz + 440 Hz	Continuous
Station busy	480 Hz + 620 Hz	0.5 s on; 0.5 s off
Network busy	480 Hz + 620 Hz	0.2 s on; 0.3 s off
Audible ring return	440 Hz + 480 Hz	2.0 s on; 4.0 s off
Off-hook alert	Multifrequency howl	1.0 s on; 1.0 s off
Recording warning	1400 Hz	0.5 s on; 15.0 s off
Call waiting	440 Hz	0.3 s on; 9.7 s off

9.4 COMMON CHANNEL SIGNALING (CCS)

In 1976, deployment began in the public U.S. telephone network on an improved system of segregating the signaling mechanism over its own dedicated supervisory data network. CCS is a signaling method in which a single channel conveys, by means of labeled messages, signaling information relating to a multiplicity of circuits, or other information, such as that used for network management. This migration to out-of-band signaling is only the beginning of a major network evolution with regard to signaling, which will culminate in the deployment of ISDN in the early 1990s and BISDN in the late 1990s [9.28].

9.4.1 Needs for Improved Signaling

Until 1976, signaling in the U.S. plant was almost entirely on a per-line or per-trunk basis, as discussed in the previous sections. The disadvantages of in-band signaling were not overwhelming when the only "users" of the network were humans. When sophisticated high-speed, high-capacity equipment is interconnected (such as a PBX or a computer) that attempts to interact with the network with the objective of providing an advanced service, the limitations of a traditional signaling system become evident. These limitations have been the motivation for the development and introduction of CCS in the telephone plant. Another motivation for the establishment of a common channel was to reduce the possibility of fraud [9.23].

9.4.2 CCS Architecture

CCS involves a separate high-speed network (the "common channel") to transfer supervisory signaling information in an out-of-band fashion; this separate network carries signaling information from a large number of different users, in a multiplexed fashion, employing packet-switching technology. The separation of the signaling from the information channel, as well as the greater repertoire of command message formats, allows a more methodic migration of the network to an advanced architectural configuration; this follows from the fact that changes can be made without the high cost associated with physical replacement or modification of hardware.

Because supervisory instructions are coded as messages, instead of some sequence of tones, and because of the higher bandwidth available for signaling, more detailed information about a call, in terms of desired network treatment, call origin, *et cetera*, can be exchanged across the network. In turn, this implies more sophisticated services. The talk-off problem is totally eliminated with the separate signaling facilities. Another advantage of CCS is that signals can be sent in both directions simultaneously, and during the conversation, if necessary. This last feature is very valuable for some advanced services. Business-case analyses also proved-in CCS on the merits of saving network equipment and trunks with faster signaling.

CCS allows access to many points in the network, not just switches. Advanced voice and data services may depend on remote data bases, processors, and facilities. Thus, CCS provides (1) direct local office–to–local office signaling connectivity; and (2) local office–to–service node signaling connectivity.

CCS is reliable and fast. It replaces both the SF and the MF signaling equipment and methods; addressing digits are converted to data messages (packets). All modern digital switches can be equipped with CCS capabilities. The CCS interface is an electronic device that interprets incoming data messages and transfers the translated signal to the common control-call processor.

Three signaling modes need to be considered:

1. *Associated Mode*. In this mode, the messages relating to a particular signaling relation between two adjacent points are conveyed over a link directly interconnecting these signaling points.
2. *Nonassociated Mode*. In this mode, the messages relating to a particular signaling relation are conveyed over two or more links in tandem passing through one or more signaling points other than those which are the origin and the destination of the messages.
3. *Quasiassociated Mode*. This mode is a subcase of the nonassociated mode. Here, the path taken by the messages through the network is predetermined and, at a given point in time, fixed.

A direct plant implementation of fully associated CCS would require a point-to-point signaling link between any two switches. Switches equipped with CCS interfaces can be interconnected with direct data links if the traffic volume of these signaling messages is high enough. Most switches, however, are connected to data communication packet switches. The nonassociated signaling is then implemented over a network that employs signal transfer points (STPs) operating as packet switches; this topology obviates the need for a large number of point-to-point links (see Figure 9.11). Signals are sent between offices by two or more links and one or more STPs in tandem. Connectionless packet-switching techniques are employed. The function of the STPs is to route signaling messages between the various constituent links, without altering the message. Thus, the only functions are the level 2 error detection-correction task on signaling message content and the level 3 network routing function.

Until the advent of ISDN, out-of-band functional signaling will be limited to the trunk side of the plant. The SS7 signaling network only applies at the trunking

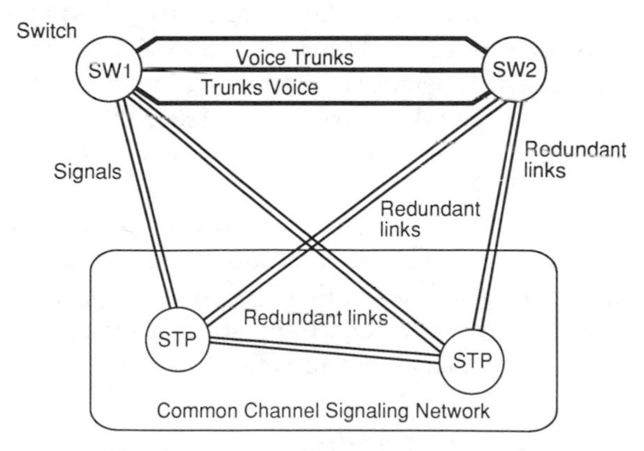

Figure 9.11 Links to the CCS network.

level. With ISDN's D channels, a complementary (but not identical) capability will be extended to the end-user.

The STPs are packet switches that handle the routing of the signaling messages. As such, these nodes provide for concentration and ensuing efficiency: Few switching offices in the inter-LATA or intra-LATA network have trunk groups large enough to justify direct connection of the signaling channels between the offices in question ("associated" or "direct connected" links). STPs may be redundant to ensure availability and reliability. While normally operating in a load-sharing mode, one STP can take over if the other fails. In most cases, signaling messages are routed over one or two STPs.

CCS was introduced in the United States in 1976, with the first implementation serving Madison, Wisconsin, and Chicago, Illinois. The network was comprised of 20 STP nodes (2 in each of 10 DDD switching regions), the links interconnecting these STPs, and the selected switching machines in that region. The format employed by AT&T was initially a variant of CCITT Signaling System 6 (SS6), specified for international use in an analog environment. SS6 was developed in the early 1970s and was optimized toward carrying signaling information related to call setup and management [9.23].

Not all switches in the United States are currently connected to a carrier's CCS system. This is an evolutionary deployment; while inroads have been made in the 1980s, ubiquitous deployment must await the 1990s. (Recall Table 4.9.)

9.4.3 CCITT Signaling System Number 7

The goal of CCITT's Signaling System Number 7 (CCITT SS7) is to provide an international standard for CCS suitable for SPC exchanges, computers, and PBXs. It operates on digital networks with DS0 channels (64 kb/s) and it accommodates low-speed analog links if desired. While CCITT SS6 lived up to its promise of providing improved call management and facilitating some new services, it was not totally adequate to a digital environment. Hence, the need to define a more comprehensive CCS system. This need gave rise to CCITT SS7. SS7 will be the building block of ISDN [9.29, 9.30].

CCITT SS7 meets the requirements of call control signaling for telecommunication services such as POTS, ISDN, and circuit switched data transmission services. SS7 can also be used as a reliable transport system for other types of information transfer between exchanges and specialized centers in telecommunication networks (e.g., for management and maintenance purposes). Hence, CCITT SS7 supports [9.23, 9.31]:

- Public switched telephone networks;
- ISDN;
- Interaction with network data bases and service control points for advanced intelligent network services;

- Public land mobile networks; and
- Operations administration and maintenance of networks.

Some of the areas where SS7 is applicable include international, domestic, local, and long-distance public switched networks and signaling over a privately deployed CCS network for large private voice networks. In addition to the point-to-point terrestrial links application, SS7 can operate over point-to-point satellite links. However, it does not include the special features required for use in point-to-multipoint operations, typical of satellite technology. The system includes redundancy of signaling links and it includes functions for automatic rerouting of signaling traffic to alternative paths in case of link failure.

CCITT SS7 supports the associated and quasiassociated signaling modes described above. The message transfer part does not include a resequencing mechanism to handle the dynamic message routing arising in the general nonassociated environment.

The fundamental principle of the SS7 structure is the division of functions into a message transfer part and separate user parts. The term "user" refers to any functional entity that utilizes the transport capability provided by MTP (in this context, the user is not the ultimate customer of the network).

The two functional parts are:

1. *Message Transfer Part* (MTP). The MTP contains the necessary mechanism to ensure reliable transmission of functional signaling messages, with maximum network availability.
2. *User Parts* (UP). The UP provides criteria for service management. Individual user parts are being introduced for a number of services; for example, operating procedures for interexchange signaling in the telephone network (TUP = telephone user part; ISDN-UP = ISDN user part), and for digital data networks (DUP = data user part).

The Blue Book CCITT SS7 comprises the following functional blocks:

- MTP;
- TUP;
- ISDN-UP;
- DUP;
- Signaling connection control part (SCCP); and
- Transfer capabilities (TC).

See Table 9.6 for a more complete list of 1988 CCITT standards. Figure 9.12 positions the standards in the physical context of the signaling network. The Bellcore specification of the SS7 protocol is found in TR-NPL-000246.

Figure 9.13 shows the partition of the protocols, as well as some users of MTP, including the ISDN-UP, the TUP, and the SCCP. Note that these user parts do not correspond precisely to the OSIRM layers described in Chapter 1 and

Table 9.6
CCITT SS7 Specifications

Function	*Recommendations*
Key aspects SS7	
MTP	Q.701–Q.704, Q.706, Q.707
TUP	Q.721–Q.725
Supplementary services	Q.730
DUP	Q.741
ISDN-UP	Q.761–Q.764, Q.766
SCCP	Q.711–Q.714, Q.716
Transaction capabilities	Q.771–Q.775
Operations maintenance and administration part (OMAP)	Q.795
Other aspects of SS7	
Signaling network structure	Q.705
Numbering of international signaling point codes	Q.708
Hypothetical signaling reference connection	Q.709
PBX application	Q.710
Test specifications	Q.780–Q.783
Monitoring and measurements	Q.791

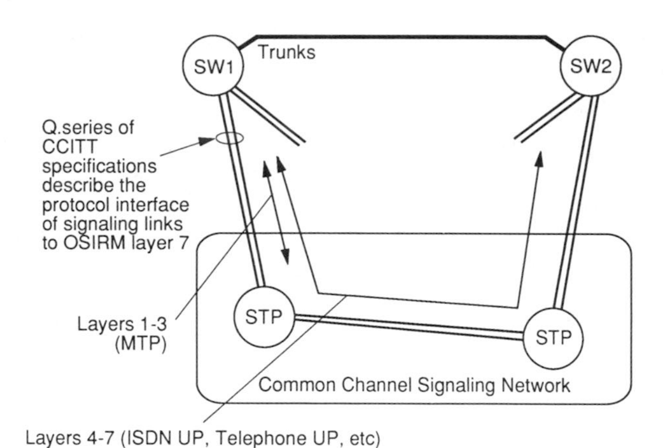

Figure 9.12 SS7 interfaces.

covered in much more detail in Chapter 13, although a substantial amount of harmonization and alignment has taken place in the past few years. The lower three layers follow the model, while user parts are still bundled. Figure 9.14 shows this relationship. (Pedantically, CCITT SS7 defines level 1 = layer 1; level 2 = layer 2; level 3 = layer 3; and level 4 = user parts covering layers 4 through 7.)

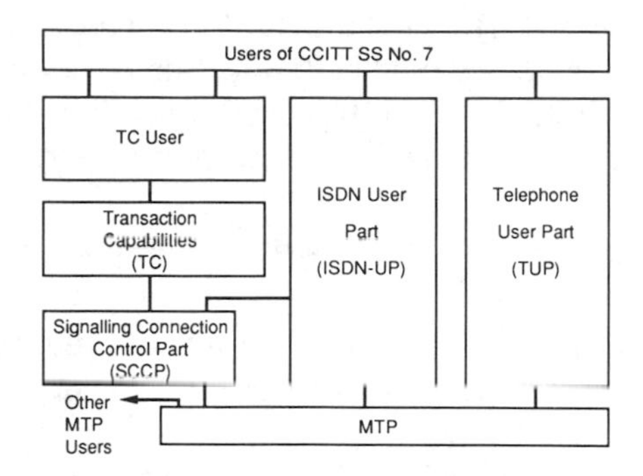

Figure 9.13 SS7 functional parts.

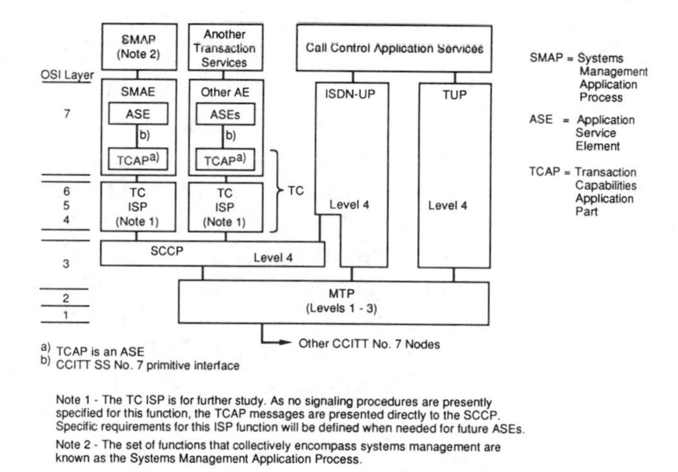

Figure 9.14 Relationship to OSIRM model.

Message Transfer Part

Level 1 specifies the physical, electrical, and functional characteristics of a signaling link (type A, B, or C); this deals with, for example, the channel plug characteristics, voltages required to identify a bit, *et cetera*.

Level 2 defines the functions and procedures for the transfer of signaling messages over a link in an error-free and reliable manner. This has to do with message blocking and retransmissions. Messages are transmitted using signal units that follow the framing conventions of the HDLC procedure. This type of modern bit-oriented protocol allows transparency and variable signal units. This is accomplished with a single (reserved) flag character 01111110, and zero insertion. Zero insertion assures that a sequence 01111110 in the data stream will not be confused with a synchronizing HDLC flag: The transmitting station over a link will insert a 0 after five consecutive 1s in the data stream; the receiving station will apply the reverse process.

Three types of message, or signal units, are transacted by the signaling network, in a link-by-link fashion [9.32, 9.33]:

1. *Fill In Signal Units* (FISUs). These are transmitted over the link when no other signal units are to be sent; basically, they maintain the link connection in the absence of real traffic.
2. *Link Status Signal Units* (LSSUs). These consist of status information (establishing or deestablishing a connection), checking alignment, and related functions.
3. *Message Signal Units* (MSUs). These really carry the content of the signaling message. Figure 9.15 shows the format of the MSU. (The interested reader may consult Q.703, Q.704, Q.713, Q.723, Q.763, and Q.773.)

Level 3 defines networking (transport) functions common to all individual links in the end-to-end circuit. It provides the means to multiplex several logical links on a single physical link. In the signaling environment, this implies the ability to carry call control information about several calls on one channel. CCITT Recommendations Q.701 to Q.707 specify the MTP.

The collection of these three layers constitutes the MTP because they have to do with the physical movement of signaling information. This provides a service similar (but not identical) to an X.25 network (MTP provides a connectionless packet service).

User Parts

As described in Chapter 1, the OSI model specifies four upper layers: transport, session, presentation, and application support. SS7 user parts correspond to these higher layers, though the clean segmentation suggested by the OSI model is not followed in the upper four layers because when SS7 was being developed, the higher OSI layers were not yet fully defined (they are now); ongoing efforts are aimed at migrating SS7 protocols to the OSI model. These higher layers have to do with the content and coding of the signaling messages. The original SS7 architecture had monolithic user protocols. Some of the message types are:

- Call establishment messages for alerting, call proceeding, connect, and setup;
- Call disestablishment messages for disconnect and release;
- Intracall information messages for suspending, resuming, and other user information; and
- Miscellaneous messages.

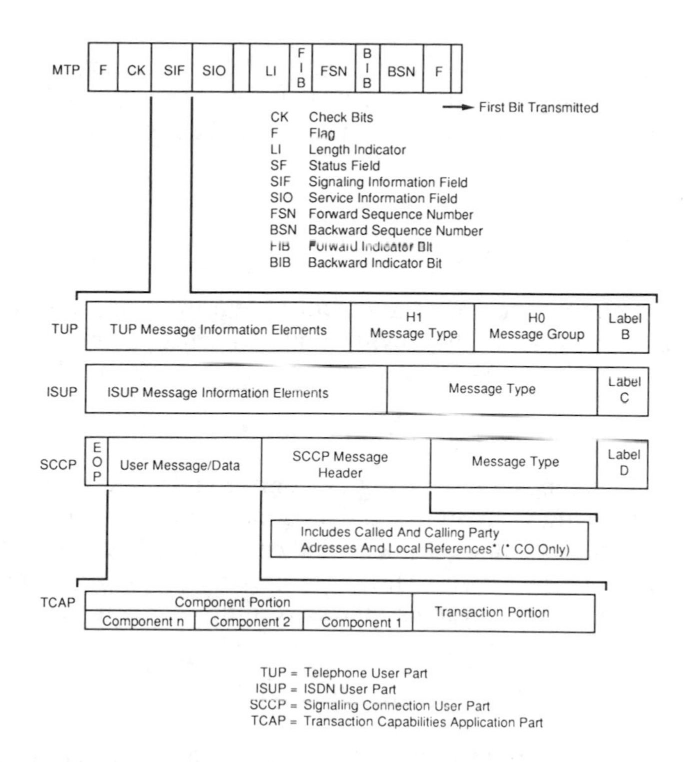

Figure 9.15 CCITT SS7 signaling message structure.

The parameters to be included in these messages for the purpose of call management are a function of the type of call: voice, data, circuit switched, channel switched, video, ISDN call, *et cetera*. This results in the definition of a number of user parts. A brief summary of the user parts is provided below. Detailed information can be found in the appropriate standards described in Table 9.6.

Signaling Connection Control Part

The SCCP provides the means to (1) control logical signaling connections in a CCITT SS7 network and (2) transfer signaling data units across the CCITT SS7 network. It provides a routing function that allows signaling messages to be routed to a signaling point. In addition to translation functions, SCCP also provides a management function. The combination of MTP and the SCCP is called the "network service part." This network service part corresponds with layers 1 through 3 of the OSIRM, as depicted in Figure 9.14.

Telephone User Part

The TUP defines the telephone signaling functions necessary to use CCITT SS7 for international telephone call control signaling. The Q.721–725 series of recommendations defines the signaling messages, their encoding, and cross-office performance.

Data User Part

This user part defines the protocol to control interexchange circuits used on data calls, as well as data call facility registration and cancellation.

ISDN User Part

The ISDN-UP encompasses signaling functions required to provide switched services and user facilities for voice and data applications in ISDN. (It is also suited for applications in dedicated telephone and circuit switched data networks, and in mixed analog-digital networks.) This series of recommendations defines the ISDN signaling messages, their encoding, and cross-office performance.

Transaction Capabilities

TCs provide the means to exchange operations and replies via a dialogue. It provides the means to establish noncircuit-related communication between two nodes in the signaling network. TC consists of two elements: (1) transaction capabilities application part (TCAP) and (2) an intermediate service part (currently being studied further).

User Parts Summary

The TUP and DUP recommendations describe monolithic signaling message groups (forward address messages, forward setup messages, and backward setup messages). The TUP will be used outside the United States. The DUP has actually been supplanted by the newer ISDN-UP because the rigid format of the DUP is not ideally suited to the ISDN environment, particularly for end-user to end-user signaling. The ISDN-UP will be supported in the United States. The ISDN-UP represents interoffice mappings of the content of the functional messages of Q.931 for the purpose of call-connect, call-terminate, *et cetera*. MTP provides the means (the "pipe") for those messages to be moved between appropriate switches. Reference [9.31] provides a detailed overview of the entire SS7 system.

9.5 NETWORK PLANT ARCHITECTURES OF THE LATE 1980s AND EARLY 1990s

The underlying structure of the telephone network and its architectural goal for the next few years is discussed here [9.6, 9.34–9.41].

9.5.1 AT&T

The evolving network architecture for the late 1980s and early 1990s is comprised of three building blocks: (1) intelligent processors that go beyond (and may be physically separate from) the digital switch (these processors are called *network control points* or *network elements*); (2) common channel signaling; and (3) digital end-to-end path connectivity. This arrangement has been called "stored program controlled network" by AT&T [9.23].

AT&T's SPC/CCIS toll network can illustrate how these building blocks interact. In spite of the fact that CCIS only applies to the trunk side of the plant, it provides substantial benefits by allowing, among other things, faster interoffice signaling for call setup and trunk allocation. This is achieved by facilitating direct network element–to–network element communication. Action control points (ACPs) are, in AT&T's version of this architecture, user offices that provide access and interface to generally centralized network control points (NCP). These NCPs are computer or data base machines that contain detailed instructions on how to treat each specific call, on a per-call basis. These instructions can be viewed as features describing some network-based vertical service. NCPs can be located anywhere in the network because their services can be obtained over the CCS system. This network architecture will employ SS7 as its signaling method.

The best example of this prototypical architecture, which will be used by the carriers in the future to complement ISDN (the digital transport and the signaling network), is 800 service. Here, the SPC switch, on detection of an 800 number, interrogates the NCP via the signaling network as to the real telephone number corresponding to the dialed 800 number; the resulting number could even be a function of the time of day. The NCP returns to the switch—again over the signaling network—the desired physical number, from which the switch can complete the call. As presently defined, the CCS network can transfer data that have small content (such as a phone number); some services being contemplated would require transfer of larger blocks of data from the NCP to the switch. Some performance problems may arise unless the CCS protocols and network are appropriately modified (work is now under way to examine these issues). The first example of this type of architecture was introduced in the U.S. network in the early 1980s [9.23].

Figure 9.16 provides a topological view of the integrated signaling-service architecture. A user accessing the network will, at some point near entry, be treated

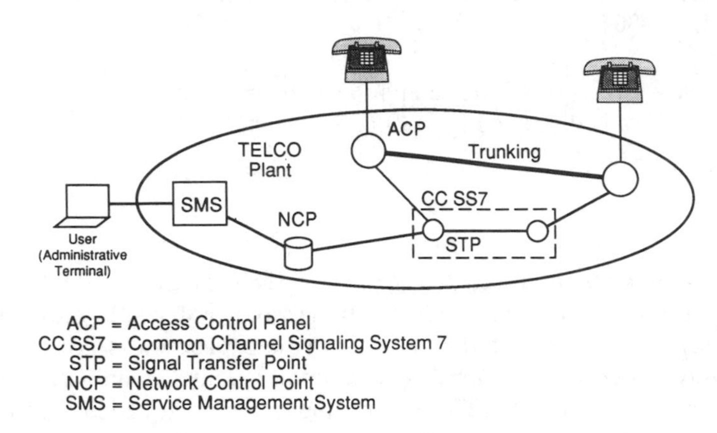

Figure 9.16 An "intelligent" network.

by a ACP that determines the required action to be taken. The NCP is consulted over the signaling network, possibly through several STPs internal to the latter network, and its disposition recommunicated to the ACP for call completion. The NCP provides centralized translation, routing, information control, and service logic. CCIS-equipped switching offices (in reality, the ACP) will query the NCPs over the signaling network. New services can be added to the network without major redevelopment; the changes can be made in software and even controlled by the end-user, as shown in Figure 9.16, via the service management system (SMS), which acts as a "service compiler."

Customer-dialed calls indicating to the SPC a need for a special service are routed to the closest ACP; the local office will forward the called number and the calling line identity (*automatic number identification*—ANI information) over the signaling network. The routing to the ACP allows any call to receive the requested service even if the local switch is not directly connected to the CCS network. Currently the ACP recognizes special calls by the digits that are dialed (800-, 900-, 555-, *et cetera*).

Initially, the CCIS network was used to improve call management (particularly call setup) and to provide improved maintenance. Later CCIS facilitated the introduction of some new services: notably call forwarding, conferencing, call storage, 800 service (possibly with end-user administration), and automatic calling card service. In the latter, a credit card call is first routed to a TSPS (traffic service position system) point where the call is temporarily halted while the signaling network carries a query to an appropriately built NCP for authorization. If the credit number presented to the NCP is valid, the NCP returns an authorization message to the local switch to connect the call [9.23].

9.5.2 BOCs

The BOCs are pursuing advanced intelligent network architectures [9.36, 9.39, 9.42]. Five factors influence the development of this network: (1) external driving forces; (2) user needs; (3) effect of evolving technologies; (4) new transport architectures such as ISDN, BISDN, and advanced intelligent network (AIN); and (5) advanced data and voice services required by the business climate of the 1990s.

Driving forces include competition, the legal and regulatory environment, BOC's own business strategies, and technology. All these driving forces affect the embedded BOC network.

User needs include compatibility, control (services, responsiveness), bandwidth, reliability, cost control, connectivity, and the ability to manage intelligent CPE (data PBXs, LANs, *et cetera*), applications, and networking. Applications range in utilization and required transmission speed. For example, voice has high utilization and relatively low speed (64 kb/s); alarms and telemetry have low utilization and require low speed; file transfer requires low utilization but high speed; video requires high utilization and high speed. Users need standard interfaces, integrated access, ease of service management, flexible bandwidth, and economical services.

Technology factors encompass advancements in networking and in the transmission, hardware, and software fields.

Advances in transmission include fiber optics and digital switching; advances in hardware include VLSI, ULSI (ultra large scale integration), integrated optics, new semiconductor substances; software innovations include artificial intelligence and expert systems. Each of these technologies impacts a segment of the BOC network. This plant is comprised of a transmission segment, a switching segment, signaling and control, and operations. At the transmission level, the plant has seen increased digitization of the feeder and interoffice plant, using digital pair gain multiplexers in the former and fiber trunking in the latter.

The "last mile" will see introduction of fiber to the home in the next few years. Fiber technology is such that the bandwidth derivable from an embedded fiber has doubled every couple of years; for example, 405 Mb/s systems were typical in 1984; 810 Mb/s systems were available in the mid- to late 1980s; and 1.7 Gb/s are now being introduced (as discussed in Chapter 7). At the switching level, the plant has seen the increased introduction of digital facilities that have a distributed and modular architecture; intelligent peripherals may complement or retrofit some of the switches. Next-generation switches may include integrated circuit, packet, and channel switching that handle voice, data, and video. Switching will take place in a semiconductor fabric, closer to the hardware level than at present.

The target architecture and its underlying architectural modules (ISDN, BISDN, and AIN) will enable the BOCs to provide new services in the 1990s to satisfy evolving business requirements. Some examples of new services are LAN interconnection, multimedia voice-data-videoconferencing, and user-defined network interfaces. This architecture will provide the user with standardized interfaces, digital transport, distributed processor control, flexible digital signaling, common service structure, and an integrated operations and network management.

The Advanced Intelligent Network

The AIN is a service control architecture proposed by Bellcore for the regional Bell companies for the 1990s and beyond. AIN will enable the production of service-independent platforms that can be used by regional companies to introduce new services. The AIN architecture provides a flexible structure, distributed intelligence, sophisticated signaling, and the ability to deploy new services rapidly. Services are defined in service logic programs, which use a standardized syntax to define the service interactions between various AIN components, in support of the ultimate user-perceived service [9.36, 9.39].

The advantages of AIN include:

- Regional company control of service offerings: The regional companies aim at greater control over what new services they will offer, when they will offer them, and where they will offer them.
- Rapid service introduction: The regional companies aim at offering new services more rapidly to enable them to respond to the needs of their customers.
- Diversity of services: The regional companies aim at offering a wider range of new services to meet their customers' needs.
- Multivendor environment: The regional companies aim at purchasing equipment from a wide range of vendors and having this equipment work together in their networks.

These goals will be addressed through a series of evolutionary releases of AIN. Three releases of AIN are currently being defined, releases 0, 1, and 2. Release 0 is scheduled for initial deployment in the 1990–91 time frame; release 1 for initial deployment in 1993; and release 2 for initial deployment in 1995. One of the major aspects of AIN release 2 is the increased support of data services.

Reference [9.43] provides an excellent overview of the entire field of intelligent networks.

9.5.3 International

More than six countries are introducing intelligent networks, including the United Kingdom (operational), Germany (operational), Spain (to be operational at the end of 1990), Italy (operational in 1992), Belgium, Taiwan, and Korea [9.44].

REFERENCES

[9.1] "Central Office Equipment Market: 1989 Database," Northern Business Information, New York.

[9.2] "User-to-User Signaling with Call Control," Bellcore TR-TSY-000845, Issue 1, December 1988.

[9.3] W. Stallings, *ISDN: An Introduction*, MacMillan Publishing, New York, 1989.

[9.4] K. Habara, "ISDN: A Look at the Future through the Past," *IEEE Communications Magazine*, November 1988, pp. 25 ff.

[9.5] *Telephone News*, May 7, 1990 (based on a Phillips Telecommunications Research report, Potomac, MD).

[9.6] G. Bregant, R. Kung, "Service Creation for the Intelligent Network," XIII International Switching Symposium, Stockholm, Sweden, May 27–June 1, 1990.

[9.7] R. Parkinson, "Outlook on Digital Communications," *Journal of Data and Computer Communications*, Winter 1989, pp. 13 ff.

[9.8] R. Smith, "Lean—But Still Mean," *TE&M*, January 15, 1990, pp. 37 ff.

[9.9] Siemens-provided information.

[9.10] "An Overview of Central Office Switches," DataPro Report CA30-020-801, May 1988.

[9.11] S. Salamone, "Lightwave Deemed Vital to International Broadband Net," *Lightwave*, August 1989, p. 8.

[9.12] E. Munter, M.R. Wernik, "Broadband Public Network and Switch Architecture," XIII International Switching Symposium, Stockholm, Sweden, May 27–June 1, 1990.

[9.13] F. Dolenc, "Switch Architecture for Intelligent Networks," XIII International Switching Symposium, Stockholm, Sweden, May 27–June 1, 1990.

[9.14] G.J. Grimes, L.J. Haas, "An Optical Backplane for High Performance Switches," XIII International Switching Symposium, Stockholm, Sweden, May 27–June 1, 1990.

[9.15] S.R. Treves, A. Fioretti, "A Novel Distributed Photonic Switch," XIII International Switching Symposium, Stockholm, Sweden, May 27–June 1, 1990.

[9.16] B. Schaffer, "ATM Switching in the Developing Telecommunication Networks," XIII International Switching Symposium, Stockholm, Sweden, May 27–June 1, 1990.

[9.17] D.J. Bastien, D.P. Worrall, "Implications of New Technologies on Future Networks," XIII International Switching Symposium, Stockholm, Sweden, May 27–June 1, 1990.

[9.18] S. Scully, "Fiber-to-the-Curb Gains Status," *Lightwave*, June 1989, p. 1.

[9.19] D. Minoli, "Engineering PBX Networks Part 3: Signaling," DataPro Report MT30-315-301, April 1987.

[9.20] J.H. Green, *The Dow Jones-Irwing Handbook of Telecommunications*, Dow Jones-Irwing, Homewood, IL, 1986.

[9.21] R.F. Rey, ed., *Engineering and Operations in the Bell System*, 2nd Ed., AT&T, Murray Hill, NJ, 1984.

[9.22] CCITT Recommendation Q.931/I.451, "ISDN User-Network Interface Layer 3 Specification," CCITT Blue Book, Geneva, 1988.

[9.23] R. Eward, *The Deregulation of International Telecommunications*, Artech House, Dedham, MA, 1985.

[9.24] "Signaling, LATA Switching Systems Generic Requirements," Section 6, Bellcore, TR-TSY-000506, Issue 2, July 1987.

[9.25] E.F. O'Neil, ed., *A History of Engineering and Science in the Bell System—Transmission Technology (1925–1975)*, AT&T, Murray Hill, NJ, 1985.

[9.26] E.B. Carne, *Modern Telecommunications*, Plenum Publishing, New York, 1985.

[9.27] R.L. Freeman, *Reference Manual for Telecommunications Engineering*, Wiley-Interscience, New York, 1985.

[9.28] P. Distler, F. Faller, "Towards an ISDN Signaling System Paving the Way for the Future," XIII International Switching Symposium, Stockholm, Sweden, May 27–June 1, 1990.

[9.29] D. Minoli, "Common Channel Signaling System Number 7," DataPro Report MT30-320-201, November 1988.

[9.30] W. Stallings, "Demystifying SS7 Architecture," *Telecommunications*, March 1989, pp. 41 ff.

[9.31] Recommendations Q.700–Q.716, CCITT Blue Books, Geneva, 1989.

[9.32] M.G. Walker, "Get Inside CCITT Signaling System No. 7," *Telephony*, March 10, 1986.

[9.33] W.C. Roehr, "Inside SS No.7: A Detailed Look at ISDN's Signaling System Plan," *Data Communications*, October 1985.

[9.34] E.G. Sable, H.A. Bauer, J.Z. Jacoby, J.B. Sharpless, "Evolution of Intelligence in Switched Networks," XIII International Switching Symposium, Stockholm, Sweden, May 27–June 1, 1990.

[9.35] K. Shulz, G. Glaeser, W. Klein, H. Thomas, "Strategy For An Implementation of the Intelligent Network," XIII International Switching Symposium, Stockholm, Sweden, May 27–June 1, 1990.

[9.36] G.F. Oram, L.A. Sych, "Service Location Tradeoffs in Intelligent Networks," XIII International Switching Symposium, Stockholm, Sweden, May 27–June 1, 1990.

[9.37] P. Bagnoli, E. Cancer, E. Guarene, "The Introduction of the Intelligent Network in Italy, A Strategic Objective and a Challenge for the 1990's," XIII International Switching Symposium, Stockholm, Sweden, May 27–June 1, 1990.

[9.38] J. Delory, G. Marx, "Data Management in the Intelligent Network," XIII International Switching Symposium, Stockholm, Sweden, May 27–June 1, 1990.

[9.39] R.J. Baseil, B. Gotz, "Integrated Service Management for Intelligent Networks," XIII International Switching Symposium, Stockholm, Sweden, May 27–June 1, 1990.

[9.40] M. Yoshimi, T. Omiya, S. Esaki, "Service Control Concepts and Architecture for Service Execution Environment of Intelligent Network," XIII International Switching Symposium, Stockholm, Sweden, May 27–June 1, 1990.

[9.41] L. Davidson, E. Valentine, "Intelligent Networks Implementing Advanced Concepts in an Existing Switching System," XIII International Switching Symposium, Stockholm, Sweden, May 27–June 1, 1990.

[9.42] P. Bloom, P. Miller, "Intelligent Network/2," *Telecommunications*, January 1987.

[9.43] W.D. Ambrosch, A. Maher, B. Sasscer, *The Intelligent Network—A Joint Study by Bell Atlantic, IBM, and Siemens*, Springer-Verlag, New York, 1989.

[9.44] M. Torabi, AT&T Bell Labs, Holmdel, NJ, personal communication, May 1990.

Chapter 10
Private Branch Exchanges: Technology and Networking

10.1 OVERVIEW

Private branch exchanges (PBXs) are stored program control CPE that provide a switching function. PBXs allow the user to switch on-net calls at a given location without requiring the services of the CO. Modern PBXs provide many useful ancillary functions, such as camp-on, distinctive ringing, and attendant station. In addition to voice applications, PBXs have also been used for data transmission. Although, in practice, low speeds up to 19.2 kb/s have been typical for traditional systems, 64 kb/s for ISDN-configured systems should become routine in the 1991–92 time frame; higher speeds may also become more widespread. Signaling to the network remains a major issue because, until ISDN (Q.931) becomes widely available, PBXs must principally rely on "stimulus" signaling rather than on "functional signaling."

PBXs have gone through four generations, with the fourth generation being the richest in terms of features, applications, and interfaces. Many of the modern features found on a PBX are a result of the computerized capabilities of SPC, originally developed for CO switches, as well as of the switching fabric. The PBX can be analog (switching through a mechanical matrix) or digital (switching through TDM techniques); contemporary technology is exclusively digital. PBXs offer the customer flexibility and control and a flexible numbering plan for intrapremises communication. Features easily obtainable with PBXs include restriction of stations from dialing specified numbers (lines that provide the scores of sporting events, weather information *et cetera*); routing of outgoing calls to use the least expensive service available; and generation of detailed management billing and traffic reports.

PBXs are similar to telephone switches, except that they do not include many of the operational and administrative functions. PBXs often omit line protection and redundancy found in CO switches. PBXs are generally smaller than a CO switch in terms of served stations. PBXs do terminate stations, but the loops are much shorter so many of the loop management functions found in CO local switches are omitted. On the other hand, local switches and tandem switches have substantial trunk interface and signaling requirements; PBXs are involved with a smaller number (both in terms of quantity and variety).

This chapter explores a number of important technology and networking issues affecting the PBX environment.

10.2 FOUR GENERATIONS OF PBXs

As indicated, four generations of PBXs have evolved in the past 30 years; some researchers are already postulating a fifth generation. Rapid evolution has resulted in PBXs that have numerous features and capabilities. At one time, PBXs handled primarily voice; today they permit voice and data integration within a building environment. Applications span office automation equipment, dispersed PCs, and even data processing applications, as long as the bandwidth requirement is small and sessions are short. The early PBXs were scaled-down CO switches. The PBXs of the 1920s established paths between callers through a central matrix of operator-run "cordboards." All PBX lines and trunks terminated in switchboard jacks; the PBX attendant connected two stations or a station and a trunk manually with a jumper cord circuit. The attendant monitored lamp signals to recognize requests for connections. The next improvement was the dial PBX, commercialized in the early 1930s, which automated intercom calling and outgoing call placement using step-by-step technology. The most well-known models were the 701 and 740 types. During the 1950s and 1960s, step-by-step PBXs were enhanced to provide new features. In the 1950s, new PBXs based on crossbar technology started to appear. These systems used consoles rather than switchboards. Enhancement of this technology continued into the early 1970s. The more well-known models were the 756 (handling up to 60 lines); the 757 (up to 200 lines); and the 770 (up to 400 lines) [10.1].

Second-generation systems emerged in the 1970s and started to incorporate software (e.g., SPC) that switched calls automatically, thus replacing operators and electromechanical switches. These systems also began carrying some data traffic in addition to the voice traffic. Fully digital PBXs appeared in 1975. These second-generation systems were engineered to handle relatively low levels of traffic, and the data capabilities were very primitive. Data were handled through add-on techniques such as port doubling, submultiplexing, alternative-use ports, and dedicated data ports. Since the early 1980s, third-generation PBXs have carried both voice and data transmissions without blocking the voice calls (second-generation systems gave callers high numbers of busy signals because of the high holding time of the data calls). The third-generation systems incorporated integrated voice and data as part of the original design, rather than as an afterthought as in the second-generation systems.

Direct digital switching is a capability that has become available with the third-generation. Second-generation systems used modems to translate digital bit streams from terminals into analog waves. The new systems reverse this process by digitizing analog voice signals so that they can interchangeably travel with computer data along the PBX's digital bus. Third-generation and fourth-generation PBXs can also exploit economies made available by VLSI components, thus permitting more PBX intelligence at a lower cost. The third-generation was sold on the basis of money-saving voice features such as call accounting (for business cost control) and automatic route selection (to exploit less expensive long-distance services).

Fourth-generation systems may handle data communication in a more efficient way. "Dynamic bandwidth allocation" is an innovation of fourth-generation PBXs that can increase data-handling efficiency. With this method, a data device that must send a file can request and receive as much bandwidth as is needed from the PBX's bus, up to its maximum capacity. The ability to alter the available bandwidth means the PBX can adapt to fluctuating needs. A user who occasionally sends files can do so without requesting a permanent system reconfiguration, thereby conserving PBX capacity. Anticipating increased volumes of data transmissions, some PBX vendors have integrated LANs into their architectures. While most third-generation PBXs can transmit at 56 to 64 kb/s, cable-based LANs carry data at speeds that range from 5 to 50 Mb/s. Fourth-generation PBXs use various methods that attempt to bridge these differences. A change in the way offices use data and information has motivated this trend: A few years ago, most computer applications involved CRT-based transaction processing applications, which required relatively slow speeds and small bandwidth. Transaction processing requires direct user interaction with the system; a speed of 64 kb/s can easily keep pace with a person typing or reading information on a screen; however, this speed may not be adequate for today's file transfers with PC-based systems and graphics applications. Currently, fourth-generation PBXs are in the development phase; the technology has proven difficult to implement and deliver reliably.

Table 10.1 is a synopsis of the different generations of PBX equipment. Table 10.2 depicts some key systems for the United States at the time of this writing.

Three techniques provide integrated voice-data transmission through a PBX:

1. Integration using data over voice (DOV) methods that apply a data signal over a voice circuit. The circuit typically achieves a speed of 19.2 kb/s. Separate modems are needed at each end of the circuit.
2. Integration using separate voice and data pairs. Each of the pairs requires two or four wires to the station. Separate ports for voice and data are required at the PBX. This approach is now being replaced with the one discussed

Table 10.1
Five Generations of PBX Equipment

- First-generation PBX systems: hardwired electromechanically switched. (1930–1950s).
- Second-generation systems: computer-controlled, programmable switches that offered substantially more features, but limited capacity due to the blocking nature of the devices. Voice-data integration on second-generation products is generally the result of capability that has been added to the original design. (1960s and 1970s).
- Third-generation devices: nonblocking architecture capable of creating a completely digital system with digital telephones to convert vocal conversation from analog to digital at the source; third-generation systems integrate voice and data by design, not as an add-on capability. They also employ a distributed architecture in which modules offload some of the work of the central processing unit. (1975–1983).
- Fourth-generation products integrate cable systems and LANs into the PBX and thus take advantage of both LAN and PBX technologies. (1983–present). This author suggests that a fourth-generation PBX is an ISDN-configured PBX.
- Fifth-generation: uses radio connectivity from a "micro cellular system" that can reach a user anywhere in a building. Will also handle data. Another view: an ISDN-ready PBX.

Table 10.2
Major PBX systems available in U.S. (Partial List)

PBX Systems	*Maximum Capacity (Lines)*
Northern Telecom Meridian 1	60,000
AT&T Definity	32,000
Intecom IBX	32,000
Ericsson MD110	20,000
NEC NEAX 2400	20,000
Rolm 9751 (Release 9005)	20,000
Fijitsu F9600	9,600
Hitachi HCX5000	3,000
Mitel SX-2000 (MS2004-01)	2,500
Siemens Saturn	800

next. Note that both the DOV and the separate pair are compatible with standard telephone sets.

3. Integration is digitized voice and data over a common transmission envelope on a single copper pair using echo cancellation or time compression techniques.

As discussed above, in second-generation PBX systems, all signals traveled to the PBX in analog form and were then converted to digital format for switching; the resulting signal was then converted back to analog. In many applications, the existing telephone wire could be used for both data and voice support. The early third-generation systems evolved on this basis; they actually were enhanced second-generation systems that had been designed to support traditional telephone sets and to convert the analog voice signal at a line card interface codec using PCM. Space division implies the use of two physically separate loops: one for voice and one for data. This also requires two ports at the PBX. Voice is carried in an analog mode over its pair; data use FDM on its pair. Some PBXs actually require three pairs (but they still use two PBX slots): The data link has one pair for the transmitting side and one pair for the receiving side. Time division requires two pairs (but only one port). The voice and data signals for the transmitting side are combined and digitized. The resulting bit stream is contained in a digital frame of various lengths. A similar process occurs for the receiving side, thus implying two pairs. Because of this predicament, these third-generation systems required an additional series of components to support true end-to-end data communication. The flexibility that had been expected did not generally evolve because of the additional components and extra wire pairs required by some systems, as well as load-balancing problems and switch capacity constraints. Therefore, these systems came with a high data communication support price tag, and have limitations (protocol differences, data rate limitations, high per-port costs). Most of these systems work on an end-to-end digital basis and either do not support traditional telephone sets, or support them only on a limited basis [10.2].

Newer PBXs support both data and voice streams from the station to the PBX using time-compression multiplexing or echo cancellation. Time-compression multiplexing and echo cancellation require only one pair (and one PBX port) by combining the voice and data and digitizing it to fit a data frame that can handle both the transmitting and receiving sides on a single pair. The goal is to achieve duplex communication over the twisted-pair medium. Duplex communication is preferable for data communication to the integrated workstation, and two wires are more economical than four in terms of material and labor. In time division, bursts of data are interleaved in time, alternating transmission in one direction and then the next. Full-duplex operation on two wires is sometimes called "ping-pong" because the alternate ends of the loop transmit bursts of digital signals at twice the nominal (64 kb/s) transfer speed, in rapid succession. The echo cancellation scheme requires that the transmitting and receiving stations at each end of the two-wire loop be connected by means of a hybrid transformer; each end of the loop has an echo canceler with an adaptive digital filter, which ensures that the sum of the filter's output and the received signal is completely uncorrelated with the transmitted signal. (These concepts were also discussed in Chapter 4.)

Studies indicate that data usage on PBX systems is relatively small. If network managers were inclined to elect a digital voice-data PBX, they should consider (1) the relatively higher cost of the integrated PBX *versus* a voice-only PBX and (2) the fact that to handle data the digital PBX must be equipped with extra data line cards. Regardless of the choice, modems generally are still required to interface the data channels with the outside world because telephone lines are still analog. This problem will be resolved with the deployment of ISDN. Vendors are now starting to develop PBXs that will handle ISDN interfaces, as discussed in more detail below.

10.2.1 Building Wiring

Twisted-pair wire costs $30 to $40 per kilofoot, while coaxial cable can range from $300 to $1000 per kilofoot for Teflon-coated material. Fiber cable costs approximately $300 per kilofoot, but the tapping connectors can be rather costly, and the tapping problem for fiber dictates a star configuration rather than a ring configuration, increasing the cable length requirement. Clearly, twisted-pair LANs would seem an economical way to move small quantities of data, rather than integrated voice and data PBX. In addition, twisted pair is consistent with voice usage.

10.2.2 CENTREX

CENTREX is the name of a set of communication services designed to provide low-cost telephone service for medium- to large-scale business customers. The service originates in the telephone company's CO rather than in an on-site system such as a PBX; hence, the origin of the name "central exchange." CENTREX service is experiencing a technical renaissance, and now appears to be a viable alternative to PBXs, particularly in conjunction with ISDN.

The early history of CENTREX is somewhat confused. The term was first applied to centralized PBX equipment, rather than to a service. Originally, a CENTREX-CO existed in the central office and a CENTREX-CU existed on the customer's premises. One of the original concepts was to eliminate the PBX churn, i.e., the short life of PBXs on customer premises, by providing a large centralized switch [10.1, 10.3].

After World War II, a substantial push occurred to provide dial service to businesses, particularly customers with large numbers of phones. CENTREX was designed to eliminate the use of operators to process calls, and large PBX users having many operators eagerly accepted the idea. Traditional CENTREX had several basic operating features, including direct inward dialing (DID), direct outward dialing (DOD), station-to-station dialing, automatic identified outward dialing (AIOD), and attendant console. Digital CENTREX offers nearly all of the main features that can be provided by a PBX, including automatic route selection,

INTERNETWORKING PBXs

corporate users that have office facilities in several buildings within a city, several cities statewide or nationwide, have at times found useful the creation private network of PBXs (referred to hereafter as "PBX networks") to interconnect each site into a cohesive closed user group network (see Figure 10.1). PBXs also have remote switching modules endowed with some, but not all, PBX capabilities. A number of interconnection techniques exist. The choice cted by distance and cost (in addition to internodal traffic); in turn, the choice the type of internode service and grade of service that the network user es. Interconnection can be direct or via tandem intermediary nodes 10.6]. This section explores some of the issues associated with these networks. The approach of employing telephone company-based facilities is more comhan the approach of utilizing totally private transmission facilities. For this

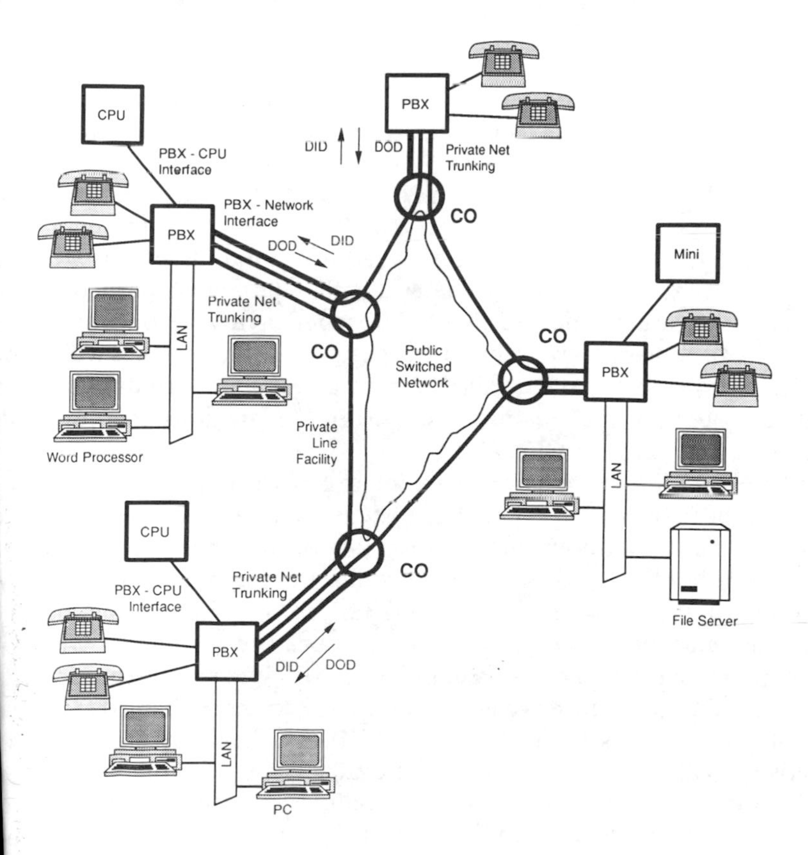

A typical PBX network.

call transfer, and speed dialing. The service is ideal for organi
number of business locations in one urban area, such as multib
chains with many stores in one city, city governments, *et ceter*

CENTREX provides service to a total of approximately
is 10% of all business telephones in the United States. New PE
the preoccupations of divestiture resulted in a loss of custom
1980s. However, since divestiture, some BOCs have been en
to regain their customer base. One enhancement allows the
their own rearrangements via CPE terminals; another allo
transmission of voice and data over the same line. LAN
explored (e.g., ISDN, CO LANs). The number of CENT
growing at a rate of 5% annually (300,000 lines). This com
of about 4 million station lines on PBX systems (in 1987) del
suppliers (which hold 80% of the market).

In some states, regulatory constraints are being rem
vary from region to region but generally include adjustmen
based on end-user line size, the distance from the CO, t
contract, and the optional features selected. These and oth
resulted in an increase of customers over the past two yea
competitive with PBXs.

CENTREX service is generally provided through
AT&T's No. 1, 1A, 5ESS, and NTI's DMS switches. A
more than 50% of all CENTREX lines in the United Stat
analog switches. As BOCs upgrade switching equipment
ISDN, more features and functions will become availabl
CENTREX services. New architectural designs for the sw
interface with computers that can interrogate special serv
intelligent network architecture discussed in the previo
provide new impetus to CENTREX.

CENTREX service advantages for a general tele
outweigh those of customer-premises systems. Any techn
hardware and software are the responsibility of the loca
work organization. Updates are done without any capit
user. The system provides nonblocking service, requir
data communication user (BOC staff maintains the eq
uses minimal space on the customer premises, receives u
from the BOC, and can serve users with from 2 to more
or multiple locations ("city-wide CENTREX"). In m
effectively make their own moves and changes via a c
or, in some cases, via an end-user PC connected wi
computer system, which then downloads this informa
into the CO switch.

10.3

Larg
or in
of a
terco
Some
of th
is aff
affect
receiv
[10.5,

mon t

Figure 10.

reason, it is important to understand what telephone company signaling is in the context of PBXs, how the network deals with it, and how this impacts PBX network users. This section applies some of the signaling techniques discussed in the previous chapter.

10.3.1 Topologies

If the distance between nodes is very short (typically 5000 to 50,000 ft) or if the designer is willing to pay the cost, the interconnection may consist of a dedicated umbilical fiber link. In this case, the users of the PBX network will probably enjoy full service capabilities regardless of the location because the PBX can transact the full repertoire of necessary signaling information across the large unconstrained digital bandwidth, employing a pertinent hardware-dependent scheme. If the distance is long or the user is unwilling to pay the cost of specially dedicated facilities, then the user will have to employ carrier-provided trunks. Three approaches are available: (1) use of aggregated T1/DS1 links, (2) use of individual two- or four-wire voice trunks, or (3) use of the switched public network (particularly when the internodal traffic is low). The traditional telephone company network (whether for dedicated lines or over switched facilities) has not carried end-user to end-user signaling beyond basic alerting; this, in turn, limits the type of functionality and services derivable by remote users of the private PBX network. User-to-user signaling under ISDN should remedy this situation, as discussed later.

Situations exist, particularly in the case of a large number of PBX nodes, in which the direct networking of these nodes with point-to-point inter-PBX (tie) links becomes impractical. In this case, the designer finds it useful to deploy one or more tandem switches, to which the PBXs are in turn connected. When the number of network nodes is small, nodes are interconnected with direct point-to-point trunks; as the number of nodes increases, the number of links and the complexity of the arrangement increase quadratically. To keep costs down, a hierarchical topology is often employed. When a user needs to communicate intra-PBX, only the switching services of the PBX node (the first level) are needed. When the user needs to communicate inter-PBX, then the switching services of the first level switch plus the services of the second level (tandem) switch are necessary. In this scenario, end-node PBXs connect directly to end-user stations, and they usually switch traffic between their own users internally. Internode traffic is routed up the switching hierarchy to a tandem node that services internode traffic. See Figure 10.2(b). If the designer were faced with such a situation, signaling techniques that closely resemble or are identical to the signaling methods employed in the telephone company plant would be necessary. In the most sophisticated case, the designer may opt to employ a private common channel signaling system.

Full interconnection, which provides point-to-point links between each node and every other node, requires approximately $N^2/2$ internode links, given N PBXs

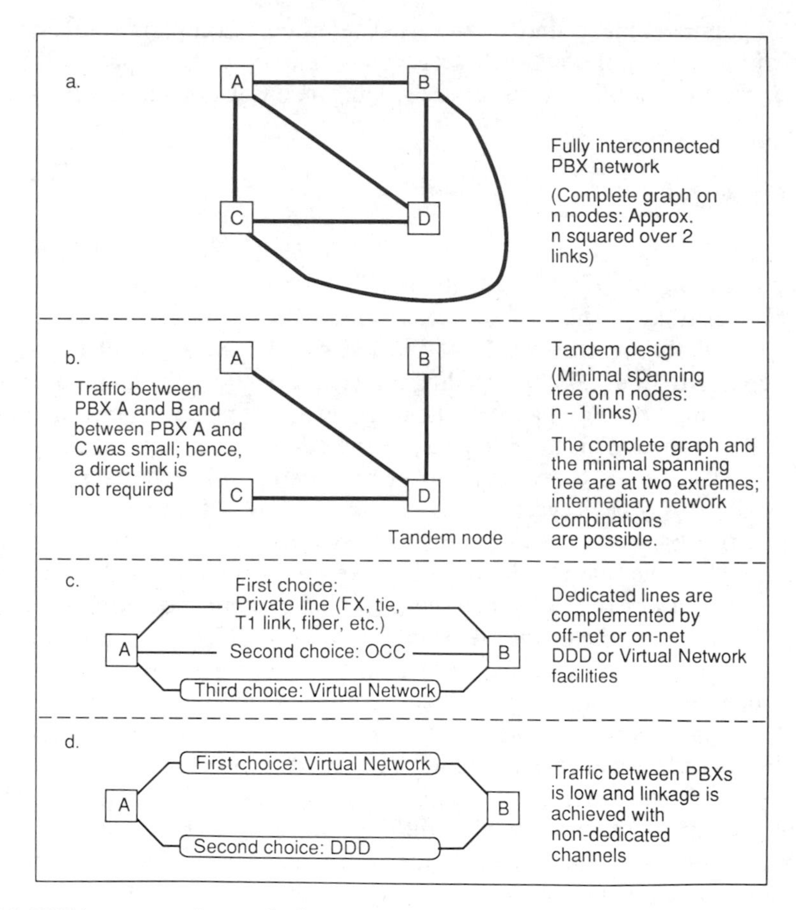

Figure 10.2 PBX interconnection methods.

in the network. This approach would be followed when the nodal traffic matrix is fairly balanced, the individual matrix entries are not trivial (namely, they involve at least several erlangs of traffic at the busy hour), the number of PBXs is small, and the airline radius of the network is not excessive (say, up to 100 miles). Full interconnection becomes expensive and complex for networks with many nodes (generally more than six) because of the large number of links and the associated cost.

Partial interconnection, which provides less than the full number of point-to-point links between nodes, is feasible when the amount and distribution of traffic can be characterized fairly accurately in advance. Nodes that exchange a large amount of traffic are connected directly to one another, while the traffic between nodes with low point-to-point requirements is routed through intermediary (tandem) nodes. For both on-net and off-net applications, a dedicated path is not always necessary; a low traffic node can connect to the rest of the system via a dial-up arrangement or virtual network service, as shown in Figure 10.2(d). Over the past couple of years, effort have been aimed at building standardized T1 interfaces for computer connections, trunking, or tie line purposes.

Drawbacks of tandem networks include long dial sequences with intermediary dial tones and, in the absence of digital trunking, accumulated distortion due to transmission limitations in subsequent stages. Out-of-band signalization of recent PBXs and T1 interfaces, as well as ISDN signaling, should facilitate tandem design.

Several of the more recently designed PBXs utilize a fully distributed architecture [Figure 10.2(a) and other intermediary graphs between the "complete graph" and the "spanning tree" graph]. In this case, each node is a stand-alone entity, at an equal level in the hierarchy with all other nodes. Distributed nodes may be fully or partially interconnected [10.5–10.7].

Many PBXs can be equipped to serve as tandem switches for "pass-through" purposes. Large customers may even find that a telephone company-grade tandem switch (such as AT&T's 4ESS or NTI's DMS 200) is required. Tandem switches are generally four-wire switches that can deal with trunks (four-wire compared to two-wire subscriber lines). The digital switches can interface directly with DS1 lines without requiring channel banks to convert the signal into analog conversations. Private voice application tandem switches afford the same type of functionality as telephone company machines, while at the same time providing a number of features unique to the private user. In particular, private tandem switches aim at maximizing the efficiency of trunk management, for cost control reasons. In large sophisticated networks employing tandem machines, several other advanced cost control and security features are incorporated as a service to the end-user, including least cost routing (LCR), queueing, and code blocking (denying access to 555-1212, for example). Tandems are invoked by dialing an access code in a PBX application where the PBX is directly trunked to the tandem switch. Transmission issues have to be considered when designing a PBX network with tandems, particularly for remote access (where an off-premises user can dial a local telephone company number to access the tandem in order to use the private network to make a long-distance call). High-attenuation problems may be encountered.

CCS is the rule for public toll tandem applications, as discussed in Chapter 9, and is increasingly being employed in private systems. In a multitandem network, CCS should be considered by the designer to achieve more cost-effective circuit utilization. The added cost of the CCS equipment and the separate data network should, however, be considered against in-band signaling to derive the truly optimal solution. (The data network need not necessarily be a physically separate network; it could employ a slot in a DS1 facility which is then removed from the trunk pool.)

10.3.2 Traditional PBX Signaling Summary

The network aspect of signaling is of interest to a PBX user because (1) PBXs may have to coexist or interact with that signaling system, particularly under ISDN; (2) a private network is a microcosm of a larger network: The methods and techniques employed on a large scale in the latter can apply to the former on a reduced scale, particularly in a case involving private tandem nodes; and (3) advanced services, either in the public network or via the PBX (or, in conjunction), are possible only with a sophisticated signaling system. In addition to the network aspects of signaling, issues pertaining strictly to PBX networking deserve attention because these will ultimately determine how cohesive and transparent the PBX network will look to the corporate users; PBX tie trunks and large private switched voice networks require essentially all the signaling capabilities of the telephone network.

PBXs interface to remote entities through a variety of signaling interfaces, depending on what the remote entity is. All switching systems are equipped with circuits to interconnect the system with stations, trunks, and other service-maintenance-telemetry circuits. PBX applications are considered special services and employ the ground-start format for CO trunks and automatic or ringdown signaling for nondial tie trunks. E&M is employed for PBX dial tie trunks or carrier systems. The M-lead is for the sending of signals (an analogy with the mouth), and the E-lead is for receiving signals (an analogy with the ear); a caller's M-lead communicates with the called party's E-lead and *vice versa*. Dial-repeating trunks are used with those PBXs that are capable of routing tie trunk calls without attendant intervention, but require dial-pulse address information. The E&M-lead interface for tie trunks is virtually identical to that used in interoffice trunks and is able to pass dial pulse of DTMF addressing information.

Dial pulses or DTMF tones are used for addressing between PBXs and their serving offices. The CO furnishes dial tone to the PBX when it is ready to accept addressing information, coded in dial pulses or DTMF. Traditionally, PBXs have been designed to terminate on the line side of a CO switch; therefore, they must use line signaling methods (loop or ground-start signaling with dial or DTMF pulses). This is beginning to change with digital interfaces and ISDN in particular. (For comparison, note that CO switches connect to each other on the trunk side, and so can employ trunk-signaling or, if equipped, CCS.)

Most PBXs connect to the CO for off-net access over trunks equipped with two-way signaling. This can lead to glare, particularly with loop-start signaling. In this type of simple signaling, the only indication that the PBX has regarding an incoming call is the ringing signal, with a pulse every six seconds (20 Hz). During those first six seconds (from the time the CO switch seizes the trunk and the first pulse), the PBX is oblivious to the possibility of an incoming trunk seizure, and may itself grab the same circuit for outgoing traffic [10.8].

In addition to segregating the incoming trunks (DID trunks) from the outgoing trunks (DOD trunks), we can use ground-start signaling. In this mode, the CO switch grounds the tip side of the line immediately upon seizure by an incoming call. By detecting the tip ground, the PBX is immediately able to recognize line seizure before the ringing pulse is actually received. When a local user of the PBX dials an access code (typically a digit 9), the PBX will seize an idle DOD trunk and pass the station's addressing information to the CO.

Signaling capabilities and compatibilities are important factors to consider when purchasing a PBX. Compatibility with CO line equipment is important to maximize the functionality of the PBX for off-net applications; the equipment available at the CO may be a function of the CO and may not always be current digital technology, though the conversion to total digital switching is expected as early as 1990 for some BOCs, and in the early 1990s for all BOCs. ISDN access will greatly streamline PBX-CO and PBX-PBX signaling.

10.3.3 Details of Traditional PBX Trunk Signaling

In this section, we take a closer look at traditional PBX signaling methods.

10.3.3.1 PBX-CO Signaling

CO-PBX Trunks with Loop-Start Signaling. In this configuration, the tip lead at the CO as well as at the PBX has a 0-V potential in an idle state; the ring lead has a −48-V potential at the CO and at the PBX. When the PBX wishes to place an outgoing call, the transmitting-receiving circuit at the PBX (more specifically at the PBX trunk interface card) is closed by inserting a resistor. The current flow is detected by the CO, signaling the switch for a request for dial tone. On the reverse path, the CO signals the PBX of an incoming call by modulating (adding) to the ring lead voltage, normally at −48 V dc, an ac component alternating at 20 Hz. Thus, the ring lead presents to the PBX a voltage that varies from −98 to +2 V; this alternation is interpreted by the PBX's ring detector as an incoming call, which is acknowledged by the PBX by attaching a 600-Ω load to the transmitting-receiving circuit, in turn completing the process by generating a current flow to the CO switch.

CO-PBX Trunks with Ground-Start Signaling. In this configuration, the ring lead has −48 V at the PBX because it is connected to a battery at the CO; the tip lead has −48 V on it, applied by the PBX. In an idle state, no current flows because the circuit is open in two places. The tip lead is open at the CO and the ring lead is open at the PBX. The PBX signals the request of a dial tone by momentarily grounding the ring side of the line (namely, changing the −48 V normally present to 0 V). When the PBX grounds the ring lead, the change in potential is detected by the CO switch. The switch will, in turn, signal the PBX by grounding the tip lead with a 0-V potential that can be detected by the PBX. After the ground start has been thus signaled to the PBX, the PBX will remove the ground previously applied to the ring lead and it will close its side of the ring lead circuit. The CO will also remove the ground and close the loop so that current can flow. At this stage, the PBX will be able to send dial pulses. For an incoming call, the tip lead from the CO is connected to ground, while the ring lead receives the 100-V ac ring voltage, again modulated on its −48-V potential [10.9].

DID CO to PBX Signaling. DID trunks are employed to reduce the number of channels between the PBX and the CO; they are one-way trunks. A PBX perceives the DID trunk as one of its single-line phones and can interpret four-digit dialing. Outside callers dialing a station on a DID-configured PBX inform the CO switch to signal the PBX for service. An available DID trunk is activated by the CO by causing a current to flow; the PBX detects the current flow but, rather than responding with a dial tone (which it would have done if it really considered this loop to be a true station-to-PBX loop), it gives the CO a wink-start signal. The wink informs the CO that the PBX has been able to allocate resources (notably, registers) to receive the incoming addressing information. The wink signals the CO switch to repeat the last four digits dialed by the distant user, from which the PBX can complete the connection. (The wink is a quarter-of-a-second reversal of the polarity of the current that the CO had previously caused to flow. The CO had placed a load across the PBX line, and the PBX forced a current through it, as if the loop was a station line with a phone on it which had gone off-hook. At the end of the 250 ms, the polarity is restored to its initial state). After the pulses have been received by the PBX and the desired station has answered the call, the PBX must signal the CO to initiate billing; the PBX does this by reversing the polarity of the current as it did for the wink-start signal, but at this juncture it will leave the polarity reversed for the entire duration of the call. When the call is finally disconnected, the DID facility returns to its idle state.

10.3.3.2 PBX-to-PBX Tie Trunk Signaling

Signaling over traditional PBX-to-PBX tie trunks is generally of the E&M type. In this configuration, two "wires" are used for signaling only, as well as two or four "wires" for the voice path. The two signaling leads, the E-lead and the M-lead, are not carried end-to-end as separate transmission facilities, but are multiplexed at the E&M interface card with the voice path; logically, they can be considered as independent of the voice. The E&M interface converts the voice and the signaling information into a stream that is compatible with the transmission medium.

When the caller's PBX puts a request signal on the M-lead for a dial tone from the remote PBX, the remote PBX will detect the request on its E-lead and, in response, will provide battery (-48 V) on its M-lead. The remote E&M interface card will appropriately encode this event so that the receiving E&M interface can render this -48-V signal to the local PBX, on the E-lead. The caller's PBX thus receives an acknowledgment that the remote PBX is able to receive addressing information (this is similar to the wink on DID trunks discussed above). In turn, an auditory dial tone will be returned to the end-user on the voice path to initiate entry of addressing information. Because the E&M signaling is essentially dichotomous, it cannot handle DTMF addressing over the E&M-leads; hence, while rotary dial pulses are passed over the E&M-leads, DTMF addressing digits are transmitted directly in the voice band.

Three common signaling methods used by various E&M systems are [10.9]: (1) immediate-start E&M signaling, (2) wink-start E&M signaling, and (3) delay-dial E&M signaling.

Immediate-Start E&M Signaling. In this mode, the caller's E-lead is at -48 V in an idle state and the M-lead is grounded. The same is true for the called party's leads. A request for service is indicated by changing the state of the M-lead from 0 to -48 V. The called party's E-lead would register a change in its state from -48 to 0 V. The far end is thus responding to the request for service with this change on the E-lead; at the same time, it provides a dial tone to the calling party over the voice path. The addressing information pulses are placed on the M-lead of the sender and are received as inverted pulses on the called party's E-lead. The remote station is signaled with a ringing voltage, and an auditory copy of the ring is sent back to the originating PBX over the voice path. When the destination answers the call, the remote PBX connects the voice path to the remote station. The remote PBX informs the local PBX of what has occurred by changing the state of its M-lead from 0 V to -48 V; this condition is maintained for the duration of the call. When either party hangs up, the closest M-lead is brought back to the idle state of 0 V grounded. The other end will receive this notification on its E-lead and will release all resources associated with the call, as did the first PBX.

Wink-Start E&M Signaling. This is similar to wink start on DID trunks, but does not involve the reversal of the loop current; instead, a wink (momentary change in the status of the signal) is applied to the appropriate lead (E- or M-lead, depending on the direction). The wink occurs when the remote PBX is ready to receive the addressing digits; the local PBX can start accepting and transferring the addressing information from the caller after the wink is over.

Delay-Dial E&M Signaling. In delay-dial signaling, the status of the caller's E-lead is changed immediately when the called PBX recognizes a request for service and is changed back to its normal state as soon as a register is found to accept the addressing information. As soon as the E-lead returns to the idle state, the caller can start entering the addressing information. In the wink-start mode, dialing can begin only after the wink is over; a register may be found sooner than the 250 ms required to complete the wink, which, in turn, expedites the call setup process [10.9].

10.3.3.3 Signaling on T1 Links

Most modern digital PBXs are capable of a direct DS1 interface, and ISDN PRIs are becoming available. If available and cost-effective from a transmission point of view, T1 facilities and associated signaling will be superior to analog facilities with ground- or loop-start signaling. Digital signaling methods and related equipment are relatively simple and inexpensive. As discussed in Chapter 3, VF-level signaling is achieved by literally robbing some real bits every 6th and 12th frame of each DS0 channel. This is also called DS1 A,B signaling.

Interest in T1 signaling in a PBX environment arises for the following reasons:

- T1-based digital loop carrier systems servicing the feeder part of the loop (and sometimes right up to the customer building) have become quite prevalent. Until the general availability of ISDN, the digital feeder plant has to be interfaced to an analog distribution plant and existing station equipment that requires traditional electrical currents to carry out their signaling.
- Even prior to the establishment of DS1-rate PBX interfaces, a large PBX network user might have used a "T1" line.
- With the establishment of DS1-rate PBX interfaces (traditional D4 interfaces, as well as vendor interfaces such as AT&T's Digital Multiplex Interface—DMI), the signaling aspects of T1 have to be considered. A number of PBXs now have T1/DS1 interfaces that can be used to interconnect two PBXs in a network [10.10]. These interfaces are a migration step toward the ISDN primary rate PBX interface of the early 1990s.

During the 1980s, various proprietary (digital) networking techniques were developed for inter-PBX signaling by manufacturers of large PBXs. Most vendors of medium to large PBXs now offer digital networking capabilities. This approach employs DS1 lines to carry on-net voice, data, and signaling between the PBX nodes in a network. The first generation of DS1 connections to PBXs performed the analog functions of a D3 channel bank, thus providing tie line connections with 24 digital PCM channels. As was the case with the analog tie lines that the DS1 link replaced, these DS1 interfaces used on-hook and off-hook signaling bits and in-band tones in each channel to establish connections and pass numbering information between networked PBXs. The need to implement a uniform set of PBX features across the network soon stretched the limits of in-band signaling to deliver the more complex information between network nodes. Most current DS1 networking options use some type of CCS implementation to carry digital messages between PBX nodes; this occurs either over a dedicated separate data connection or one that is shared with voice and data channels. Messages needed for call processing and management of a group of connected channels are generated by the software of each networked PBX [10.11].

The majority of DS1-based inter-PBX common-channel systems in use at the beginning of 1990 were proprietary. In addition, these proprietary schemes are closed because no published specifications exist; therefore, users, network integrators, and other PBX vendors are not able to develop common internetworking solutions. AT&T's, NTI's, and NEC's schemes are the most widely deployed today. In spite of the closed nature, the vendor's CCS for DS1 internetworking offers a fairly high degree of feature transparency between networked PBX nodes. This allows PBX users to have the same calling and data features across a network of PBXs as they have available at their local PBX. To internetwork two PBXs from two different vendors, a consistent protocol at each of the seven OSIRM layers would be needed, plus a standardized language above layer 7 to manage the remote PBX resources. We need a consistent layer 1 (physical), layer 2 (data link control), and layer 3 (packetization) set of protocols, plus upper layer consistency for end-to-end integrity, session, character set representation, and application-support functions. In addition, as indicated, we need to standardize the functional procedures (for example, how to complete the transfer of a call, how to connect to the central attendant, how to obtain remote traffic data, *et cetera*).

10.3.4 ISDN and Its Effect on User-to-Network Signaling

In ISDN, a separate user-to-network signaling channel is provided for both the basic (2B + D) and the primary (23B + D) interfaces, as discussed in Chapter 4. In fact, ISDN may find its first commercial success in applications between the PBX and the network (the PBX may, if desired, continue to use pre-ISDN methods to the stations on the PBX). The D channel for out-of-band signaling employs a three-layer protocol stack for user-to-network signaling. This allows functional signaling, where messages are exchanged between the CPE and the network [10.12, 10.13]. Note that the signaling path terminates at the exchange terminator (ET) at the CO; the "protocol partner" of the user is at the ET. The ET will remap the signaling requests over the SS7 network (for both user-to-network and user-to-user signaling).

In addition to PBX-to-network functions, ISDN also provides methods of transferring end-to-end user-to-user signaling using the D channel in conjunction with SS7 [10.14]. Fields are now provided in the setup message that allow a certain amount of user signaling bytes to be carried end-to-end during the call setup phase. Clearly, this can be beneficial to a PBX network arrangement over the public switched network (for example, in a software-defined network) to pass administrative information. An application might be to display the extension of the calling party from the distant PBX; this can already be done now if a proprietary umbilical link is employed between some of the more well-known PBXs, but not across a normal link facility such as a T1 trunk or an analog trunk. User-to-user signaling (UUS) with call control (UUSCC) is a service that allows user equipment to exchange user information in the call request phase, call confirmation phase, or the call clearing phase of a call. This feature is applicable for point-to-point, on-demand, circuit switched, and packet-mode calls. In all these phases, the UUS is transmitted by inclusion of the information in the appropriate end-to-end call control message [10.14]. During any of the three call phases, the calling party may send an integral number of octets (up to 128) of user information to the called party. For a circuit switched call request, the calling party requests UUS for the call by sending to the switch a setup message that contains the user-user information element (the setup message was described in Chapter 4). The user-user information element may, but is not required to, contain user information. If the UUS request is accepted by the switch and the circuit switched call is interswitch, an SS7 user-to-user information (UUI) parameter, with any user information sent by the calling party, is contained in the outgoing SS7 initial address message. Figure 10.3 depicts the format of the setup message to request UUS for a circuit switched call. Five other ISDN messages that can be used in UUS are connect, disconnect, release, release complete, and progress.

ISDN signaling, as applied to the PBX-to-PBX environment, provides the first three layers of the required protocol. The upper layers and the messaging remain to be defined. CCSS7 defines upper layers that are primarily geared to carrier applications and are not totally conformant with the OSIRM. TCAP defines message formats, but again is geared to the public network. Q.931 covers the protocols for basic call setup and supervision over a primary rate ISDN interface and is used for PBX-to-CO signaling. This specification, however, does not properly cover voice and data signaling capabilities that are required to achieve sophisticated PBX networks. Examples of needed supplementary services are call transfer mechanisms, call waiting, call forwarding, camp-on to extensions, centralized attendant services, data services, *et cetera*. The definition of these and other services will eventually be included in Q.932 ("Generic Procedures for the Control of ISDN Supplementary Services").

ECMA Committee TC32/TG6 and ANSI/ECSA T1S1 were working on supplementary services at the time of this writing. The ECMA work was more mature at that time than that of T1S1, and may be the key driver of the eventual standard. The two groups investigated different approaches: a carrier-sponsored approach supports networking based on CCSS7; the other, a manufacturer-sponsored approach, supports extension of Q.931 [10.11]. The CCSS7 approach would build on the TCAP protocol (discussed in Chapter 9) to define the message set needed to facilitate supplementary services; TCAP already provides 800 and credit card calling services. In particular, a subset of TCAP, a layer 7 protocol, called *remote operations service element* (ROSE), could in theory satisfy the needs for supplementary services in both the inter-PBX and the PBX-to-CO environment.

Direction: user → network

Information Element	Value	Inclusion Conditions
Protocol discriminator	Q.931	Must include
Call reference	Flag = 0, CR value = any valid value not in use	Must include
Message type	SETUP	Must include
Bearer capability	Speech, 3.1 kHz audio, 64 kbps digital or 56 kbps digital	Must include
User-user	Transparent to SPCS; may include user information	C/O (Included if the calling party wishes to request UUS for the call; may also transfer user information; N/A for packet)
Other elements	As defined in Bellcore's TR-TSY-000268, Issue 2	As required in Bellcore's TR-TSY-000268, Issue 2

Part a: Calling party request for UUS in the SETUP message

Direction: network → user

Information Element	Value	Inclusion Conditions
Protocol discriminator	Q.931	Must include
Call reference	Flag = 0, CR value = any value not in use	Must include
Message type	SETUP	Must include
Bearer capability	Speech, 3.1 kHz audio, 64 kbps digital, 56 kbps digital or packet	Must include
Packet layer binary parameters	Fast select: Restriction or No Restriction on response	C (Included for PMD call, mapped from Fast Select facility in Incoming Call packet)
Called party number	Local number in ISDN numbering plan, digits = called party number	Must include
User-user	Transparent to SPCS; may include user information	C(Included if request for UUS received from calling party and passed by SPCS/PHF;may also transfer user information)
Other elements	As defined in Bellcore's TR-TSY-000268, Issue 2	As required in Bellcore's TR-TSY-000268, Issue 2

Part b: SPCS delivering request for UUS in the SETUP message

C = Conditional
O = Optional

Figure 10.3 User-to-user signaling.

Movement of the inter-PBX links to an ISDN primary rate-based implementation would afford the network designer a number of advantages including (1) the ability to use ISDN links if the dedicated DS1 link experiences an outage; (2) the ability to overflow to ISDN trunks if the traffic exceeds the capacity of the in-place network (perhaps if the busy hour is very high, compared to the average, making sizing of the network based on the busy hour inefficient); and (3) the ability to cover thin routes without employing dissimilar technologies.

Critics of the CCSS7 approach claim that CCSS7 was designed for public networks and contains more management and maintenance features than needed in the private PBX network context. A firm resolution will not come until 1992, with PBX product implementation in the 1993–94 time frame.

While waiting for formal standards (which some vendors are claiming is taking too long), in 1990 some manufacturers started to announce bilateral interface specifications (for example, Alcatel and Siemens). Some industry observers argue that these moves may preempt the efforts of the standards-setting bodies by flooding the market with a *de facto* standard. Work by Alcatel and Siemens on defining a protocol, based on the ISDN D channel protocols, started in 1988; by early 1989, enough common areas existed between their proprietary standards (ABC and CorNet) to define a common protocol. The protocol, initially called Inter-PBX Networking Specification (IPNS), specified more than 50 features, including call forwarding, call transfer, and three-way conferencing [10.15].

One of the main *proprietary* protocols available in the early 1990s is the Digital Private Network Signaling System (DPNSS), specified by British Telecom. This standard is widely deployed in the United Kingdom and is supported by vendors such as Mitel, Philips, and Siemens. Originally specified in the mid-1980s, the protocol was more feature-rich at the time of this writing than IPNS. A further delay in specifying an Q.931-based protocol may cement the hold of X.25-based protocols for inter-PBX signaling [10.15].

The ultimate objective of inter-PBX signaling is to allow end-user to end-user signaling at any point during the conversation or session. This is not yet possible under the currently specified ISDN protocols, but this facility may be made available at some point in the future.

10.3.5 Issues Affecting Inter-PBX Signaling

This section provides some product information for PBX interfaces as of January 1, 1990, and explores issues affecting inter-PBX signaling for 1990–92 and beyond. The reasons for providing this information are:

1. To give the reader an idea of the product status, six years after the (rather simple, i.e., OSIRM lower layer) ISDN protocols were codified in the Red Book; this interval may be representative of other standards-based product delivery cycles.
2. The large and medium PBX product families are going to be around for years to come; major investments are tied up with hardware development.
3. In reading this material, in retrospect, we are able to assess how well the vendors delivered on their promises.

In the discussion below, the nature of the physical, link, and network protocols used can be discerned, in addition to a reference to the messaging method used (upper layers and message formats proper). A distinction is also being made between (1) DS1-based inter-PBX links, (2) ISDN PRI-based inter-PBX links, and (3) PBX-to-CO links. (This discussion is based on published reports—the reader should contact the specific vendor of interest for the latest information).

AT&T's Definity System 75/85 uses the Distributed Communications System (DCS) as the DS1-based signaling protocol between networked nodes. Twenty-three channels are available for information transfer. This signaling approach uses a separate proprietary data link based on the HDLC link layer protocol. Messages are transported with an implementation of the X.25 protocol. DCS uses a proprietary message set between System 75/85 nodes. DCS has been available for several years. AT&T is designing implementations of ISDN PRI-based PBX networking conformant with Q.931 and ECMA's recommendations for supplementary messages. AT&T's PBX networking capabilities will be migrated to ISDN primary rate for both on-net and off-net communication. Direct SS7 support on PBXs may also evolve. PBX-to-CO ISDN PRI has been available since 1987 on System 85 (however, the slots did not at first provide Clear 64 [10.16]). DSC should not be confused with Communications Protocol (DCP), which is similar to the ISDN BRI. DCP is used on System 85 for signaling between the PBX and the stations.

Ericsson's MD110 employs DS1 links to connect distributed nodes in a star configuration. One 64 kb/s D channel per DS1 link is used for signaling. Twenty-two channels on each DS1 link are available for information transfer (one channel is used for synchronization tasks). The link layer protocol is proprietary. The signaling mechanism to support the distributed nodal architecture uses a proprietary message set and protocols over each DS1-associated D channel. Ericsson's existing implementation of the DPNSS was to be incorporated as part of an ISDN-based inter-PBX networking option for the U.S. market during 1990. DPNSS features will be added to Q.931 primary rate connection capabilities, available at the same time. ISDN CO connections have already been demonstrated in the field, and were to be available in the first quarter of 1990. DPNSS protocols were developed in the United Kingdom. In the early to mid-1980s, British Telecom, along with a committee of PBX manufacturers, started to work on a common-channel standard for PBX networking over digital links. The committee developed a standard designed to provide a set of interconnection capabilities for voice and data. Although ISDN standards were already under development (with the first set available in 1984), these were not used. The group used instead available OSI/CCITT standards such as traditional CEPT framing (CCITT G.703 and G.732) at the physical level, HDLC at the link level, and DPNSS-characteristic protocols at layers 3 through 7 (British Telecom Network Requirement 188). This so-called "pre-ISDN standard" diverges from ISDN's standards at layer 2 and higher. Product implementations started to appear in 1986. The technology has achieved a fair degree of penetration in the United Kingdom due to multivendor support and as a definable migration strategy to ISDN.

By first quarter 1990, Fujitsu's F9600 was expected to incorporate a signaling capability for inter-PBX networking called ISDN Network Transparency (INT). INT uses one 64 kb/s D channel per DS1 link, and employs ISDN LAP-D; this leaves 23 channels for information transfer. INT will use Q.931 basic call establishment messages and protocols. "Supplementary service" messages are based on the Australian's PTT extensions, which are similar to the evolving ECMA recommendations. PRI-based networking, providing feature transparency between Fujitsu PBXs, was expected to be available by the end of 1989 for the F9600 and selected Omni products. A primary rate CO interface for the F9600 and for selected Omni PBXs was to be available in 1990.

Intecom's IBX DS1 nodal connection capability, called InterExchange Link (IXL), uses one 64 b/s signaling channel per DS1 link. IXL supports 21 bearer channels per DS1 link. The link layer protocol is proprietary; IBX uses a proprietary message set and protocol between IBX nodes. PRI-based nodal interconnection will be adopted, according to the vendor, as ISDN networking standards become available. Networking based on Q.931 and Q.932 is desirable because these would provide a uniform interface for PBX-to-PBX and PBX-to-CO applications. The latter is currently under development and was to be available in mid-1990.

Mitel's SX-2000 DS1 nodal connection capability, called Mitel SuperSwitch Digital Network (MSDN), uses one 64 kb/s D channel per DS1 link. MSDN supports 23 bearer channels per DS1 link. HDLC is used at the link layer. The message set and protocols used for private digital networking are an implementation of DPNSS discussed above; Mitel-specific messages have been added, in a transparent fashion. MSDN offers common-channel networking functionality consistent with Q.931 and the ECMA extensions. For PRI-based PBX internetworking, Mitel was developing a Q.931-based capability in 1990 for future delivery. A primary rate CO interface is available in the United Kingdom using the British Digital Access Signaling System 2 (DASS2).

NEC's NEAX 2400 offers a digital networking signaling capability based on the Common Channel Interoffice Signaling #7 (CCIS #7). One 64 kb/s signaling channel is used per DS1 link, leaving 23 channels for information. The link layer protocol is proprietary. NEC uses a proprietary message set based on a subset of the CCITT recommendation for SS7. NEC has stated the intention to migrate its CCIS #7 proprietary system to the ISDN standard, although apparently no plans appear to exist to publish the specifications for the former. The strategy is to implement the same standard for inter-PBX and PBX-to-CO interfaces. A primary rate CO interface for the NEAX 2400 was expected to be available in the United States by the first quarter of 1990.

Northern Telecom PBXs (SL-1 and SL-100) use a nodal connection capability called Electronic Switched Network (ESN). One 64 kb/s signaling channel is used per DS1 link. The link layer protocol used is LAP-D. ESN uses a proprietary message set and protocol to network SL-1 and SL-100 nodes, above layer 2. Northern Telecom has been a proponent of PBX networking based on ROSE, which can be viewed as a subset of the public SS7. Northern's PBX-to-CO ISDN networking and PBX-to-PBX networking will be based on a common implementation. The PBX-to-CO ISDN primary interface for the SL-1 has been available since 1988.

Siemens' Saturn PBX DS1 internetworking system, called Corporate Network (CorNet) uses one 64 kb/s D channel per DS1 for signaling. The link layer utilizes ISDN LAP-D. CorNet uses CCITT Q.931 basic call establishment messages and protocols. "Supplementary service" messages are defined by Siemens, based on evolving ECMA recommendations. CorNet was available on Saturn PBXs as of the end of 1989. A primary rate CO interface for Saturn PBXs was available as of the end of 1989. See Table 10.3, which is based partially on Reference [10.17].

10.3.6 Computer-to-PBX Interfaces

The growing use of intelligent processors and advanced software is creating new information service requirements in many companies. Computer-to-PBX interfaces are being defined to meet aspects of this requirement. These interfaces will provide users with integrated voice-data applications. For example, an incoming call with

Table 10.3
Major PBXs with ISDN Capabilities

	Product	*Basic Rate*	*Primary Rate (to CO)*	*ISDN Stations*
Large PBXs (10,000 or more lines)				
AT&T	5ESS	1988*	1988*	1988*
AT&T	System 85	1Q89	1988*	1988*
NTI	Meridian SL-100	1988*	1988*	1988*
NEC	NEAX 2400	1990	3Q89	3Q89
InteCom	IBX S/80	1990	3Q89	1990
Ericsson	MD110	1990	4Q89	1990
FPN	Personal eXchange	1990	1990	1991
Medium PBXs (between 1000 and 9999 lines)				
Fujitsu	Focus 9600	1990	3Q89	3Q89
NTI	Meridian SL1	1988*	1988*	1988*
Mitel	SX-2000	1990	4Q89	4Q89
Hitachi	LDX	1990	2Q89	1988*
Fujitsu	Focus 960	1991	1990	1990
Harris	20-20	1991	4Q89	1988*
Small PBXs (less than 1000 lines)				
Siemens	Saturn III	1990	3Q89	3Q89
Alcatel	Alcatel One	1990	1990	1990
AT&T	System 75	1991	1Q89	1Q89
Mitel	SX-200	1991	1990	1990
Telenova	Telenova I	1990	4Q89	1988*
SRX	System One	3Q89	4Q89	4Q89

*or earlier

ANI delivery can be routed to a company service representative while at the same time a computer record on the incoming customer can be simultaneously displayed at the representative's CRT.

A number of working relationships have been developed to enable computer-to-switch communication, but for the most part these have been proprietary in nature. For example, in 1990, IBM announced a CallPath Services Architecture (CSA) to interconnect IBM mainframes to AT&T, Northern Telecom, Rolm, and Siemens AG PBXs. This interface is also applicable to CO equipment. CSA supports integrated voice and data applications for both incoming and outgoing call traffic (earlier interfaces supported only incoming traffic). Advanced voice-data application software utilizing these interfaces should begin to become available in 1991 [10.18]. Other vendors have announced interfaces, including Digital Equipment Corporation, with its Computer Integrated Telephony, connecting VAX minicomputers and about 12 PBX systems.

Vendor-independent protocols for switch-to-computer interfaces were under discussion in ECMA and ECSA/ANSI T1S1 committees at the time of this writing. These open interfaces allow synergistic interworking of PBX-based private networks, the public network, and computer systems and pertinent applications. The standards, based on the OSIRM, should establish commonality and vendor-independence. This effort is known as *switch-to-computer application interface* (SCAI). Standards for SCAI may be available in the 1991–92 time frame.

10.4 NETWORK DESIGN CONSIDERATIONS

Modern PBXs allow refined measurements that can aid network design. These PBXs have complex software to manage hunt groups, pickup groups, intercom groups, automatic route selection (ARS) patterns and tables, and many other service and routing features. These modules have hooks that permit data collection at various points in the system. The important point to realize, however, is that we need more information, not just large amounts of data (although, of course, information is built on data). All too often, large amounts of data that are more or less useless and do not meet the needs of the organization are collected. This is a leftover from the early 1960s when computer power was scarce and researchers were willing to accept partially processed data and complete the task by hand [10.19, 10.20].

The optimal monthly report for a three-node PBX network could be one page long and look like the following:

PBX-1 to PBX-2 bundle: no additional trunks needed with confidence 90%; projected change to new status: 3 months.
PBX-1 to PBX-3 bundle: 1 additional trunk required with confidence 80%; projected change to new status: 6 months.

PBX-2 to PBX-3 bundle: 1 trunk can be decommissioned with confidence 95%; projected change to new status: 4 months.

This report gives action items (secure new resources; decommission existing resources); gives the level of confidence of the recommended actions (nothing is absolutely certain because a number of assumptions go into a calculation); and forecasts for the length of time the new design should be valid before additional changes are required (for example, if the addition of a trunk would be good for the duration of one month, at which time it has to be decommissioned, perhaps the installation should not be pursued). Of course, other types of reports not related to engineering (cost allocations, NPA/LATA frequency, collect call details, and other general communication management information) would have to be of the appropriate length and scope.

Some proponents would argue that paper reports should not be used at all as an analysis tool (and only be used after the fact for archiving and documentation), and that we should rely on on-line data display and management. The focus of AI and expert systems follows this philosophy of data reduction. A well-engineered network design tool and a set of raw (or semi-raw) data are at two ends of the spectrum. The design tool will process the data and present final results that can be easily displayed in one, or few, screens of a CRT. The issue here is one of designing a tool for a CRT, where at most only a few pages can be scrolled, which forces the developer to design a tight, condensed, and highly focused system.

The advent of fourth-generation data base management languages (FOCUS, RAMIS, DB2, NOMAD, *et cetera*, both mainframe and PC versions) makes these new tools within easy reach, even for the nonprogrammer. What might have taken 100 lines of COBOL programming several years ago can now be done with perhaps as little as 10 lines of 4GL (4th Generation Language).

10.4.1 Measurement Purposes

The typical goals of network measurement efforts in a PBX environment are aimed at facilitating the following:

- PBX system and PBX network design and refinement, which encompasses facilities usage optimization (facilities include hardware, software, and transmission equipment); facilities planning; facilities procurement; capacity information for eventual system upgrade or replacement;
- Cost control: underutilized facilities, or misused facilities (for example, LCR with DDD as a first choice rather than last);
- Abuse control: unnecessary or unauthorized off-net calls (555-1212, 900-, *et cetera*);
- Early identification of system faults, malfunctions, impairments, or suboptimal operation;

- Network cost allocation to various corporate departments;
- Productivity studies (for example, telemarketing staff productivity; corporate-wide cost of communication per employee; some companies estimate a $12,000 to $18,000 per year per employee for total telecommunication expenditures; corporate communication department efficiency measured in network cost per minute, *et cetera*); and
- Business information (distribution of company business by LATA; users of the 800 service; *et cetera*).

10.4.2 Issues

Some issues associated with measurements are the following:

- Of the several possible goals listed above, which are the specific objectives of data collection? If the answer is all of the above, then a substantial amount of data will have to be gathered and processed. The amount and type of data needed will vary greatly depending on the use to which it is put. A determination of the requirements should come first.
- Will standard data collected by PBXs be sufficient for the task at hand, or will additional data and measurements be required (either as a special PBX software-driven collection task, or via an ancillary apparatus)?
- Is the design effort concerned with a small number of "high volume switches" (say, a dozen or less) or does it involve hundred of sites, with PBXs ranging in power, sophistication, application, and size? Some customers have 300, 500, and even thousands of branch locations (for example, brokerage firms, insurance firms, General Motors, *et cetera*).
- How should traffic data be collected and processed in multilocation situations? In the case of many branch locations, centralized treatment of the data may be desirable because the remote locations do not have the personnel, skills, or tools to carry out the task. Data can be stored and processed at each site, stored at each site but processed centrally, or stored centrally and processed centrally. In terms of storage, we can keep the data off-line (paper files) or on-line (either with PBX facilities or on a separate micro, mini, or mainframe). Of course, on-line storage is mandatory for any kind of sophisticated design effort. In terms of processing, customized design tools are available or commercial software. Design software that is highly automated, requiring minimal intervention between the original raw data input and the production of focused reports to which we alluded earlier, is generally desirable. The establishment of a separate data communication network for the collection of the voice-PBX data may not be cost-effective. Many organizations already have a DP data network in place for traditional purposes; PBX data may be combined with that network, which may involve some hardware accommo-

dation to accept RJE-type traffic over the interactive link (channel splitters, dual-port modems, protocol converters, *et cetera*). [10.21].

- Data from different PBXs may have different formats; these have to be reconciled prior to conglomeration (this may include some CENTREX data—185 character records on magnetic tape). Also, data for all types of calls (inbound, within-the-PBX, intra-LATA, inter-LATA, international, on various trunk bundles, to voice storage units, *et cetera*) are required. Data volume may become an issue.
- Not all data may be available to the PBX because they may reside on another computer (departmental breakdown, tariff tables, facilities inventory, *et cetera*). These data may have to be appended to the traffic data before processing can commence.
- How long an interval is required for the calculations to have validity? For example, we cannot measure blocking for a single day of the year and then use that value as representative of any day (see the section below on sampling). We may be required to measure blocking for, say, 45 days, and then be able to generalize. The same is true for traffic arrival parameters, busy hour, *et cetera*. This implies that data for several days (or even weeks and months) must be kept on-line and accessible. Clearly, with tens of thousands of records for each day, the volume of the data accumulated over a period of time goes up very quickly. Fortunately, in general, the voluminous raw data are not kept, but are processed and reduced. For example, we are interested in calculating the mean call length over the period of a month (this horizon having been selected to achieve a good degree of confidence). Given all the calls for day i, we can calculate the mean m_i and store it and the number of calls n_i for day i. This would be a total of 60 numbers for the entire month. The final mean is calculated as:

$$A = (n_1 \times m_1 + n_2 \times m_2 + \ldots + n_{30} \times m_{30}) \div (n_1 + n_2 + \ldots + n_{30}).$$

  Additionally, we must assess the necessary daily windows as to when the measurements must be carried out; not all components of a system are busy at the same time, and this would necessitate different measurements. For example, the busy hour for an inter-PBX trunk bundle may not coincide with the busy hour of the PBX itself (consider a company on the East Cost calling a subsidiary on the West Coast: station-to-station traffic internal to the East Coast PBX may occur around 11:00 AM, while the busy hour for the West Coast bundle may occur around 2:00 PM).
- Generally, near real-time access of the data for fault detection and other time-sensitive tasks is desirable. This, in turn, dictates the type of storage required.

- The scope of the desired reports is needed. Are the data for traffic engineering only, or will they also serve some of the other management purposes listed above?
- The cost-effectiveness of the undertaking must be studied. Classical optimization may look at percentages, but we also need to look at absolute values. For example, collection of large quantities of data at a given collection and processing cost may indeed reduce the size of the trunk bundle by 25%, and thus the cost of the bundle by 25%; however, if the trunk bundle cost $12,000 a year (four tie lines New York–to–Philadelphia), saving 25% corresponds to yearly savings of $3000: How much manpower and computer power did it take to reach that conclusion? Percentages may look good, but the absolute value is also important.

10.4.3 Measurements

The relevant window of measurement must be specified following the identification of the goals to be achieved by the measurements. The windows of measurement generally employed in traffic engineering for the PBX are as follows [10.22]:

- *Average Busy Season* (ABS): the average of the three highest months' (not necessarily consecutive) average usage values through the busy season. This then equates to an average of 65 daily busy hour usage values.
- *Ten High Day* (THD): the average of the ten highest daily busy hour values that occurred in the busy season.
- *High Day* (HD): the highest of the daily busy hour values that occurred during the busy season.
- *Extreme Value* (EV): the highest hour usage value that occurred during the busy season.

The windows of measurement generally employed in traffic engineering for the inter-PBX trunking are [10.22]:

- *Average Network Busy Hour* (ANBH): the trunk bundle usage in the hour in which the network of which this trunk bundle is a constituent is busiest.
- *Average Consistent Busy Hour* (ACBH): the usage in the time consistent clock hour in which this trunk group is the busiest.
- *Average Group Busy Hour* (AGBH): the average of the usage values in the hours in which this trunk bundle is the busiest (also called the *bouncing busy hour*—BBH).

Some typical measurements are peg count, all busy, last busy, overflow, queue logs, and traffic usage. Three *key* types of data are: peg count, overflow, and usage. Not all traffic users have a need for all types of measurements. Load balancing calculations would require usage data; certain types of traffic separation studies require only peg counts. However, the combination of peg count, overflow, and usage is generally a good starting set.

Peg Count. This is the number of attempts to use a trunk or bundle of trunks, whether or not that attempt is successful. (Some equipment manufacturers report peg count as the number of attempts accepted by the circuit group—this is then a type of "carried peg count").

Overflow. This is the number of attempts to use a trunk or bundle of trunks that were unsuccessful.

Usage. This is the occupancy of a circuit or group of circuits expressed in erlangs or CCS.

Other data include:

Queue Logs. Queue logs provide a measure of resource utilization; these logs can be used to assess the need for additional computer ports, modems in the modem pools, and interswitch trunks.

SMDR. Important and readily available data are the station message detail recording (SMDR) data, which consist of call information produced by the PBX and other switching equipment. These are also called *call detail records* (CDRs). In effect, SMDRs are usage data; however, they are principally intended for billing and accounting purposes, so that traffic information must be extracted and adapted. One of the issues dealing with SMDR is that of the size of the data file; a medium-size organization may generate from several hundred thousands to several million records a month. The solution here is to use sampling theory, rather than accounting theory (sampling theory is a modern and sophisticated discipline); accounting theory says that one has to look at 100% of the data to get the answer, sampling theory says that 5% of the data, or even 1%, suffices to obtain an answer that can be trusted 99% of the time (99% confidence). See the sampling section below for more details on this issue.

External Data. Additional data may be required or desired to feed the more refined models described above; this may be collected through other facilities of the PBX, or even adjunct equipment (response time monitors for data paths, front-end processor statistics for a mainframe, *et cetera*).

MIS Reports. Typical traffic reports would include, among others [10.22]: a THD summary, a monthly summary of traffic usage, a call volume and service indicator report, and a trunk group usage summary.

Only a few years ago, this wealth of data was difficult to obtain. The telecommunication manager should now take full advantage of the available information. On-line traffic processing subsystems of modern PBXs allow such tasks as (1) data base management for system inventory applications, (2) data validation (by user-supplied rules in the data base, the data can be subject to validation tests to identify items that are spurious or in error), (3) traffic data archiving, (4) data analysis (traffic engineering), and (5) report generation (the final packaging of the data in report form).

10.4.4 Typical PBX Measurements and Reports

Some typical statistics, in addition to or independent of SMDR, as available on major PBXs are shown in Table 10.4 [10.23]. These data, when properly compiled, combined, analyzed, and interpreted can be useful in designing the PBX network, particularly in identifying possible bottlenecks.

PBX-assembled data are typically stored in the following ways:

Detailed data for any multiple of 15-min intervals
Detailed data for 15, 30, or 60 min
Every half hour
By group, on system hard disk
Peak day data in internal buffer
Last hour in internal buffer
Previous day peak in internal buffer
Detailed data for given period (say, 10 h)
On disk
External device

Table 10.4
Design Data on Major PBXs

Data on CO trunk groups	*Occupancy or Peg*
Tie trunks	Attempts and seizures
Extension hunt groups	ATB (all trunks busy) count, duration, overflow
Service circuits	Average or maximum wait in queue
Feature usage	Exact time in queue
Matrix blocking (for blocking PBXs) and balance	Count of calls dropped from queue at timeout
Dial-tone delay	
Trunk queues	Average or maximum calls in queue
Console queue	Queue overflow during hour
Other:	Information for consoles and ACD (Automatic Call Distributor)
Carrier usage	Calls answered
Processor occupancy	Calls abandoned by attendant
Automatic alternate routing	Average waiting time
Invalid log-in attempts	
Modem pool utilization	
Data switching use	
Data by route	
Calls unanswered, abandoned	

Nonvolatile RAM
Redundantly on disk and tape.

Data are typically made available to the user and displayed as follows:

Printed, on request
Printed, automatic
TTY
Console CRT (on request)
Maintenance CRT or TTY
To RS-232-C port (for PC access or other)
Customer administration center system
Nine-track tape.

Data are available for analysis as system-generated reports, user-customized reports, menu-driven fields, and by request by a data base management system. The data are typically kept in the system:

Until polled
One hour
One day for detailed data
One day for peak
Several days
Several weeks
Until displayed or printed
Until manually cleared
User specified
Until disk is full
Stores until disk is full, then writes over
No limit except disk space
Until power lost (to maximum capacity of solid-state drive).

In addition to endogenous PBX traffic statistics, most PBXs provide SMDR information, as discussed above. PBX-compiled SMDRs typically include the information shown in Table 10.5.

Processing of SMDRs is typically as follows:

PBX buffers call records
  Built-in record storage
    For all active calls
    Up to some maximum (tens or hundreds)
  Nine-track tape
  Paper tape
  Disk drive
  PC
Real-time CDR display

Table 10.5
Typical Information in SMDRs Compiled by PBXs

Access code
Called number
Calling extension
Billing-authorization code
Trunk group
Specific trunk used
  Tie trunks records
  Toll calls records
    FX calls
    WATS calls
  Local calls records
  Intra-PBX calls records
  Local extensions records
    Incoming CO/WATS
    Incoming tie trunks
Queue waiting time
Answer time
  Variable timer per trunk group
Hang-up time-duration
Total trunk holding time
Other information
  Facilities restriction level
  Ineffective call attempts
  Interexchange carrier
  Data call information
  On-hook versus off-hook queue
  IDDD calls
  Operator calls
  911, 800, and 900 calls

  Historical search
PBX-supplied processing packages
  Recording and reporting
  Analysis on main or standby processor
  Analysis on applications processor
  Analysis on external computer
  Output to RS-232-C port
  Output to modem for remote access
  Output to X.25 channel.

CENTREX also offers SMDR capabilities. Usually the SMDR data are derived from the billing data available at the Revenue Accounting Office (RAO); these data are first preprocessed by telephone company software. The expense associated with SMDR data acquisition is sometimes an issue. Customer-premises hardware (PCs or minis) and software (purchased, licensed, or leased) is needed to process real-time SMDR information [10.24].

A PBX-network manager with multiple locations served by different PBXs could encounter different SMDR record formats unless the data are normalized by a common process. Another issue is that of integrating the data when their use has a hybrid PBX-CENTREX arrangement (PBXs at major sites and CENTREX at smaller sites). Consistency between the PBX format and the switch format is required for comprehensive processing of the data.

10.4.5 Conclusion on Vendor-Provided Data

Many of the more refined measurements must be devised by the user because they may not be automatically available through the standard reports. In particular, histogram tables characterizing the actual distribution (and eventual "goodness of fit test results") are not supplied automatically (much of the theoretical traffic models rely on the distribution of call arrival and call service being Poissonian). Perhaps the PBXs of the future will provide this information more routinely.

10.4.6 Issues of Data Integrity

Traffic data can become corrupted for several areas, particularly for large multinode networks. Efforts should be made to establish some validation criteria to identify erroneous data before they enter the design (or billing) stream. Reference [10.22] provides a list of seven general areas of concern, including the PBX; the scanner-converter equipment, or transmission to it; the traffic acquisition-accumulation computer, or transmission to it; the traffic data processing computer, or transmission to it; and the data bases associated with each step of the collection or analysis process.

10.4.6.1 Basics of Sampling Theory

Most people are likely to be aware of the outstanding results achievable with applied sampling theory, a branch of statistics. It is used in political polls and presidential elections, making it possible to predict who will win an election four to eight hours prior to the closing of the West Coast polls in the United States.

Sampling theory can be applied to communication to achieve substantial monetary savings in certain situations. Sampling is a way to choose and study a small subset of cases from a large population and to generalize the results to the entire population. The generalization can be carried out with high statistical confidence. The theoretical results are fully developed and are generally easily understood.

As indicated, sampling theory states that to obtain results about a large population, we need only do the following [10.25]:

- Determine the population homogeneity.
- Use the theory to obtain the necessary sample size.
- Obtain the scientifically specified sample for which appropriate measurements will be made.
- Adjust the results to fit the whole population using the formula.

Homogeneity tests the uniformity of a given population. If all objects, people, or situations are likely to be similar in principle, then homogeneity exists, and an "unstratified sample" would be used. If the objects or entities can be classified into areas likely to have some intrinsic difference, then a "stratified sample" would be used. This means that a sample must be taken from each of the classes defined.

For example, if we wish to determine the characteristics of traffic habits of asynchronous data users throughout the United States, we may want to define an "urban population" and a "rural population." On the other hand, stratification of the population by male and female users may not be necessary because both categories can be assumed to have the same traffic characteristics.

Sample size is determined by a set of formulas and depends on the degree of accuracy demanded. If the accuracy must be 100%, then the entire population must be consulted; this could be very expensive. If 95, 99, or 99.9% accuracy is acceptable, then only a small subset of the population needs to be consulted. This subset varies from 1% for very large situations in which the universe is in the one-million-population range, to 5% for a 10,000 population range, to 20% in a 500-population range.

Once the size of the sample is determined, the sample (namely, the individual representatives to be examined) must actually be picked. We are not free to select representatives, otherwise the results would be biased. Generally, we must choose a truly random subset (in a stratified environment, a random subset from each population is required). With the example above, we would not, despite the convenience, be able to select the "urban population" from New York City only; some from New York, some from Los Angeles, some from Seattle, and some from other cities would have to be included.

Before discussing the application to SMDR or other PBX traffic data, we list two concrete examples, based on experience, in which sampling techniques would have saved money had they been applied.

An international record carrier wanted to classify the 800,000 telex calls a day carried by its switch into national destination percentages and to calculate blocking factors to each nation. The type of transaction records produced by the telex switch are very similar to SMDR records.

To achieve this goal, the company executed a daily batch program requiring eight hours every night and several disk drives—an inefficient 20-year-old program that parsed every one of the 800,000 transactions. Clearly, the firm could have applied sampling techniques and processed only 8000 records (a 1% sample based on the size of this population) and used a program estimated to require only 0.05 min to complete. The sampling techniques would have saved machine cycles (which could have been applied to more "revenue-related" programs, such as billing) and storage capacity.

In the second example, an operations manager at a financial services firm wanted to determine the type of communication problems experienced by the more than 200 major remote branches. A project was undertaken to contact each of the 200 branch managers—at considerable manpower and travel expense. Again, a sample of approximately 40 users (20% sample based on the size of the population) would have been statistically sufficient.

10.4.6.2 Sampling Applications to SMDR

Some SMDR applications must be done with the classical accounting techniques of exact cost allocations per user (department or even station). Many other applications, particularly those pertaining to networking and traffic engineering, can be done with sampling methods.

In fact, even cost allocation can be done by sampling if the departmental users are willing to accept the fact that a small statistical uncertainty in the bottom line number may exist (the advantage might be a faster availability of the data or smaller DP charges associated with having to sort and store millions of records—particularly if these large charges are by an outside time-sharing service). Here, we might first quantify the total cost (TC). We would then select 1000 random calls for each month using sampling techniques and calculate the total cost for these calls. We would allocate these 1000 calls by department and obtain the fraction of the cost based on the 1000 total. Those percentages can be extrapolated to the entire TC to obtain total individual departmental costs. If done right, this could be ±10% with 95% assurance.

If the cost has to be allocated down to the individual station, then the statistical technique would not be applicable, unless each station makes enough calls such that the sample is meaningful. (The number of calls per station would have to be around 300 to 400 per month before the technique can be applied. For a 1000-station organization, this would be 400,000 SMDRs a month, and 20,000 would have to be sampled.)

To a person unaccustomed to sampling methods, the approach may at first look superficial and suspicious because we are familiar with accounting methods. This method is similar to the first exposure to base 7 when all we have learned is base 10: It looks strange. In reality, sampling is extremely powerful. Why count 100,000 objects when we can guarantee the same result by only counting 1000 of them? One could look at 100,000 calls; count the ones blocked, B, and then calculate blocking with the formula $b = B/100{,}000$; we obtain the same data by looking at 1,000 properly selected calls. The same argument holds for any other measurement in which the number of events (population) is large enough. (How *not* to use sampling: If we have three tie bundles, do not measure blocking on one and extrapolate the same number to all three—if we have 100 bundles, then we can measure blocking on, say, 50 properly selected bundles and then extrapolate to the 100.)

The use of sampling for networking and engineering is related to another: If some of the variables are only known within $\pm 10\%$, why enumerate millions of data records to obtain a theoretical accuracy on this given variable of ± 0 instead of going through 1000 records to obtain a value $\pm 10\%$, as is the case for other data items. The added accuracy for one variable will be washed out by the uncertainty in other variables. Also the ± 0 theoretical error is probably unlikely because of some intrinsic error with the raw data itself: Some of the SMDR data may have been lost or corrupted during transmittal (even a modern data link control protocol has a certain undetectable error rate); the station file on which the SMDR data ride may have an inaccurate entry; the WATS bundle may not have as many trunks as thought because one was erroneously canceled in the recent past, *et cetera*. This is not to say that we should be careless with or disinterested in the data; be aware of the need for data consistency: All data must have the same relative accuracy.

10.4.6.3 Example

To provide a flavor for the sampling method, we provide one simple example: obtaining average call holding time for the purpose of engineering a PBX network.

Note that the real holding time for a PBX resource is the sum of the dialing time, the ringing time, the conversation time, and the call disconnect time. SMDR data would only capture the conversation time; the holding time for a trunk would be the above terms, except for the dialing time. These components would have to be included in a precise calculation of the holding time, rather than just considering the conversation time, as is customary. Other aspects involve false starts and premature abandonment. Rotary dial times are around 10 seconds and dial tones vary between 7 and 4 seconds. Reference [10.26] provides a number of useful statistics about holding time.

We assume first that the calls are fairly homogeneous (namely, we do not have a mixture of voice and data calls, which are known *a priori* to possess different holding times) so that we do not need a stratified sample, but can get away with a simple random sample. Assume further that the calls are known to have negative exponential distribution (we are assuming that this is known from a previous study, or it is just assumed for convenience as some people do—if we actually wish to determine or test a distribution, we can make use of a statistical methodology called "goodness of fit" [10.27]).

Suppose that we want to determine the mean holding time h within $\pm 10\%$ accuracy and with 95% confidence. Because the standard deviation σ_h of the calls equals the mean h (a property of the "memoryless" distribution), the standard deviation of a sample of N random calls is $h/\sqrt{N}$. By the "law of large numbers," the mean of a large sample will be asymptotically normal (Gaussian). In a normal distribution, 95.5% of the population is contained within ± 1.96 standard deviations from the mean. Hence, the sample size required to achieve the stated accuracy goal will be:

$$(1.96) \times (h/\sqrt{N}) = 0.1 \times h,$$

or, solving for h, $h = 385$. This means that by looking at only 385 properly selected calls we are able to obtain the average holding time h within 10% with 95% confidence.

This method is implemented, given a SMDR file of, say, 100,000 records, by programming a simple pseudorandom number generator ranging from 1 to 100,000 (a congruence relation from any book on simulation—probably less than 10 computer instructions. The reason we can use this simple method to pick records is because we did not need a stratified sample.). The generator will spill out a list of 385 numbers, say, 32567, 22873, 87263, 16004, *et cetera*. The holding time calculation program will now obtain the 32567th record and note its holding time; then look at the 22873rd record and accumulate its holding time; *et cetera*. When all 385 records have been looked at, the accumulated total is divided by 385 and the desired number is obtained. The entire program could be 30 lines of code and run in a very short period of time, granted that the SMDR file is suitably indexed.

10.4.7 Busy Hour Issues

Day-to-Day Traffic Variations

Because of hourly and daily variations, the steady-state equilibrium position may not be reached in a real system; as a result, the average blocking may be higher than predicted. One practical approach to address this problem is to set a permissible blocking factor given specified percentages of increase over the nominal value. This, however, does not allow for the fact that some tie bundles in the PBX network system are subject to more volatility than others. (Consider, for example, a bundle from a large brokerage firm order room to their back office DP center, and a bundle from the order room to an exchange: The bundle to the exchange is more likely to be affected by large variations in traffic patterns.)

CCITT recommends that the blocking averaged over the 30 highest days should not exceed 1% for international carrier trunking, and that the blocking at busy hour for the five highest days should not exceed 7%. This method may be acceptable to carriers, but is not advisable for end-users. The better approach is to use the call-defection formulas to calculate the probability of blocking.

A more refined approach is to consider the offered traffic A as a function of time t (and even route). Traffic tables have been calculated on this basis, assuming A to have normal or gamma distributions.

The ratio α of the variance to the mean is a convenient way to study the day-to-day and within-busy-hour variations. For very stable traffic, the effect of day-to-day variation may be negligible, so that the distribution is close to a Poisson process ($\alpha = 1$). Studies on toll traffic show α to be from 1 to 2 (this "peakiness" applies to the offered traffic, as well as to the day-to-day variations in average load); of course, traffic collected by a carrier tends to be more stable than one might experience in a private PBX network because a large number of users are involved here, and a result from probability theory predicts that the effects of individual users will tend to neutralize each other, as long as there are enough of them.

The effect of variations within the busy hour has not received extensive attention until recently; modern measurement techniques on the PBX make it possible to determine such variations simply and economically. These variations may be taken into account for design of large PBXs systems in which short deviations from the postulated equilibrium may have considerable effect on the call handling capability of a high-occupancy processor. After all, no fundamental reason exists to use an hour as the design interval; in fact, we would be quite disconcerted if a congestion state lasted an entire hour; using a smaller interval (now made possible by the measurements) may cut this congestion state to a shorter period.

Design Level

The busy hour method may not always be the most cost-effective way to engineer a PBX network because it may either be volatile or be substantially higher than the traffic for other hours. Consider a tie trunk bundle between New York and Chicago, with the average traffic arrival as follows (in erlangs):

Hr 1: 10
Hr 2: 20
Hr 3: 20
Hr 4: 30
Hr 5: 20
Hr 6: 20
Hr 7: 20

The mean traffic is 20 E. Using $BH = 30$ and an erlang B (over)simplification for $b = 0.005$ and full availability, we would need 44 trunks. At approximately \$1000 each, this corresponds to \$44,000 per month. On the other hand, if we chose to engineer for 20 erlangs with an overflow to DDD at the busy hour, we would require 32 trunks in the tie group for \$32,000, plus 10 erlangs to DDD for one hour for 22 days. At approximately \$0.50 per minute on DDD, we have $\$0.50 \times 10 \times 60 \times 22 = \$6600$. The total would now be \$38,600.

Now consider another average traffic profile that still has a mean of 20 erlangs and a busy hour of 30 as follows:

Hr 1: 7
Hr 2: 7
Hr 3: 6
Hr 4: 30
Hr 5: 30
Hr 6: 30
Hr 7: 30

If we engineered for 20 erlangs (32 trunks) and put the overflow to DDD, we would have $\$0.50 \times 10 \times 60 \times 4 \times 22 = \$26,400$ plus the \$32,000 for the tie line bundle, for a total of \$58,400. Here, designing for the entire 30 erlangs, or 44 trunks, would be cheaper.

What is the difference between these two cases? Define the following quantities:

FME = fraction of time the average A is equaled or exceeded
FMA = fraction of time the maximum M ($M = BH$ value) is achieved
$\delta = \text{FME} - \text{FMA}$
design level $= M \times (1 - \delta) + A \times \delta$.

For the examples above, we have:

	First Example	Second Example
FME	0.86	0.57
FMA	0.14	0.57
δ	0.72	0.00
Design Level	22.8	30.0

Note that the FME is always greater or equal to the FMA, so that $\delta \geq 0$. When the FMA is low, only a few hourly intervals are equal to the BH value; it may not pay to secure dedicated transmission resources that will be employed (i.e., for traffic overflows) only a few times a month; conversely, if the FMA is high, the traffic tends to equal the BH for several hours. Figure 10.4 depicts three examples, all having the same average and busy hour traffic, but with different characterizations for FME and FMA [10.28]. The quantity δ is a measure of the spread:

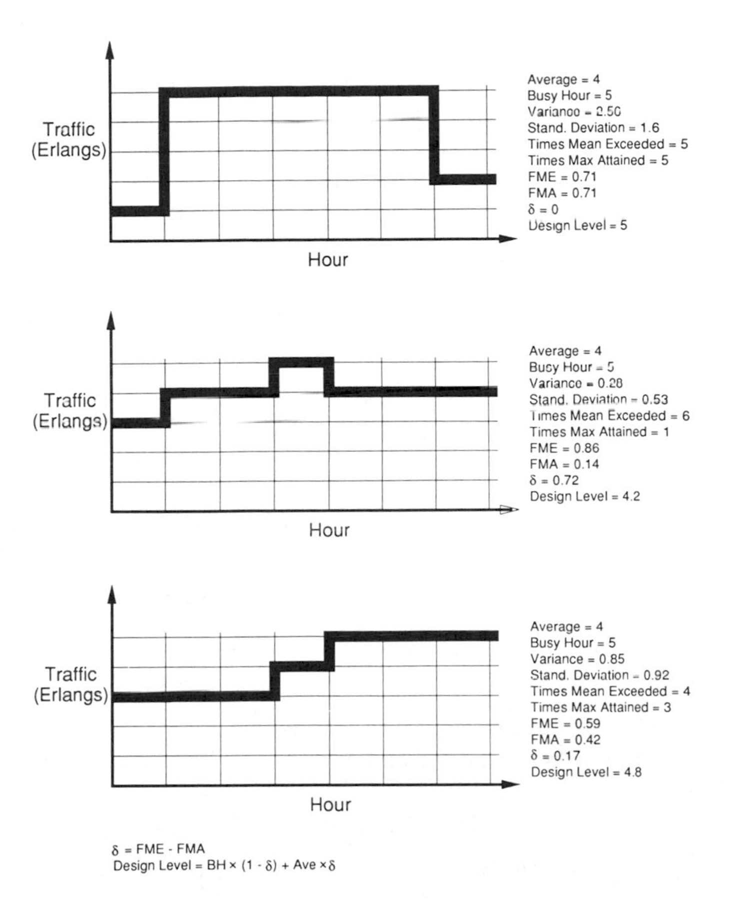

Figure 10.4 Link level.

When δ is close to 1, then most of the traffic clusters around the average; when δ is close to 0, the traffic clusters around the maximum value (the busy hour value).

An empirical design criterion is then to engineer the system at traffic level $M \times (1 - \delta) + A \times \delta$ ($M = BH$ and the other terms are as defined above).

This approach is related to the mean and standard deviation relationship that bounds the amount of probability in the "tails" (Chebyshev's inequality), but keeps the data values within the maximum value point. This empirical method is not necessarily optimal in the sense that it produces the lowest cost. If the lowest cost is desired, an optimization process similar to the arguments provided in the preceding few paragraphs can be carried out by using actual facility costs. The method provides one specific design level that is a reasonable compromise between the

"mean" and the "maximum" traffic values. A more sophisticated approach to traffic smoothing is discussed in [10.29].

REFERENCES

[10.1] J.W. Falk, Bellcore, personal communications, June 1990.

[10.2] D. Minoli, "Issues in Voice/Data Integration," *Teleconnect*, May 1986.

[10.3] P.D. Shea, "Centrex Service," *Transactions of the American Institute of Electrical Engineers*, Vol. 80, Part I, 1961, p. 474.

[10.4] J.R. Abrahams, "Centrex Versus PBX: The Battle of Features and Functionality," *Telecommunications*, March 1989, pp. 27 ff.

[10.5] D. Minoli, "Engineering PBX Networks Part 3: Signaling," DataPro Report MT30-315-301, April 1987.

[10.6] D. Minoli, "Engineering PBX Networks Part 1: Design Models," DataPro Report MT30-315-120, September 1986.

[10.7] T.C. Bartee, ed., *Data Communications, Networks, and Systems*, Howard W. Sams, Indianapolis, IN, 1985.

[10.8] J.H. Green, *The Dow Jones-Irwing Handbook of Telecommunications*, Dow Jones-Irwing, Homewood, IL, 1986.

[10.9] M.L. Gurrie, P.J. O'Connor, *Voice/Data Telecommunications Systems—An Introduction to Technology*, Prentice-Hall, Englewood Cliffs, NJ, 1986.

[10.10] E.E. Mier, "PBX Trends and Technology Update: Following the Leaders," *Data Communications*, September 1985.

[10.11] R.L. Koenig, "How to Make the PBX-to-ISDN Connection," *Data Communications*, May 1989, pp. 91 ff.

[10.12] D. Minoli, "ISDN Bodes Improvements," *ComputerWorld*, January 13, 1986.

[10.13] D. Minoli, "ISDN Stands at the Threshold," *ComputerWorld*, January 20, 1986.

[10.14] "User-to-User Signaling with Call Control," Bellcore, TR-TSY-000845, Issue 1, December 1988.

[10.15] P. Sharma, "The Trouble with the PABX," *Telecommunications*, March 1990, pp. 29 ff.

[10.16] P.R. Strauss, "An Update on U.S. Services and Standards," *Data Communications*, May 1989, pp. 129 ff.

[10.17] P.F. Kirvan, "ISDN PBX Systems—Next Generation or Next Generality?," *Telecommunications*, March 1989, pp. 33 ff.

[10.18] S. Girishankar, "The IBM Connection," *CommunicationsWeek*, May 14, 1990.

[10.19] D. Minoli, "Net Optimization Takes Practical Insight," *ComputerWorld*, May 6, 1985.

[10.20] D. Minoli, "Network Models Inadequate?," *ComputerWorld*, May 13, 1985.

[10.21] R.R. Fields, "SMDR—Designing a Resource Management Tool," *Business Communications Review*, January–February 1985.

[10.22] R.W. Lawson, *A Practical Guide to Teletraffic Engineering and Administration*, Telephony Publishing Corporation, Chicago, 1983.

[10.23] *The BCR Manual of PBXs*, BCR Enterprises, Hinsdale, IL.

[10.24] S.B. Duggan, "SMDR Can Be a Telco's Ace in the Hole," *Telephony*, July 4, 1988, pp. 26–28.

[10.25] D. Minoli, "Sampling: The Smaller View," *ComputerWorld*, July 1, 1985, pp. 61, 70.

[10.26] J.C. McDonald, *Fundamentals of Digital Switching*, Plenum Press, New York.

[10.27] M. Fisz, *Probability Theory and Mathematical Statistics*, John Wiley and Sons, New York, 1963.

[10.28] D. Minoli, "Engineering PBX Networks Part 2: Gathering Support Data," DataPro Report MT30-315-201, September 1986.

[10.29] D. Minoli, "A New Design Criterion for Store-and-Forward Networks," *Computer Networks*, Vol. 7, 1983, pp. 9–15.

Chapter 11
Data Communication: Issues Affecting the 1990s

11.1 INTRODUCTION

Data communication involves the reliable transmission of properly coded signals, generated by a terminal, computer, or other device, to a remote terminal, computer, or other device. The following chapters explore a number of key issues that will affect the industry in the 1990s. In a nutshell, these issues are use of broadband transmission facilities (Chapter 12), open interconnection standards (Chapter 13), LANs (Chapter 14), MANs (Chapter 15), network management (Chapter 16), and network security (Chapter 17). This chapter opens the treatment of data communication with a survey of a number of important factors, not the least of which is the strong market presence of *de facto* architectures. These architectures have arisen out of necessity because of the void created by the unavailability of commercially robust, end-to-end, and user-friendly standards in the 1980s, and the continued trend toward decade-long discussions for the development of each single standard.

Transfer of information between the two end systems can be undertaken in a "connectionless mode" or a "connection-oriented" mode. In the connection-oriented mode, communication is initiated through a call request phase, which establishes an end-to-end path. After an information transfer phase that lasts as long as needed, the communication path is taken down via a call clearing phase. This process is typical of circuit switched connections, as described in Chapter 9, and of packet switched virtual circuits using the CCITT X.25 protocol. In the connectionless mode, the end-to-end route is not decided *a priori*, but each packet can take an independent route that may be a function of real-time parameters such as traffic congestion, link outages, node overload, or availability of inexpensive "first choice" transmission links. Commercially tariffed connectionless communication is only now beginning to be considered, after some fruitless starts with the "datagram" service in the early 1980s. The ARPANET has used datagrams from its inception, and some ISO standards have become available (as discussed in Chapter 13). Commercial carrier applications of connectionless systems should become available in the MAN environment. The SS7 system described in Chapter 9 uses connectionless mode communication; connectionless methods are also employed in LANs.

IBM's SNA is one commercial example of early implementation of a layered architecture, as discussed in Chapter 1. Most large data communication installations, particularly in the business sector, have SNA networks to some extent. When the traffic is low and highly dispersed, a packet switched architecture may be superior to SNA and similar architectures. Packet switching is based on vendor-independent standards. Packet networks need packet assemblers-disassemblers (PADs) because X.25 terminals never really appeared on the market.

This chapter explores some important aspects of the issues identified in this introduction. In particular, it discusses digital encoding methods, error detection, packet-switching technology, and the connection-oriented mode *versus* the connectionless mode of transmission. In addition, it discusses an important proprietary network architecture that has been the *de facto* standard for the past 18 years and which may, in fact, continue to be so for the foreseeable future.

11.2 DIGITAL ENCODING METHODS

The continued progress toward an all-digital wide area network (of the type afforded by digital loop carriers, ISDN, SONET, *et cetera*) calls for an understanding of digital coding techniques. The bipolar method described in Chapter 3 and the 2B1Q method discussed in Chapter 4 are well-known examples of the available methods; other methods also exist. In this section, some of these encoding methods are discussed briefly to aid the reader in discerning the scope of the discipline (other techniques discussed in Chapter 4 are not revisited here).

Baseband LANs (using either twisted-pair, coaxial cable, or fiber), digital PBXs, and digital access to the public telecommunication network over a digital local loop employ some of the advanced coding techniques. The determination of the type of technique to employ must be based on the signal spectrum, signal synchronization capability, signal interference and noise immunity, error-detection capability, and cost and complexity. Detailed information can be found in Reference [11.1], on which this presentation is partially based.

Some of the more important digital coding methods are as follows [11.2]:

Nonreturn to Zero-Level (NRZ-L). 1 = high level and 0 = low level. This is the traditional means of encoding digital signals and is the simplest code to implement. It is used for common short-distance physical interfaces such as EIA RS-232. The limitation of this scheme is the presence of a dc component and lack of a self-synchronization capability. A long string of 1s produces a constant high voltage; consequently, any drift in clocks between the sender and the receiver cannot be detected by looking at the signal alone. The dc component arises from the fact that all the bits involve a high value, without pulses in the opposite direction to cancel the signal (when the coding scheme has no dc components, a copper transmission channel in which repeaters are coupled by way of transformers can be designed, which is advantageous).

Nonreturn to Zero-Mark (NRZ-M). 1 = transition at beginning of interval and 0 = no transition. This is a type of differential coding: The signal is decoded by comparing the polarity of adjacent signal elements, rather than the absolute value of the signal itself. This type of coding is more reliable in the presence of channel noise because we can sense the polarity.

Nonreturn to Zero-Space (NRZ-S). 1 = no transition and 0 = transition at beginning of interval. This is identical to NRZ-M, except for the coding rule. Clock synchronism remains a problem. NRZ-type codes are at this time the most common techniques for data communication; the reader may be familiar with the NRZI (NRZ inverted), typical of IBM's SDLC links.

Return to Zero (RZ). 1 = pulse in first half of bit interval, the other half is no pulse; and 0 = no pulse. With this method, we have a maximum rate twice that of the bit rate; this is advantageous because the bandwidth is greater than that achieved with NRZ. However, the dc component and lack of self-synchronization limitations remain.

Biphase Level (Manchester). 1 = transition from high to low in middle of interval and 0 = transition from low to high in middle of interval. Biphase schemes were developed to compensate for the three major limitations of the NRZ codes. More channel bandwidth is required to carry a biphase signal because at least one signal transition per bit (and sometimes two) is required. The advantages these schemes afford are: *Synchronization*: the receiver can synchronize on the sender because a predictable transition exists during each bit interval; *dc components*: none, the majority of the energy is between one-half and one the bit rate; and *error detection*: the absence of an expected signal transition can be employed to detect errors. Manchester codes are gaining ground in communication at the expense of the entrenched NRZ; magnetic tape recording and fiber optic systems employ this code.

Biphase Mark. Always a transition at beginning of interval; 1 = transition in middle of interval and 0 = no transition in middle of interval. Same benefits discussed above; it is a differential code.

Biphase Space. Always a transition at beginning of interval; 1 = no transition in middle of interval and 0 = transition in middle of interval. Same benefits discussed above; it is a differential code.

Differential Manchester. Always a transition in middle of interval; 1 = no transition at beginning of interval and 0 = transition at beginning of interval. Same benefits discussed above; it is a differential code. Used in baseband token-ring LANs on coaxial cable.

Delay Modulation. 1 = transition in middle of interval and 0 = no transition if followed by 1; transition at end of interval if followed by 0. This approach, known as the *Miller code*, requires at least one transition per two-bit interval, while there is never more than one signal transition per bit. Consequently, this scheme has synchronization capabilities, but requires less channel bandwidth than the biphase method.

Bipolar. 1 = pulse in first half of bit interval, alternating polarity from pulse to pulse, and 0 = no pulse. This is the scheme employed by T1 systems; following this discussion, we should be able to appreciate its properties more clearly. Bipolar has no dc components (so that the T1 line can have transformer-coupled repeaters). Bipolar does not afford self-synchronization from the incoming signal, which is why the elaborate synchronization process undertaken by the T1 hardware is required. Bipolar does provide some error-detection capabilities because contiguous 1s must have opposite pulses. The energy content of the bit stream concentrates around 722 kHz; nonetheless, the power spectrum of this signal covers a broad frequency range with significant spectral lines from 20 kHz to 1 MHz [11.3]. The bipolar signal can be transmitted approximately one mile over 24-gauge twisted pair; for additional mileage, regenerative repeaters powered over the line are required, as discussed in Chapter 3. Manchester-type codes provide overall superior performance, and could one day supplant the older bipolar scheme, particularly when more of the plant will be fiber-based and the ancillary equipment (channel banks, but also switches) becomes more integrated with fiber technology.

11.3 CYCLIC REDUNDANCY CHECKING

Cyclic* redundancy checking (CRC) is used to perform error detection (and subsequent correction) in all modern communication systems. (Error detection was introduced in Chapter 1.) Error treatment is typically done at the data link control layer of the OSIRM, although it can also be done at the transport layer. The data field within the message is used to compute the CRC value, which is included as the last field of a transmitted frame. When the frame arrives at the destination, the receiving system uses an identical process to calculate a CRC from the received

*Some authors use "cyclical" instead of "cyclic." The term "check sum" is also employed. While check sum characters can be used at the transport layer (ISO 8072/8073) for error-detection purposes, the algorithmic process for calculating the check sum is not the same as that used in CRC. The term *frame check sum* (or sequence) (FCS) is also used, particularly in the context of LANs.

data field and then compares its independently calculated CRC with the one that arrived with the frame. If the two CRCs do not match, an error has occurred and the communication session proceeds under the protocol's error recovery provisions. If the two CRCs do match, the message has a high probability of having arrived intact. Using a CRC does not guarantee 100% error detection, but can come close; the degree of protection achieved with CRC depends on the CRC method used.

A typical data link frame is shown in Figure 11.1. In a character-oriented data link control protocol (such as binary synchronous communication, widely available in the 1960s and 1970s), the beginning of the header is indicated by a SOH character; the text field begins with an STX character, which also terminates the header field. The text field is terminated by an ETX (end of text) or ETB (end of text block). The CRC field follows the ETX-ETB character. In a more modern bit-oriented protocol, the PDU is delimited by a starting flag, from which all field positions are determined, at prespecified starting points away from the flag.

In CRC calculations, a frame's bit pattern is treated as shorthand for a corresponding polynomial's coefficients. Assuming 8 bits per message byte, a given 11-byte message has 88 bits that uniquely define an 88-term polynomial (having terms with degrees 87 through 0). By definition, these polynomial coefficients all have a value of 0 or 1. The polynomial thus derived is called $F(x)$. To compute a frame CRC, a selected polynomial $P(x)$ divides the polynomial $F(x)$ derived from the frame using the rules of modulo 2 arithmetic. This yields a remainder polynomial $R(x)$ with coefficients that have the same shorthand representation as the bits of the frame's CRC value [11.4].

The calculated CRC value is transmitted immediately following transmission of the original frame. The CRC is transmitted with the high-order bit first as a single unit, even if it consists of more than 8 bits. (All the CRC bits are transmitted, even high-order zero bits.)

During message transmission, bytes are sent sequentially, in the same order as they are in the message; however, bits of a byte are traditionally transmitted low-order bit first. For example, the ASCII character A, equivalent to a binary 01000001, reverses to become 10000010. To perform the CRC calculation, the bits must be reversed on a byte-by-byte basis to obtain the polynomial that actually represents the message's bit pattern. After a message's bit pattern is determined, several binary digits—which are usually 0s or 1s and are equal in number to the degree of the polynomial divisor—are added to the beginning of the message bit pattern. This preconditioning step introduces an initial value that is equivalent to an intermediate remainder during the subsequent CRC division. After this, several

Flag or SOH	Header	Data Field	CRC

Figure 11.1 Level 2 PDU, including CRC.

binary 0s equal in number to the degree of the polynomial divisor are appended to the message bit pattern. This padding step allows the divisor polynomial to be applied against every bit position of the original message polynomial.

The polynomial obtained from the original message bit pattern and the adding and padding steps is divided using a selected CRC divisor polynomial under modulo 2 arithmetic division rules. The resulting quotient is discarded, and the remainder is subject to further processing under the rules of the CRC generation procedures. In most cases, the value is not further processed; in some cases, it is subject to a bit-wise inversion in which all binary zeros are transformed into binary ones and *vice versa* [11.4].

The receiving system calculates the CRC on the arriving data characters and includes the arriving CRC characters as though they were part of the original message. It also omits the CRC postconditioning procedure—thereby producing a residue from the process, instead of a CRC value. If a specific final remainder does not exist after the original message bytes and the CRC bytes are processed, the message is considered to be in error. The specific value that must remain depends on the CRC procedure used.

The most commonly used CRC polynomials, their associated preconditioning values and postconditioning procedures, and receiving station specific final remainders are shown in Table 11.1. Because the remainder for degree 16 divisor polynomials is degree 15 or less, the use of one of the first two polynomials (i.e., CRC-16 or CRC-CCITT) results in a 16-bit remainder. This allows for the detection of all errors spanning 16 bits or less and about 99.995% of other errors. The CRC-32 polynomial is the CRC divisor in the IBM Token-Ring and IBM PC Network as well as in CSMA/CD Ethernet LANs. CRC-32 provides a 32-bit CRC value that can receive messages that are too large for the SDLC/HDLC CRC, though the 32-bit CRC uses a similar preconditioning and postconditioning process. Reference [11.4] provides a detailed treatment of this subject, with a number of examples.

11.4 PACKET-SWITCHING TECHNOLOGY

Traditional packet switching is a technology of the mid-1960s that attempted to solve the networking problems of the 1960s. At that time, bandwidth was scarce

Table 11.1

CRC type	*Quotient Polynomial*	*Preconditioning value*	*Postconditioning*	*Final residue*
CRC-16	$x^{16} + x^{15} + x^2 + 1$	0x0000	None	0x0000
CRC-CCITT	$x^{16} + x^{12} + x^5 + 1$	0xFFFF	Bit inversion	0xF0BB
CRC-32	$x^{32} + x^{26} + x^{23} + x^{22} + x^{16} + x^{12} + x^{11} + x^{10} + x^8 + x^7 + x^5 + x^4 + x^2 + x + 1$	0xFFFFFFFF	Bit inversion	0xDEBB20E3

and networks attempted to maximize the efficiency in transport. Now with the widespread availability of fiber, with an intrinsic traffic-carrying capacity that has been doubling every two to three years for the past few years, the need to maximize efficiency at the expense of end-to-end delay and switching node complexity is no longer an imperative. Also, in the 1960s, BER left a lot to be desired, with 10^{-6} stretching the technical limit. Fiber now routinely provides 10^{-9}, and BER can improve that further. The error-prone circuits necessitated complex error checking and recovery procedures at each node of a network.

The X.25 packet standards described below assume that the transmission media is error-prone. To guarantee an acceptable level of end-to-end quality, error management is performed at every link by the fairly sophisticated but resource-intensive link protocol HDLC. HDLC provides (1) core functions, including frame delimiting, bit transparency, error checking with CRC, and error recovery, and (2) other functions. Frame relay service (Chapter 4) only supports core functions in the network.

Packet switching (private, public, and equipment) represents about 4% of the data communication revenues in the United States ($1 billion per year compared to a total of $25 billion per year). Hence, it deserves some discussion. Proponents claim major gains in the future, while others remain skeptical. Ultimately, many types of communication, including T1 frames (Chapter 3), SNA/SDLC, LANs, MANs, and BISDN, use packet technology in the sense that data are transferred in frames and frames can be viewed as packets (although not X.25 packets); people, however, tend not to classify these as packet switched systems at this time.

11.4.1 X.25 Packet Protocols

Packet networks follow a layered architecture. The lowest layer, the physical layer or layer 1, is concerned with physical connectivity to the network. The next layer, the link layer or layer 2, deals with error control and flow between two adjacent points. Layer 3, or the network layer, deals with end-to-end networking aspects, including routing. This is consistent with the OSIRM. CCITT Recommendation X.25 is a standard interface protocol between packet switched DCE equipment and packet-mode DCE. The recommendation was first adopted in 1976 and was significantly revised in both 1980 and 1984; minor revisions took place in 1988. Related protocols are shown in Table 11.2.

The physical level deals with the representation of data bits, timing aspects, and physical contact between DTE and DCE (X.21, in particular). The link level provides the functions of link initialization, flow control, and error control (LAP and LAP-B). The network layer (X.25 Packet Layer Procedure—X.25 PLP) deals with support of multiple data streams independently of others and with the initialization of the communication interface after a serious fault.

Unfortunately, X.25 was written from a DCE perspective and can be interpreted in many ways. Different DTE interpretations may be incompatible with

Table 11.2
CCITT Packet Protocols

X.21 bis	This standard is used on public data networks of DTE that is designed for interfacing to synchronous V-series modems.
X.21	This standard is used to interface between DTE and DCE for synchronous operation on public data networks.
X.25	This standard is used to interface between DTE and DCE for terminal operation in the packet mode on public data networks.
X.29	This standard is used to specify the procedures for the exchange of control information and user data between a PAD and a packet-mode DTE or another PAD.
X.28	This standard is used to specify the DTE-DCE interface for start-stop DTEs accessing a PAD in a public packet switched network.
X.3	This standard is used to specify PAD in a public network.
X.121	This standard is used to specify any international numbering plan for public data networks.
X.75	This standard is used to specify terminal and transit call control procedures and data transfer systems on international circuits between packet switched data networks.

Note: Additional ISO-created standards are identified in Chapter 13.

DCEs. The use of X.25 requires verification of implementation compatibility; this process is called *conformance testing*.

11.4.2 Data Link Level

The link level protocols used in X.25 are the link access procedures (LAP and LAP-B). These procedures are similar and compatible with HDLC [11.5]. LAP-B provides an essentially error-free data channel despite the unreliability of the physical medium. At the receiving end of the link, the information is delivered in units (or packets) to the packet level without loss or duplication and in the same sequence of transmission.

The basic transmission unit at the link level is the frame. Specific bit patterns (flags) delimit the frames and also provide a means for link level synchronization. A LAP-B frame structure consists of a link level header, an information field (if any), and to detect transmission errors, a 16-bit CRC code. The frame structure is similar to that of Figure 11.1; the header consists of an address field and a control field. The header contains information on whether the frame is a command or a response and differs with frame type (that is, information frames, supervisory frames, or unnumbered frames). The information field is present only in information frames (I frames) and frame reject frames (FRMR). In a FRMR frame, the information field indicates the reason for the FRMR response, while the information field of an I frame normally contains one layer 3 packet to be transferred across the data link.

For each direction of transmission, every information frame to be transmitted across the interface is sequentially numbered in modulo 8 (or 128) arithmetic. This number, denoted as the send sequence number N(S), is contained in a special field within the information frame. Furthermore, information and supervisory frames contain a receiving sequence number N(R) to acknowledge the reception of I frames up to and including the I frame with send sequence number N(R) − 1.

To avoid congestion, the maximum number of unacknowledged I frames for each direction of transmission is restricted to the window size *k*. The lower window corresponds to the oldest unacknowledged I frame and the upper window edge is the lower window edge plus *k* in modulo 8 (or 128) arithmetic. As acknowledgments are received, the lower window edges are moved and the transmission station can send additional I frames. The receiving station may acknowledge more than one information frame ("piggybacking"). The transmitting station maintains a copy of each packet it is sending and restarts a timer *T* whenever an I frame is sent. When the receiving station acknowledges receipt of one or more information frames, the packets associated with those information frames are removed from the storage buffer. If the acknowledgment for an I frame is not received before the timer expires, the sending station will query the receiving station with an appropriate supervisory frame. Should the acknowledgment still be missing after *N* attempts, the transmitter will take alternative recovery actions. When a station receives information frames faster than the rate acceptable by its higher level entities, it may indicate a busy condition to the transmitting station with a "receiver not ready" supervisory frame. The sending station will stop the transmission of information frames until it receives a "receive ready" supervisory frame from the receiving station.

When an "out of sequence" frame is received, a reject frame is sent to the transmitting station. The transmitting station will retransmit all information frames starting from the earliest unacknowledged frame. A frame reject is transmitted following reception of an invalid frame that cannot be corrected by retransmission when one of the following conditions occur: (1) an invalid receive sequence number, (2) an invalid control field, or (3) a noninformation frame with an information field is received.

Because the FRMR frame indicates a serious link condition, the data link must be reinitialized transfer. Link level parameters may be adjusted to enhance the efficiency of link operations.

(For comparison with other layer 2 protocols, the LAP-D data link control procedure was discussed in Chapter 4; the SS7 procedure was discussed in Chapter 9; and LAN procedures are discussed in Chapter 14.)

11.4.3 Packet Level

The X.25 PLP (CCITT X.25 PLP 1980, 1984, ISO 8208) defines rules to use logical channels to multiplex user sessions and utilize the link bandwidth more efficiently, in addition to data transfer flow control. The X.25 PLP provides a number of capabilities including:

- Multiplexing: the ability to support multiple data streams;
- Transfer of addressing information: the ability to transfer addressing information including OSI network service access point (NSAP) addresses (see Chapter 1);
- Segmenting and reassembly: the ability to divide a data unit into smaller packets for transfer over a network or LAN and to reassemble packets into the original data unit;
- Flow control: the ability to control, for each data stream, the flow of data between transmitting and receiving DTEs;
- Transfer of expedited data: the ability to transfer a small amount of data outside the normal flow-control procedures;
- Error control: the ability to detect errors at the packet level; and
- Reset and restart: The ability to reinitialize communication paths at the packet level in the event that nonrecoverable error conditions are encountered.

The allocation of logical channels can be either static in the case of a permanent virtual circuit (PVC) or dynamic. A PVC emulates a point-to-point private line. The network delivers packets to the destination in the order of transmission by the origination DTE. The definition of which channels are reserved for virtual calls and for PVCs is determined at subscription time. In the 1984 version of X.25, these channel ranges may be dynamically changed using registration procedures.

X.25 allows up to 4096 logical channels on an individual interface. Therefore, many individual low-speed terminals can be multiplexed on a single higher speed digital channel at each DTE-DCE interface, using the logical channel number for identification. Theoretically, up to 7 (or 127) outstanding data packets could each carry 4096 octets on each of these 4096 logical channels. In practice, however, the number of logical channels and the window size supported are determined by available buffer storage, processing power, speed of the access line, *et cetera*.

The operation of the packet level in X.25 is as follows: After link setup and packet level initialization, the calling DTE must establish a virtual circuit between itself and the destination. The calling DTE chooses one idle logical channel from those available for outgoing virtual calls. To minimize the possibility of call collision, the available logical channel with the highest number is selected. If the called DTE informs the network that it accepts the call, the virtual circuit is established and both DTEs may now use the full-duplex virtual circuit to exchange packets. At each interface, transmission of data packets is controlled separately for each direction of transmission. Data packets are sequentially numbered in modulo 8 (or 128) arithmetic and the corresponding packet send sequence number P(S) is contained in each data packet [11.5].

11.4.4 PADs

Non-X.25 synchronous and asynchronous terminals can interface to X.25 networks through a PAD. The PAD performs the necessary X.25 functions on behalf of the terminal. Other CCITT recommendations specify the physical and logical aspects of this asynchronous interface, as listed in Table 11.2. Recommendation X.3 specifies a set of parameters used by the PAD to control asynchronous terminals. Users of asynchronous terminals may, in turn, change individual parameters according to procedures specified by Recommendation X.28. The remote X.25 DTE may also modify the parameters according to the procedure identified by Recommendation X.29. X.25 packets that control the PAD are called *data qualified packets* and have their Q bit (bit 8 in the first octet of a data packet header) set to 1 by higher level entities. The Q bit is always 0 for data packets containing user data, and all data packets in a complete packet sequence must have the same Q bit setting. Terminal handling, data forwarding, and other PAD functions associated with asynchronous terminals are controlled by approximately 40 X.29 parameters.

11.4.5 Recent Activities in X.25 Standards

The purpose of this section is to highlight some recent advancements in the X.25 packet-switching environment.

CCITT Blue Book 1988

Publication of CCITT standards has traditionally occurred over a four-year cycle, as discussed in Chapter 2. In 1984, CCITT published the Red Book on a wide range of topics, including the X.25 standards. The Plenary Assembly session to revise the 1984 X.25 standard took place in November 1988. The 1988 technical changes to X.25 were rather small. In general, we have seen an effort at harmonizing with the ISO standards. CCITT X.25 PLP specifies a virtual circuit service (virtual calls and PVCs). A compatible version of the packet standard was issued by ISO in ISO 8208. Preparations were also being made as of this writing to carry longer addresses in the DTE field to facilitate interworking with ISDN (E.164). The information discussed below is based on Reference [11.6].

No changes were anticipated at the physical and link levels. At the packet level, a new facility for redirecting calls, identified as *call deflection*, is being made available. In 1984, a new call redirection facility was made available whereby the network can redirect all calls destined to a given address. This redirection could be based on the fact that the destination was out of order or busy, or based on time of day or other criteria. The 1988 facility extends this capability. The destination subscriber can clear incoming calls to another party on a call-selective basis. The clear request packet will contain the call deflection information as to the desired alternative party.

The address extension facilities have been modified to be consistent with ISO address length. Previously, a provisional 32-decimal, 16-octet field had been recommended. The address length will now be 40 decimals and 20 octets. The address recommendations were added to Addendum 2 of ISO 8348, and have been adopted by CCITT into X.213 (naming and addressing is covered further in Chapter 13).

Also, a preferred binary encoding has been defined for network service access point addressing. The ISO defines three properties that NSAP addresses must possess (ISO 7498/AD3) [11.7]: (1) global nonambiguity, (2) global applicability, and (3) route independence.

In addition to defining these properties of NSAP, ISO has developed an NSAP addressing scheme to provide these properties, as described in ISO 8348/AD2. The ISO specification, however, only defines the abstract notation and semantics of the NSAP addresses. The actual encoding of the address information is left for specification by the specific network layer protocol standard, for example, X.25. The 1984 X.25 standard had an address extension feature, providing a flag that indicated that a partial NSAP address was carried; that feature has been removed.

11.4.6 Connectionless-Mode Lower Layers: Relationships with ISO Efforts

At any layer of the OSIRM, except the physical layer, two basic forms of operation are possible, as highlighted in the introduction to this chapter: the connection-oriented mode and the connectionless mode. Connection-oriented service involves a connection establishment phase, a data transfer phase, and a connection termination phase; a logical connection is set up between end systems prior to exchanging data (see Figure 11.2). These phases define the sequence of events ensuring successful data transmission. Sequencing of data, flow control, and transparent error handling are some of the capabilities inherent with this service. The similarity of these phases with those described in Chapter 9 for a telephone call are not coincidental: traditional telecommunication services are connection-oriented.

In a connectionless service, each PDU is independently routed to the destination; no connection-establishment activities are required because each data unit is independent of the previous or subsequent one. Connectionless-mode service provides for unit data transfer without regard to the establishment or maintenance of connections, as shown in Figure 11.2. The service element required to perform this data transfer is called an (N)-UNITDATA (or (N)-UNIT-DATA) service element, and it provides the function of data transfer of discrete data units. Clearly, each unit of data must contain at least the addressing information and the data. In the connectionless-mode, transmission delivery is uncertain, due to the possibility of errors. To the uninitiated, this may appear contrary to the goal of communication: We would like assurance that the message gets to the other end. In reality, connectionless-mode communication simply shifts the responsibility for the

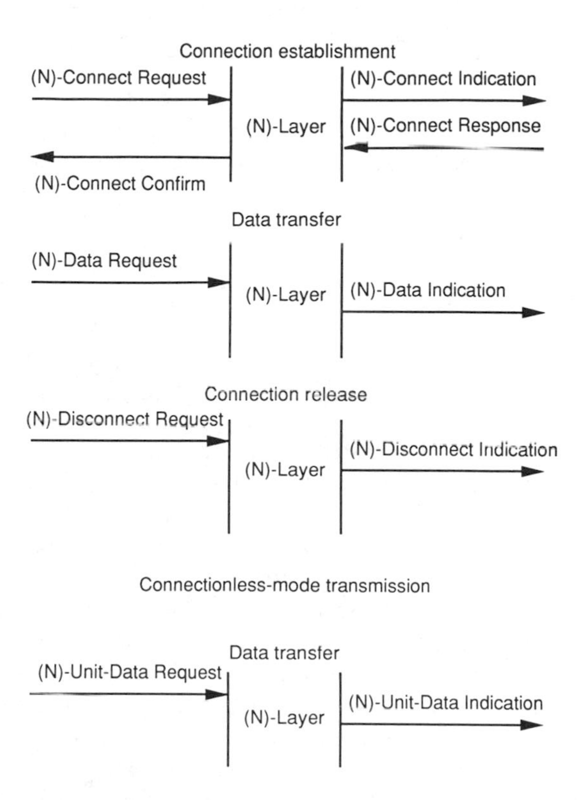

Figure 11.2 Connection-oriented mode *versus* connectionless mode communication.

integrity to a higher layer, where the integrity check is done only once, instead of being done at every lower layer. Alternatively, the recovery mechanism may lie within each unit of data. Table 11.3 compares a connection-oriented network layer service to a connectionless-mode service.

Each mode of transmission has a niche where it represents the best approach. For example, file transfers of the order of gigabytes may benefit from a connection-oriented mode lower layer service, while point-of-sale inquiries may be best served by a connectionless-mode service. Generally, connectionless-mode data transmission may be ideal for (1) broadcast or multicast of information; (2) inward data collection, which involves the periodic sampling of many sources (such as in process control); (3) transient processes (in military, aviation, and meteorological systems, in which frequent and abrupt dissociation from peer processes often occurs); and (4) OSI-defined functionality, for example, directory inquiry (discussed in Chapter 13).

Table 11.3

	Connection-oriented	*Connectionless*
Typical protocol	ISO 8208 (X.25 PLP)	ISO 8473 connectionless network protocol (CLNP)
Packet treatment	Packet layer sets up logical channel Each packet has a logical channel identifier Same virtual circuit for duration of the call	Packets are sent independently Each packet has complete addressing information Packets can take totally different routes
Services		
Packet sequencing	Yes	No
Flow control	Yes	No
Acknowledgments	Yes	No
Protocol type	Complex	Simple

Traditionally, CCITT has pursued a connection-oriented philosophy, while ISO has shown interest in the connectionless mode. Connectionless communication at the lower layers of the OSIRM is now a well-entrenched technique and is found, for example, in LANs and MANs. While the original OSIRM described in ISO 7498 was connection-oriented, ISO saw the need to provide connectionless service by issuing an addendum to that protocol (ISO 7498/AD1). Considerable effort is under way in ISO to update the Connectionless Addendum, and a parallel process is under way in CCITT SG VII, although CCITT has been reluctant to accept connectionless-mode data transmission concepts into its version of the OSIRM (which is CCITT X.200). ISO has issued a standard for network layer service, ISO 8348, which, in addition to the connection-mode, also contains connectionless service (in AD1). However, with regard to X.25 itself, ISO has decided not to pursue the connectionless service (which was known in the past as "datagram service").

ISO has shown considerable interest in connectionless *internetworking* protocols, as discussed in more detail in Chapter 13. X.75 is also a connection-oriented internetworking service. Interest is now growing for connectionless communication at upper layers. The standards community generally accepts that:

- All levels (except the physical layer) should support connectionless-mode data transmission; and
- The "crossovers" of connectionless-mode transport over a connection-mode network and connection-mode transport over a connectionless-mode network must be facilitated.

11.4.7 Data Link Layer and Physical Layer Standards

This section identifies for reference some key data link and physical layer standards. HDLC standards from ISO are shown in Table 11.4, while physical layer standards

Table 11.4
HDLC Network and Higher Layer Standards

ISO 3309 (1984)	HDLC frame structure
ISO 3309 AD1	Start-stop transmission
ISO 4335 (1987)	HDLC: Consolidation of elements of procedures
ISO 4335 AD1	Unnumbered information (UI) frames and SREJ (selective reject) extension
ISO 4335 AD2	Enhancement of the XID (exchange identification) function utility
ISO 4335 AD3	Start-stop transmission
ISO 7478 (1987)	Multilink procedures
ISO 7776 (1986)	HDLC: Description of the X.25 LAP-B compatible data link procedure
ISO 7809 (1984)	HDLC: Consolidation of classes of procedures
ISO 7809 AD1	UI extensions
ISO 7809 AD2	Description of optional functions
ISO 7809 AD3	Start-stop transmission
ISO 7809 AD4	List of standard DLL protocols that utilize HDLC classes of procedures
ISO 8471	HDLC balanced classes of procedures
ISO 8885 (1987)	HDLC: General-purpose XID frame information field content and format
ISO 8885 AD1	Additional operational parameters for negotiation
ISO 8885 AD2	Start-stop transmission

Table 11.5
HDLC Physical Layer Standards

ISO 2110	25-pin DTE-DCE interface connector and pin assignments
ISO 2593	34-pin DTE-DCE interface connector and pin assignments
ISO 4902 (CCITT X.20)	37-pin DTE-DCE interface connector and pin assignments
ISO 4903 (CCITT X.21)	15-pin DTE-DCE interface connector and pin assignments
ISO 8481	DTE-to-DTE physical connection using X.24
ISO 8482	Twisted-pair multipoint interconnections
ISO 8877	ISDN interface connector at S and T reference points and pin assignments
ISO 9578	Communication interface connectors used in LANs

*Comparison with common U.S. standards**

Standard	*Mechanical Equivalent*	*Electrical Equivalent*	*Functional Equivalent*
EIA RS-232-D	ISO 2110	CCITT V.28	CCITT V.24
EIA RS-422-A	ISO 4902	CCITT V.11	CCITT V.24
EIA RS-423-A	ISO 4902	CCITT V.10	CCITT V.24

*Refer to Reference [11.8] for a more comprehensive comparison.

are shown in Table 11.5 (standards associated with the network layer and upper layers are discussed in details in Chapter 13). Reference [11.7] is an excellent tutorial reference on HDLC; for physical layer interfaces, Reference [11.8] is recommended.

11.5 SYSTEMS NETWORK ARCHITECTURE (SNA) CONCEPTS

SNA is IBM's computer network architecture for large mainframes [11.9]. SNA has been available since the mid-1970s and is continually being evolved to keep up with varying requirements. SNA employs a layered architecture. Traditionally, the system has operated in a hierarchical master-slave relationship, with the mainframe acting as the master and the terminals acting as slaves. Starting in the late 1980s, SNA was enhanced to allow peer-to-peer communication. SNA has generally utilized conditioned dedicated voice-grade lines. DDS and, more recently, T1 digital facilities have also been utilized; some establishments (particularly banks) have made limited use of packet switched networks.

Traditional SNA is a centralized (tree) network, with a mainframe host computer acting as the network control node controlling the front-end processors (FEPs), the cluster controllers, and the terminals—in that order. The above-mentioned controllers and terminals are referred to as the network's domain. SNA establishes a logical path between network nodes, and it routes each message with addressing information contained in the protocol. SNA uses the SDLC protocol, although use of packet-switching protocols is also allowed on a limited scale. SNA also allows host-to-host communication; each host controls its own domain.

Currently, type 3270 SNA communication protocols are the *de facto* standards for mainframe communication and are implemented by almost one-third of the world's installed base of display terminals. More than 7 million devices are directly attached to SNA networks, including 1 million printers, 500,000 cluster controllers, and 1 million protocol converters [11.10]. In 1989, an estimated 750,000 3270-compatible terminals and 100,000 controllers were sold, generating $3 billion of revenue [11.11].

11.5.1 Architecture

SNA is a layered architecture consisting of seven layers, each performing a separate function to ensure the reliable flow of data. These seven layers are not precisely the same as the layers in the OSIRM, although good agreement is found at the lower three layers.

SNA's layer 1 is the physical connection between the communication circuits and terminal equipment.

Layer 2 is called the *data link control*; its function is to detect and correct line errors. SNA employs the SDLC protocol for this function. The framing consists of a number of bytes of overhead, including the start of the frame, the address, control fields, the data field (having varying lengths), and a two-byte CRC.

Layer 3 is the *path control*. Its two functions are to establish a path through the network and data flow control through the network, and it regroups messages to avoid congestion. Long messages are segmented into smaller ones to help eliminate errors that would cause excessive retransmission. Short messages are combined into longer ones so that the network will not be cluttered with large amounts of small transmissions.

Layer 4 is called *transmission control*. Its function is to check the speed and buffer management between logical units (LUs); it prevents the transmitting LU from sending more data than the receiving LU can handle.

Layer 5, called *data flow control*, uses two methods of preparing messages for transmission. The two methods are chaining for one-way transmission and bracketing for two-way transmissions.

Layer 6 consists of the *function management data services*; it handles three functions: (1) activating and deactivating internal links; (2) network operator services, which allow operators to send commands and receive responses; and (3) management services, which are responsible for testing and troubleshooting the network.

Layer 7 is the *network addressable units* (NAU) services layer; it formats host data for display at CRTs and for printers.

Superficially, the architectures of OSIRM and SNA appear to be similar, and indeed they are in a number of ways. Fundamental differences appear when investigating the underlying assumptions in the initial implementation of SNA products.

Although the OSIRM chronologically followed SNA, both were being defined in the same time period. Like OSIRM, SNA uses a collection of layer functions, as listed above, with a number of its layers somewhat equivalent to the OSI layers. The SNA layers containing transmission control through NAU services are system-resident protocols supporting the communication activities of local applications. These layers are independent of the underlying layers that have the responsibility for communication services between systems. Communication services are provided at the three lowest layers. The initial backbone service used a bit-oriented protocol managed by the network control program (NCP) executing in a FEP. Private leased lines could be utilized efficiently through multiplexing many sessions on the same physical circuit. The NCP manages all switching nodes, configuring them and selecting routing options before any traffic flows; switching nodes are passive, following instructions issued by the NCP. Failures of network components cause the NCP to issue directives initiating error recovery and reconfiguration procedures. In traditional SNA, LUs cannot communicate directly with each other, even on a powerful PC; they must use the mainframe or FEP as a relay point. See Figure 11.3, which depicts the original hierarchical structure of SNA.

When SNA was being defined, IBM viewed the mainframe as its major computer product. The choice of a mainframe-driven hierarchical communication architecture followed from these considerations. These initial assumptions were appropriate when SNA was first defined; however, in the meantime, users and

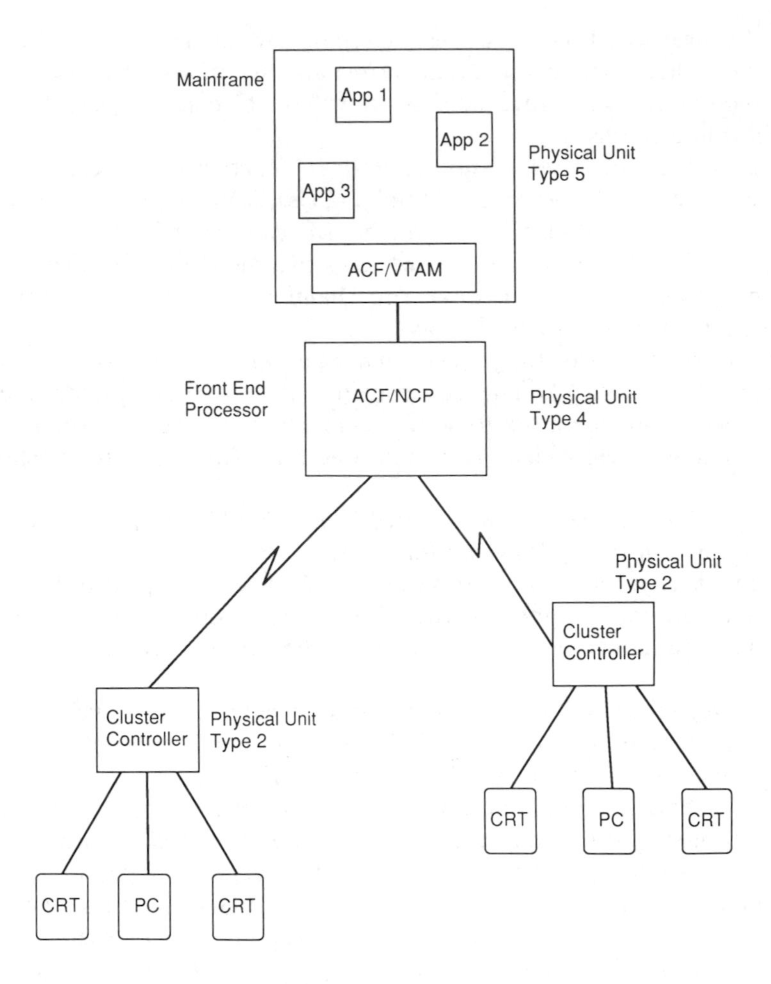

Figure 11.3 Traditional SNA environment.

networks have evolved. When SNA was specified, most communication was between terminal devices and a mainframe; few users were actively pursuing computer-to-computer communication and even fewer were concerned with internetworking between different computer architectures. The first products were influenced by this terminal-to-mainframe orientation. Terminal-to-mainframe environments are characterized by many remote terminals communicating with a few mainframe sites. These traffic patterns are suited to hierarchical management of communication and routing: Routing decisions can be made so that traffic is evenly spread across all the links and congestion is minimized. Terminal devices depend on a computer to direct and control communication activities. Normally, terminal devices cannot communicate directly with each other. They allow remote users to access mainframe resources for program execution or data base inquiries.

Changes are now being made to SNA to facilitate distributed data processing, as discussed in a subsequent section.

11.5.2 Key Mainframe Facilities

Multiple Virtual Storage (MVS) is the general name given to the flagship operating system used on large IBM mainframes. There are currently three variants of MVS: MVS/SP, MVS/XA, and MVS/ESA. MVS is to an IBM mainframe what DOS is to a PC. The nucleus is the portion of MVS code in which very critical, low-level routines reside. Software code that resides in the nucleus usually cannot be altered without shutting down a computer and shutting out all users of that computer.

The virtual telecommunication access method (VTAM) is the software module that mainframe applications use to send and receive messages over an SNA network; VTAM implements layers 2, 3, and 4 of SNA. NetView is the IBM software package that provides VTAM network management capabilities.

NCP is the operating system for IBM FEPs. Advanced Communication Function (ACF) is a nomenclature given to IBM's mainframe communication modules in the early 1980s to differentiate between the old releases of VTAM, NCP, and SSP, which came with the MVS operating system at no extra charge, and the then-new releases of those products, which are available for a fee.

11.5.3 SNA Physical Unit Types

In SNA, a node embodies the set of responsibilities that governs physical attachment to the network. Whatever fulfills these responsibilities is said to implement the physical unit (PU), which can be hardware, software, or firmware. A variety of PU types exist:

- PU type 4 is the node type of a FEP; this type of physical unit may eventually be phased out. A node type 4 is subordinate to a node type 5.
- PU type 5 is the node type of a mainframe itself, or of a node that contains a system services control point. Implementation of a PU type 5 emulates a mainframe. This type of physical unit may eventually be phased out.
- PU 2.1 is IBM's strategic node type. A 2.1 node can be directly connected to any other PU 2.1 node. Future IBM products largely will be type 2.1 nodes.

The PU name is the mnemonic name given to a node. This name is made known to VTAM or NCP and sometimes to the node itself [11.12].

A few years ago, the term "physical unit" was dropped in favor of the use of the term "node" instead. PU type 2, now retroactively called PU 2.0, is the specification for a peripheral node. Examples of PU 2.0 implementations include 3174 cluster controllers and most minicomputers. Two type 2.0 PUs cannot attach to each other; instead, they must be connected to a larger node PU type 4 or 5, hence, this node type is being phased out. Other SNA nodes include: (1) A network node (NN) is a type 2.1 node that is capable of routing traffic between two other computers; a network node can perform intermediate routing of sessions between two peripheral nodes. (2) A peripheral node (PN) is a type 2.1 node that contains LUs, one of which is a session partner.

Other key SNA concepts include the following:

- Network identification (NETID) is a character string, up to eight characters in length, that uniquely identifies the network in which a user seeking SNA services resides. The new releases of VTAM requires that every network be given a name. To enable cross-network identification, node names can be qualified with the name of the network (NETID) in which they reside. This should not be confused with NETNAME which, with VTAM, is analogous to an area code.
- Control point (CP) is the intelligence that manages a node and provides network services to attached LUs.
- Subarea is a division of an SNA network that is under the control of, and addressed by, a FEP or a host access method such as VTAM. Each PU 4 and PU 5 is a subarea.
- SSCP is the entity that controls and manages the resources of a data communication network that are owned by a host.
- Systems support program (SSP) is the set of utility programs that assembles the NCP system generation and loads, dumps, and debugs the NCP. (SSP as used here is short for the mainframe ACF/SSP, and is not to be confused with the SSP operating system of the System/36) [11.12].

11.5.4 SNA Logical Units

An LU is the addressable entity in an SNA network to and from which a program or a user can send and receive messages. A LU could also be described as the definition of a generic device. A LU can be implemented in either hardware, software, or firmware. Whatever fulfills the SNA responsibilities of the LU is said to implement the LU [11.12]. The LU software is resident on the mainframe (now also on other devices) as a telecommunication access method used by the various applications. The initial LUs reflected the hierarchical, terminal-based philosophy because these LUs were based on devices such as 3270 terminals, remote job entry (RJE) devices, and remote printers. LU-to-LU sessions were always managed by the mainframe-resident LU. One result of this approach is that two remote LUs cannot directly communicate with each other.

In the traditional SNA environment, applications resided only on a mainframe and were responsible for managing interactions with remote partners. Application management procedures for remote interaction, error recovery, and other situations had to be constructed without depending on the terminal-based partner for assistance. Institutionalizing this assumption in applications makes the applications harder to adapt to interworking with other computers [11.13]. For example, in a terminal environment, an error may result when repeating the transmission of the entire file because terminals have limited capabilities (such a file would probably be small because terminals have very limited buffers). However, if another computer is involved, partial file contents can be saved on disk, thus allowing recovery to begin from the point of error. These files may be large, and retransmitting the entire file is inefficient.

LUs are embedded in proprietary software, making the units difficult for users to modify. As a result, application designers may pragmatically choose the LU that comes closest to their needs; this means, however, that at that point many factors such as data streams, delivery, and recovery procedures are determined by terminals, rather than by application needs.

Several LU types have been defined over the years to satisfy different session requirements, as listed below. For example, LU 1 describes printer and keyboard-printer devices; LU 2 describes display stations with IBM 3270 data stream support; LU 0 allowed product-specific combinations of SNA protocols (this protocol, which was used in the 1970s, provided no standardized presentation services). With the advent of distributed data processing, the need arose to provide any-to-any connectivity through a single LU:

Logical Unit Type 6.0. This was the first attempt to provide program-to-program communication. LU 6.0 was made available in CICS/VS 1.4 and facilitated only CICS-to-CICS communication. CICS (customer information control system) is a well-established data base, data communication module that runs under the MVS operating system and is optimized for end-user applications.

Logical Unit Type 6.1. This LU allowed three types of communication: (1) CICS-to-CICS; (2) IMS-to-IMS (information management system), where IMS is one of two major data base access systems provided by IBM for the MVS environment (the other is DB2); and (3) CICS-to-IMS. However, communication with printers continued to be in a master-slave relationship (LU 0, LU 1, LU 2, and LU 3); this implied that a uniform LU type across the transaction did not exist.

Logical Unit Type 6.2. This is the SNA protocol now used for program-to-program communication between any two processes. This was originally called "logical unit convergent" because it is the LU type around which the entire IBM product line will eventually converge. LU 6.2 is IBM's strategic direction for program-to-program communication. Some observers call the LU 6.2 a *de facto* standard for communicating across a variety of dissimilar hardware platforms [11.14].

LU 6.2 was first available on CICS/VS 1.6. LU 6.2 defines a *peer-to-peer* protocol for managing the conversation between the two end entities. A *session* in the SNA sense is a connection between two LUs. Such a session is initiated with a BIND command and terminates with the UNBIND command. A *transaction* is a unit of work done between the application program on one end of the session and the application program or device at the other end of the session. The application communicates within the host to VTAM over the VTAM application programming interface (API). While LUs are connected by sessions, transaction programs are connected by conversations. When a program needs to converse, it issues an allocate request. If a session already exists and is not busy with another conversation, that session is used to initiate the conversation. Otherwise, a new session is initiated, after which the two transaction programs initialize LU 6.2 functions to send and receive data that constitute the transaction. When this is completed, one of the programs can deallocate the conversation.

LU 6.2 is based on *conversational-mode* communication between transaction programs and can be half- or full-duplex. In a conversational environment, each command received by an end application is responded to in real time. CICS is an example of a conversational system. In half-duplex, responses follow a strictly temporal sequence; in full-duplex conversations, applications respond asynchronously—but in real time.

The introduction of peer-to-peer communication is more difficult when the end application systems are queued.* IMS does not currently support LU 6.2 communication, and it may be in the 1991–93 time frame before such capability becomes available.

The LU 6.2 protocol consists of a series of verbs (with parametric information) and return codes. The four types of verbs are:

1. Basic conversation mode, for lower-level interface to the LU;
2. Mapped conversation verbs, for user-written transaction programs, giving an interface suitable for high-level languages;
3. Type-independent conversation verbs, providing functions that span both mapped and basic conversations; and
4. Control-operator verbs, giving functions for initiating and controlling SNA protocols to establish and terminate sessions across the network.

LU 6.2 Base and Optional Functions

The LU 6.2 functions and protocols are organized into functional subsets: the "base function set" and "optional function sets." The base function set includes

*A queued environment implies that requests are processed as time allows and must wait in queue for resources. IMS is an example of a queued-transaction system.

basic conversation verbs, mapped conversation verbs, and control-operator verbs. The principal optional function sets are syncpoint, program initialization parameters, security, and performance options.

Products implementing LU 6.2 can communicate to each other using the base set; any two products supporting functions in the same optional set can communicate between them using that full function set. Figure 11.4 depicts the base and several towers of LU 6.2 implementations.

The VTAM has implemented LU 6.2 basic conversation verbs, as shown in Table 11.6.

Syncpoints represent a mechanism for acknowledging completion of subatomic tasks within the transaction life cycle. LU 6.2 base supports either a no-sync point operation or a simple confirm operation in which a data phase is followed by a confirm phase.

11.5.5 SNA Directions

The SNA implementations of the 1970s and early 1980s exhibited limitations as more sophisticated applications requirements involving distributed computer-to-computer communication began to appear. In particular, the incorporation of microprocessor workstations opened many new networking alternatives that SNA could not exploit effectively. The main characteristic of the initial SNA imple-

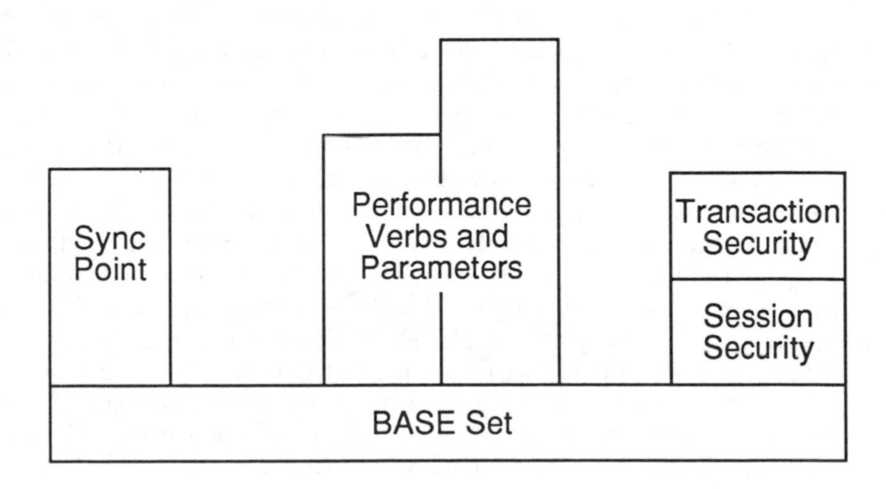

Figure 11.4 LU 6.2 base and tower functional subsets.

ALLOCATE	associate a conversation with a session from the pool
CONFIRM	transmit any data in the send buffer and request confimation
CONFIRMED	acknowledge confirmation
DEALLOCATE	end conversation
FLUSH	transmit any data remaining in the send buffer
PREPARE_TO_RECEIVE	transmit any data remaining in the send buffer with a indication of Change Direction
RECEIVE_AND_WAIT	basic Receive
REQUEST_TO_SEND	signal for Change Direction Indicator
SEND_DATA	place data in the send buffer and, if enough data is waiting, transmit it
SEND_ERROR	indicate error

Table 11.6 Basic Conversational Verbs Implemented by VTAM.

mentation is the choice of hierarchical organization and central control at every level, as discussed in previous sections. Creating an internetworking environment is more difficult when working with a centralized philosophy. Centralized control of communication traffic is appropriate for terminal-to-mainframe operations, but direct interaction between microprocessors still requires a relay through the central control site. Increasing traffic adds to the loading of the relay point, and contributes as well to network congestion and decreasing application performance.

Many applications were customized to work with a terminal-based partner (usually a 3270 data stream device). More sophisticated control procedures must be designed to allow each partner to take on the appropriate role for the application. The task is further complicated by the need to develop more sophisticated application software on each participating system, rather than on just one. Data movement will involve more complex data objects than terminal-oriented character streams. Additional translation software within each system will be needed for each new object. Moving large amounts of data for local processing is another aspect that demands new software development. Interworking may require that much of the existing terminal-based software be replaced. Many current applications must still continue supporting terminal-based interactions as well as newer interworking partners; changing or replacing applications is a large endeavor.

Further complications resulted from the fact that the SNA "standard" was loosely followed even within IBM. Each major product line is incompatible with the other at hardware, operating system, and communication levels. Many SNA products have slightly different formats and procedures, making it difficult to interconnect different computer systems.

IBM is making fundamental changes in SNA to address the restrictions caused by hierarchical implementation throughout the layers. IBM has moved to rectify these problems at several levels by introducing new LUs and PUs, network topologies, and control procedures. Software products for the different computer lines will augment the interworking environment as well.

Low Entry Networking: Advanced Peer-to-Peer Networking

Low entry networking (LEN) is a relatively new technical direction within SNA that provides for the routing of traffic between any two nodes, automatic definition of routes, and directory services so that traffic can find its way to the intended destination. It was introduced in 1986. The enhancement to SNA that makes LEN possible is the PU 2.1, or simply, type 2.1 network node. *Advanced peer-to-peer networking* (APPN) is the term for products operating with LEN services. Initially, only the System/36 had APPN products, although other computer product lines are being added. VTAM/NCP LEN support was made available in 1987. LEN allows a dynamic routing capability in System/36-based products. Intermediate System/36 nodes can route traffic for other systems as needed. APPN systems communicate changes in operational status or configuration with each other to make dynamic routing possible. Note that with APPN the central system no longer controls the routing because the routing can be changed by the nodes. These new features provide the capability to open IBM networking toward a distributed environment.

A drawback of APPN is an internal SNA compatibility problem because older hierarchal networks cannot coexist with this new SNA implementation. The advantages of distributed communication in a small office are lost if mainframes cannot be accessed through the same facilities. IBM is adding mainframe support of PU 2.1 so that both approaches can operate concurrently in the same domain, as shown in Figure 11.5.

APPC/LU 6.2—Advanced Program-to-Program Communication

APPC is the IBM marketing term for LU 6.2, as discussed above and its interface. Basic interworking services are supplied by LU 6.2, which serves as the major foundation for interworking products. LU 6.2 breaks the management responsibilities in ways that suit each partner's needs. Any LU 6.2-based application can initiate a session with a remote partner and negotiate the roles each will assume. LEN allows two LU 6.2 elements to establish an SNA session without the mediation of a mainframe. On the older hierarchical network, LU 6.2 sessions are still mediated by the mainframe [11.13, 11.15].

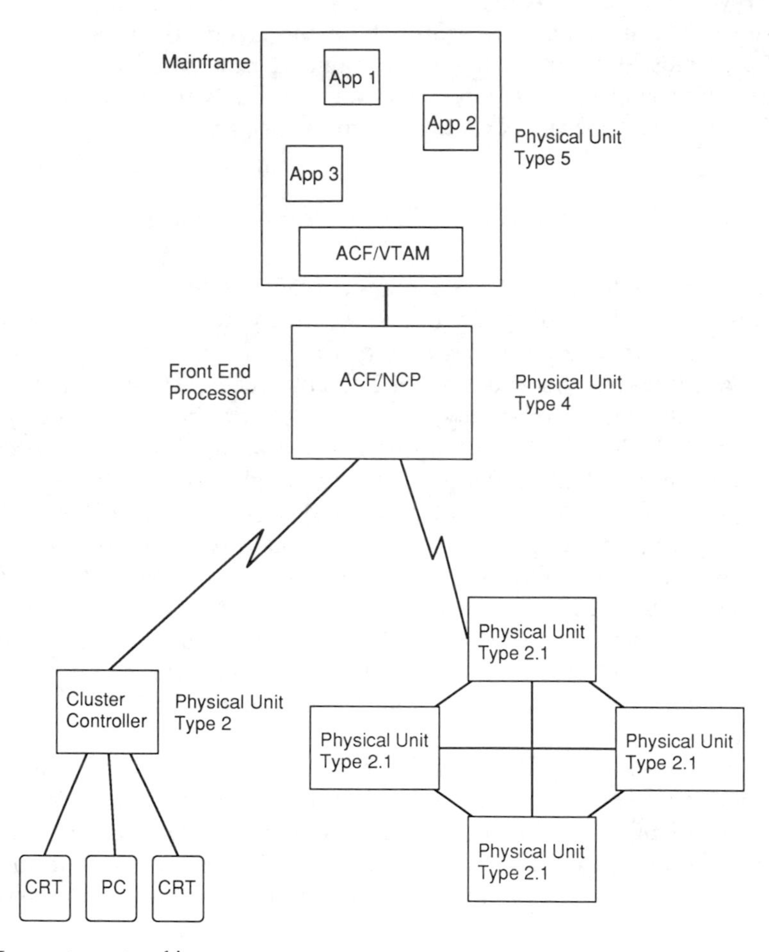

Figure 11.5 Low entry networking.

A favorable feature of LU 6.2 is the definition of a standard application interface. The interface is defined by a set of verbs that directs the communication services and controls a session with another application, as discussed above, and in Table 11.6. A standard set of verbs offers important advantages such as a consistent application development environment that should facilitate implementation and a consistent interface across the incompatible product lines. The interface also adds to the insulation between applications and networking services, so that applications need not be concerned with networking details because they deal in verbs across a standard interface. Future SNA changes will have less impact on applications, a welcome situation for many users.

Some Recent Enhancements in Local Connectivity

A new generation of cluster controllers, the 3174 controller, was introduced in 1986. It can be used as a gateway between a token-ring attached PC running the APPC/PC program and a host-based APPC program, giving the user the benefits of transparent program-to-program communication. Prior to token-ring connectivity for 3174 controllers, users had to choose between speed and distance. Local controllers located within 400 ft of the channel could operate at channel speed if they were channel-attached. Remote controllers that exceeded the channel distance specification were required to connect to the mainframe via a telephone line (operating up to 56 kb/s), a pair of modems, and a FEP. Token-ring connectivity extends high speeds to remote controllers located beyond the channel distance limitation, but within token-ring limitations (for larger distances, MANs or channel extenders can be used); in addition, users are no longer required to purchase a FEP for remote workstation connections unless they also need other FEP functions (the 3174 Model 01L with Token-Ring Gateway performs the function formerly done by the FEP). Another advantage is that each PC no longer requires a coaxial adapter card and a coaxial cable to attach to the controller, but can connect to the twisted-pair LAN. Figure 11.6 depicts a typical topology for local connectivity.

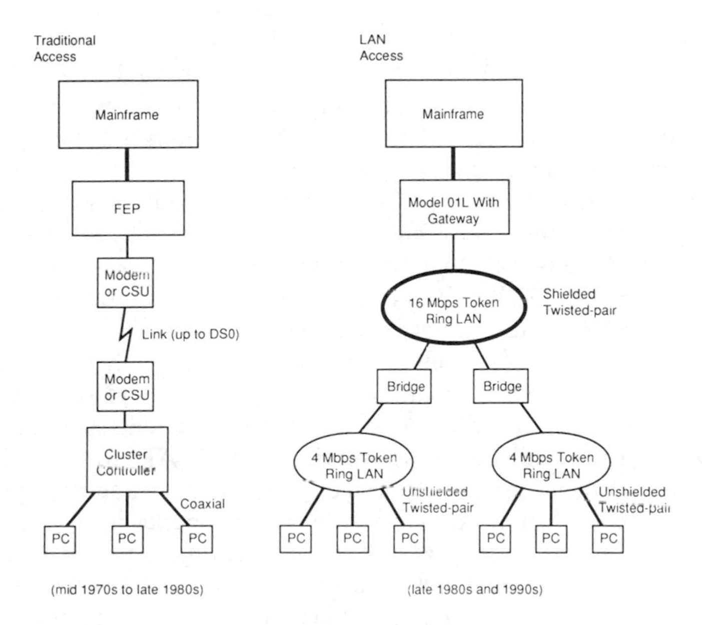

Figure 11.6 Contemporary host access in a SNA environment.

Late in 1988, IBM announced its new 16 Mb/s Token-Ring Network. Industry observers point to IBM's desire to see the 16 Mb/s LAN as a replacement for the current 4 Mb/s LAN as the desktop LAN used for user-to-host communication [11.16]. At least in the short term, the 16 Mb/s LAN may become a backbone LAN interconnecting various 4 Mb/s LANs; eventually, it may be supplanted by the 100 Mb/s LANs described in Chapter 15. Unlike the 4 Mb/s LAN, which can use unshielded twisted-pair cabling, the 16 Mb/s LAN requires the installation of shielded twisted-pair cabling; in addition, new network interface cards installed in PCs are required.

11.5.6 Systems Application Architecture

Systems application architecture (SAA) was introduced in 1987. Its goal is to facilitate writing programs that can be distributed and ported to different computing environments. The SAA goal is to be achieved by creating a common, standardized set of operations and interfaces. The intention is for these universal functions to be "distributable" across interconnected IBM systems. SAA is an attempt to achieve interoperability with the disparate types of IBM's product families. Each SAA environment may not employ the identical code for a basic operation, but because an equivalent function exists in each environment by definition, programs using only the basic functions can be easily converted and recompiled. SAA is an architecture that binds several existing architectures within the IBM arena, including SNA and DIA/DCA (document interchange architecture/document content architecture) office systems architecture (see Figures 11.7 and 11.8).

SAA contains about two dozen "standards" within four groups of SAA elements: common communication, common programming interface, common user access, and common applications.

Common communication support is a set of about 12 standards that define peer-to-peer open communication. Elements include ACF/VTAM (Advanced Communication Function/Virtual Telecommunication Access Method, release 3.2 or higher), NPSI (X.25 NCP Packet Switching Interface, release 4.3 or higher), and SNADS (SNA Distribution Services). Capabilities include the support of SDLC, X.25, and token-ring links at the lower layers and LU 6.2 at the upper layers. *Common programming interface* is a set of about 12 standards that define standardized high-level programming languages. Elements include ANSI Cobol 1985, ANSI Fortran 1987, and other services such as API formats. *Common user access* defines a consistent user interface across SAA products over possibly different operating environments and hardware. Elements include screen layouts, keyboard layouts, and others. *Common applications* are programs that use the facilities of the other three elements listed above to create applications that can be easily ported from microcomputers (such as PS/2) to minicomputers (AS/400) to mainframes (3090, 4381, *et cetera*).

REFERENCES

W. Stallings, "Digital Signaling Techniques," *IEEE Communications Magazine*, December 1984, Vol. 22, No. 12.

W. Stallings, "Encoding Techniques in Local Area Networks," *Journal of Data and Computer Communications*, Spring 1990, pp. 23 ff.

] J. Kane, "An Introduction to T1 and Its Role in Meeting Growing Business Needs," *Communications News*, December 1984.

4] W.D. Schwaderer, "Cyclic Redundancy checking," *Journal of Data & Computer Communications*, Spring 1989, pp. 4–25.

1.5] M.H. Sheriff *et al.*, "X.25 Conformance Testing—A Tutorial," *IEEE Communications Magazine*, January 1986, Vol. 24, No. 1.

11.6] K. Dally, OMNICOM Inc., personal communication, March 25, 1988.

[11.7] W. Stallings, *Handbook of Computer-Communications Standards*, Macmillan Publishing Company, New York, 1987.

[11.8] U. Black, *Physical Level Interfaces and Protocols*, IEEE Computer Society Press, New York, 1988.

[11.9] A. Guruge, *SNA Theory and Practice*, Pergamon Infotech Limited, United Kingdom, 1987.

[11.10] M. Canipe, "The Evolution of the 3270 Market," *The McData Link Magazine*, Vol. 1, No. 1, June 1988.

[11.11] L. Coulson, "The Demise of System Compatibility," *The McData Link Magazine*, Vol. 1, No. 4, Spring 1989.

[11.12] *An Updated SNA Glossary*, Data Communications, June 1989.

[11.13] *OSI and SNA Compared: The Integration of Network Standards*, Communication Solutions, San Jose, CA, 1987.

[11.14] L. Lord, "In Search of Peer-to-Peer for the 3174," *The McData Link Magazine*, Vol. 1, No. 4, Spring 1989.

[11.15] A. Meijer, *Systems Network Architecture, A Tutorial*, John Wiley and Sons, New York, 1988.

[11.16] P. Sikorski, "Sixteen-Mb/s Token-Ring Networks and the 3174: A New Partnership," *The McData Link Magazine*, Vol. 1, No. 4, Spring 1989.

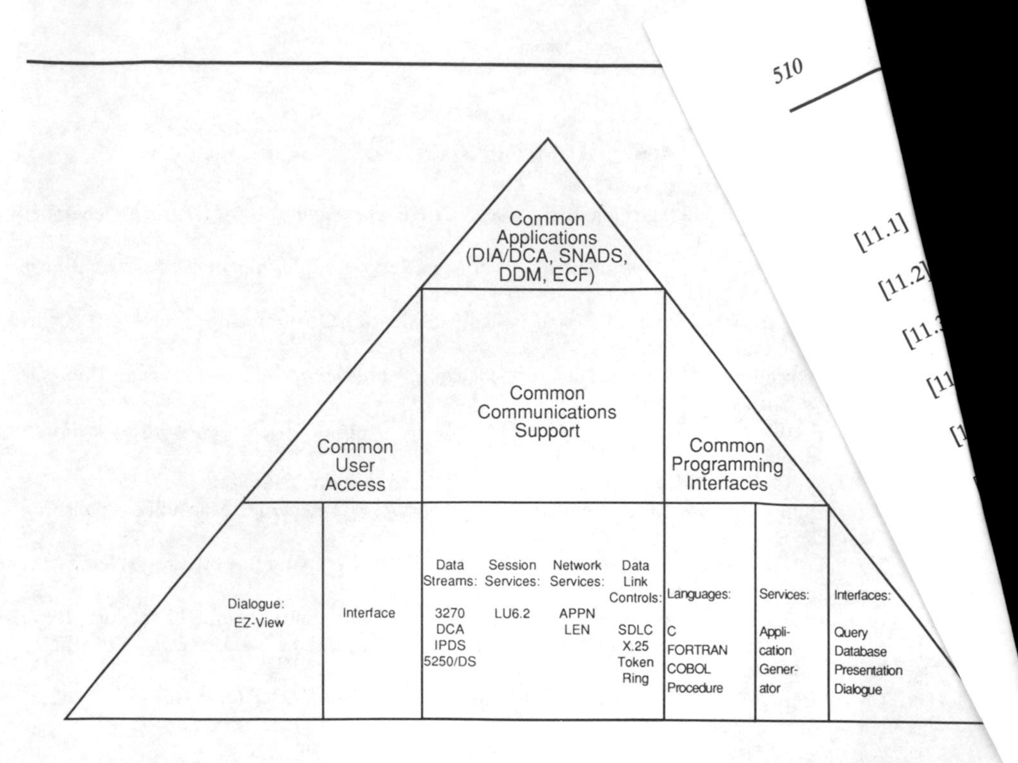

Figure 11.7 SAA relationships.

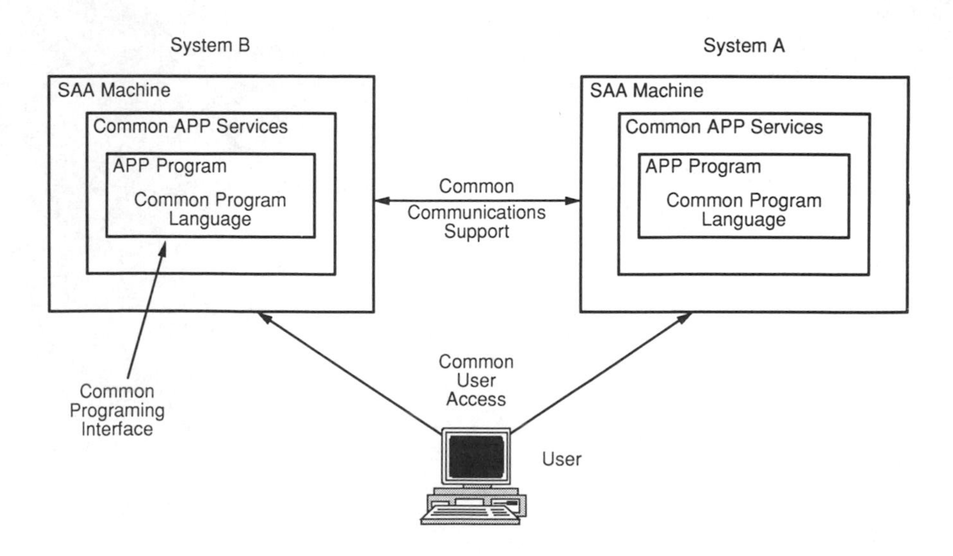

Figure 11.8 SAA from a user's perspective.

Chapter 12

Channel-to-Channel Computer Communication

12.1 OVERVIEW

Distributed data processing has become more practical in the past five years. A relatively new type of hardware can extend the mainframe's channel across town or across the nation, so that high-speed peripherals can be located remotely, closer to the place at which the user needs them. In this context, the channel is the high-speed input-output facility of a large mainframe. Although the introduction of PCs in the work place has been sustained, the mainframe continues to be the central resource where the company's data base, programs, and utilities reside. The host provides the processing power for sorting the data and preparing major corporate reports. In addition, a mainframe connection allows multiple workstations to run simultaneously the same complex application program and accomplish tasks co-operatively with it. Consequently, for many institutions, an efficient connection to one or more large mainframes remains a key goal [12.1].

Channel extenders allow peripherals, such as high-speed laser printers, storage devices, tape units, and FEPs, to be remote. The technology has been available for a few years, but the applications are only now starting to be realized [12.2]. Key factors in this development are the decreased costs of carrier-provided T1/DS1 lines and private-installation multimode fiber cable and related fiber equipment for data communication usage. Channel attachment minimizes the processing time needed for communication between users and applications; bypassing the FEP and attaching directly to a mainframe channel results in higher speed links to terminals, LANs, and other mainframes. As more corporations disperse their operations, networking using channel extension is becoming a common design approach.

Remote channel attachment serves four basic purposes: (1) connecting distant host processors in a high-speed network, (2) connecting distant terminals requiring subsecond response time, (3) connecting distant high-speed printers, and (4) connecting remote tape or disk drive units for disaster recovery applications. The economic advantages of using channel extension are reported to be substantial by those institutions that have deployed them. The cost to extend a mainframe channel ranges between $10,000 and $90,000. Speed-preserving channel extenders can extend the channel over fiber links to about 6000 ft (typical device cost: $10,000 to $25,000 per end). Other extenders that employ a speed-reduction link, such as a carrier's T1/DS1 link, can provide a remote channel over hundreds or thousands of miles (typical device cost: $35,000 to $90,000 per end).

Large corporations are expanding mainframe input-output channels to prepare for the increased bandwidth expected to accompany the emergence of new data-intensive business applications in the 1990s.

This chapter assesses the importance of channel extension as a data communication technique. The purpose and types of mainframe channels are described first. Features of channel extenders are then identified, and typical applications of the technology are described. Some of the engineering issues pertaining to the internal architecture of channel extenders are also highlighted. Brief descriptions of two related products, the channel switch and the channel distribution network, are also provided.

12.2 DATA COMMUNICATION ENVIRONMENT NEEDS FOR THE EARLY 1990s

A computer system consists of a number of distinct components: processors, memories, and peripherals. These components must share and exchange information consisting of either instructions or data. To do this, a physical means of transferring the information has to be provided, and a mutually accepted set of rules governing the transfers has to be established. This combination of protocols and physical medium constitutes a data path. The two key data paths of a computer are the bus and the channel. The bus normally interconnects computer components, such as the CPU (central processing unit) and the ALU (arithmetic and logical unit), memory, cache memory, and the input-output processor. The channel normally connects the computer with external peripherals, such as disk and tape drives, printers, and communication devices. Typically, the bus speed is higher than the channel speed (10 Mb/s compared to 500 Mb/s for the former). In some computer systems, the bus and the channel are the same. Buses, channels, LANs, MANs, and wide-area networks (WANs) are all components of the continuum of technology used to interconnect computers and computing equipment over a wide range of distances and speeds. Each of these technologies has a characteristic, but not rigidly defined, set of features, as shown in Table 12.1. Buses tend to be viewed as connecting hardware cards and submodules to form a computer system, while channels, LANs, and WANs interconnect systems. However, the distinction between buses, channels, and LANs is becoming more blurred as time passes [12.3].

Table 12.1
Interconnecting Technologies

Feature	*Bus*	*Channel*	*LAN*	*MAN/WAN*
Distance covered	A few tens of feet in a backplane or among cabinets	200–400 ft within a building	A few miles within a building or campus	Hundreds or thousands of miles
Type of devices interconnected	Parts of a computer system	Peripherals of a computer	Computer systems, servers, and workstations	Computer systems and terminals
Address granularity	Fine (a memory location)	Fine (peripherals)	Coarse (different computers and terminals)	Coarse (different computers and terminals)
Data rate	High (10 to 500 Mb/s)	High (24 to 36 Mb/s)	Medium (1 to 100 Mb/s)	Low (1 kb/s to 1.5 Mb/s today; 45 Mb/s and SONET rates are emerging)
Ownership of facility	Private	Private, but may also employ carriers' links	Private, but may also employ carriers' links	Typically carrier based
Cable	10 to 100 wires, parallel transmission	10 to 30 wires, parallel transmission	1 channel, serial transmission	1 channel, serial transmission
Topology	Bus	Bus	Bus, ring, star, tree	Star, mesh
Error rate of cable	Very low	Low	Medium (depends on communication medium used)	Higher (depends on communication medium used)
Error detection	Usually not implemented	Implemented by channel extender hardware	Always implemented	Always implemented today; perhaps not with frame relay, ATM
Device relationship	Hierarchical	Hierarchical	Peer-to-peer	Master-slave and peer-to-peer
Response time	Media inherent	Very low 10–30 ms	0.5–3 s	0.5–10 s
Data unit	Word	Byte	Message or file	Message or file

12.2.1 Application Environment

Channel extension is applicable to two key topological environments: terminal-to-host connection (THC) and host-to-host connection (HHC).

Terminal-to-Host Connection

In the 1970s and early 1980s, connectivity in a data communication environment meant mainly the interconnection of dumb terminals and centralized host processors. (See Chapter 11.) This connection to the mainframe has been traditionally accomplished by linking a mainframe input-output channel through a communication processor. These links have also been used to connect remote peripherals and remote mainframes to the central mainframe at relatively low data rates (9.6 to 56 kb/s). In this environment, the FEP solution allows remote users to move screen images of data interactively; file transfers are not, however, handled very efficiently. The increased use of PCs and workstations, particularly those operating in a peer-to-peer environment (such as in a LU 6.2 configuration), has created the need to move data files on a routine basis. It requires approximately 5 to 10 min to transmit a 1-Mbyte file over a 56 kb/s DDS link between a PC and the mainframe (including protocol overhead). This delay is due partially to the transmission speed of the line; higher speed facilities are becoming increasingly available, as discussed in Chapter 3. The delay may also be due to the fact that data are still being sent to the PC in a 3270-emulation mode, rather than through a sophisticated file transfer protocol. Finally, this delay may be due to the bottlenecks interjected by the FEP and related access software.

Figure 12.1 depicts the traditional environment with terminal emulation hardware and associated communication processors to mediate access to the mainframe. As indicated, communication links operating at 9.6 kb/s have been standard, with some use of DDS at 56 kb/s; until recently, the maximum data rate between two FEPs was 256 kb/s. These speeds are becoming inadequate for today's business applications. The additional software these processors require [such as NCP, NCP Packet Switched Interface (NPSI), and other native and emulation software] also

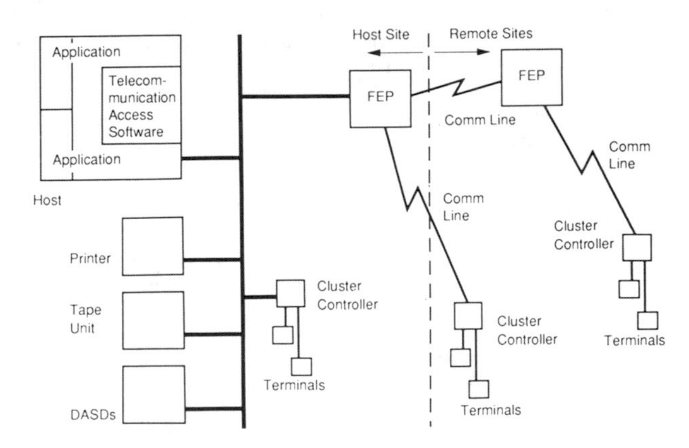

Figure 12.1 Traditional teleprocessing environment.

increases the total transmission delay. That added delay is enough to make the desirable goal of subsecond response times on terminals difficult to achieve.

In the late 1980s, a new wave of front-end communication processors was announced; all aimed at increasing the back-end throughput while interfacing with new higher rate communication lines and LANs at the front end. The T1/DS1 interfaces are being pushed by the vendors, but potential buyers should make sure that the traditional communication processor (the FEP) has the horsepower to handle the DS1 throughput at the line card level and still provide reasonable overall performance. Many processors, in fact, may become bogged down at T1/DS1 rates [12.4]. Even if the processor handles T1 and possibly T3/DS3 rates (45 Mb/s) effectively, the communication protocol used may not. Common WAN protocols in use today were usually designed for low to medium speeds and, therefore, may not be efficient at higher speeds. Users with applications requiring high data rates (large file transfers, high-resolution graphics, CAD/CAM, *et cetera*) may need to bypass the communication processor and achieve direct host-to-channel communication. This is the point at which the channel extender becomes useful.

Examples of FEPs include the IBM 3705, 3725, and the latest in this family, the 3745, as well as Amdahl's 4725 and the NCR Comten 5620XP. The 3745 employs cache memory, data streaming in channel adapters, and direct memory access. It can accommodate up to 16 DS1 lines, 512 data communication lines with speeds up to 256 kb/s, and 8 token-ring networks operating at 4 Mb/s. In spite of these features, this type of communication is still relatively slow for some applications (for example, the deployment of remote direct access storage devices—DASDs—or laser printers such as IBM's Model 3800). FEP connections are optimized for terminal-oriented transaction processing. In today's computing environment, a broader set of communication requirements has emerged; in particular, diverse computer networks and peripherals need to be connected in a more symmetrical peer-to-peer manner.

In addition, LANs are being increasingly interfaced to the mainframe (as discussed in the previous chapter) as a more efficient way to interconnect a local population of peer-to-peer PCs or 3270-emulation PCs. Traditional cluster controllers with their cabling constraints are often being replaced with LANs, particularly in new installations. LANs operate at speeds ranging from 1, 4, 10, and 80 to even 100 Mb/s (for new backbone products, including FDDI, discussed in Chapter 15). Thus, the mainframe link must be able to accommodate these higher LAN transmission speeds.

Recognizing the growing need to make the mainframe easier to access as a node on a network, mainframe manufacturers now offer built-in connectivity on some of their computers. These integrated communication controllers enhance LAN-to-mainframe connectivity by opening the mainframe system to a broad range of users. This was accomplished by enabling the built-in controllers to interface directly to LANs and to other computers, without having to go through FEPs. But these built-in connectivity devices are not available for the entire family of mainframe computers. Therefore, they do not meet the connectivity challenges posed by "closed" mainframe architectures [12.1]. The solution here is to use channel extenders, which facilitate the connection of devices beyond the traditional 200-ft limit imposed by the mainframe channel. T1/DS1 circuits over copper, fiber, or microwave can be used to link to peripherals located away from the host site. Channel extenders can be grouped in two topological classes: local area extension (typically in a multibuilding or campus environment) and wide area extension.

Host-to-Host Connection

Collocated IBM computers can communicate with one another via a channel-to-channel adapter (CTCA). The connection between mainframe and control unit (or other peripherals) is, according to standard specifications, a cable that limits the distances to 200 ft (in some situations, 400 ft). Direct channel attachment is implemented with an interface known as a *channel adapter* on the processor end and as a *device adapter* on the peripheral end. The two adapters are connected using the bus-and-tag cable, a large round cable (about 1 in.) with paddle-shaped connectors on the ends. Signal degradation and protocol requirements impose distance limitations on these connections; even with fiber optic channel extenders, the cable length is limited to several thousand feet [12.5]. Channel extenders provide host-to-host data transfers at high volumes and high speeds, allowing users to access many hosts located at different sites as if they were a single host, bypassing the communication processor. Connecting the mainframes using a T1/DS1 facility has been done to some degree by FEPs for a number of years. However, FEPs have not, in the past, used the DS1 link efficiently; while the host is delivering the data at 3 Mbytes per second, the parallel-to-serial conversion in the FEP is such that the real traffic throughput has typically been around 256 kb/s [12.2].

12.3 MAINFRAME INPUT-OUTPUT MECHANISMS

Channel attachment is an IBM term for the high-speed physical interface between the host processor and other devices. Traditional IBM System 370 computers have employed a number of channels to communicate with the outside world: the byte multiplexer channel and the block multiplexer channel. (Another channel type, the selector channel, which was used for tape drives, is now obsolete.) In a large host processor setup, processors in a multimodule complex communicate with one another over channel connections using a block-oriented protocol. These connections are referred to as a *block channel attachments*. Data streaming permits data transfer rates up to 3 Mbytes per second on block multiplex channels. Slower devices—peripherals and FEPs—communicate using a byte-oriented protocol over a channel attachment interface at rates up to 2 Mbytes per second [12.5]. All printers, terminal controllers, disk, and tape controllers that are colocated with the data center connect with the processors via channel attachments.

The channel is a parallel facility, carrying 8 information bits and 1 control bit. The channel typically operates at 3 Mbytes per second (3 Mb/s connections for each of the 8 bits of the byte) with a serial data rate of 24 Mb/s (27 Mb/s if we include the control bit). The actual bus has 32 cable conductors, which handle the input-output in arrays of 32-bit words. Remote channel extender devices take the 32 concurrent inputs and align them in a serial stream for transmission.

Byte Multiplexer Channel

This channel transfers bytes interleaved between memory and several input-output devices (byte mode) or transfers a byte string to one device (burst mode). The channel is composed of a number of subchannels, each of which can control one device and has one unit control word (UCW) implemented in control storage. The channel is microprogram-controlled, and only has one active UCW. During multiplex mode operation, subchannel UCWs are fetched into the active UCW, updated, and returned to control storage. In some implementations, all storage access for byte multiplexer channel operation interferes with processor operation.

Block Multiplexer Channel

By interleaving blocks of data, this type of channel allows concurrent operation of several burst-mode devices on the channel's single data path. This channel has a high data rate and operates only one device at a time, but it can have several active channel programs. Devices capable of block multiplex operation logically disconnect from the channel when not prepared for data transfer and can reconnect when the channel is no longer busy transferring data to another device via the interrupt system. (Channel extenders typically support the block multiplexer data transfer protocol.) The byte multiplexer channel can be connected to a number of slow- and medium-speed devices and is capable of operating with a number of input-output devices simultaneously. The block multiplexer channel provides a connection to a number of high-speed devices, but all input-output transfers are conducted with an entire block of data, as compared to a byte multiplexer channel, which can transfer only one byte at a time. The computer system may have a number of channels and each is assigned an address. Similarly, each channel may be connected to several devices and each device is assigned an address. In the System 370 environment, any channel can be assigned any valid channel address without concern for priority. In the System 370 Extended Architecture environment, up to four channel paths are available to any attached input-output device.

During any input-output operation, one of the available channel paths to any specific input-output device is selected (channel path selection is a hardware function).

In the byte multiplex mode, several relatively slow-speed input-output devices can operate concurrently. In block multiplex operation, channels can operate either in the high-speed transfer mode or data streaming mode. In the data streaming mode, a traditional block multiplex channel can transfer at up to 3 Mbytes per second. Each byte multiplexer is capable of operating with an aggregate data rate in the range of 90 to 300 kbytes per second for data transfer burst sizes of 4 bytes or more. Configurations consisting of control units with faster input-output interface tags and larger data transfer burst sizes can achieve the higher performance.

Computers typically have more than just one channel of each type. Table 12.2 shows the configuration of a typical computer family.

While IBM and the PCM (plug compatible manufacturers) have had 3 Mbytes per second channels (raised to 4.5 Mbytes per second in the late 1980s), other mainframes may have channels operating at higher rates, as seen in Table 12.3. An IBM announcement of plans to increase the channel speed for the 3090 family of computers was made in 1987. National Advanced Systems (NAS) responded to the 3090 announcement with a 6 Mbytes per second channel speed for the Alliance series of processors. By 1989, Amdahl and IBM both offered 4.5 Mbytes per second channels; both vendors may increase these rates in the future. In 1990, IBM announced a 10-Mbyte fiber-based channel. Channel speed may be increased by IBM to 18 Mbytes by the mid-1990s according to some indications, presumably with the help of fiber optics [12.6, 12.7].

Table 12.2
Computer Channel Operating Rates

Computer System	*Byte Multiplexer Channels*	*Block Multiplexer Channels*
IBM 4381 Model 24	0 to 4	20
IBM 3083 CX	0 to 4	8 to 16
IBM 3083 EX	0 to 4	8 to 16
IBM 3083 BX	0 to 4	8 to 24
IBM 3083 JX	0 to 4	8 to 24
IBM 3081 GX	0 to 4	16 to 24
IBM 3081 KX	0 to 4	16 to 24
IBM 3084 QX	0 to 4	48
IBM 3090 120 E	0 to 4	16, 24
IBM 3090 150 E	0 to 4	16, 24
IBM 3090 180 E	0 to 4	16, 24, 32
IBM 3090 200 E	0 to 4	32, 40, 48, 64
IBM 3090 300 E	0 to 4	32, 40, 48, 64
IBM 3090 400 E	0 to 4	64, 80, 96, 120
IBM 3090 600 E	0 to 8	64, 80, 96, 128

Table 12.3
Mainframe Channel Operating Rates

Computer system	*Model Number*	*Channel or input-output speeds (bytes per second)*
Amdahl	5990 Series	3M, 4.5 M
Amdahl	5890 Series	3M, 4.5 M
Control Data	Cyber 180 930 family	3M
Control Data	Cyber 180 840 family	3M
Control Data	Cyber 180 992 family	3M, 12 M
Control Data	Cyber 180 994 family	3M, 12 M
Honeywell Bull	DSP 7000 Series	1.25M, 2.5M
Honeywell Bull	DSP 8000 Series	17.8 aggregate
Honeywell Bull	DSP 88 Series	3M
Honeywell Bull	DSP 90 Series	17.8 aggregate
IBM	9373 Model 20	1.5M to 1.9M
IBM	9375 Model 60	1.9M to 3M
IBM	4380 Series	3M
IBM	3090	3M, 4.5M
NAS	AS/VL Series	6M
NAS	AS/XL Series	3M, 6M
NCR	9800 Systems	14M aggregate
NCR	V-8800 Systems	2M
Unisys	A1, A4	3.4 M, 4.5M (A 6K)
Unisys	A2, A3	3.4 M, 4.5M (A 5K)
Unisys	A9, A10	3M
Unisys	V Series	8M, 16 M (V 380/V 530)
Unisys	A 12	3M
Unisys	A 15, A 17	8M
Unisys	1100/90 Systems	35.2 M aggregate

Modern Input-Output Processors

Modern processors (such as the IBM 3090) employ a channel subsystem (CSS) to handle all input-output operations for the CPU. The CSS controls communication between a configured channel, a control unit, and a device. The input-output configuration data set (IOCDS), selected at system initialization, identifies channel, control unit, and device configurations to the channel subsystem. The input-output configuration program creates the IOCDS, which is stored on disk drives attached to the processor controller. During initialization, the IOCDS information is used to build the necessary control blocks in the hardware system area of central storage. In addition, the CSS contains a channel control element that interacts with central storage, the central processors, and the channels. In operation, the channel control element initiates and ends channel operations, provides central storage access control, and sets priorities for input-output operations.

Even newer systems using a separate processor to handle input-output still employ channels to communicate with the outside world. Modern computers also have integrated input-output channels. These are normally in an integrated input-output processor that contains and controls channels. The channels can be configured for either byte or block multiplexer operation. Channel units are increasingly becoming small programmed processors to permit extension of the channel functions. For example, the Unisys A, B, and V systems use microprogrammed data link processors.

12.4 CHANNEL EXTENDER USAGE

Table 12.4 describes a number of common channel extension applications. One of the key reasons users install channel extenders is to minimize the risks by relocating important peripherals away from the mainframe computer. Channel extenders have been used extensively for "hot" disaster recovery sites to transfer backup data bases. By having a remote host or storage device at a separate site, the potential disaster that could result from a host failure is minimized; for many financial services companies, such as banks and brokerage firms, any downtime or loss of data can be a serious problem. Other applications include remote tape vaults for routine off-site storage.

Table 12.4
Common Applications of Channel Extenders

Application	*Explanation*
Distributed data processing	Different portions of the data can reside in different host's disk drives, spread throughout the country. With channel extenders and T1/DS1 links, all hosts have quick access to all the information.
Instantaneous data backup	Large amounts of data can be dumped in real-time to a secure location mirroring all data base operations. This eliminates the need to generate, manage, and ship tapes, which is a labor-intensive activity. It also eliminates the current bottlenecks of occasional dumps of small selected data sets (or portions thereof) over slow communications lines.
Dual center transaction processing	With appropriate programming, each transaction can be maintained in two or more hosts for fully mirrored operation with negligible delays.
Darkened data center	All the primary mainframe equipment is placed in a remote, secure, and darkened facility, possibly in a suburban area. T1/DS1 or fiber channel extenders can be used to connect remote users. Minimum personnel would be needed.
LAN integration	Channel-to-channel attachment offers the possibility of using computers as true servers, while at the same time retaining the hierarchical control within the mainframe system.

Another reason to install channel extenders is economics: A remote location service level can be as effective as if the mainframe computers were at that remote location. This is particularly attractive when trying to relocate the data center to a suburban location. Real estate costs are soaring in many cities; thus, many companies are moving their data centers to lower cost facilities. For example, a number of New York banks have built data centers in New Jersey. Using channel extenders, a host could be located in New Jersey while terminals accessing it from New York would appear to be local.

Adding to the impetus for increased user demand for channel extender equipment are the declining cost of T1/DS1 facilities and the promises of ISDN [12.8].

12.5 NETWORK TOPOLOGIES FOR USING CHANNEL EXTENSION

Topologically, channel extension can be accomplished in two ways. The first way involves a simple point-to-point architecture; the second way employs a high-speed backbone LAN approach.

In the first approach, shown in Figure 12.2, each remote device (host, printer, cluster controllers) has its own point-to-point facility connecting it to a channel of the host; each host channel connects to a channel adapter, which communicates with the device adapter at the remote end.

In the backbone LAN approach, shown in Figure 12.3, each processor in a location is connected to a high-capacity LAN through channel adapters. The LAN is then connected to other LANs supporting hosts, peripherals, or servers at other locations. The LAN approach offers flexibility in terms of its ability to network many peripherals connected to various hosts in addition to its host-to-host interconnections. Applications and processors can also share the T1/DS1 transmission facilities required by these networks. The LAN approach has certain drawbacks, in spite of its greater level of interconnectivity. These drawbacks include increased

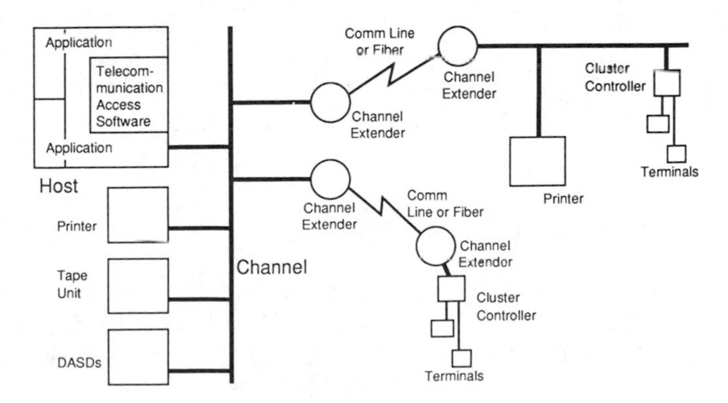

Figure 12.2 Typical configuration showing point-to-point channel extender usage.

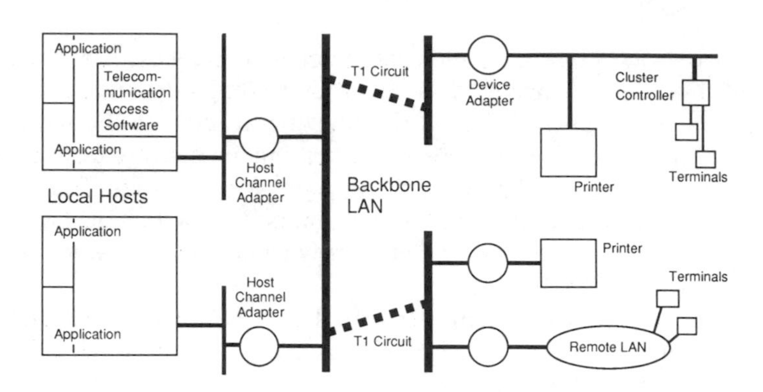

Figure 12.3 Typical configuration showing channel extender in a LAN configuration.

cost compared to a straight point-to-point solution (for those cases where the number of interconnection points is small), and the increased management complexity.

A diversity of communication links can be used with channel extenders: the T1/DS1 and DS3/T3 carriers and private microwave or fiber. Some installations have used a communication-link resource manager that allocated only a portion of the available communication bandwidth to the channel extender. Some researchers have also studied coaxial-based CATV links provided by local CATV companies. CATV lines have already been used in security systems to monitor remote sites; sufficient bandwidth may be available to provide a full-duplex T1 speed link for the controllers on the same cable. However, the BER on these types of CATV facilities is normally too high for efficient high-throughput usage. The effects of transmission link errors can be significant. As the BER approaches 10^{-5}, the frequency of retransmissions causes the internal controller message queues to increase and, consequently, the round-trip delay increases. When the BER reaches 10^{-3}, few messages are successfully transmitted. An alternative is to employ FEC methods on noisy links, but this also degrades the throughput. A sophisticated channel extender employs flow control algorithms that throttle back the number of messages accepted from the host CPU, but the overall throughput will also be degraded.

12.6 CHANNEL EXTENDER EQUIPMENT

The channel extension equipment must perform the following functions: meet the channel data transfer block or byte multiplex protocol, serialize the parallel data, provide for possible buffering where speed transparency is not maintained, interface to the communication link (T1/DS1 or fiber).

Users seeking to extend the mainframe channel access have the following three options:

1. Use channel extenders that allow connection of remote peripherals;
2. Use channel extenders that provide connection to another mainframe's channel and, in turn, connect the peripherals to the remote host; and
3. Convert the channel to a LAN or LAN backbone. These LANs could use the IEEE protocols, or the FDDI protocols. (These products are beginning to appear, although they are slightly higher in cost than a dark-fiber system [12.9].)

While direct channel attachment sounds simple, such interfaces might require extensive programming, particularly with a system of diverse hosts. This could involve modifying the operating system or the communication access method software (such as VTAM). This programming can be difficult and at times risky. The host vendor may also object to these modifications, and might require that the user back out all changes before modifications or troubleshooting can be done.

Two categories of channel extenders exist, each with a niche of applications. Products such as the IBM 3088, Data Switch's 1488, and similar fiber optic extenders provide high-speed data transfer among computers located in a relatively small geographic area [12.10]. These products are intended for 308X and 3090 mainframes. This approach is ideal for data shops with multiple hosts, including multivendor situations.

The second category is for systems using carriers' T1/DS1 links. Channel-to-T1 couplers are now marketed by about a dozen vendors [12.8]. Getting a remote peripheral to run as though it were connected to a channel when T1/DS1 lines operate at 1.544 Mb/s and the channel operates at 1.5 to 4.5 Mbytes per second is not necessarily a problem. While the channel runs faster, it is supporting up to eight different device controllers, and the channel speed is the burst speed required to keep all those devices running. A sophisticated channel extender will perform a type of address filtering to transfer only the traffic destined for the remote locations [12.5]. In data transfers, a message passed from one processor usually requires a response. To increase the efficiency of T1/DS1 links by cutting down the flow of information across a line, this equipment typically employs buffers at each end of the link to withhold acknowledgment that individual words or blocks of words have been received. Another efficiency enhancing strategy is to use large data block sizes to minimize the number of responses or acknowledgments [12.11]. Buffer size is an important consideration when selecting any of the channel extender products. As the data arrive at channel speed rate, a channel extender unit, repacking the data for a T1/DS1 link without appropriate buffers, gets congested relatively quickly. It then signals back to the host to throttle the incoming stream. This can be expensive in terms of computer processor inefficiency. A better method would be for the host to dump these data into a buffer and then go on to other processing tasks [12.2].

DS1 data transfers over these channel systems are significantly slower than 3 and 4.5 Mbyte per second channel speeds, but are significantly faster than using FEPs. Despite inefficiencies with T1/DS1 links because of the parallel-to-serial data conversion required, many corporations are installing fiber lines in campus environments for VTAM-to-VTAM communication at DS1 rates [12.11]. A few vendors offer a T3/DS3 channel extension system. The 45 Mb/s serial speed is more than adequate for both the 3 and 4.5 Mbyte per second channels (the latter operating at 36 Mb/s, the former at 24 Mb/s).

A nonchannel alternative option to increase the remote throughput in a large IBM computer environment would be to replace the existing 3705s and 3725s with the 3745. However, the cost can be substantially higher than a pair of extenders. Running the FEP lines through a T1 multiplexer or a T1 resource manager does not achieve the same results as using the channel extenders because, in this case, the data flow is still through the bottleneck of the FEP (see Figure 12.4).

12.6.1 Equipment Categories

Three related types of channel extender products are (see Figure 12.5):

1. Channel extenders, as discussed above.
2. Point-to-point (static) channel matrices are switches that can connect the IBM channel to a number of possible devices. Typical applications are for

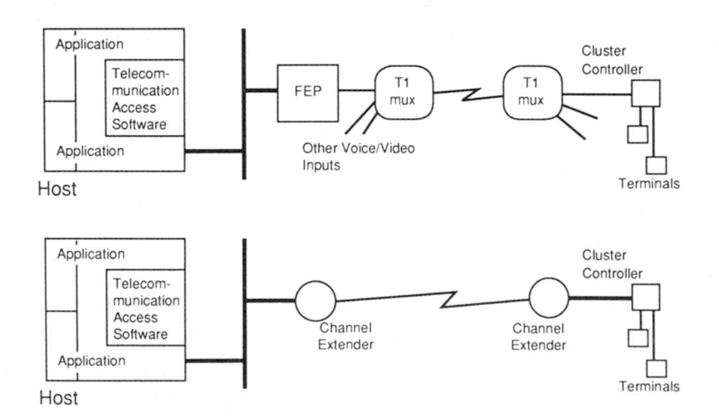

Figure 12.4 Alternative connection methods: top, utilization of T1 multiplexers; bottom, preferred solutions using channel extenders.

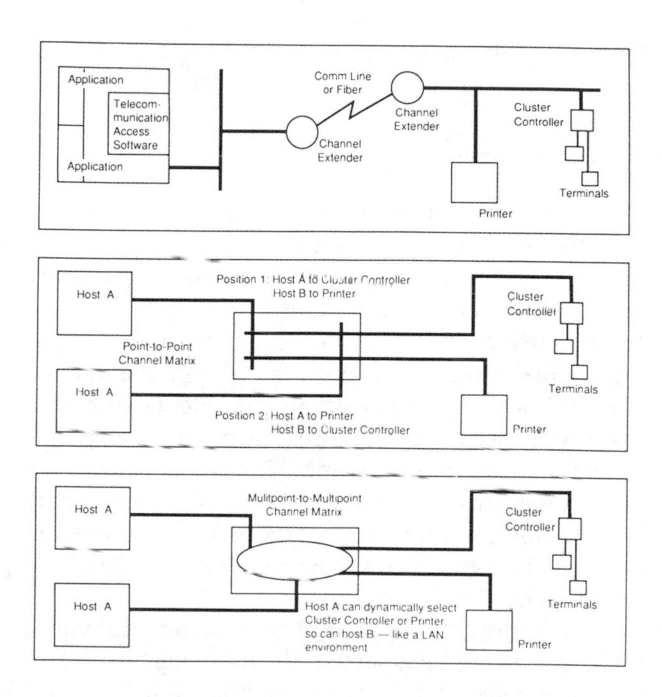

Figure 12.5 Comparison among various types of channel equipment.

putting spare FEPs on line; the matrix eliminates the need to disconnect and reconnect cables. The equipment can be thought of as a patch panel on the channel side.

3. Multipoint-to-multipoint (dynamic) channel extender matrices are dynamic channel switches that provide true mainframe-to-mainframe internetworking. The mainframes, however, need to be in relative proximity to each other for this technology to apply directly. Otherwise, channel extenders can be used in conjunction with the matrix.

IBM and DEC computers can also be interconnected. As already discussed, the IBM parallel connection port is a channel, which is a combination of a processor and a parallel port. The processor executes programs in its own right and is able to transfer data from the memory of the host computer to the device attached to the channel, without involving the CPU. The DEC parallel port, however, is a simpler device, which transfers data either directly to the CPU or via a direct memory access device to the memory of the host [12.12]. DEC systems cannot easily emulate an IBM channel. Fortunately, other computers and channel extender devices can. Hence, achieving the equivalent of channel-to-channel communication by linking the IBM channel to certain super micros and minis by a channel extender device is possible. This device can then be linked to DEC systems via a conventional parallel port.

12.6.2 Channel Extender Design Issues

A prototypical channel extender system is described in this section to familiarize the reader with some of the internals and some of the issues associated with this type of hardware [12.13]. Many channel extenders follow the same principles and need to deal with the same design constraints.

The design objective for this type of equipment is to develop a communication controller that is channel-attached to a host in order to communicate with another host over a distance. Each unit must emulate the remote computer to its channel-attached host, so that the two controllers are transparent, except for some reduction in throughput because of the communication link's limitation of approximately 1.5 Mb/s. The pair of them must typically provide a channel-to-channel connection between two host computers with no changes to VTAM or other computer software. Also the channel extenders must be capable of sustaining significant utilization of the transmission line.

Channel-to-channel communication is handled one message at a time. When two host computers are channel-connected, the receiving host acknowledges receipt of each message before the next message is transmitted. This assures reliable, orderly, and fast communication. This technique works well between hosts interconnected by a few hundred feet of cable. As cable lengths are increased, however, the maximum attainable throughput is reduced because of the cable delay experienced by the outgoing message and the returning acknowledgment. With two controllers and a communication link separating the two hosts, the delay is much greater.

Even at the speed of light, a signal travels only 186 miles in 1 ms. This speed is actually only achievable in free-space. In most media, the speed will be reduced. The worst is plain copper; if 1.544 Mb/s links were truly T1, then we would be severely limited. In transmission engineering, T1 uniquely identifies copper as the underlying media, as discussed in Chapter 3. Fortunately, many DS1 channels today, particularly long-haul applications, are not T1, but are fiber-based. In any event, a terrestrial link of 1800 miles (say, New York to Denver) would impose a minimum one-way delay of 10 ms; if we include media end repeater processor time, the delay can be as high as 20 ms. The round-trip delay for confirmation of the delivery of a channel data unit may thus be 40 ms. In satellite links, the one-way delay is on the order of 250 ms. Generally, running channel extender hardware over satellite-based links is not advisable.

To mask the effects of delay, each channel extender unit must simulate the host computer attached to its companion. Consequently, an acknowledgment can be sent back across the channel as soon as the last byte of the message has been checked and stored in the controller's memory. The message or its acknowledgment does not need to traverse the communication link between the two controllers before an acknowledgment can be returned to the originating host computer. After the message has been stored in its memory, the controller prepares the message for transmission and sends it across the communication link to its companion; from there, the message is passed to the target host computer. In every interaction between a controller and a host computer, the controller simulates the host attached to its companion controller and reacts and responds using VTAM channel-to-channel protocols.

Prior to sending the data over the communication link, the controller splits the message received from the originating host computer into small blocks with formatting characters. Splitting is done because the host computer may send the communication controller a message as large as 64 kbytes, although we can demonstrate that the optimum size message for this application is less, typically 1 kbyte (depending on channel speed and link error rate). Included in the format for each block are beginning and ending flags that delineate the formatted message. Additional fields are also included: address, command, sequence number, message length, and check sum.

The address field contains both a "from" and "to" controller address. The command field identifies the type of message. Examples of message type include data, acknowledgment, statistics, and controller link-control messages. A unique sequence number is inserted by the transmitting controller in each message for identification purposes. This sequence number allows the receiving unit to determine if messages are missing and to establish the correct order for passing data messages to the host computer. The transmitting unit is requested to retransmit any message that is missing. Often data messages get out of sequence if they are retransmitted. When this occurs, the receiving controller retains undelivered data message in its memory until the missing messages are received and then it presents them in the correct sequence to the target host computer. Any data message that is determined to be in error when received from the transmitting controller is discarded by the receiving unit and is treated as a missing data message [12.13].

The size of the sequence number field is specified so that it does not limit the number of messages that may be transmitted between acknowledgments. As a minimum, the size of the sequence number field has to be such that a unique number can be assigned to each message transmitted during the time required for a message to travel across the communication link to the other end, plus the time required for an acknowledgment to be returned to the message originator. Because acknowledgments are not returned for each message and we must allow for the possibility of an acknowledgment being damaged or lost, the size of the sequence number field must be appropriately selected to accommodate these factors.

A message length field specifies the number of data bytes contained in the message. This may be added to the format to increase the controller's capability to detect message errors. The CCITT CRC code is typically used by the transmitting unit to generate a 16-bit check sum. The use of this check sum in combination with format, sequence number, address checks, and message length ensures that a message has been received correctly by the receiving unit. The check point mode (CPM) variant of the HDLC data link protocol can be used between controllers. CPM allows all messages that are received error free to be retained by the receiving unit. Only missing or flawed messages must be retransmitted. Some data link protocols require that all messages received after a missing or flawed message be retransmitted, reducing link utilization. A CPM acknowledgment identifies the sequence number of the next message expected, and it specifically lists by sequence number each message requiring retransmission.

12.6.3 Equipment Features

Several means exist to differentiate products that fall under the channel extender umbrella [12.14]. First, while some products use telecommunication facilities with various degrees of distance limitations, others can connect only hosts that are within a few miles. Another way to evaluate channel extension products is to determine whether they interconnect hosts only, or allow interconnection of a group of high-speed peripherals (laser printers or disk drives). To date, the most popular application has been for host-to-peripheral channel extension. Channel extension equipment is not cheap: A device is needed at each end of whatever is being connected. In addition, nontrivial software charges may apply.

The higher end channel extender products incorporate microprocessors, memory buffers, and control logic to operate as channel extenders both locally and—via T1/DS1 or T3/DS3 facilities or equivalent—over wide geographic areas. These sophisticated channel extenders can perform a number of communication functions by recreating the block multiplexed channel of an IBM mainframe at a distance.

Almost all vendors claim that their products are transparent to VTAM. Earlier versions of VTAM presented problems to designers and users of channel extenders and adapters because of time-outs and block sizes. More recent VTAM releases have overcome many of these problems [12.8]. For example, version 3.2 brought a number of enhancements to SNA networks as follows [12.15]:

- Support for multiple type 2.1 peripheral nodes;
- Improved T1/DS1 performance;
- Direct support of X.25 links; and
- Dynamic reconfiguration of SNA networks.

Users have several options to consider when choosing a channel extension product. Key considerations are:

- Does the product need host-based software?
- Does the product need modifications to VTAM or the operating system? (Modifications should be avoided whenever possible.)
- How much control can a remote operator have?
- Can multiple hosts be connected?
- How many and what type of peripherals can be supported?
- Does this product operate with an integrated network management system, such as Netview?

An important feature to consider is the ability to support multiple device types. This avoids the expense and performance inefficiency of multiple layers of protocol conversion software on the mainframe. A channel adapter should be configured readily with a wide range of adapter boards to support standard applications. Where time is a critical element to the data flow, such as in engineering and scientific environments, a high-speed interface of at least 3 Mbyte per second transmission in a data streaming mode is desirable (not limiting transmission to a predetermined amount).

12.6.4 Relationship with Fiber Optic Modems and Multiplexers

Fiber optic multiplexers do not normally attach directly to a channel, but rather accept traditional data input, presented in some standard interface such as RS-232-C. This input is multiplexed over a dedicated point-to-point fiber link, much the same as traditional TDM. The input rates of the devices are generally 9.6, 19.2, and occasionally 56 kb/s. The number of ports can vary from 4, 8, 16, and 64 to occasionally more. The typical data rate achieved on the fiber link is a relatively modest DS1 or DS2, which does not employ the full potential of the fiber. These devices are functionally similar to a T1 multiplexer. More than 30 vendors are in the fiber optic data link and module market. The price range is generally $3000 to $10,000. The difference between a fiber optic multiplexer and a fiber optic modem is generally one of size and sophistication: Modems handle a single port, may operate at lower speeds (19.2 kb/s is typical), and cost less ($100 to $500).

12.6.5 Relationship with Digital Switching Matrices (DSM)

Channel switches should not be confused with DSM hardware now being used by some data centers to interconnect mainframe digital ports on the FEP to the digital side of a bank of modems, while providing automated intermediate electronic patching functions (this eliminates bulky 25-pin EIA patch panels). The difference between a DSM and a channel switch-channel extender is that the DSM operates at FEP-type data speeds, which are normally 4.8, 9.6, or 19.2 kb/s (as selected by the FEP or modem clock), while the channel switches operate at the channel speed

(3 Mbytes per second and higher); they also apply to different parts of the data center. Different technologies, applications, scopes, and functions apply to the two types of equipment. Table 12.5 and Figure 12.6 depict some of the differences. Channel switches often incorporate a channel extension function; in this context, these can be viewed as high-end channel extenders.

12.7 POTENTIAL FUTURE DEVELOPMENTS

Some observers believe that channel extenders may represent a future threat to FEPs. Figure 12.7 depicts a potential evolution of channel extender connectivity in the early 1990s. One early product along these lines was available in 1989; it allowed mainframe channels to be connected to a FDDI network [12.16].

In the future, we may wish to extend the CPU bus, not just the channel, to the 150 Mb/s range. Bundled T3/DS3 lines (operating at 45 Mb/s) are being considered to achieve this goal. The promise of broadband ISDN, however, is better than bundled DS3s. Broadband ISDN will make available bandwidths of 150 to 600 Mb/s. SONET links, operating up to 2.4 Gb/s, may become generally available in the mid-1990s to the end-user.

Table 12.5
Comparison of Switches and Extenders with DSMs

	Channel extender	*Channel switch*	*Digital Switching Matrix*
Location	Channel side	Channel side	FEP side
Speed per second	3 Mbytes	3 Mbytes	9.6 kb
Purpose	Extend channel	Extend and switch channel	Facilitate modem sharing
Size	1 × 1	$n \times n, n \leq 16$	$n \times n, n \leq 256$

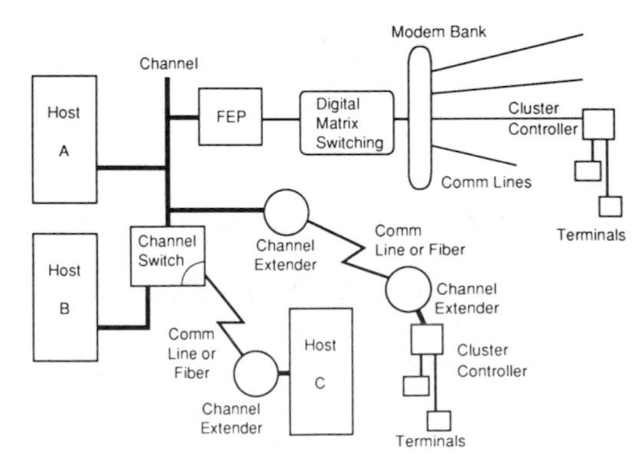

Figure 12.6 Placement of channel switches, channel extenders, and a digital switching matrix in a computer installation.

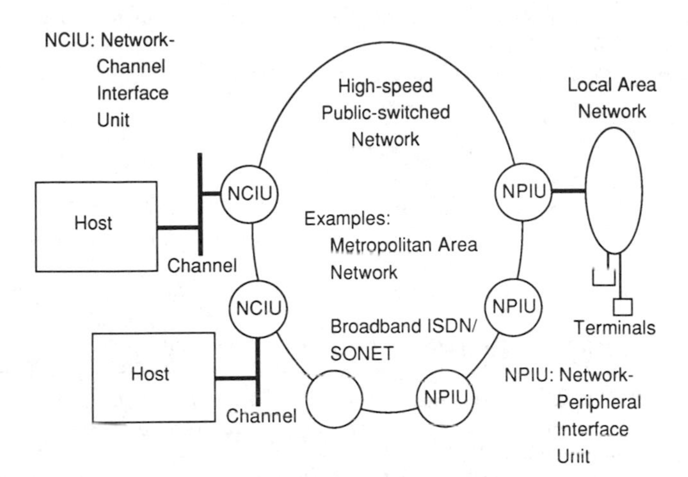

Figure 12.7 Possible future remote peripherals configuration.

Until recently, most channel adapters were designed with proprietary interfaces, which has restricted users to connecting only those products offered by the channel-adapter vendor. Thus, the user's choice of those computing tools was limited. Both local and remote units must be purchased from the same vendor and be of the same model type. In addition, most of the adapters have been special-purpose connectivity products capable of handling only one type of connection and have not supported general attachment for multiple devices. The industry needs a multisource channel adapter. A nonproprietary open-architecture channel adapter can bring less limiting solutions to users and serve as a connectivity "platform." An open-architecture channel adapter is a generic interface device that attaches directly to the mainframe channel at channel speed. Such adapters could be useful in the next few years.

Finally, the speed of the channel will increase in the future. The first step was the announcement by IBM in 1990 of a fiber-based channel architecture, increasing the speed to 10 Mbytes/s over 62.5/50 multimode fiber. This is the first major redesign since 1964. In 1991 the channel speed was raised to 17 Mbytes/s over single-mode fiber. The distance was also extended to 60 km. The *enterprise systems connection* (ESCON) architecture provides the capability to reconfigure channels dynamically. The equivalent of a multipoint-to-multipoint dynamic channel extender matrix, called ESCON Director, was introduced.

REFERENCES

[12.1] M.W. Cerruti, M. Voce, "Zap Data Where it Really Counts: Direct-to-Host Connections," *Data Communications*, July 1987.

[12.2] "Newsfront," *Data Communications*, October 1987, p. 74.

[12.3] W.K. Dawson, R.W. Dobison, "Buses and Bus Standards," *Computer Standards and Interfaces*, Vol. 6, 1987, pp. 403–425.

[12.4] J.J. Hunter, "Product Focus: Communications Processors," *NetworkWorld*, November 2, 1987.

[12.5] J.B. Pruitt, L. Wood, "Extending Mainframe Reach," *Computer Decisions*, July 1988.

[12.6] *Computer Economics*, Vol. 9, No. 11, November 1987, p. 3.

[12.7] *Computer Weekly*, August 4, 1988, p. 6.

[12.8] J. Bryce, "Extending Mainframe Access," *ComputerWorld*, October 26, 1987.

[12.9] C. Fox, "Fibronics Boosts Fiber LAN Line," *CommunicationsWeek*, September 4, 1989, p. 20.

[12.10] J.B. Iida, "Channel Extenders Help Remote Hosts Communicate," *Information Week*, July 13, 1987.

[12.11] R. Moran, "Opening I/O Floodgates," *Computer Decisions*, July 1988.

[12.12] H. Woodsend, "DEC and IBM: An Affair Begins," *Communications* (GB), June 1987.

[12.13] E.W. Miller, "T1 Speed Connections within SNA," *Telecommunications*, September 1988.

[12.14] K. Korostoff, "Changing the Channel," *ComputerWorld*, November 16, 1987.

[12.15] D. Strom, *PC Week*, July 21, 1987.

[12.16] "Fibronics to Announce FDDI-to-IBM Link," *ComputerWorld*, January 30, 1989, p. 85.

Note: This chapter is based on the following report produced by the author: "An Overview of Channel Extenders," DataPro Report C14-010-101, March 1989.

Chapter 13
Data Communication Standards

This chapter describes the state of affairs in the standardization of key data communication functions, including file transfer, transaction processing, messaging, virtual terminals, and internetworking. Approximately 100 key standards are required by a sophisticated contemporary network to undertake data communication (see Table 13.1). Issues pertaining to standards associated with LANs are described in Chapter 14; standards associated with MANs are discussed in Chapter 15; network management standards are discussed in Chapter 16; and network security standards are discussed in Chapter 17.

This chapter is, by necessity, only an overview of the subject matter; the CCITT 1988 Blue Book (in reality, 61 books) alone spans 20,000 pages of material, and this does not include standards from ISO, ECSA, ANSI, Bellcore, *et cetera*. In addition, this chapter does not aim at strict formalism, but focuses on trends that may affect computer communication in the 1990s. References [13.1], [13.2], and [13.3] are good texts that adopt a more formal approach; however, ultimate formalism can only be achieved by consulting the original standards directly. Reference [13.4] is also an excellent source of information.

13.1 THE OPEN SYSTEM INTERCONNECTION REFERENCE MODEL

The OSIRM, introduced in Chapter 1, imposes order and structure on the process of communication. The OSI seven-layer environment provides the necessary functionality for communication between applications processes (APs) that reside in discrete remote end systems. An AP represents the functionality to perform the information processing required for a particular user task. Examples of APs are airline reservation applications, payroll processing, inventory management, and sales tracking.

Table 13.1
Key OSIRM Communications Standards

Standard	Description
ISO 7498 (CCITT X.200)	Reference model of open systems interconnection
ISO 7498-1	Basic reference model
ISO 7498-1 AD1	OSIRM Connectionless-mode transmission
ISO 7498-2	OSIRM Security architecture
ISO 7498-3	OSIRM Naming and addressing
ISO 7498-4	OSIRM Network management framework
ISO 8072 (CCITT X.214)	Transport layers service definition
ISO 8072 AD1	Connectionless-mode transport layer service definition
ISO 8073 (CCITT X.224)	Transport layer protocol
ISO 8073 AD1	Network connection management subprotocol
ISO 8073 AD2	Class 4 operation over connectionless network service
ISO 8208 (CCITT X.25 PLP)	X.25 packet level protocol (connection-oriented network layer protocol)
ISO 8208 AD1	Alternative logical channel number allocation
ISO 8208 AD2	Extensions for private and switched use
ISO 8326 (CCITT X.215)	Session layer service definition
ISO 8327 (CCITT X.225)	Connection-oriented session layer protocol
ISO 8348	Network layer service definition (connection mode)
ISO 8348 AD1	Connectionless-mode network service definition
ISO 8348 AD2	Network layer addressing
ISO 8348 AD3	Additional features of the network service
ISO 8473	Protocol for providing connectionless-mode network service (CLNP)
ISO 8473 AD3	Provision of the underlying service assumed by ISO 8473 over subnetworks which provide the OSI data link service
ISO 8571-1,-2,-3,-4,-5	File transfer, access, and management
ISO 8602	Protocol for providing connectionless-mode transport service
ISO 8613 family (CCITT T.411-T.419)	Office document architecture and interchange format
ISO 8648	Internal organization of the network layer
ISO 8649 (CCITT X.217)	Association control service element (ACSE) service
ISO 8650 (CCITT X.227)	Association control services element protocol
ISO 8802-1 and -2	LAN logical link control
ISO 8802-3 and Addenda	LANs' access methods and physical layer for carrier sense multiple access
ISO 8802-4	LANs' access methods and physical layer for token-passing bus
ISO 8802-5	LANs' access methods and physical layer for token ring
ISO 8822 (CCITT X.216)	Presentation layer service
ISO 8822 AD1	Connectionless-mode presentation service
ISO 8823 (CCITT X.226)	Presentation layer protocol
ISO 8824 (CCITT X.208)	Specification of abstract syntax notation one (ASN.1)
ISO 8825 (CCITT X.209)	Specification of basic encoding rules for abstract syntax notation one (ASN.1)

Table 13.1 continued
Key OSIRM Communications Standards

ISO 8831	Job transfer and manipulation service
ISO 8832	Job transfer and manipulation protocol
ISO 8878	Use of X.25 to provide connection-oriented network service
ISO 8880-1	General principles of protocol combinations to provide support of network service
ISO 8880-2	Provision and support of the connection-mode network service
ISO 8880-3	Provision and support of the connectionless network service
ISO 8880-4	Interconnection of OSI environments
ISO 8881	Use of X.25 PLP in LANs
ISO 8886 (CCITTX.212)	Data link service definition
ISO 9040	Virtual terminal service definition—basic class
ISO 9040 AD1	Extended facility set
ISO 9040 AD2	Additional functional units
ISO 9041	Virtual terminal protocol—basic class
ISO 9041 AD1	Extended facility set
ISO 9041 AD2	Additional functional units
ISO 9066-1 (CCITT X.218)	Reliable transfer service definition
ISO 9066-2 (CCITT X.228)	Reliable transfer protocol specification
ISO 9072-1 (CCITT X.219)	Remote operations service element (ROSE) service definition
ISO 9072-2 (CCITT X.229)	Remote operations service elements protocol specification
ISO 9542	End system to intermediate system routing exchange protocol
ISO 9545 (CCITT X.207)	Application layer structure
ISO 9545 AD1	Application layer structure, connectionless operation
ISO 9548	Connectionless session protocol
ISO 9574	Provision of OSI connection-oriented mode network service by packet terminal equipment connected to an ISDN
ISO 9576	Connectionless presentation layer protocol
ISO 9579	Remote database access (RDA)
ISO 9594 -1 through -10 (CCITT X.500)	Directory services (see Table 13.5 for more detail)
ISO 9595	Common management information service
ISO 9596	Common management information protocol
ISO 9804 (CCITT X.237)	Service definition for commitment, concurrency and recovery (CCR)
ISO 9805 (CCITT X.247)	Commitment, concurrency and recovery (CCR) protocol
ISO 9834 family	OSI registration authorities
ISO 10021 (CCITT X.400 and CCITT F.400 series)	Message handling system (MHS) (See Table 13.6 for more detail)
ISO 10022 (CCITT X.211)	Physical layer service definition
ISO 10026-1 and -2	Transaction processing service definition

Table 13.1 continued
Key OSIRM Communications Standards

ISO 10026-3	Transaction processing protocol specification
ISO 10028	Definition of the relaying functions of a network layer intermediate system
ISO 10030	End system to intermediate system routing information exchange protocol for use in conjunction with ISO 8878
ISO 10035	Connectionless application layer
ISO 10038	MAC bridges
ISO 10039	MAC service definition
ISO 10164 family	Systems management (see Chapter 16)
ISO 10165 family	Structure of management information (see Chapter 16)
ISO 10589	Intermediate system to intermediate system intra-domain routing exchange protocol
ISO N1972	Connectionless session service

Notes:
(1) Not all standards listed in this table were at the IS status at the time of this writing; some were DIS, DAD, or PDAD.
(2) Table is not exhaustive. In particular, secondary addenda and proposed addenda are not always shown.
(3) Standards are identified by function and not always by the formal title.
(4) ISO standard shown above should be properly labeled as ISO/IEC xxxxx, as alluded to in Chapter 2; the IEC nomenclature in only implicitly assumed.

To achieve orderly communication, many functions need to be performed. We can group these functions into natural layers that share task affinity and logical proximity [13.1]. As discussed in Chapter 1, OSIRM layers are hierarchical in the sense that a given layer calls for the services of the layer immediately beneath it; it cannot ask directly for the services of a layer several levels away or jump into the middle of another layer (in some implementations, some layers may be "null"—in formal terms, however, these implementations do not conform strictly with the OSIRM). The model includes precise definitions of the services provided by each layer to its next higher layer. Services defined for a given layer are, in turn, employed by the layer immediately above it. Each layer passes down to the lower layer (or up to the next layer, at the receiving end) blocks of data requiring transmission, manipulation, or service. These layers normally attach a characteristic header that contains appropriate information (such as the real network address, block sequence number, *et cetera*). The headers are physically nested, with the lower layer headers being outermost and higher layer headers being innermost. Through the use of these well-defined headers, the protocols between the remote open systems are implemented.

To realize the OSIRM, the ISO has formulated standards—specifications about how information is coded and passed between the partners in the communication. As discussed above, the OSI model is characterized by a standardized description of the services provided at each layer, as well as the peer-to-peer protocols for communication between layers across the network. The OSIRM and the service definitions are only structures for discussing the tasks involved in communicating between open systems. Only the peer-to-peer protocols need to be implemented by vendors in order to conform practically with the OSIRM and achieve open communication. In some situations, however, the formal layer interfaces are also implemented by vendors, consistent with the appropriate ISO and CCITT layer services recommendations.

The OSIRM is described by document ISO 7498, which was adopted as an international standard in 1984. It consists of several parts and addenda. Document 7498-1 describes the basic reference model. Additions to the reference model have been made for connectionless-mode transmission (ISO 7498-1, Addendum 1), security (ISO 7498-2), naming and addressing (ISO 7498-3), and network management (ISO 7498-4). This model is also described in CCITT X.200, adopted in 1988, although some differences exist (for example, CCITT did not support connectionless service at the beginning of 1990, although an effort on the "alignment for connectionless was under study"—at the time of this writing, the disposition of the other addenda was unknown because X.200 does not include any of them). Note that at this point in time many, but not all, OSI standards are collaborated on by both ISO/IEC and CCITT.

The layering (clustering of functions) in the OSIRM is as follows, in descending hierarchical order (see Figures 13.1 and 13.2):

Layer 7: Application layer;
Layer 6: Presentation layer;
Layer 5: Session layer;
Layer 4: Transport layer;
Layer 3: Network layer;
Layer 2: Data link layer; and
Layer 1: Physical layer.

(Note that Figure 13.1 does not show the structure imposed by the addenda; Figure 16.1, for example, depicts the network management extensions.)

In the OSIRM environment, communication between APs is modeled by communication between application entities (AEs). An AE represents the communication functions of an AP. Multiple sets of communication functions may exist in an AP; hence, an AP may be represented by multiple AEs. Each AE is a set of communication capabilities whose components are application service elements (ASEs). An ASE is a coherent set of integrated functions; some examples will be discussed later. ASEs can be configured to provide a variety of functionality to

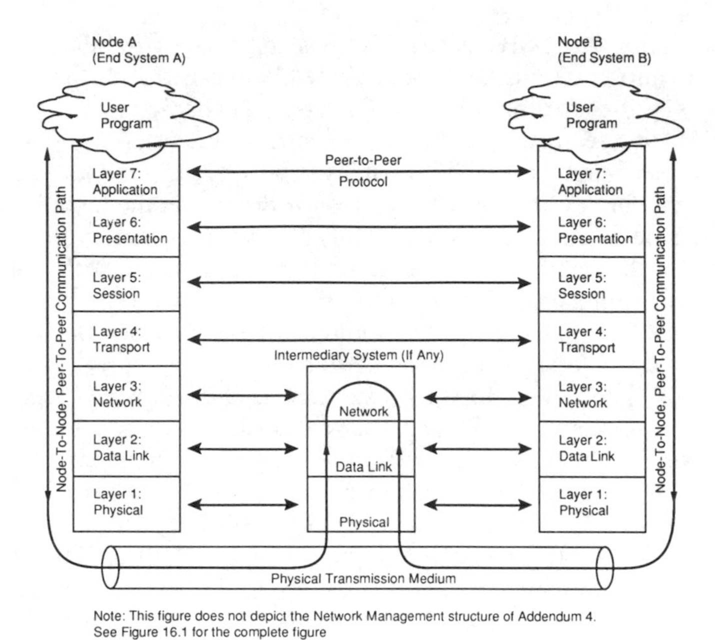

Figure 13.1 The OSIRM framework.

meet many operational contexts that are required for diverse information systems. Figure 13.3 depicts the possibilities described in this paragraph.

Table 13.1 lists, for easy access, the key OSIRM standards discussed in this chapter, ordered by ISO designation; the table is based partially on Reference [13.5]. Table 13.2 lists some additional standards that are of interest in the MAP/TOP context discussed toward the end of the chapter.

The *application layer* is concerned with communication activities on behalf of the ultimate APs running on the computer. The application layer provides ASEs for management and activation of resources. Examples of ASEs include file transfer, transaction processing, virtual terminals, and message handling services. The application layer provides access between the user APs in the local systems and the OSI environment. The application layer provides facilities for a semantic exchange between APs: association control; commitment, concurrency, and recovery; remote operations; and reliable transfer.

The application layer (ISO 9545/CCITT X.207) functionally consists of two parts: those common to all applications and those specific to a particular application. The common application part consists of the common application service elements (CASEs), which supply a standard set of services to the various ASEs. CASEs have been or are currently being standardized by ISO and CCITT. The

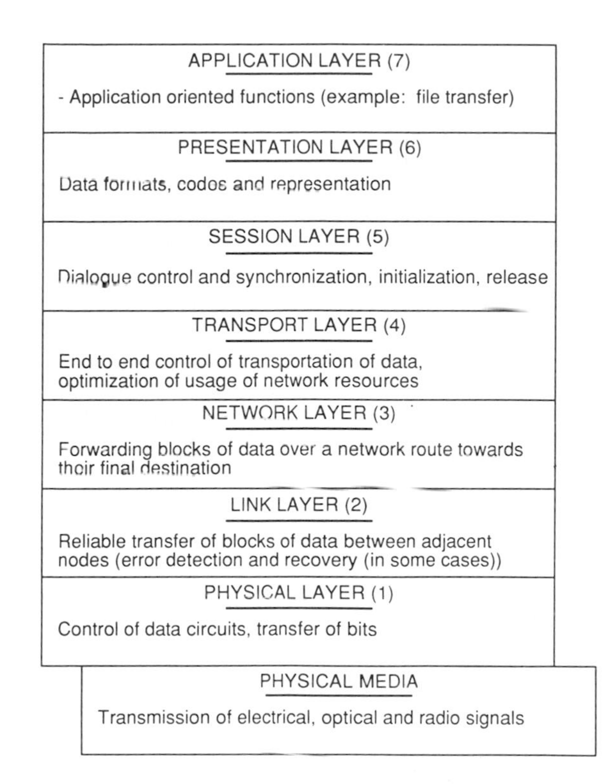

Figure 13.2 Basic functions performed by the layers of the OSIRM.

second functional part contains the specific application service elements (SASEs), which are designed to meet the unique requirements of a particular application.

Two important CASEs are the *association control service element* (ACSE) and the *commitment, concurrency, and recovery service element* (CCR). Association control initiates contact with another application within the OSI environment. All of the lower layer mechanisms are activated by an attempt to create the application association; if all lower layer operations are successful, the association is established. CCR provides coordination services for a group of AEs operating on different systems; functions that applications would use within a local system are extended to coordinate the modification of shared data, to commit to actions, and to manage application problems.

The ACSE provides the functions for conceptual establishment of an association so that the communication can take place between the corresponding APs. The association control process exchanges AE titles. An AE may derive name and address information from the directory (see Section 13.2.1.2). An exchange of application context names also occurs to determine ASE functionality and the version to be performed during the association. The context defines the set of

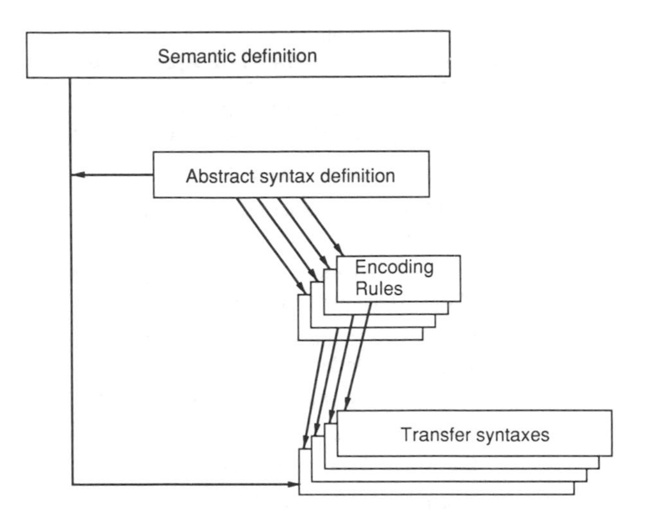

Figure 13.4 The relationship between ANS.1 and BER.

and terminate a session between end-users of the open communication facilities. This allows the end-users to negotiate and manage the type of dialogue needed to support information flow requirements and to synchronize and resynchronize their respective activities by inserting distinguishable signals in the information stream. Also, the session protocol has the capability to recover from prematurely terminated conversations. The session layer provides the basic functions for dialogue coordination between the communication APs. It establishes the session connection, manages the exchange of data, provides recovery for data lost in the local receiving system, and gracefully terminates the communication.

The *transport layer* has the responsibility of guaranteeing that the received data blocks, as delivered by the network layer, are correctly assembled. Hence, this layer provides a universal end-to-end transport service in conjunction with the underlying network layer. It provides transparent transfer of data between the communicating entities, relieving these entities of the details of how reliable transfer of data is achieved across the network. The transport layer compensates for the differences in network services provided at the lower layers, regardless of the type or quality of the network employed. The transport layer includes such functions as multiplexing a number of independent message streams over a single connection and segmenting data into appropriately sized units that can be handled by the network layer. (The session layer recovery deals with both end-system and network failure, while transport layer recovery deals only with network failure.)

The transport layer defines a number of "transport classes." Class 0 provides no additional reliability other than that which is intrinsic to the network layer services that have been employed; Class 1 provides basic error recovery; Class 2 allows multiplexing (to reduce network connection costs); Class 3 provides additional error recovery capabilities; and Class 4 provides comprehensive error detection and recovery and additional multiplexing capabilities. These transport classes provide a range of functionality: Class 0 is the simplest, for use in highly reliable (packet switched or circuit switched) networks; Class 4 is a high-quality service, with error detection and recovery services for possibly unreliable networks (particularly for node failure), and is for use over connectionless-mode networks, discussed in Chapter 11. If we know that the underlying physical medium or network is basically error-free (for example, with fiber), then a Class 0 service may suffice. A twisted-pair LAN in an electronically noisy environment (for example, on a factory floor) is not as intrinsically reliable as a LAN with fiber links in an office environment; therefore, Class 4 may be preferable in this case.

The *network layer* routes data over a series of physical links, moving the data toward their destination. It provides functional and procedural means to exchange data units between two transport layers and provides these layers with independence from routing and switching considerations. In the connection-oriented mode, the network layer also provides the means to multiplex several "logical" links onto a single physical link (i.e., the ability to carry information from several users on one channel). In addition to routing, it provides congestion control functions.

Two types of network layer services exist as indicated above and in Chapter 11: connection-oriented (also called connection-mode) network transmission and connectionless (also called connectionless-mode) network transmission. Connection-oriented communication provided by the network layer service is reliable in the sense that acknowledgments, retransmissions, and in-sequence delivery are provided; however, it requires a somewhat complicated protocol with overhead. The connectionless communication provided by the network layer service is such that no acknowledgments, retransmissions, or in-sequence delivery is provided. It is, however, simpler and, for some applications, more efficient than connection-oriented communication. Connection-oriented service requires that a formal connection be established between two ends before the transmission of information can start. This is typical of traditional telephonic connections and of PVC or switched virtual circuit (SVC) packet connections. (In the packet environment, the connection is virtual: A route over which data packets of a given connection are forwarded is specified by a logical association.) Connectionless service involves communication in which each data unit travels independently and is unacknowledged (at the network layer). The path to be taken by the data units may be established in advance or as the message arrives at each network node. As an analogy, consider the delivery of letters dropped in a mailbox for a common destination—each letter may well take a different route to its final destination, independent of the other letters. Connectionless communication is typical of a LAN environment. The OSI connection-oriented network layer service is specified in ISO 8348; the connectionless mode is specified in ISO 8348, Addendum 1.

At the *data link layer* (DLL), data are segmented into frames (blocks) that can be easily managed; they are then moved in and out of the storage (buffers) typically included in data termination or processing equipment. The DLL functionality is concerned with transmission of data in a reliable fashion over a single point-to-point communication link, which is accessed through the layer 1 cables and facilities. The control procedure specifies the headers and trailers of the blocks. Communication links may be subject to noise, corruption, and timing jitter, which are unacceptable for data transmission (but may be tolerated for voice transmission).

The DLL has traditionally guaranteed data integrity over a link; the functions required to achieve this include framing, error detection, and, optionally, error recovery and flow control. As discussed in Chapter 4, newer communication services, such as frame relay, may only implement a subset (a core) of this functionality. The same is true in LANs, under the LLC procedure 1, discussed in Chapter 14. This shift in perspective in DLL services has arisen because (1) telecommunication links today are less noisy than they used to be, particularly when compared to older analog transmission carrier systems; (2) the network may relinquish the responsibility (or the user may wish to acquire the capability) to perform error correction on an end-to-end (i.e., end system to end system) basis, rather than on a link-by-link basis; (3) in some situations, end-to-end error management may be more cost-effective than link-by-link; and (4) some user applications simply do not require link-by-link error correction.

Examples of network layer and DLL standards were discussed in Chapter 4 (ISDN) and Chapter 9 (CCSS7).

The *physical layer* is concerned with the physical, mechanical, electro-optical, functional, and procedural characteristics necessary to establish, maintain, and disconnect the physical link of a communication network. For example, it describes the size and shape of the data cable receptacles to connect the computer to the data communication equipment, the voltage levels to represent bits, the number of pins, *et cetera*.

Status

Many upper layer standards were completed in the 1987–90 time frame and others are being developed. Many lower layer standards were completed between 1985 and 1988. Some of the more important available upper layer standards (most of which are finalized) include the following subset from Table 13.1:

- Message handling system (MHS) (ISO 10021/CCITT X.400 and CCITT F.400);
- File transfer (ISO 8571-1, -2, -3, -4, -5);
- Directory services (ISO 9594/CCITT X.500);

- Transaction processing (ISO 10026-1);
- Virtual terminal (ISO 9040, ISO 9041, and addenda);
- ACSE (ISO 8649/CCITT X.217, ISO 8650/CCITT X.227);
- Presentation layer service and protocol (ISO 8822/CCITT X.216, ISO 8823/CCITT X.226); and
- ASN.1 (ISO 8824/CCITT X.208, ISO 8825/CCITT X.209).

13.1.1 Application Layer Standards

This section provides a brief summary of the status of major application layer standards. Some of the key ASEs are briefly described.

13.1.1.1 X.400—MHS

MHS (CCITT X.400 and F.400 families) is a store-and-forward electronic messaging architecture that allows end-users and application processes to send and receive messages. A message consists of a message envelope and message content. The message content may contain facsimile, graphics, text, voice, or binary data structures. The 1984 standards have been enhanced in the 1988 recommendations, including the new feature of message store. ACSE and ROSE are employed by MHS. MHS can be viewed as a platform to deliver a variety of services, including E-mail, electronic data interchange, and others. Many vendors are committed to developing X.400 products. This protocol is likely to be widely accepted. A more detailed description of this standard is provided in Section 13.2.4.

13.1.1.2 FTAM

FTAM (ISO 8571 family) provides the ability to transfer and access files remotely across multiple computer systems. A virtual file store concept is used that gives FTAM independence from the local implementation of end systems. A more detailed description of this important standard is provided in Section 13.2.2. Many vendors are committed to developing FTAM products. This protocol is likely to be widely accepted, although it may require a substantial amount of work to implement.

13.1.1.3 X.500—Directory

The directory (CCITT X.500 family) is a specialized repository of information (a data base) that contains data used to facilitate communication among addressable entities in a network. X.500 specifies a common protocol to retrieve and update information stored in the directory. This service will provide support for users who require information regarding establishment of communication with another user.

About six vendors had committed by early 1990 to bring X.500 products to the market. A more detailed description is provided in Section 13.2.1.

13.1.1.4 Transaction Processing

The transaction processing (TP) (ISO 10026 family) ASE supports transaction processing among distributed transaction programs. The function of TP is to handle computer-to-computer transactions. TP will provide support for remotely submitted transactions and distributed transactions. This protocol is less mature than SNA LU 6.2 but appears to be progressing within ISO. This protocol will probably not displace IBM's LU 6.2 as a product in the near future.

The status of TP was that of a draft international standard (DIS) at the time of this writing. It was expected to become an international standard (IS) in May 1992. TP employs CCR (ISO/DISs 9804.2/CCITT X.237 and 9805.2/CCITT X.247), which was being reworked to be more functional and provide better performance, and ACSE. Though TP is stabilizing, the impact of the CCR changes may be substantial. Some of the issues being addressed in 1990 are:

- Compatibility with ROSE (ISO DIS 9072-1/CCITT X.219);
- Compatibility with remote data base access (RDA) (ISO DP 9579);
- Separation of data transfer mechanisms from TP; and
- Separation of data base from protocol, allowing product heterogeneity (separate data base and protocol stack vendors), and encouraging data base vendors to implement products supporting the TP standard.

In 1992, we should see the first basic implementation providing conversation between two programs; in 1993, the first full implementation of commitment protocols should appear; from 1994 on, several vendor products should become available.

ISO TP and LU 6.2 are mappable; a sample of these mappings is shown in Table 13.3. However, some differences exist in the means of providing contention resolution. Currently, ISO TP is conversational. OSI-based queued message service is just beginning to be studied.

13.1.1.5 Virtual Terminal

The OSI VT (ISO 9040 and 9041) standard allows a device-independent way of representing data and control actions. The VT service is intended to support communication between (1) two human terminal users, (2) two computer systems, or (3) a human terminal user and a computer system. The VT is a service provided by a pair of virtual terminal application entities (VTAEs) that provides access to

Table 13.3
Comparison of TP and LU 6.2

TP	*IBM's LU 6.2*
Dialogues	Conversations
Associations	Sessions
Coordination	Synchronization
None	None
Handshake	Confirm
Commitment	Two-way commitment
TP-BEGIN-DIALOGUE	ALLOCATE
TP-DATA	SEND-DATA
TP-GRANT-CONTROL	RECEIVE-AND-WAIT
TP-HANDSHAKE (request)	CONFIRM
TP-HANDSHAKE (response)	CONFIRMED
TP-END-DIALOGUE	DEALLOCATE
TP-REQUEST-CONTROL	REQUEST-TO-SEND

terminal-like facilities across the network. The object-oriented approach taken by the VT standard introduces abstract data objects described by object type definitions. Terminals and hosts using a "virtual" description of this data interact by means of a shared communication area, known as a *conceptual communication area*. The responsibility for mapping the virtual description data to the real terminal belongs to the local end system. VT is an ASE.

The VT service provides syntactic translation and communication between the abstract views of the terminal object as maintained within the open systems. The service is provided by the two VTAEs connected by an application association; ACSE and presentation layer services are utilized by VT.

ISO 9041 is the VT protocol for character-oriented terminals. Currently, VT standards support a variety of terminal types categorized into two classes: (1) the basic class, which supports terminals such as TTYs, DEC VT100s, and page-mode terminals; and (2) the forms addendum to the basic class, which supports forms processing for 3270-type terminals. These are described in addenda to the two standards as ISO 9040/AD1 and ISO 9041/AD1.

VT basic class for character-oriented terminals is now complete and has international standard status; work in the NIST OSI Implementors' Workshop (discussed in Chapter 2) was under way in 1990 on phase II agreements for VT.

Future work in this area will include support for terminals that display graphics using vector or raster-based technology, and windowing. On a related topic, work is ongoing in ISO/IEC JTC1/SC21/WG5 on a new terminal standard, called terminal management (TM). The scope of this new work item is to manage and correlate dialogues. For example, voice, data, and video requiring end-user resources such as telephones, printers, and workstations can be coordinated by TM. TM will also be responsible for managing window mechanisms in OSI. TM has a model available and preliminary service was expected in 1990. Many issues, such as windowing and multimedia synchronization, need to be addressed. VT will probably not become an important standard because it cannot compete with the *de facto* IBM 3270 protocol, both in terms of sophistication and ubiquity.

VT is not related to X-Windows; the latter is an industry *de facto* standard for fairly sophisticated graphics and windowing interfaces.

13.1.1.6 Other Application Layer Standards

Some secondary but equally important standards include the following:

- *Job Transfer and Manipulation* (JTM) (ISO 8831 and ISO 8832) is a service and protocol that enables users to define, submit, and receive results of processing on remote end systems. It allows users to inquire about the status of the processing environment and to suspend, cancel, or resume processing.
- *CCR* (ISO 9804/CCITT X.237, ISO DIS 9805/CCITT X.247) is a service and protocol to ensure consistent data integrity across distributed systems. It provides reliable completion of distributed actions, identifies error conditions, and facilitates recovery from such fault events.
- *ROSE* (ISO 9072) is the capability for a request-response function when one AP (the requester or client) requests that a particular operation be performed. The other AP (the responder or server) attempts to carry out the operation and then reports the outcome of the attempt to the requester.
- *Reliable Transfer Service* (RTS) (ISO 9066-1/CCITT X.218, ISO 9066-2/CCITT X.226) is a procedure for use of the ACSE and presentation service to achieve the assured transfer of a single application protocol data unit (APDU) or a series of APDUs. RTS is an application layer standard and should not be confused with transport Class 4, which is a layer 4 protocol. RTS ensures that no message is lost in the event of network or end system failure.
- *Remote Data Base Access* (RDA) (ISO 9579) is a mechanism to allow interconnection of applications and data base systems from different manufacturers, under different managements and having differing complexity levels, using different technologies. More precisely, RDA defines a standard environment in which data manipulation language statements from a particular device can be formatted, sent over an OSI network, received by a data base server, and decoded [13.7]. Like other AEs, RDA provides facilities to control an association between the client and server and to negotiate parameters specific to an RDA association. These services are provided through the ASCE. Work on the standard started in Europe in 1985. Additional work is required to achieve a transparent distributed data base management. Work will continue through the middle of 1991.

- *OSI Upper Layers Architecture* (ULA). The work on ULA is concerned with the standardization of protocols and services in the OSI session, presentation, and application layers. It is also concerned with establishing a structure for the application layer to provide a framework and an appropriate basis for the development of application layer standards. Other related projects are also studied by the ULA group, such as naming and addressing, registration authorities, and connectionless services in the upper layers (see below).

13.1.1.7 Commercial Trends for Upper Layer Standards

The movement to open standards is gathering momentum in the vendor community. The Enterprise Networking event held in June 1988 aimed at demonstrating interoperability between systems from different vendors in the FTAM and MHS areas. Most of the major computer and data communication vendors participated in the event. Increased support for standards implementation is also evidenced by growing vendor participation in the OSI Network Management Forum and in the Corporation for Open Systems. Approximately 50 vendors supported the OSI Network Management Forum in early 1990.

Many vendors are planning to bring out X.400 products in the 1989–92 time frame; most products are based on the 1984 version of the standard, but some products based on the 1988 version should also start to appear. X.400 is likely to be accepted widely by the user community, at least for E-mail interconnection. Many vendors are also committing to provision of FTAM products in the 1990–92 time frame. FTAM is likely to be accepted in the industry, although it requires a substantial amount of work to implement. The complexity arises from having to map real system files onto the FTAM format. On the other hand, as mentioned above, VT will probably not become an important standard. The thrust is in favor of public and standardized directories. However, the potential market for directories cannot be assessed yet. For network management, real products will probably not be available before 1991–93. The market is highly segmented with vendor-proprietary products. Note, too, that TP will probably not displace IBM's LU 6.2; any possible impact would come after 1992.

13.1.1.8 Connectionless Communication Issues for Upper Layers

Some of the issues pertaining to lower layer connectionless-mode communication were discussed in Chapter 11; Chapters 14 and 15 examine the topic in the context of lower layers of LANs and MANs, respectively. Some additional internetworking issues pertaining to "crossovers" of (1) connectionless-mode transport over a connection-mode network and (2) connection-mode transport over a connectionless-mode network are discussed below in the internetworking section.

This section briefly addresses connectionless-mode communication at the upper layers.

Currently, many of the OSI upper layer protocols are based on the connection-mode operation. The transport layer was the highest layer (in 1990) in the OSI protocol suite to provide a connectionless service (ISO 8072/AD1 and 8602). Standards are now emerging for connectionless data transfer at the session, presentation, and application layers.

ISO has produced an addendum to the existing connection-oriented session layer service definition (ISO 8326). The connectionless session service is currently a separate document (ISO N1972), and the connectionless session protocol is an international standard (ISO 9548). The connectionless presentation layer service definition is being processed as an addendum to the existing connection-oriented presentation layer service definition (ISO 8822). A separate connectionless presentation layer protocol has also been developed (ISO DIS 9576) and should become an international standard by the end of 1990. The connectionless-mode application layer service is described in an addendum to the ACSE service specification (ISO 8649). A protocol for connectionless application layer operation was available in DIS form (ISO DIS 10035) at the time of this writing [13.7].

13.1.1.9 Lower Layers

The protocols discussed above rely on lower layer functionality provided by other CCITT and ISO standards. Because the emphasis of the current work (and of this chapter) is on "communication services" provided principally at layer 7, these lower layer issues are not pursued here in detail, except for the internetworking issues covered below.

Some of the key standards at each layer are shown in Figure 13.5 and Table 13.4, according to various applications. The specific selections are called *protocol stacks* or *protocol suites*. Traditionally, carrier-provided networks have offered capabilities up to and including the network layer (i.e., layers 1 through 3), while the CPE end systems implement the upper layers (layers 4 through 7).

Internetworking Issues

With the proliferation of communication networks, the importance of these networks being capable of internetworking has increased. (Although some subtle differences in meaning may be found in some quarters, in this section, we employ the terms *internetworking*, *interconnect*, and *interoperate* interchangeably.) Internetworking is accomplished through the use of relays. The OSIRM defines an (N)-relay as "an (N)-function by means of which an (N)-entity forwards data received from one correspondent (N)-entity to another (N)-entity." Internetworking two

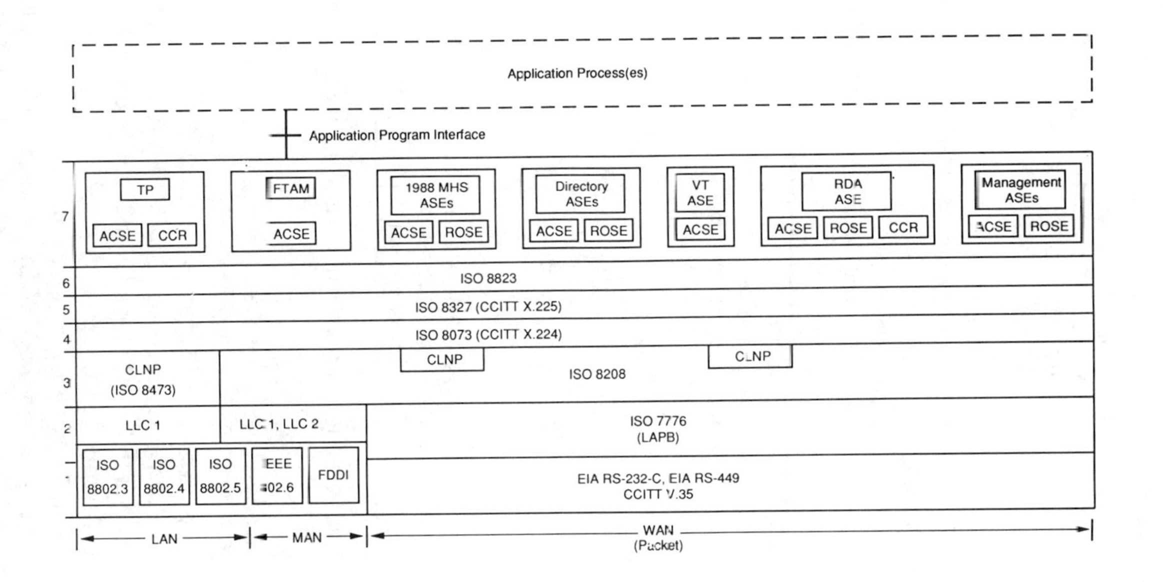

Figure 13.5 Typical protocol suites.

Table 13.4
Commonly Used Protocols in Data Communication Environments

WANs (connection-oriented mode)	
Layer 7	As described in the paragraphs above
Layer 6	ISO 8823/CCITT X.226, ISO 8822/CCITT X.216
Layer 5	ISO 8327/CCITT X.225, ISO 8326/CCITT X.215
Layer 4	ISO 8073/CCITT X.224, ISO 8072/CCITT X.214
Layer 3 (packet)	ISO 8208
Layer 2	ISO 7776 (LAP-B)
Layer 1	EIA RS-232-C, EIA RS-449, CCITT V.35, X.21, X.21bis
LANs	
Layer 5 and higher	As listed for WANs
Layer 4	ISO 8073 AD2 (TP4 over connectionless)
Layer 3	ISO 8473 (CLNP)
Layer 2	LLC1 and LLC2 (ISO 8802-2)
Layer 1 and part of layer 2	ISO 8802-3 (CSMA/CD bus)
	ISO 8802-4 (Token bus)
	ISO 8802-5 (Token ring)
	IEEE 802.6 (MAN)
	ANSI FDDI
ISDN Information Channel	
Layer 3 and higher	As listed for WANs
Layer 2	ISO 7776 (LAP-B)
Layer 1	CCITT I.430 or I.431
ISDN Signaling Channel	
Layer 4 and higher	CCSS7
Layer 3	CCITT Q.931 (it can also carry X.25 PLP)
Layer 2	CCITT Q.921 (LAP-D)
Layer 1	CCITT I.430 or I.431

Note: This table depicts one (or few) suites. These are in no way the only suites possible.

networks is relatively simple when the networks are similar. For example, if both networks are IEEE 802.3 LANs, then the task can be achieved by a bridge, which is a relatively simple device. [Bridges operate at the DLL; in a LAN context, bridges operate at the media access control (MAC) sublayer of the DLL—for the purpose of this chapter, a bridge spans layer 2; detailed functionality of bridges in a LAN context is described in Chapter 14.]

Complexities arise when trying to interconnect incompatible heterogeneous networks, such as connection-mode X.25 packet-switched networks and connectionless-mode LANs. The incompatibility arises from the differences in the services provided by the subnetworks, as depicted in Figure 13.6.

Internetworking can be defined as the provision of the network layer service over an interconnection of different real networks. The network layer is responsible

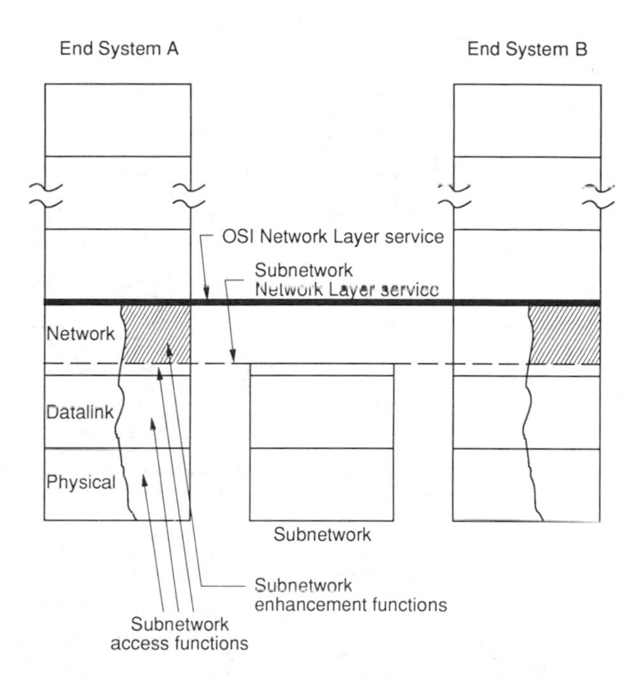

Figure 13.6 Differences in the network layer's subnetwork services.

for resolving some of the differences among diverse network types and presenting to the transport layer a uniform image (by providing subnetwork-dependent convergence functions described later); as such, the network layer plays a crucial role in internetworking two networks. In turn, the transport layer presents a uniform image of the communication infrastructure to the rest of the upper layers.

Figure 13.7 depicts five general strategies to accomplish internetworking. Figure 13.7(a) depicts an approach in which two homogeneous subnetworks can be connected by a physical layer relay; this is typically a repeater, which amplifies the signal as needed (pass-through function) or, at most, interfaces the two types of physical media, such as fiber and twisted-pair cable (conversion function). The repeater sends all traffic across the interface with no interim storage or protocol conversion. Repeaters are technology-dependent.

Figure 13.7(b) shows a DLL relay, which is commonly known as a *bridge*, already mentioned. The function of a bridge is to forward frames and possibly to map across two different DLL protocols. In this arrangement, multiple homogeneous networks can be connected. In a LAN environment, bridges operate at the MAC sublayer of the DLL. They do not provide LLC sublayer processing; hence, these protocols must be similar in both subnetworks for the LAN bridge to be effective.

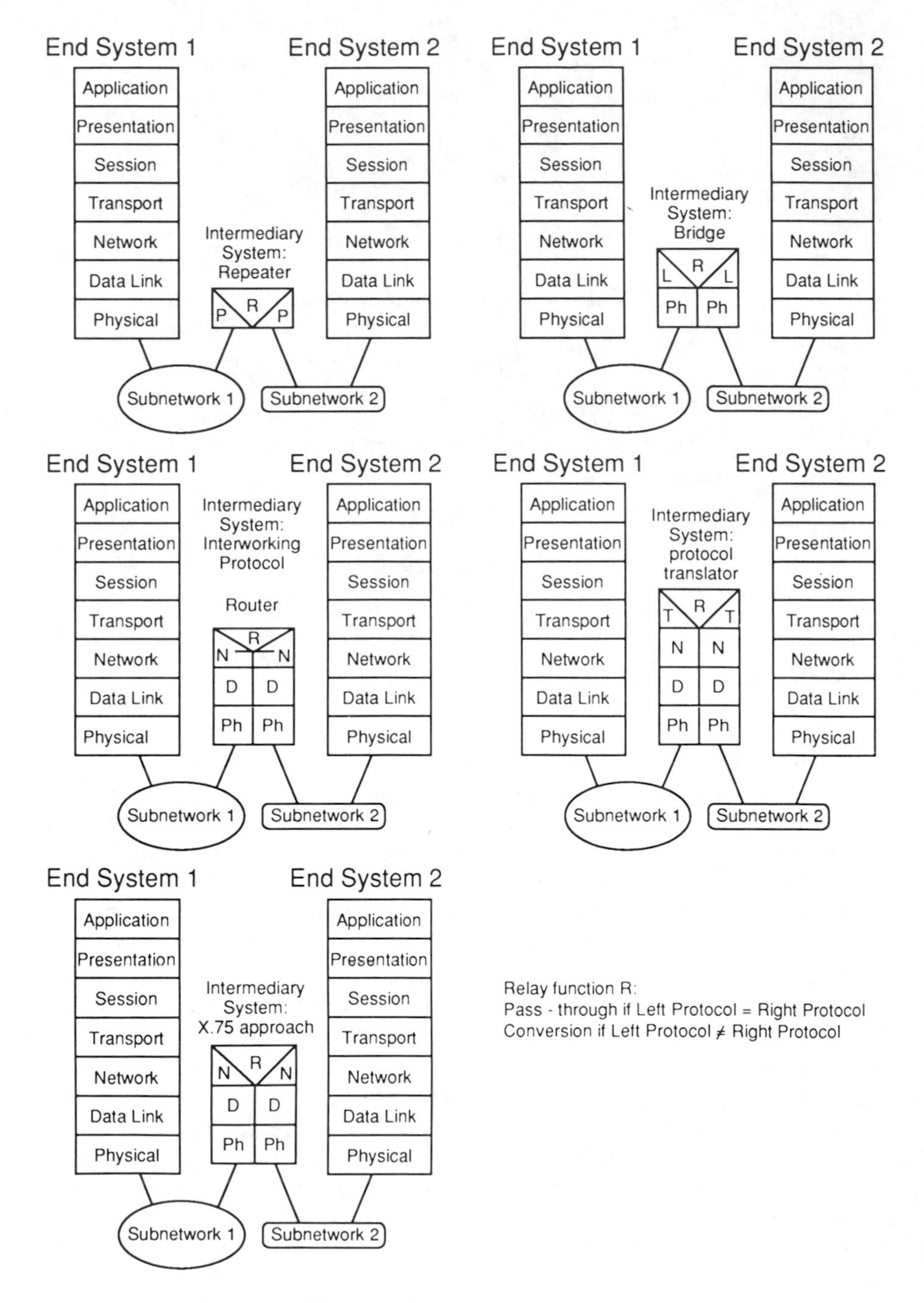

Figure 13.7 Possible internetworking strategies.

Figure 13.7(c) depicts a network layer relay approach. It involves a protocol that sits "above the standard network layer," but is still considered part of the network layer; an alternative view is to consider this protocol as operating in a sublayer of the network layer (this view is discussed more precisely below). An interconnection device operating at the network layer is called a *router*.

The internetworking device must operate the internetworking protocol and the two network access schemes used in the subnetworks. The end systems also operate the internetworking protocol and the appropriate network access protocol. Figure 13.8 depicts the configuration schematically, as an amplification of Figure 13.7(d). Data units are transmitted from the sending end system to the relay system across subnetwork 1 using that network's access protocol. The relay takes the internet-level data unit and retransmits it across subnetwork 2; a single transport connection exists between the two end systems. Hence, data are relayed transparently at the intermediate system's network layer. The router may be provided by a carrier in the network; the implication, however, is that the end systems must deploy the appropriate network layer protocols.

Figure 13.7(d) depicts a transport layer relay (referred to by industry as a "protocol converter" or "gateway," although the latter is a layer 7 relay by definition). This relay assumes that the two subnetworks are different and that the two end systems have different transport protocols. Each subnetwork uses a specific network access protocol. The gateway performs a layer-by-layer termination of the protocols on one side of the protocol "wall" and initiation of other protocols

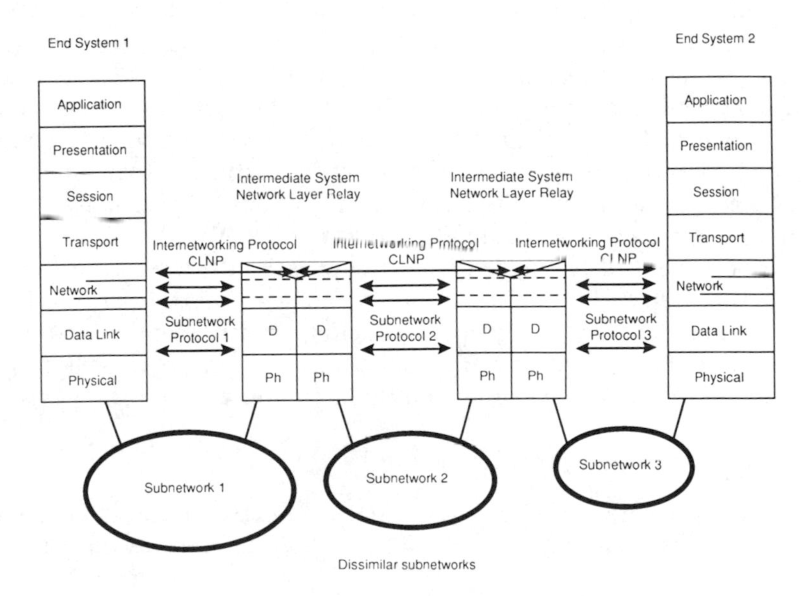

Figure 13.8 Internetworking using network layer routers.

at the other side. The sending end system establishes a transport layer connection with the relay; the device includes a relay mechanism that converts between the transport protocols of the sending and receiving end systems. The problem is complicated by the variety of technologies and protocols that must be considered for the various subnetworks. A particularly active area in the mid- to late 1980s was the efficient interconnection of connection-oriented networks (such as the traditional WANs) with connectionless-mode networks (such as LANs). Europe and the United States have emphasized different transport layer protocol classes and associated network service paradigms. In the United States, transport protocol Class 4 (TP4) with a connectionless network service appears to be common, while in Europe, transport protocol Class 0 (TP0), together with a connection-oriented network service, predominates. The former is due to the success in the United States of LANs in general, and TCP/IP† as a protocol in particular, while the latter has developed because of the common availability of X.25-based public data networks in Europe [13.8]. When users of TP4 over a connectionless LAN need to communicate with users of TP0/X.25 PLP (or even TP4/X.25 PLP), a transport layer relay is required in order to translate between the two transport layer protocols and service types.

Because of the operational limitations of gateways, including issues related to performance, network management, and sensitivity to the end system configuration, network layer internetworking is a technically preferable way to achieve the interconnection.

Figure 13.7(e) depicts the method employed in X.75. X.75 specifies a protocol for exchange of packets between networks that allows a series of internetworked X.25 network layer virtual circuits to be "spliced" together. To the end systems on different networks, they appear to be a single virtual circuit connection end-to-end (actually, the virtual circuits terminate at the relay node, which maintains the mapping information required to connect the distinct virtual circuits).

ISO has developed a family of standards that supports internetworking among current and emerging network architectures at the network layer (i.e., as depicted in the environment of Figure 13.8). One approach advocated is to use transport layer Class 4 over the *connectionless network layer protocol* (CLNP, ISO 8473), as discussed more below. To complete the protocol set, adaptive routing protocol standards were beginning to emerge in 1990 as support for reliable transmission of routing information between the intermediate router systems.

Network Layer Relay Approaches

To provide a common service across a variety of interconnected subnetworks, ISO 8648, "Internal Organization of the Network Layer," divides the network layer

†TCP/IP (Transmission Control Protocol/Internet Protocol) is a DARPA-based pre-OSIRM protocol set that is widely deployed in commercial systems. In OSIRM nomenclature, TCP is a transport layer protocol; IP is a network layer protocol. TCP/IP is discussed in Chapter 14.

into three sublayers: SNACP (Sub-Network Access Protocol), SNDCP (Sub-Network Dependent Convergence Protocol), and SNICP (Sub-Network Independent Convergence Protocol), as shown in Figure 13.9.

The SNACP is the protocol that operates between a network entity in the subnetwork and a network entity in the end system. SNACP is responsible for interfacing with the underlying DLL. The SNACP can be viewed as the locally operating subnetwork protocol; i.e., it is the native protocol of the subnetwork. The level of service that such a protocol contributes toward meeting the requirements of the network layer service is specified by the capabilities of the subnetwork, and it may not be sufficient to meet the expected end-to-end service. The SNACP entity in the end system directly utilizes services of the subnetwork and performs data transfer, connection management, and quality of service selection functions. Clearly, SNACP is dependent on the subnetwork type. An example of SNACP is X.25 PLP.

SNDCP is needed to provide a service mapping between SNICP and SNACP if a direct interface is not possible. To achieve communication across subnetworks, these functions are required in both the intermediate system and the end systems. Relay and routing functions are needed in intermediate systems. The SNDCP is designed for a particular type of protocol and is used to enhance a SNACP to provide a particular network layer service to transport entities; another way of

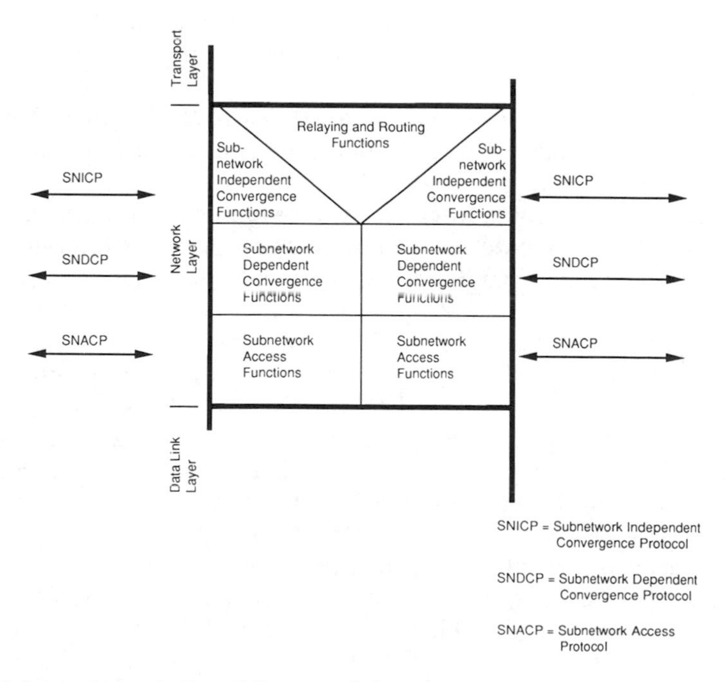

Figure 13.9 Internal organization of the network layer.

saying this is that it corrects for subnetwork deficiencies. SNDCPs, which are either explicit protocols or a set of rules for manipulating the subnetwork service, "provide the underlying service" required by the SNICP; they relax the dependency of SNICPs on the property of the subnetworks [13.9].

SNDCP either provides the network layer service or, as an exception, provides the service required for the SNICP. Therefore, because, in general, the service required by a SNICP is always satisfied by any subnetwork, SNDCP is not required if SNICP is being used (only exceptionally will a SNDCP be required with a SNICP; it is expected that a SNDCP will be used to realize the network service directly) [13.1].

SNICP is concerned, among other things, with providing global network services to the transport layer. A SNICP supplies the network layer service to the network service users; the service is either the connection-mode service (ISO 8208) or the connectionless-mode service (ISO 8473). A SNICP is intended for use over a variety of subnetwork types and requires a minimal subnetwork infrastructure. A primitive service is defined for the underlying service, which all known subnetworks are known to meet with ease (any useful features over and above this minimal service, actually available from a subnetwork, are ignored when a SNICP is in use) [13.1].

A protocol within the network layer may fulfill one of these roles in a particular configuration; the same protocol may fulfill the same or a different role in different configurations. Each role may not be fulfilled individually by a separate discrete protocol: In some cases, a single network layer protocol may provide all the necessary functions.

Two approaches are used to implement SNACP, SNDCP, and SNICP functions in order to achieve internetworking, depending on the network layer services available in each subnetwork.

In the first approach, internetworking is accomplished by relaying the services of one subnetwork directly onto the corresponding services of the other subnetworks. Network layer relays map the services offered by one subnetwork onto the other. This implies that either all of the subnetworks provide equivalent services or that each subnetwork must be enhanced to a common level of service. This "hop-by-hop" enhancement strategy of interconnection is described in ISO 8648, in which each subnetwork is enhanced by the addition of SNDCPs (the middle sublayer of the network layer) up to the level of the ISO 8348 network layer service. In this mode, an end-to-end connection is established for the duration of the session [13.10]. This technique uses a SNDCP over a SNACP. This hop-by-hop enhancement approach is depicted in Figure 13.10.

The enhancement approach enhances each subnetwork in the system individually in terms of network layer services, as needed, up to the OSI network layer service, as shown in Figure 13.10. This requires the deployment, over each interconnected network, of an SNDCP enhancement protocol tailored to that

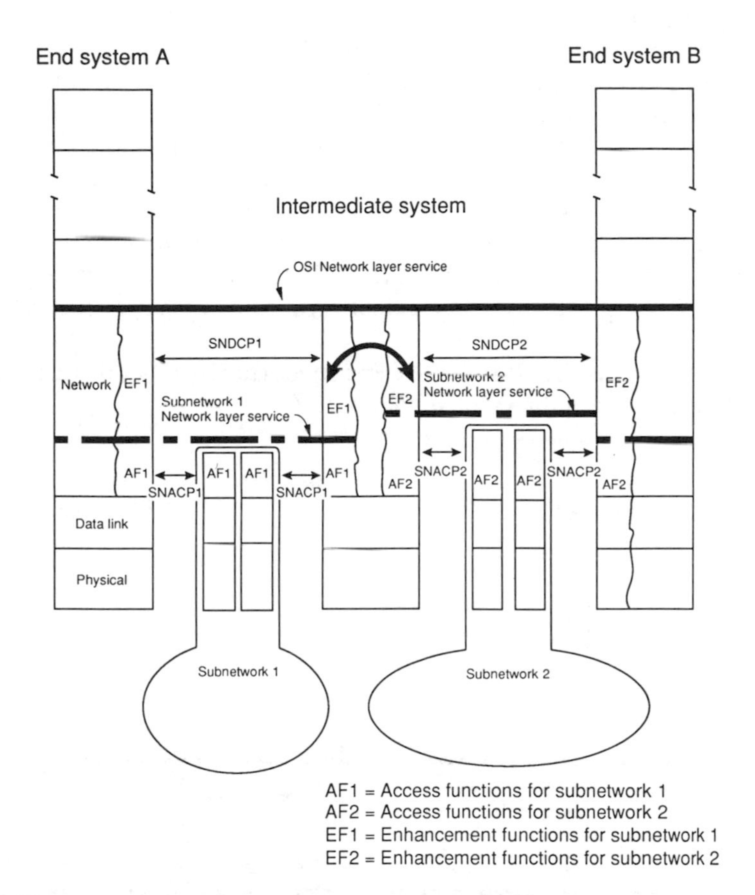

Figure 13.10 Hop-by-hop enhancement for internetworking.

particular subnetwork, followed by the provision of the necessary mappings between different but equivalent implementations of the OSI network layer service in intermediary relay devices between the subnetworks.

In the second approach, the subnetworks are interconnected by using a single end-to-end protocol to convey the information over a combination of subnetworks. This approach uses a SNICP over a SNACP. This internetworking protocol, which operates in a sublayer above the network-specific protocols in both LAN and WANs, performs addressing and routing functions necessary for end-to-end communication independent of the subnetwork-specific functions. Internetworking protocols (also called "internet" protocols) are network layer protocols that end systems agree to support in order to achieve communication among themselves. Here, no subnetwork enhancement is necessary and the protocols can operate directly over the DLL in the sense that no formal SNACP or SNDCP exists. (However, the service interface between any SNICP and *data link control* (DLC) must exist; this is normally implementation dependent and is not standardized [13.11].) The connectionless internetworking protocol is specified in ISO 8473 (also known as CLNP). The ISO CLNP performs the SNICP functions for the connectionless mode of operation. A connection-oriented SNICP is X.25 PLP.

This approach involves the operation of the same enhancement protocol across all the interconnected subnetworks. Hence, we have a single SNICP. This protocol must support every feature of the OSI network layer service that will not be automatically supported by every subnetwork over which it operates (see Figure 13.11).

Operation of the CLNP as a SNICP is supported by a family of SNDCPs that has been defined for a number of subnetworks:

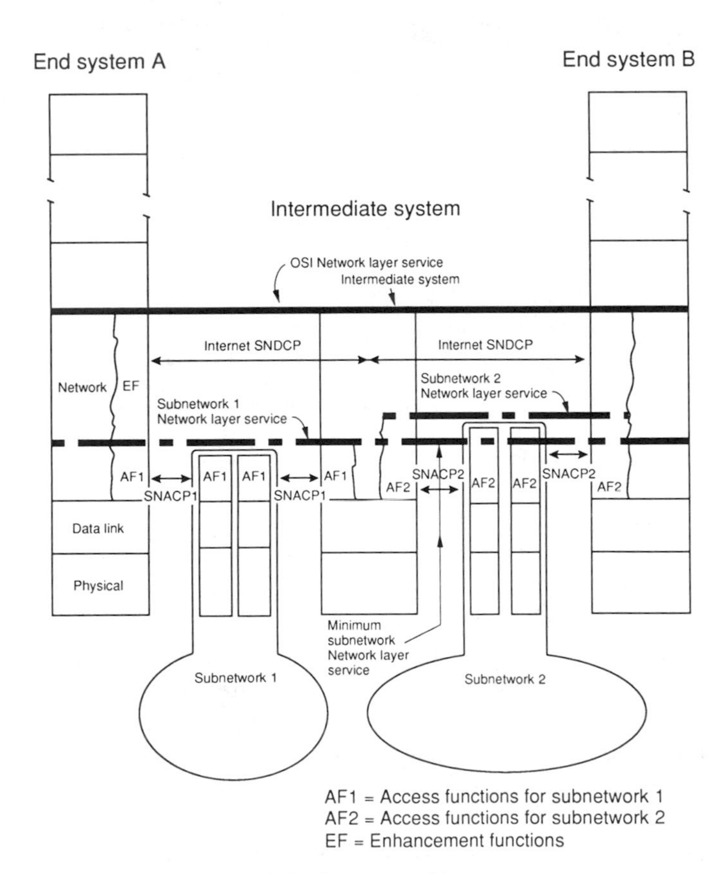

Figure 13.11 Common internet approach for internetworking.

- SNDCPs included in ISO 8473 are the interface between CLNP and ISO 8208/CCITT X.25 PLP subnetworks; and the interface between CLNP and ISO 8802-2 LAN subnetworks with both LLC1 and LLC2 (see Figure 13.12); and
- An addendum to ISO 8473 will also specify a SNDCP that interfaces between CLNP and point-to-point links with ISO 7776 (LAPB); and between CLNP and CCITT Q.921 (LAPD) protocols.

A version of X.25 (DTE to DTE) can reside over LAN LLCs; in this case, X.25 PLP is a connection-oriented SNICP, the LLCs are SNACPs, and the SNDCP is defined in ISO 8881.

CLNP is simpler than X.25 PLP; therefore, a number of proponents are advocating this approach to internetworking, which requires that CLNP and the appropriate SNDCP be available in each end system attached to the subnetworks (in addition to also being in the subnetworks) and that TP Class 4 be used.

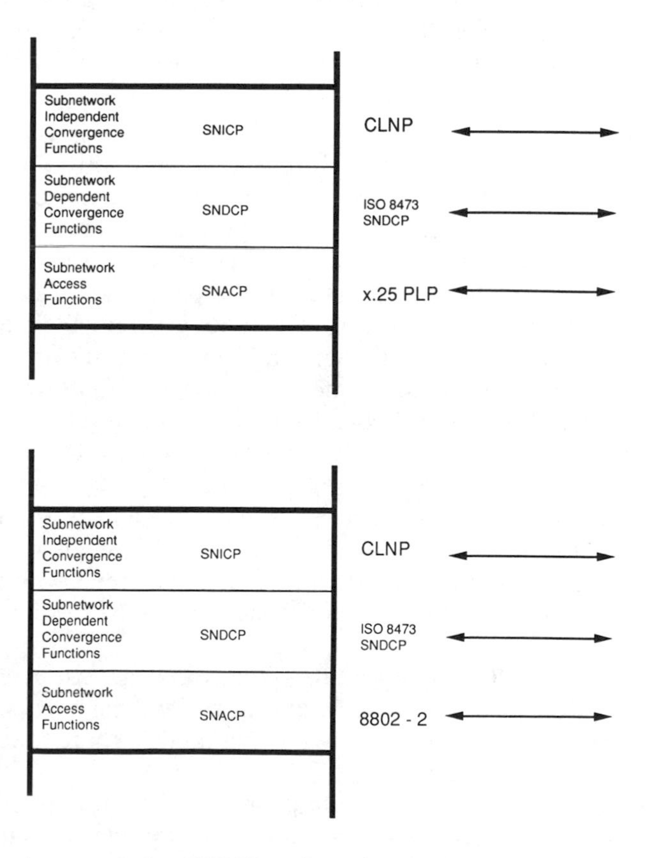

Figure 13.12 Two basic connectionless SNDCP configurations for network layer internetworking.

Both internetworking approaches discussed in the previous paragraphs, in combination with the transport layer protocol, aim at providing reliable data transfer; however, the layers where the end-to-end reliability functions are performed are different. In the first case, both the internetwork routing and the end-to-end reliability functions are performed in the network layer; hence, a transport protocol Class 0 could suffice. In the second case, the connectionless internetworking protocol focuses on dynamic routing of datagrams, and leaves the responsibility for end-to-end reliability to the transport layer protocol; for this reason, the more reliable transport protocol Class 4 is needed to guarantee data integrity between end systems.

Routing Protocols for Interworking

Two efforts were under way in 1990 to define adaptive routing protocols to be employed in conjunction with the internetworking protocols described above: One is from an OSI perspective, the other is from a TCP/IP perspective. The routing protocols allow dynamic information to be propagated, so that the network PDUs (described by ISO 8473) can be moved efficiently.

Routing Protocols from an OSI Perspective. To support large routing domains, provision is made in the ISO approach for hierarchical intra-domain routing (see Figure 13.13). A large domain may be administratively divided into areas. Each system resides in exactly one area. In the protocol, routing within one area is referred to as *level 1 routing*; routing between areas is referred to as *level 2 routing*. Level 2 intermediate systems keep track of the paths to destination areas. Level 1 intermediate systems keep track of the routing within their own area. For a network PDU destined for another area, a level 1 intermediate system sends the network PDU to the nearest level 2 intermediate system in its own area, regardless of the destination. Then the network PDU travels via level 2 routing to the destination area, where it again travels via level 1 routing to the destination end system.

Level 1 intermediate systems are systems that deliver and receive network PDUs from other systems; they also relay network PDUs from other source systems to other destination systems. They route directly to systems within their own area and route toward a level 2 intermediate system when the destination system is in a different area.

Level 2 intermediate systems are systems that act as level 1 intermediate systems in addition to acting as a system in a subdomain consisting of level 2 intermediate systems. Systems in the level 2 subdomain route toward a destination area or another routing domain.

As of early 1990, an ISO draft proposal (DP) for an intermediate system–to–intermediate system (IS-IS) intra-domain routing exchange protocol was being developed (DP 10589, February 1990). This proposal provides a protocol

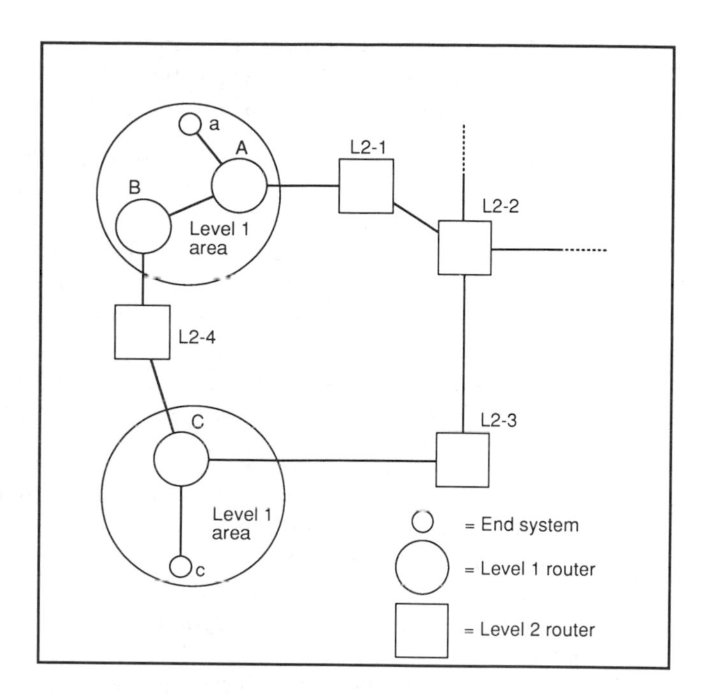

Figure 13.13 ISO hierarchical routing.

for use in conjunction with the protocol for providing connectionless-mode network service (ISO 8473). It permits intermediate systems within a routing domain to exchange configuration and routing information to facilitate the operation of the routing and relaying functions of the network layer. The intra-domain IS-IS routing protocol is intended to support large routing domains consisting of combinations of many types of subnetworks (including point-to-point links, multipoint links, X.25 subnetworks, and LANs). To support these large routing domains, provision is made in DP 10589 for intra-domain routing to be organized hierarchically, as discussed above.

These protocols do not handle the routing mechanism. They only provide information for internetworking dispatch of packets or datagrams to the correct subnetwork or gateway. The goals of handling large domains and incompatible subnetworks made it necessary for the standard to solve a number of key problems [13.12]:

- Handle up to 100,000 network SAPs, which correspond with end systems, and up to 10,000 intermediate systems (routers);
- Encapsulate the different subnetwork protocols within a common backbone;
- Permit determination of the correct route for a message in a reasonably short time;

- Detect breaks in routes between subnetworks; and
- Keep intermediate systems (routers) supplied with up-to-date information on network topology, routes, and addresses in a reasonably efficient way.

Timely advancement of this standard to DIS status is an important milestone in making feasible the interconnection of various connectionless LAN and MAN subnetwork architectures into reliable WANs. The IS-IS intra-domain protocol is the "middle" protocol of the three that will comprise the ISO routing protocol set. The first, end system–to–intermediate system (ES-IS) routing exchange, is already an international standard, while the third, the IS-IS inter-domain routing exchange, is still the subject of debate in ANSI X3S3.3 [13.9].

The ES-IS routing exchange protocol for use in conjunction with the protocol for providing connectionless-mode network service (ISO 8473) was standardized in March 1988 and is known as ISO 9542.

Routing Protocols from a TCP/IP Perspective. Work is also being conducted on an internet draft specifying an integrated routing protocol, based on the OSI intra-domain IS-IS routing protocol, which may be used as an interior gateway protocol to support TCP/IP as well as OSI-based systems. This allows a single routing protocol to be used to support pure TCP/IP environments, pure OSI environments, and dual environments.

The TCP/IP protocol suite has been growing in importance as a multivendor communication architecture. With the anticipated emergence of OSI-based systems, we can expect the coexistence of TCP/IP and OSI to continue for the foreseeable future. Routers are needed to support TCP/IP traffic and OSI traffic in parallel.

Two main methods are available for routing protocols to support dual OSI and IP routers. One method is to use completely independent routing protocols for each of the two protocol suites. The second method is to use a single integrated protocol for routing both protocol suites. The second approach is preferable. The proposed protocol is based on ISO DP 10589 (intra-domain IS-IS protocol) with IP-specific functions added. The integrated IS-IS provides a single routing protocol that will simultaneously provide efficient routing for TCP/IP and for OSI.

The protocol described in the draft internet protocol ("Use of OSI IS-IS for Routing in TCP/IP and Dual Environments," January 1990) allows for mixing of TCP/IP-only, OSI-only, and dual (TCP/IP and OSI) routers, as defined below.

An IP-only IS-IS router is defined to be a router that (1) uses the IS-IS protocol for routing TCP/IP packets, as specified in the draft protocol, and (2) does not otherwise support OSI protocols (for example, such routers would not be able to route OSI CLNP packets).

An OSI-only router is defined to be a router that uses the IS-IS protocol for routing OSI packets, as specified in ISO DP 10589 discussed above. OSI-only routers conform to OSI standards and are implemented independently of the internet protocol.

A dual IS-IS router is defined to be a router that uses the IS-IS internet protocol for routing both IP and OSI packets.

The proposed approach, which describes an interior gateway protocol, does not change the way in which IP packets are handled: The TCP/IP-only and the dual routers are required to conform to the requirements of the internet gateways (Internet RFC 1009, June 1987). Similarly, the approach does not change the way OSI packets are handled (ISO 8473).

Internetworking Efforts, 1990–92. Rapid progress is being made on routing protocols for use in conjunction with connectionless transport. The ES-IS protocol is an international standard. In 1990, the IS-IS protocol was approaching DIS status. Proposals have been made to ANSI X3S3.3 to adopt TCP/IP internet protocols as U.S. national standards, hinting at a possible coalescing of technologies in the 1992–95 time frame. Joint efforts between the OSI and internet communities will pool the considerable available resources, and should enhance the move toward high-speed data networking [13.9].

13.2 TECHNICAL DETAILS OF UPPER LAYER STANDARDS

13.2.1 Naming, Addressing, and Directory

13.2.1.1 Name and Addressing Issues

In a data communication (or telecommunication) network, the requirement to identify and locate entities is of axiomatic importance. Three important elements of this requirement are:

1. The name of the entity or resource of interest;
2. The address of the entity or resource of interest, indicating where the resource is; and
3. A route on how information can flow from the user to and from the resource of interest.

Entities requiring a name include physical objects (computers, file servers, printers, *et cetera*), as well as logical entities (processes, services, agents, *et cetera*). The function of mapping names onto addresses and eventually onto routes is a basic communication function. Typically, user-friendly names are used to denote entities in human-readable form; these names screen the user from tedious nomenclature and from the details of the network. While names tend to be static, routes depend heavily on the details of the network, particularly when these routes are selected dynamically. In highly internetworked environments, in which entities may be widely distributed, and may be members of subcommunities, the addressing and routing tasks are fairly complicated.

Routes are described in terms of subnetwork facilities. These facilities are typically identified by numeric addresses. Thus, an address serves as a link between a name and a route. Addresses, therefore, separate the name-to-route mapping process into two distinct steps, each of which may take place at different locations and may be distributed over many cooperating entities. Name-to-address mapping is best suited to the application layer because that is the interface where a user may access the network. A directory service may be employed to perform this task. Address-to-route mapping is best suited to the network layer because that is where access to internetworked communication facilities is provided.

In the OSIRM, names are used where human users interact at the application layer; addresses are used below this layer, both at the interface between adjacent layers and between peer entities. ISO and CCITT have also specified 17 categories of entities that require naming, including country, locality, organization, person, application, and process [13.13]. Naming, addressing, and directory functions interact to support communication.

Addressing

An OSI address is comprised of a set of nested (N)-SAP addresses, introduced in Chapter 1. Figure 13.14 depicts an OSI address, composed of a network layer SAP (NSAP) address, transport layer SAP (TSAP) address, session layer SAP (SSAP) address, and presentation layer SAP (PSAP) address. The NSAP portion of the address is the most important because it uniquely identifies an NSAP, which can be intuitively thought of as identifying a network layer service provider. Selectors are used to identify (N)-SAPs within a particular open system; hence, they exist only above the network layer. A TSAP address is used to access a session layer entity and consists of an NSAP address plus a transport selector. An SSAP is used to access a presentation layer entity and consists of a TSAP plus a session selector. A PSAP is used to access an application layer entity, and consists of an SSAP address plus a presentation selector. The NSAP addressing scheme is described in the network layer service standard ISO/IEC 8348:1987/AD2:1988, which is equivalent to CCITT X.213 NXA, and is summarized here.

An addressing authority is postulated to control a collection of addresses called an *addressing domain*. Within each addressing domain, all NSAP addresses are mandated to be unique and are allocated by the addressing authority. At the national level, the addressing authority may be a domestic standards body or some

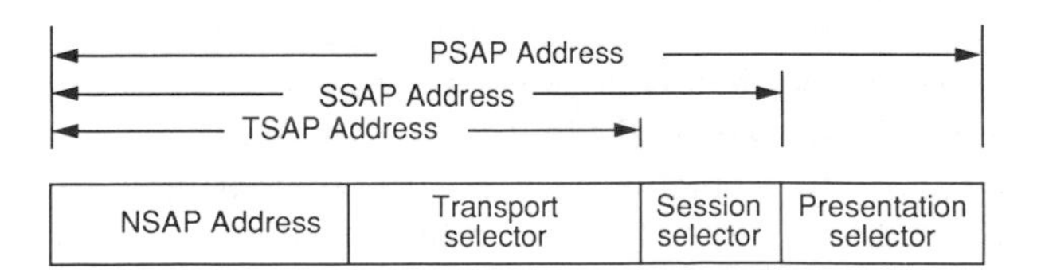

Figure 13.14 OSI address.

regulatory entity. The addressing authority may further delegate the responsibility for portions of the domain to suitable organizations. The authority need not be a formal body and may be a "distributed" mechanism—the only requirement is the uniqueness of the NSAPs, regardless of how the uniqueness is achieved.

The complete set of NSAP addresses allocated within the OSI environment is known as the *OSI global network addressing domain*. This global domain is hierarchical, being divided into a number of addressing domains that, in turn, can be further divided into subdomains. NSAP addresses must be globally unambiguous.

The NSAP addressing scheme of ISO 8348 AD2/CCITT X.213 provides a number of allocation schemes and, while facilitating the hierarchical nesting of allocation authorities, guarantees global nonambiguity [13.13]. Figure 13.15 depicts the structure of the address.

The NSAP can be up to 40 decimal digits‡ and is structured hierarchically. The NSAP is composed of an initial domain part (IDP), followed by a domain-specific part (DSP). The format of the IDP is described in ISO 8348/AD2, while the format of the DSP is a national or local matter. Various combinations of IDP formats and lengths (with specific DSPs) are possible.

The IDP is, in turn, composed of an authority and format identifier (AFI), which is always two decimal digits, followed by an initial domain identifier (IDI), with a length that depends on the IDI format specified by the AFI, of up to 17 decimal digits.

The AFI identifies the allocation authority for the IDI and DSP. AFIs are administered by ISO and CCITT. For example, CCITT has been allocated a number of AFIs to identify numbering schemes used in existing public networks (addressing schemes in these existing networks are called *subnetwork-dependent*

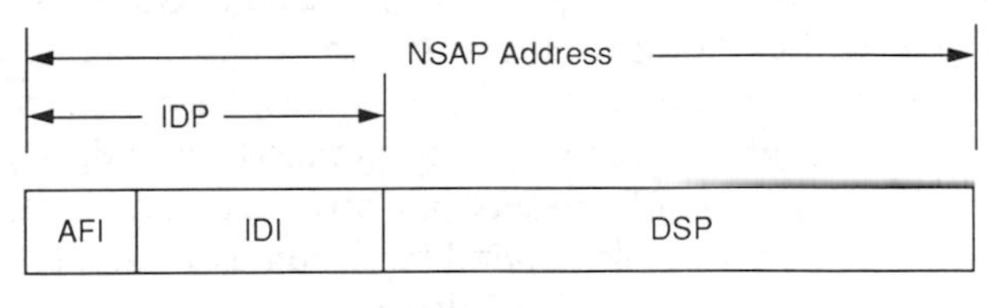

Figure 13.15 Structure of an NSAP address.

‡At the time of this writing, the U.S. position presented to CCITT was to make the NSAP a 20-octet field; 20 octets allow up to 40 decimal digits when represented with a four-bit "half-octet" code. The binary view of the NSAP, contrasted to the decimal view, facilitates inclusion of binary LAN and MAN addresses.

addressing schemes). For example, AFI = 36 is allocated to CCITT for the X.121 public packet data network address scheme; AFI = 42 is allocated to CCITT for the CCITT E.163 public switched telephone network address scheme; and AFI = 44 is allocated to CCITT for the CCITT E.164 (CCITT I.331) addressing scheme for the ISDN era. ISO has also allocated AFIs (= 38, 46), for ISO subnetwork-independent addressing schemes [13.13].

The IDI, which has a maximum length of 15 decimal digits, identifies the initial addressing domain of the authority responsible for allocating within the domain specified by the AFI. Examples of existing IDIs include:

- CCITT X.121, public data network numbering (IDI = 14 digits);
- CCITT E.163, public switched telephone network numbering (IDI = 12 digits); and
- CCITT E.164, ISDN numbering (IDI = 15 digits).

The network addressing authority (or an appropriate delegate) identified in the AFI-IDI is responsible for specifying the form of the DSP. Only one of the following four configurations is possible [13.13]:

- Binary octets;
- Decimal digits;
- Special graphics characters (only allowed if the AFI specifies a local format for the NSAP); and
- Characters from a national character set, as determined by the addressing authority (only allowed if the AFI specifies a local format for the NSAP).

The foregoing discussed the various NSAP addressing schemes that are available. In particular, subnetwork-dependent schemes allocate NSAP addresses by deriving them from the underlying subnetwork addresses; these addresses are known as *subnetwork point of attachment* (SNPA) addresses. Subnetwork-independent schemes for NSAP address allocation are desirable, if possible, especially when international connectivity among networks of complex topologies is involved. Because they are independent of the underlying subnetwork addressing scheme, they remain unaffected by a change of subnetwork provider or a change in subnetwork topology. However, subnetwork-independent schemes necessitate a directory inquiry to map the NSAP onto an SNPA address.

Naming Issues

The OSIRM recognizes the need for naming of communication entities (ISO 7498-3). User-friendly names should not contain elements that depend on underlying subnetwork addresses. Because a user-friendly name and an address are necessary, a directory service is required to perform the necessary mapping (directory functions are described in the next section).

The technique used by the OSI directory, MHS, and management for user-friendly naming is *attribute-based naming*. The resulting identifiers are called *distinguished names*, and a directory service is used to support the process. The distinguished name is an attribute list, namely, an unordered set of zero or more distinct attributes that is used to describe characteristics of the named entity. In turn, an attribute is a tuple consisting an attribute type and an attribute value. The attribute type defines the category of information, while the value is an instance of that category. An application entity title (which is a distinguished name) is mapped onto a PSAP address by an application layer directory. The NSAP address portion of the PSAP is mapped into an SNPA address in the subnetwork, via a yet-to-be-specified network-based network layer directory.

ISO and CCITT have defined 41 standard attribute types. Some examples are countryName, commonName, postalName, organizationName, *et cetera* [13.13]. Attribute types can also be multivalued to define distribution lists.

13.2.1.2 The X.500 Directory

The directory is a specialized repository of information that contains data used to facilitate communication among addressable entities in a network. CCITT X.500 specifies a common protocol to retrieve and update information stored in the directory. This service will provide support for users who require information regarding communication with another user. Vendors and users are increasingly in favor of public standardized directories.

The directory contains addressing information. This basic addressing capability is required by applications, management processes, OSI layer entities, and telecommunication services. The relationship between the directory and security is fairly obvious. Secure communication is dependent, in part, on each communicating party's knowledge of the other's identity; this requires assurance that the claimed identities are the actual identities. Authentication is a security service that provides the assurance that the claimed identity is correct. The directory is a resource in which user credentials are kept. The directory stores electronic credentials that can be unwrapped to authenticate users. Each user is an object that has an entry in the directory information base. Local copies of global authentication information cannot be stored in every computer engaged in open communication; this is where the public directory comes into play.

Among the capabilities the directory provides are those of "user-friendly naming," whereby objects can be referred to by names that are suitable for citing by human users (though not all objects need have user-friendly names); and "name-to-address mapping," which allows the dynamic binding between objects and their locations, as described in the previous section. The latter capability allows OSI networks, for example, to be "self-configuring" in the sense that addition, removal, and changes of object locations do not affect network operation. The directory is not intended to be a general-purpose data base for all possible network information, although other such data bases may be built in a fashion similar to the directory.

Table 13.5 depicts the family of the CCITT X.500 (ISO 9594) standards.

Overview of the Directory

The users of the directory, including people and computer programs, can read or modify the information, or parts of it, subject to having permission to do so. Each user is represented in accessing the directory by a directory user agent (DUA), which is considered to be an application process, as shown in Figure 13.16.

As is typical with existing communication directories, a considerably higher frequency of "queries" than updates is likely. The rate of updates is expected to be governed by the dynamics of people and organizations rather than, for example, by the dynamics of networks. Also, instantaneous global commitment of updates is not necessary: Transient conditions in which both old and new versions of the same information are available are generally acceptable.

The X.500 series of recommendations refers to the directory in the singular, and reflects the intention to create, through a single, unified, name space, one

Table 13.5
Standards for the Directory

X.500 (ISO 9594-1)	The Directory, Part 1: Overview
X.501 (ISO 9594-2)	The Directory, Part 2: Information Framework
X.509 (ISO 9594-8)	The Directory, Part 8: Authentication Framework
X.511 (ISO 9594-3)	The Directory, Part 3: Abstract Service Definition
X.518 (ISO 9594-4)	The Directory, Part 4: Procedures for Distributed Operation
X.519 (ISO 9594-5)	The Directory, Part 5: Access and System Protocols Specification
X.520 (ISO 9594-6)	The Directory, Part 6: Selected Attribute Types
X.521 (ISO 9594-7)	The Directory, Part 7: Selected Object Classes
ISO 9594-9	The Directory, Part 9: Directory Information Tree Structure and Naming
ISO 9594-10	The Directory, Part 10: Replication and Knowledge Management

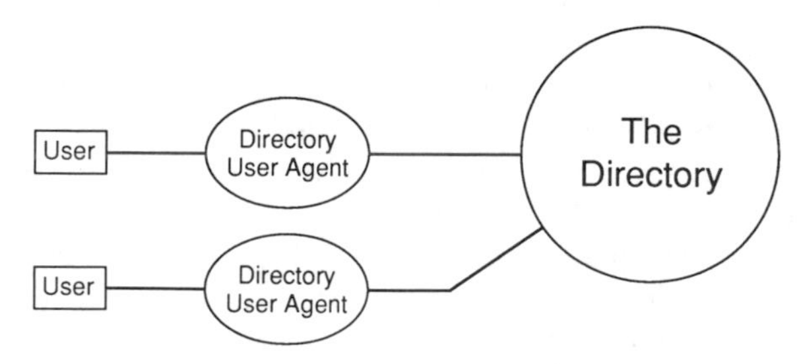

Figure 13.16 Access to the X.500 directory.

logical directory composed of many systems and serving many applications. Whether or not these systems choose to interwork will depend on the needs of the applications they support. Applications dealing with nonintersecting worlds of objects may have no such need. The single name space facilitates later interworking should the needs change.

The information held in the directory is collectively known as the *directory information base* (DIB). The DIB and its structure are defined in Recommendation X.501. The DIB is made up of information about the objects. It is composed of directory entries, each of which consists of a collection of information on one object. Each entry is made up of attributes, each with a type and one or more values. The types of attributes present in a particular entry are dependent on the class of object that the entry describes. The entries of the DIB are arranged in the form of a tree, known as the *directory information tree* (DIT), in which the vertices represent the entries. Entries higher in the tree (nearer the root) will often represent objects such as countries or organizations, while entries lower in the tree will represent people or application processes. The services defined in the X.500 recommendation operate only on a tree-structured DIT.

Every entry in the DIB has a distinguished name, which uniquely and unambiguously identifies the entry. These properties of the distinguished name are derived from the tree structure of the information. The distinguished name of an entry is made up of the distinguished name of its superior entry, together with specially nominated attribute values (the distinguished values) from the entry. Some of the entries at the leaves of the tree are *alias entries*, while all other entries are object entries. Alias entries point to object entries, and provide the basis for alternative names for the corresponding objects.

The directory enforces a set of rules to ensure that the DIB remains well-formed in the face of modifications over time. These rules, known as the directory schema, prevent an entry from having the wrong types of attributes for its object class, attribute values that are the wrong form for the attribute type, and even subordinate entries of the wrong class. The directory also provides a well-defined set of access capabilities, known as the *abstract service*, to its users. This service provides a simple modification and retrieval capability. This can be expanded with local DUA functions to provide the capabilities required by the end-users. The directory is likely to be distributed, perhaps widely distributed, along both functional and organizational lines. Models of the directory have been developed to provide a framework for the cooperation of the various components to provide an integrated whole. The provisions and consumption of the directory services require that the users (actually the DUAs) and the various functional components of the directory cooperate with one another. In many cases, this will require cooperation between application processes in different open systems, which, in turn, requires standardized application protocols to mediate this cooperation.

The directory has been designed to support multiple applications drawn from a wide range of possibilities. The nature of the applications supported will govern which objects are listed in the directory, which users will access the information, and which kinds of access they will carry out. Applications may be very specific, such as the provision of distribution lists for E-mail, or generic, such as the "interpersonal communication directory" application. The directory provides the opportunity to exploit commonalities among the applications:

1. A single object may be relevant to more than one application; perhaps even the same piece of information about the same object may be relevant. To support this, a number of object classes and attribute types are defined, which will be useful across a range of applications. These definitions are contained in Recommendations X.520 and X.521.
2. Certain patterns of use of the directory will be common across a range of applications.

All services are provided by the directory in response to requests from DUAs. Some requests allow interrogation and modification of the directory. The directory always reports the outcome of each request that is made of it. The form of the normal outcome is specific to the request and is evident from the description of the request.

A number of aspects of the eventual directory service are not currently provided by the X.500 standards. The corresponding capabilities will, therefore, need to be provided as a local function until such time when a standardized solution is available. These capabilities include addition and deletion of arbitrary entries, thus allowing a distributed directory to be created; the management of access control (i.e., granting or withdrawing permission for a particular use to carry out a particular access on a particular piece of information); the management of the directory schema; the management of knowledge information; and the replication of parts of the DIT. Currently available directory interrogation services described in X.500 are as follows.

Read. A read request is aimed at a particular entry and causes the values of some or all of the attributes of that entry to be returned. In cases for which only some attributes are to be returned, the DUA supplies the list of attribute types of interest.

Compare. A compare request is aimed at a particular attribute of a particular entry and causes the directory to check whether a supplied value matches a value of that attribute. For example, this can be used to carry out password checking, a process in which the password, held in the directory, might be inaccessible for reading, but accessible for comparing.

List. A list request causes the directory to return the list of immediate subordinates of a particular named entry in the DIT.

Search. A search request causes the directory to return information from all of the entries within a certain portion of the DIT that satisfy some filter. The information returned from each entry consists of some or all of the attributes of that entry, as with read.

Abandon. An abandon request, as applied to an outstanding interrogation request, informs the directory that the originator of the request is no longer interested in the request. The directory may, for example, cease processing the request and discard any results achieved thus far.

The directory modification services described in X.500 are as follows:

Add Entry. An add entry request causes a new leaf entry (either an object entry or an alias entry) to be added to the DIT.

Remove Entry. A remove entry request causes a leaf entry to be removed from the DIT.

Modify Entry. A modify entry request causes the directory to execute a sequence of changes to a particular entry. Either all of the changes are made, or none of them, and the DIB is always left in a state consistent with the schema. The changes allowed include the addition, removal, or replacement of attributes or attribute values.

Modify Relative Distinguished Name. A modify relative distinguished name (RDN) request causes the relative distinguished name of a leaf entry (either an object entry or an alias entry) in the DIT to be modified by the nomination of different distinguished attribute values.

A *directory system agent* (DSA) is an OSI application process that is part of the directory. The DSA's role is to provide access to the DIB to DUAs or other DSAs. A DSA may use information stored in its local data base or interact with other DSAs to fulfill requests. Alternatively, the DSA may direct a requester to another DSA that can help carry out the request. Local data bases are entirely implementation dependent.

The DUA interacts with the directory by communicating with one or more DSAs. A DUA need not be bound to any particular DSA. It may interact directly with various DSAs to make requests. For some administrative reasons, direct interaction with the DSA, which will carry out the request, is not always possible. The DUA may also be able to access the directory through a single DSA. For this purpose, DSAs will need to interact with each other (see Figure 13.17).

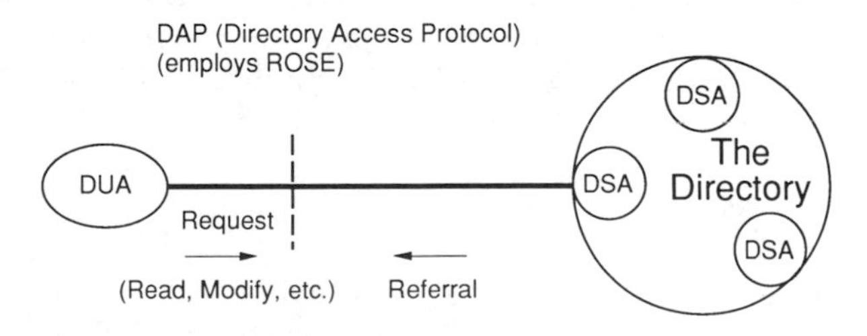

Figure 13.17 X.500 functional model of the directory.

13.2.2 Standardization of File Transfer

The ISO 8571 FTAM standard defines a basic file service and specifies a file protocol available within the application layer [13.14–13.17]. FTAM was an international standard as of late 1987 [13.18]. FTAM allows the open interconnection of a wide range of systems of different complexity that need to transfer and manipulate files. *File transfer* refers to a function that moves an entire file between open systems; *file access* refers to the inspection, modification, or replacement of part of the file's content; *file management* refers to the creation and deletion of files and the inspection or manipulation of the attributes associated with the file.

FTAM uses the virtual filestore concept to allow differences in style and specification of real filestores (file systems existing in real systems) to be absorbed into a mapping function from the open system onto the real end system. By screening the details of the local operating systems from the FTAM users, the need to modify existing systems is minimized. This, in turn, implies that FTAM users need not be concerned with the idiosyncrasies of the end systems. Part 1 of the FTAM, ISO 8571-1, gives the general concepts; part 2, ISO 8571-2, defines the virtual filestore; part 3, ISO 8571-3, defines the file service; and part 4, ISO 8571-4, defines the file protocol.

The field of application of FTAM is very wide; some simple applications include:

- Transmission of a complete file or part of a file to a remote system via a public data network;
- Remote data base access (making a series of inquiries from a remote data base system); and
- File manipulation (e.g., a workstation wishing to establish a connection with a central filestore to manipulate its working files).

A key capability is the need to access remotely a single record from a file without having to retrieve the whole file. This feature may prove to be a positive force in favor of FTAM commercial deployment. During the past few years, more than 200 computer manufacturers, semiconductor manufacturers, word processing vendors, process-control vendors, communication carriers, and industry and government users have participated in a NIST workshop to advance the development of commercial products implementing FTAM [13.19]. Many vendors reportedly consider "FTAM the most critical layer 7 protocol right now" [13.20].

13.2.2.1 The FTAM Protocol

In this section, some of the basic aspects of the standard affecting implementation are discussed. In addition to ISO 8571, FTAM relies on the following standards:

ACSE Service/Protocol (ISO 8649/ISO 8650); Presentation Service/Protocol (ISO 8822/ISO 8823); Abstract Syntax Notation and Basic Encoding Rules (ISO 8824/ISO 8825); and Session Service/Protocol (ISO 8326/ISO 8327). For a pedagogical description of FTAM the reader may consult Reference [13.1].

13.2.2.2 FTAM Part 2: Virtual Filestore Definition

Part 2 of the FTAM standard is devoted to the definition of the *virtual filestore*. This includes a definition of a set of virtual file attributes; a hierarchical model of the file's content; and the definition of a number of constraint sets and document types (which provide some restrictions on the general model to make it suitable to generally available commercial file structures).

A filestore is an organized collection of files, including their attributes and names, residing in an open system. The OSI file service is based on the virtual filestore; a virtual filestore is an abstract model for describing files and filestores residing in the local system environment (LSE). The virtual filestore defines one common intermediate representation of file operation commands (for example, open file, close file, *et cetera*). Each real system converts the virtual FTAM language to or from real commands (for example, F-OPEN, F-CLOSE). Attributes describe features of the file such as name, length, *et cetera*. The OSI file service definitions are given in terms of the characteristics and data of the virtual filestore rather than any real local filestores that contain the actual files; each open system is responsible for mapping the virtual filestore descriptions and operations into local file description and operation (see Figure 13.18).

A file consists of data units that are delivered to the presentation layer, which will invoke the services of lower layers needed to carry out the transfer to the remote location. These data units carry a relationship, typically a hierarchical tree structure; this relationship is employed to access desired data units. Each node of the hierarchical tree structure contains an identifier, possibly 0 or 1 data units, and

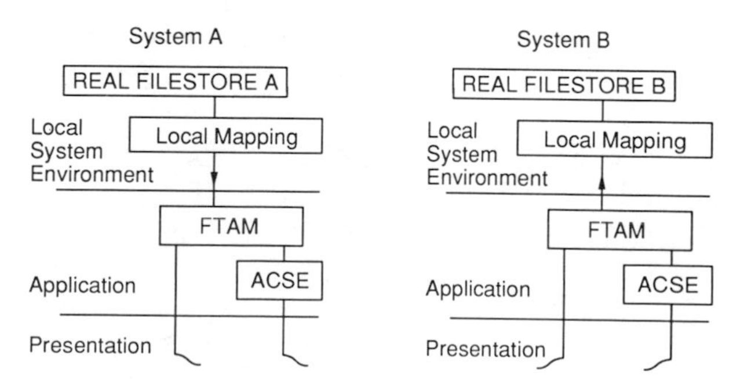

Figure 13.18 FTAM environment.

a subtree. The entire node is called a *file access data unit* (FADU). Thus, a FADU can be defined (recursively) as a sequence consisting of a FADU identifier, an optional data unit, and a subtree sequence of FADU. This mechanism allows the definition of any number of sophisticated file structures. The access context determines which part of the structure is actually transmitted over the network. Five contexts are defined in FTAM: entire FADU, DUs only, root DU only, given level DUs only, and FADU structure only. These can be considered user views of the file data. For example, for a flat file, the structure consists of two levels: only data units are in the leaf nodes and the root node has no data unit (see Figure 13.19). In this example, the file content consists of a FADU; the root node does not contain any data units. The FADU subtree consists of three descendants, each of which is also a FADU.

The general model described above needs to be considered with the help of two other constructs. One has a number of constraint sets that restrict the access structure, while not constraining the actual data types in the file. One also has a number of document types that completely define the type of the file's contents by specifying a constraint set and the possible data types in the file [13.1].

Constraint set definitions are concerned with restricting the structure of a file. Some examples include (1) a sequential flat constraint set, (2) an ordered flat constraint set, and (3) an ordered hierarchical constraint set, among others. Document type definitions go beyond the constraint set definitions because these not only constrain the structure (by enumerating a constraint set), but also the data types forming the data units. Thus, document type is a powerful and flexible tool for defining a file structure in closed form.

As an example of a document type, consider the unstructured text document type. This file satisfies the requirements of the unstructured constraint set, with a single data unit consisting of zero or more instances of an ASN.1 GraphicString data value; the GraphicString is defined as an unbounded sequence of printing characters from any character set registered in the ISO Register of Coded Character Sets Used with Escape Sequences [13.1].

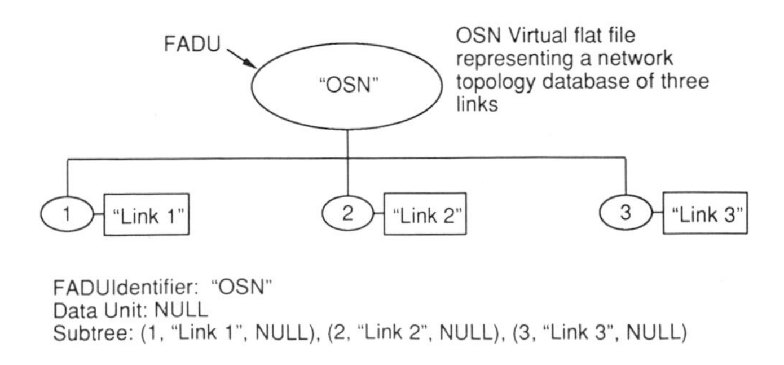

Figure 13.19 An example.

The files in the virtual filestore have data content and a number of attributes. Some typical attributes are filename, permitted actions, storage account, contents type, date and time of creation, identity of creator, and file size, among others.

For example, the permitted actions attribute allows the following actions: read FADU, insert FADU, replace FADU, extend data unit at root FADU, erase FADU, read attribute, change attribute, delete file. The contents type attribute contains a single document type name or a single constraint set name with a single abstract syntax name.

After an FTAM file transfer implementation has been established, all variable attributes of that implementation have fixed values (or are not supported); the only other variable is the actual file contents. In the simplest implementation, all attributes except the filename attribute are fixed.

FTAM employs abstract syntaxes (a part of presentation contexts). Abstract syntaxes refer to the syntactic information that is architecturally passed between the application layer and the presentation layer. FTAM has a number of syntaxes. The FTAM protocol control information header is delivered to the presentation layer as depicted in Figure 13.18, according to the abstract syntax defined in FTAM Part 4 [13.17]. This syntax is an ISO-registered syntax. The file structure information (defining the data types that must be transferred to describe the file structure remotely) is delivered to the presentation layer according to the abstract syntax defined in FTAM Part 2 [13.15]. This syntax is also an ISO-registered syntax. The third syntax is for the description of the file user data to the presentation layer.

13.2.2.3 FTAM Part 3: Service Definition

Part 3 of FTAM is the service definition. It specifies the interactions between the FTAM ASE and the virtual filestore, and between the FTAM ASE and the remote user. Part 3 specifies the allowed sequence of interactions and the parameters associated with such interactions. Typical primitives are F-CLOSE, F-OPEN, F-ERASE, F-CREATE, F-DELETE, *et cetera*. Most of the primitives are confirmed primitives; namely, the service action is completed only when a confirm primitive is received following the request primitive. The complete sequence, as described in Chapter 1, is: (1) request, (2) indication, (3) response, and (4) confirm.

Part 3 provides key information for FTAM users because it specifies the effect of the defined service primitives on the virtual filestore.

FTAM operates through a series of nested regimes: (1) the FTAM regime, (2) the file selection regime, (3) the file open regime, and (4) the data transfer regime. For example, the FTAM regime uses the primitive F-INITIALIZE to negotiate FTAM capabilities. The selection regime uses primitives such as F-SELECT and F-CREATE. Many of these service primitives involve the transfer of a number of parameters, which have to be specified appropriately for the implementation to work in a heterogeneous environment.

13.2.2.4 *FTAM Part 4: Protocol Specification*

Part 4 defines the messages that are transferred across the network, namely, the peer-to-peer protocol. It specifies the detailed form of the protocol messages that carry the semantics of the request and response primitives. ASN.1 is used for this specification.

13.2.2.5 *Implementation Issues*

Implementing FTAM is not a trivial exercise. First, the protocol is quite rich, with many options; the ability to describe any real file structure with the FADU construction makes it general, but complex. In addition, the services of the presentation layer can be somewhat inefficient in terms of overhead bytes and required protocol exchanges.

From a practical perspective, the major issue in the implementation of FTAM appears not to be the development of the software to support the peer-to-peer protocol (which after all will be the same for all vendor implementations following ISO 8571), but the mapping of the FTAM primitives to and from the real file system. This means that we cannot simply buy an off-the-shelf implementation of FTAM and drop it in the desired systems. A lot of detailed programming work will be required to develop the local mapping.

13.2.2.6 *FTAM Implementors Agreements*

In 1983, at the request of industry, the National Bureau of Standards organized the NBS Workshop for Implementation of OSI to bring together users and suppliers of OSI-based products. (NBS is now NIST, as discussed in Chapter 2). The workshop accepts as input the specifications of emerging standards and produces as output agreements on the implementation and testing details of these protocols. This activity is the key to expediting the development of OSI standards and products and to promoting interoperability between products developed by various vendors. Reference [13.21] records stable implementation agreements of OSI protocols, including FTAM; the agreements were completed in December 1987. The agreements are based on final or near-final standards. SIGs composed of qualified technical leaders from participating vendors produce the technical documentation contained in the implementors agreements.

The agreements provide detailed guidance for the implementor and eliminate ambiguities in interpretations. The agreements define FTAM functions for minimal functionality and for several implementation profiles that are tailored to different classes of user requirements. Thus, these agreements serve as the basis for product development. The systems implemented according to these agreements will probably interoperate successfully. The FTAM phase 2 agreements cover transfer of and access to files between the filestores of two end systems and also management of a remote filestore.

(The phase 1 implementation specification was based on the second ISO draft proposal of April 1985; phase 2 is based on the international standard. No backward compatibility exists with the NBS Phase 1 FTAM.)

13.2.3 Abstract Syntax Notation One (ASN.1)

The presentation layer (PL) of the OSIRM is responsible for determining how data exchanged by users will be represented while traversing the network. The users are actually AEs. The presentation layer does not deal with the human-machine interface. Different computers represent data in different ways. Typically, IBM computers have used EBCDIC, while other computers have used ASCII. To transfer information between two such computers, in a heterogeneous environment, a common representation must be used during the actual transfer phase. Integers, floating point numbers, and other digital data are stored in a variety of ways; a common format must be agreed on before the information can be transacted. This common representation may not be visible to the ultimate users. The presentation layer provides services and protocols for reaching the necessary agreements. The presentation layer protocol within the OSIRM context is ISO 8823 (CCITT X.226); the service is ISO 8822 (CCITT X.216).

In the 1990s, new user information may have to be encoded. For example, the X.400 E-mail standard allows storage of voice, facsimile, graphics, *et cetera*. The presentation layer, therefore, must solve the more general problem of facilitating data representation of all kinds. To accomplish this, the presentation layer standards define a number of abstractions, as follows [13.22]:

Types and Values. The type and value of information must be invariant under transfer, while the form (the bit representation) may change. Types can include floating point numbers, integers, characters, and other complex structures.

Grouping Types and Values. To undertake information exchange, two entities must reach agreement (via the presentation layer) on how values of all relevant types will be represented. This could be a very large set. Thus, the presentation layer provides a way to group types that will be used into abstract syntaxes.

An abstract syntax is the specification of data or protocol control information by means of notation rules that are independent of the encoding technique used to represent them. The syntax defines the constituent types, but not how to represent values for those types. A syntax can be simple or complex. A syntax may contain a single type "integer." Another syntax may contain the types "integer" and "Boolean." Another syntax could contain the type "integer" and a record type containing "character" and "Boolean" subtypes. The range of values for each of the types is not specified in these examples, nor is an attempt made to describe if "integer" in the first syntax is the same as "integer" in the third syntax.

Merely describing the types to be used by the AEs is not enough. The responsibility of the presentation layer is to determine how the values of the types at hand should be represented. This is accomplished by agreeing on a transfer syntax for each abstract syntax. Thus, a transfer syntax is a set of rules for encoding values of some specified group of types (namely, of some abstract syntax). In practical terms, it describes the actual bit-level representation that results when applying the syntax rules to a particular value. For a transfer syntax to be usable with a given abstract syntax, it must be able to encode the values of all types contained in that abstract syntax.

Several different transfer syntaxes are capable of encoding the types contained in an abstract syntax. For example, an abstract syntax may simply contain the type "character"; then one transfer syntax can be ASCII (characters are encoded following the ASCII rules); another transfer syntax can be EBCDIC (characters are encoded using the EBCDIC rules).

The application layer determines which abstract syntaxes are needed. All information dealt with by AEs must be defined within one or more abstract syntaxes (for example, the types of information contained in an E-mail message). Before any information can be transacted, end systems must agree on which abstract syntax and transfer syntax must be used. This is accomplished through the establishment of a *presentation connection.* This logical connection is initiated by the application layer with a P-CONNECT request primitive to the application layer. The AP sends a connect presentation (CP) PDU to a remote peer, describing the names of the transfer syntaxes to be supported with each abstract syntax (clearly, this connection must be supported by appropriate connection activities at the lower layers, including a physical path between the two peers). The P-CONNECT request would specify the names of the abstract syntaxes to be supported; the CP PDU pairs transfer syntaxes with each abstract syntax specified.

For example, the AE may need ASyntx#1 and ASyntx#2. The P-CONNECT conveys this request to the PL. The PL may then issue a CP specifying ASyntx#1 : {TSyntx#17, TSyntx#77, TSyntx#96}; ASyntx#2: {TSyntx#9}.

On receipt of a CP request, the remote PL will issue a P-CONNECT to the AE at that end, including among the parameters the names of the abstract syntaxes contained in the CP PDU. The AE will, in turn, respond with a P-CONNECT containing the names of the abstract syntaxes it is willing to use in the dialogue with the remote end. Clearly, this must be a subset of the abstract syntaxes presented to it.

To complete the process, the remote PL, having received a P-CONNECT from its own AE, will issue a connect presentation–accept (CPA) PDU, indicating the proposed transfer syntaxes it has picked for each selected abstract syntax. The origination PL must accept these selections and, on receipt of the CPA, it will issue a final P-CONNECT to its AE. At this juncture, (1) a presentation connection has been established, (2) the AEs have negotiated which abstract syntaxes (i.e., a set of types) to use for this communication, and (3) the PLs have negotiated a transfer syntax (i.e., a representation) of each of the abstract syntaxes to be employed by the AEs.

To facilitate application of OSI principles, a formal notation has been created to describe the constituent parts of the negotiation, as shown in Figure 13.20. The goal of this notation is to describe abstract syntaxes. The notation is called Abstract Syntax Notation One.

ASN.1 is similar in some respects to programming languages such as C, PL/1, and Ada. However, because its goal is to define abstract syntaxes, ASN.1 has no executable statements (for example, no GOTO, ADD, *et cetera*); rather it includes only facilities to describe types and values.

13.2.4 Message Handling Systems and Services (X.400)

Message handling systems and services enable users to exchange messages on a store-and-forward basis. The following discussion is based on the 1988 specification [13.23]. The MHS is designed in accordance with the principles of the OSIRM for CCITT applications (Recommendation X.200) and uses the presentation layer services and services offered by other, more general ASEs. A message submitted by one user, the originator, is conveyed by the message transfer system (MTS),

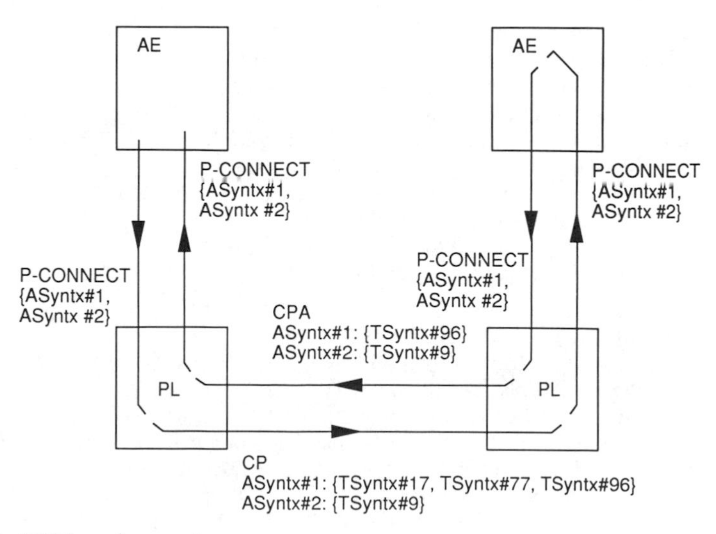

Figure 13.20 ASN.1 environment.

the principal component of a larger MHS, and is subsequently delivered to one or more additional users, i.e., the message's recipients. An MHS comprises a variety of interconnected functional entities. Message transfer agents (MTAs) cooperate to perform the store-and-forward message transfer function. Message stores (MSs) provide storage for messages and enable their submission, retrieval, and management. User agents (UAs) help users access the MHS. Access units (AUs) provide links to other communication systems and services of various kinds (e.g., other telematic services, postal services). The X.400 series of recommendations specifies the overall system and service description of message handling capabilities (see Table 13.6). The discussion that follows is based on the X.400 standard.

A MHS can be constructed using a variety of subnetworks (layers 1 through 3). The message transfer service provided by the MTS is application independent. An example of a standardized application is the interpersonal messaging (IPM) service. End systems can use the MTS for specific applications that are defined bilaterally. MHSs provided by administrations (telecommunication carriers) belong

Table 13.6
Standards supporting MHS

MHS Standards:

MHS System and Service Overview (CCITT X.400; ISO 10021-1)
MHS Overall Architecture (CCITT X.402; ISO 10021-2)
MHS Conformance Testing (CCITT X.403)
MHS Abstract Service Definition Conventions (CCITT X.407; ISO 10021-3)
MHS Encoded Information Type Conversion Rules (CCITT X.408)
MHS Message Transfer System: Abstract Service Definition and Procedures (CCITT X.411; ISO 10021-4)
MHS Message Store: Abstract Service Definition (CCITT X.413; ISO 10021-5)
MHS Protocol Specifications (CCITT X.419; ISO 10021-6)
MHS Interpersonal Messaging System (IPMS) (CCITT X.420; ISO 10021-7)
Telematic Access to IPMS (CCITT T.330)
MHS Naming & Addressing for Public Message Handling Services (CCITT F.401)
MHS The Public Message Transfer Service (CCITT F.410)
MHS Intercommunication with Public Physical Delivery Services (CCITT F.415)
MHS The Public Interpersonal Messaging (IPM) Service (CCITT F.420)
MHS Intercommunication Between IPM Service and Telex (CCITT F.421)
MHS Intercommunication Between IPM Service and Teletex (CCITT F.422)

Other OSIRM Standards:

OSI Basic Reference Model (CCITT X.200; ISO 7498)
OSI Specification of Abstract Syntax Notation One (CCITT X.208 ISO 8824)
OSI Association Control: Service Definition (CCITT X.217; ISO 8649)
OSI Reliable Transfer: Model & Service Definition (CCITT X.218; ISO 9066-1)
OSI Remote Operations: Model Notation & Service Definition (CCITT X.219; ISO 9072-1)
OSI Association Control: Protocol Specification (CCITT X.227; ISO 8650)
OSI Reliable Transfer: Protocol Specification (CCITT X.228; ISO 9066-2)
OSI Remote Operations Protocol Specification (CCITT X.229; ISO 9072-2)

to the group of telematic services defined in F Series recommendations. Various other telematic services and telex (Recommendations F.60, F. 160, F.200, F. 300, *et cetera*), data transmission services (X.1), or physical delivery services (F.415) gain access to and intercommunicate with the IPM service or intercommunicate with each other, via AUs. *Elements of service* are the service features provided through the application processes. The elements of service are considered to be components of the services provided to users and are either elements of a basic service or of optional user facilities. They are classified as either essential optional user facilities or as additional optional user facilities.

Functional Model of the MHS

The MHS functional model serves as a tool to aid in the development of standards for the MHS and to describe the basic components that work together to provide MH services. The model can be applied to a number of different physical and organizational configurations. A functional view of the MHS model is shown in Figure 13.21. For the purpose of the model, a user is either a person or computer process. Users are either direct users or indirect users (i.e., users that engage in message handling through another communication system that is linked to a MHS). A user is referred to as either an originator (when sending a message) or a recipient (when receiving a message). Message handling elements of service define the set of message types and the capabilities that enable an originator to transfer those types of messages to one or more recipients.

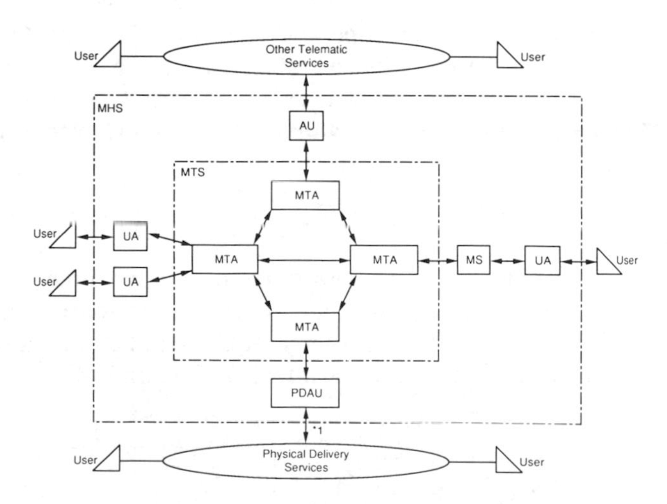

*1 Message input from PD Services to MHS is for further study.
Flow from PD Services to the PDAU shown is for notifications.

Figure 13.21 Functional model.

An originator prepares messages with the assistance of her or his UA. The UA is an application process that interacts with the MTS or MS to submit messages on behalf of a single user. The MTS delivers the message submitted to it to one or more recipient UAs, AUs, or MSs, and can return notifications to the originator. Functions performed solely by the UA and not standardized as part of the message handling elements of service are called *local functions*. A UA can accept delivery of messages directly from the MTS, or it can use the capabilities of the MS to receive delivered messages for subsequent retrieval by the UA.

The MTS contains a number of MTAs. Operating together, in a store-and-forward manner, the MTAs transfer messages and deliver them to the intended recipients. Access by indirect users of MSs is accomplished by AUs. Delivery to indirect users of MHS is accomplished by AUs, such as in the case of physical delivery, by the physical delivery access unit (PDAU).

The MS is an optional general-purpose capability of MHS that acts as an intermediary between the UA and the MTS. The MS is depicted in the MHS functional model shown in Figure 13.21. The MS is a functional entity whose primary purpose is to store and permit retrieval of delivered messages. The MS also allows for submission from and alerting to the UA. This collection of UAs, MSs, AUs, and MTAs make up the message handling system.

Structure of Messages

The basic structure of messages conveyed by the MTS is shown in Figure 13.22. A message is made up of an envelope and a content. The *envelope* carries information that is used by the MTS when transferring the message within the MTS. The *content* is the information the UA wishes delivered to one or more recipient UA. The MTS neither modifies or examines the content, except for conversion.

Application of the MHS Model

Users access UAs for message processing purposes, for example, to create, present, or file messages. A user interacts with a UA via an input-output device or process (e.g., keyboard, display, printer, *et cetera*). A UA can be implemented as a set of computer processes in an intelligent terminal. A UA and MTA can be collocated

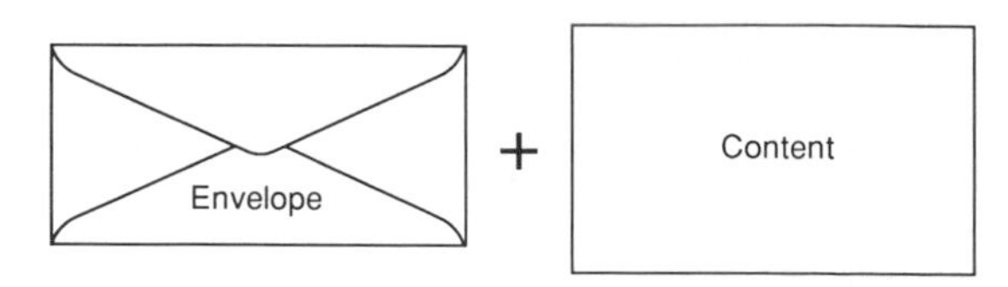

Figure 13.22 Basic message structure.

in the same system, or a UA-MS can be implemented in physically separate systems. In the first case, the UA accesses the MT elements of service by interacting directly with the MTA in the same system. In the second case, the UA-MS will communicate with the MTA via standardized protocols specified for MHS. An MTA may also be implemented in a system without UAs or MSs. Some possible physical configurations are shown in Figures 13.23 and 13.24. The different physical systems can be connected by means of dedicated lines or switched network connections.

Administrative Management Domain

An administration or organization can play various roles in providing MHSs. An organization in this context can be a commercial or noncommercial enterprise.

The collection of at least one MTA, zero or more UAs, zero or more MSs, and zero or more AUs operated by an administration or organization constitutes a *management domain* (MD). An MD managed by an administration is called an *administration management domain* (ADMD). An MD managed by an organization other than an administration is called a *private management domain* (PRMD). The relationship between MDs is shown in Figure 13.25.

In one country, one or more ADMDs can exist. An ADMD is characterized by its provision of relaying functions between other MDs and the provision of MTS for the applications provided within the ADMD. An administration can provide access for its users to the ADMD in one or more of the following ways:

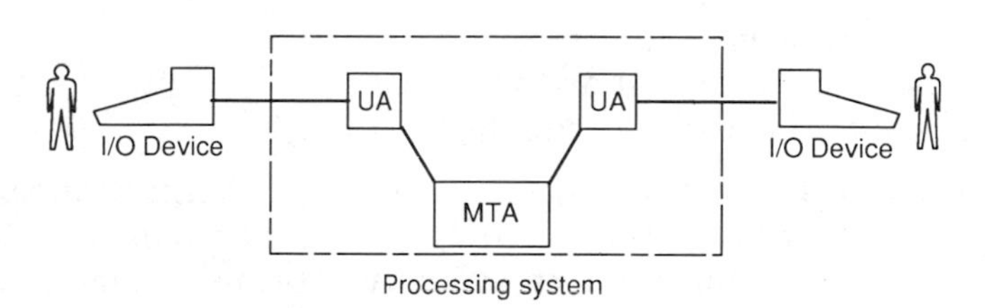

Figure 13.23 Co-resident UA and MTA.

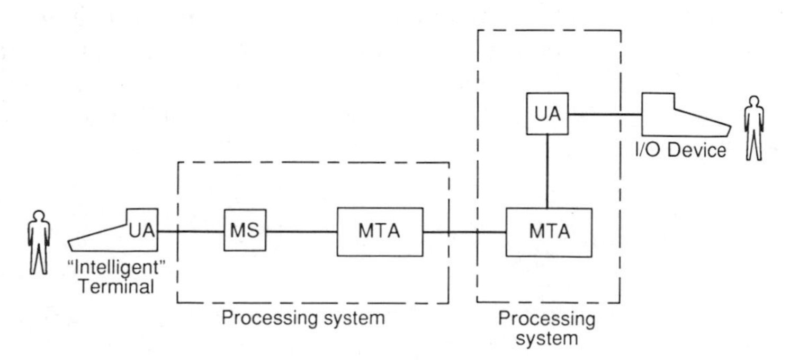

Figure 13.24 Stand-alone UA and co-resident MS-MTA and US-MTA.

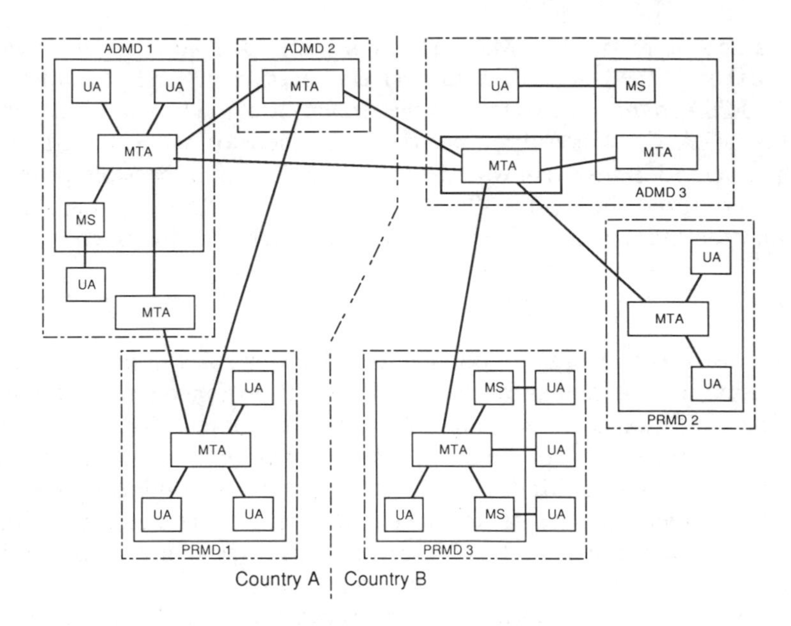

Figure 13.25 Relationships between MDs.

- User to administration-provided UA;
- Private UA to administration MTA;
- Private UA to administration MS;
- Private MTA to administration MTA; and
- User to administration-provided AU.

Administration-provided UAs can exist as part of an intelligent terminal with which the user can access MHS. They can also exist as part of administration resident equipment that is part of the MHS; in which case, the user obtains access to the UA via an input-output device.

In the case of a private UA, the user has a private stand-alone UA that interacts with the administration-provided MTA or MS, using submission, delivery, and retrieval functions. A private stand-alone UA can be associated with one or more MDs, provided that the required naming conventions are preserved. A private MTA as part of a PRMD can access one or more ADMDs in a country, following national regulations.

Private Management Domain

An organization other than an administration can have one or more MTAs, UAs, AUs, and MSs forming a PRMD interacting with an ADMD on an MD-to-MD

(MTA-to-MTA) basis. A PRMD is characterized by the provision of messaging functions within that MD.

A PRMD is considered to exist entirely within one country. Within that country, the PRMD can have access to one or more ADMDs, as shown in Figure 13.25. However, in the case of a specific interaction between a PRMD and an ADMD (such as when a message is transferred between MDs), the PRMD is considered to be associated only with that ADMD. A PRMD will not act as a relay between two ADMDs.

In the interaction between a PRMD and a ADMD, the ADMD takes responsibility for the actions of the PRMD that are related to the interaction. In addition to ensuring that the PRMD properly provides the MTS, the ADMD is responsible for ensuring that the accounting, logging, quality of service, uniqueness of names, and related operations of the PRMD are correctly performed. As a national matter, the name of a PRMD can be either nationally unique or relative to the associated ADMD. If a PRMD is associated with more than one ADMD, the PRMD can have more than one name.

Direct interaction between PRDMs (without ADMD) is allowed under the X.400 standard, using the same protocols as those used for the interaction between the PRMD and the ADMD.

The Message Store

Because UAs can be implemented on a wide variety of equipment, including PCs, the MS can complement a UA implemented, for example, on a PC by providing a more secure, continuously available storage mechanism to take delivery of messages on the UA's behalf. The MS retrieval capability provides users who subscribe to an MS with basic message retrieval capabilities potentially applicable to messages of all types. Figure 13.26 shows the delivery and subsequent retrieval of messages that are delivered to an MS, and the indirect submission of messages via the MS. One MS acts on behalf of only one user, i.e., it does not provide a common or shared MS capability to several users. When subscribing to an MS, all messages destined for the UA are delivered to the MS only. The UA, if on line, can receive alerts when certain messages are delivered to the MS. Messages delivered to an MS are considered delivered from the MTS perspective. When a UA submits a message through the MS, the MS is, in general, transparent and submits it to the MTA before confirming the success of the submission to the UA. However, the

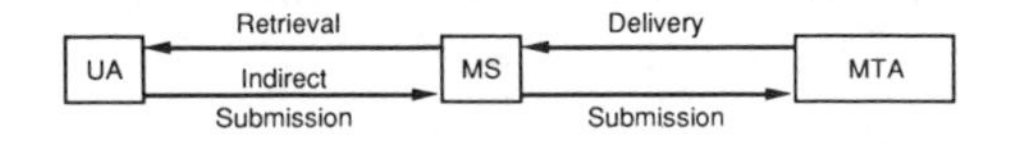

Figure 13.26 Submission and delivery with an MS.

MS can expand the message if the UA requested the forwarding of messages that exist in the MS. Users are also provided with the capability to request the MS to forward selected messages automatically on delivery. Particularly because of this new feature, implementations based on the full 1988 CCITT X.400 MHS do not interwork directly with implementations based on the 1984 version of the standard.

Physical Configurations of the Message Store

The MS can be physically located with respect to the MTA in a number of ways. The MS can be collocated with the UA, collocated with the MTA, or stand-alone. From an external point of view, collocating the MS with the MTA offers significant advantages that will probably make it the predominant configuration.

Organizational Configurations of the Message Store

Either ADMDs, PRMDs, or UAs can operate MSs. In the case of administration-supplied MSs, the subscriber either provides the UA or makes use of an administration-supplied UA via an input-output device. In either case, all the subscribers messages are delivered to the MS for subsequent retrieval. The physical and organizational configurations described above are examples only and other equally valid cases can exist.

The Message Transfer Service

The MTS provides the general, application-independent, store-and-forward service.

Submission and Delivery

The MTS provides the means by which UAs exchange messages. The two basic interactions between MTAs and UAs or MSs are:

1. The submission interaction is the means by which an origination UA or MS transfers to an MTA the content of a message and the submission envelope. The submission envelope contains the information that the MTS requires to provide the requested elements of service.
2. The delivery interaction is the means by which the MTA transfers to a recipient UA or MS the content of a message plus the delivery envelope. The delivery envelope contains information related to delivery of the message.

In the submission and delivery interactions, responsibility for the message is passed between the MTA and the UA or MS.

Transfer

Starting at the originator's MTA, each transfers the message to another MTA until the message reaches the recipient's MTA, which then delivers it to the recipient UA or MS using the delivery interaction. The transfer interaction is the means by which one MTA transfers to another MTA the content of a message plus the transfer envelope. The transfer envelope contains information related to the operations of the MTS plus information that the MTS requires to provide elements of service requested by the originating UA. MTAs transfer messages containing any type of binary coded information. MTAs do not interpret or alter the content of messages except when performing a conversion.

Notifications

Notifications in the MTS include the delivery and nondelivery notifications. When a message cannot be delivered by the MTS, a nondelivery notification is generated and returned to the originator in a report signifying the situation. In addition, an originator can specifically ask for acknowledgment of successful delivery through use of the delivery notification element of service on submission.

Three users of the MTSs are:

1. *User Agent*: The UA uses the MT service provided by the MTS. A UA is a functional entity by means of which a single direct user engages in message handling. UAs are grouped into classes based on the type of content of messages they can handle. The MTS provides a UA with the ability to identify its class when sending messages to other UAs. UAs within a given class are referred to as *cooperating UAs* because they cooperate with each other to enhance communication among their respective users.
2. *Message Store*: The MS uses the MT service provided by the MTS. An MS is a functional entity associated with a user's UA. The user can submit messages through it, and retrieve messages that have been delivered to the MS.
3. *Access Unit*: An AU uses the MT service provided by the MTS. An AU is a functional entity associated with an MTA to provide for intercommunication between the MHS and another system or service.

The Interpersonal Messaging Service

The IPM service provides a user with features to assist in communication with other IPM service users. The IPM service uses the capabilities of the MTS for sending and receiving interpersonal messages.

IPM Service Functional Model

Figure 13.27 shows the functional model of the IPM service. The UAs used in the IPM service (IPM-UAs) comprise a specific class of cooperating UAs. The optional AUs shown (TLMA, PTLXAU) allow for teletex and telex users to intercommunicate with the IPM service. The optional PDAU allows IPM users to send messages to users outside the IPM service who have no access to the MHS. The MS can optionally be used by IPM users to take delivery of messages on their behalf.

Structure of IP Messages

The IPM class of UAs creates messages containing a content specific to the IPM. The specific content that is sent from one IPM UA to another is the result of an originator composing and sending a message, called an *IP message*. The structure of an IP message as it relates to the basic message structure of the MHS is shown in Figure 13.28. The IP message is conveyed with an envelope when being transferred through the MTS.

Figure 13.29 shows an analogy between a typical office memo and the corresponding IP message structure. The IP message contains information (e.g., to, cc, subject) provided by the user that is transformed by the IPM-UA into the heading of the IP message. The main information that the user wishes to communicate (the body of the memo) is contained within the body of the IP message.

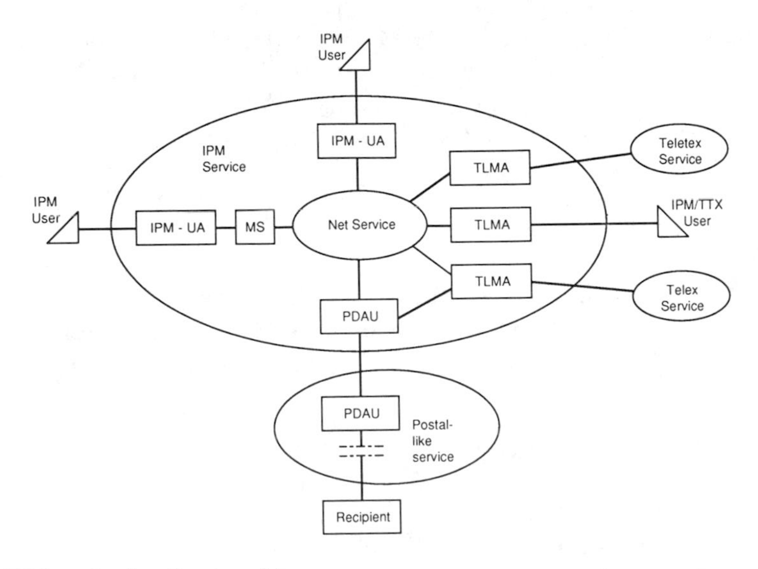

Figure 13.27 IPM service functional model.

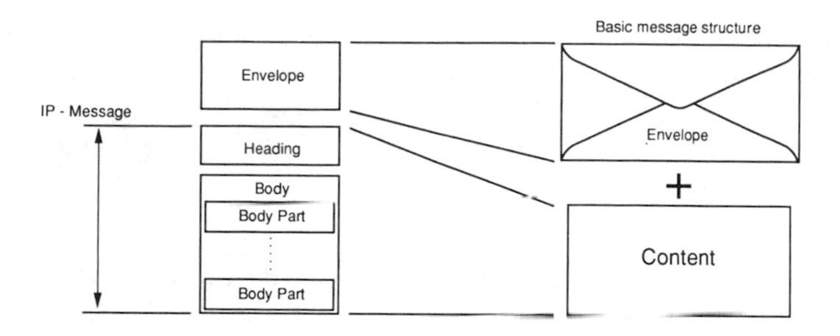

Figure 13.28 IP message structure.

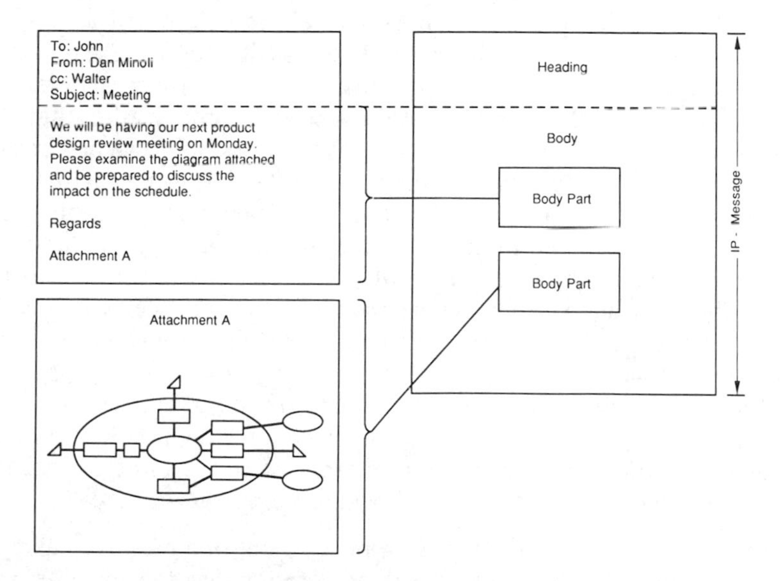

Figure 13.29 IP message structure for a typical memo.

In the example shown, the body contains two types of encoded information: text and facsimile, which are referred to as *body parts*. In general, an IP message body can consist of a number of body parts, each of which can be of a different encoded information type, such as voice, text, facsimile, and graphics.

IP Notifications

In IPM service, a user can request a notification of receipt or nonreceipt of a message by a recipient. These notifications are requested by an originator and are generated as a result of some recipient or intermediate MTA action (such as reading or not reading the message). In certain cases, the nonreceipt notification is generated automatically by the recipient's UA.

MHS Use of a Directory

Users of the MHS can be identified by a name, called the *directory name*. A directory name must be looked up in a directory to find out the corresponding originator-recipient (O-R) name. (This is the same directory discussed in the previous sections.) The O-R name is an information object by means of which a user can be designated as the originator or a user can be designated as a potential recipient of a message. O-R names also apply to distribution lists. An O-R name distinguishes one user (or distribution list) from another and can also identify its point of access to MHS. Thus, an O-R address contains information that enables the MHS to identify uniquely a user for delivery of a message or return of a notification. The prefix O-R recognizes the fact that the user can be acting as either the originator or recipient of the message or notification in question. An O-R address is a collection of information called *attributes*. Recommendation X.402 specifies a set of standard attributes from which O-R addresses can be constructed (standard attributes mean that their syntax and semantics are specified). In addition to standard attributes and to support existing messaging systems, domain-defined attributes exist that have syntaxes and semantics that are defined by MDs.

As described earlier, the directory defined by the X.500 series of recommendations provides information about communication resources useful in the use and provision of a variety of telecommunication services. The directory capabilities used in message handling fall into the following four categories:

1. *User-friendly Naming*: The originator or recipient of a message can be identified by means of the X.500 directory name, rather than the machine-oriented O-R address. At any time, the MHS can obtain the latter from the former by consulting the directory.
2. *Distribution Lists* (DLs): A group with its membership stored in the directory can be used as a DL. The originator simply supplies the X.500 directory name of the list. The MHS can obtain the names of the individual recipients by consulting the directory.
3. *Recipient UA Capabilities*: MHS capabilities of a recipient (or originator) can be stored in the O-R directory entry. At any time, the MHS can obtain (and then act on) those capabilities by consulting the directory.
4. *Authentication*: Before two MHS functional entities (two MTAs, or a UA and an MTA) communicate with one another, each verifies the identity of the other. This can be done by using authentication capabilities of MHS based on information stored in the directory.

Besides the above, one user can directly access the directory, for example, to determine the O-R address or MHS capabilities of another. The recipient's directory name is supplied to the directory, which returns the requested information.

13.3 MAP/TOP STANDARDIZATION EFFORTS

The need to allow open communication within and between very different departments such as engineering, manufacturing, publishing, procurement, sales, and finance is the key to running an organization efficiently. Sharing computing resources and exchanging information electronically, however, often involves a diverse set of computers, network technologies, and application environments. As a result, information systems engineering requires integration of an increasing number of computers, each supporting a wide variety of applications specializing in different aspects of information processing. International standards concerning interoperability are now emerging, but interoperability has historically been a cumbersome, expensive, and inefficient process [13.24]. The benefits of using information technology products based on an open architecture include lower costs, a wider selection from multiple vendors, increased functionality, simpler implementation, more effective management, and ensured compatibility through test methodologies.

Integration problems associated with multivendor networks tend to inhibit the desired free exchange of information. To get closer to the goal of universal interoperability, implementation profiles are necessary. An implementation profile selects a specific standard (where more than one exists) and defines implementation options and guidelines necessary to achieve true interoperability. The Technical and Office Protocol (TOP) Version 3.0 Specification is an implementation specification that defines a family of implementation profiles pertaining to user requirements in the engineering and administrative environments. TOP grew out of efforts at the Boeing Company. TOP networks will be able to transfer and accept data with any of the Manufacturing Automation Protocol (MAP) networks, linking engineering and office environments to the factory floor.

MAP grew from the operating difficulties of a very large organization, General Motors. By the end of the 1970s, GM found itself with 20,000 robots and more than 40,000 intelligent devices in its manufacturing operations, and only 15% of them were capable of communicating beyond the limits of their own "islands of automation." The crucial factor inhibiting greater integration was the lack of a suitable communication system. Up to half the cost of introducing automation was being absorbed by the communication component. GM decided to do something about it, and gathered together representatives from its own division and from its suppliers to set up a task force. Their charter was to develop the specification for an independent computer network protocol capable of creating and sustaining a true multivendor environment on the plant floor. Progress since then has been rapid. GM, which has introduced MAP into new or refurbished plants, is now planning to extend it to other manufacturing facilities [13.25]. GM's activities were matched by a parallel exercise, led by an equally influential organization, Boeing, and aimed at establishing a specification for multivendor communication serving organization functions other than those on the plant floor. TOP does for the office and engineering environments what MAP does for the plant. Both MAP and TOP are standards-based specifications. The two specifications complement each other but are not identical.

13.3.1 MAP/TOP Network Services

The MAP/TOP specification provides a range of services designed to satisfy the information technology needs of the user in several areas, including information exchange, network operations, and application interfaces. The information exchange process includes two distinct aspects: (1) a data transfer aspect supported by a data transfer service and (2) an interpretation aspect supported by common data interchange formats. Data transfer services are those services that provide for actual transfer of data or information across the network. These include services traditionally associated with the application layer of the OSIRM. Specifically, MAP/TOP 3.0 defines the following data transfer services [13.24]:

- *Remote File Access*: Provides real-time access and management of remote files for the purpose of transferring their contents. The application level functionality is provided by the ISO FTAM protocol.
- *Electronic Mail*: Provides an electronic analog to normal postal service and is an example of store-and-forward technology. The application level functionality is provided by the CCITT X. 400 protocol.
- *Remote Terminal Access*: Allows intelligent workstations to operate in a terminal mode for access to centralized application processes. Also provides a method for low-cost terminals to access distributed applications through terminal servers. TOP specifies a basic service corresponding to simple character-oriented terminals equivalent to those supported by TELNET services in the TCP/IP environment (discussed in more detail in Chapter 14). The application functionality is based on the ISO VT protocol.

These services can be provided over any of the following subnetwork types:

- CSMA/CD LAN subnetwork specified by ISO 8802-3 based on Ethernet;
- Token-ring LAN subnetwork specified by ISO 8802-5 based on IBM's token ring;
- Token bus LAN subnetwork specified by ISO 8802-4 (used by MAP); and
- X.25 packet-switching network-subnetwork specified by CCITT X.25 packet-switching protocols for layers 1, 2, and 3.

Data transfer services such as E-mail or remote file access do not guarantee that information can be properly interpreted when it reaches its destination. Using FTAM or X.400 to move a file from a remote system to a local system does not mean the local text editor, graphics program, *et cetera*, can interpret the information. True interoperability is achieved through the use of standardized data structures. These can then be interpreted by any conforming implementation and, if necessary, translated into the proprietary local context for local processing. In MAP/TOP, these standards data structures are defined by "interchange formats." MAP/TOP 3.0 specifies interchange formats for [13.24]:

- *Office Documents*. Provides the ability to exchange general-purpose compound documents containing text, raster graphics (Group IV fax), and geometric graphics. It is known as the *office document interchange format* and is based on the ISO Office Document Architecture/Office Document Interchange Format (ODA/ODIF) (ISO DIS 8613/CCITT T.411 through T.419) Standard.
- *Product Definition Data*. Provides the ability to exchange product data information among applications and users. Product definition data include all elements of data pertaining to the analysis, design, manufacture, and test of a product and is necessary for CAD-CAM applications. The Initial Graphics Exchange Standard (IGES) Version 3.0 forms the basis for this interchange format known as the *product definition interchange format*.
- Two-Dimensional Graphics. Provides the ability to exchange two-dimensional computer graphics picture information in final or revisable form. It is known as the *computer graphics metafile interchange format* and is based on the Graphics Metafile Standard.

At each of the seven layers of the OSIRM, the MAP and TOP specifications selected appropriate subsets of protocols, options, and parameters to provide the required functions and services. Figure 13.30 depicts the MAP/TOP protocol suite.

13.3.2 MAP/TOP Network Management

Support services designed to facilitate the day-to-day operations of communication networks are also specified in MAP/TOP. These include network management and directory services. The MAP/TOP 3.0 network management specification is based on ISO and IEEE work and is supplemented as necessary to provide a basic level of functionality. TOP network directory services are a powerful efficiency tool for the operation of distributed networks. TOP allows "user-friendly addressing" of objects (printers, application entitles, host computers, terminals, people, mailboxes, lists, *et cetera*) by maintaining a logical separation of the object name from the object address. The network directory properly associates an object name with its current address. A user of the directory can, therefore, address an object by

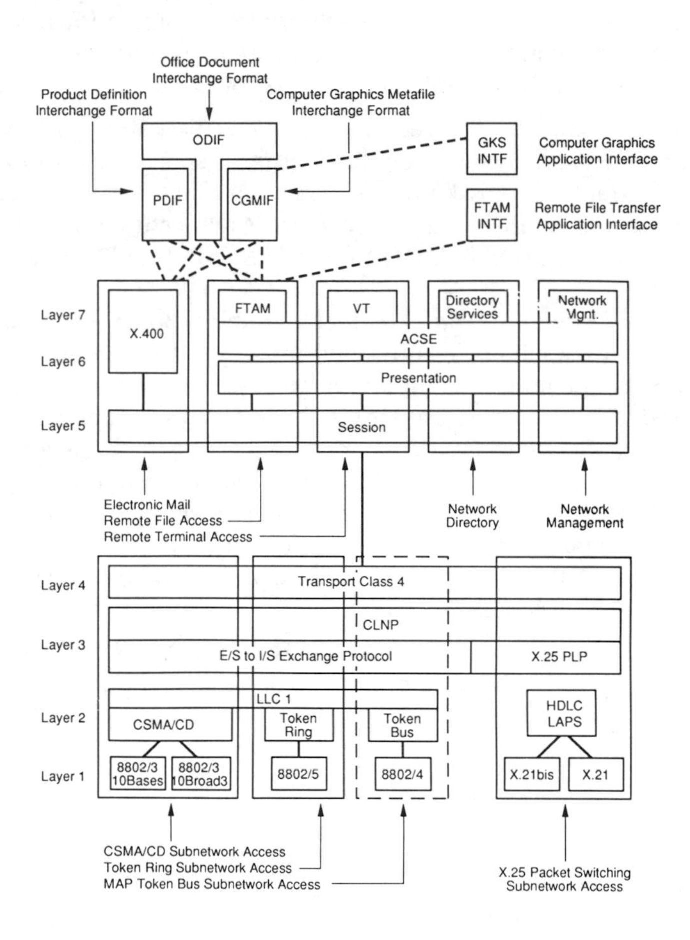

Figure 13.30 TOP Version 3.0 building block overview.

referring to its common name without explicit knowledge of its address. This functionality is based on the ISO/CCITT X.500 directory services standard.

13.3.3 MAP/TOP Application Program Interface

Another key capability provided by MAP/TOP 3.0 is that of application interfaces. Interfaces are not defined by standards addressing the peer-to-peer communication requirements of distributed systems. In TOP, application interfaces are used to isolate application programs from their computing systems environment. This allows the development of programs utilizing MAP/TOP network services that are, however, independent of the system in which they are installed. In other words, application interfaces provide for the portability of applications programs. A significant savings is realized by the user if applications programs do not have to be rewritten when a change to the computing systems environment is made.

MAP/TOP 3.0 defines two application interfaces [13.24]: (1) the FTAM application interface, which defines a standard library of access functions for using the remote file access service, and (2) the graphical kernel system (GKS) application interface, which defines a set of functions for creating and displaying two-dimensional graphics.

13.3.4 MAP/TOP Releases

The first MAP document, which was published in October 1982, covered general aspects of the network and its implementation. After being expanded to include more in-depth information, it was issued as version 1.0 in April 1984. Version 2.0, which took into account new standards activity and GM's work on the upper layer protocols, appeared in February 1985. Version 2.1, released in March 1985, added internet and file transfer. TOP Version 1.0 was published in November 1985. After the appearance of MAP 2.1 and TOP 1.0, the specifications were frozen for two years to allow implementations to be brought to the marketplace. A MAP version 2.1A was released correcting some of the errors found. MAP 2.2, published in August 1986, introduced Enhanced Performance Architecture, carrier-band bridges, and an introduction to naming and addressing.

The version 3.0 implementation releases of both TOP and MAP were issued in mid-1987. A joint MAP/TOP Steering Committee has been formed to ensure that the two specifications develop in unison. Both MAP 3.0 and TOP 3.0 are major upgrades from their previous versions. A MAP/TOP Users Group with approximately 2000 members in the United States from 750 different organizations has been formed. International interest has grown at a similar rate with active groups in Canada, Europe, Japan, and Australia. These groups have joined together to form the World Federation of MAP/TOP Users Groups to encourage international acceptance of the MAP/TOP specifications and to promote development of conformant products [13.25].

In the late 1980s, vendors complained about the frequent releases of new MAP/TOP specifications and their inability to develop products conforming to the specifications. An agreement was made at that time to try to keep the key parts of the document stable for six years. This would time a next major release for the 1993 time frame [13.26].

MAP 3.0 supports two different physical media. The most sophisticated, but more expensive, is broadband cable. Broadband is typically used as the backbone of a large plant-wide network. Broadband has multiple-channel capability, which allows it to carry voice, video, and other signals in addition to data. The other MAP media is carrier-band. This is a single channel cable that is slower, and less expensive, than broadband.

For time-critical applications, MAP supports an alternate architecture called MAP Enhanced Performance Architecture (MAP-EPA). This is a dual-architecture device incorporating a full-MAP version (with seven full layers) on one side and a "collapsed architecture" (consisting only of layers 1, 2, and 7) on the other. A MAP-EPA node can use either communication path. The full-stack side lets it communicate with other full-MAP nodes in a normal fashion, and the collapsed-stack side gives it faster access to similar nodes.

TOP 3.0 is the counterpart of MAP for the business and engineering sectors. It is designed to provide integrated business, engineering, and publishing systems. The primary medium for TOP is baseband cable, but as an option it also supports broadband with CSMA/CD subnetwork access. Other options for TOP include token-passing rings on shielded twisted pairs and a token-passing bus on a broadband cable. The token-passing bus is not a recommended option for most applications.

In addition to supporting FTAM for remote file transfer, TOP provides E-mail based on the CCITT X.400 protocol for MHS. TOP also supports remote terminal access using the basic VT services, which provide communication facilities for transfer of data between terminals and applications based on host computers in a multivendor environment.

13.4 OTHER OSI-BASED EFFORTS

MAP and TOP are not the only specifications based on the OSIRM. Governments are also pursuing OSI-based solutions to the communication problem. Both the United States and the United Kingdom, for example, have a specification called GOSIP (Government OSI Profile). GOSIP is intended to permit communication and interoperation of end-user and intermediate-level systems throughout the government. It provides computer systems within and between individual agencies with peer level, process to process, and terminal access capabilities.

REFERENCES

[13.1] K.G. Knightson *et al.*, *Standards for Open Systems Interconnection*, McGraw-Hill, New York, 1987.

[13.2] M.T. Rose, *The Open Book, A Practical Perspective on OSI*, Prentice Hall, Englewood Cliffs, NJ, 1989.

[13.3] W. Stallings, *Handbook of Computer-Communication Standards, The Open Systems Interconnection Model and OSI-Related Standards*, Howard W. Sams & Company, Indianapolis, IN, 1987.

[13.4] W. Stallings, *Data and Computer Communications*, 2nd Ed., Macmillan Publishing Company, New York, 1988.

[13.5] A.L. Chapin, "Computer Communications Standards," *Computer Communication Review*, June 1989 and June 1990 (and possibly future mid-year issues).

[13.6] H. Folts, "Open Systems Standards," *IEEE Network*, January 1989.

[13.7] OMNICOM, Open Systems Communication, December 1989.

[13.8] L.H. Landweber, M. Tasman, "An ISO TP-4-TP0 Gateway," *Computer Communication Review*, April 1990, pp. 16 ff.

[13.9] E.W. Geer, Bellcore, personal communication, May 1990.

[13.10] G.W. Poo, W. Ang, "OSI Protocol Choices for LAN Environments," *Computer Communications*, January/February 1990, pp. 17 ff.

[13.11] M. Seyed-Naghavi, Bellcore, personal communication, June 1990.

[13.12] OMNICOM, Open Systems Communication, April 1990.

[13.13] A. Patel, V. Ryan, "Introduction to Names, Addresses and Routes in an OSI Environment," *Computer Communication*, Vol. 13, No. 1, January/February 1990, pp. 27 ff.

[13.14] "Information Processing Systems—Open Systems Interconnection, File Transfer, Access, and Management—Part 1: General Description," ISO 8571.

[13.15] "Information Processing Systems—Open Systems Interconnection, File Transfer, Access, and Management—Part 2: The Virtual Filestore", ISO 8571.

[13.16] "Information Processing Systems—Open Systems Interconnection, File Transfer, Access, and Management—Part 3: The File Service Definition," ISO 8571.

[13.17] "Information Processing Systems—Open Systems Interconnection, File Transfer, Access, and Management—Part 4: The File Protocol Specification," ISO 8571.

[13.18] D. Minoli *et al.*, "Application of OSI/FTAM to Telephone Companies' Operations Systems Networks," ENE 88 Conference Record, SME, Dearborn, MI.

[13.19] *EDP Weekly/Computer Age*, June 29, 1987.

[13.20] L. Mantelman, "Upper Layers: From Bizarre to Bazaar," *Data Communications*, January 1988.

[13.21] "Stable Implementation Agreements for Open Systems Interconnection Protocols," Version 1, Edition 1, NBS Special Publication 500-150, December 1987.

[13.22] D. Chappell, ENE '88 Conference ASN.1 tutorial notes.

[13.23] "Message Handling and Directory Services—Operations and Definition of Service," F.400–F.422 and F.500, Fascicle II.6 of the CCITT Blue Book, 1988–89.

[13.24] T.A. Haug, "TOP—A Giant Stride Toward Universal Interoperability," *Gateway*, May/June 1988.

[13.25] B. Thacker, "The Computer Integrated Organization," Department of Trade and Industry, United Kingdom, 1988.

[13.26] A.H. Grossman, Bellcore, personal communication, June 1990.

Chapter 14
Local Area Networks and Their Management

LANs are designed to operate as high-speed low-cost data systems over a limited distance, usually linking terminals, PCs, and resources (servers) in a building or in a group of buildings within a few miles of each other. Many LANs also interconnect hosts, either via host communication ports or by way of channel extension, as discussed in Chapter 12. Wide area access to national and international networks is accomplished using a LAN-resident gateway.

With the ever-increasing power of PCs and workstations, LAN usage in the corporate environment has become more prevalent. By the end of 1991 there were 6 million LAN connections in the U.S.; that number is expected to exceed 13 million by 1995. Stand-alone systems and individual PCs, once considered adequate for most jobs, no longer meet user needs. Users now require connection to their coworkers on a flexible, high-speed network.

While LANs are becoming the standard intracompany communication apparatus for large as well as medium-size firms, the software of many LAN operating systems lacks comprehensive network management tools needed to monitor users, hardware resources, and data files. Given the importance of LANs and the continued deployment expected in the 1990s, the need to understand LAN technology and LAN management issues is critical. This chapter covers both topics, from a pragmatic perspective; Chapter 16 addresses the more general topic of network (WAN) management, particularly from an OSI standards perspective.

14.1 LAN BASICS

LANs are customer-owned facilities used to communicate within a building or a campus. In some limited instances, building management (through joint tenant services) provides LAN access to the tenants of the building; in this case, the tenants do not own the LAN facilities. The transmission speed of LANs varies from 1 Mb/s at the low end to 4, 10, 16, 50, and 100 Mb/s at the (current) high end, depending on the system and the technology used [14.1]. LANs fit in a continuum of communication speeds available to end-users, as shown in Table 14.1.

While data transmission is the major application, some LANs also carry voice and video, particularly analog-based broadband LANs. Typically, the underlying transport medium is coaxial cable, although beginning in the mid- to late 1980s, low-end twisted-pair systems and high-end fiber-based systems have become increasingly common.

14.1.1 Topologies

Three major LAN topologies are star, ring, and bus [14.2, 14.3], as discussed below:

1. A star network is joined at a single point, generally with central control.
2. In a ring, the nodes are linked into a continuous circle on a common cable, and signals are passed unidirectionally around the circle from node to node, with signal regeneration at each node. A ring with a central control is known as a "loop."
3. A bus is essentially a single line of cable to which the nodes are connected directly by taps. Like the ring, it is normally employed for distributed control, but it can also be based on central control. Unlike the ring, however, a bus is passive, namely, the signals are not regenerated and retransmitted. Both MAP and TOP (discussed in Chapter 13) have adopted the bus topology.

Other configuration variations include: (1) the star-shaped ring, (2) the double ring, and (3) the tree. The first variation represents a wiring methodology to

Table 14.1
Continuum of Communications Bandwidths

Communication System	*Approximate Bandwidth (b/s) (orders of magnitude)*
Voice-grade lines with modems	10^3
ISDN	10^4
Fractional T1/DS1 facilities	10^5
T1/DS1 facilities	10^6
LANs	10^6
T3/DS3 facilities	10^7
Backbone LAN-current MAN	10^7
Broadband ISDN	10^8
"Dark fiber" private line	10^9

facilitate physical management: at the logical level, the transmission facility is a ring; at the physical level, it is a star centralized at some convenient point. The second variation provides two rings: one for transmission and one for reception; this method is used, for example, in the fiber backbone networks discussed in the next chapter. The tree is used most often in CATV applications.

14.1.2 Bandwidth

The *nominal capacity* of a communication link depends on the physical medium (wire, coaxial cable, optical fiber, *et cetera*) used to transmit the information. Different media have different transfer capabilities, which are usually defined in terms of the frequency range or, alternatively, by the number of bits per second. Coaxial cable has a bandwidth of 300 to 400 MHz; the twisted-pair wires used in the telephone system have a bandwidth of 1 to 2 MHz; optical fibers have bandwidths in the 2- to 6-GHz region. (These bandwidths are typical for repeatered cable; for unrepeatered cables, the bandwidth is affected by distance.) During the 1980s, LANs generally used coaxial cable similar to that used to carry CATV; more recently, twisted-pair and fiber systems have emerged, as alluded to earlier.

The *actual capacity* of a LAN depends on a number of other factors besides the medium. While the media clock rate can support 10 Mb/s, the DLL may have a throughput of only 7 Mb/s. In addition, the higher layer protocols impose additional overhead: In some cases, the internetworking protocol only supports an effective throughput of 4.5 Mb/s, and the transport layer protocol typically only passes 3 Mb/s. The application may, in turn, achieve a throughput of 1 Mb/s. Some practitioners claim an effective throughput of 8 Mb/s using TCP/IP on a 10 Mb/s Ethernet.

Bandwidth can be allocated using two traditional techniques: FDM and TDM. FDM divides the available bandwidth into separate channels, each adjusted to give the frequency range required to deal with the particular type of information to be transmitted. TDM allows several nodes to access the network channel on a rotating basis; each node, in turn, is given control for long enough to transmit part of its information, and all the information is reassembled from the data stream at the destination.

A communication system based on FDM is known as *broadband*; a system in which the signals are carried on a network at their original frequency is called *baseband*. One in which all the signals are modulated on a single carrier frequency is called *carrier band*. The main advantage of a broadband system is its multichannel capacity; however, it is more expensive than a carrier-band system. In baseband transmission, the unmodulated data are pulsed in a single channel directly onto the transmission medium. Baseband LANs do not employ multiplexing to derive multiple channels from the medium. These LANs employ random access techniques to achieve medium access control (broadband LANs employ multiplexing).

In a broadband network, the signals can be transmitted within an allotted frequency range only in one direction along the cable (because of amplification considerations). The two methods of dealing with this are: (1) to establish separate sending and receiving channels by FDM and to control these by some kind of relay station; or (2) to provide the sending and receiving channels by using two cables. With the first method, the data are sent up-cable, usually on the lower frequency channel, to a head-end remodulator, as shown in Figure 14.1. The head-end then amplifies the signals, modulates them to the higher frequency of the other channel, and retransmits them down-cable to the addressee node. All signals put on the cable travel toward the head-end; all signals taken off the cable travel away from the head-end. With a double-cable system, the nodes send on one cable and receives on the other (see Figure 14.2). A head-end is not needed, but cable and tap costs will be higher.

14.1.3 Access to the Network

Two ways exist for ensuring that no node waits too long to gain access to the network and that no more than one node at a time gains control of the LAN

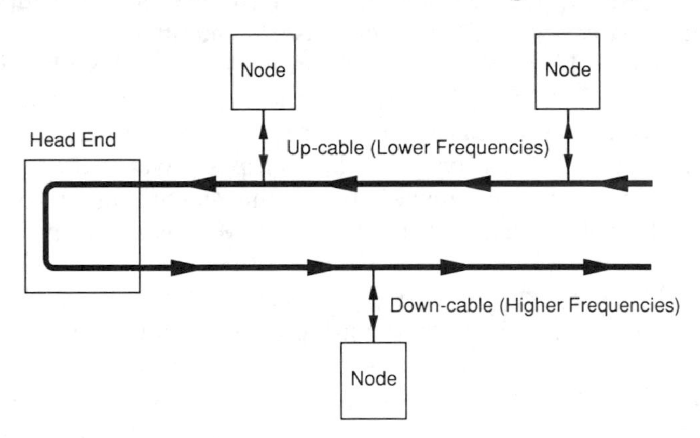

Figure 14.1 Single LAN cable.

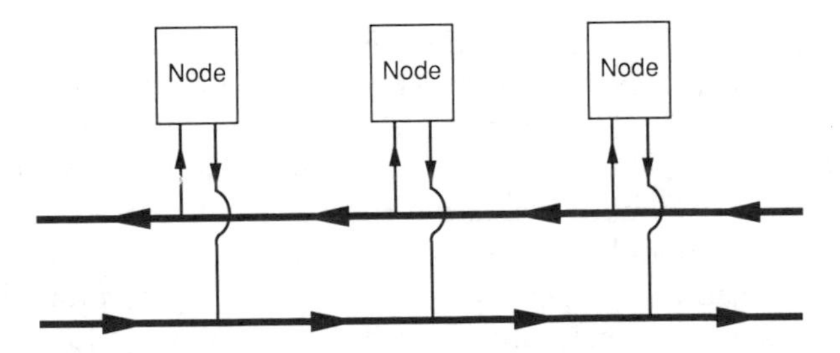

Figure 14.2 Dual-cable system.

channel. The first is by the *contention method*; the second is by a *variant of polling* (in theory, we could also use reservation, but this method has not been implemented extensively in the LAN context).

arrier Sense Multiple Access with Collision Detection (CSMA/CD). If node α has essage to send, it checks the network until it senses that it is traffic-free ("listen-re-talk") and then transmits. However, because all the nodes on the network the right to contend for access (hence, "multiple access"), node α keeps ring the network in case a competing signal has been transmitted simulta- with its own ("listen-while-talk"). If a second node is indeed transmitting, gnals will collide. Both nodes will detect the collision by a change in the rgy level, stop transmitting, and wait for a random time (of the order nds) before attempting to gain access again. Contention techniques inistic: They cannot guarantee access to the network within a given

t-of-polling method uses the concept of a *token*. The token is a ket that is circulated around the network from node to node, equence, when no other transmission is taking place. Each ess of the predecessor and the successor; the sequence by lated to the nodes is set by network management and does conform to the physical bus or ring location. The token for transmission, and no node can send data without it. nitors the network to detect any packet addressed to it; the token (the token does not have an address). When ode, and the node has nothing to send, then the node hannel access delay) to the next node in the sequence. passed on after the node has completed transmitting en must be surrendered within a specific time so that network resources.

M layer 1 and layer 2 standards are typically defined Ns. Layers 1 and 2 are defined by the IEEE 802 tworking" protocols defined at layer 3 (such as ISO nnection-oriented transport protocols, we can then ite up to layer 7 (including FTAM and MHS). availability of OSI-based products (which should cussed in Chapter 13), the use of TCP/IP has been . TCP/IP is a well-entrenched *de facto* standard, products constitute a $6.5 billion business yearly; 2000 networks with more than 180,000 hosts and 1 million users; also, more than 2000 private networks exist, not on Internet, that use TCP/IP. Nonetheless, the availability of international standards will probably exert a strong force toward eventual migration. We describe the standards at layer 1 and layer 2 that are peculiar to LANs below (see Figure 14.3).

The effort to formulate standards for LANs was begun by the IEEE in 198 as Project 802. Originally expected to standardize the Ethernet technology, whic was brought to the market by a joint effort among Xerox, Intel, and Digit Equipment Corporation (DEC), the project grew in scope to include two toke based standards as well. Ethernet was designed in 1973 at the Xerox Palo A Research Center as a network for minicomputers (the name "Ethernet" der from the long-discarded notion that electromagnetic energy is transmitted thro a fluid substance that permeates the universe, called "ether"). Ethernet (X Corporation) products started to reach the marketplace in the early 1980s.

The charter of Project 802 encompassed the physical and link lay OSIRM. The range of transmission speed was set at 1 to 20 Mb/s. Because are based on a shared medium, unlike point-to-point links intrinsic in the O the link layer had to be split into two sublayers, the MAC and the LLC. D IEEE 802 MAC standards represent different protocols used for sharing th while the LLC sublayer provides a protocol–media-independent interface layers. The MAC procedure is part of the protocol that governs acce transmission medium, independent of the physical characteristics of the but it does, however, take into account the topological aspects of the su in order to provide services to the LLC sublayer. Token-based LANs collisions inherent in Ethernet by requiring each node to defer transm it receives a "clear-to-send" message, called a token. Token technolo adopted by large vendors, such as IBM, while the Ethernet technolo brought to the market by DEC and many smaller vendors [14.4].

Contention access uses CSMA/CD techniques and is represe Ethernet bus architecture (IEEE 802.3). Noncontention methods us ically in a ring configuration (IEEE 802.5). Token bus is described i

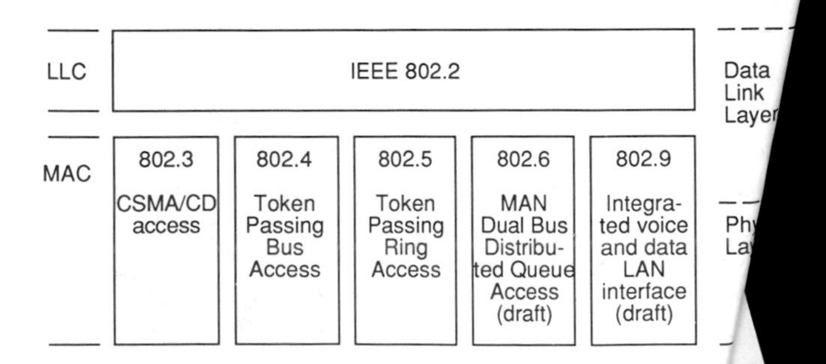

Figure 14.3 LAN standards.

14.1.4.1 Logical Link Control

IEEE Standard 802.2-1985 (now technically superseded by ISO 8802-2:1989) provides a description of the peer-to-peer protocol procedures that are defined for the transfer of information and control between any pair of DLL SAPs on a LAN. As described in Chapter 13, each layer provides a specific service to adjacent layers. A number of different SAPs are defined in the IEEE 802 protocol (see Figure 14.4; in particular, note the multiple L-SAPs). The SAP can be considered to be an address within a station that identifies a particular application or service. Service primitives are used to exchange information between the different layers. This exchange of information is the service provided by these sublayers.

The LLC procedure is that part of the protocol that governs the assembling of DLL frames and their exchange between data stations, independent of how the transmission medium is shared. The LLC sublayer supports medium-independent data link functions and employs the MAC sublayer service to provide services to the network layer. The protocol is important because it provides a uniform interface between higher layers and the MAC protocols of 802.3, 802.4, 802.5, and 802.6. Effectively, it provides transparency to the network layer with respect to the underlying LAN media and, thus, transparency to application software (E-mail, word processing, *et cetera*).

LLC connects two peer L-SAPs in the two end systems (end stations). LLC 1 provides an unacknowledged connectionless service. It allows the sending and receiving of frames between SAPs without the prior establishment of a connection between the two communicating endpoints; no call setup or call termination phase is required. No guarantee of delivery or sequentiality is provided by LLC 1, but this can be accomplished at the transport layer if desired or needed. Unacknow-

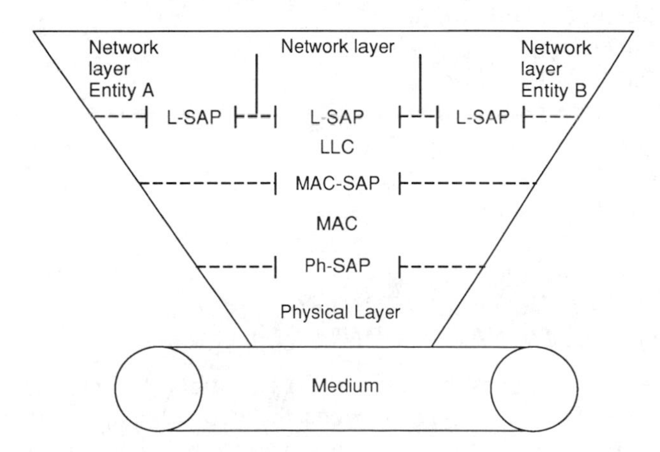

Figure 14.4 Lower layer LAN SAPs and protocols.

ledged connectionless service may be point-to-point, multicast, or broadcast. LLC 2 is a connection-oriented service, similar to virtual circuit service. LLC 2 provides a point-to-point connection between SAPs; it allows flow control and error recovery. A call setup procedure is required to establish a logical connection between the two communicating endpoints prior to exchanging frames containing data. Sequential delivery of frames is guaranteed: Data frames contain sequence numbers and the frames must be acknowledged by the receiver [14.5]. Class I LLC service only supports LLC 1; Class II LLC service supports LLC 1 and LLC 2.

LLC defines two sets of primitives (refer to Chapter 1 for a refresher on primitives). Primitives between the LLC sublayer and the higher layers are prefixed by "L"; primitives between the LLC and the MAC sublayers are prefixed by "MA."

Unacknowledged connectionless LLC 1 service supports only the higher layer–to–LLC sublayer L-SAP primitives necessary for one instance of data transmission, without prior connection establishment. Hence, the service primitives are simply *LDATA.request* and *LDATA.indication*. Connection-oriented LLC 2 requires several types of higher layer–to–LLC sublayer LSAP primitives to undertake such activities as logical link establishment and deestablishment, flow control, *et cetera*. The primitives are:

- Link connect activities:
  LCONNECT.request
  LCONNECT.indication
  LCONNECT.confirm
- Data transfer activities:
  LDATACONNECT.request
  LDATACONNECT.indication
  LDATACONNECT.confirm
- Link disconnect activities:
  LDISCONNECT.request
  LDISCONNECT.indication
  LDISCONNECT.confirm
- Link reset activities:
  LRESET.request
  LRESET.indication
  LRESET.confirm
- Link flow control activities:
  LCONNECTIONFLOWCONTROL.request
  LCONNECTIONFLOWCONTROL.indication

The MAC-SAP primitives between the LLC sublayer and the MAC sublayer provide for the transfer of data between the two sublayers; the interface allows passage of frames between the two sublayers. The primitives are *MADATA.request*, *MADATA.indication*, and *MADATA.confirm*. The interface and primitives are independent of the MAC specifics (whether 802.3, 802.4, or 802.5, described below).

Figure 14.5 depicts the format of an LLC frame, which, in turn, is enveloped (as described in Chapter 1) inside the information field of the lower MAC sublayer frame, as shown in the lower portion of Figure 14.5. The LLC standard includes both the destination and source address fields within the header, as shown in Figure 14.5. These addresses are associated with the SAP of the LLC sublayer, and their values are defined to be unique only within a given station address (so that after a frame has been routed to the target station, it undergoes additional internal routing to reach the proper network layer entity). A network layer entity, which is a user of LLC services, can be reached through an address formed by the concatenation of the LLC SAP (i.e., L-SAP) with a given (physical) station address, as shown in Figure 14.6. The station address identifies the SAP associated with

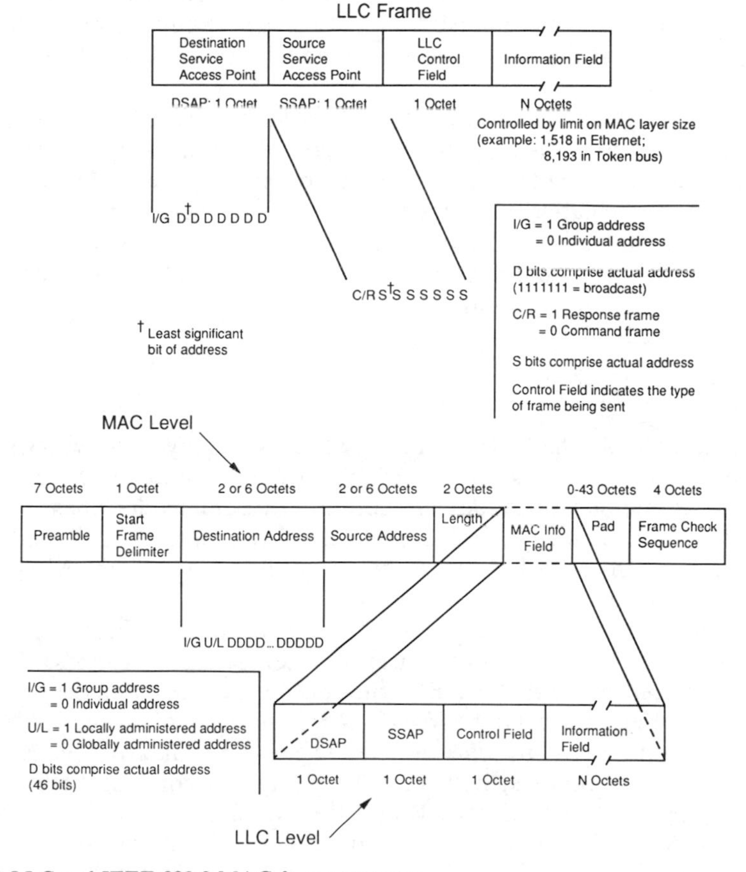

Figure 14.5 LLC and IEEE 802.3 MAC frame structure.

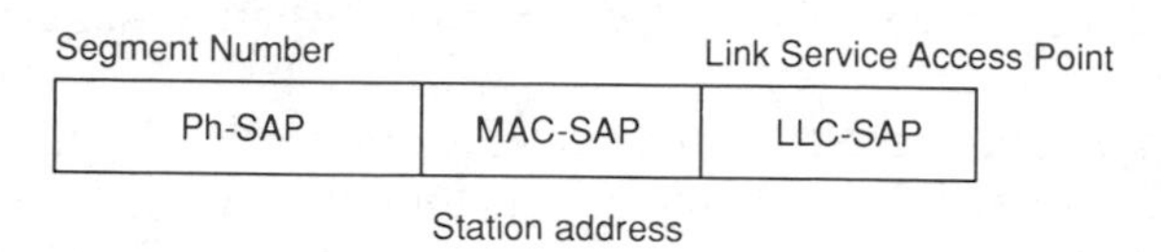

Figure 14.6 LAN address.

the MAC entity within each system. This address has a one-to-one mapping with the physical address or physical SAP. (The physical address is not explicitly carried in each frame, but is implied by the physical connection of the station to the network [14.6].)

The maximum size of the LLC data field is determined by the maximum frame size imposed by the MAC layer because the LLC PDU is encapsulated inside the MAC PDU. For example, the 802.4 MAC has a maximum size of 8193 octets; an 802.3 Ethernet MAC has a maximum of 1518 (excluding the preamble and the start delimiter); FDDI has a limit of around 4500; and the IEEE 802.6 MAN has a limit of 9188 [14.7].

A minimum frame size is also required for correct MAC protocol operation, and is specified by the particular implementation of the standard. If necessary, the MAC data field is extended by extra pad bits in units of octets after the LLC data field but prior to calculating and appending the frame check sequence (or "sum") (FCS). The size of the pad, if any, is determined by the size of the data field supplied by the LLC and the minimum frame size and address size parameters of the particular MAC implementation.

14.1.4.2 MAC: CSMA/CD

IEEE Standard 802.3-1985 (now technically superseded by ISO 8802-3:1989) provides a medium access method by which two or more stations share a common bus transmission medium. The standard applies to several media types and provides the necessary specifications for a baseband LAN operating at 1, 5, and 10 Mb/s. The specification also describes the service primitives between the MAC sublayer and the physical layer.

Only very minor MAC-level differences exist between IEEE 802.3 and Ethernet systems (consisting of a slightly different interpretation of fields). Most commercial systems on the market are Ethernet-based. People commonly refer to the technology with the nomenclature *IEEE802.3/Ethernet*. This is the LAN technology most commonly found in office LANs. The 802.3 standard defines a "logical bus"; many implementations, however, have physical configurations that topologically may not be a bus. Ethernet, and the initial 802.3 standard based on it, operate at 10 Mb/s over a coaxial-cable bus, with all stations connected to one transmission line. The IEEE 802.3 group has subsequently built on its original effort by developing new standards, particularly with reference to twisted-pair LANs. The four variants of the 802.3 LANs are as follows [14.8]:

- *10BASE-T*, a baseband LAN operating at 10 Mb/s over twisted pair, allowing a distance between the station and hub of 100 m. Topologically, this LAN has a star rather than a bus configuration (as of 1990 this draft became a full standard [14.9]).
- *1BASE5*, a baseband LAN operating at 1 Mb/s over twisted pair, allowing a distance between station and hub of 250 m. Topologically, this LAN has a star rather than a bus configuration.
- *10BASE5*, a baseband LAN operating at 10 Mb/s over coaxial cable, with a maximum bus length of 500 m. This LAN is the closest to the original Ethernet. The standard was approved in 1983. In the original 802.3 standard, a station (a computer, printer, file server, *et cetera*) was connected with the coaxial bus using heavy-gauge shielded and twisted copper wires. Shielding and pairing is employed to minimize electromagnetic and radio-frequency interference.
- *10BASE2*, a baseband LAN operating at 10 Mb/s over thin coaxial cable (type RG-58) with a maximum bus length of 185 m. Regular coaxial cable is approximately 1 cm thick; this thickness makes it difficult to bend the cable in tight office environments. Thin cable performs almost as well as regular coaxial, except that signals attenuate more rapidly.

The original 802.3 coaxial LAN may in the future become a backbone for tying a number of smaller twisted-pair networks.

10BASE-T

The primary objective of the 10BASE-T IEEE specification for the medium access unit is to allow Ethernet LAN users to utilize the existing telephone twisted-pair cable (typically, 24 American wire gauge solid copper wire with unshielded PVC insulation, containing four or more pairs) as the physical medium for 10 Mb/s transmission over segments up to 100 m. The 10BASE-T Ethernet requires two pairs of wire to each service location; because most telephones operate on a single pair, and most modern buildings in the United States are cabled with four-pair cable, usually spare pairs are available in each work area [14.9].

The 10BASE-T medium access unit must be able to coexist with other signals that may be present on the other pairs in the telephone cable (including 1BASE5 LANs, token-passing LANs, analog voice, and ISDN). Another objective in developing this latest standard was backward compatibility with existing Ethernet hardware at the attachment unit interface (the standard 15-pin D connector). Users can replace the coaxial-based medium access units with the 10BASE-T medium access units, hence, protecting the investment in Ethernet controllers. In addition to specifying the requirements for the transmitter, the receiver, and the attachment unit interface electronics contained within a 10BASE-T medium access unit, the specification also places performance constraints on the link segment between the access units (these specifications are critical to the proper functioning of the Ethernet LANs built on this type of cables) [14.9].

The 10BASE-T twisted-pair Ethernet standard received final approval in September 1990. Twisted-pair Ethernet products were popular even before the standards were finalized because of their cheaper, and often already-installed, wire [14.10].

14.1.4.3 MAC: Token-Passing Bus Access Method

IEEE Standard 802.4-1985 (now technically superseded by ISO 8802-4:1989) deals with all elements of the token-passing bus access method and its associated physical signaling and media technologies. The goal of the standard is to achieve compatible interconnection of stations in a LAN. The access method coordinates the use of the shared medium among the attached stations. It specifies the electrical and physical characteristics of the transmission medium; the electrical signaling used; the frame formats of the transmitted data; the actions of a station on receipt of a data frame; and the services provided at the conceptual interface between the MAC sublayer and the LLC sublayer above it. The specification also describes the service primitives between the MAC layer and the physical layer.

This standard corresponds to the type of LAN used most frequently in process and manufacturing control applications, and adopted by the MAP/TOP consortium.

14.1.4.4 MAC: Token-Passing Ring Access Method

IEEE Standard 802.5-1985 (ISO 8802-5) specifies the formats and protocols used by the token-passing ring medium at the control sublayer and physical layer. It also specifies the means of attachment to the token-passing ring access method. The protocol defines the frame format, including delimiters, addressing, and FCS, and includes timers, frame counts, and priority stacks. It also defines the MAC protocol and provides finite-state machines and state tables, supplemented with descriptions of the algorithms. It identifies the services provided by the MAC sublayer to the LLC sublayer and the services provided by the physical layer to this MAC sublayer. These services are defined in terms of service primitives and associated parameters. Also defined are the physical layer functions of symbol encoding and decoding, symbol timing and latency buffering, and the 1 Mb/s and 4 Mb/s twisted-pair attachments of the station to the medium. The specification also describes the service primitives between the MAC layer and the physical layer.

This standard is representative of IBM's token-passing ring LAN. The 802.5 standard is for the 4 Mb/s rate and did not yet include, as of 1990, the 16 Mb/s rate supported by IBM's token ring.

14.1.4.5 MAC Layer Summary

Figure 14.7 compares the format of the three standardized MAC layers. Figure 14.8 depicts the physical layout of a typical LAN.

14.1.4.6 OSI Upper Layers

Table 14.2 depicts a typical LAN protocol suite when OSI protocols are used at the upper layers (note, however, that many commercially available LANs do not yet employ OSI-based communication at the upper layers).

14.1.4.7 802.10 LAN Security Standard

Security has emerged as a critical LAN issue, as discussed in more detail later in the chapter. A new IEEE committee (IEEE 802.10) is now devoted to specifying DLL security standards.The Standard for Interoperable LAN Security (SILS) will resolve interoperability problems independent of the encryption algorithms employed by devices on the LAN. This standard will comply with the OSI Security Architecture (ISO 7498-2), discussed briefly in Chapter 13 and covered in more detail in Chapter 17. The 802.10 standard will include provisions for authentication, access control, data integrity, and confidentiality. Authentication will prevent stations from reading packets destined for a different station; access control limits the use of resources (such as files, servers, gateways, *et cetera*) by unauthorized users; data integrity guarantees that the data are not modified before they reach

Ethernet	Preamble 56	SD 8		Destination Address 48	Source Address 48	PT 16	MAC Data Unit (≤1500)	FCS 32		
IEEE 802.3 MAC	Preamble 56	SD 8		Destination Address 48	Source Address 48	Length 16	MAC Data Unit (≤1500)	FCS 32		
IEEE 802.4 MAC	Preamble (≥1 octets)	SD 8	FC 8	Destination Address 48	Source Address 48		MAC Data Unit (≤8174)	FCS 32	ED 8	
IEEE 802.5 MAC		SD 8	AC 8	Destination Address 48	Source Address 48		MAC Data Unit	FCS 32	ED 8	FS 8

SD = Start Delimiter (=10101011 in 802.3)
ED = End Delimiter
FC = Frame Control
AC = Access Control
FS = Frame Status
PT = Protocol Type

Field lengths shown in bits
except for Information Field shown in octets

Figure 14.7 Comparison of MAC Layers

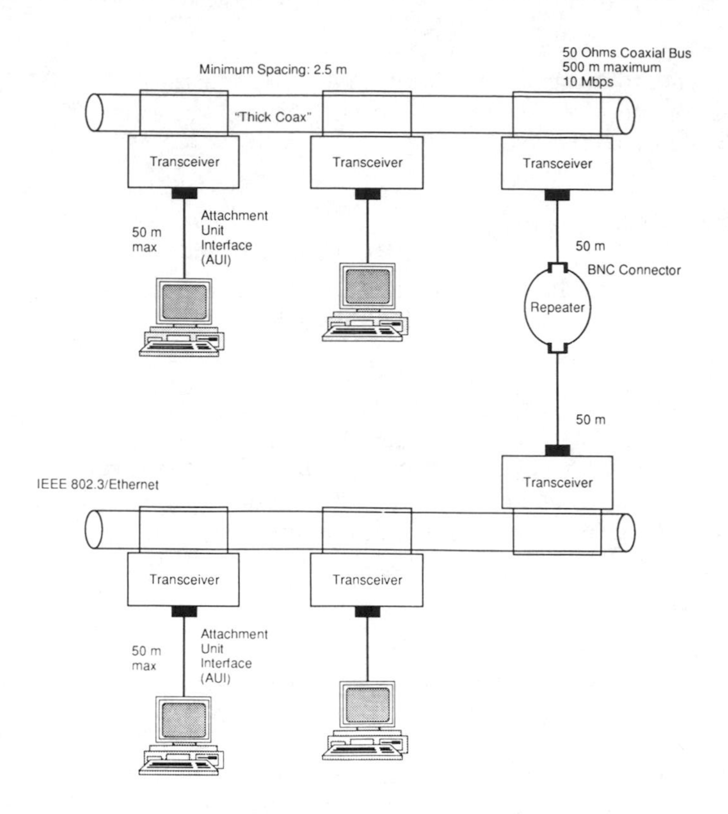

Figure 14.8 LANs connected by a repeater.

	LAN environment	WAN environment
Layer 7	FTAM, TP, VP, MHS, etc	
Layer 6	ISO 8822 & 8823 with ISO 8824 & 8825 (ASN.1/BER)	
Layer 5	ISO 8326 & 8327	
Layer 4	ISO 8072 & 8073 Class 0, 1, 2, 3, or 4 as needed	
Layer 3	Connectionless Network Protocol, ISO 8473	Connection-oriented X.25 PLP
Layer 2	LLC MAC: CSMA/CD, Token ring, Token bus	LAP-B
Layer 1	PMD: IEEE 802.3, .4, .5	Telco channels

Table 14.2 Typical OSI-Based Communication

the intended user; and confidentiality uses encryption to mask the information to users without the appropriate key (some of these issues are treated later in this chapter from a practical perspective).

The standard consists of four parts: (1) "the model," (2) "key management," (3) "system management," and (4) "secure data exchange." In mid-1990, the first three were in draft form and the fourth had been sent for balloting [14.11].

14.1.5 Internetworking LANs Using OSI Methods

A number of interconnection scenarios exist (see also Figure 14.9):

- Two similar collocated LANs internetworked directly via a local relay device;
- Two dissimilar collocated LANs internetworked directly via a local relay device;
- Two similar remotely located LANs internetworked via a backbone similar or dissimilar LAN-MAN;
- Two dissimilar remotely located LANs internetworked via a backbone similar or dissimilar LAN-MAN;

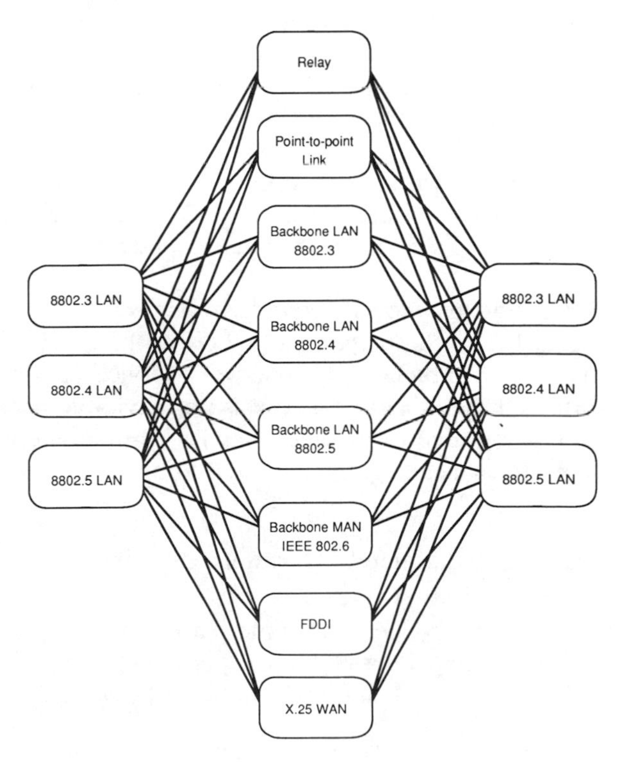

Figure 14.9 LAN interconnection needs.

- Two similar remotely located LANs internetworked via a point-to-point link or WAN; and
- Two dissimilar remotely located LANs internetworked via a point-to-point link or WAN.

Two "open" LAN internetworking approaches are available*: one through ISO protocols means, as described in the previous chapter; the other via TCP/IP *de facto* standard means. The TCP/IP method is the most commercially prevalent method.

As discussed above, in a LAN environment, the LLC sublayer provides a connection-oriented mode and a connectionless mode service. The connection-oriented mode data link procedure (LLC Type 2) provides sequencing, error recovery, and other link layer functions associated with a traditional protocol, such as HDLC. However, because this capability is restricted at the DLL, it does not provide end-to-end integrity. The transport layer protocol Class 4 can provide this end-to-end service. The connectionless procedure (LLC Type 1) assumes that the higher layer is responsible for resequencing and error recovery, which means that unnumbered information frames can be sent at any time by any station, without first establishing a connection. LLC 1 is preferred in the LAN environment because of its simplicity and efficiency.

Two strategies are being advanced by proponents for interconnecting LANs using ISO methods, as depicted in Figure 14.10. The first involves X.25 PLP; the second involves ISO CLNP 8473.

X.25 PLP can operate with either LLC 1 or LLC 2, as specified in ISO 8881. High data rates require fast routers, implying the need for simpler protocols; networks with data rates of several gigabits per second are already being planned [13.12].

In the second strategy, the subnetwork service required by the ISO CLNP is provided by LLC 1, and the mapping of CLNP to LLC 1 is relatively simple. Hence, LLC 1 is used to provide the subnetwork service required by ISO 8473. (A connection-oriented mode LLC 2 service is not advisable because in this case a SNDCP protocol is required to provide a service mapping between ISO CLNP and LLC 2, with performance degradation implications.)

14.1.6 LAN Interconnection Methods: Hardware Perspective

From a practical perspective, LANs are internetworked using repeaters, MAC-level bridges, and routers, as described in Chapter 13 for general internetworking application. The functions of this hardware are as follows:

---

*Vendor proprietary approaches, application layer gateways, and other closed solutions are not treated here.

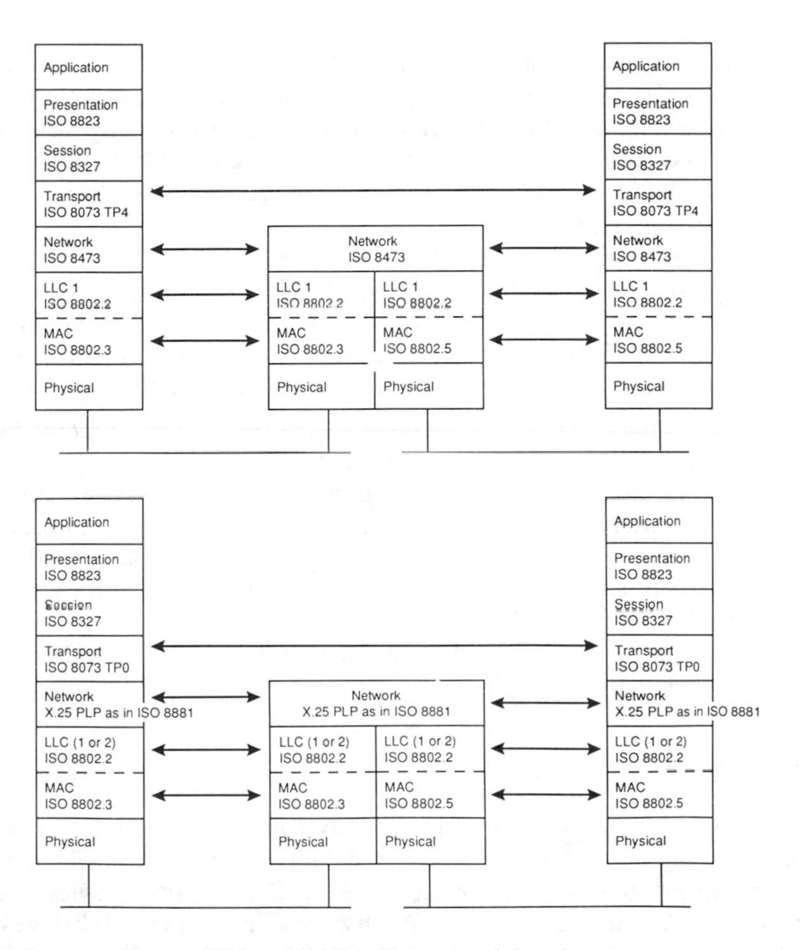

Figure 14.10 Interworking two OSI-based LANs with a network-based router.

- *Repeaters*: Relay hardware at the physical layer. Although they could, in theory, connect two types of media, in practice, they copy electrical signals from one medium to a similar medium. Typically, sites use repeaters to connect a physical Ethernet cable on each floor of a building to a backbone cable. The disadvantage of repeaters is that they also amplify the noise. Figure 14.8 depicted an example of LAN interconnection using a repeater.
- *Bridges*: Relay hardware at the MAC sublayer of the DLL. The major protocol difference between IEEE 802-based LANs is in their MAC sublayers. In the LAN context, a bridge is a functional unit that interconnects two subnetworks that use a single LLC procedure but may use different MAC procedures; bridges can map some of the MAC protocols to one another. They are technology dependent in the sense that, for example, an Ethernet

bridge connects two physical Ethernet cables and forwards from one cable to another nonlocal packet (bridges store and then forward complete packets, while repeaters forward at the bit level, repeating and amplifying the electrical level). The technology dependence can also be described in the sense that we generally cannot decouple, in hardware products, the physical layer from the MAC sublayer.

- *Routers* (also known as "internet"† router): Relay hardware at the network layer. A router is a functional unit that interconnects a LAN to a WAN or MAN, or perhaps two LANs that have different LLC services (such as LLC 1 and LLC 2, discussed above). (In this case, they are hybrid systems.)
- *Gateways*: Relay hardware at the transport layer or above.

The interconnection of two different LANs, say, an 802.5 network with an 802.3 network, is not a trivial task. This difficulty arises from the fact that the functionality offered by these networks is not the same, implying that rigorous adherence to a MAC sublayer routing and relaying is impossible. Instead, in practical situations, we must move up to the LLC or network layer (and sometimes even higher) to obtain the desired functionality end-to-end. (See Reference [14.13] for some specific examples of differences.)

The next few sections provide additional details on these LAN interconnection methods.

14.1.6.1 Bridges

Bridges provide a means to extend the LAN environment in physical extent, number of stations, performance, and reliability; they route packets across extended networks comprised of multiple LANs. Bridges also provide a means to interconnect MAC-dissimilar LANs. As discussed in Chapter 13, bridges apply to the DLL; in a LAN environment, bridges typically apply to the MAC layer and, to be precise, are called *MAC-layer bridges*. In the following, the term *bridge* is utilized to mean MAC-layer bridge. By definition, a bridge is independent of the higher level protocols (see Figure 14.11). Algorithms have been developed that allow bridges to build their routing tables automatically by observing the source addresses of packets passing the bridge [14.14, 14.15].

MAC-level bridges have become increasingly popular among network users, planners, and managers. Bridging of LANs is desirable for a variety of reasons, including improved performance (because each local network segment contributes

---

†A lower case "internet" refers to a generic network of interconnected subnetworks. "Internet" capitalized refers to the specific network structure supported by the Internet Advisory Board, discussed in Chapter 2 (Internet is a collection of more than 2000 packet switched networks located principally in the United States, but also includes systems in other parts of the world; interconnection is via the TCP/IP).

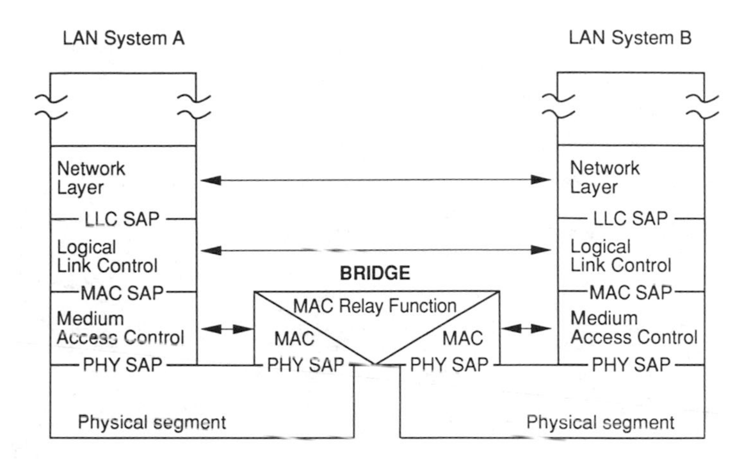

Figure 14.11 A LAN MAC-level bridge.

its own bandwidth), signal quality (compared to attaching all stations at a location to the same local segment), and availability (because the breakdown of one segment does not cause all stations on all segments to lose service) [14.6].

Bridges perform three basic functions: (1) frame forwarding, (2) learning of station addresses, and (3) resolving possible loops in the topology. Bridges provide frame-relay functions within MAC; each network retains its independent MAC, such as contention or token passing. By contrast, a router terminates both LLC and network layer protocols. Unlike network layer routers, bridges possess no capabilities for flow control, either at the LLC sublayer or the network layer [14.6]. Hence, with finite buffer capacity and no guarantee of time-bound access to adjacent segments, bridges are subject to congestion. When this occurs, the bridge discards frames (the responsibility for integrity is relegated to the end station's transport layer protocol).

LAN bridges can be classified as *pass-through* and *converting*. Pass-through LAN bridges connect similar LANs and provide functions such as repeating, packet filtering, and packet queueing; frames are passed without any conversion. Converting bridges interconnect dissimilar LANs and must therefore convert the header and trailer of incoming frames.

LAN bridges can be further classified as *local* or *remote* (also known as "half") bridges. Local bridges internetwork collocated LANs. Remote bridges internetwork LANs through a subnetwork. Remote bridges between two similar LANs encapsulate the MAC frame into a frame specific to the intermediate subnetwork; remote bridges between two dissimilar LANs perform a frame conversion and a frame encapsulation function (the encapsulation is specific to the intermediate subnetwork). See Figure 14.12. Bridges do not require a network-wide addressing plan.

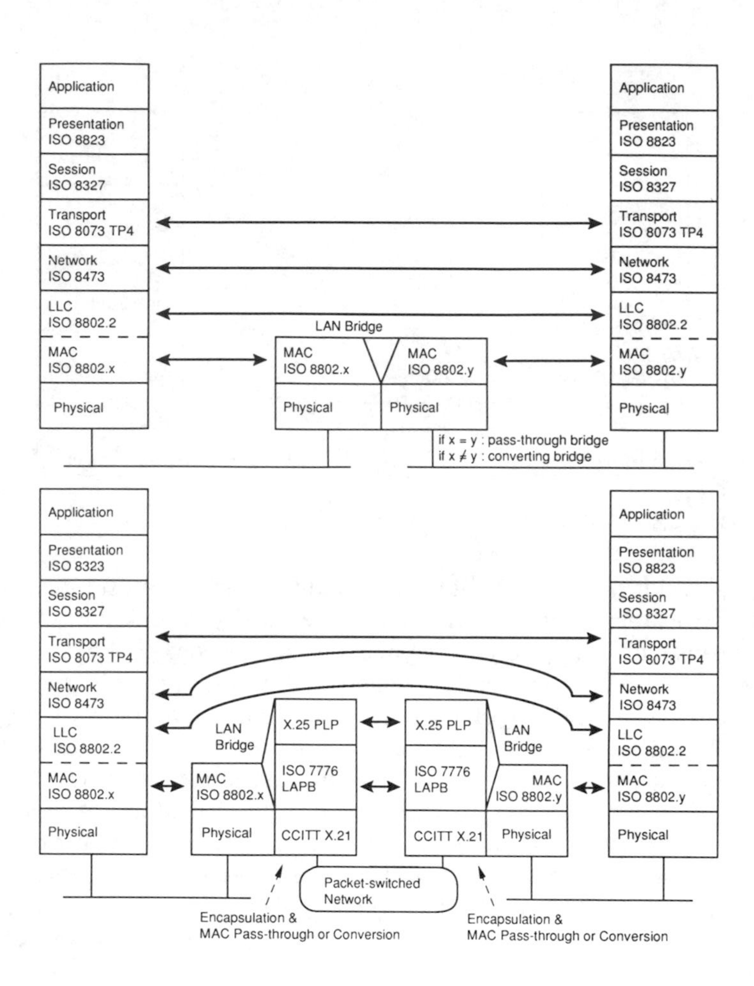

Figure 14.12 LAN bridges (OSI example).

Demand for bridges is leading to a multiplicity of offerings by an increasing number of vendors; products include both local and remote applications [14.16].

Bridges can further be classified as [14.17]:

- *Transparent Routing,‡ Full Broadcast*: Bridges are transparent to the stations in the sense that installing a new bridge does not require any modifications to stations or existing bridges. Stations on one LAN communicate with stations on another LAN as if they were on a single LAN. Stations provide the

‡"Routing," as a descriptor of the technique employed by a MAC bridge to forward frames to an internet LAN, should not be confused with a router device, which, as explained, operates at a higher OSIRM layer.

destination with a station address, not the address of the intermediate bridges. "Full broadcast" refers to the fact that all traffic of each attached LAN is sent to all other LANs (no filtering or routing is performed). These are the most basic bridges, providing a repeater function coupled with packet queueing.

- *Transparent Static Routing*: Transparent operation, as described above. "Static routing" refers to the mechanism of how the bridges deal with all traffic of an attached LAN and forward it to the destination LAN; the data base is updated manually by the LAN administrator. The destination LAN is determined by a routing data base maintained in the bridges. If the destination LAN is not the same as the originating LAN, the frame is forwarded; if the destination is the same, the frame is discarded (if no entry is found, every LAN except the originating LAN is flooded with a copy of the frame).
- *Transparent Adaptive Routing*: Transparent operation, as described above. "Adaptive routing" refers to the fact that the bridge data base is automatically updated from the information the bridge "learns" from passing frames and from other bridges. Adaptive routing bridges automatically initialize and configure themselves. Adaptive routing can be isolated, distributed, or centralized. Isolated adaptive routing utilizes information locally available to the bridge. Distributed adaptive routing utilizes local as well as shared information from other bridges. Centralized adaptive routing utilizes information that resides at a central location (the central location, in turn, collects information from all internetworked bridges).
- *Source Routing*: In this environment, bridges are not transparent to the stations. The stations must supply the routing information required to forward the frames to the intermediate bridges. The routing information consists of the sequence of bridges and LANs that the source deems necessary to traverse (both bridges and LANs have unique identifiers, in this context). The source routing bridges only forward those frames that contain its identifier and the identifier of one of the LANs connected to the given bridge. Source routing bridges are simpler that transparent bridges because they do not need to maintain the routing data base (in effect, each LAN station maintains the data base).

Transparent routing is advocated by companies such as DEC; source routing is advocated by IBM.

In a simple bridge, the address table is static, based on *a priori* knowledge of station addresses. Filtering performed for the purpose of access control also relies on static tables. These bridges have the advantage of simplicity and speed. The major disadvantage is lack of flexibility. For example, simply moving a station from one network to another requires changing the configuration of all bridges in the system [14.18].

A learning bridge modifies its address tables dynamically as each packet is transacted. Station locations relative to the location of the bridge are also learned so that the traffic is not forwarded onto a link unless it is required.

By not processing LLC or network layer protocols, a bridge has the following advantages compared to a router:

1. Performance is higher because fewer layers must be processed [14.7];
2. No incompatibilities between LLC services must be mediated (particularly between connection-oriented and connectionless LLC services); and
3. No incompatibilities between network layer services (as discussed in Chapter 13) must be mediated.

14.1.6.2 IEEE 802.1 Bridge Standards

The IEEE 802 Committee has been developing standards to ensure that LANs extended by bridges exhibit consistent characteristics. The internetworked LANs should retain as many properties intrinsic to a LAN as possible, including properties such as low delay; low occurrence of frame loss, misordering, and duplication; and high throughput [14.19, 14.20].

The IEEE 802.1 Committee has evaluated several bridging schemes, including transparent bridging and source routing. The committee selected transparent bridging as the standard for interconnecting 802 LANs [14.19, 14.21]. The IEEE 802.1 Part D MAC bridge specification ("Transparent Spanning Tree Scheme") describes transparent local bridges that interconnect 802 LANs. IBM had been lobbying in the late 1980s for source routing. In 1990, IBM submitted a proposal to annex source routing to the transparent bridging standard. The proposed source routing transparent (SRT) bridges would offer source routing as an option, maintaining interoperability with transparent bridges [14.11] (previously, source routing was being studied under the IEEE 802.5 banner). To support both types of routing, the SRT proposal adds a routing information indicator bit to the source address; if the bit is set to 1, the bridge looks for a routing information field; otherwise, the bridge routes packets using transparent routing.

The transparent bridging scheme that the IEEE committee worked on throughout the late 1980s became a standard in May 1990 (actual publication was to follow by the end of 1990). ISO may adopt the standard soon after IEEE approval.

14.1.6.3 Transparent Spanning Tree Routing Bridges

In a transparent bridge, no participation by an end station is necessary in order for the station to make use of the services provided by a bridge. A transparent bridge automatically initializes and configures itself and runs with no intervention from the network manager. Transparent bridges are backward-compatible with existing IEEE 802 LANs. Figure 14.13 depicts an example of a logical (and physical) spanning tree architecture connecting several LANs using bridges.

A forwarding data base is maintained in the bridge by observing the source address of frames received on each side of its ports. Frames are forwarded or discarded based on a comparison of the frames' destination address to the information contained in the forwarding data base. The bridge's forwarding data base contains a list of group and individual station addresses and information that relates these addresses to one or more of the bridge's ports. If the destination address is not found in the forwarding data base, the frame is transmitted on all bridge ports, except the one on which it was received (this action is called "flooding"). A bridge typically examines both the source and the destination addresses of packets. The source address is used to maintain a list of active users known to be on either side of the bridge. The destination address is examined to determine if the frame must be forwarded [14.13].

The forwarding of a frame by a bridge involves receiving or copying a frame from one local network segment, processing it internally, and transmitting it on an adjacent segment (the source and destination addresses are not modified by the bridge). When two stations on a LAN exchange frames, the original DLL FCS appended by the transmitting station is received and checked by the destination

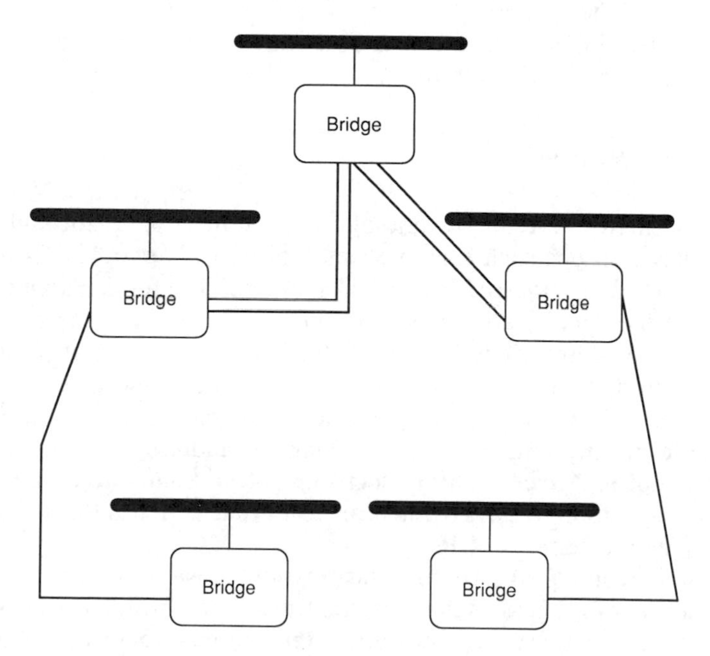

Figure 14.13 Bridges connected with a spanning tree backbone (note redundant links in some cases).

station. When a bridge copies a frame, having recognized a forwarding address, it ascertains that the frame is error-free by examining the FCS. Some implementations perform software check sum computations only on frames traveling between different LANs. Because transparent bridges make multiple LANs appear as one to higher layer protocols, they have the responsibility of passing the original FCS through to the destination station whenever possible. When transparent bridges interconnect similar LANs (for example, two Ethernets), they do not modify the frames they forward and, consequently, should always preserve the original FCS. When interconnecting different LANs (for example, 802.3 to 802.5), transparent bridges must translate between different frame formats. As a result, the FCS cannot be preserved; this introduces some risk of undetected user data corruption [14.22].

The key features of transparent spanning tree routing are [14.23]:

- Stations on bridged LANs communicate with remote stations as if they were local. As a result, no additional protocols or route maintenance are required by the end stations. Frames forwarded by bridges maintain their original FCS, thus ensuring data integrity.
- Although the physical topology may be arbitrary, a spanning tree algorithm is operative at all times to prune the physical topology into a tree.
- Bridges need not be configured prior to installation.
- Group-addressed and multicast frames are transmitted throughout the network along the spanning tree.
- Bridges learn and store information on the location of stations based on the source address in the data frames.

14.1.6.4 Source Routing

IBM has specified a bridging technology for routing frames through a multiple-ring local network (one such example was shown in Chapter 11). This technology is now also included in the IEEE 802.1 D standard. The term *source routing* stems from the specification by a frame's transmitter, or source, of the route the frame is to follow; the source identifies the route. The source associates the route with the destination station and includes the route in the frame header. In this environment, freed from the demands of address-table maintenance and lookup, bridges perform only simple string matching for address recognition; this makes them capable of high-speed routing decisions [14.6]. Source routing is acquired by the originating station by employing a protocol that searches through the network for the destination station [14.18].

In source routing, the routing information consists of a series of identifiers of the local network segments that a frame transmitted from one station to another station must utilize. Local networks allow the source to determine the route by a number of different means. One typical method is *dynamic discovery*. The bridge makes a routing decision based on the contents of the MAC frame header. This header consists of a routing information field, which immediately follows the source address field in the frame. The routing information field describes the path across one or more bridges to the LAN containing the destination station [14.24]. The frame format described in MAC standards includes both a destination and source address field. The source address must always be an individual address; the destination address can be either an individual or a group address.

Communication between two stations in a bridged network requires the existence of a route between the individual local segments on which the two stations reside and also a means for a frame to follow that route. A route is comprised of an ordered sequence of local network segments on which the frame must be transmitted, as shown in Figure 14.14. Bridges are responsible for assuring that a frame follows an appropriate route to its destination; they must, therefore, decide which frames to forward.

In a bridged network, individual station addresses must still be unique across the internetworked system, not just in the local segment. When routers are used, only the network layer address must be unique; local network station addresses need only to be unique within the local segment. When stations communicate through a network layer router, they use the router's station address in the destination address field [14.6].

Source routing must be accomplished while satisfying the following requirements [14.24]:

- The ability to maintain the original frame's FCS across ring boundaries. End-to-end FCS protection is desirable across bridges connecting two rings to preserve data integrity across the bridge.
- The ability to transmit forwarded frames using the original source address of the sender. A MAC layer bridge should not be visible in the MAC address fields; therefore, a bridge must transmit a frame with the source address of the originating station, not its own.
- The ability to transmit frames with a higher priority than those of lower priority.
- The ability to set the address recognized indicator and the frame copied indicator bits if it has been determined that the frame should be transported across the bridge to another ring.

The MAC layer for source routing is depicted in Figure 14.15. Key features of source routing are [14.23]:

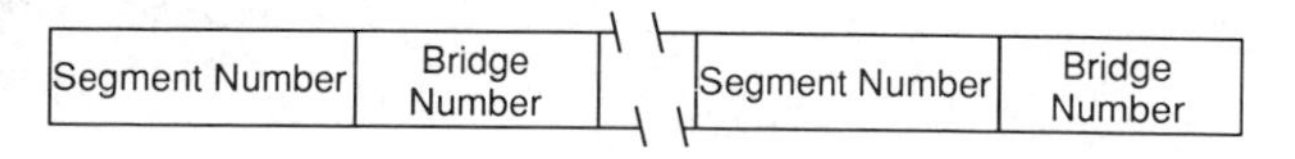

Figure 14.14 Local route number.

- Frames are routed through the multiring network by having the originating station specify the route. The routing information is included in each frame.
- The presence of the routing information is indicated by the setting of the "group" or "multicast" bit in the source address field. The routing information consists of a routing control field and a list of route descriptors, as shown in Figure 14.15. Each route descriptor comprises a ring number and a bridge number. The ring number is 12 bits long and must be unique throughout the bridged LAN; the bridge number is 4 bits long and must be unique among parallel bridges (i.e., a set of bridges connecting the same pair of rings). The routing information field allows up to 14 hops (bridges). An option allows extension to 28 hops.

14.1.6.5 Intelligent Multiport Bridges

A new generation of bridging devices—intelligent multiport bridges—was emerging in 1990. These Ethernet bridges offer comprehensive network management capabilities and are positioned as an alternative to using cascaded two-port bridges for connecting several LAN segments. The per-port forwarding-filtering rate ranges from 10,000 to 25,000 packets per second, with five to ten active ports. The high-performance multiport bridges fall somewhere between stand-alone intelligent de-

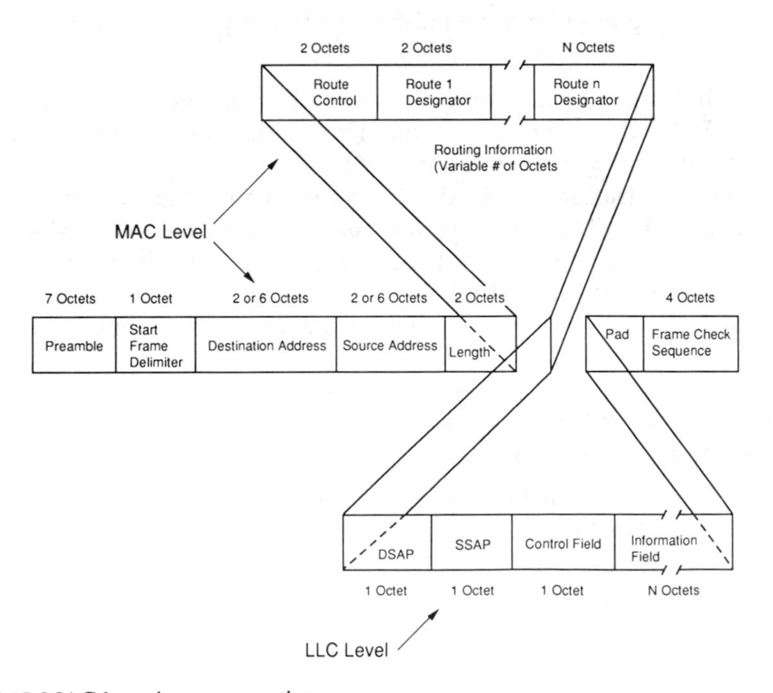

Figure 14.15 MAC layer in source routing.

vices and closed wiring hubs used to link multiple LANs. They cost between $10,000 and $25,000 [14.25].

14.1.6.6 Routers

A router is also called an *intermediate system*, as discussed in Chapter 13. A router differs from a bridge in the sense that bridge operation is transparent to the communication end stations, while a router is directly addressed by the end systems requiring its service. A router offers more sophisticated services than a bridge: It can select one out of a group of paths to forward a packet. This selection is based on a variety of parameters, including transit delay and congestion on primary routes.

Routers operate at the network layer; they can be used to connect among LANS, MANs, and WANs. Use of routers is based on a sublayering of the network layer; the internetworking protocol applicable to the upper sublayer must be the same across the end systems and the intermediate (router) systems. Routers employ the hierarchical address of the network layer described in Chapter 13. Routers have unique addresses and the stations must adhere to a uniform addressing plan. Routers attached to a LAN only operate on frames addressed to that router. They, in turn, strip the MAC and LLC headers and wrap the PDU of the originating station into the DLL format appropriate for the outgoing network. (Segmentation of the PDU may also be required if the frame size of the outgoing network exceeds the size of the originating LAN's PDU.)

In the OSIRM, the function of the DLL is to relay frames (packets), with error detection, across one hop. Networks typically consist of multiple links. The network layer is designed to allow multiple-hop operation, route identification, route selection, and packet lifetime control [14.26]. In a LAN environment, the issue of one hop or multiple hops is a matter of perspective. A DLL header designed for single-hop operation carries only one pair of addresses; a network layer header carries an additional pair of addresses, possibly explicit routing information, and antiloop fields [14.26]. Bridges, by strict definition, must operate totally within the DLL and its fields [14.21, 14.27]. Table 14.3 provides a comparison between bridges and routers. Figure 14.16 depicts an example of a typical internet. Services such as frame relay and SMDS are being increasingly used.*

14.1.7 TCP/IP

14.1.7.1 Background

As discussed at length in Chapter 13 and above, methods and technologies are required to interconnect separate and distinct physical networks in general, and

*Routers typically employ DDS and T1 links. By 1990, routers employing frame relay and SMDS communication facilities were beginning to enter the market. Now a variety of such equipment is available.

Table 14.3
Comparison of LAN bridges and LAN routers

Bridge	*Router*
Operates at the MAC-level; store-and-forward	Operates at the LLC and network layer*; store and forward
Commercialized in 1984†	Available since mid-1970s in ARPANET
Cable systems connected to a bridge are electrically independent	Subnetworks connected to a router are independent
Independent of higher layer protocols (MAC-level bridge can be used in a variety of practical situations, regardless of higher layer structure)	In practical terms, the network layer tends to identify the communication architecture (i.e., if IP is used there, then almost certainly it is in an Internet protocol environment). Hence, in practical terms, routers tend to depend on an entire higher-layer protocol suite‡, ¶
Connect networks with same LLC and same upper layers	Connect subnetworks with the same transport and some upper layers†, *
"Protocol transparent"† (a) Physical and MAC layers can be different on the inbound and outbound side of the relay	"Protocol specific"† (a) Physical, MAC, LLC, and network layers can be different on the inbound and outbound side of the relay
(b) LLC through application layer are the same on the inbound and outbound side of the relay	(b) (in theory) transport through application layers are the same on the inbound and outbound side of the relay (see point above about network layer)
Local and remote connectivity	Local and remote connectivity
(Typically) unique address space	Separate address space in different administrative domains
Flat addressing; addresses to scan proportional to number of active users	Hierarchical addressing; addresses to scan proportional to number of subnetworks in internet
End stations are not explicitly aware of their presence, except in source routing	Routers are explicitly addressed by the end stations
Adapt well (automatically) to changing environment, particularly with learning and aging techniques*	Typically require manual intervention to accommodate new network technologies, due to complexity
Variety of routing methods (static routing, source routing, adaptive routing). Explicit protocols to disseminate real-time routing information between bridges not typically used	Static (configured by system manager) or dynamic routing. Utilize explicit protocols to disseminate real-time routing information between routers
Integrity problems (availability, frame loss, misordering, duplication, error rates, et cetera)	Higher level of quality of service¶
Make simple forward-filter decisions based on flat (nonhierarchical) address fields	Complex internetworking arrangements including mixed-technologies: WANs, MANs, et cetera. Hierarchical addressing is common. Multiple types of network layer services and protocols to be reconciled

Table 14.3 continued
Comparison of LAN bridges and LAN routers

Bridge	*Router*
Cannot tolerate loops; spanning tree structure may be used	Complex topologies are accommodated
No flow control	Flow control
Limited variability in features of subnetworks served	Wide variability in functions and performance of higher protocols and protocol implementations, from multiple vendors†
Relatively simple packet processing; fast (all other variables being equal, a bridge will sustain several times the forwarding rate of a router†)	Relatively complex packet processing; performance becomes an issue
Negatively impacts security because it is protocol transparent and can transfer any traffic from one segment to another	Security controls can be instituted at off-net routing points
Simplicity of installation, configuration, and operation	More demanding installation, configuration, and operation
Best suited to implement an extended local network system	Best suited to interconnecting LANs and WANs

*Some lack of precision exists in this term when it is used in the LAN context. In the OSIRM context, a router (network layer relay, to be more precise) operates only at the network layer. In the LAN context, *in practical terms,* it typically includes LLC and the network layer.

†See Reference [14.28] for additional details.

¶See Reference [14.18] for a detailed discussion.

‡If the term were used formally (i.e., dealing strictly with the network layer), this would not be the case. "The choice of a Network layer protocol is tied to the choice of a protocol stack, or sometimes to the choice of a network vendor" [14.28].

LANs in particular, and make them appear to the ultimate user as a single coordinated system. These methods and technologies are known as internetworking (or, by way of jargon, "internet"). Internetworking accommodates multiple and diverse underlying network hardware by adding both connections and a new set of conventions. Internetworking technology hides the details of network hardware and permits computers to communicate independent of their physical network connections [14.29]. The OSI view of internetworking was discussed in Chapter 13 and above. However, OSI techniques are relatively recent and are not yet well-supported commercially. This section describes a more classical approach, but one that still has major commercial relevance today.

Government agencies were the first to realize the importance of internetworking technology. The U.S. government has funded research over the past

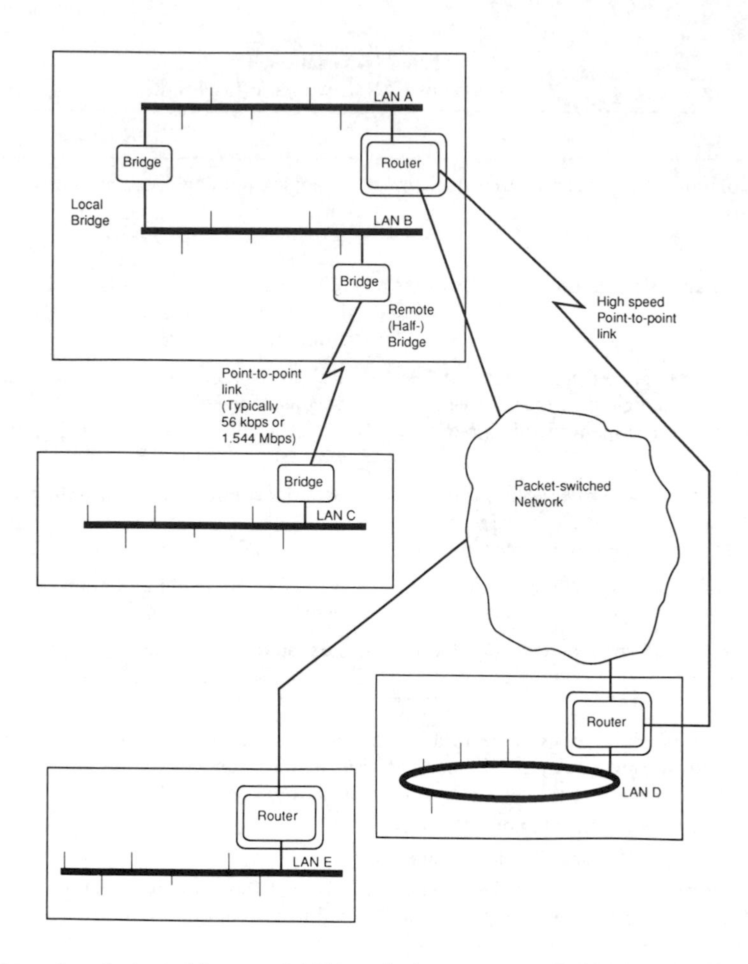

Figure 14.16 Examples of a typical internet LAN in today's corporate or industrial environment.

quarter century to make possible national internetworked communication systems. In the early 1970s, the U.S. Defense Advanced Research Projects Agency (DARPA, also known as ARPA) funded work to develop network standards specifying the details of how computers communicate, as well as a set of conventions for interconnecting networks and routing information. These protocols are commonly referred to as TCP/IP. TCP is a layer 4 protocol, and IP is a layer 3 protocol. TCP/IP is flexible and robust; in effect, TCP takes care of the integrity and IP moves the data.

The TCP/IP architecture and protocols acquired their current form in the late 1970s. For the ARPANET, the transition to TCP/IP technology was completed in 1983. The decision on the part of DARPA to make an implementation available at low cost encouraged the academic community to install the technology. Internet protocols were developed to run under Berkeley UNIX, which became the impetus that has made TCP/IP so popular in the 1980s. More than 150 vendors support TCP/IP, including IBM [14.29, 14.30]. TCP/IP supports both LANs and WANs.

As of the beginning of 1990, TCP/IP was the most widespread vendor-independent suite of communication protocols in commercial use, achieving *de facto* standard status. Eventually, communication suites will migrate to the OSI-based architectures described in Chapter 13; however, due to the penetration of TCP/IP, we can expect products based on it to still be around in the mid-1990s.

Host systems are computers connected to networks (a host can be connected to more than one network or have more than one connection to one network). Networks are connected to one another by a "gateway" (also known as a "IP router"). Dynamic routing relies on protocols to convey the routing information around the internet. When a topology change occurs in the internet, the routers directly involved in the change are responsible for propagating this information to the other routers in the system.

Gateways in the Internet are grouped for administrative purposes into autonomous systems. Gateways within an autonomous system communicate with each other using one of a number of dynamic routing protocols, known generically as *interior gateway protocols* (IGPs). This communication is required to update the routing information dynamically in each gateway router to reflect real-time conditions of the topology. Connections between gateways of different autonomous systems require exterior gateway protocols (EGPs); these facilitate exchange of routing information [14.18] (see Figure 14.17).

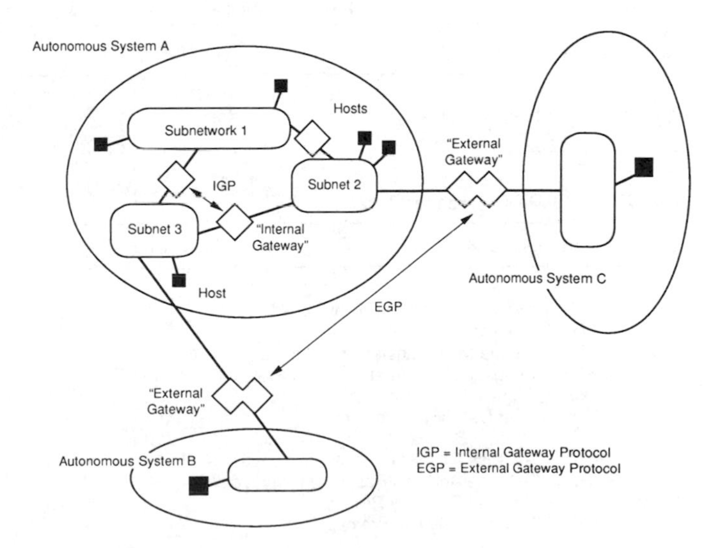

Figure 14.17 Internal and external gateways in Internet.

An Internet gateway must implement the IP, the Internet Control Message Protocol (ICMP), one or more internal gateway protocols, and, optionally, the EGP if the gateway connects autonomous systems. An IP gateway must be able to fragment IP datagrams to match the maximum size supported by an outbound link. A gateway is not required to reassemble a fragmented packet, unless the packet carries the address of that gateway [14.18].

The basic TCP/IP-based suite is shown in Table 14.4, for both LAN and wide area applications (the Internet suite has more than 94 protocols). TCP is connection-oriented, while IP is connectionless. The IAB, discussed in Chapter 2, provides the focus for much of the research and development underlying the Internet.

The IP is employed to send datagrams through the internet. Each datagram is treated independently. The IP does not maintain any physical or virtual connections. It provides two basic services: addressing and fragmentation and reassembly of long datagrams. IP does not provide end-to-end integrity such as retransmission, acknowledged delivery, and flow control. For these services, the system must rely on TCP, which resides at the transport layer. To provide its services, IP employs four fields [14.18]:

1. *Type of Service*: Parameters set by the end station specifying, for example, expected delay characteristics, expected reliability of path, *et cetera*.

	LAN environment	WAN environment (non-SNA)
Layer 7-Layer 5	Application specific protocols such as TELNET (terminal sessions), FTP and SFTP (file transfer), SMTP (e-mail), SNMP (management), DNS (directory)	
Layer 4	TCP, UDP, EGP/IGP	TCP, UDP
Layer 3	IP, ICMP, ARP, RARP	IP, ICMP X.25 PLP
Layer 2	LLC CSMA/CD, Token ring, Token bus	LAP-B
Layer 1	IEEE 802.3, .4, .5	Telco channels

SFTP = Simple File Transfer Protocol
FTP = File Transfer Protocol
SMTP = Simple Mail Transfer Protocol
SNMP = Simple Network Management Protocol
DNS = Domain Name Service
UDP = User Datagram Protocol
ICMP = Internet Control Message Protocol
ARP = Address Resolution Protocol
RARP = Reverse Address Resolution Protocol
EGP = External Gateway Protocol
IGP = Internal Gateway Protocol

Table 14.4 TCP/IP-based Communication: Key Protocol.

2. *Time to Live*: Parameter used to determine the datagram's lifetime in the Internet system.
3. *Options*: Parameters to specify security, time stamps, and special routing.
4. *Header Check Sum*: A two-octet field used by IP to determine datagram integrity.

Internet provides two broad types of services that all applications use: connectionless packet delivery service (layer 3 service) and reliable stream transport service (layer 4 service). TCP/IP makes no explicit assumptions about the reliability of the underlying networks. The ISO 8473 Connectionless Networking Protocol is analogous to IP in the sense that it acts as a router, routing datagrams received from above or below either upward to higher layers or down across network links, according to destination addressing information. TCP is similar to the ISO transport Class 4 protocol (both use positive acknowledgment schemes with timer-driven retries; both can reorder data that arrive out of order) [14.31].

From the point of view of a user, the TCP/IP internet appears to be a set of application programs that uses the network to carry out useful communication tasks [14.29]. The most widespread Internet application programs are E-mail, file transfer, and access to hosts on remote networks. As seen in Figure 14.6, a TCP/IP LAN application involves: (1) a user connection over a standard LAN environment (IEEE 802.3, 802.4, 802.5 over a LLC utilizing CSMA/CD, token ring, or token bus); (2) software the PC or server can use to implement the IP and TCP protocols; and (3) programs running on the PCs or servers to provide the needed application (the application may use other layer 7 protocols for file transfer, network management, *et cetera*).

Reference [14.29] provides a detailed assessment of TCP/IP; the interested reader is encouraged to consult it.

14.1.7.2 Network Management

SNMP (Simple Network Management Protocol) is a relatively new specification of a protocol for the management of internets [14.32]. SNMP has evolved from the Simple Gateway Monitoring Protocol (SGMP), which is a mature, successful, and widely deployed protocol in Internet. This existing protocol was changed based on operational experience and was generalized to provide management (not just control) of facilities beyond gateways. SNMP was built via the cooperative effort of researchers, users, network managers, and vendors [14.33]. It was declared a draft standard by the IAB in March 1988. It was elevated to recommended standard status in April 1989, with the release of the CMOT (Common Management Protocol Over TCP/IP) specification. As of April 1990, SNMP is a full Internet standard. SNMP allows transfer of data pertaining to fault, performance, and configuration management of TCP/IP systems in general, and LANs in particular.

As of early 1990, more than 30 vendors had developed products based on SNMP, and a total of 70 vendors had committed to products in the near future [14.34]. Some observers were already labeling it a *de facto* standard for LAN management; in the absence of interoperating and fielded OSI-based network management systems, these observers see it as the only choice for management of multivendor TCP/IP internets [14.33]. SNMP was originally developed to meet the management needs of the largest collection of heterogeneous networks in the world, the Internet [14.34].

The desire to ease eventual transition to OSI-based management protocols led to the definition, in the ASN.1 language, of an Internet-standard Structure of Management Information (SMI) and Management Information Base (MIB), which are OSIRM constructs, covered in more detail in Chapter 16 [14.35].

SMI is a policy statement describing how the managed objects (hosts, terminals, routers, bridges, gateways, *et cetera*) are defined. Such definition basically involves agreement on a standard language for management. One of the basic definitions in the SMI pertains to the representation of data elements. The 1990 version of the SMI (RFC 1065) specified that only the integer and the octet string types within ASN.1 are supported (in the future, support for other types may be added) [14.34].

Conceptually, the MIB represents the data base of management information and parameters. The management parameters are stored in a tree-structured data base (similar, for example, to a Unix file). The MIB specification for use in SNMP is described in RFC 1066. All mandatory variables listed there must be supported, either statically (with null or default value) or dynamically (i.e., actually employed by the device in question).

While SMI and MIB define and logically store the set of parameters pertaining to managed entities, the SNMP actually provides a means of communicating information between the management station and the managed devices. Reference [14.34] provides a more detailed discussion of some protocol transition implications.

14.1.8 IEEE 802.9

The IEEE 802.9 effort is aimed at defining a standard for integrated voice and data LAN (IVDLAN) interfaces. Backbone integration has shown its value in terms of cost savings and improved network management. The proliferation of PCs in the working environment has created a need to integrate on-premises computer and voice networks. Proposals that attempt to provide this convergence by adapting existing LANs or PBXs have been advanced, but have met with little success because of the differing service needs of voice and data traffic. The interface defined in IEEE 802.9 takes a new approach to the problem by combining ISDN and LAN interfaces to provide voice and data services, and videoconferencing and colorfax in the future [14.36]. The standard defines the MAC and physical layer compatible with IEEE LAN standards (802.3, 802.4, and 802.5) and ISDN. These functions are provided over a single interface. The detailed operation of the IVDLAN interface was not finalized at the time of this writing. A draft of IEEE 802.9 was expected by late 1990.

The utilization of the 802.9 interface requires the use of a TA and an AU.[†] The TA allows a PC and a telephone to be adapted to the interface, as shown in Figure 14.18. The AU permits access to ISDN via an integrated services PBX (ISPBX), and to a backbone LAN through a bridge. The AU also acts as a concentrator for a number of IVDLAN interfaces, each of which is a point-to-point link. (The standard, however, only defines the interface.)

802.9 follows the OSIRM; however, the specification only deals with the physical layer and the MAC sublayer operation. The interface will support isochronous traffic and asynchronous traffic (for example, computer data). Isochronous channels provide a guaranteed bandwidth and minimal delay. Asynchronous channels require an arbitration scheme to resolve contention among multiple users. The MAC sublayer must support several LLC protocols including LAP-D and APM (additional packet mode). APM is a superset of LAP-D protocols in ISDN that is aimed at providing X.25-like capabilities within ISDN [14.36].

Layer 1 has been subdivided in IEEE 802.9 into three sublayers (see Table 14.5): (1) physical medium dependent (PMD) (lowest), (2) physical signaling (PS) (middle), and (3) multiplexing (MUX) function (highest). The PMD sublayer defines the electrical and mechanical interface to the medium at hand; this includes voltages, line code, and connectors. The PS sublayer specifies the medium-independent operation of the physical layer; this includes frame synchronization, scrambling, and order of bit transmission. The multiplexing function is responsible for combining LAN and ISDN interfaces.

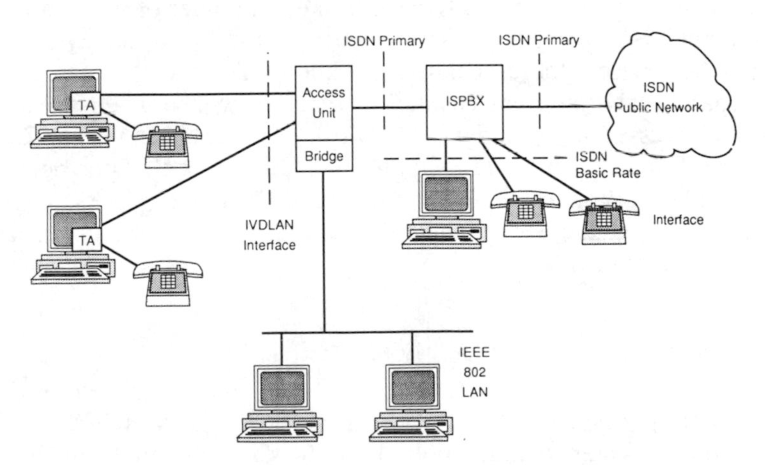

Figure 14.18 IEEE 802.9 environment.

[†]The concept of access unit discussed here is totally unrelated to that described in Chapter 13.

Table 14.5
Proposed IEEE 802.9 IVDLAN Channels and Protocols

	C channel	P channel	P_D channel	D channel	B_1 channel	B_2 channel
Upper Layers	←------	------	As needed	------	------	------→
Layer 3	Null	X.25 PLP	Q.931	Q.931	X.25 PLP	Null
Layer 2	Null	LLC 1 or 2 IVDLAN MAC	LAP-D IVDLAN MAC	LAP-D	LAP-B	Null
Layer 1	←------	------	MUX	------	------	------→
	←------	------	PS	------	------	------→
	←------	------	PMD	------	------	------→

Layer 1 Definitions

This definition includes the medium and physical connectors; data rate and line code; and TDM frame structure.

A point-to-point connection exists between each IVDLAN and the AU. The medium specified for the IVDLAN interface is unshielded twisted-pair cable. Two twisted pairs are needed: one for the transmitting side and one for the receiving side. The connector is the RJ-45 eight-pin plug-socket (widely used in telephony and ISDN). The total transfer rate over the IVDLAN interface is initially defined at 4.096 Mb/s, full-duplex. Each twisted pair carries a TDM frame repeating every 125 ms and consisting of 64 octets. In each frame, a number of octets are reserved for each channel. Six channels are supported over the MUX sublayer to provide voice and data services; these channels are known as C, P, P_D, D, B_1, and B_2. The AU separates the different channels associated with a given service. These channels are discussed in more detail below. An overhead field occupies 5 octets in each 125-ms frame. Each of the B and D channels is allocated one octet in every frame. The balance is shared between the C and P channels; the C channel will probably use a larger number of octets than the P channel for some isochronous applications, as discussed below.

Layer 2 Definitions

The C channel specified in IEEE 802.9 provides circuit switched facilities for isochronous data transfers that are multiples of 64 kb/s, ranging from C_1 at 64 kb/s to C_{30} at 1.92 Mb/s. C channels are routed to the PBX for local or off-net switching. Circuit switched facilities can be used for voice, video, or data and do not need an intrinsic protocol suite at or above Layer 2 (the user may implement any desired protocol over this system-transparent channel).

Digitized video traffic is also transferred over a circuit switched C channel. Digitized voice traffic is transferred between TAs and the AU via a 64 kb/s B channel (B channels are also circuit switched). The B_1 channel may use LAP-B at layer 2 and X.25 PLP at layer 3. The B_2 channel is transparent at and above layer 2. The C and B channels may also be used as bearer channels to access other networks, including, for example, a packet network.

The circuit switched channels require a signaling mechanism for call management. This is accomplished in the 802.9 IVDLAN over a D channel, which is similar in protocol suite to the ISDN D channel (LAP-D at layer 2 and Q.931 at layer 3).

IVDLAN also provides a connectionless packet service over a P channel. The P channel is routed to and from conventional LANs (802.3, 802.4, 802.5, and FDDI) at the AU. The P channel uses LLC 1 or LLC 2 and AMP protocols over the IVDLAN link layer MAC. The size of the P and C channels may be interchanged. The C channel capacity could be selected on a per-call basis; the remaining bandwidth is then available for the P channel (a minimum capacity is always available to the P channel for management activities; this is expected to be 1.024 Mb/s). The maximum capacity is 3.584 Mb/s [14.36].

If desired, the D channel can be carried within the P channel over the IVDLAN MAC. This is identified as a P_D channel (in practice, only one of the signaling stacks would be implemented). Higher layer protocols can be selected by the user as needed.

An 802.9 characteristic MAC sublayer is defined for the P channel for LAN applications. It provides access control via contention resolution between the integrated voice and data terminals for AU resources. Traditional MAC functions are frame delineation and specification, error detection, link level protocol identification, addressing, and priority indication. MAC PDUs (which typically exceed the 64-octet length of the TDM frame at layer 1) are segmented into groups of octets at the transmitter before submission to the P channel interface. The groups of octets must be reassembled with the layer 2 protocol at the receiving end. In turn, an 802.2 LLC PDU is carried within the MAC frame. The LLC entity only receives SAP-defined services from the MAC; therefore, it is oblivious to its peculiarities. Hence, a LAN service is derivable over the P channel.

Future

LAN service is provided over the P channel. LANs are accessed at the AU through a bridge function (although not all the interworking issues are fully resolved). Interest exists in the study of 16 Mb/s, which would allow 14.336 Mb/s for the P channel for CAD-CAM workstations and desktop publishing systems incorporating graphics. IEEE 802.9 products could begin to appear in 1992.

14.2 PRACTICAL ASPECTS OF LAN MANAGEMENT

LANs are commonly susceptible to malfunction: Connectors come loose, cables snap, file servers develop problems and software may be affected by bugs. In the late 1980s, only about 10% of the people who had LANs were estimated to have some sort of diagnostics tools. The cost of these tools is dropping and people who are buying LANs must consider buying testing equipment before the whole LAN management issue becomes a corporate liability. A variety of diagnostic tools, ranging from sophisticated network monitors to inexpensive testers, is available to facilitate fault management. Configuration, accounting, and performance management of LANs is also critical. Security is recognized as one of the major network management concerns in LANs. Beyond physical transport security, LAN managers are facing increased threats to establish information security, from unauthorized internal as well as external access. In contrast to centralized DP environments, as LANs proliferate, so does general dissemination of information, compounding information security management responsibilities.

Most work groups probably started as a few PCs sharing resources, but, with increased processing power, many have expanded to include thousands of PCs connected to hundreds of servers and other shared resources. Multiple software protocols and interfaces will become more commonly included in a single server or gateway interface. Network management will be very important in this setting.

A pragmatic definition of LAN management is "whatever cures whatever ails the LAN: if one can't figure out how to assign user names, network management is assigning user names; if the cable is damaged and one can't figure out where, then network management is figuring out where cables are damaged" [14.37]. LAN management and software covers a wide range of applications and functions. In broad terms, management functions in a LAN can be broken into two classes: user support and system support. User support capabilities insulate users from the complexities of the LAN and reduce the amount of training and interaction that must be undertaken by the LAN manager. These tools include front-end systems, document organizers, and remote user support. System support capabilities give the network manager information about the LAN. System support utilities provide data on user activity, server usage, network traffic, system configuration, *et cetera*. These tools complement the deficiencies in the network operating system [14.38].

The OSI network management standard has defined functional areas pertinent to management activities [14.39]. This nomenclature, although originally applied to OSI network management standards, is now becoming commonly accepted for all types of network management, whether OSI-based or not. The five categories are described in Table 14.6. These standards as well as issues pertaining to the management of WANs are discussed more fully in Chapter 16. The treatment below will attempt to relate the methods, techniques, and tools available to the LAN network manager to these categories.

A network cannot be managed effectively if we cannot establish in real time what is happening on the LAN, at least with respect to critical tasks. When LANs were first being installed, the LAN manager's major task was to establish the connectivity and make the network operational. Experience shows that the demand for LAN management tools typically lags about one year behind the initial LAN installation. Today's large and complex LANs require sophisticated software-based management and performance-optimization tools. Management must not only prevent deliberate misuse of the LAN but it must also monitor utilization and frequency of network access per node [14.40]. The common complaint from LAN managers is that the number of quality network management systems is minimal.

Table 14.6
A Contemporary Definition of Network Management

Fault Management	Encompasses fault detection, isolation, and the correction of abnormal operation. Faults may be persistent or transient. Faults will manifest themselves as particular events; error event detection provides a capability to recognize faults. Fault management is the discipline of detecting, diagnosing, bypassing, repairing, and reporting on network equipment and service failures.
Accounting Management	Enables charges to be established for the use of communication resources, and for costs to be identified for the use of those resources.
Configuration Management	Identifies, exercises control over, collects data from, and provides data to communication systems for the purpose of initiating and providing continuous reliable connectivity. Configuration management is concerned with maintaining an accurate inventory of hardware, software, and circuits and the ability to change that inventory in a smooth and reliable manner in response to changing service requirements.
Performance Management	Enables evaluation of the behavior and effectiveness of resources and related communication activities. Performance management is concerned with the utilization of network resources and their ability to meet user service level objectives.
Security Management	Supports the application of security policies. Security is an important issue: unauthorized or accidental access to network control functions must be eliminated or minimized. Security management controls access to both the network and the network management systems. In some cases, it may also protect information transported by the network from disclosure or modification.

MIS departments are challenged with accepting the power and flexibility of the PC while retaining their traditional corporate control over software and sensitive data. Many network management systems in use today were designed to address a communication environment with four primary network components: the mainframe computer, carrier-provided circuits, data communication equipment, and terminals. The main function of diagnosis has been to isolate a problem to one of these four general areas and then to the specific failing element, if possible. These monitoring systems do not work well in a LAN environment. Fortunately, a new class of software utilities is becoming available to help LAN managers to maintain the LAN hardware and software resources [14.41]. In terms of functionality, LAN management software has come a long way since then. Not only are more products available, but the products are easier to use and much more powerful, with capabilities ranging from audit trail and software metering to system (disk) management, front-ending, and traffic monitoring [14.42].

Some of the available tools address shortcomings of NetWare, NetBIOS, and MS-Net operating systems typical of the LAN environment. In terms of products, LAN management tools can be grouped in the areas shown in Table 14.7. In the discussion that follows, these tools will be mapped to the five industry-accepted classes shown in Table 14.6, although in a few cases, the mapping is a matter of judgment (particularly for the user support tools).

A network manager preferably wants to monitor the entire network with one tool and to have one consolidated view of it. At present, many vendors, interfaces, protocols, and devices exist, making network control and management a challenge. If the network happens to contain a mix of IBM equipment, other mainframes or minicomputers that run TPC/IP, gateways and bridges, and work groups that share files and printers, then network managers may be forced to use several tools to get information on the total network. Consolidating this information into one overview is difficult. This results in added costs for training and for support of these multiple network management systems.

Table 14.7
LAN Management Tools

User support
Front ends
Document managers
User support
Software meters
System support
User listing
Usage reporting
Performance monitors
Fault recovery

A trend faced by network managers that makes network management a more difficult task is the erosion of their control over the networks they must support. The proliferation of low-cost intelligent local devices has left many managers with little or no control over what is plugged into their networks. Desirable network management features are [14.43]:

1. Expert systems for diagnosis of common LAN problems;
2. A mechanism that allows LAN alerts for the network administrator to be generated automatically when problems arise;
3. On-line LAN help facilities; and
4. Central LAN backup system.

Expert Systems. At times, more network data are available than managers can efficiently use. Network management expert systems that could filter the data and infer the underlying problems would be useful. Network management expert systems for all the categories of Table 14.6 have evolved for WANs in the late 1980s, as surveyed in Reference [14.44]; however, these systems were generally not available for LANs as of early 1990.

Automatic Alerts. The LAN management system should issue automatic alerts when a problem (security or performance) arises. Such a system should also alert the user directly about problems they may be responsible for (for example, a printer out of paper). Some of the LANs' network operating systems track statistics that indicate the status of the LAN and of a number of LAN services. However, notifications of problems are not generally proliferated automatically to the appropriate individuals.

On-Line Help Facilities. A network manager should be able to query a LAN server about static answers to the most-asked technical questions. An enhancement would be an expert system that leads the unsophisticated user to a solution in easy-to-follow steps. The file could contain product name and configuration information such as interrupt usage, input-output address, and port location. Installation software for new products could check the file to self-configure or recommend a configuration [14.43].

Backup. A secure and reliable way to do file backup is needed. Security issues may arise out of this backup process. To minimize the vulnerability, the backup services must come from the LAN vendors; interworking between the LAN operating system and backup software should include the ability to encrypt data files based on the identity of the owner.

Microsoft's OS/2 LAN Manager represents an emerging *de facto* work group networking standard that includes extensive network management support. It is gaining acceptance among network vendors and applications developers.

14.2.1 Fault Management

Fault management is the discipline of detecting, diagnosing, bypassing, repairing, and reporting network equipment and service failures. Three basic steps are required for LAN fault management and troubleshooting. The first step is an understanding of the particular LAN that is being managed [14.45]. What are its characteristics? Is it a token ring, bus, or star? What type of cable is used? How is the cable laid out and how long are the runs? How many servers are there and how are they configured? What is the network operating system and version? Which applications software is being used? The second step requires a logical procedure for sectionalizing the problem. By moving, replacing, and testing cable, servers, and workstations, a manager can narrow the problem areas. The third step involves applying the proper tools to diagnose problems that are not immediately obvious.

Fault management in a LAN is typically undertaken with hardware tools. LANs fail in three areas: hardware, software, and cabling. Of the three, cabling is the most common problem, based on empirical observation. For this discussion, cabling includes both the cable proper and the interface card.

14.2.1.2 Physical Network Issues

Cards. Network interface cards, cables, and other low-level hardware all have an impact on the operation, performance, speed, and throughput of the LAN. If the interface card and cable are working properly, we cannot expect speeds greater than the rated throughput. However, if an interface card is working improperly or a cable is not correctly terminated, errors may occur causing the LAN to run significantly slower due to higher layer retransmissions of mutilated and lost packets.

A number of ways exist to determine if interface cards and cables are working properly. Most LAN manufacturers include a basic diagnostic utility program integrated with their hardware. While these programs can detect severe errors, they may not show subtle errors. These utilities may report that they see other workstations on the LAN and that, therefore, the LAN is "working" although a problem may exist [14.46]. Other tools are available from the network interface card manufacturers: many of them have programs for their boards. There are also third-party products from companies that offer diagnostic tools [14.45]. Some available products monitor traffic on the LAN and indicate trouble spots. These diagnostic programs will generally provide much better information about the physical LAN than the simple diagnostics included with the interface cards.

Another problem is presented by a mixture of different vendors' Ethernet cards that may not work together. Diagnostic programs included with vendors' cards may help indicate any incompatibilities. With the use of more sophisticated hardware diagnostic programs, an experienced installer should be able to determine easily compatible card mixes. Once the data are read from the server, they must be transmitted over the LAN to the station that requested it. Unless the network is optimized, a fast server may not be of benefit; hence, performance management and fault management must be done in a cooperative fashion. Typical products evaluate the network's capacity, reporting what percentage of the network's bandwidth is being used (this function is similar to a traffic monitor); also these products detect physical problems on the network. In addition to bad network interface cards, such physical problems include impaired cables, hubs, and repeaters. The cable test generates packets, then measures the success rate of the transmissions. Some of the products will inform the manager of how many reconfigurations were necessary because of lost tokens and how many packets collided, lost bits, or were dropped. This information is then used to formulate fault scenarios.

Wiring. For cable problems, a LAN manager can perform continuity, loopback, and reflectometry tests (cable installation companies can also do that on behalf of the manager). This will help ensure that the cable system is working correctly. The most common problem with LAN cabling is broken cable or improper termination. Most connectors do not handle movement well, and may break. Regular inspection of the cable ends and connectors will help identify defective terminations. Various tools exist for diagnosing cable problems [14.47]:

- *Ohmmeters*. An ohmmeter is a simple tool that provides impedance measurements. An ohmmeter is used to locate open or shorted cable. If the impedance reading matches the rated impedance of the cable, the cable should be fine. If the reading does not match the rated impedance, then a short, a crushed cable, or a cable break exists somewhere along the cable run. Ohmmeters cannot be used in fiber-based LANs. Ohmmeters are priced at well under $100.
- *Outlet Testers*. Sometimes the problem is not with the cable but with the electrical outlet used by some LAN equipment. For example, if the outlet is not grounded properly, power may be introduced through the power supply into the workstation, and then through the network interface card onto the a copper-based (twisted-pair or coaxial) LAN cable. An outlet tester costs around $10. Even if power is not introduced in the cable, improper grounding could result in noise.
- *Coaxial Connectors, T Connectors, and Terminators*. Particularly when working on a bus network, extra terminators are a basic requirement. They are used to isolate sections of the cable for testing.
- *Oscilloscopes*. An oscilloscope allows a signal to be examined over the cable. An oscilloscope helps detect the existence of noise or other disturbances on the wire, such as repetitive voltage spikes. Again, this applies to a copper-based LAN.

- *TDRs*. A time-domain reflectometer operates by sending an electrical pulse over the LAN cable, monitoring for signal reflections. On a good cable, no reflections will exist; the cable will be clean, with no breaks or shorts. If a break or short exists in the cable, however, the time it takes for the pulse reflection to return gives the TDR a very accurate indication of where the fault is located. Many TDRs can locate cable breaks to within a few feet or less. TDRs have traditionally been relatively expensive instruments; a TDR with an oscilloscope can cost $5000 or more. A new generation of TDR equipment, which became available in the late 1980s, is less expensive: Although still accurate to within a few feet, these new instruments now cost under $1000. They are also much more compact than their predecessors, often measuring about the size of a paperback book or smaller; they are also easier to use [14.47].

Troubleshooting Fiber LANs

For fiber-based LANs, the manager needs different equipment from that which has been used with traditional LANs. While offering significant advantages over conventional techniques, the design and production of fiber-based systems require test equipment capable of more than just functional checkout. A comprehensive design verification is needed, including precise determination of bandwidth, sensitivity, and linearity. These measurements can be performed only with fiber optic test equipment that has sophisticated parametric capabilities.

An optical LAN connects devices using optical fiber. Although optical LANs are available in three distinct topologies—star, bus, and ring—an estimated 80% of all optical LANs today are the ring type [14.48]. Most of the LANs (95%) use graded-index fiber; the balance is step-index and plastic-clad fiber. The use of single-mode fiber for LANs is still very low. The transmission speed of optical LANs is increasing each year. Recently, 10 to 50 Mb/s speeds have gained attention; backbone optical LANs have also been introduced with transmission speeds of 100 Mb/s and higher. This, however, is no major technical breakthrough because fiber can easily carry 2.4 Gb/s for field-deployed technology, and a higher rate in the laboratory. Ninety-five percent of optical LANs use the 0.85-mm wavelength and light sources that are LEDs (the other 5% use the 1.3- and 1.55-mm bands). The main receiving elements used in the 0.85-mm band are silicon photodiodes (Si PDs). Seventy percent of the receiving elements are Si-PIN PDs (where PIN is positive-intrinsic-negative) and 30% are Si avalanche photodiodes (APDs).

In troubleshooting optical LANs, optical loss is a major measurement item; a stabilized light source and optical power meter are key elements of this process. Each optical fiber and circuit part used in optical fiber communication produces an optical loss, which is one of the factors that limits the range of the optical communication system. A stabilized light source, a mode scrambler, and an optical-power meter are used to measure this optical loss. The optical output stability of the stabilized light source is important for increasing optical-loss measurement precision. Two devices are used as the light-emitting element of a stabilized light source, as discussed in Chapter 7: LDs and LEDs. Each has unique properties for measurement purposes. The LD has a high output level, narrow light emission spectral width, large optical output change *versus* ambient temperature change, and high coherence. The LED, on the other hand, has a low output level, wide spectral width, small optical output change *versus* ambient temperature change, and low coherence. Because of these different properties, the LD stabilized light source is used in high-loss measurements and the LED stabilized light source is used in low-loss measurements that demand greater precision [14.48]. The demands for an optical-power meter differ between a high-capacity, long-haul communication system and optical LANs. The repeating zone is longer in low-loss optical fiber used for long-distance communication; thus, the optical-power measurement is made over a wide dynamic range. This is not true for optical LANs; the requirement is for simple operation.

A wide variety of test equipment has come to the market in the late 1980s. This instrumentation assists the technician in diagnosing and maintaining a fiber-based wiring system. Instruments vary in complexity from pocket-size power measuring units to console-type testers. Only instruments that are optical in nature, however, can make performance measurements on the optical parts of any electro-optical system. Fortunately for electronics test engineers, the few optical instruments used outside research laboratories are relatively simple in design. Instrumentation can be divided into three general categories: (1) power meters (optical-loss test sets), (2) optical time-domain reflectometers (OTDRs), and (3) optical bandwidth test sets (OBTSs).

A power meter is used to measure the optical power from a length of fiber in much the same way that one is used to measure electrical power. It is used when performing a one-way loss measurement. The loss can occur in the fiber, the connectors, splices, jumper cables, or other system areas. Some power meters also have a built-in transmitting source. Two sets, both having transmitting and receiving capabilities, are used together to make measurements in both directions without the movement of personnel or equipment. The individual power meter is a single unit consisting of an optical receiver and an analog or digital readout. A light source, typically at the point of origination, supplies the power that is detected at the end or test access point. The meter displays the power detected in decibels. The wavelength range often given indicates the various wavelengths that the meter is able to detect. The resolution parameter indicates the smallest step that the meter will display.

OTDRs are used to characterize a fiber in which an optical pulse is transmitted through the fiber and the resulting light scattered and reflected back to the input is measured as a function of time. They are useful for estimating the attenuation coefficient as a function of distance and identifying defects and other losses. These devices operate on basically the same principles as a copper-based TDR. The difference is one of cost: OTDRs typically range from $10,000 to $20,000.

The OBTSs consist of two separate parts: the source, with an output data rate that varies according to the frequency of input current applied to the source (specified by frequency range parameter), and the detector, which reads the changing signal, determines the frequency response, and then displays a bandwidth measurement. The instrumentation is calibrated using a test fiber and then the results of the actual measurement are compared to the calibrated value to display the bandwidth value.

14.2.1.3 LAN Hardware

Fault management of LAN hardware other than cables and cards is another problem. Until recently, LAN hardware troubleshooting was limited to LAN built-in diagnostics. Most LAN troubleshooters still use manual, step-by-step procedures for diagnostics. Many tools are available for Ethernet; not as many are available for token-ring LANs. Ethernet can be checked by sending traffic to a particular node and waiting for a response; that response can be compared to the response of another node. If the node is sending packets and not receiving any, then a fault exists. We can also count or sample packets on a network. A typical problem is loss of information (mutilated or incomplete packets). Hardware and software diagnostic tools have been developed to assist the manager in determining if this kind of problem exists. A number of networks perform loopback tests, which involve a workstation sending a message to itself. If the packet is not received intact, then a problem exists [14.47].

A number of products have appeared that compensate for the simple diagnostics included by the manufacturer in the network hardware. This equipment is installed as a node on a network (monitoring mode) or it can be connected only when a specific problem exists (testing mode). In the monitoring mode, the equipment checks network parameters such as preamble length, alignment and CRC faults, frame lengths, and collision rates. In the testing mode, the equipment allows the troubleshooter to run interactive diagnostics on different parameters, notifying the troubleshooter when a given level exceeds a given threshold. This hardware monitoring equipment ranges in price from $300 to $5000, depending on sophistication.

As internetworking becomes commonplace, problems in diagnostics become even more critical. Undertaking troubleshooting across router and bridge links is a challenge. Even though LAN vendors have not yet been able to offer adequate network management products for a single LAN, users are already demanding more sophisticated products that can manage and integrated multiple LANs over a geographically dispersed area, possibly in a MAN arrangement. Network management limitations continue to be the single most frequent reason why users limit the size and scope of LAN implementations, according to experts [14.40]. A whole new set of tools is necessary to monitor the traffic between networks. Some vendors are already building these tools into bridges and routers.

Currently, fault management revolves around diagnostic tools and software used by network administrators to monitor and diagnose hardware. The next logical step is expert systems to maintain and diagnose networks [14.44].

Troubleshooting Software with Protocol Analyzers

Protocol analyzers are important high-end tools for fault management. The analyzer is a specialized workstation that collects, analyzes, and displays the data circulating in a LAN cable. An operator using an analyzer literally sees everything on the cable: PDUs, messages, files, passwords, IDs. An analyzer can display the information and store it to disk for future study [14.12]. They assist in diagnosing LANs, breaking out information on protocols ranging from TCP/IP, OSI, and DECnet to AppleTalk and NFS. Protocol analyzers can solve a wide variety of network troubleshooting problems; however, they require a fair level of technical expertise to assess the results. Additionally, protocol analyzers are expensive, generally ranging from $15,000 to $30,000, depending on the number and type of protocols supported, although some cheaper models are beginning to appear. Both token-ring and Ethernet models are available. While dozens of vendors build traditional protocol analyzers for X.25 and BSC/SDLC networks, only a handful of vendors had LAN analyzers at the time of this writing.

One of the problems with network analyzers, however, is speed. Protocol analyzers are not designed for full-packet capture in a heavily loaded environment. Most analyzers barely keep up with network utilizations of 30 or 40%; very few can handle 60 to 70% utilization on Ethernet. They need to provide the expanded packet buffers; with some models, the buffers now are in the 8-Mbyte range. LAN protocol suites now include full OSI session and presentation layers, which means the sophisticated analyzers must support layers one through six of the OSIRM [14.49].

Typically, the protocol analyzer is needed in one of two circumstances: (1) when network performance is degraded and must be monitored in order to identify the cause or (2) when explicit problems are uncovered by the network monitoring or management software that need to be resolved. In troubleshooting networks, the technique of "stepwise refinement" is a pragmatic course of action, which protocol analyzers make possible. The steps are [14.50]:

1. Begin with a generic test to observe the current state of the network. A protocol analyzer works well here because it collects all packets on one channel and several types of errors on other channels.
2. Look for high utilization. Sixty percent or higher is considered to be the region at which Ethernet performance begins to degrade. Also look for high error rates.
3. Try to isolate high-traffic or high-error producing stations on the network (very often, they will be the same).
4. Look for high counts of broadcast or multicast packets. In an internetwork situation with bridge, router, or gateway problems, large numbers of broadcast or multicast can slow down network performance with spurious traffic.
5. Look for high FCS-alignment or short packet counts, indicating media or connection problems. In general, elimination of physical media-related factors is advisable before proceeding to investigate software as a potential source of network problems. Network managers should ascertain that name tables or name servers are defined correctly and working properly; another factor is to make sure that multiple nodes are not contending for the same Ethernet address. Only after the obvious problems are eliminated is it worthwhile to use more sophisticated protocol analysis.

Empirical observation shows that 90% of the fault problems are identified in this straightforward fashion. When these steps fail to isolate the problem, the protocols must be analyzed. Protocol analysis can be particularly helpful in identifying whether the application or transport mechanisms are causing problems. An advisable way of proceeding is to start from the lower levels of the protocol suite under consideration and work up the hierarchy until the problematic layer can be isolated through a process of elimination. The data needed for this task include the collection of PDUs for the various layers and components of the protocol, beginning with the lower layers of Ethernet and proceeding all the way to the application environment. Most protocol problems occur at the interfaces between layers or between software systems. Typical problems are values that are out of range or out of order, fields that are too long or too short, missing or incomplete data, and garbled information due to incorrect reformatting [14.50].

The extensive capabilities of analyzers make them a potential security threat, allowing a hostile user access to anything that runs on the network. LAN cables are "party lines": Every frame reaches every station on the LAN (although in bridged LANs, frames may not leave the local LAN for which they are destined). By agreement, each LAN workstation looks only at frames addressed specifically to it (as well as at broadcast frames, which are addressed to everyone, and at frames sent to group addresses). Protocol analyzers, however, violate this agreement. They read every frame, or whichever frames the operator tells them to read.

14.2.1.4 Fault Recovery

Protection of data against hardware failure is an important issue. This aspect of management can be either fault management or security management. In general, fault recovery may also include operations recovery: Recovering a LAN or server to a stable fault-free state goes beyond mere data recovery. This section, however, focuses on data recovery.

Reliable software designs that ensure against data loss in the event of hardware failure are becoming more prevalent in the LAN market. Several methods of fault recovery are available that provide for [14.51]:

- *Disk Bad-Track Handling*. Very few disks have no flaws; software that detects these flaws and deals with them transparently is important.
- *Mirrored Disks*. This involves writing data to two separate disk drives so that both drives will contain the same data. In the event that one of the drives has an error, the alternative drive will continue operating without interruption to the LAN system. For example, Novell's NetWare offers duplexed drives, with two drives each having separate controllers. While this is an effective approach to protect a disk from failure, it does introduce a higher hardware cost.
- *Transaction Commit, Concurrency, and Recovery (CCR)*. This feature provides an application with the ability to protect data files from application failure. By grouping several input-output requests into a single transaction, the operating system will not write the transaction to a disk the application has terminated or issue a commit command.
- *Transaction Logging*. A powerful data management method is transaction logging, which is real-time backup. With this feature, after the data are written to the disk, the information is "echoed" to the server's local tape unit. As each input-output is performed on files, the data are written to the tape drive. In the event of a system failure, the state of these files may be recovered by repeating the input-output requests for that file from data written to the tape.

14.2.2 Configuration Management

Configuration and name management is concerned with maintaining an accurate inventory of hardware, software, and circuits and the ability to change that inventory in a smooth and reliable manner in response to changing service requirements. Configuration management affects network design, performance issues, and even security.

One of the most basic issues for the LAN administrator is maintaining system configuration maps. A *system configuration* is the list of system parameters showing who has access to given network software and given data bases. While most LAN operating system software gives the administrator the ability to add, delete, and modify system configuration parameters, it typically provides little functionality in monitoring the system configuration.

A configuration management technique to improve the performance of a network is to use a bridge or router to break a single network into two separate but logically connected LANs. Network hardware and cabling has a specified maximum bandwidth. Each time another user is added to a network, more workstations share the bandwidth. Eventually, if the network is heavily used, that bandwidth may become saturated, and network performance will suffer. Bridges and routers can be a solution to this problem. Both bridges and routers link two LAN systems into a single system, as discussed earlier in this chapter. The two separate LANs can still communicate transparently, but only internetwork traffic (traffic intended for the LAN on the far side of the bridge) actually gets passed over the bridge to the other side and circulates in that network's cabling system. Each LAN's local traffic remains local and each network will use up about half the bandwidth compared to a single network. A number of bridge and router strategies exist, as discussed earlier. Bridges at the MAC layer are usually external devices that work transparently to any LAN software. MAC-layer bridges usually link similar hardware systems; the most common are Ethernet-to-Ethernet bridges. Routers are protocol-dependent, as defined in Table 14.2; hence, they only work with one network operating system. Under these network operating systems, several different network cards can be plugged into the bus of the file server at the same time. These different interface cards each run two separate networks; the file server software will link them so that the different networks can still communicate.

When bridges and routers are used to attempt to improve performance, careful attention to LAN traffic patterns must be given before deciding where to place this equipment. If the equipment were placed where cross-bridge traffic is still high, the performance will improve only minimally.

While many types of cabling and network interface hardware are available, only two basic logical architectures are supported on PC LANs today: (1) peer-to-peer and (2) client-server architectures [14.51]. Peer-to-peer architectures require no dedicated file server because any node on the network may share its local hard disk with other nodes on the network. This type of architecture is used, for example, in the IBM's PC LAN program. This approach is often used in smaller LAN installations because it requires no additional hardware and generally has a lower cost per-node. Security problems are clearly of concern with a peer-to-peer architecture because of the lack of centralized data storage, which can, in turn, be physically and logically protected. These networks have lower performance and require greater administrative effort to configure and maintain security definitions. The client-server architecture depends on services provided by dedicated file and print servers. The higher cost of such LAN installations, because of the additional hardware, is compensated by the higher performance and added security. The centralized disk storage architecture provides a mechanism for control of user access and of backup operation. During the configuration selection phase of selecting a network operating system, the network manager should classify security needs into three categories: (1) minimal or no access control, (2) medium access control, and (3) maximum access control. For installations requiring little or no security control, any network operating system (NOS) on the market is adequate. The lowest costing security implementations can be configured by use of peer-to-peer LAN software. Higher security requirements will require selection of a NOS with more specific and sophisticated security. See Table 14.8, adapted from Reference [14.51]. More information on security is provided later in this chapter. (Unfortunately, space does not permit a full definition or description of a network operating system.)

14.2.2.1 Document Managers

Document managers assist end-users in discovering files and documents without requiring either the full DOS name or the exact location of the directory. They are included under configuration management in this discussion. With document managers, users need to know only basic information to locate a file. With these

Table 14.8
LAN Configurations as a Function of Security

Minimum security LAN:

- Peer-to-peer architecture
- DOS disk format
- Bootable workstations (local storage)
- No directory or file access control
- Shared printers across the network

Added-security LAN:

- Dedicated file server
- Non–MS-DOS disk format
- Diskless workstations (remote boot)
- Access control down to lowest level possible (file)
- Encryption of passwords
- Security monitoring and accounting
- Network encryption devices
- Printers attached to secured file server
- Automatic logout after dormant periods
- No remote login
- Reduced system privileges
- Fault-tolerant design

tools, a user can call a file by author, subject, project name, or creation date [14.52]. Users can search on any field to find a file anywhere in the network. Several of these LAN management tools also load the software that created the file to facilitate work on the document by the recipient. Others interwork with E-mail packages. Document-file organizers are very useful when many people need access to the information. These systems also act as front ends because they insulate the users from the complexities of the directory structure. A handful of products is available. The 1990 prices ranged from $500 to $1000.

14.2.2.2 Real-Time User Trackers

The first step in collecting network information is determining who is using it as a function of time. In some instances, all users must be logged off (for example, before doing a backup); this real-time tool will be of assistance. Trackers are included under configuration management in this discussion. While a listing of active users is generally available from the NOS, these tools have value-added features not otherwise available to the LAN network manager. These value-added features include graphical floor plans, identification beyond bridges, and sorting capabilities. The 1990 prices ranged from $100 to $300.

14.2.2.3 User Support

Many problems the user attributes to the LAN are, in fact, problems having to do with the user's unfamiliarity with the various applications that can be employed. User support tools allow the LAN manager to gain access to the remote workstation or PC; the keyboard and screen can be acquired remotely so that the manager can determine the possible source of the problem. In some systems, the manager is automatically informed of the user's hardware. In addition, some of these systems maintain statistics on who was helped, the nature of the assistance, and the duration of the session; this can be useful in billing back the support or in designing training programs. These systems range in price from $300 to $1500.

14.2.2.4 Front Ends

Nontechnical end-users should not necessarily be expected to understand DOS, the LAN software protocol suite, and other details such as version number. Front ends are included under configuration management in this discussion. Front ends are menu-driven interfaces that insulate the users from the native environment. The menus will convert from the user-friendly commands to the necessary DOS or LAN operating system commands. Many front-end systems provide a template that the network manager can customize for different LAN user groups; these menus can typically be produced at the manager's terminal and distributed over the network. A front end may perform additional functions. Some front ends provide basic usage tracking. While not as detailed as a sophisticated audit trail tool, these systems can report how much time a user spends in a particular application. Others may blank out the screen after a period of inactivity, or even log the user off. Others have a software meter that monitors the usage of applications for consistency with the license restrictions. Front-end systems range in price from $100 to $700 per server. About a dozen products were available in 1990.

14.2.3 Accounting Management

Accounting management enables charges to be established for the use of communication resources, and for costs to be identified for the use of those resources. Generating a report detailing each user's access problems is an important tool for the LAN manager for at least three reasons: (1) This can be the basis for chargeback activities. (2) It provides a hardcopy list of all activities for that reporting period; this can also be employed for security monitoring. If subtle security problems arise later (e.g., discovery of sabotage of data), this type of report may aid in determining what and who is responsible for the problem. (3) This report can be a basis for LAN usage statistics and can aid in making plans for network expansion. Such statistics reporting features, however, are lacking in all but a few of the currently available network operating systems [14.51, 14.53].

14.2.3.1 Software Meters

PC software vendors are particularly concerned about product licensing; they wish to maintain control over the number of copies of their software. Obviously, they do not wish to sell one copy of their product to a 50,000-employee Fortune 500 company and have all the employees use copies of the software. On the other hand, it would be quite inefficient for the LAN manager to buy as many copies of the software as there are employees. At any one point in time, the possibility that thousands or even hundreds of people will need a given application is unlikely. While some applications come with built-in meters, others do not. The LAN manager may consider a meter as a way to save money. A meter works on the same principle as a lending library. When a user starts an application, it is checked out of the license library; that copy of the program is returned to the library when the user is done. When all copies have been loaned, the meter will return a temporary denial. Usage statistics will be kept to assist with the "traffic engineering" of the software library. Meters range in price from $200 to $2000 (1990 prices).

14.2.3.2 Audit Trail Management Tools

Audit trail systems are a key component of security management, although the function can be considered part of accounting management. It informs the manager as to who is doing what, when, and where in the LAN. For example, it can track which users access a given resource, server, or file, and when. Additionally, audit systems can assist in billing management, by providing the data needed to charge back usage. For an audit trail system to be effective, it must provide the LAN manager with streamlined and useful information (rather than mountains of raw data). The manager may wish to audit only certain users, operations on files with certain extensions or in certain subdirectories, only certain types of operations, or certain servers [14.52]. All file and directory creations, deletions, and renames may need to be reported. A system error log report listing all system error messages to alert the LAN manager of potential problems may be advantageous. A sophisticated audit tool must allow for this management flexibility. These tools range in price from $300 to $700.

14.2.4 Performance Management

Performance management enables the manager to evaluate the behavior and effectiveness of resources and related communication activities. Performance management is concerned with the utilization of network resources and their ability to meet user service level objectives. Proactive management can optimize the network's performance. In the past, when smaller LANs (typically with fewer than 25 users) were common, performance optimization could be accomplished more easily. Today, at a minimum, a dedicated network manager is needed to manage either large LANs or the connectivity of multiple LANs into a seamless network for as many as several hundred users. LAN performance management also requires a distinction between the backbone LAN facility and smaller sub-LANs in order to monitor and maximize total network utilization. A company may have either a small number of large LANs or many small LANs connected via a backbone LAN. In either case, users require functionality across multiple-bridge LANs to gain access to information that is not on their own LAN [14.40].

The first step in improving a LAN's performance is knowing what needs to be improved. End-users can be a source of information about what needs improvement; however, analytical tools are better. It is not difficult to measure the performance of distinct stand-alone LAN components objectively. For example, the speed of a file server can be estimated by its CPU type and disk speed; the speed of a LAN transmission medium can be measured in raw Mb/s figures, which are intrinsic with the underlying technology; user workstations can be measured by how fast they run the NOS software (to build or decompose enveloped protocol data units). When all of these components are combined, however, the performance of the LAN is by no means trivial to compute. Simultaneously improving the performance of the file server, the LAN hardware (the network interface cards and cables), and the workstation should assure improvements in the entire LAN system. The situation is more complicated when only some of the LAN components are changed. For example, by changing settings on the file server, the LAN may seem slower, while by improving workstation performance, the entire LAN may seem faster. A reliable methodology is required by the network manager to study these performance issues. Performance management allows the manager to: (1) establish a benchmark to measure the current network performance against future measurements; (2) provide means to recognize mismatched network or applications software; and (3) establish analytical measures to compare possible network performance improvement strategies.

The LAN manager can improve the relative speed at which files are saved, the amount of time spent changing menus, and the amount of time it takes to load programs or data. Several performance measurement programs are now available to determine how well the LAN can move data between file servers and workstations. Some systems send records to a file on the file server. Several workstations on the LAN can run the measurement program simultaneously and the record size can be adjusted to show when the most data are being transferred. Depending on the network software, either the workstation packet buffer size, the number of buffers, or the file server cache block size or packet buffers can be adjusted. By running such programs several times, the manager should be able to reach "best case" settings, although this can be a rather long process and requires extensive record-keeping. Other systems are easier and more complete. These tools are a more rigorous approach to measuring LAN performance. By simulating several user applications, such as word processing, spreadsheet, and data base entry, and running automatically from several workstations, these programs measure LAN performance in terms of the length of time it takes to do a specific set of tasks. Because these programs keep their own statistics, they are much easier to use. Each workstation's execution time is reported to a text file, and the results are also provided in a graph [14.46].

14.2.4.1 Disk Usage Monitoring

Utilities are needed to monitor disk usage for those operating systems not providing detailed information. This will assist the LAN manager in assessing whether a user is monopolizing the server and facilitates configuration management in terms of sizing file server needs. Additionally, the manager will be able to forecast the need for new facilities at future times. Disk usage statistics include how many files are in a given directory or volume, the owner of the file, the size of the file, and the access chronology. Exception reporting for users colonizing more than a specified threshold of space is available with some products. Also, some products allow the end-user to check his or her disk usage. Other products provide partial NetWare security reports by listing users, access privileges,and group membership. These tools generally range in price from $100 to $200.

14.2.4.2 Traffic Monitoring Tools

To undertake performance management, a network manager needs a complete traffic matrix of the traffic patterns of the LAN. With such information, the manager can subsequently look at the LAN configuration to determine, for example, if a server is being used too heavily or if the network should be partitioned using bridge technology. A protocol analyzer provides detailed information about packets and related PDU headers that populate the LAN transmission medium, as discussed above; it can also compile traffic matrices. The drawbacks of protocol analyzers are: (1) they are complex and need a certain sophistication on the part of the LAN manager to use effectively and (2) they are relatively expensive. Software-based traffic monitors will collect some of the needed traffic statistics. In addition to traffic collection, some of these systems can send probes to diagnose nodal problems. Some log errors and some situations in which a user-selected threshold is exceeded will issue an alarm. A typical performance tool monitors traffic and it records how much data are sent and received from every network node, documenting the packet size, frequency, and type (data or system packet). For system packets, these systems typically distinguish between commands and internal operations messages. The data should be collected into sequential ASCII files or a spreadsheet file. These systems range from $200 to $8000.

14.2.4.3 Issues Affecting Performance

As indicated, file server performance is critical. The file server is shared by all network users, supplying them with files and applications. It is a major factor in actual LAN throughput. File server hardware can range from a PC with a small hard disk for minimal use to a high-performance, high-cost 80386-based system with large and fast external disk drives. Commonly available file servers (such as Novell, 3Com, and Banyan) are either optimized IBM PC/AT compatibles or proprietary microcomputers. In both cases, they are made up of basically two elements: (1) the CPU (together with RAM and operating system software) and (2) the disk drive subsystem. If either of these two elements is not performing well, it will affect the overall performance of the file server. Many LAN NOSs for printer sharing or program loading may perform relatively well with an 8088-based file server. If the server supports manipulation of large amounts of data, as in a multiple-user accounting or data base system, then a dedicated file server based on a high-speed processor is a better solution. Advances in disk technologies have led to 150-Mbyte drives and controllers capable of delivering a throughput of as much as 500,000 bytes per second. While these components are not inexpensive, the cost is amortized over many users.

The disk server CPU and the support hardware (RAM) clearly affect performance. If the file server is running software for the 8088 CPU, then RAM cannot normally exceed 640 kbytes. Some NOSs can take advantage of expanded RAM cards to work above 640 kbytes. File servers with 80286 or 80386 processors can address more RAM, which will improve file server performance. These systems also make more effective use of RAM with file caching and directory hashing, which allows data normally stored on disk to be loaded into RAM. This allows faster execution, faster table interrogation, and expeditious data movement [14.46].

The disk drive subsystem is also important in its own right. Many subsystems use a proprietary disk file structure to achieve higher performance than DOS. This disk format can make as much as a 50% difference in the actual throughput. Also, in general, the larger the disk capacity, the quicker the disk drive can move data to the file server. The IBM PC/AT disk subsystem is relatively slow, primarily due to the method used by the disk controller to talk to the CPU. With the advanced *small computer system interface* (SCSI) controllers and high-performance drives though, data can be delivered much more rapidly. A PC/AT with a SCSI disk controller may now be able to move 250 kbytes/s or more. Many vendors now use the SCSI disk drive interface.

One bottleneck in the process, however, occurs when the data come from the disk so fast the CPU cannot keep up. A solution is to use a disk coprocessor that can capture the remaining data until the CPU catches up. Another bottleneck occurs when the data or program files on the drive become fragmented. Most new files are laid out on the disk in a contiguous order, but as time goes by, portions of the file may be rewritten to different sections of the disk. This results in the disk drive taking longer to collect all the parts of a file. The easiest way to solve this problem is to undertake periodically a complete backup, reinitialize the disk drive, and restore the files from the backup tape. Interface cards can also affect performance. Besides the fault management issues mentioned earlier, memory management is crucial to speed and performance. Factors such as DMA *versus* shared memory, and onboard processors and buffers can mean large differences in the actual throughput of two cards on the network. The performance difference between Ethernet cards can be as high as 50% [14.46].

Another element affecting the performance is the network workstation. For example, a high-performance file server on a 10 Mb/s LAN will show the inefficiency of an IBM PC workstation with limited RAM: The workstation is now the bottleneck because it cannot accept or display data as fast as the file server and the network hardware can supply it. At times, upgrading the workstation is cheaper and more practical, rather than upgrading the LAN. Adding more RAM or a coprocessor could improve the grade of service without a single change to the network. The protocol software can also affect workstation performance. A full seven-layer OSI suite could require considerable resources to run. Even at the network layer, packet sizes, transfer buffers, and other workstation network software settings can have a major effect on the network performance.

14.2.5 Security Management

Security management supports the application of security policies. It controls access to both the network and the network management systems and may also protect information transacted by the LAN from disclosure or modification (network security in general is covered in Chapter 17).

No product or product family on the market provides the total solution for maximum LAN security. However, relatively good LAN software and hardware products are available that will meet the security requirements of most installations. The secure LANs of today are configured as a combination of products from several vendors.

Choosing a LAN software or hardware configuration that will support the needed security requires an understanding of LAN architectures, the safeguards provided by the products, and the specific security requirements of the specific target environment. With the NOSs now available, centralized data storage is achieved via common access to the file server. In this environment, critical files may reside on a central device that is accessible by all workstations within any work group. Unlike mainframe data base files, however, many considerations must be addressed by the LAN manager to ensure proper security. Control of access to LAN data poses several security problems that are not found in mainframe installations [14.51]:

- PC LAN users are usually more sophisticated than "dumb" terminal users. Because the PC operator must acquire some knowledge of DOS and its commands, an understanding of internal security structures is more common.
- In the mainframe environment, only computer operators have access to tapes and hard disks; in a LAN environment, every PC (except for the diskless workstations) stores data. Protecting this data from theft or destruction becomes a more difficult task.
- Utilities are readily available to do bypass copy protection, to expose disk substructures, or to perform sophisticated file and disk copying. Use of these utilities expose all data on the local workstation or LAN file server to security risks.

Evaluation of the company's (or even work group's) security requirements is important in making the decisions pertaining to LAN configurations: Low-security

LANs are generally less expensive and allow a wider selection of software and hardware, while high-security requirements may force the selection list to only a few options. Additional hardware and more expensive software is generally required for the more secure LAN installations. Implementation of a very secure LAN is considerably more costly than a single work group LAN.

LAN security issues fall into three major areas [14.51]:

1. *Physical Access*. Security in any data processing environment starts with controlling access to the equipment. Although intrinsically distributed in topology, LAN security requires that the file servers and printers be installed in secured rooms. Access to the LAN's cabling system is also a concern because of the potential to "tap" into the network, insert new nodes, or monitor network data traffic. Access to the PC workstation also must be considered. Even without the network or server available, the local hard disk of the workstation can pose a security risk for loss of data.
2. *Logical Access*. Physical access techniques aim at keeping unauthorized users off the network. Logical access aims at keeping authorized network users away from unauthorized files. Access to the data is partially the responsibility of the NOS (the applications must also provide their own protection). The logical control for information access is carried out via the NOS. Password access to servers, input-output rights to directory or file structures, and user accounting features represent typical support features provided by a NOS. The level of security required by any site will dictate the LAN manager's final choice of LAN software.
3. *Administrative Control*. An important but often neglected aspect of LAN security is the role of the LAN manager. This individual is responsible for physical and logical access control, in addition to undertaking fault recovery procedures, performing backups, and monitoring for potential security infractions.

14.2.5.1 Physical Access

Physical access security for LAN installations affects the three major components of a LAN: (1) the workstations, (2) the servers, and (3) the cabling.

Workstations may represent the highest security risk in a LAN: unchecked access to workstations and their local storage provides an avenue for theft of sensitive information. An authorized user may download information from the server to the workstation; once the information is stored on the local disk, usually no security processes are in place to prevent an unauthorized user from obtaining it, except by stringent physical controls. This involves installation of one or more add-on items of software and hardware. For example, unauthorized use of a workstation may be prevented by attaching keyboard lock devices; these can be software-

or hardware-based. With such tools, only the authorized user of that PC may activate the keyboard. Also, adding physical restraint equipment to the PC will prevent it from being removed from its work area. Such restraints were common for $1000 typewriters, but are not too common for $10,000 PCs.

Because of the decreased physical size of most file servers, this equipment is subject to theft. The file server should be placed in the most secure location possible; this could be a controlled MIS computer room or a special room designed for secure equipment. Even if physical access is strictly controlled to prevent theft of the server, loss of information could occur through misuse of the server console. The storing, archiving, and vaulting of the system's backup tapes must also be considered carefully. On-site storage should be under the same security restrictions as the file server, but, if possible, the backup tapes should be in a separate location for disaster recovery purposes. An off-site storage facility must also meet the same security requirements as the central site. Other areas of concern include communication gateways that can be accessed through dialup modems. Security problems, including virus infections, often arise because of the poor control over these calls. A separate port controller may be considered if security is critical.

Cable is one of the first and easiest places where a LAN security infraction can occur. Copper-based systems can be tapped easily: Tapping into a twisted-pair LAN does not even require direct contact with the cable; an electromagnetic pickup antenna can be used. Such a device can be purchased for under $20. In spite of the myth, fiber can also be tapped, but it does require somewhat more sophisticated equipment and slightly more skill. A perpetrator can use a razor blade-like knife to scrape off the fiber's cladding. The cut is deep enough to penetrate the cladding, but not the fiber itself. Once the cladding is compromised, the fiber cable can be tapped. Also, to extend a fiber link, a connection to the cable has to be available. These connections are a fiber system's potential weak point. Devices to tap a fiber connector cost around $275 [14.54]. The physical cabling is at risk if unauthorized personnel are allowed to tap it or to attach a network monitor and protocol analyzer. With these devices, not only can user passwords be determined by observing packet traffic on the cable system, but sensitive information can actually be captured.

Electromagnetic signal leakage outside the building is another vulnerability for LAN security. Coaxial and twisted-pair cabling and the devices that connect cabling (connectors, amplifiers, tap boxes) will leak a certain amount of signal. Depending on the quality of the antennas used, these signals can be decoded from a quarter-of-a-mile to several miles away. Fiber cable is more secure in this respect.

14.2.5.2 Logical Access Security Issues

Logical access control is provided principally by the LAN's NOS. A LAN's NOS should manage access to LAN resources (terminals, servers, *et cetera*); the applications should manage access to its data. The security management task is the maintenance of the "access environment," which includes all aspects of access control, including the methods of control, monitoring the effectiveness of the control, and reporting and saving the audit trails for later analysis.

A common way to attack LAN security is through a PC on the LAN. LAN security schemes must include ways of controlling access to the network. Two basic forms of access control are user authentication (usually in the form of a password, but also with biometrics) and file and program security.

Passwords. A user must own a legitimate password to gain access to the system. Theoretically, an unauthorized user will not have a valid password and, thus, will be excluded from the system. Unfortunately, passwords are notoriously weak. Users tend to employ simple-to-remember household names; they are written down, sometimes right next to the terminal; they may be shared with other users; they do not get changed for long periods of time; and not enough companies have rigid controls to revoke all passwords from all systems when people are terminated or resign. Some systems make users' passwords expire every 30 days. Password length and randomness are also critical. A number of password generator products are on the market.

One of the more reliable password mechanisms is the one-time password system. This approach employs convenient user-owned devices that generate a one-time password for the user to enter into the system and a LAN-based software counterpart. Both the remote device and the LAN employ the same algorithm to generate the next legitimate password. Access depends on the user's possession of the device. Naturally, the user must safeguard the portable device. In addition to the security risk, losing one of these devices can be expensive: The portable generators cost around $200 each.

Biometric access control is purported to be the best way to verify that the user is a valid user. Biometrics uses unique body characteristics for identifying an individual; characteristics that cannot be stolen or forgotten. Biometric controls work from the actual physical presence of the user. Devices that read a user's fingerprints or thumbprints can be used. The fingerprint reader records users fingerprints and stores them [14.55]. Another product uses infrared light to scan the retinal patterns. Biometric controls are used mostly in government installations. These devices are expensive, however, and they are still not perfect. In particular, voice-prints are currently an unreliable means of identifying people because the signal processing techniques used are very crude; some of these systems that can be trained by a particular user do not even recognize that user when he or she has a cold.

At the time of this writing, the most reliable method was a combination approach: a password generator or biometric scheme used in conjunction with a memorized password. Many LAN managers, however, currently rely only on simple passwords. This approach may suffice in a number of cases, particularly for office automation environments. Note, however, that a trade-off exists between security and productivity and user friendliness [14.56].

File and Program Security. Once a user has been authenticated and has gained entry into the system, security concerns turn to what the user can access. This is termed *access control* in the evolving international security standards discussed in Chapter 17. The effectiveness of any access control scheme depends on how granular a control the LAN manager has over resources and objects. In a LAN environment, resources include servers, disk volumes, directories, and files. A user is assigned access rights to one of these objects and that determines the operations that can be performed. A minimum of READ or WRITE access control is given to the user, while some operating systems extend control to UPDATE, ADD, and DELETE. In terms of implementation, access control is provided by a combination of the features of the NOS and the specific security utility. While the NOS will allow access control at the directory or file level, the security utility must provide control at the record and even field level [14.51].

The manager needs to delimit the data a user can look at, and even limit the applications and utilities that can be accessed. Most NOSs offer a form of file security that allows the LAN manager to determine to which volumes, directories, and even individual files a given user can have access. Access to programs must also be controlled. One way to prevent unauthorized users from running given programs is to put them in a directory that the file security system does not allow these users to read. Another approach is to use an application control system [14.54].

Access monitoring allows the LAN manager to track security problems in real time. Most NOS software lacks this ability, which is common in mainframe systems. Being able to detect a security infraction as soon as it happens is often critical to determining the problem and quickly resolving it. With this information, the manager can temporarily suspend the user account in question, review security detail information, and make a determination as to the source of the problem.

Table 14.9 provides a summary of security approaches for the LAN environment.

Debate in some quarters has occurred as to whether access control must be provided by the network or by the applications. The answer will probably depend on the environment. Military or financial environments may require the highest possible level of security, and access control in both the network and the application may enhance this security. In other cases, the security requirements may not be as rigid.

Diskless PCs are another means of physical security on a LAN. Diskless PCs have no floppy or hard disk drives. Each user stores data on the server's hard disk. Diskless PCs impede users from stealing corporate information or software. Some institutions, particularly the government, also eliminate printers. By eliminating the disk drive, diskless PCs make it difficult to introduce viruses onto the network.

Table 14.9
Security Protection for Various LAN Environments

General office applications—low or no access control
Peer-to-peer architecture
Password protection
Sensitive data and proprietary software—medium access control
Client-server architecture
Password protection
Access control to directories
Classified or secure data—high access control
Client-server architecture
Password format control
Diskless workstations
Security monitoring
Access reporting

Taking a different stance, some argue that diskless PCs reduce user flexibility and, in some cases, productivity.

14.2.5.3 Administrative Issues

The administrative procedures and responsibilities are critical to security. The LAN manager must generate the security policies, perform backup and restore procedures, and implement the specific access architecture to support the desired level of security. The manager must educate users on the scope of the security policies; for example, the simple practice of logging out from the network each time a user leaves his or her desk is a key component for maintaining security. Periodically, the LAN manager should run diagnostic utilities to ensure that disk information is intact. Also, a review of the actual data should be performed in order to detect possible sabotage or corruption of the information [14.51].

Maintaining the data integrity of any data processing system requires procedures that provide insurance against disk failures. Usually, backup security involves periodically generating a backup copy of the information. Typically, magnetic tape is used for backup because it is inexpensive and can be stored economically. Some NOSs can duplicate directory structures, do read-after-write verification, and on-the-fly disk sector error recovery. However, if the server crashes, the data still need to be reconstructed from a separate backup (possibly from tape). The removable nature of tape reels and cartridges represents another security risk. When tape backup is performed, the tape data are in one of two formats: image or file-based. The image tape format is an exact disk image; restore operations are direct, consisting of restoration of the complete disk. File-based tape methods sequentially store entire files in directory order on the tape. With a file-based tape, a certain amount of search is needed to restore files from the tape, exposing all files. In considering security aspects of backup and restore utilities, the image format is both faster in execution and more secure because of the limited operations support for these formats—many backup software packages require that image backup be restored to the same drive from which they were backed up.

Fault recovery tools automate the task of rebuilding servers. This is done by gathering critical file server information on hard disks or floppies, including the bindery, login scripts, directory rights, system autoexec file, and printer definitions [14.52]. Products typically range from $200 to $1500.

14.2.5.4 Protocol Analyzer Issues for Security

A protocol analyzer device, described earlier, in the hands of the wrong people can be a security threat. If an infiltrator can gain access to a LAN port or is able to tap the cable, the analyzer can reveal useful information. An analyzer can capture the entire dialogue taking place over the LAN, and it can display passwords in an easily readable form. Appropriating passwords is easy with analyzers, but passwords may not always be useful to the infiltrator in a properly designed LAN. For example, the stations from which a user can log in can be restricted. Thus, although the infiltrator may have the manager's password, he or she cannot log in as the supervisor unless the actual manager's terminal is used. In addition, audit trail utilities may report logins and logouts, with special attention paid to the manager's ID. This is the reason reliable security measures that go beyond basic password protection are needed.

While protocol analyzers can present problems for LAN security, they can in turn be used to monitor the network for infractions. One simple technique involves looking for stations that are not supposed to be on the network. The manager can set the display to depict unknown stations. This is done by declaring an easily readable name for each LAN station. If a program claims to lock certain files, for example, the analyzer can be used to test that claim. With some analyzers, the network manager can write programs in C for specialized functions such as monitoring compliance with security procedures. For example, such a program might look through the data to find stations that are logged on to a file server but show no activity for long periods of time. This may indicate a station where the user has walked away without logging off, which is a violation of security policies in most institutions [14.12].

14.2.6 NetView and LANs

IBM's NetView allows users to identify and correct problems in a traditional teleprocessing network. Announced in 1986, it allows key nodes of a large network to send alarms and alerts to the mainframe. In addition to the control and problem determination features, NetView provides a network billing system, designed to charge back costs for a voice system, and a traffic engineering line optimization system to design voice networks. NetView/PC is an interface to NetView for non-SNA devices. The interface does not provide any functionality; a LAN manager program must generate the information for NetView/PC to format it and send it to NetView. Technically, NetView has the potential to carry out centralized network management for a large collection of local work group LANs; the problem is one of jurisdiction.

In spite of a number of limitations of NetView, comprehensive alternatives do not appear to be available for the foreseeable future. Compared to AT&T's Unified Network Management Architecture, NetView had almost a two-year head start. According to industry watchers, if real standards are not finalized soon, NetView will become a solid *de facto* standard [14.37].

A number of vendors have announced NetView interfaces for their products. Most networking vendors will probably have to support NetView to some extent. However, NetView has a centralized architecture. Many LAN network managers prefer a more decentralized approach. Some industry watchers guess that the big LAN vendors will try to develop superior, low-cost proprietary management systems for their LANs, while providing gateways to NetView [14.37].

Architecturally, IBM has developed an approach to network management that separates the management of a network from the transport of data through the network. Network components acquire two types of responsibilities in this framework: transport and management [14.57]. The approach defines four entities within the management framework: (1) focal point, (2) service point, (3) entry point, and (4) target. These entities interact to take advantage of the capabilities that each network component offers to the rest of the network.

A focal point offers the functions required to manage the network from a central location. The focal point manages all of the remotely and locally attached network components. This would be NetView, in IBM's product parlance. The service point makes a network component visible to the network management system residing at a focal point. The service point handles the interaction between the focal point and the network component, especially with regard to transporting management information. This is NetView/PC, in IBM's product context. The entry point is a network component from the perspective of transporting information through the network. It assumes the responsibility of a service point, providing network management information about itself to the network manager.

The target is a network component that does not have its own direct access to the network management system. It is a device or subsystem that provides only information transport; the management capabilities are provided to it by a service point (see Figure 14.19).

The focal point–server point–target approach is applicable to all network components regardless of their structures or transport characteristics; however, this approach is particularly well suited when the LAN subsystem is considered to be a target and the management server of the LAN is coupled with the functions of a service point, as seen in Figure 14.19. LAN Manager provides the LAN-specific operation at a service point and reporting to a centralized focal point. The NetView/PC program provides the service point function that is linked to NetView, the focal point in a mainframe. In this fashion, the hierarchy of network management is extended into the LAN.

LAN Manager can provide centrally accessible control for the management servers distributed to the rings through LLCs, with the LAN reporting mechanism residing with each server. Statistical or problem information can be forwarded to the LAN Manager; the LAN Manager also has a local operator interface, allowing active management of the LAN. Because communication with the LAN Manager uses the LLCs, an implementation of the LAN Manager common to both the token-ring network and the PC network is possible. The alert interface of the NetView/PC program forwards selected error information, compiled by the LAN Manager, to the NetView program. The alert is displayed at the NetView program console along with probable causes and recommended actions. The NetView/PC program also provides other interfaces to the service point application, such as the service point command facility (SPCF), which receives commands from the local point for response by the service point application. The SPCF will allow remote LANs and other systems to be incorporated into an automated operations strategy,

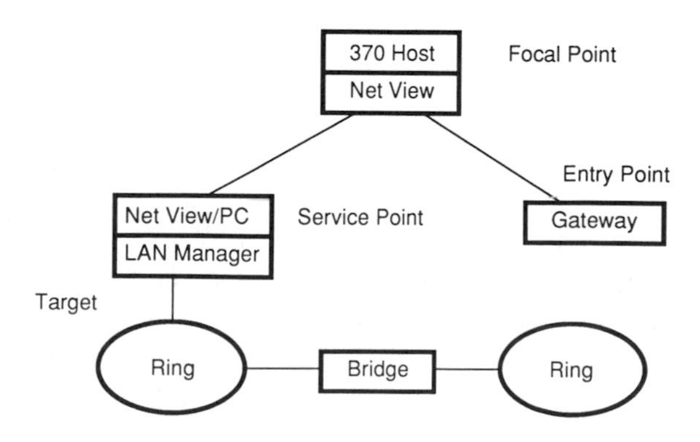

Figure 14.19 IBM's hierarchy for LAN management.

whereby human responsiveness is augmented and enhanced with programmed control [14.57, 14.58].

REFERENCES

[14.1] B.G. Kim, *Current Advances in LANs, MANs, and ISDN,* Artech House, Norwood, MA, 1990.

[14.2] B. Thacker, "The Computer Integrated Organization: Some Business and Technical Issues for OSI-MAP/TOP Solution," Department of Trade and Industry, United Kingdom, and Society of Manufacturing Engineers, 1988.

[14.3] D. Minoli, "Where the Action Is—Some Trends in Local Area Networking," *LAN Magazine*, October 1986.

[14.4] J. Mollenauer, "What to Expect from IEEE P802.6," *Journal of Data and Computer Communications*, Summer 1989, pp. 48 ff.

[14.5] G. Kessler, "Logical Link Control," *LAN Magazine*, January 1988, pp. 65–67.

[14.6] R.C. Dixon, D.A. Pitt, "Addressing, Bridging, and Source Routing," *IEEE Network*, January 1988, Vol. 2, No. 1, pp. 25 ff.

[14.7] J.B. Rickert, "Evaluating MAC-Layer Bridges, Beyond Filtering and Forwarding," *Data Communications*, May 1990, p. 117 ff.

[14.8] G. Stix, "Telephone Wiring: A Conduit for Networking Standards," *IEEE Spectrum*, June 1988, pp. 38 ff.

[14.9] D. Zwicker *et al.*, "Ethernet over Twisted Pair: Not As Easy As It Looks?," *Telecommunications*, June 1990, pp. 23 ff.

[14.10] T. Haight, "A Twisted Future," *Communications Week*, June 18, 1990, p. 1.

[14.11] "Standards Watch," *Data Communications*, May 1990, p. 52.

[14.12] M. Hurwicz, "The Sniffer Threat," *LAN Magazine*, April 1988, pp. 90–93.

[14.13] H. Salwen *et al.*, "Examination of the Applicability of Router and Bridging Techniques," *IEEE Network*, January 1988, Vol. 2, No. 1, pp. 77 ff.

[14.14] W.D. Sincoskie, C.J. Cotton, "Extended Bridge Algorithms for Large Networks," *IEEE Network*, January 1988, Vol. 2, No. 1, pp. 16 ff.

[14.15] P. Kelly, "Connecting LANs with Bridges," *Telecommunications*, June 1990, pp. 31 ff.

[14.16] L. Bosack, C. Hendrick, "Problems in Large LANs," *IEEE Network*, January 1988, Vol. 2, No. 1, pp. 49 ff.

[14.17] J. Boroumand, Bellcore, personal communication, June 1990.

[14.18] W.M. Seifert, "Bridges and Routers," *IEEE Network*, January 1988, Vol. 2, No. 1, pp. 57 ff.

[14.19] F. Backes, "Transparent Bridges for Interconnection of IEEE 802 LANs," *IEEE Network*, January 1988, Vol. 2, No. 1, pp. 5 ff.

[14.20] L. Zhang, R. Perlman, "Comparison of Two LAN Bridge Approaches," *IEEE Network*, January 1988, Vol. 2, No. 1, pp. 44 ff.

[14.21] Glasgal product literature, Northvale, NJ.

[14.22] J. Hart, "Extending the IEEE 802.1 MAC Bridge Standard to Remote Bridges," *IEEE Network*, January 1988, Vol. 2, No. 1, pp. 10 ff.

[14.23] M. Soha, R. Perlman, "Comparison of Two LAN Bridge Approaches," *IEEE Network*, January 1988, Vol. 2, No. 1, pp. 37 ff.

[14.24] M.C. Mamner, G.R. Samsen, "Source Routing Bridge Implementation," *IEEE Network*, January 1988, Vol. 2, No. 1, pp. 33 ff.

[14.25] C.M. Brown, "An Alternative for Interconnecting Multiple LANs," *Data Communications*, May 1990, p. 60.

[14.26] R. Perlman *et al.*, "Choosing the Appropriate ISO Layer for LAN Interconnection," *IEEE Network*, January 1988, Vol. 2, No. 1, pp. 81 ff.
[14.27] C. Cargill, M. Soha, "Standards and Their Influence on MAC Bridges," *IEEE Network*, January 1988, Vol. 2, No. 1, pp. 87 ff.
[14.28] E. Benhamou, "Integrating Bridges and Routers in a Large Internetwork," *IEEE Network*, January 1988, Vol. 2, No. 1, pp. 65 ff.
[14.29] D.E. Comer, *Internetworking with TCP/IP, Principles, Protocols, and Architecture*, Prentice Hall, Englewood Cliffs, NJ, 1988.
[14.30] A. Brenner, "OSI Model Update," *LAN Magazine*, June 1987, pp. 48 ff.
[14.31] D. Wallace, "How to Interwork Between TCP/IP and OSI," *Telecommunications*, April 1989, pp. 46 ff.
[14.32] J.D. Case, M.S. Fedor, M.L. Schoffstall, J.R. Davin, "A Simple Network Management Protocol, Request for Comments 1098," DDN Network Information Center, SRI International, April 1989.
[14.33] J.D. Case, "Managing Your INTERNET: The Simple Network Management Protocol," INTEROP89, San Jose, CA, October 4–6, 1989.
[14.34] G. Bennett, "The Simple Network Management Protocol," *Telecommunications*, February 1990, pp. 21 ff.
[14.35] J.D. Case, "The Simple Network Management Protocol for TCP/IP-based Internets," Advanced Computing Environments' Tutorial Material, Mountain View, CA.
[14.36] C. Gallanger, "Beyond ISDN: A LAN Interface for Integrated Voice and Data," *Telecommunications*, April 1989, pp. 75 ff.
[14.37] M. Hurwicz, "NetView Now," *LAN Magazine*, April 1988, pp. 76–79.
[14.38] D. Minoli, "Managing Local Area Networks: Fault and Configuration Management," DataPro Report NM50-300-401, August 1989.
[14.39] "Information Retrieval, Transfer, and Management for OSI: OSIRM Part 4—OSI Management Framework," ISO/IEC JTC1/SC21, Revision of DIS 7498-4, October 1988.
[14.40] M. Pyykkonen, "Local Area Network Industry Trends," *Telecommunications*, October 1988, pp. 21–28.
[14.41] C. Zarley, "Software Helps Managers Track LANs," *PC Week*, January 1989, p. C/11.
[14.42] P. Schnaidt, "Smorgasboard: 20 Ways to Feed a Hungry LAN Manager," *LAN Magazine*, June 1988, pp. 84–95.
[14.43] B. Enyart, "Network Managers' Wish Lists Keep LAN Vendors on Their Toes," *PC Week*, March 13, 1989, pp. 33, 34.
[14.44] E. Ericson, L. Ericson, D. Minoli, *Expert Systems Applications to Integrated Network Management*, Artech House, Norwood, MA, 1989.
[14.45] J. Schwartz, "Fixing a LAN," *LAN Magazine*, March 1988, pp. 70–75.
[14.46] J. Diehl, "Network Tune-up," *LAN Magazine*, May 1988, pp. 120–123.
[14.47] M. Mohanty, "Troubleshooting Tools," *LAN Magazine*, March 1988, pp. 64–69.
[14.48] T. Ooka, "Accurate Loss Measurement for Optical LANs," *Telecommunications*, July 1988, pp. 48–56.
[14.49] M. Hurwicz, "New Smells," *LAN Magazine*, December 1988, pp. 86–87.
[14.50] E. Tittel, "The LANalyzer," *LAN Magazine*, December 1988, pp. 89–93.
[14.51] R. Watson, "Fortifying a LAN," *LAN Magazine*, October 1988, pp. 51–56.
[14.52] P. Schnaidt, "The Arsenal: 36 Ways to Arm a LAN Manager for Network Battle," *LAN Magazine*, December 1988, pp. 69–84.
[14.53] D. Minoli, "Managing Local Area Networks: Accounting, Performance, and Security Management," DataPro Report NM50-300-501, June 1989.
[14.54] M. Mohanty, "Defending a LAN," *LAN Magazine*, April 1988, pp. 84–88.
[14.55] D. Greenfield, "Sensible Paranoia," *LAN Magazine*, April 1989, pp. 84–88.

[14.56] T. Peak, Bellcore, personal communication, March 1990.

[14.57] M. Willett, R.D. Martin, "LAN Management in an IBM Framework," *IEEE Network*, March 1988, Vol. 2, No. 2, pp. 6–12.

[14.58] D. Minoli, "Interworking LANs," *Network Computing,* October 1990, p. 96.

Chapter 15

Metropolitan Area Networks and Fiber Distributed Data Interface

A metropolitan area network (MAN) is a standardized, high-speed network providing LAN-to-LAN and LAN-to-WAN connections for public or private communication systems in noncontiguous real estate within metropolitan-range distances [15.1]. In addition, MANs may also provide voice and video services.

Two MAN technologies are emerging: fiber distributed data interface (FDDI) and distributed queue dual bus (DQDB). The former is more suited to private applications in a campus environment, and needs dedicated facilities and rights of way; the latter provides a system for public communication over a larger geographic area, can use shared telecommunication facilities, and does not require the customer to own rights of way. Both of these technologies of the 1990s are discussed in this chapter. In addition, a proposed BOC service that could exploit MAN technology, Switched Multi-Megabit Data Service (SMDS[SM]), is discussed.

15.1 FDDI BACKGROUND

In the 1980s, a lack of standards hampered the growth of the short- to medium-distance fiber optic market, including LANs and private point-to-point premise communication. FDDI standards that are aimed at rectifying this situation are now available from ANSI. In October 1982, ANSI Committee X3T9.5 was chartered to develop a high-speed data networking standard. The standard that evolved specifies a packet switched LAN backbone that transports data at high throughput rates over a variety of fibers. FDDI grew out of the need for high-speed interconnections among mainframes, minicomputers, and associated peripherals (the general charter of the ANSI X3T9 parent committee is computer interfaces) [15.2].

FDDI is intended for use in a high-performance general-purpose multiple-station network. FDDI specifications encompass a token-passing network employing two pairs of fibers operating at 100 Mb/s. FDDI technology provides a fiber backbone for interconnecting multiple LANs that is independent of the protocols used by the constituent LANs connected to the FDDI ring. In addition to this backbone application, FDDI can be used as a front-end technology, i.e., as a high-speed LAN for high-end workstations. As of the beginning of 1990, the standard had developed to the point at which products based on it were already reaching the market, and more will appear in the 1991–92 time frame. The standard covers the first two layers of the OSIRM, through the MAC sublayer [15.3]. FDDI can use ISO or TCP/IP protocols for the other layers [15.4].

The optical-based FDDI LAN was designed to enjoy the same type of serial interconnection provided by LANs, while providing the high bandwidth, inherent noise immunity, and security offered by fiber. Remember that, at the inception of the FDDI effort in 1982, fiber was used mostly for point-to-point applications, not for the "any-to-any" connectivity allowed by LANs. Although considerably higher data rates are possible over fiber, higher rates would result in significant increased costs and smaller ring capacity in terms of the number of repeaters. FDDI is meant to provide inexpensive connectivity; thus, it focused on the 100 Mb/s rate.

FDDI accommodates synchronous and asynchronous data transmission. An extension of FDDI, called FDDI II, addresses isochronous channels for real-time digitized voice and compressed video. Unlike existing open standards for LANs, in which fiber optic variants have been introduced following successful specification on conductive media, FDDI was designed from the beginning as a fiber optic network. This has involved standardization issues in such areas as duplex optical connectors, fiber characteristics, optical bandwidth, bypass relays, and cable assemblies [15.5]. The FDDI ring is designed for an overall BER of less than 10^{-9}. The network can tolerate up to 2 km of fiber between stations and can support a total cable distance of 100 km around the ring with 500 attachments (1000 physical connections and a total fiber path of 200 km) [15.6]. The intrinsic topology of FDDI is a counter-rotating token-passing ring (note the arrows in Figure 15.2).

At least part of the reason why FDDI employs a ring topology is based on the characteristics of optical communication. Bus and passive star topologies would require the optical transmission to be detected at several sources simultaneously. Although practical fiber optic taps are beginning to become available, the attenuation is still such that the number of nodes is relatively limited [15.6]. Because fiber optic transmission is best handled with a point-to-point configuration, this aspect was included in FDDI's definition. Two types of topologies are possible with point-to-point links: the active star and the ring. The problem with an active star is that is has a single point of failure that incapacitates the entire network; a similar problem exists with a single ring. Consequently, the FDDI specification calls for a dual or active star ring. Table 15.1 provides a glance at FDDI features.

Table 15.1
FDDI at a Glance

Standardized by ANSI X3T9.5

Dual counter-rotating ring

- Multimode fiber for connections up to 2 km
- Single-mode fiber for connections up to 50 km

100 Mb/s on each ring

4500 octets per frame (including header and trailer)

Up to 500 stations

100-km total perimeter path (per fiber)

4B/5B NRZI encoding ("1" generates a transition, "0" generates no transition)

Isochronous traffic included in FDDI II

MAC operation:

- A station must wait for a token before transmitting
- Each station repeats the frame it receives to the downstream neighbor
- If the destination address matches the station address, the frame is copied into the station's buffer
- The receiving station sets indicator symbols in the frame status field to indicate reception
- The transmitting station is responsible for removing its own data frames from the ring

Note that, although FDDI is somewhat similar to the IEEE 802 standards, it is not part of that family of standards.

FDDI-specified lasers operate at 1300 nm (general fiber transmission-receiver technology operates at either 850, 1300, or 1550 nm, as described in Chapter 7). While the performance increases as the wavelength increases, so does the cost. For local data communication, both in LANs and in point-to-point applications employing fiber optic modems, the 850-nm light sources are typically employed; however, this technology becomes infeasible for 100 Mb/s beyond a couple of miles. At the other end of the range (1550 nm), the system becomes expensive and may not be necessary, particularly because special dispersion-shifted fibers would be required. In the early days, the committee designing FDDI was also looking at short-wavelength implementations. Sometime around 1985, they realized that to meet all the requirements, particularly the 2-km station-to-station spacing, the system would have to be specified at the longer wavelength of 1300 nm. Based on these considerations, the FDDI committee selected the 1300-nm technology; this allows the use of less expensive LEDs and the resulting distance and data rates are within the range desired for LANs and LAN backbones. The issue of fiber size was settled sometime after the wavelength decision [15.7].

The FDDI standards directly address the need for reliability. This need arises from the fact that a backbone transports a large number of user sessions and its loss would be a serious outage. FDDI incorporates three reliability-enhancing methods. First, a failed or unpowered station can be bypassed optionally by an automatic optical bypass switch. Second, wiring concentrators are used in a star wiring strategy to facilitate fault isolation and correction. Third, two rings are used to interconnect stations so that a failure of a repeater or cable link results in the automatic reconfiguration of the network.

15.1.1 FDDI Standards

As of the end of 1990, FDDI consisted of four adopted or draft proposals. The MAC is specified in X3.139-1987, approved on November 5, 1986. The *physical* (PHY) layer standard is contained in X3.148-1988. The *physical medium dependent* (PMD) is specified in X3.166, the latest revision of which is dated March 1, 1989. The revisions were being edited into the standard, which was expected to be available in mid-1990. The *station management* (SMT) is under development; revision 6.2 was to be voted on in early 1990. Letter ballot work is continuing in 1991. Table 15.2 provides a more detailed listing of standards activities [15.8].

Table 15.2
Status of FDDI Documents as of June 1, 1990

Document	Status
MAC:	
ANSI X3.139-1987	Standard is published
ISO 9314-2:1989	Standard is published
Enhanced:- MAC-2	Resolving letter ballot
Rev. 3.1 (5/25/90)	
PHY:	
ANSI X3.148-1988	Standard is published
ISO 9314-1:1989	Standard is published
Enhanced:- PHY-2	In public review
Rev. 3.1 (5/25/90)	
PMD:	
Rev. 9 (3/1/89)	ANSI standard approved
X3.166-1989	(Await final ISO text for publishing)
ISO IS 9314-3	ISO standard approved
	(Published 1990)
SMT:	
Rev. 6.2 (5/18/90)	At X3T9 Letter Ballot
HRC:	
Rev. 6.0 (5/11/90)	At X3 for 2nd public review
→ X3.186–199x	
DP 9314-5	DP letter ballot issued
SMF-PMD:	
Rev. 4.2 (5/18/90)	At X3T9 for further processing
→ X3.184–199x	(Forward to X3 for 2nd public review)
ISO IS 9314-4	Ready for publication

Extensions to FDDI are also under development and are covered later in this chapter.

The MAC specifies access to the medium, addressing, data checking, and frame generation-reception. PMD specifies the optical fiber link and related optical components. PHY specifies encoding-decoding, clocking, and data framing. SMT specifies the control required for proper operation of stations on the ring; services such as station management, configuration management, fault isolation and recovery, and scheduling procedures are provided.

The token-passing protocol used in FDDI is somewhat similar to the IEEE 802.5 standard for a 4 Mb/s twisted-pair ring [15.9], although some differences exist. When no station is transmitting, the token (a small control packet) circulates around the ring. When a station needs to transmit data, it will wait until it can detect a token. It then removes the token from the ring and transmits the data frame. The station then places the token back on the ring for the next station to use it. The data frame will circulate around the ring and will be captured (copied) by the intended recipient; the data frame returns to the transmission station, which will remove it from the network. This process is depicted in Figure 15.1.

15.1.2 FDDI Station Types

Two station types are allowed in FDDI: Class A and Class B. Class A stations (also known as "dual attachment stations") connect to both the primary and secondary rings of the network. Data flow in opposite directions on the two rings. A Class A station can act as a wiring concentrator to interconnect several Class B (single-attachment) stations. Wiring concentrators give the network administrator a common maintenance point for a number of stations (see Figure 15.2).

The reconfiguration of the ring in case of failure is shown in Figure 15.3. Here the link between two Class A devices is truncated. The two stations are able to detect the failure and patch the network by channeling data through part of the secondary ring.

In Figure 15.4, we see a break in the cable between a Class A and a Class B device. Here communication continues over the primary link as the Class A device detects the failure and makes appropriate internal modifications. The Class B device, however, remains detached; a Class B device trades lower cost against the fault-tolerance of the more sophisticated Class A device.

The secondary ring can also be used to carry traffic data; this gives the fully configured FDDI system 200 Mb/s of effective throughput. When the two rings merge to support a backup configuration, the network data rate drops to 100 Mb/s.

15.1.3 Applications

Although optical fiber has seen widespread introduction in the telecommunication environment (long-hauls, interoffice, feeder plant, *et cetera*), it has not yet done

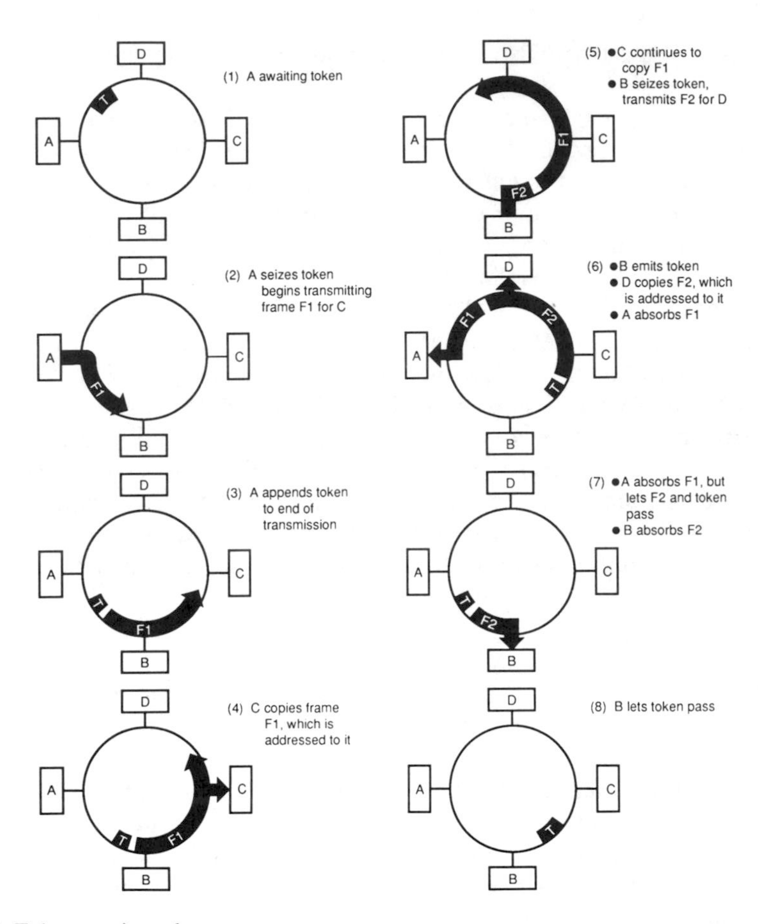

Figure 15.1 Token-passing scheme.

as well in the LAN arena. Three possible reasons are: (1) increased technical complexity compared to passive copper and coaxial cable, (2) cost considerations, and (3) lack of a workable standard. FDDI will solve the third problem and, in the process, also begin to resolve the second problem.

FDDI allows (1) the building of larger capacity LANs or LAN backbones to serve new data needs (file transfer, graphics, *et cetera*) and some voice needs; or (2) the interconnection of LANs in MANs (see Figure 15.5). The initial application of FDDI as a "back-end" interconnect for high-powered computing devices and peripherals required a high degree of fault tolerance and data integrity. As developers proceeded, they realized that FDDI could also serve high-speed "front-end" applications. These front-end applications include terminal-to-terminal and terminal-to-server communication, typical of a LAN. In large networks (say, 3000 to 10,000 terminals, particularly when workstations are involved), the aggregate

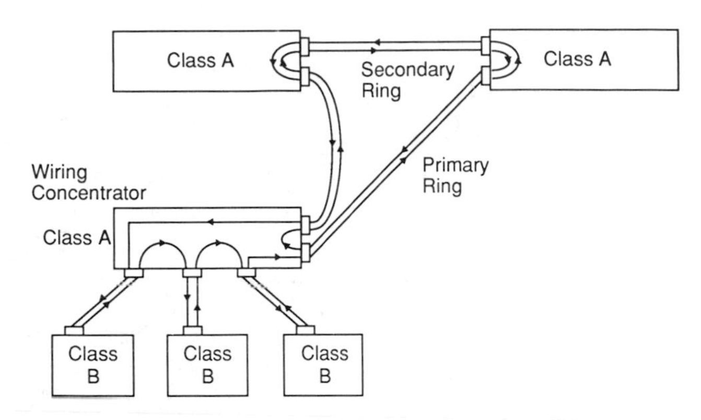

Figure 15.2 The FDDI environment.

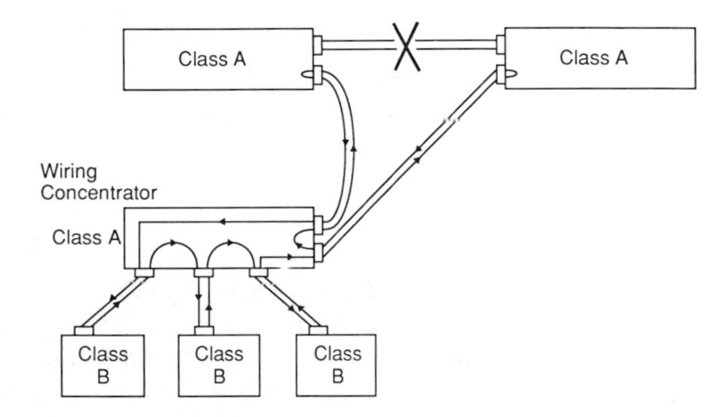

Figure 15.3 Ring rearrangement under failure.

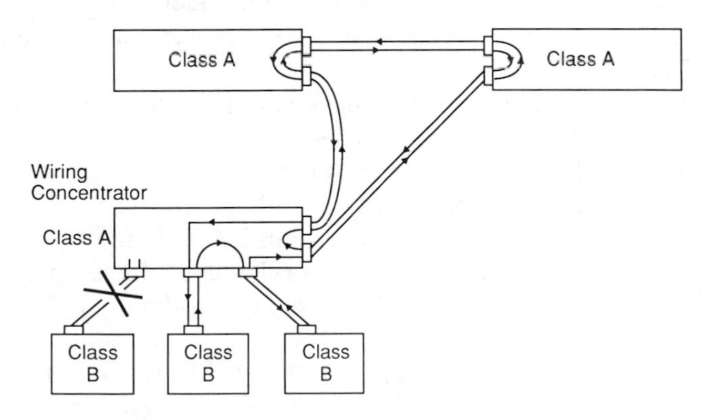

Figure 15.4 Failure of Class B stations.

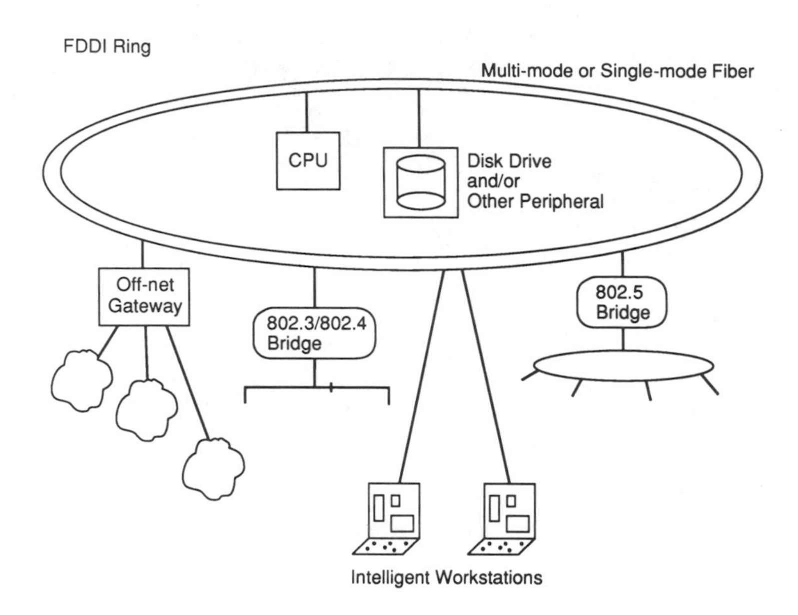

Figure 15.5 Typical applications of FDDI.

demands for network resources can overwhelm a 4 or 10 Mb/s LAN; this is where FDDI becomes important. Most LAN suppliers acknowledge that today the trend in corporate-premises networking is toward higher speed backbones to link departmental LANs, and to higher speed processing in bandwidth-intensive applications such as CAD-CAM graphics and desk top publishing [15.10, 15.11]. Desktop systems that are likely to be linked via high-speed networks now find applications in real-time simulations, graphic transfers in a CAD-CAM environment, supercomputer terminals, medical imaging, and video. FDDI will both influence the rate of acceptance of fiber-based LANs, as well as be influenced by the penetration of fiber LANs.

A high-speed FDDI ring is ideal as a backbone for "departmental" LANs, typically operating at lower speeds. High capacity on LANs can be achieved in two ways: by using multiple channels with each at relatively low speed or by using one channel at a relatively high data rate (this is the FDDI approach). The multiple-channel approach can be used with a broadband bus LAN. One drawback of this approach is that bridges must be provided between channels; the architecture must be designed properly to avoid high rates of interchannel traffic and bottlenecks at the bridges.

The one area for which FDDI proper is not well-suited is for broadband LANs carrying full analog video (6 MHz). A broadband LAN can easily carry multiple channels of video as well as data; digitizing a 6-MHz TV channel results in 45 to 90 Mb/s (45 Mb/s is a slightly compressed version); this can easily swamp a FDDI backbone. BISDN, particularly in a HDTV environment, will require around 150 Mb/s of dedicated bandwidth per channel, after compression (uncompressed HDTV requires on the order of 1.5 Gb/s). Planners are now discussing delivery of up to three channels per domicile, reaching into the 600 Mb/s range. FDDI is a shared-medium technology; uncompressed video applications or local loop applications may not be able to use the FDDI standards. Compressed video is more easily handled by FDDI II. FDDI II aims at supporting up to sixteen 6 Mb/s isochronous channels. However, FDDI II is not as well stabilized in the standards process compared to FDDI.

NASA is planning to use a FDDI backbone in the $16 billion space station Freedom [15.12].

15.2 FDDI SPECIFICATIONS

Layer 1 (physical layer) is specified in two documents: the FDDI PMD and the FDDI PHY (see Figure 15.6). The physical layer provides the medium, connectors, optical bypassing, and driver receiver requirements; it also defines encoding-decoding and clock requirements as required for framing the data for transmission on the medium or to the higher layers of FDDI.

When the FDDI committee realized that considerable discussion would be needed with regard to fibers, connectors, and other hardware, they decided to break the standardization of the OSI physical layer into the two pieces. In this way, the relatively noncontroversial issues, such as coding and other matters that IC chip manufacturers need to know to begin design, could be put in a formal document and approved independently of other items pertaining to the standard [15.7].

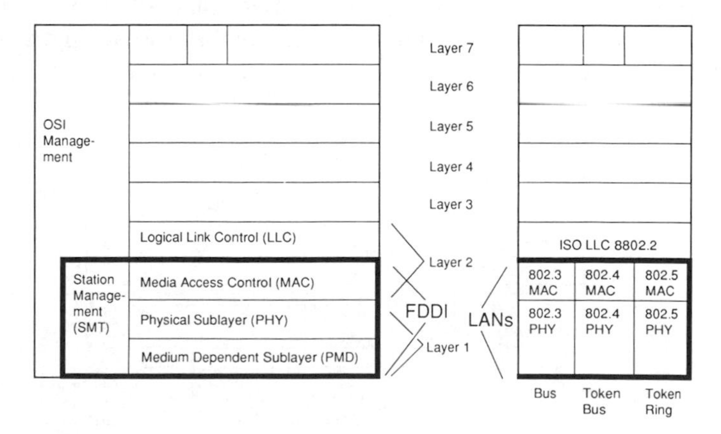

Figure 15.6 FDDI protocols (compared to LANs).

The specification *single-mode fiber PMD* (SMF-PMD) is shown in Figure 15.7. Multimode fiber has a high attenuation, as discussed in Chapter 7. The original FDDI PMD limited multimode links between nodes or repeaters to 2 km. To overcome this limitation, X3T9.5, following the commercial introduction of a number of vendor-proprietary converters to extend this distance, standardized a version of FDDI for use in single-mode fiber. Single-mode fiber allows the signal to travel up to 50 km between nodes or repeaters. A node can be directly attached to a single-mode fiber or a converter can be used between multimode and single-mode fiber [15.8]. Single-mode FDDI equipment was appearing on the market as of early 1991.

The DLL is also divided into two sublayers:

1. A MAC portion that provides fair and deterministic access to the medium, address recognition, and generation and verification of frame check sequences. Its primary function is the delivery of frames, including frame insertion, repetition, and removal.
2. A LLC portion provides a common protocol to provide data assurance services between the MAC and the network layer. The LLC specification, however, is not part of the FDDI standard, as depicted in Figure 15.6.

Each of these sublayers is now discussed in detail.

15.2.1 Physical Medium-Dependent Specification

PMD defines the optical interconnecting components used to form a link. It describes the wavelengths for optical transmission, the fiber optic connecter, the functions of the optical receiver, and (as an option) the bypass switch that can be incorporated into the station. It specifies the optical channel at the bulkhead of a station. The source is defined to radiate in the 1300-nm wavelength. PMD also describes the peak optimal power, optical rise and fall times, and jitter constraints.

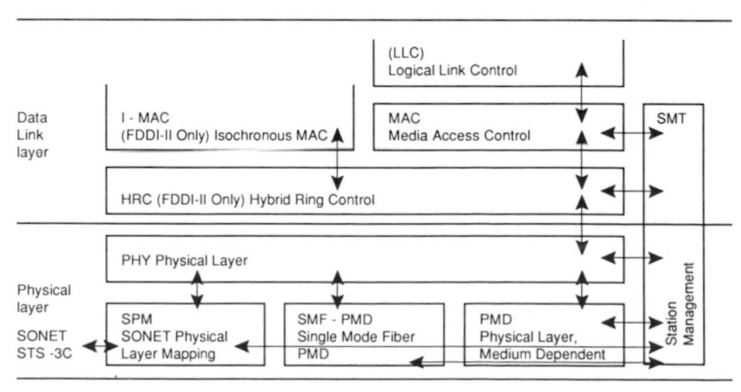

Figure 15.7 Structure of FDDI and FDDI II standards.

The minimum rise-fall time is 0.5 ns [15.7]. The standard includes the following specifications:

1. Services:
    1.1 PMD to PHY services
    1.2 PMD to SMT services;
2. Media attachment;
3. Media signal interface;
4. Interface signals; and
5. Cable plant interface specification.

As discussed earlier, multimode fibers were initially specified for distances up to 2 km. The dimensions of the optical fiber are specified in terms of the diameter of the core of the fiber and the outer diameter of the cladding layer. Fiber specification is 62.5/125 μm (core diameter/cladding diameter). The applicable standards for the fiber are EIA-455-48 (core), EIA-455-27 or EIA-455-48 (cladding), and EIA-455-57 (aperture) [15.7].

As with all layer 1 specifications, PMD defines the duplex connector to be used for FDDI access. The intention is that the primary and secondary ring connections to each Class A station be attached simultaneously using the duplex-connector and a dual fiber cable. The connectors can be used for both Class A–to–Class A as well as Class B–to–Class A (wiring concentrator) links. The bypass relay is used to connect the optical inputs (at the primary and secondary rings) directly to the optical output in case of a station or link failure; this allows the ring to maintain continuity.

In the 1300-nm region, the dispersion due to multimode interference is at minimal. The combination of physical parameters selected will ensure the 10^{-9} desired BER. LEDs (either surface- or edge-emitting) are implicitly assumed in PMD. However, PMD does not specify that the emitter has to be a LED; it could also be a laser, as long as—at the optical port—the optical interface parameters are met [15.7]. At some future point, some manufacturers may include lower cost local-loop-grade laser emitters, or even long-haul-grade LDs in a FDDI package by adjusting the optical output at the optical port to conform with the standard. For the foreseeable future, however, all manufacturers pursuing FDDI products are employing LEDs.

15.2.2 Physical Sublayer Specification

PHY represents the upper sublayer within OSI layer 1. It defines the encoding scheme used to represent data and control symbols; it also describes the method for retiming transmission within the node. The standards include the following specifications:

1. Services:
   1.1 PHY to MAC services
   1.2 PHY to PMD services
   1.3 PHY to SMT services;
2. Facilities:
   2.1 Coding
   2.2 Symbol set
   2.3 Line states.

Digital data needs to be encoded in some form for proper transmission, as discussed in Chapter 11. The type of encoding depends on the nature of the transmission medium, the data rate, and other factors such as noise, reliability, and cost. Given the fact that fiber is inherently an analog medium, a digital-to-analog technique is required. Intensity modulation is the norm for fiber: A binary 1 can be represented by a pulse of light and a binary 0 by an absence of optical power. The disadvantage of using this method in the simplest form is its lack of synchronization: Long strings of 1s or 0s create a situation in which the receiver is unable to synchronize its clock to that of the transmitter. The solution is to first encode the binary data in such a way as to guarantee the presence of signal transitions, even if no transitions are present in the incoming digital signal. After this encoding is done, the signal can be presented to the optical source for transmission using intensity modulation. A typical encoding scheme is Manchester encoding (see Figure 15.8).

Differential Manchester is only 50% efficient because each data bit is represented by transitions in signal. Two transitions allow a degree of robustness in the presence of noise, as would be the case in coaxial cable. Because fiber is less

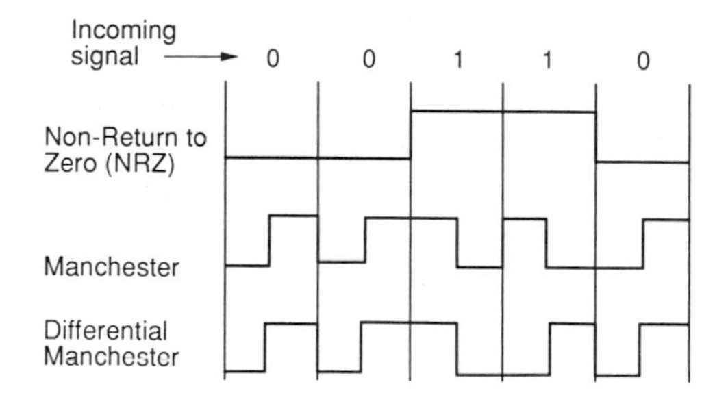

In Manchester code there is a transition at the middle of each bit period. The mid-bit transition serves as a clock and also as data: a high-to-low transition represents a 1, and a low-to-high transition represents a 0.

In Differential Manchester the mid-bit transition is used only to provide clocking. The coding of a 0 (1) is represented by the presence (absence) of a transition at the beginning of the bit period.

Figure 15.8 Comparison of coding schemes used in LANs or fiber.

susceptible to noise, two transitions are not required to identify a bit with a good degree of confidence. To avoid having to use a 200-MHz signal, FDDI specifies the use of a code referred to as 4B/5B group encoding. The result is that the 100 Mb/s throughput is achieved in FDDI with a 125-MHz rate, rather than the 200-MHz rate needed in differential Manchester; this helps keep the cost and complexity of the equipment down. One drawback of the group encoding pertains to clock recovery: Because differential Manchester has more pulses in its stream, it is easier to extract the clock in that scheme than in the former. One of the key responsibilities of this FDDI sublayer is that of decoding the 4B/5B NRZI signal from the network into symbols that can be recognized by the station, and conversely. The synchronization clock is derived from the incoming signal; the data are then retimed to an internal clock through an elasticity buffer. In this scheme, 4 bits of data are translated into a 5-baud value transmitted over the network, giving an 80% efficiency factor. This group-encoding scheme employed in FDDI is a departure from differential Manchester codes normally specified in LAN standards.

To understand how the FDDI scheme achieves synchronization, note that encoding occurs in two stages. In 4B/5B, the encoding is done 4 bits at a time; in effect, each set of 4 bits is encoded as 5 bits. Then, each element of the 4B/5B stream is treated as a binary value and encoded using NRZI. In this code, a binary 1 is represented with a transition at the beginning of the bit interval; no other transitions occur. The advantage of NRZI is that it employs differential encoding: The signal is decoded by comparing the polarity of adjacent signal elements rather than the absolute value of a signal element. This scheme is relatively robust in detecting transitions in the presence of noise or other distortions; the NRZI encoding will improve reception reliability [15.13]. Table 15.3 shows the symbol encoding used in FDDI.

Encoding 4 bits (16 combinations) with 5-bit patterns (32 combinations) results in patterns that are not needed. The codes selected to represent the sixteen 4-bit patterns are such that a transition is present at least twice for each 5-bit code; given a NRZI format, no more than three 0s in a row can be allowed, because, with NRZI, the absence of a transition indicates a 0. The remaining symbols are either declared invalid or are assigned special meaning as control symbols as shown in Table 15.3.

PHY also provides line states for establishing the station's links with its neighbors (upstream and downstream) to determine the integrity of the station's links to these neighbors. These line states are used to exchange a handshake with a neighbor. A node receiving a line state on its primary input can respond by sending the proper line state on the secondary output. The line states are composed of a repetition of one or more I symbols [15.6].

Another item that needs to be resolved in this sublayer is the issue of timing jitter. Jitter is the deviation of clock recovery that can occur when the receiver attempts to recover clocking as well as data from the received signal [15.13]. The

Table 15.3
4B/5B Codes Used in FDDI

Function or 4-bit group (4B)	*Group Code (5B)*	*Symbol*
Starting Delimiter (SD)		
First symbol of sequential SD pair	11000	J
Second symbol of sequential SD pair	10001	K
Data Symbols		
0000	11110	0
0001	01001	1
0010	10100	2
0011	10101	3
0100	01010	4
0101	01011	5
0110	01110	6
0111	01111	7
1000	10010	8
1001	10011	9
1010	10110	A
1011	10111	B
1100	11010	C
1101	11011	D
1110	11100	E
1111	11101	F
Ending Delimiter		
Used to terminate data stream	01101	T
Control Indicators		
Logical ZERO (reset)	00111	R
Logical ONE (set)	11001	S
Line Status Symbols		
Quiet	00000	Q
Idle	11111	I
Halt	00100	H
Invalid Code Assignment		
These patterns shall not be transmitted	00001	Void or Halt
because they violate consecutive	00010	Void or Halt
code-bit zeros or duty cycle requirements	00011	Void
Some of the codes shown shall	00101	Void
nonetheless be interpreted as a Halt if	00110	Void
received.	01000	Void or Halt
	01100	Void
	10000	Void or Halt

clock recovery will deviate in a random fashion from the transitions of the received signal; if no countermeasures are used, the jitter will accumulate around the ring. In LANs, the IEEE 802.5 standard specifies that only one master clock will be used on the ring, and that the station that has that clock will be responsible for eliminating jitter, using an elastic buffer. If the ring as a whole runs ahead of or behind the master clock, the elastic buffer expands or contracts accordingly. This centralized clocking method is not practical for a 100 Mb/s ring: At this speed, the bit time is only 10 ns compared to a bit time of 250 ns at 4 Mb/s, making the effect of distortion more severe. Consequently, FDDI specifies the use of the distributed clocking scheme. In this environment, each station uses its own autonomous clock to transmit or repeat information onto the ring. Each station has an elastic buffer; data are clocked into the buffer at the clock rate recovered from the incoming stream, but they are clocked out of the buffer at the station's own clock. This distributed system is considered to be more robust than the centralized method and to minimize jitter; as a consequence of reclocking at each station, jitter does not limit the number of repeaters in the ring, as is the case in LANs where the master clock is used.

15.2.3 Media Access Control Specification

Layer 2, DLL, of the OSIRM is traditionally divided into two sublayers in a LAN context: LLC and MAC. FDDI only defines MAC, which controls the flow of data over the ring. The token-passing protocol incorporated in FDDI controls transmission over the network. MAC defines packet formation (headers, trailers, *et cetera*), addressing, and CRC. It also defines the recovery mechanisms. This standard defines the following specifications:

1. Services:
   - 1.1 MAC to LLC services
   - 1.2 PHY to MAC services
   - 1.3 MAC to SMT services;
2. Facilities:
   - 2.1 Symbol set
   - 2.2 PDUs
   - 2.3 Fields
   - 2.4 Timers
   - 2.5 Frame counts
   - 2.6 Frame check sum.

The FDDI packet format is shown in Figure 15.9 [15.3, 15.6]. Packets are preceded by a minimum of 16 IDLE control symbols; the packet itself is characterized by a start delimiter (SD) composed of the J and K control symbols. This is followed by a frame control field that identifies the type of packet. The destination address, which follows, identifies the recipient of the frame; the source address is also included to identify the station that originated the packet. The address field can be up to 48 bits in length. The variable information field follows, along with a frame check sequence field of 32 bits. The check sequence covers the frame control field, the two addresses, and the information field. An end delimiter, which consists of the T symbol, is transmitted. The maximum length of the packet is limited by the size of the elastic buffer in the physical sublayer and the worst-case frequency difference between two nodes; the upper bound is 9000 symbols or 4500 bytes [15.6, 15.9]. Figure 15.9 also shows the format of the token.

In an idle condition, MAC connects to an internal source of IDLE control symbols to be transmitted over the ring. When a SD is detected from the ring, MAC switches to a repeat path; the packet is monitored and copied if it is meant for this destination. The packet is simultaneously repeated on the ring for relaying. The MAC can also inject its own packet if it has the token. Packets are removed only by the origination station. The MAC repeats the packet only until the sender address field is detected; if the destination recognizes this as its own destination, it will insert IDLE control symbols back onto the ring (the fragmented packet is ignored and removed by any station holding a token for transmission).

Stations that desire to transmit must first obtain a token (this is the unique packet shown in Figure 15.9). The procedures for obtaining the token and the amount of time allowed for data transmission (to retain fairness) are specified in the Timed Token Protocol (TTP), similar to IEEE 802.4. A station obtains the token by performing the stripping function on the incoming token. Only the SD field is repeated onto the ring; the station will inject its own information at this juncture. When the packet is sent, the station immediately issues a new token. TTP guarantees a maximum token rotation time. The token-based operation is exemplified in Figure 15.1, in Section 15.1.2. TTP allows two types of transmission: synchronous and asynchronous. In the synchronous mode, stations obtain a pre-

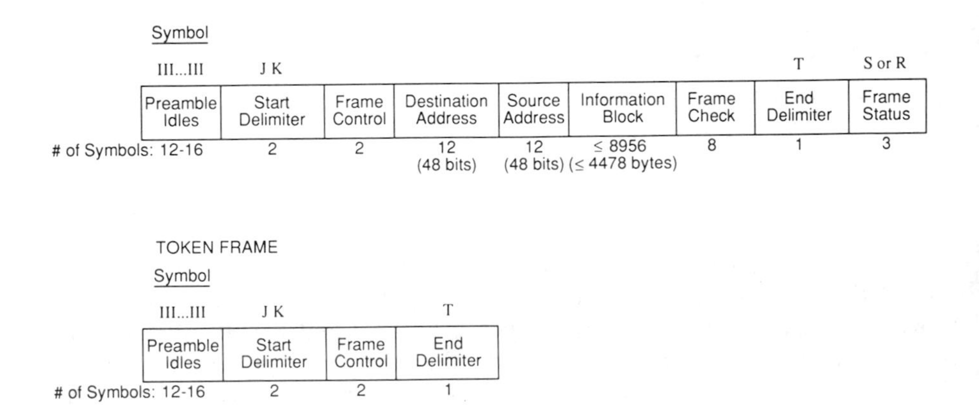

Figure 15.9 FDDI frame formats.

defined amount of transmission bandwidth on each token rotation. The balance of the bandwidth is shared among stations using the asynchronous service; these stations can send data when the token arrives earlier than expected. Any unused capacity left over from synchronous capacity is available to asynchronous traffic, which may be subdivided into up to eight levels of priority. The amount of time allowed for asynchronous transmission is bounded by the difference of the token's actual arrival time and the expected arrival time. In essence, each station keeps track of how long it has been since it last saw the token; when it next sees the token, it can send synchronous traffic or any asynchronous traffic for which time remains available.

15.2.4 Station Management

As mentioned earlier, the FDDI specification also includes SMT; this is outside the OSIRM itself (see Chapter 16), but it can be part of OSI network management (OSINM) standards, at least in the long term. SMT interfaces to the bottom three FDDI sublayers; it provides intelligence for operation of the individual sublayers in an FDDI node, including error detection and fault isolation.

SMT would define the interaction of stations and nodes on the network. In addition to network management, SMT is the means by which FDDI establishes and maintains connections between stations [15.14]. The debate in the standards committees has centered on how much remote network management should be included in the standard. One view is that management aspects are already covered in OSIRM and OSINM, leading to duplication of standards and possible incompatibilities. A second view is that the SMT must include a closed set of management capabilities, and that the standard may be delayed until consensus on this approach is reached. If SMT, describing configuration and control is not complete, "FDDI" products on the market cannot be vendor independent. A number of manufacturers at the beginning of 1990 claim that their products are FDDI-conformant; these vendors are using their own version of station management [15.15].

15.2.5 Higher Layers

From LLC upward, FDDI is generally intended to fit traditional protocol stacks, ISO or TCP/IP. Upper layer protocols are needed for any type of productive network activity (such as transferring files over an FDDI system). Progress has been made toward TCP/IP implementations over FDDI [15.8, 15.16]. Three methodologies have been proposed, particularly with reference to the treatment of the IP subnet number; convergence should occur in the near future. Other efforts are also under way to migrate upper layer LAN stacks to ISO-based standards, particularly under the MAP/TOP thrust. These market demands are forcing investigations into an upgraded third and fourth layer [15.10].

15.3 1990–92 ACTIVITIES

FDDI was originally expected to be completed by the end of 1989; however, the full standard may not be completed until early 1992, with stable SMT conformant products appearing in 1992–93 and penetration into the market to occur thereafter. The delay has arisen because of debate late in 1989 in the SMT portion of the four-part standard. Active dialogue is taking place on the features that should be included in the SMT. The previous chapter emphasized the need for good management capabilities in a LAN environment, and this issue cannot be underestimated.

SMT

In 1989 and early 1990, SMT was the focus of the X3T9.5 task group. SMT was the only major part of the FDDI standard that as of June 1991 was not yet published as a full standard. The issues pertain to the management information base (discussed briefly in Chapter 15 and in more detail in Chapter 16). Initially, SMT was conceived as a mechanism for managing the processes in the various FDDI layers, providing services such as station insertion and removal, connection management, ring configuration management, and fault isolation. To perform these tasks, SMT must maintain a store of information, including resource availability and timer settings. Eventually, new protocols were introduced to SMT so that external entities might examine this MIB information. More recently, further protocols were introduced so that the external entities might be able to change the values in the MIB. This "write" capability raises numerous issues, which are holding up the final resolution of the standard.

In February 1990, the FDDI station management *ad hoc* working group advanced the SMT document to the X3T9.5 task group. The next steps were to return the letter ballot by July 1990 and then resolve the many expected comments (this letter ballot was expected to return a substantial number of "No" votes, as many contentious issues remain) [15.8].

In the meantime, manufacturers were emphasizing the stable part of the specification; some of these vendors plan to put SMT in loadable software modules that can be exchanged when the standard is finalized. Much development can take place at the physical and link layer level.

FDDI-SONET Mapping

A FDDI-SONET mapping working group, reporting to X3T9.5, is working on a mapping of the FDDI physical layer for SONET STS-3c transmission. A SONET mapping for the FDDI symbol stream has been making progress toward standardization. FDDI standards and T1X1.5 (SONET standards) are converging on a mutually agreeable mapping, and technical stability was expected by the end of 1991. The FDDI-SONET *ad hoc* working group must still define precisely how the mapping is used to provide an equivalent to the multimode PMD or SMF-PMD, but this should be a straightforward exercise [15.8]. The mapping standard will be included in ANSI T1.105 Addendum B.

FDDI II

FDDI II was bogged down in discussions of signaling at the time of this writing. The hybrid ring control (HRC), which is essentially complete, standardizes how up to sixteen 6 Mb/s isochronous channels may be superimposed on the ring. However, it says nothing about how stations should signal each other to allocate these channels (i.e., set up calls); this is where the 1990 efforts were directed.

While much of the FDDI mainstream has lost interest in FDDI II, new proposals continue to be made to standardize a fiber distributed voice-video-data interface. These efforts, which were attracting some attention in 1990, are variously known as FDDI Follow-On LAN (FFOL), FDDI III, Future Integrated Services LAN (FISLAN), and Fiber Distributed Video/Voice Data Interface (FDVDI) [15.8].

FDDI over Twisted-Pair Wire

In mid-1990, two companies announced a technology to transmit 100 Mb/s over shielded and unshielded twisted-pair wire. The technology is motivated both by the lower cost of electrical transceivers, compared to electro-optical transceivers, and by the embedded base of twisted-pair wire. In 1990, optical fiber cable was nearly as cheap as twisted-pair copper wire, but the FDDI electro-optical adapter cards cost around $10,000 each. To achieve real savings, however, a standard is required (the Pronet-80 LAN from Proteon Inc. has offered 80 Mb/s on twisted-pair wire for some time, but it uses a proprietary method [15.17]). The X3T9.5 FDDI committee agreed to address the need and feasibility of new PMDs for alternative media, including twisted-pair wire and lower cost fiber (for example, plastic fibers). A standard may be available by late 1992 (based on technology developed by Crescendo Communications, Inc.). The cost of FDDI connections had decreased to about $2,500 per terminal by 1992.

15.3.1 Some FDDI Vendors

One of the more active vendors in this area is Fibronics, which began development work early; the vendor now has a product line including Ethernet and token-ring interfaces, an IP router, and a channel-extender. Other vendors include Fibermux

Corporation, Digital Equipment Corporation (DEC), Sun Microsystems, Sumitomo Electric USA, and Advanced Micro Devices Inc. (these last three manufacture FDDI chips) [15.18, 15.19]. Several vendors manufacture FDDI-ready bridges, including Canoga, DEC, and Fibronics.

Some observers had doubts about customer acceptance of a product built on an incomplete standard, but interest in FDDI chips and products was already strong in the 1989–90 time frame [15.20]. (Approximately 100 FDDI networks were deployed in the United States at the beginning of 1990 [15.8].) In addition to implementing the layer 1 and 2 standards, any such chips need to maintain the throughput dictated by FDDI. When FDDI is used as a backbone technology, the bridge or router needs to be able to do efficient packet filtering, based on address destination. To be viable in an FDDI environment, the addresses must be kept in RAM. Typically, 4000 address comparisons per microsecond are required. Content-addressable memory is the best VLSI technology for this task.

15.4 PUBLIC MANs

The media access protocols generally used in a LAN cannot accommodate networks covering large geographic areas without a loss of efficiency. MAN protocols are tailored to larger networks. MANs must transmit faster than LANs because these networks typically will interconnect sets of LANs, high-speed computer links (including channel-to-channel transmission), and perhaps voice and video. For this type of connection, 100 Mb/s is just the beginning. Both fiber and coaxial cable are suitable as the underlying media, with the preference being for fiber (see Figure 15.10).

LANs are typically privately owned facilities, dedicated to one user. The geographic reach of MANs in most cases requires them to be public in order to be cost-effective. A public MAN will interconnect the premises of many organi-

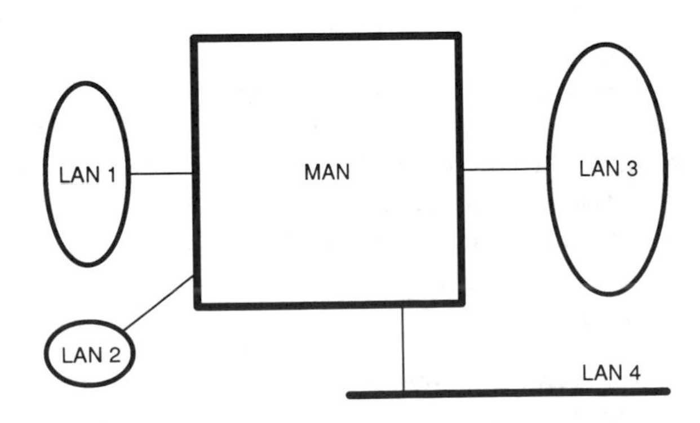

Figure 15.10 A MAN.

zations. An exception to this rule would be in a large campus environment that covers dozens of buildings belonging to a university, medical center, or industrial park. A private MAN will be made up of dedicated circuits distributed over the campus or city.

For premises communication, little incentive is provided to use packetized LAN technology for voice communication because PBXs do a fairly good job, and the advantages of multiplexing over a common medium do not compensate for the cost associated with the multiplexing equipment. Over metropolitan distances, however, the economies of integration begin to make sense. PBXs in various buildings throughout the city may be interconnected effectively over shared media rather than over dedicated DS1 or DS3 links.

Some of the applications of MANs include interconnection of LANs; interconnection of mainframes (channel-to-channel); interconnection of PBXs for private network applications (discussed in Chapter 10); medical imaging transfers; graphics and CAD-CAM; and digital (compressed) video for teleconferencing use. Two requirements that determine the type of protocols that can be used are: (1) the treatment of stream traffic (voice and video) and (2) security. Delay requirements for voice are stringent because of echo problems [15.21–15.23]. To meet the voice restrictions, a maximum delay of 2 milliseconds is allowed in a MAN. Because none of the existing IEEE 802 standards meet this constraint, a new standard is needed—this is IEEE 802.6.

A private network (such as a WAN) interconnects only the premises of one company and is as secure as the telephone network. Namely, to tap into the information stream, a potential intruder needs to identify physically—from among numerous cables—the cable carrying the information in question (this cable may be underground or on top of a pole). In a LAN-like environment, every user's data passes through every node in the network. In public MANs, the solution is to separate the MAN into a transport network and an access network. The transport network carrying the totality of the data remains within the reach of the carrier. The bridging equipment is carried as far as a vault or pole outside of the customer's premises; alternatively, it can be in a CO, with dedicated fiber links connecting the customer to the CO. The user is connected to the transport network by an access network that extends outward from their premises and attaches to the transport segment via a bridge; only that customer's data are available over the access network. The length of the access network is from zero to several miles.

Because the bridging occurs at the MAC level on the carrier's property, the connection between the access network and the transport network is secure; it is also transparent to the user (see Figure 15.11). Because the bridge operates at the MAC level, it is invisible to the layers above the MAN or downstream from the LANs. Data circulating in the transport network destined for other users are ignored by the bridge: The bridge transfers data onto the access network only when it is addressed to the user connected to the given access network. The bridge

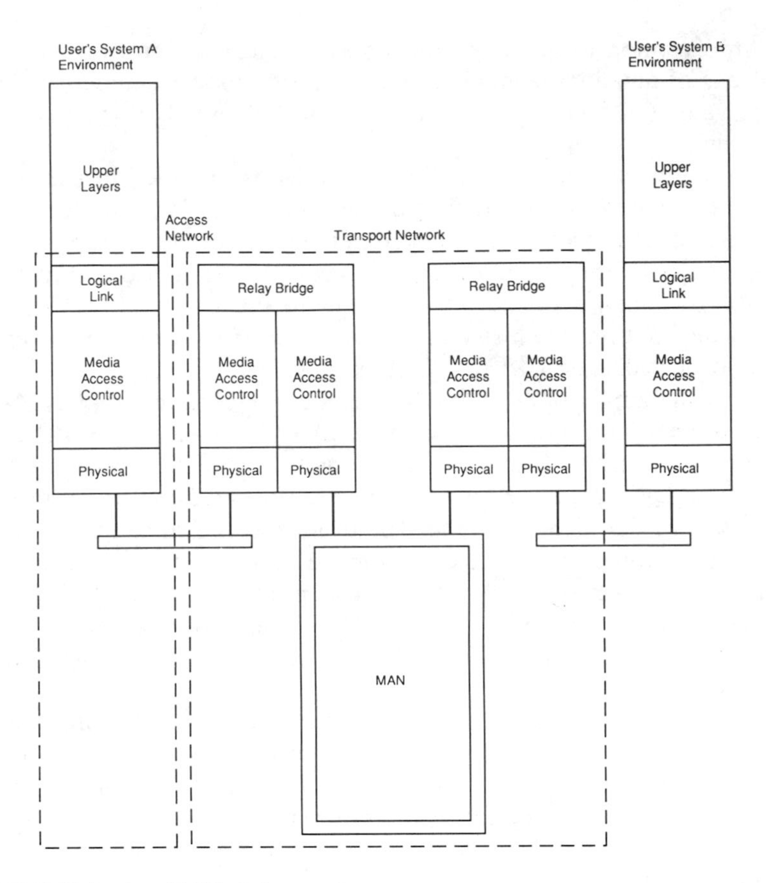

Figure 15.11 Bridging in a MAN environment.

also performs a *closed user group* function for data originating within a subnetwork LAN connected to the MAN: The bridge forwards data onto the transport network only if it is addressed to a MAN destination, and it ignores data destined to locations on the local network.

In addition to the transmission facilities, MAN systems require major network management support, including security management, accounting management, and fault management (this is discussed in more detail in the next chapter).

15.4.1 IEEE Project 802.6

Many carriers now support the MAN concept. Work in defining MAN standards is coordinated by IEEE Project 802.6. The widespread deployment of the technology requires standards for development of equipment and for interconnection. The purpose of MANs, as embodied in IEEE 802.6, is to provide integrated services such as data, voice, and video over a large geographical area (at least 50 km in diameter). The use of media access bridging is a relatively new development; although it does not conform precisely to the OSIRM, it offers performance advantages relative to bridging done at the network layer. In addition, it requires less knowledge of network details on the part of each node [15.21].

The access network need not be based on the IEEE 802.6 protocol; it can, in fact, be an extension of the user's own bus- or token-based LAN or LANs.

The P802.6 group started to work on the MAN issue in 1982. Early supporters included the satellite and CATV industry. The satellite providers wanted to provide inexpensive high-speed links between the earth stations and the local customers; CATV wanted to deliver data over their TV networks. A number of protocols and media were analyzed between 1982 and 1987 [15.21]. In late 1987, the committee achieved a consensus, and a dual-bus architecture proposed by Telecom Australia was selected as the system on which to base the MAN standard. The system is formally known as Distributed Queue Dual Bus. DQDB technology was invented by students at the University of Western Australia [15.1]. In May 1987, a company called QPSX was founded and the technology was further developed.

The ninth draft of IEEE 802.6, published in August 1989, modified the length of the slot to 48 octets for the payload to interwork more directly with the ATM standards (discussed in Chapter 4). IEEE 802.6 became a standard in December 1990.

The 802.6 standard defines a high-speed shared medium access protocol for use over a dual unidirectional bus subnetwork (taken together the two buses provide bidirectional connectivity). The standard specifies the physical layer and the DQDB layer required to support the following:

- A LLC sublayer by a connectionless MAC sublayer service provided to support a LLC sublayer in a manner consistent with other IEEE 802 LANs;
- Isochronous service users by a connection-oriented service that may be used to transport isochronous data (for example, PCM digitized voice). (*Isochronous* refers to the time characteristics of an event or signal recurring at known, periodic time intervals); and
- Connection-oriented data service users by a connection-oriented service that may be used to transport bursty data, for example, signaling.

The applicability of the DQDB technology is shown in Figure 15.12. IEEE 802.6 specifies the DQDB subnetworks required to provide telecommunication services within a metropolitan area "up to and exceeding 50 kilometers in diameter" [15.24]. Typically, a MAN would consist of interconnected DQDB subnetworks; the in-

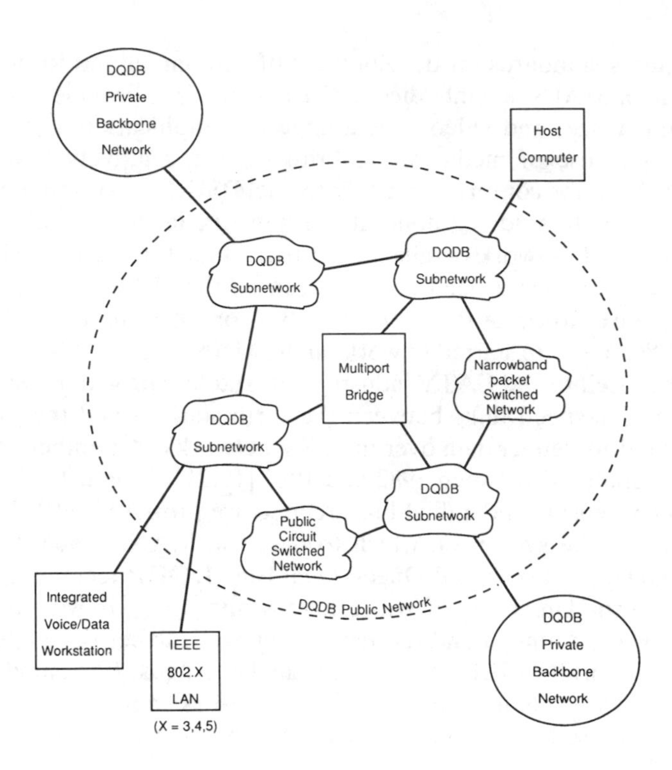

Figure 15.12 DQDB as a public network.

terconnection of the subnetworks can be via bridges, routers, or gateways. The various connectionless, connection-oriented, and isochronous data streams share the total capacity of the subnetwork flexibly and equitably.

Currently, the DQDB operates on fiber at 1.5 or 45 Mb/s in each direction (if necessary, coaxial cable and microwave segments of adequate bandwidth can also be used). The dual bus is physically star-wired at a central location. The cable can be brought back to a CO for physical maintenance. The scheme differs from the earlier ring systems: The network, although cabled as a ring, is logically a bus. In practical terms, this means that a designated node in the network does not repeat the incoming data at the other nodes; this node serves as the logical beginning and logical end of the two buses. The end node generates the slot framing for each fiber cable (this is similar to the master node in a ring system). See Figure 15.13. Packets in this environment do not have to be specifically removed from the network at the end of their journey.

The advantage of the dual bus topology is its fault tolerance. If a node or line segment fails, continuity can be achieved by logically designating the failed portion as the end of the bus and letting the previous bus end become a regular

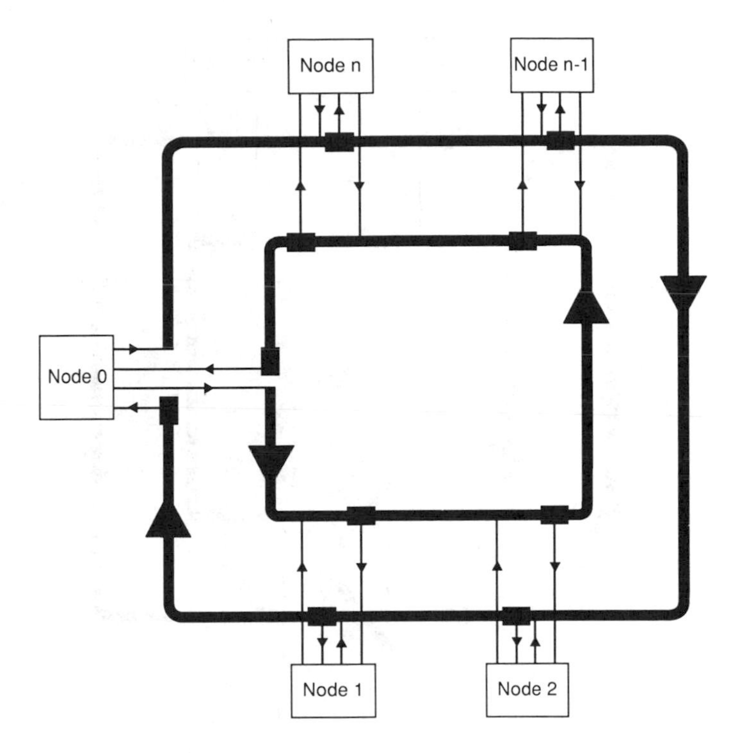

Figure 15.13 Dual bus architecture.

node. The nodes on either side of the failure can take up the bus-end function, as shown in Figure 15.14. In addition to this mechanism, bridging relays protect against the failure of nodes.

The bus uses a kind of distributed reservation method. The bus controller maintains a counter for each direction of transmission. The counter is incremented for each slot request arriving from the direction of the destination, and is decremented for each vacant slot toward the destination. If the counter is zero when a station wishes to transmit, it can use the next vacant slot. If requests are pending, the counter will have a nonzero value. The controller determines the value of the counter (which continues to count for future requests) and separately counts down as vacant slots come by. These unfilled slots satisfy the requests already pending when the station decided to transmit. When the value has reached zero, the next vacant slot is available for use. Greedy terminals are kept under control by limiting the terminals to only one outstanding reservation.

The distributed queue access control is used for bursty traffic. The operation of the distributed queue protocol provides three key advantages over existing LAN protocols:

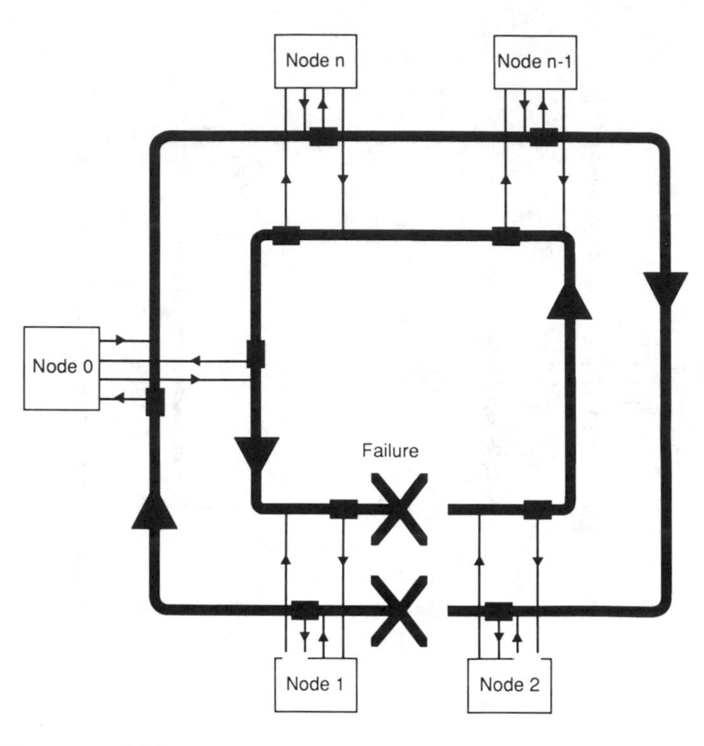

Figure 15.14 Treatment of failures.

1. An ordered queue is established for medium access such that capacity on the network is never wasted.
2. The access discipline allows fine granularity and control in the allocation of the bandwidth.
3. The performance on the network is substantially better than that achievable on a token ring (such as in the FDDI scheme). Access delay for a token ring is typically around 0.3 milliseconds at 25% network loading and 0.5 milliseconds at 50% loading. The access delay remains flat around a few microseconds for 10% loading all the way to 90% loading [15.25].

Figure 15.15 shows the slot structure as defined in Reference [15.24]; however, this is conceivably still subject to change.

Figure 15.16 depicts the mapping of MAC-level data units into the smaller queue-arbitrated segments. For additional technical information on IEEE 802.6, see References [15.24], [15.25], and [15.26].

15.4.2 IEEE 802.6 Timetable

Once the standard is approved within the IEEE (December 1990), the standard is expected to be submitted to ISO for international ratification. This process takes

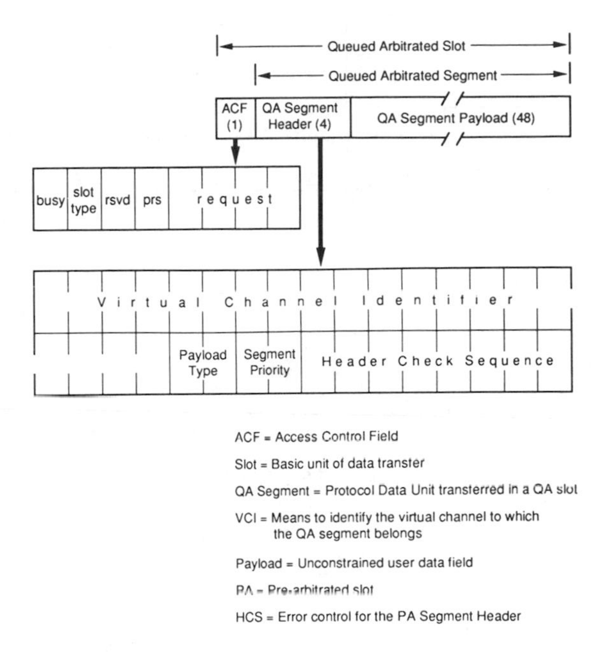

Figure 15.15 DQDB layer PDU formats as of late 1989.

a year or more, even if no difficulties are identified by other countries in applying the technology to their environments. International standard status for 1992 is an optimistic estimation [15.1, 15.21].

15.4.3 Switched Multi-Megabit Data Service

The demand for switched high-speed data services, as opposed to dedicated lines, is now emerging. SMDS is a Bellcore-proposed high-speed (DS1-DS3) connectionless packet switched service that provides LAN-like performance and features over a metropolitan area. SMDS is a public connectionless packet-switched service that provides for the exchange of variable length data units up to a maximum of 9188 octets. SMDS is defined as a technology-independent service, whose early availability via MAN technology is envisioned for the 1991–93 time frame. SMDS is expected to be one of the first switched broadband offerings and will eventually be supported by BISDN [15.27].

User Access to SMDS

A technical advisory has been published by Bellcore on SMDS. The advisory defines the subscriber-network interface (SNI), which complies with IEEE 802.6 (see Figure 15.17). SMDS is offered via a SNI that is dedicated to an individual

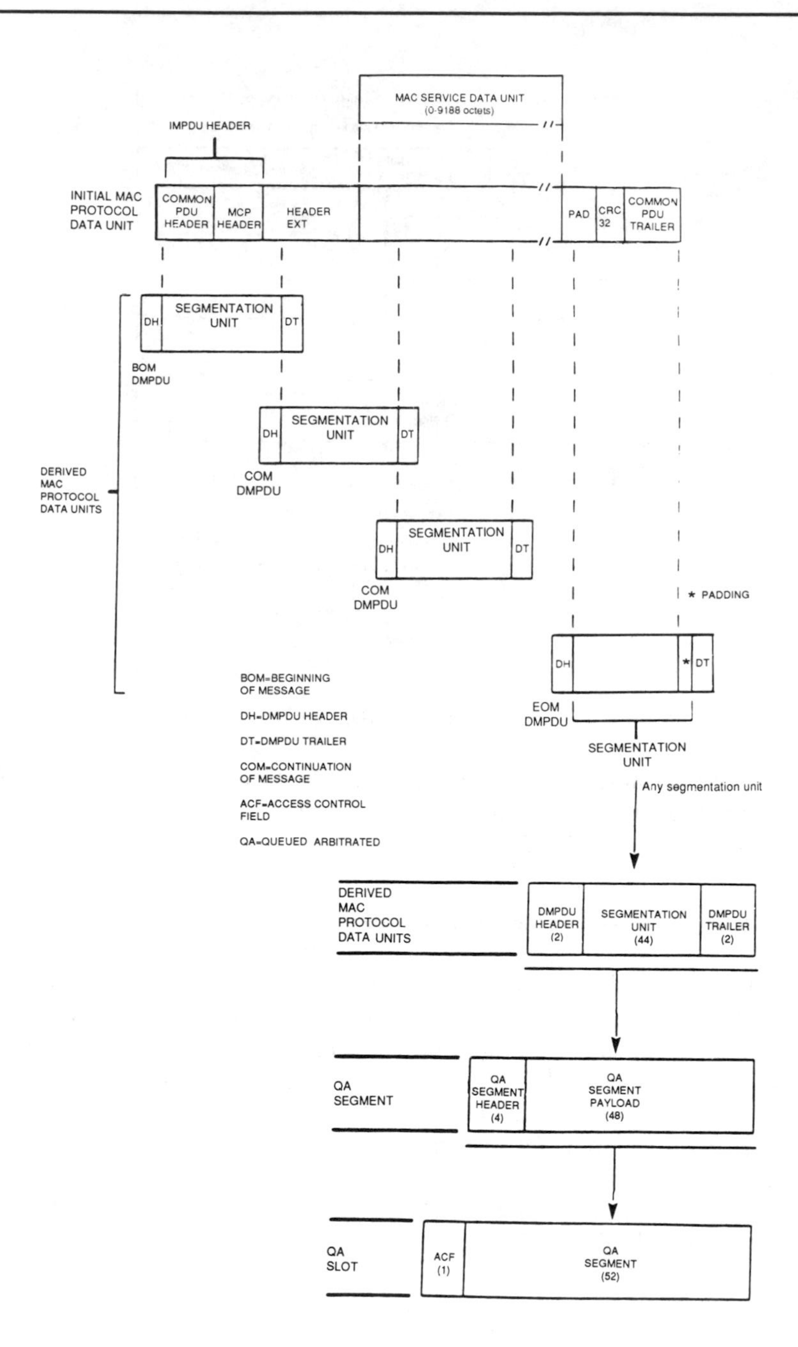

Figure 15.16 MAC data unit mapping.

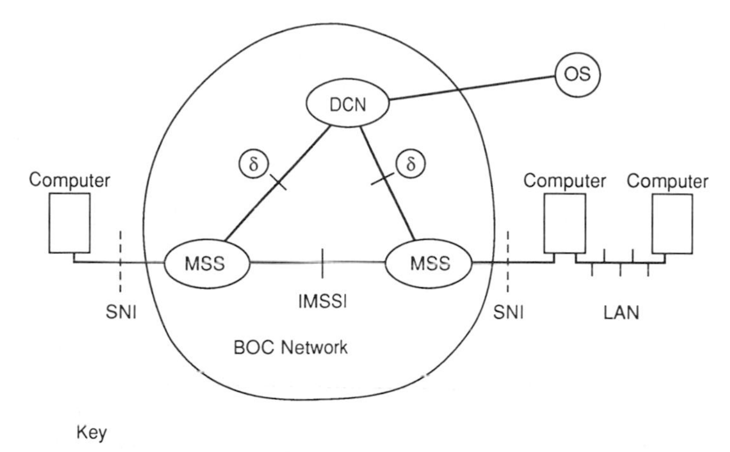

Key

SNI: Subscriber-Network Interface
IMSSI: Inter-MAN Switching System Interface
(δ) Generic Interface for Operations
MSS: MAN Switching System
OS: Operations System
DCN: Data Communications Network
LAN: Local Area Network

Figure 15.17 Network support for SMDS.

customer. The network validates that the source address associated with every data unit transferred using SMDS is an address that is legitimately assigned to the SNI from which the data unit originated. For customers desiring to maintain higher degrees of security and privacy for their traffic, features are also provided that allow customers to form "logical private networks." Two such features, destination address screening and source address screening, allow a customer to restrict the communication across a SNI. SMDS uses addresses similar to telephone numbers with extensions. SMDS also includes a capability for group addressing (group addressing is a feature analogous to the multicasting feature of LANs). When a SMDS user sends a data unit that has a group address as its destination, SMDS will deliver copies of the data unit to a set of destination addresses identified by that group address. In addition, a SMDS user could also designate multiple addresses (up to 16) to a single SNI.

To facilitate early availability, SMDS currently will (1) provide a maximum bandwidth in the DS3 range, rather than in the SONET's STS-3 range, and (2) only provide data capabilities (i.e., no isochronous traffic supported).

SMDS provides different "access classes" at service initiation; each access class provides for different traffic characteristics by placing limits on the level of sustained information flow and the burstiness of the information flow. The network

[15.16] D. Katz, "A Proposed Standard for the Transmission of IP Datagrams over FDDI," Request for Comments 1103, DNN Network Information Center, SRI International, June 1989.

[15.17] T. Haight, "A Twisted Future," *Communications Week*, June 18, 1990, p. 1.

[15.18] *Communications Week*, September 4, 1989, p. 24.

[15.19] *Lightwave*, June 1989, p. 3.

[15.20] S. Salamone, "FDDI Component Demand Is Surpassing Expectations," *Lightwave*, August 1989, p. 1.

[15.21] J. Mollenauer, "What to Expect from IEEE P802.6," *Journal of Data and Computer Communications*, Summer 1989, pp. 48 ff.

[15.22] D. Minoli, "Packetized Speech Networks, Part 1: Overview," *Australian Electronics Engineer*, April 1979, pp. 38–52.

[15.23] D. Minoli, "Issues in Packet Voice Communication," *Proceedings of IEE*, August 1979, Vol. 126, No. 8, pp. 729–740.

[15.24] "Proposed Standard: Distributed Queue Dual Bus (DQDB) Metropolitan Area Network (MAN)," Unapproved 9th Draft, IEEE, New York, August 7, 1989.

[15.25] R.M. Newman *et al.*, "The QPSX MAN," *IEEE Communications Magazine*, April 1988, pp. 20 ff.

[15.26] J.F. Mollennauer, "Standards for Metropolitan Area Networks," *IEEE Communications Magazine*, April 1988, pp. 15 ff.

[15.27] C.F. Hemrick *et al.*, "Switched Multi-Megabit Service and Early Availability Via MAN Technology," *IEEE Communications Magazine*, April 1988, pp. 9 ff.

[15.28] "Switched Multi-Megabit Data Service," *Bellcore Digest of Technical Information*, May 1988, pp. 1 ff.

[15.29] T. Haight, "Ungermann-Bass Makes SMDS Move," *Communications Week*, September 1990, p. 21 ff.

Chapter 16
Network Management Standards and Issues

Networks are increasingly essential to the day-to-day activity of the business enterprise. Several factors point to increased operational complexity in the contemporary communication environment. More users are buying PCs and intelligent workstations and connecting them with LANs and with gatewayed LANs. Peer-to-peer networking is on the rise, as distributed data bases and E-mail become more widely used and accepted. The declining cost of desktop and departmental computing power is increasing the number of intelligent systems that need to be connected. In this environment, the task of identifying and solving problems, achieved via network monitoring and control, has become both more difficult and more important.

A network management system allows communication management to examine the way in which an important corporate resource is being used, and it provides the necessary information for adjusting the operation of that resource so it more closely meets the needs of the organization. Network management involves the application of information technology to measure and control the effectiveness of the organization's telecommunication technology [16.1]. As networks have grown and become more important within businesses, control centers have grown in complexity. A center with 6 to 12 terminals monitoring various pieces of the network is not uncommon. One of the major benefits of end-to-end network management is the integration of these terminals into one complete control device. Network management integration has become a hard requirement for many users [16.2].

This chapter provides an assessment of the WAN network management field, particularly from the perspective of integration of management information from multiple-user networks. Chapter 14 provided a pragmatic view of LAN management, with emphasis on what can be done today; this chapter addresses the OSI standards efforts and the potential benefits that can be derived in the near future as these standards reach fruition. The changing user environment and the new requirements are described. The newly defined network management categories being studied by the international bodies are identified. The need for standards to achieve interoperability is discussed, followed by a summary of the accomplishments to date by the ISO. This is followed by an evaluation of the current industry status, particularly in terms of *de facto* standards. NetView, Net/Master, Unified Network Management Architecture, and Enterprizewide Management Architecture are briefly discussed.

16.1 THE CHANGING ENVIRONMENT AND THE NEW DIRECTIONS

Divestiture has transferred many of the network monitoring and control functions formally undertaken in an integrated fashion by the Bell System into the hands of the actual end-users. These users now typically need to deal with several vendors; one estimate showed that a typical communication link can consist of services and equipment provided by up to 14 vendors [16.3]. With larger networks being used by Fortune 500 firms to ensure continued competitiveness in a global market through information-based business practices, network management in general, and network control and monitoring in particular, have become critical tasks. Customers have data and voice equipment on their premises, use the LECs for intra-LATA services and connections, may go across the country with an interexchange network (possibly made up by several carriers), and may connect with PTTs for international services. Where multidrop lines are involved, several intra-LATA exchange carriers may provide service. Managing such links is clearly a challenge; multiply that by a factor of 10 or 100 lines and the problem becomes serious. Another factor driving network management is that of network reliability, particularly for networks based on a high degree of integration over T1/DS1 or T3/DS3 facilities [16.4].

A trend faced by network managers that makes network management a more difficult task is the erosion of their control over the networks they must support. The growth in decision-support systems and in the LANs has left many managers with decreased control over what is plugged into their networks. Even at the backbone level, many Fortune 1500 companies have segmented corporate networks; some companies have up to 50 separate internal networks, for example, several synchronous networks, one or more E-mail networks, one or more packet switched networks, possibly some private packet switched networks, and several dozen or hundred LANs. Multiple network management access points or stations need to be supported. Network information and statistics-gathering must be centralized and must be messaged into comprehensive reports and presented in easily understandable graphics. The integration of the network management function is a key prerequisite to any large network [16.5].

In the mid- to late 1980s, network management catapulted to the status of a strategic issue. Network designers have come to realize that much more is involved in making data communication a cost-effective undertaking than simply optimizing tariffs to achieve reductions of a few percentage points in transmission costs. The day-to-day operations, including outage resolution, service restoration, documentation, "moves," and changes, can often totally obliterate *prima facie* savings if the technology put in place is difficult to manage or a plethora of systems and solutions exists.

The network management field received an impetus when IBM formally entered the market in 1986. IBM created much of the language used in current discussions of network management and introduced the first "strategic" network management products, NetView and NetView/PC. At the commercial level, other vendors have resisted establishing yet another *de facto* standard. This has resulted in accelerated efforts in the area of standards [16.1].

The three basic objectives of any network management and control system are maintaining network availability, ensuring network performance, and providing user support. Traditionally, network management has included (1) monitoring and control, (2) administration, and (3) planning and design [16.6].

A network control system should provide comprehensive diagnostic and control facilities for monitoring, testing, and reconfiguration that are accessible from the network control center and from any other point in the network with proper security safeguards [16.7]. Management and control tools (and, hence, the network management standards) must support:

1. Activities that enable communication managers to plan, organize, supervise, control, and account for the use of communication facilities;
2. The ability to respond to changing requirements;
3. Facilities to ensure predictable communication behavior; and
4. Facilities that provide for information integrity and for the authentication of sources and destinations of data.

A network manager preferably wants to monitor the entire network with one tool and to have one consolidated view of it. At present, a diversity of vendors, interfaces, protocols, and devices exists, making network control and management an ever-changing challenge. If the network contains a mix of equipment, then today's network managers are forced to use several tools to get information on the total network; consolidating the information into one overview is difficult. In addition, network control and monitoring is not inexpensive: Personnel account for about a third of the five-year life-cycle cost of owning a network. By contrast, circuits constitute, on average, only 20% of the five-year life-cycle cost [16.1]. The direct cost in lost business resulting from network failure can be substantial, particularly in the financial services industry. Incorrect diagnosis of a network fault or attempts to restore service before adequate repairs are completed often adds to the cost by lengthening the downtime. While many users have resorted to redundant facilities for the purpose of rerouting, these solutions tend to be expensive. A proactive network control and monitoring system is needed that, under preventive maintenance algorithms, can identify early symptoms before the service is totally interrupted. Artificial intelligence techniques are being explored to facilitate this goal [16.5].

All network management functions are related to both physical and logical network management. Generally speaking, network management becomes more complicated as it becomes more logical in nature. This is particularly true when we compare applications software complexity *versus* basic equipment monitoring achieved with purely physical network control and monitoring [16.8].

Table 16.1 depicts the changing environment of network management [16.9]. These changes have to do not only with the regulatory climate, but also with the technology and the user environment. Stand-alone systems and individual PCs, once considered adequate for most jobs, no longer meet user needs. Users now require connection to their coworkers on a flexible, high-speed network. LANs are becoming ubiquitous, as discussed in Chapter 14. Growth in the T1/DS1 arena is also very strong, continuing the trend of a compound annual growth rate approaching 30% through the early 1990s.

The network management systems in use today were designed to address mainframe communication environments with carrier-provided circuits. These monitoring systems do not work well in a new networking environment. A typical monitoring process consists of four key steps [16.10]: (1) identify critical performance variables; (2) establish methods to collect, measure, and analyze performance; (3) determine the nature of the corrective action; and (4) periodically evaluate the monitoring process. For a network size of any consequence, some element of centralized control must exist. This could be in the form of a central manager-operator with distributed network intelligence, or in the form of a master-slave configuration, such as that of the IBM NetView environment. Limits exist on the degree of distribution that can be handled effectively in network manage-

Table 16.1
The Changing Role of Network Control and Monitoring

Traditional	*New*
Voice only	Voice, data, and video
Single vendor	Multiple vendors
Vendor-supplied expertise	In-house or contracted expertise principally, vendor-supplied expertise secondarily
Limited staff	Large staff with increased technical skills
Limited tools	Large use of on-line tools
Low-level administrative function	Complex business and technical function
Expense item (utility)	Competitive factor (generation of revenues and reduction of costs)

ment, even more so than in the traditional debate over distributed *versus* centralized data processing in the MIS environment. The debate also considers to what extent network management investments should be made in systems that do not conform with the evolving international standards. In addition, network management is more than just a technical issue; often it is organizational and political, especially in cases of voice-data integrated DS1 or DS3 links.

16.2 NEW DEFINITIONS AND SCOPE

Consistent with the changing role, the scope of network management is also evolving. Traditionally, network management consisted of seven areas [16.11, 16.12]:

1. Fault isolation;
2. Operational control;
3. Capacity planning;
4. Configuration management;
5. Accounting and billing;
6. Management of change; and
7. Hardware and software evaluation.

This definition is changing, expanding, and reflecting the growing importance and technology behind network management, although the traditional needs continue to exist. Two new functions have been added to network management: asset and systems management. Asset management will help corporations model networks and find the cheapest way to run operations. Systems management refers to managing not just the communication links but also the applications.

The OSINM standards have defined functional areas pertinent to management activities [16.4]. This nomenclature originally applied only to OSINM standards, but it is now becoming commonly accepted for all types of network management, whether OSI-based or not. This common nomenclature will be adopted for the rest of this discussion. The functional areas defined in OSI are discussed below in some detail.

Fault Management. Fault management encompasses fault detection, isolation, and the correction of abnormal operation. Faults cause communication systems to fail in attempts to reach their operational objectives. Faults may be persistent or transient. Faults will manifest themselves as particular events; error event detection provides a capability to recognize faults. Functions included in fault management are:

- trace and identify faults;
- accept and act on error-detection notifications;
- carry out sequences of diagnostic tests;
- correct faults; and
- maintain and examine error logs.

Thus, fault management is the discipline of detecting, diagnosing, bypassing, repairing, and reporting on network equipment and service failures.

Accounting Management. Accounting management enables charges to be established for the use of communication resources, and for costs to be identified for the use of those resources. Accounting management functions include:

- Inform users of costs incurred or resources consumed;
- Enable accounting limits to be set and tariff schedules to be associated with the use of resources; and
- Enable costs to be combined where multiple resources are used to achieve a specified communication objective.

Configuration Management. Configuration management identifies, exercises control over, collects data from, and provides data to communication systems for the purpose of initiating and providing continuous reliable connectivity. Configuration management functions include:

- Set the parameters that control the routine operation of the communication system;
- Associate names with managed objects and sets of managed objects;
- Initialize and terminate managed objects;
- Collect information on demand about current condition of the communication system;
- Obtain announcements of significant changes in the condition of the communication system; and
- Change the configuration of the system.

Configuration management is concerned with maintaining an accurate inventory of hardware, software, and circuits and the ability to change that inventory in a smooth and reliable manner in response to changing service requirements. This aspect of network management is concerned with the network directory and is partly related to CCITT's X.500 standard. Configuration and name management ensures consistency and validity of operating parameters, naming and addressing tables, and software images and hardware configurations of managed systems.

Performance Management. Performance management enables the evaluation of the behavior and effectiveness of resources and related communication activities. Functions include:

- Gather statistics about the system;
- Maintain and examine logs of the system's history;
- Determine system performance under natural and degraded conditions (real or self-induced); and
- Alter modes of operation of the system for the purpose of conducting performance management activities.

Performance management is concerned with the utilization of network resources and their ability to meet user service level objectives.

Security Management. Security management supports the application of security policies. Security is an important issue: Unauthorized or accidental access to network control functions must be eliminated or minimized. Security management functions include:

- the creation, deletion, and control of security services and mechanisms;
- the distribution of security-relevant information; and
- the reporting of security-relevant events.

Security management controls access to both the network and the network management systems. In some cases, it may also protect information transported by the network from disclosure or modification. The ISO 7498-2 document, addendum to the basic OSIRM, covers the security issue in more detail.

Some extend these OSI functions even further. For example, some have proposed adding the following three categories: capacity management, provisioning management, and administration management [16.13, 16.14]. Capacity management includes demand forecasting, network design, and network engineering. Provisioning management includes service ordering, installation, and preservice testing. Administration management includes equipment inventory, corporate directory functions, data base administration, and generation of management reports.

Others have argued that an ideal network management system should be able to provide, in an integrated end-to-end fashion, the following functions in addition to the five OSI functions just defined [16.2]:

Capacity Planning Management. In network planning, the network manager consolidates usage trends and performance data to develop options and models for potential network changes. Capacity planning is the routine fine-tuning of a network, such as adding or rearranging trunks. Planning also must accommodate periodic events such as seasonal rearrangements and annual reoptimization of the network, as well as scheduled moves and changes.

Strategic Planning Management. This allows evaluating the network as part of business planning, for example, for new applications, significant growth plans, acquisitions, and reorganizations. End-to-end control is particularly important to ensure that all aspects of the network are considered. Contingency planning is the ability to have backup or disaster recovery options available in the event of a major disruption of the network, including an estimate of associated costs; this activity can also be considered part of strategic planning.

Operations Support Management. Operations support includes management of all the "people" functions needed to staff and operate the network management center or centers. Four major aspects of operations support exist. The first is development of local methods and procedures for trouble logs, maintenance fixes, shift changes, and handoffs. The second aspect is work and information flow analysis, which examines the people and system interactions necessary for an effective operations center. Third, force management analysis is used to determine how many people are needed and the types of expertise required for a particular network management center. The last is preparation of training and development plans for network management systems users. (Some view this as not being a true functional category because every function category needs to address the issue of management staff, unless everything is completely automated [16.15].)

Programmability Management. User programmability allows the user to tailor the management system based on the unique characteristics of the network to be managed. Customization can be supported with parameters for key system characteristics, flexible report capabilities, customizable scripts, and custom programming options.

16.3 EVOLUTION OF VENDOR APPROACHES

As the need for management became clear to industry vendors in the early to mid-1980s, they began to include some type of network management capability in their product offerings. Their products ranged from inexpensive protocol analyzers and manual testing systems to help diagnose problems to high-level teleprocessing monitors, focusing on the administrative aspects of keeping track of real-time network configurations, under the contingency of failure.

Rather than manage each piece of equipment locally, centralization of the management staff and control systems in one or a few locations is generally desirable. To manage devices remotely, the centralized automated management systems must have a pathway over which they can communicate with these devices. The information collected from the network is stored on one or more network management systems. These systems typically implement a user interface that allows operators to monitor the network status and issue instructions to alter device operations [16.1]. Modem vendors, including AT&T, Codex, IBM, and Paradyne/AT&T were the first to offer control tools for networks built over voice-grade private lines. These vendors augmented their basic modem technology with built-in intelligent test capabilities that communicated with a main controller located at a central network control center. The modems communicate with the central control unit over an in-band secondary channel (derived either with FDM or TDM techniques). The speed of the telemetry subchannel is between 75 and 150 b/s. The drawback of these systems is that the in-band technique relies on the same physical channel being monitored. The centrally located component of the network control systems have also grown and evolved. Early systems were manually operated and cumbersome. More modern computer-based systems are easier to use.

IBM's modems used a virtual in-band diagnostic channel. The monitoring and test data are interleaved with the user data stream. This technique allowed IBM to integrate its modem-based network diagnostics with the diagnostics for mainframes and FEPs. Using this method, the network control data became available for real-time processing by host-resident traffic and performance analysis software. These early vendor-specific systems enabled the network manager to monitor the operation of nodes (usually cluster controllers) and paths in the network. They allow diagnostic testing on the network components and can fairly rapidly sectionalize problems on the link. Microprocessor-based test units are now built into many multiplexers, switches, and other network components, in addition to modems [16.7]. The network manager has access to all these network components and to information on their performance from a network control center. Information management tools, such as rudimentary report generators, are available with many network management systems to organize data from and about network components into detailed reports for network and corporate management. Table 16.2 depicts a chronology of recent and near-term network management developments.

Vendor-packaged solutions have been demonstrated to perform relatively well in a number of specific user environments, particularly when the functions are limited to hardware diagnostics (e.g., line monitoring, response-time mea-

Table 16.2
Network Management Timetable

Year	*Vendor and/or other relevant event*	*Environment*
1986	SCI's Net/Master, IBM's NetView and NetView/PC	NCCF and NPDA widely deployed
1987	AT&T's UNMA NetView Version 1 Release 2	NetView becomes de facto standard (for SNA networks)
1988	HP's OpenView DEC' DECManage Formation of OSI Network Management Forum	Continued proliferation of vendor-proprietary systems, particularly in T1-multiplexer environment
1989	NetView Version 1 Release 3 Early OSINM/drafts	
1990	Additional OSINM standards NetView Version 2 Release 2	
1991	First wave of OSINM products	Migration to two or three standards
1992	NetView Release 4 (Projected)	Standardization of control center console
1993	OSINM standards mature	
1994	OSINM products appear based on full standard	Network management reaches a rational stage

surement) [16.8]. However, to reach the next level of synergy, standards are necessary.

16.4 NEED FOR NETWORK MANAGEMENT STANDARDS

The concurrent development of management systems by many equipment providers (and even some carriers), as shown in Table 16.3, leads to a variety of products. Many of these products perform some aspect of management on the respective vendor's own equipment, but most cannot communicate with other vendors' equipment [16.16]. While this predicament is at best tolerable for networks dominated by a single provider, it is unworkable for heterogeneous environments. The increased availability of "open systems" operating under CCITT/OSI standards is driving an increased number of users to a multivendor environment.

The growing number of commercial network management packages became a major problem as networks grew in scale, scope, and diversity. Rather than helping, network management tools were adding to the complexity. The integration of separate tools and displays has become a major issue in network management, control, and monitoring efforts. Table 16.4 depicts the multitude of monitoring and control products made available by a single vendor: The need for integration is obvious. Customers today want more information about and control over their communication systems. They want end-to-end management of their entire networks, regardless of whether they reach into a PBX, a CO, a long-distance carrier, or even a private network. Integration of network management functions is deemed critical by a majority of users [16.17]. Customers also want access to carriers' operations and maintenance information [16.18].

Starting in the mid-1980s, but more vigorously at present, communication providers began to realize that it benefits their customers, as well as themselves, to focus their efforts on interoperable network management systems. Two avenues to achieve this interoperability are available: conform to a published *de facto* standard, or develop an open international standard that is not dominated by any one vendor. Cohesiveness can be accomplished only with a set of standardized protocols and services. Work is now under way to develop network management standards to integrate the data generated from a multitude of carriers and equipment into a single, cohesive network control system. A product conforming to a published standard is guaranteed to interact with other products developed to the same standard, or at least to a common-denominator level of functionality [16.16]. A product developed to conform to an open system will do the same, but have the advantage of not giving unfair treatment to a vendor. All the interfaces and transactions of such a system are documented and published so that others can design their systems to interact with it.

When IBM announced NetView in early 1986, it was the first vendor to respond to the need for integration. IBM provided a plan for reducing the number

Table 16.3
Management Systems Developed by Equipment Providers

Company	*Network Management Product*	*Security*	*Accounting*	*Fault*	*Performance*	*Configuration*	*Integrated Tools*
				(Management)			
AT&T	ACCU-MASTER	Y	Y	Y	Y	Y	Y
Avant-Garde	Net/Alert Plus	Y	N	Y	Y	N	Y
SCI	NET/MASTER	Y	Y	Y	Y	Y	Y
Codex	9800 Series	Y	N	Y	Y	Y	Y
Diglog	Network Analysis and Management System (NAMS)	N	N	Y	Y	Y	Y
Digital Communications Associates (DCA)	Series 300	Y	Y	Y	Y	Y	Y
Digital Equipment Corp. (DEC)	Current tools	NS	NS	Y	Y	Y	N
	Enterprise Management Architecture (EMA; forthcoming)	Y	Y	Y	Y	Y	Y
EMCOM	NCS70	N	N	Y	Y	Y	Y
Hewlett-Packard (HP)	OpenView	Y	Y	Y	Y	Y	Y
IBM	NetView	Y	Y	Y	Y	Y	Y
	NetView/PC			Y	Y	Y	Y
Infotron	Integrated Network Manager	Y	N	Y	Y	Y	Y
Network Equipment Technologies (N.E.T.)	Model 4010 and Integrated Network Command System (INCS)	Y	Y	Y	Y	Y	Y
Novell	NetWare Care	NS	NS	Y	Y	Y	Y
Proteon	OverVIEW	Y	NS	Y	Y	Y	Y
Racal-Milgo	CMS 2000	N	N	Y	Y	Y	Y

Table 16.3 continued
Management Systems Developed by Equipment Providers

Company	Network Management Product	Security	Accounting	Fault	Performance	Configuration	Integrated Tools
			(Management)				
Sytek	9100 Network Management Center	N	N	Y	Y	Y	Y
3Com	3←Open	Y	Y	Y	Y	Y	Y
Timeplex	TIME/VIEW	Y	Y	Y	Y	Y	Y
Ungemann-Bass (U-B)	Access/One	Y	Y	Y	Y	Y	Y
Wang	Distributed Management Facility	Y	Y	Y	Y	Y	Y

Table 16.3
Management Systems Developed by Equipment Providers

Company	Network Types	Console	Centralized/ Distributed (C/D)	Standards Based/ Proprietary	Multivendor Integration
AT&T	STARLAN, T1	Workstation	C	UNMA Based on OSI. Proprietary Network Management Protocol	Limited to Net View/ PC
Avant-Garde	SNA; DECnet. Others	Workstation	C	Proprietary	Yes
SCI	SNA	Sys/Master	C	Proprietary	Yes (includes NetView/ PC)
Codex	Various	Apollo Workstation	C	OSI Based	Yes
Diglog	Various	VAX Computers	D or C	Proprietary	No
Digital Communica-tions Associates (DCA)	T1	Personal Computer (PC) Based	C	Proprietary	No (NetView/PC Link Forthcoming)
Digital Equipment Corp. (DEC)	DECnet	VAX or MacroVAX	D	Proprietary	Yes (NetView/PC, Appletalk)
	DECnet, SNA. Others	VAX to PCs	D	OSI Based	Yes
EMCOM	Various	IBM PC	D	Proprietary	Yes
Hewlett-Packard (HP)	Various	HP IBM or Other PCs	D	OSI Based	Yes
IBM	SNA	IBM 3270	C	Proprietary	No
	Various	IBM PC		Proprietary	Access to NetView for Other Vendors' Alarms
Infotron	T1	Apollo Workstation	C	Proprietary	No
Network Equipment Technologies (N.E.T.)	IDNX/T1, SNA	N.E.T. Console	C	Proprietary	Yes (Includes NetView/PC)

Table 16.3 continued

Company	Network Types	Console	Centralized/ Distributed (C/D)	Standards Based/ Proprietary	Multivendor Integration
Novell	NetWare	PCs	D or C	Proprietary	Yes
Proteon	Token Ring. TCP/IP	Workstation	C	SNMP Based: OSI Based (Forthcoming)	Yes
Racal-Milgo	T1, SNA	PC Based	C	Proprietary	Yes (includes NetView/ PC)
Sytek	Various	Sun Workstation	C	Proprietary	Limited to NetView/PC
3Com	OS/2	Bridge Network Control	D or C	OSI Based	Limited
Timeplex	T1. Others	Sun Workstation	C	Proprietary	No
Ungemann-Bass (U-B)	XNS, TCP/IP, Others	U-B Consoles	C	Standards Based	Yes
Wang	DECnet, SNA, WangNet	Wang Workstation	D or C	Proprietary but *OSI-Like*	Yes (to DEC and IBM)

Table 16.4
An Example of Lack of Integration — A Vendor's Plethora of Products

Element Management Systems	Network Element
ACCUMASTER™ Trouble Tracker	System 85, 75, DIMENSION PBX
Customer Controlled Reconfiguration	ACCUNET® T1.5 Service
Customer Test Service	DATAPHONE® Digital Service
DATAPHONE II System Controller	Modems, Multiplexers
DATAPHONE II ACCULINK™ Network Manager	Modems, Multiplexers
STARKEEPER Network Management System	DATAKIT™ VCS, ISN
Customer Network Control Center	EPSCS
Centralized System Management VMAAP	System 85, 75, Dimension PBXs
Multi-Function Operations System	5ESS Switches, PBXs
Service Management System/MISR	SDN
Order Management Service	Provisioning
Customer Network Controller	DACS
MACSTAR™ End Customer Management System	5ESS® Switches
Advanced Communications Package	5ESS Switches
Basic Communications Package	1A ESS, 5ESS Switches

of separate consoles needed to manage an SNA network. However, as of 1990, only a few non-IBM products can be directly managed by an SNA host. IBM's major network management strategy has been to provide integration of non-IBM products through the NetView/PC interface. As of 1989, this strategy was, at least in terms of announcements, the most extensive effort at vendor support. NetView is constructed on a centralized architecture. Many users retain the belief that

NetView is a CPU-intensive system [16.12]. Some favor a matrix management structure whereby companies adopt a centralized approach but distribute people to operate the network.

AT&T, Hewlett-Packard, DEC, and others responded to the IBM entry into the market by endorsing the rapid definition and adoption of network management standards under the OSI banner [16.1]. These three vendors have also introduced their own architectures, tools, and services for consistent, unified network management. Of these, AT&T's Unified Network Management Architecture (UNMA) is the most well publicized. Table 16.5, based partially on Reference [16.17], provides a sense of the penetration of various network management products.

16.5 INITIAL CONVERGENCE

While various attempts have been made in the past by different groups to develop network management standards and *de facto* standards, four major developments appear to be reaching critical mass. Each of these developments addresses the needs of a specific network area. The good news is that the data communication industry appears to be converging on a small number of network management standards, and away from the numerous proprietary offerings of the past. The bad news is that four approaches are still more than one cohesive, integrated system. The major clusterings of activities are as follows.

IBM Networks (typically, corporate business-support networks). IBM's presence in the market is strong enough to have allowed IBM to develop its own network management solutions. IBM was rated by Datamation as the number one vendor by market size in each of the following four areas: mainframes, minicomputers,

Table 16.5
Various Organizations' Network Management and Control Directions

Product	*1989 Penetration (%)*
Vendor-based	
NetView	25
UNMA	10
Net/Master	2
EMA	1
Other vendors	7
Multivendor-based	18
OSI Standards-based	3
Other	
In-house	10
No decision	24

PCs, and software provision [16.19]. As already discussed, IBM has developed NetView, a complete management system focusing on SNA networks. It has became a *de facto* industry standard [16.20]. NetView allows for integration of other systems through a relatively "open" interface, in the sense that a published specification is available; however, a major software development effort is required from the vendors desiring to communicate with it. New releases have aimed at addressing a number of perceived shortcomings of NetView. In particular, automation facilities have been expanded. Rexx, an upper-level SAA language, has been added to ease the difficult task of writing commands in CLISTs. Logging functions have been enhanced. The ability to monitor alerts has been expanded to any associated distributed host [16.17].

OSI Networks (also referred to as "enterprise networks"). For several years, the ISO's OSI standards have been promoted as the basis for interconnectivity between all devices in heterogeneous networks (as discussed in Chapter 13). The basic model adopted in 1984, however, needed to be extended to allow network management functions [16.4]. Such companies as DEC, AT&T, and Hewlett-Packard have announced that their network management strategies are based on the OSI CMIS/CMIP standards (see Table 13.1). Recent major events—including CMIS/CMIP attaining international standard status and the definition of how the components of CMIP work with TCP/IP—have made OSI a viable solution for future network management products [16.16].

Internets. In the mid- to late 1980s, a "rediscovery" of TCP/IP (discussed in Chapter 14) occurred. TCP/IP is a product-proven solution to interoperate heterogenous systems while waiting for the full-scope OSI products. TCP/IP-based systems have been widely available and well supported in the past five years. TCP/IP networks include WANs (such as NSFnet Backbone, regional and consortia networks, statewide networks), and LANs. TCP/IP networks include end systems (such as hosts and terminal servers) and intermediate systems (such as routers and gateways). SNMP (also discussed in Chapter 14) is a relatively new specification for the management of Internets [16.21]. SNMP was declared a standard by the IAB in April 1990. Some observers already label it a *de facto* standard for LAN management; in the absence of interoperating and fielded OSI-based implementations, these observers see it as the only choice for management of multivendor TCP/IP Internets [16.22, 16.23].

Network Work Groups (also known as "PC LANs; unlike "traditional" networks, work groups are not centered around a company mainframe, but are comprised of smaller computing devices such as PCs and high-powered workstations). Microsoft's OS/2 LAN Manager represents a *de facto* work group networking standard that includes extensive network management support. It has gained rapid acceptance among network vendors and applications developers. Most work groups probably started as a few PCs sharing resources but, with increased processing power, many have expanded to include thousands of PCs connected with hundreds of servers and other shared resources.

A true comprehensive network management standard is still in the future. From a user's perspective, a multiple-standard solution is far less than ideal. The desirable solution is to integrate these "standards" into a single management system so that all the components can communicate and exchange essential management information [16.16].

16.6 TARGET: OSI NETWORK MANAGEMENT STANDARDS

Integrated network management depends critically on standardization. It requires the definition of common information standards that devices and network management systems can use to exchange status, statistics, and action commands. A description of these efforts follows.

The basic OSIRM, ISO 7498, discussed in Chapter 13, provides a description of the activities necessary to undertake open communication. The information retrieval, transfer, and management for OSI forms Part 4 of ISO 7498 [16.4]. Part 4 provides a description of the framework and structure of OSI management, supplementing the basic model. The document reached international standard status in October 1988. Areas of OSI management standardization include: (1) the services and protocols used to transfer management information between open systems and (2) the abstract syntax and semantics of the information transferred in management protocols. The model of OSI management is defined in the specifications in terms of:

- The OSI management structure;
- The supporting functionality required by OSI management;
- The MIB;
- The flow of control among processes; and
- The flow of information between entities.

Approximately two dozen ISO documents were available or defined as of early 1990. (The numbering and organization of these documents has recently been restructured.)

In addition to the framework described in Reference [16.4], other parts of the standard are Common Management Information Service, ISO 9595, International Standard, November 1989; Common Management Information Protocol, ISO 9596, International Standard, November 1989; Systems Management Overview (SMO), ISO 10040, Second Draft Proposal, November 1989; Structure of Management Information (SMI), ISO 10165, Draft Proposals. The five functional areas described above are addressed in the following five documents:

1. Configuration management (N3311);
2. Fault management (N3312);
3. Performance management (N3313);
4. Accounting management (N3314); and
5. Security management (N3315).

These functional documents were in the working draft stage as of early 1990. A set of systems management functions is defined (such as error reporting and information retrieval function, log control function, *et cetera*) that are, for the most part, in the draft proposal stage (see Table 16.6 [16.24]). Although many tasks need to be undertaken in a network management environment, a basic set of common functions is required by all these tasks. These common elements are separated and described as a support protocol, known as Common Management Information Service Element (CMISE). CMISE provides a structure for network management application messages that are further defined in the specific management functional areas of layer 7 [16.2].

The OSI management environment is that subset of the total OSI environment (OSIE) that is concerned with the tools and services needed to control and supervise interconnection activities. The OSI management environment includes both the capability for managers to gather information and to exercise control, and the capability to maintain an awareness of and report on the status of resources in the OSIE [16.4]. OSI management is effected through a set of management processes. These processes are not necessarily located at one local system but may be distributed in many ways over a number of systems. In cases where management processes that are not co-resident need to communicate with one another in the OSI environment, they communicate using OSI management protocols. A *managed object* is the OSI management view of a resource within the OSIE that is subject to management, such as a layer entity, a connection, or an item of physical communication equipment. Thus, a managed object is the abstracted view of such a resource and represents the resource's properties as seen by management. A managed object is defined in OSI in terms of attributes it possesses, operations that may be performed on it, notifications that it may issue, and its relationships with other managed objects. This is distinct from, but related to, any definition or specification of the resource represented by the managed object as an element of the OSIE. The set of managed objects within a system, together with their attributes, constitutes that system's MIB. The MIB is that information within an open system that may be transferred or affected through the use of OSI management protocols.

The logical structure of management information is standardized; the MIB definition does not imply any form of physical or logical storage for the information, and its implementation is a matter of local concern and outside the scope of OSI standards. Management information may be shared between management pro-

Table 16.6
Network Management Standards and Their Status as of Early 1990

Standard	*ISO number*	*Status*	*Disposition*
OSI Basic Reference Model Part 4: Management Framework	ISO 7498 Part 4	International Standard	October 1988
System Management Overview	ISO 10040	Second Draft Proposal	Ballot closed May 1989
Common Management Information Service	ISO 9595	International Standard	November 1989
ISO 9595 Addenda 1 and 2		Draft International Addendum	
Common Management Information Protocol	ISO 9596	International Standard	November 1989
ISO 9596 Addenda 1 and 2		Draft International Addendum	
Structure of Management Information Part 1: Information Model	ISO 10165-1	Draft Proposal	Third draft May 1989
Structure of Management Information Part 2: Definition of Support Objects	ISO 10165-2	Draft Proposal	Second draft November 1989
Structure of Management Information Part 3: Definition of Management Attributes	ISO 10165-3	Draft Proposal	Second draft November 1989
Structure of Management Information Part 4: Guidelines for Definition of Managed Objects	ISO 10165-4	Draft Proposal	November 1989
Systems Management Part 1: Object Management Function	ISO 10164-1	Draft Proposal	Second draft November 1989
Systems Management Part 2: State Management Function	ISO 10164-2	Draft Proposal	Second draft November 1989
Systems Management Part 3: Relationship Management Function	ISO 10164-3	Draft Proposal	Second draft November 1989
Systems Management Part 4: Alarm Reporting Function	ISO 10164-4	Draft Proposal	Second draft November 1989
Systems Management Part 5: Event Reporting Management Function	ISO 10164-5	Draft Proposal	First draft November 1989
Systems Management Part 6: Log Control Function	ISO 10164-6	Draft Proposal	First draft November 1989

Table 16.6 continued
Network Management Standards and Their Status as of Early 1990

Standard	*ISO number*	*Status*	*Disposition*
Systems Management Part 7: Security Alarm Reporting Function	ISO 10164-7	Draft Proposal	First draft November 1989
Part ?: Confidence and Diagnostic Function	ISO 10164-?	Draft Proposal July 90?	
Part ?: Security Audit Trail Function	ISO 10164-?	Draft Proposal July 90?	
Part?: Accounting Metering Function	ISO 10164-?	Draft Proposal July 90?	
Part?: Workload Monitoring Function	ISO 10164-?	Draft Proposal July 90?	
Part ?: Measurement Summarization Function	ISO 10164-?	Draft Proposal July 90?	

cesses and structured according to the requirements of those processes. The MIB neither restricts the interpretation of management data to a predefined set, nor to whether the data are stored in a processed or unprocessed form. However, both the abstract syntax and the semantics of information that is part of the MIB are defined so that they can be represented in OSI protocol exchanges. The actual specification of the syntax, semantics, services and protocols, and the concepts applicable to managed objects are provided in specific OSI standards. The physical representation of managed objects and their physical storage are local matters and are not subject to standardization. (SNMP can now access MIB variables; an international code designator has been obtained for TCP/IP, and MIB objects are being standardized in that context consistent with OSI specifications. See Chapter 14.)

OSI management is accomplished through (1) systems management, (2) (N)-layer management, and (3) (N)-layer operation. Figures 16.1 and 16.2 depict the OSINM architecture. Systems management provides mechanisms for the monitoring, control, and coordination of managed objects through the use of application-layer systems management protocols. OSI communication concerning systems management functions are realized through a systems management application entity (SMAE). Systems management may be used to manage any objects within or associated with an open system. (N)-layer management provides mechanisms for the monitoring, control, and coordination of managed objects that relate to communication activities within an (N)-layer. (N)-layer management can effect multiple instances of communication and is effected either through systems management protocols or through the use of special-purpose management protocols within (N)-

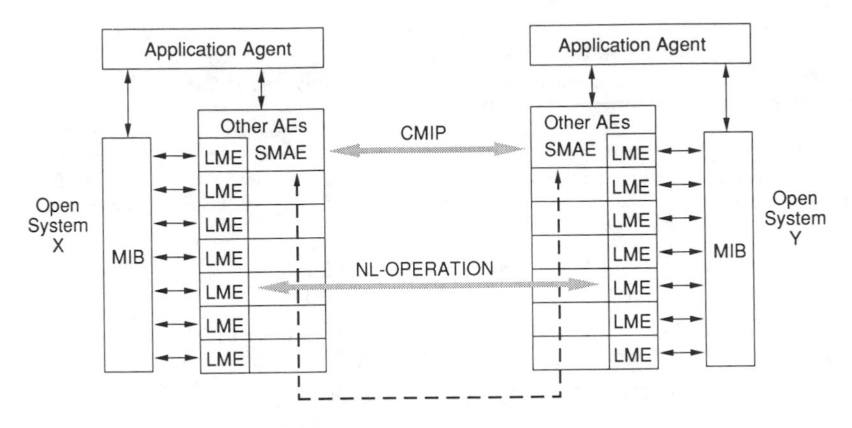

Figure 16.1 Systems and layer management model.

layer. (N)-layer operation provides mechanisms for the monitoring and control of a single instance of communication.

(N)-layer management supports the monitoring, control, and coordination of (N)-layer managed objects. (N)-layer management protocols are supported by protocols of the layers (N − 1) and below; they do not provide communication capability offered by the (N + 1) and higher layers. (N)-layer management protocols can only convey management information between peer (N)-layer management entities pertinent to the (N)-subsystems in which these entities reside. (N)-layer management protocols provide such functions as:

- Communicating parameter values associated with managed objects that relate to the operation of the (N)-layer;
- Testing the functionality provided by the (N–1)-layer; and
- Conveying error information describing faults or diagnostic information related to the operation of the (N)-layer.

Management functions may exist within the (N)-protocols in all seven layers of OSI. The management information that is carried within an (N)-protocol must be distinguishable from information that the protocol carries for other purposes. The (N)-protocol is responsible for providing this distinction. Each (N)-layer management protocol is independent of other management protocols. Standardization efforts do not mandate the development of (N)-layer management protocols for each of the seven layers.

Management information carried by (N)-protocols also exists for the purpose of controlling and monitoring a single instance of communication. Examples of

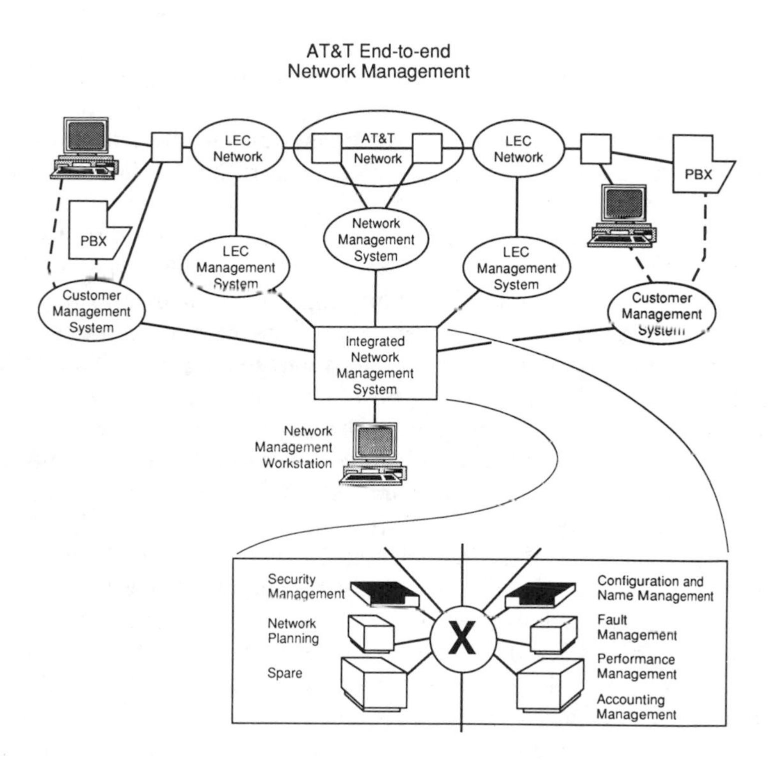

Figure 16.2 UNMA framework.

management information carried within an (N)-protocol for (N)-layer operation are:

- Parameters carried in connection establishment PDUs that apply to the specific instance of communication that is being established;
- Parameters carried in particular PDUs that can modify the environment in which this instance of communication operates;
- Error information describing faults encountered during the operation of that specific instance of communication; and
- Parameters carried in connection release PDUs that report information pertaining to the specific instance of communication that is being released.

An open naming convention is needed to identify multivendor network elements. The naming convention must be hierarchical, so that a registered code at the higher level can allow independence across vendors, and a structured approach at the lower levels can provide an integrated system with the understanding it needs. Further, data fields must be standardized across vendors. For example, the meaning of alarm severity values must be the same for all vendor equipment to allow an integrated system to filter alarms based on the dynamic input of an administrator [16.2].

Once an association between two remote systems wanting to transfer network management data is established, management operations define the dialogue required to monitor and control the managed objects. Key operations are: *Get, Set, Report-Event, Action, Create,* and *Delete*. *Get* and *Set* support inspection and modification of the information associated with the managed objects. The *Report-Event* operation is used to notify the managing system of the various events generated by the managed objects. *Action* allows the managing process to request that a certain action be performed on the managed objects. The addition and deletion of the managed objects are supported by the *Create* and *Delete* operations.

The timetable for the full-fledged availability of OSINM standards is 1991–93, as shown in Table 16.2. Products should become available in 1991–94. However, some acceleration efforts have already started. An Open Systems Interconnection/Network Management Forum (OSI/NMF) was established in mid-1988 by eight founding members (Amdahl, AT&T, British Telecom, Hewlett-Packard, Northern Telecom, STC PLC, Telecom Canada, Unisys). The OSI/NMF has grown to more than 20 voting member companies and 80 associate member companies, based in 15 countries [16.25, 16.26]. The goal of the OSI/NMF group is to accelerate the specification of interface standards that are OSI-compatible and on which vendors can develop their own network management products [16.8]. OSI/NMF is an independent, open, incorporated consortium of international network equipment vendors, service providers, and users. The member companies are investing business and technical resources to accelerate the development of, and promote the use of, OSI standards for network management. These goals are accomplished through a combination of developing specifications based on existing and emerging standards, and sponsoring trade show events in which implementations based on these specifications are demonstrated.

A key achievement of OSI/NMF, at the time of this writing, was the completion and approval of a first set of nine technical specifications designed to support an interoperable interface for network management between multivendor supplied products and services. The specifications, which are suitable for use in both LANs and WANs, provide a communication infrastructure based on the OSIRM to support the management needs of data communication and telecommunication networks developed by different suppliers.

OSI/NMF demonstrations at eight major computer and communication trade shows were planned for 1991 [16.26].

16.7 INTEGRATED SYSTEMS

Table 16.3 summarized the features of some of the available network management systems. Monitoring and control is included in the fault management and configuration management columns. In this section, the more widely deployed systems identified in Table 16.3 are discussed.

16.7.1 NetView

With NetView, IBM offers a coherent set of network management products that can be combined and integrated. Some of the features of NetView were discussed above. The NetView product line is based on earlier products such as the Network Problem Determination Application (NPDA), the Network Communication Control Facility (NCCF), and the Network Logical Data Manager (NLDM). Under the first release of NetView, these products were basically repackaged with a common interface. NCCF was introduced in the late 1970s; it provides a pipeline out of VTAM for network management information flowing from controllers and terminals to the network control terminal. NPDA provides status information about many kinds of hardware on the network. NLDM was introduced in the early 1980s; it monitors the network logically, looking for the breakdown of individual communication sessions not reported by NPDA. VTAM Node Control Application (VNCA) is a tabular listing providing the network operator with a summary of network resources on a single screen. Network Management Productivity Facility (NMPF) is a program that assists the operator in selecting diagnostic routines [16.16].

The NetView package has since been expanded by IBM. In September 1986, IBM unveiled an approach to allow "openness." With NetView/PC, SNA can collect network management information from non-IBM devices not running in an emulation mode. This allows other equipment providers with a set of protocols to send their own alert information to NetView in a native mode. Currently, most non-IBM products are handled through NetView/PC or specially configured NetView software running on an IBM or compatible. In the long term, IBM plans for NetView to support every IBM computer and communication product under the SAA family, as well as gateway to non-IBM voice and computer products.

Version 1 Release 2 of NetView became available in 1987, and Version 1 Release 3 in 1989. (Version 2 Release 2 for MVS/ESA was announced in 1990 and Version 2 Release 2 for VM/ESA should be available in 1992.) According to IDC estimates, IBM has sold NetView into 55% of the existing SNA networks. More than 55 vendors pledged to develop NetView/PC interfaces Furthermore, some vendors have actually developed NetView/PC products [16.27]. Some vendors have complained about NetView/PC, suggesting that IBM wishes to keep non-IBM products under the NetView umbrella. Developers claim that NetView/PC is difficult to program and that sending commands from NetView is also difficult. The problem is the lack of peer-to-peer treatment by NetView [16.28].

Additional features of NetView can be found below, in the discussion of Net/Master.

16.7.2 Net/Master

Net/Master is an SNA network management product being sold by Systems Center, Inc., of Vienna, VA (previously available from Cincom Systems). Net/Master was actually developed by Australia's Software Development Corporation [16.17]. Net/Master was introduced in 1984, two years before NetView. Since then, 1200 licenses have been sold into more than 700 accounts, according to IDC of Framingham, Massachusetts; nearly 22% of those accounts are Fortune 100 firms. More than 39,000 SNA sites existed worldwide at the end of 1989; hence, SCI's Net/Master has been licensed to less than 2% of the sites.

While IBM's NetView and SCI's Net/Master have the same objective, and are relatively similar in terms of the network control data provided to the user, Net/Master users generally praise the product's ease-of-use and flexibility. Net/Master's Sys/Master component with the new Expert System Foundation option provides a higher level of automation than NetView. The basis on which Net/Master is built is the Network Control Language (NCL). NCL is a full-fledged fourth-generation language. Also, because Net/Master was built from the bottom-up as a single product, it has a number of advantages compared to NetView, which was basically a repackaging of existing IBM products. According to observers, if a new release does not offer a true fourth-generation language for customizing and writing commands, if it does not integrate its different components more smoothly, and if it does not provide a more friendly user interface, then its control of SNA networks may be in jeopardy.

The capabilities of the NetView and Net/Master control-management systems are compared below.

Control Facility. (1) Accepts network operator commands and provides control from a centralized site and (2) collects and interprets information on errors and events. IBM's NetView: Command Facility (formerly Network Communication Control Facility). SCI's Net/Master: Operator Control Services.

Physical Network Management Facility. (1) Manages the physical portion of the network such as lines and modems and (2) collects and displays statistical data, alerts, and events. IBM's NetView: Hardware Monitor (formerly Network Problem Determination Application). SCI's Net/Master: Network Error Warning Systems (NEWS) (part of the Advanced Network Management component of Net/Master).

Logical Network Management Facility. (1) Manages the logical portion of the network and (2) provides for access to information such as session configuration data, session response time, and session trace data. IBM's NetView: Session Monitor (formerly Network Logical Data Manager). SCI's Net/Master: Network Tracking System.

Network Status Facility. Provides displays of all or portions of the network through hierarchical, interactive displays. IBM's NetView: Status Monitor (formerly

VTAM/NCA). SCI's Net/Master: (a) Network Control System, part of Advanced Network Management component, and (b) Net/Stat, which provides multidomain status.

Help Facilities. (1) Aid operators in problem determination with step-by-step instructions. IBM's NetView: (a) Help Desk Facility and (b) Online Help Facility (formerly Network Management Productivity Facility). SCI's Net/Master: (a) Help Desk Facility, (b) Online Help Facility, and (c) Net/Info, which is a data base of status information and messages in English text.

Automation Facilities. Provides for the automation of manual tasks for local and remote systems. IBM's NetView: Automated Operations Management. SCI's Net/Master: (a) Sys/Master and (b) Automated Operations Management.

16.7.3 AT&T's UNMA

Currently, unintegrated network monitoring and control systems exist in at least three domains: customer premises, local carrier, and interexchange carrier or PTT networks. In 1987, AT&T announced the Unified Network Management Architecture (UNMA). UNMA integrates management systems that control separate pieces of a network. Through UNMA, AT&T attempts to provide a blueprint for end-to-end management and control of voice and data networks, customer-premises equipment, the LEC network (via NetPartner NMS), and interexchange services [16.18]. AT&T's approach concentrates on managing the communication elements of the network, e.g., premises equipment, transmission facilities, and network services [16.2].

UNMA is positioned as a framework for integrated end-to-end management in a multivendor environment. The major benefits of a common user interface are more efficient and effective analyses of fault, configuration, performance, and accounting data. On a customer's premises or in the network, systems are available that manage multiple network elements of the same family. These are called *element management systems* (EMSs) in UNMA. EMSs generally support pieces of a network connection, but generally do not manage all components of a network end-to-end.

As shown in Figure 16.2, the UNMA has three tiers, as well as a standard machine-to-machine protocol and a unified user interface. The protocols are based on evolving international standards. The components of the architecture are described below.

1. *Network Elements.* The first tier includes the network elements, e.g., devices on the customers' premises such as modems, multiplexers, LANs, hosts, PBXs, and local, interexchange, or international network services including analog or digital facilities, packet switching, and virtual data networks.

2. *Element Management Systems*. The second tier is comprised of systems that manage the network elements, including operations, monitoring, control, administration, maintenance, and provisioning functions. These are typically customer-owned stand-alone systems.
3. *Network Management Integration*. The third tier ties the element management systems together and provides value-added functions and analyses, allowing users to operate, administer, and maintain a network from end to end.
4. *Network Management Protocol*. Implementation of UNMA requires a standard intersystem interface (machine-to-machine protocol) that developers of EMSs and operations systems (AT&T and other vendors) can easily support and that is powerful enough to allow for full network control. AT&T is committed publicly to OSI network management protocols. AT&T's implementation of these OSI protocols is referred to as the *network management protocol* (NMP), which links to the management and operations systems. AT&T is working with other vendors to finish the specifications and ensure that industry needs are incorporated in initial NMP implementations. AT&T has published initial protocol specifications as well as initial application messages. NMP will, according to the vendor, follow all relevant international standards that will evolve in the future.
5. *Unified User Interface*. This user-to-machine interface allows a network manager to work with end-to-end applications of the integrating system and in the native mode of EMSs in a uniform way. As a result, customers with separate management systems can consolidate functions into a common workstation and, using windowing software, they can monitor, control, and reconfigure their network elements without learning separate procedures and systems for each element. UNMA's ACCUMASTER Consolidated Workstation combines the network management functions of various EMSs into a single user interface. ACCUMASTER applications will process information from various EMSs, simplifying the operator's task.

AT&T's proprietary NMP, which is similar to the CMIP standard, promises a smooth migration to the OSI systems of the future. NMP is implemented in the higher layers of the seven-layer model and is made independent of the specific implementation of the three lower layers as long as the transport layer is appropriately implemented. Within layer 7 (application), a number of sublayers are used for network management. ACSE and FTAM are also final specifications, as discussed in Chapter 13. ROSE and CCR have undergone extensive review and were relatively stable at the time of this writing. AT&T has published its target NMP stack up to and including the CMIS, and has also published a number of fault management, configuration, and name management application messages. For example, AT&T's NMP fault management specifications define a set of alarm collection and control messages. If the CCITT/OSI final specifications differ from AT&T's NMP specifications, AT&T is committed to modifying NMP accordingly and migrating AT&T products to the new specifications. Seventeen vendors pledged as of 1989 to develop NMP interfaces for their products. As of January 1989, AT&T had published four NMP documents: TR54004, TR54005, TR54006, and TR54007 [16.27].

An initial step in the integration process under UNMA is the consolidation of multiple control and monitoring data streams into a single, windowing workstation. AT&T's ACCUMASTER Consolidated Workstation is an MS-DOS based system that displays a consolidated network view by providing separate windowing sessions to various AT&T EMSs. With ACCUMASTER Trouble Tracker, both AT&T and non-AT&T product alarm information can be integrated and displayed graphically on a network status display. This integration of alarms allows customers to be aware of problems and potential problems in the network before a user reports a failure. The ACCUMASTER Consolidated Workstation consists of an AT&T applications software package and a communication board designed for installation in an AT&T PC 6300-series PC. The applications package allows a user to open window sessions into any four of the following EMSs simultaneously, with each system retaining its own functionality:

DATAPHONE II System Controller
DATAPHONE II Dial Backup Unit ASCII Interface
DATAPHONE II ACCULINK Network Manager
ACCUMASTER Trouble Tracker
StarKeeper Network Management System
Customer Controlled Reconfiguration (CCR).

StarKeeper was one of the first EMSs to come on-line as part of UNMA. StarKeeper is an AT&T 3B processor-based system designed to centrally manage a network of Datakit VCS and ISN switches. Before StarKeeper NMS was introduced, each node required its own network administrator to monitor, control, configure, and diagnose the operations of the node. With StarKeeper NMS, one centrally located network administration can view the entire network and can interact with the various nodes. The system supports all five of the OSI NM functions.

16.7.4 DEC

Until September 1988, DEC did not have an architecture or a strategy for providing integrated network monitoring, control, and management. That strategy came with the introduction of the Enterprisewide Management Architecture (EMA). EMA will be a consolidation and an extension of DEC's existing network management strategy. This strategy will be a DECnet-based integration of discrete packages, presumably similar to the way IBM integrated various existing systems under NetView. EMA will replace DEC's proprietary protocols with OSI's CMIP protocol to transfer network management information. EMA will be based on OSI protocol where possible, according to the vendor. Plans are to add support for OSI protocols when the standards have stabilized. Stability includes the standards, necessary implementors agreements, and conformance testing tools. For example, DEC VAX computers will use the OSI CMIP to transfer network management information to EMA, instead of the DEC-proprietary protocol used before [16.29].

EMA has a number of strong points compared to alternative network management architectures. These include its ability to support two-way communication, using CMIP or any other protocol, and the option to run EMA data bases on all VAX computers, ranging from PC-class machines to high-end minicomputers. Netview's data base now runs only on IBM 370 architecture machines [16.29]. DEC's emphasis on distributed architecture with local control over the network differs from IBM's highly centralized approach. As an open architecture, EMA will allow peer-to-peer connectivity with other vendors' systems. DEC was planning to publish the interface specifications in the first quarter 1989. The specification will describe an interface through which EMA will accept data from, and send data to, remote devices using CMIP. At the same time, DEC will publish specifications for programming interfaces that will support network management applications written by third parties, and "presentation" interfaces that will provide standard menus or graphics for interacting with EMA [16.30].

Currently, DEC is using proprietary protocols to transmit network management data between DEC computers in its existing network management systems.

REFERENCES

[16.1] J. Herman, "What Is Network Management and Why Is Everyone Talking about It?," *Business Communications Review*, February 1989, pp. 81–83.

[16.2] J.B. Brinsfield, W.E. Gilbert, "Unified Network Management Architecture—Putting It All Together," *AT&T Technology*, Vol. 3, No. 2, 1988, pp. 6–17.

[16.3] D. Minoli, "Numbers Add to Net Hassles," *Computerworld*, June 3, 1985, p. 53 ff.

[16.4] "Information Retrieval, Transfer, and Management for OSI: OSIRM Part 4—OSI Management Framework," Revision of DIS 7498-4, ISO/IEC JTC1/SC21, October 1988.

[16.5] E. Ericson, L. Ericson, D. Minoli, *Expert Systems Applications to Integrated Network Management*, Artech House, Norwood, MA, 1989.

[16.6] J.F. Donohue, "Systems Integrators Jump into Network Management," *Mini-Micro*, May 1987.

[16.7] J. McDowell, "Building a Network Management Strategy," *Business Communications Review*, July–August 1988, pp. 41–47.

[16.8] M. Pyykkonen, "Network Management: End-User Perspectives," *Telecommunications*, February 1989, pp. 23–24.

[16.9] D.E. Harper, "The Future of Telecommunications: Part II," *Telecommunications*, February 1989, pp. 49–56.

d directory. A standard for the directory, along wit
ted aspects, is now available; this issue is briefly sur
the U.S. government and by the financial industry i
rity management standards is described.

ems that comply with the communication protocol
and CCITT, as described in Chapter 13, in ac-
openness is a desirable goal to achieve, it has a
ty. In an age when information is considered a
ole. Privacy laws have appeared in many coun-
mmunication manufacturers to demonstrate the
their systems and networks. Specific security-
ployed for given threats [17.2].
e needed to reduce the cost of security prod-
y and evaluation. Standards allow for com-
, thereby increasing competition, which
Unless security is inexpensive and conven-
n, leaving many applications unprotected.
be added to the basic OSIRM to protect
n 2 of the OSIRM (ISO 7498) deals with
ecture, standards work is under way and
are studying the standardization of ap-
n applications [17.3].
twork management international stan-
urity management as one of the five
he framework, document security is
nd the network management system.
secure (because it can, for example,
anagement system should provide
il mechanisms, key management,
urity (provide encryption capabil-

ke the cost of obtaining or mod-
ng it, or make the time required
a is lost. Security measures will
e specific threats against which
m network is potentially vul-
of these ways may be usable
because the result does not

[16.10] D.A. Ameen, "Systems Performance Evaluation," *Journal of Systems Management*, March 1989, pp. 33–36.

[16.11] N. Streatfield, "Host-Independent Network Management," *EDP Performance Review*, Vol. 15, June 1987, pp. 1–6.

[16.12] S. Kerr, "The Politics of Network Management," *Datamation*, September 15, 1988, pp. 50–54.

[16.13] D. Tow, "Corporate Data Network Management: The Challenge," *Proceedings of NCF'88*, October 3–5, 1988.

[16.14] R.E. Caruso, "Network Management: A Tutorial Overview," *IEEE Communications Magazine*, March 1990.

[16.15] D. Tow, Bellcore, personal communication, March 1990.

[16.16] A. Ben-Artzi, "Network Management Today," *Connect*, Winter 1989, p. 53.

[16.17] *Applied Networks Report*, February 1989, Vol. 2, No. 2, pp. 1–10.

[16.18] G.D. Gex, "NetPartner System Turns Customers into Network Managers," *AT&T Technology*, Vol. 3, No. 2, 1988, pp. 6–17.

[16.19] "The Datamation 100s," *Datamation*, Special Issue, June 1988.

[16.20] "OSI and SNA Compared, The Integration of Network Standards," Communications Solutions Inc. Report, San Jose, CA, 1987.

[16.21] J.D. Case, M.S. Fedor, M.L. Schoffstall, and J.R. Davin, "A Simple Network Management Protocol," Request for Comments 1098, DDN Network Information Center, SRI International, April 1989.

[16.22] J.D. Case, "Managing Your INTERNET: The Simple Network Management Protocol," INTEROP 89, San Jose, CA, October 4–6, 1989.

[16.23] J.D. Case, "The Simple Network Management Protocol for TCP/IP-based Internets," Advanced Computing Environments' Tutorial Material, Mountain View, CA.

[16.24] "Computer Communication Standards", *Computer Communication Review*, June 1989.

[16.25] OSI/NM Forum Press Release, January 24, 1989.

[16.26] A.H. Grossman, Bellcore, personal communication, June 1990.

[16.27] "IBM NetView/PC Gets Vendor Support," *Communications Week,* April 2, 1990, p. 20.

[16.28] "NetView/PC: Users Are Just Saying 'No'," *Data Communications*, January 1989, pp. 49–52.

[16.29] "DEC Steers New Course toward OSI Net Management," *Data Communications*, September 1988, pp. 49–52.

[16.30] D.M. Weston, "Network Management: Issues, Products, and Challenges," SRI International Report D89-1318, February 1989.

the aid of a network-bas
security management-orien
veyed. Finally, progress by
the area of security and secu

17.2 THE ENVIRONMENT

Open systems are computer syst
standards specified by the ISO
cordance with ISO 7498. While
problem: inherent lack of securi
commodity, security plays a vital
tries that require computer and co
integrity, security, and privacy of
preserving mechanisms must be emp

Computer security standards a
ucts and to allow for interoperabilit
patibility among vendors' products
eventually lowers the cost of products.
ient, it will be used only as an exceptio
Security controls and mechanisms must
the exchanges of information. Addendu
security. In addition to this security archi
several national and international groups
proaches to information security for certai

Another key standard, the evolving n
dard discussed in Chapter 16, describes se
major areas of network management. In t
discussed both in the context of the network a
Namely, (1) the management system must be
reroute traffic, add users, *et cetera*); (2) the
tools to monitor and control security (audit tr
et cetera); and (3) it should ensure network se
ities, digital signature capabilities, *et cetera*).

Security controls and mechanisms should m
ifying data greater than the potential value of do
to obtain the data so great that the value of the da
usually increase the cost of the system; therefore, t
protection is required should be identified. A syst
nerable to attack in many ways. However, only som
by an attacker because of the lack of opportunity, o

justify the effort and risk of detection (see Figure 17.1). One security threat, viruses, were in the news often in the late 1980s; viruses represent a new threat to networked end systems.

The difficult task of security management and control is complicated by the proliferation of networks that connect processors, peripherals, and user terminals. Physical protection can be employed when the premises are relatively small, controlled, and economical to protect. Physical protection is practical for end systems, but it is more complicated for communication links (other means are normally required to protect links, such as encryption). The difficulties of applying security controls in a network environment stem from these factors, among others [17.4]:

1. Networks and hosts connected to them are exposed to a large number of potential attackers.
2. Networks are dynamic in both topology and user community.
3. Networks frequently span administrative domains and often are administered without formal network management processes. Internetworked resources become protected at the level of the least secure system on the network.
4. Networks use a variety of communication media, some of them potentially susceptible to eavesdropping. This problem is sometimes addressed by using encryption, a solution that introduces its own problems in the area of key management and administration.
5. Networks use a variety of communication protocols. Differences among them make it difficult to apply strong network-wide security measures.
6. The scope of a network grows exponentially in terms of new network connections: a new connection may introduce a single new user to a network, or it may introduce a new network of wider extent than the first one. This situation, coupled with weak accounting management, permits new network connections to appear without knowledge of the attendant risks.

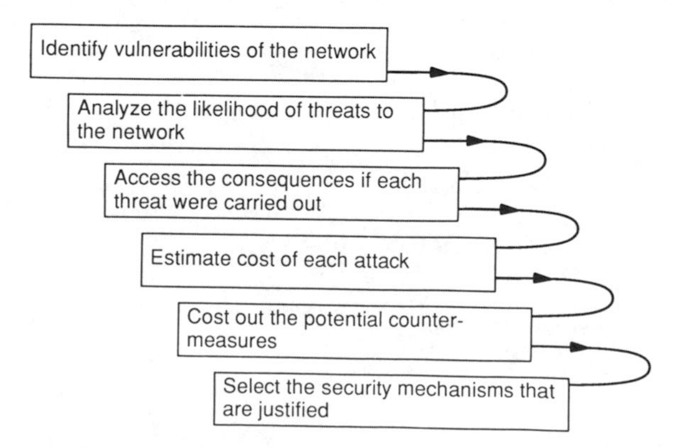

Figure 17.1 A process for managing security risks.

Figure 17.2 depicts some of the problems and safeguards for a typical network environment [17.5]. Issues pertaining to LAN security were discussed in Chapter 14.

Security and security management must be realized at many levels in a network hierarchy. At a low level in the hierarchy, individual links in a network may have to be secured against tapping; this could be done by, for example, using end-to-end or link-by-link encryption devices. At a higher level, individual computers, servers, workstations, and data bases may have to be secured; this could be done with access control mechanisms to protect against unauthorized entry and use. At yet higher levels, entire WANs may need protection, for example, in the form of a closed user group; this could be done by a combination of network security servers and localized control mechanisms. The key to achieving security in a system of any size is to analyze the security problem starting at the highest possible level; then, proceeding from there, requirements for lower level components can be specified to ensure that their collective effect is the desired level of system security and preservation of system performance [17.4].

According to the OSI specification [17.6], security management supports the application of security policies. Security is an important issue: Unauthorized or accidental access to network control functions must be eliminated or minimized. Security management functions include:

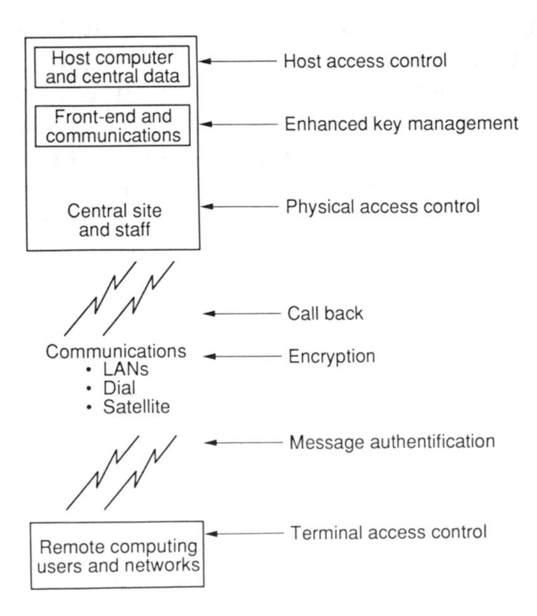

Figure 17.2 Network security countermeasures.

1. The creation, deletion, and control of security services and mechanisms;
2. The distribution of security-relevant information; and
3. The reporting of security-relevant events.

Security management controls access to both the network and the network management systems. In some cases, it may also protect information transported by the network from disclosure or modification. Key security-related terms critical to the discussion are shown in Table 17.1; Reference [17.7] provides a more complete glossary.

In a computer system network environment, security refers to minimizing the risk of exposure of assets and resources to vulnerabilities and threats of various kinds. A vulnerability is any weakness of a system that could be exploited to circumvent the policy of the system. A threat is a potential violation of the security of the system [17.2]. In an open system network, every component (software or hardware) may be the object of a threat or attack and may be vulnerable. Therefore, the protection scheme must cover the (1) the physical equipment and facilities, (2) communication and data processing services, and (3) information, data, and software.

The OSI security standard assumes, however, that the real system is secure; that is, OSI is not concerned with the security of the host seen as a discrete entity, but is concerned with security of the interface between the host and the outside world. OSI security functions are concerned only with those visible aspects of a communication path that permit end systems to achieve the secure transfer of information between them. OSI security is not concerned with security measures needed in end systems, except where these have implications on the choice and position of security services visible in OSI. These latter aspects of security may be standardized but not within the scope of OSI standards. Additional security measures are needed in end systems, installations, and organizations. Policies to provide host protection and site protection must be put in place above and beyond the OSI security services.

The U.S. government has either performed or funded much of the security research in the United States. As the demand for security reached the private sector, the need for other security standards arose. ANSI, ISO, and CCITT have initiated several security-related projects [17.3]. Security-related activities are also being pursued in the financial community, the first large commercial security market.

17.3 OSI SECURITY AND SECURITY MANAGEMENT STANDARD

In 1984, ISO/IEC/JTC1/SC21 recognized that provision must be made for security in the OSI basic reference model (ISO 7498). The OSI security architecture was

Table 17.1
Abbreviated Glossary of Security-Related Items

Access control: the prevention of unauthorized use of a resource, including the prevention of use of a resource in an unauthorized manner.

Audit: an independent review or examination of systems records and activities in order to test for the adequacy of system controls, to ensure compliance with established policy and operational procedures, and to recommend any required changes in control, policy, and procedures.

Authorization: the granting of specified rights.

Ciphertext: string of characters produced through the use of encipherment, the semantic content of which is not available.

Cleartext (also plaintext): intelligible data.

Confidentiality: the condition in which information is not made available or disclosed to unauthorized individuals, entities, or processes.

Credentials: data that are transferred to establish the claimed identity of an entity.

Cryptoanalysis: the analysis of a cipher system or of a ciphertext to derive confidential information, up to and including the cleartext itself.

Digital signature: data appended to a data unit that allows the recipient of the data unit to prove the source and integrity of the data unit and to protect against forgery.

Discretionary access control: allows users to give access to other users independent of the receiving user's attributes.

Encipherment: the cryptographic transformation of data to produce ciphertext.

Encipherment key: the sequence of symbols that controls the operations of the encipherment and decipherment process.

Mandatory access control: restricts users and data flow based on the clearance of the end-user and sensitivity level of the data.

Masquerade: the claim by an entity to be a different entity.

Multilevel security: allows data with different sensitivity levels to exist simultaneously on the same computer and allows access by users with different security clearances.

Notarization: the registration of data with a trusted third party that allows the later assurance of the accuracy of its characteristics (content, origin, time of creation, and delivery).

Password: confidential authentication information, usually composed of a string of characters.

Privacy: the right of individuals to control or influence what information related to them may be collected and stored, by whom, and to whom such data may be disclosed.

Repudiation: the denial by one of the entities involved in a communication of having participated in all or part of the communication.

Rule-based security policy: a security policy based on global rules imposed for all users. These rules usually rely on a comparison of the sensitivity of the resources being accessed and the possession of corresponding attributes of users, a group of users, or entities acting on behalf of users.

Threat: a potential violation of security, in a number of forms.

Traffic analysis: the inference of information from observation of traffic flows (presence, absence, amount, direction, and frequency).

Traffic padding: the generation of spurious instances of communication, spurious data units, or spurious data within data units.

the first of the SC21 security standards and was published in 1989 as Part 2 of the OSI basic reference model (ISO 7498-2). However, an architectural document, such as ISO 7498-2, is only the first stage in defining security services and mechanisms. It is not an implementation specification. Additional standards are required that build on the concepts of the architecture. This led to the development of related documents—security models, security frameworks, and proposals for specifying particular types of protection in the existing OSI protocol standards. While some of these documents are directed toward OSI systems, an SC21 objective is that, wherever possible, the broader open systems perspective is considered. This is being attempted in the framework documents, but whether or not this stance is possible will not be apparent for some time because currently only one security architecture (namely, the OSI Security Architecture) exists [17.8]. Discussion of some of the frameworks and current areas of study is included in Section 17.6 of this chapter.

ISO's 7498-2 OSI Security Architecture defines the general security-related architectural elements that can be applied appropriately in the circumstances for which protection of communication is required. In particular, ISO 7498-2:

- Provides a general description of security services and related mechanisms, which may be provided by the OSI basic reference model; and
- Defines the positions within the reference model at which services and mechanisms may be provided.

Thus, the security addendum establishes, within the framework of the reference model, guidelines and constraints to improve existing standards or to develop new standards in the context of OSI in order to allow secure communication and, thus, provide a consistent approach to security in OSI. Basic security services and mechanisms and their appropriate placement have been identified for all layers of the basic reference model. In addition, the architectural relationships of the security services and mechanisms to the basic reference model have been identified.

17.3.1 Network Security Threats

One must first identify the threats to the network before these threats can be managed. The threats that a network manager must consider, as identified in the OSI addendum, are: (1) theft or unauthorized removal of information or other resources; (2) unauthorized modification of information; (3) destruction of information or other resources; (4) disclosure of information; and (5) interruption of service. Most threats actually arise from the inside, from people who have knowledge of the network. The celebrated cases that are reported in the press are in the minority. Insider threats arise when legitimate users of the system behave in unintended or unauthorized ways. The most serious computer crimes have involved insider attacks that compromise the intended security of the system. Outside attacks are carried out by persons who are not authorized to use the system. Security audits are necessary to detect both insider and outsider attacks.

Threats can be classified as either accidental or intentional, and may be active or passive. Accidental threats are those threats that exist without any premeditated intent. Examples include system malfunctions, operational mistakes, and software bugs. Intentional threats may range from casual examination using easily available monitoring tools to sophisticated attacks using special knowledge of the system. An intentional threat, if realized, may be considered an attack. Passive threats are those in which no modification is made to any information contained in the system or systems and where the operation and state of the system is not changed. Wire-tapping is an example of a passive threat. Active threats are those threats that involve the alteration of information contained in the system or changes to the state or operation of the system. An authorized change to the routing tables of a packet switch is an example of an active threat.

17.3.2 Specific Threats to the Network

Denial of Service. This is a form of attack in which an entity acts so as to prevent other entities from performing their proper function. A specific target may have been chosen (an entity may suppress all messages directed to a particular destination) or the attack may be general (an entity suppresses all messages). The attack may involve suppressing traffic or it may involve generating spurious traffic so as to flood some network facility with limited storage. An entity may also generate continuous traffic, directed to a particular destination, directed via a particular route operation of the network, especially in a network where entities make routing decisions based on status received from other relay entities, such as in a dynamic packet network.

Masquerade. A masquerade is a type of attack in which an entity claims to be a different entity. A masquerade usually includes various other forms of active attack (especially "replay" and "modification of messages"). Such an attack can take place, for example, by capturing and replaying authentication sequences after a valid authentication sequence has taken place. A masquerade may be used by an authorized entity with limited privileges in order to acquire enhanced privileges; this is done either directly or by impersonating the entity that has those privileges.

Modification of Messages. Modification of messages is a threat that exists when the content of a data transmission is altered without detection and results in an unauthorized effect. For example, a message meaning "Allow John Smith to withdraw $100 at the automated teller machine and debit John Smith's account" is modified to mean: "Allow Linda French to withdraw $100 at the automated teller machine and debit John Smith's account."

Replay. A replay is a threat that is carried out when a message, or part of a message, is repeated in order to produce an unauthorized effect. A particular example of a replay threat is the repeating of a valid message that contains authentication information by an inimical entity in order to authenticate itself fraudulently.

Trapdoor. A trapdoor is planted when an entity of the system is modified to allow an attacker to produce a future unauthorized effect on a command or on a predetermined event. For example, a password validation entity could be modified so that, in addition to its normal effect, it also validates an attacker's password. Trapdoors are usually associated with security systems in which special access is permitted without the benefit of authentication.

Trojan Horse. A Trojan horse is an entity that, when innocuously introduced into the system, has a deliberately planned unauthorized effect in addition to its authorized function.

In addition to malicious security infractions, the possibility of protocol vulnerability also exists. Three types of protocol vulnerabilities are:

1. *Logical Deficiencies*. These deficiencies will result in incorrect operations and, possibly, denial of service. Common problems are: deadlocks, unboundness, inexecutable interactions, and others.
2. *Lack of Robustness*. In this case, the protocols cannot self-stabilize within a given time threshold to isolate and correct the failure without human intervention.
3. *Nonsecure Protocols*. Many network protocols are designed assuming that they will reside in a benign environment. Specialized military protocols for hostile environments are much more complex and difficult to specify.

17.3.3 OSI Security Services and Mechanisms

This section describes the security included in the OSI security architecture and the mechanisms that implement it in order to manage the security threats discussed above [17.7]. The following basic services are defined:

1. Authentication;
2. Access control;
3. Data confidentiality;
4. Data integrity; and
5. Nonrepudiation.

Authentication. The authentication services provide for the verification of the identity of a remote communicating entity and the source of data. Authentication consists of two services: peer entity authentication and data origin authentication. The peer entity authentication service is used at the establishment of a connection or at times during the data transfer phase to confirm the identities of one or more of the entities connected. This service provides confidence—at the time of usage only—that the corresponding entity is not attempting a masquerade or an unauthorized replay of a previous connection message. The data origin authentication service provides corroboration to an entity in a particular layer that the source of the data is really the claimed peer entity.

Access Control. The access control service provides protection against unauthorized use of resources accessible via the network. These may be OSI or non-OSI resources accessed via OSI protocols. This protection service is applied to various privileges of access to a resource (e.g., the reading, the writing, or the deletion of an information file or record in the file, or the execution of a program).

Data Confidentiality. These services provide for the protection of data from unauthorized disclosure. Four subtypes of service are connection confidentiality; connectionless confidentiality; selective field confidentiality; and traffic flow confidentiality. *Connection confidentiality* provides for the confidentiality of all user data on a connection. *Connectionless confidentiality* provides for the confidentiality of all user data in a single connectionless message exchange. This service provides for the confidentiality of selected fields (records) within the user data. *Traffic flow confidentiality* provides for the protection of the information that might be derived from observation of traffic flows.

Data Integrity. This service provides for the integrity of all user data or of some selected fields over a connection or connectionless message exchange; it detects any modification, insertion, or deletion of data. Four subservices are: connection integrity with recovery; connection integrity without recovery; selective field connection integrity; and connectionless integrity.

Nonrepudiation. This service can take one or both of two forms: nonrepudiation with proof of origin or nonrepudiation with proof of delivery. In the first case, the recipient of data is provided with the proof of the origin of the data; this proof will protect the recipient against any attempt by the sender to deny falsely the sending of the data or its contents. In the second case, the sender of data is provided with the proof of the delivery of the data; this proof will protect the sender from any attempt by the recipient to deny falsely the receiving of the data or its contents.

The security services listed earlier can be provided via a number of available security mechanisms. A distinction between the two groups is that security services specify "what" controls are required; the security mechanisms specify "how" the controls are to be implemented. The selection of the appropriate mechanisms is part of architectural considerations. Types of mechanisms include encryption, access control, and others. Mechanisms can also be classified as preventive, detective, and recovery.

Security mechanisms include the following:

- Encipherment;
- Digital signature;
- Access control;
- Data integrity;
- Authentication exchange;
- Traffic padding;
- Routing control; and
- Notarization.

Encipherment. Encipherment (also called encryption) is a key security mechanism that can provide confidentiality of either data or traffic flow; often it is used in conjunction with other mechanisms. Cryptography is the discipline involving principles, means, and methods for the mathematical transformation of data in order to hide its information content, prevent alteration, disguise its presence, or prevent its unauthorized use. The transformation uses confidential variables, called keys, to map between the plain text and the cipher text. Message content release and traffic analysis can be minimized by using this tool, which also affords protection against repudiation and forgery. This mechanism is required for many security services. Cryptography can be used as part of encryption, data integrity, data origin authentication, password storage, and checking.

Encipherment algorithms can be divided into two categories: symmetric encipherment and asymmetric encipherment. In a symmetric encipherment algorithm, the encipherment key is secret and the knowledge of the encipherment key implies the knowledge of the decipherment key and *vice versa*. In an asymmetric algorithm, the encipherment key is public and the knowledge of the encipherment key does not imply the knowledge of the decipherment key or *vice versa*. The two keys of such a system are referred to as the "private key" and the "public key," respectively. When symmetrical or private-key encipherment is used, the key must be kept secret. Key management is a major aspect of security management, as discussed in more detail below.

Encryption can take place at a number of layers. The focus of encryption has been at the lower two layers of the OSIRM. The two types of encryption are link-by-link encryption and end-to-end encryption. Link encryption has limitations in that data are handled in plain text in the communication processors, and it is inappropriate for broadcast networks. End-to-end encryption is more robust, but it requires increased resources, and is subject to Trojan horse attacks and covert channel infiltration.

Link encryption protects individual links. All information passed to the ph ical link is encrypted. A procedural advantage of link encryption is that it car implemented independently of the end systems; the end systems can continue to operate as they did prior to the introduction of the service, while at the same time achieving communication link protection. One disadvantage is that the encryption process must be applied at each link; at the nodes where links join, the message must be retransformed into clear text, introducing liabilities. All hosts and switching nodes of the network must be trusted to provide the necessary protection. Also, the problem of managing (storing and distributing) all the required keys is present. Additionally, link encryption is inappropriate for broadcast networks.

End-to-end encryption is an older term for what is now referred to as "encryption above the physical layer." However, the term lacks specificity in the OSI environment because no indication is provided as to the layer in which the encryption is taking place. In an OSI context, referring to the layers where the encryption occurs is desirable: network layer encryption, transport layer encryption, application layer encryption, *et cetera*. The integration of encryption protocols with the standards at each of these layers is one of the objectives of the current OSI standardization efforts. Upper layer encryption is considerably more complex architecturally than link encryption, particularly if in-band electronic key distribution is included in the algorithms.

Encipherment reduces the chances of message content release and traffic analysis; message stream modification, denial of service, and masquerading can be detected. Encipherment provides confidentiality and contributes to data integrity and authentication of peer entities. Link-by-link encryption (physical layer encryption) provides data confidentiality and traffic flow confidentiality through the use of data encryption devices attached at the termination points of each link. End-to-end encryption can also provide confidentiality and can be performed at any point from the upper boundary of the network layer up to the application layer or even the application process. Data integrity is achieved by the detection of data stream modification (seen by the CRC, which is also protected by encryption) and the subsequent use of separate data correction procedures. End-to-end encryption will also provide peer entity authentication; for encryption to be effective in this area, the initiator of the connection must use a proper preestablished key: Use of that key must have been agreed to prior to establishing the session, and the key must be transported in a secure fashion.

As for all other security services, standardization is necessary to achieve encryption (at the very least to the extent of using the same algorithms at both ends of a link). To date, encryption equipment is closely held and physically protected, particularly for military applications; nondisclosure of encryption mechanisms and algorithms is the norm. In the business arena, standardization can lead to economies of scale. Currently, only one standard is available: the Federal Information Processing Standard (FIPS PUB 46) data encryption standard (DES), also known at times as the *data encryption algorithm*—DEA. However, note that the point of application of encryption, methods of request for use, and key distribution techniques are subjects for standardization by OSI; the encryption algorithms and the form of the keys are not. The encryption algorithms used and the form of the keys are the concerns of national agencies. In the United States, the National Security Agency (NSA) provides these guidelines.

Digital Signature. This is an encryption method that provides data integrity, namely, the guarantee that data have not been altered or destroyed in an unauthorized fashion. The digital signature is data appended to (or a transformation of) a PDU that allows a recipient to prove the source and integrity of the data. In the public-key environment, the entire encrypted message is referred to as the digital signature. In the secret-key environment, the message authentication code (MAC), which is a cryptographic check sum added to the data, is called the digital signature. These security mechanisms comprise two different processes: (1) The signing process uses information that is private, i.e., confidential to the signer. This process involves either an encipherment of the data unit or the generation of a cryptographic check-value of the data unit, using the signer's private information as a private key. (2) The verification process involves using the public procedures and information to determine whether the signature was produced with the signer's private information. The basic characteristic of the signature mechanism is that the signature can only be produced using the signer's private information. When the signature is verified, it can subsequently be proven to any third party that only the unique holder of the private information could have produced the signature.

Access Control Mechanisms. These mechanisms use the authenticated identity of an entity, its capabilities, or its credentials to determine and enforce the access rights of that entity. These mechanisms ensure that only authorized users have access to the protected facilities. If the entity attempts to use an authorized resource in the proper manner, then access to the resource is granted. If the entity attempts to use an unauthorized resource, or an authorized resource with an improper type of access, then the access control function will reject the attempt and may additionally report the incident for the purpose of generating an alarm or recording it as part of a security audit trail. The access control mechanisms may be applied at either end of a connection or to a connectionless message exchange. Five basic access mechanisms involve the use of the following tools: access control lists: lists of entities together with access rights to a resource; passwords: confidential information used to authenticate the user of a resource; capabilities lists: an unforgeable identifier conferring access rights to a given user for a given resource; credentials: data passed from one entity to another that is used to establish the access rights of the requester entity; labels: tokens possessed by a user that confer specified access rights.

The implementation of access control mechanisms requires the existence of one or more of the following types of information (possibly in the directory discussed in Chapter 13):

1. Access control data bases, in which the access rights of authorized peer entities to resources are maintained; this information may be maintained by authorization centers or by the entity being accessed. The information may be in the form of an access control list or matrix.
2. Authentication information, such as passwords, the possession and subsequent presentation of which is evidence of the accessing entity's authorization.
3. Capabilities, the possession and subsequent presentation of which is evidence of the right to access the entity, or the resource defined by the capability.
4. Security labels (secret, top secret, *et cetera*) which are used to grant or deny access according to a security policy.
5. Other information, such as time of the attempted access, route of the attempted access, duration of access, *et cetera*.

Access control is based by definition on authenticated identification. Typically, a user claims the identity of a person or process operating on behalf of the person. Then the identity must be proven; in manual systems, a common proof is a credential carrying the photograph and signature of the individual. Authentication information must be validated before the user identification is accepted through comparison of known and presented information. Automated identification is based on two methods: nonencryption and encryption.

The nonencryption method relies on:

- Information the user has, such as passwords;
- Instruments the user possesses, such as a physical key or card (typically for challenge-and-response authentication); and
- Information about a physical characteristic, such as a traditional signature.

(These last two approaches require an appropriate entry device.) The host computer normally stores the comparison information in encrypted form; when the user presents the authentication information, it is encrypted and then compared with the stored cipher text. Challenge-and-response authentication eliminates the majority of the stored information because it assumes that the user possesses secret information. After the challenge is issued by the computer, the user can use devices such as portable cards or other portable devices to formulate a response; the response is compared with the expected string and the access is granted or denied. Handwritten signatures entered via a pressure-acceleration device can also be employed.

The other approach is encryption-based authentication for access. This method assures that the message came from the claimed source and was delivered unchanged. The functional similarity to a handwritten signature has caused message authentication to be known as digital signature.

Data Integrity. These mechanisms ensure that the data have not been altered or destroyed. These security mechanisms deal with two aspects of data integrity: the integrity of a single data unit or field, and the integrity of a stream of data units or fields. Different mechanisms are used to provide these two types of integrity services. Two specific tools include check sums (cryptographic remainder) and sequencing (which assures that the PDU sequence is correct).

The determination of the integrity of a single data unit involves symmetric processes, one at the sending entity and the other at the receiving entity. The sending entity appends a quantity to a data unit that is a function of the data. This quantity may be supplementary information such as a block check code or a cryptographic check-value and may itself be enciphered. The receiving entity generates a corresponding quantity and compares it with the received quantity to determine whether the data have been modified in transit. This mechanism alone does not protect against the replay of a single data unit. Detection of manipulation may trigger a recovery action at that layer or at a higher layer of the architecture. For example, the recovery may be performed by retransmission or error correction. The determination of the integrity of a stream of data units or fields varies for connection-oriented and connectionless information exchange.

For connection-oriented data transmission, protecting the integrity of a sequence of data units, namely, protecting against misordering, losing, replaying, inserting, or modifying data, requires some additional explicit ordering mechanism such as sequence numbering, time stamping, or cryptographic chaining. For connectionless data transmission, time stamping may be used to provide a limited form of protection against replay of individual data units.

Authentication Exchange Mechanism. This mechanism provides corroboration that a peer entity is the one claimed. Specific mechanisms include authentication information, such as passwords, supplied by a sending entity and checked by the receiving entity, and cryptographic means. The mechanism may be incorporated into a layer in order to provide peer-to-peer entity authentication. If the mechanism does not succeed in authenticating the entity, the connection is rejected and terminated. An entry in the security audit trail or a report to a security management center may also be generated. When cryptographic techniques are used, they may be combined with handshaking protocols to protect against replay. In many cases, authentication exchange mechanisms will be used in conjunction with nonrepudiation services by digital signature or notarization mechanisms.

Traffic Padding. This mechanism provides generation of spurious traffic or filling of PDUs to achieve constant traffic rates or message length. Traffic padding mechanisms can be used to provide various levels of protection against traffic analysis. This mechanism is only effective if traffic padding is protected by a data confidentiality service.

Routing Control. This is a mechanism for the physical selection of alternative routes that have a level of security consistent with that of the message being transacted. Routing control mechanisms ensure that the routes used by the data across the network are those that have been specified. Routes can be chosen either dynamically or by prearrangement so as to use only physically secure subnetworks or transmission links. The initiator of a connection or the sender of a connectionless message may specify routing instructions that request particular subnetworks or links to be avoided. This mechanism can direct the network service provider to establish a connection via a different route if persistent attacks on the initial route are detected. Data carrying certain security labels may be forbidden by the security policy to travel through certain subnetworks not cleared at the appropriate level. Routing control is not easy to implement in practice.

Notarization Mechanism. This security mechanism provides the assurance that the properties about the data communicated between two or more entities, such as their integrity, origin, time, and destination, are what they are claimed to be. The assurance is provided by a third-party notary, which is trusted by the communicating entities and which holds the necessary information to provide the required assurance in a manner that can be tested. Each communication channel may use digital signature, encipherment, and integrity mechanisms as appropriate to interface with the service being provided by the notary. When the notarization mechanism is invoked, the data are exchanged between the communicating entities via the protected communication channels and the notary.

Pervasive Security Mechanisms. These types of mechanisms that are not specific to any particular service; some of these security mechanisms can be regarded as aspects of security management. The major three follow.

1. *Event Detection.* Security-related event detection encompasses the detection of apparent violations of security and may also include detection of normal events, such as successful network logons. Examples of security-related events are a specific security violation; a specific network logon; a specific selected event. Security-related events can be detected by entities within OSI including the security services and mechanisms. The specification of what constitutes an event is maintained by event handling management facilities. Detection of various security-related events may, depending on implementation, trigger one or more of the following actions: local reporting of the event; remote reporting of the event; logging of the event; recovery action for the event. For each event, the relevant information for event reporting and event logging, and the syntactic and semantic definitions to be used for the transmission of event reporting and event logging, are standardized.
2. *Security Audit Trails.* A security audit permits detection and investigation of breaches of security by permitting a subsequent security audit. A security audit is an independent review and examination of system records and activities in order to ensure compliance with established policies and operational procedures, to test for the adequacy of system controls, to aid in damage assessment, and possibly to recommend any indicated change in controls, policies, and procedures. A security audit requires the analysis, recording,

and reporting of security-related information found in the security audit trails. The security audit trail is considered to be a security mechanism because the known existence of a security audit trail may serve as a deterrent to some potential sources of security attacks. Note that the analysis and report generation associated with the security is considered a security management function.

3. *Security Recovery*. Security recovery handles the requests originating from mechanisms such as event detection handling and management functions, and performs recovery actions as the result of applying a set of rules. The recovery actions may be of three types: immediate, temporary, or long term.

17.3.4 Relationship of the Services, Mechanisms, and Layers

Table 17.2 from the OSI addendum indicates how the security mechanisms, individually or in combination with others, are used to provide security services required by the network manager to ensure a trustworthy communication environment. Table 17.3 indicates the layers of the OSIRM in which particular security can be provided.

17.4 SECURITY MANAGEMENT

ecurity management is concerned with the administrative management aspects of rity, i.e., those administrative procedures and operations needed to support

Table 17.2
Security mechanisms

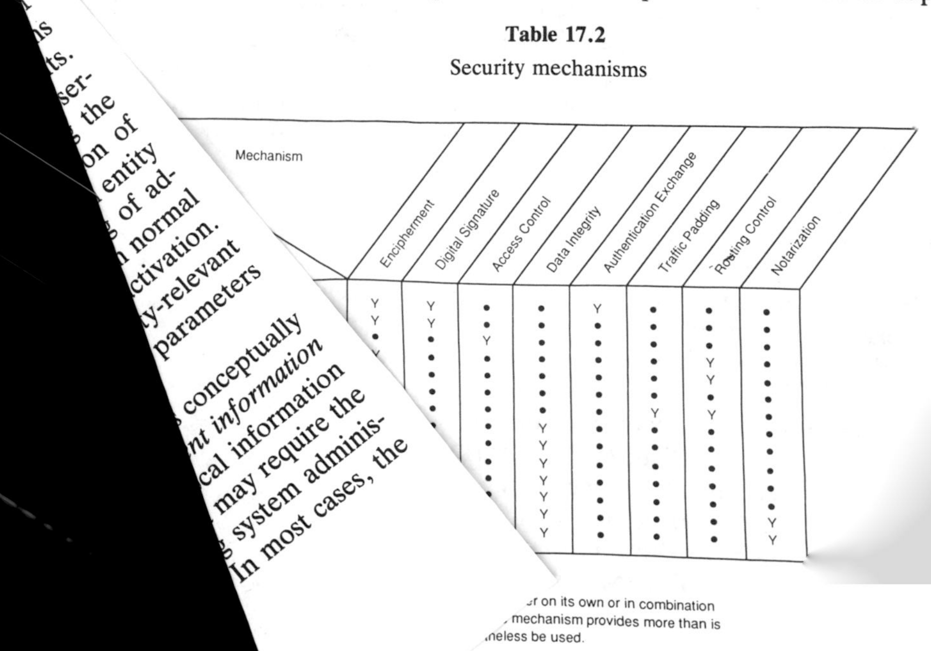

...r on its own or in combination
... mechanism provides more than is
...neless be used.

Table 17.3
Layers and Security

Service \ Layer	1	2	3	4	5	6	7
Peer Entity Authentication	•	•	Y	Y	•	•	Y
Data Origin Authentication	•	•	Y	Y	•	•	Y
Access Control Service	•	•	Y	Y	•	•	Y
Connection Confidentiality	Y	Y	Y	Y	•	Y	Y
Connectionless Confidentiality	•	Y	Y	Y	•	Y	Y
Selective Field Confidentiality	•	•	•	•	•	Y	Y
Traffic Flow Confidentiality	Y	•	Y	•	•	•	Y
Connection Integrity with Recovery	•	•	•	Y	•	•	Y
Connection Integrity without Recovery	•	•	Y	Y	•	•	Y
Selective Field Connection Integrity	•	•	•	•	•	•	Y
Connectionless Integrity	•	•	Y	Y	•	•	Y
Selective Field Connectionless Integrity	•	•	•	•	•	•	Y
Non-repudiation, Origin	•	•	•	•	•	•	Y
Non-repudiation, Delivery	•	•	•	•	•	•	Y

Legend: Y: Yes, service should be incorporated in the standards for the layer as a provider option.
•: Not provided

and control the security aspects of communication. Security management is also concerned with the security of the network management mechanism itself. Thus, management aspects of security are concerned with those operations that are outside normal instances of communication but are needed to support and control the security aspects of communication.

Entities that are subject to a single security policy, administered by a single authority, are sometimes collected into what has been called a "security domain." By contrast, networked systems are likely to be required to satisfy different security policies. Security management standards support the control and distribution o information to various end systems for providing security services and mechanism and for reporting on security services, mechanisms, and security-related eve Such management requires distribution of management information to these vices and mechanisms, as well as the collection of information concernin operation of these services and mechanisms. Examples are the distributi cryptographic keys, the distribution of information on the access rights of a prior to that entity initiating a connection to another system, the settin ministratively imposed security selection parameters, the reporting of bot and abnormal security events (audit trails), and service activation and de owever, security management does not address the passing of secur rmation in products that call up specific security services (e.g., in nnection request).

he set of security-related information needed by open system es a data base called, in OSI terms, the *security managem* B). Each end system needs to contain the necessary l to enforce appropriate security. Security managemen ertinent security information between cooperatin s so that the SMIB can be created or updated.

security-related information may be exchanged over a data communication connection. The security management protocols and the communication channels carrying the management information are especially vulnerable; hence, particular care must be taken to ensure that the security management protocols and information are protected. In other cases, the security-related information may be transferred through non-OSI communication paths, and the local open system's administrators may update the data base through local methods not standardized by OSI. The SMIB concept does not presume any storage implementation or format; this data base can be implemented as a centralized or distributed data base.

The OSI standard distinguishes three categories of security management activities: (1) system security management, (2) security service management, and (3) security mechanism management.

System security management is defined as the management of security aspects of the whole open system network. It encompasses the following activities: overall security policy management, including updates and maintenance consistency; security-related event-handling management; security audit management; interaction with other OSI management functions; interaction with security mechanism management; and security recovery management.

Security service management is defined as the management of particular security services. The following activities may be performed in managing a particular security service:

- Determination and assignment of the target protection for the service;
- Local or remote negotiations of available security mechanisms;
- Invocation of specific security mechanisms via the proper security mechanism management function;
- Interaction with other security service management functions and security mechanism management functions; and
- Selection of a specific security mechanism to be used to provide the requested security service when alternatives exist.

Security mechanism management is defined as the management of particular security mechanisms. It encompasses the following security mechanism management functions:

- Key management;
- Encipherment management;
- Access control management;
- Data integrity management;
- Authentication management;
- Routing control management; and
- Notarization management.

Management of each of these security mechanisms is addressed below.

Key Management. The security and integrity of cryptographic keys are of vital importance to the organizations that use them to secure the information. Compromise of the key used to encrypt or decrypt the information will be counterproductive. Key management involves the tasks listed in Table 17.4. Key management and distribution via the network, which employs encryption to transmit (encrypted keys), is convenient and efficient. Some nontelecommunication key loading mechanisms are: manual switches, in which rotary thumbwheel switches can be set manually to select the desired encryption at each encryption device; plug-in modules, in which the key can be inserted into an encryption device without anyone knowing what the key is; magnetic stripes, in which the key encoded on a magnetic stripe can be inserted into the encryption device, again without the key being disclosed even to the manager; and processor interfaces, in which keys are computed by a processor and loaded directly into an on-line encryption device.

Encipherment Management. The OSI standard defines encipherment management as consisting of: (1) interaction with cryptographic key management, (2) establishment of cryptographic parameters, and (3) cryptographic synchronization. The existence of an encipherment mechanism implies the use of key management and of common ways to reference the cryptographic algorithms. A common reference for cryptographic algorithms can be obtained by using a notary for cryptographic algorithms (say, the maintainer of the directory, see below) or by prior agreement between the communicating entities.

Access Control Management. Access control management consists of the management and distribution of security attributes (e.g., passwords) or updates to access control lists or capability lists. Access control management may also encompass the use of an access control protocol between communicating entities and other entities providing access control services.

Access control systems often suffer from the vulnerabilities of their password tables. The tables gather all the passwords in one data file; anyone who obtains

Table 17.4
Key Management Tasks

Key management functions within the OSI environment:

- Generation of suitable keys at intervals satisfying the level of security required
- Determination of which entities should receive a copy of each key, in accordance with access control requirements
- Secure distribution of the keys to the proper entities belonging to the systems involved

Key management functions outside the OSI environment:

- Secret cryptographic key distribution by:
  - First class registered mail
  - Special courier
  - Telecommunications carrier

this information can impersonate any of the system's users. To guard against this possibility, the access control manager must make sure that the password file does not contain the passwords, but instead contains images of the passwords under a one-way function. Such a function is easy to compute, but difficult to invert. For any password, the table entry representing the image is easy to compute. Given the image of the password, however, it is very difficult to find the input string that produced it. This reduces the value of the password table to a potential intruder because its entries are not passwords and are not acceptable to the password verification routine [17.9].

Data Integrity Management. Data integrity management consists of interaction with key management, establishment of cryptographic parameters and algorithms, and the use of data integrity protocols between communicating entities.

Authentication Management. Authentication management is defined as the distribution of descriptive information, passwords, or secret codes to entities responsible for performing authentication. It also encompasses the use of an authentication protocol between the communicating entities that provide the authentication services.

Traffic Padding Management. Traffic padding management is defined by OSI as the maintenance of the set of rules that is used for traffic padding; for example, prespecified data rates, specified random data rates; specified message characteristics such as length, *et cetera*.

Notarization Management. Notarization management is defined as the distribution of information about notaries across networks, the use of a protocol between a notary and the communication entities, and interaction with notaries.

Authentication Via the Directory

The CCITT X.500 Directory can be used to undertake authentication. Many applications require the objects taking part in a session to offer some proof of their identity before they are permitted to carry out specific actions. The directory provides support for this authentication process. (As a separate matter, the directory requires its users to be authenticated, so as to support access control.) The more straightforward approach to authentication, called "simple authentication," is based on the directory holding a "user password" attribute in the entry for any user that desires to be authenticated for a certain a privilege. At the request of the service under consideration, the directory will confirm or deny that a particular value supplied is actually the user's password. This avoids the user needing different password for every service. In cases for which the exchange of passwor in a local environment that uses simple authentication is considered to be ir propriate, the directory optionally provides means to protect those passw against replay or misuse by a one-way function (see Table 17.5).

Table 17.5
The Three Classes of Authentication Provided by Public X.500 Directories

Simple authentication	Employs the user's unique name and password to provide a level of identification assurance
Strong authentication with secret key	Employs symmetric encryption to provide a level of assurance
Strong authentication with public key	Employs asymmetric encryption to provide a level of assurance

The more complex approach, called "strong authentication" is based upon public key cryptography, where the directory acts as a repository of users' public encryption keys, suitably protected against tampering. The steps that users can take to obtain each others' public keys from the directory, and then to authenticate with each other using them, are described in detail in Recommendation X.509.

17.5 SECURITY FRAMEWORKS

The purpose of the security frameworks is to provide comprehensive and consistent descriptions of specific functional areas of security such as authentication and access control. These frameworks can be seen as a set of interrelated documents and are parts of a multipart set [17.8]. The documents are being developed by the ISO.

The security frameworks are concerned with defining the means of providing protection for systems and objects within systems; and the interaction between systems. The frameworks are not concerned with the methodology for constructing systems or mechanisms. The frameworks address both data elements and sequences of operations that are used to obtain specific security services.

In the case of access control, access may either be *to* a system or *within* a system. The data elements that need to be presented to obtain access, as well as the sequence of operations to request the access and to be notified of the access, are considered to be within the scope.

17.5.1 Security Framework Overview Document

This part of the framework standards [17.8]:

- Describes the organization of the individual security frameworks;
- Defines security concepts that are required in more than one part of the security framework standard; and
- Describes the interrelationships of the services and mechanisms identified in other parts of the framework standard.

17.5.2 Authentication Framework

This framework describes all aspects of authentication as they apply to open systems, the relationship of authentication with other security functions such as access control, and the management requirements for authentication.

17.5.3 Access Control Framework

This framework describes all aspects of access control (e.g., user-to-processes, user-to-data, process-to-process, process-to-data) in open systems; the relationship to other security functions, such as authentication and audit; and the management requirements for access control.

17.5.4 Nonrepudiation Framework

This framework describes all aspects of nonrepudiation in open systems, the relationship to other security functions, and the management requirements for nonrepudiation.

17.5.5 Integrity Framework

This framework defines the basic concept of integrity. It also identifies possible classes of integrity mechanisms. Services and the required abstract data types for the classes of integrity mechanisms are described.

17.5.6 Confidentiality Framework

This framework defines the basic concept for confidentiality in information retrieval, transferring, and data management. It identifies possible classes of mechanisms, along with services and abstract data types.

17.5.7 Security Audit Framework

This framework defines the concepts of security audit trails. It identifies the classes of events that may give rise to the creation of a security audit trail record.

17.5.8 Current Standardization Efforts

Appendix A, based on ISO/IEC JTC1/SC21, "Information Retrieval, Transfer and Management for OSI: Guide to Open System Security," a November 19 working draft, provides a detailed listing of recent ongoing security standardization efforts. Table 17.6 summarizes areas of activity.

17.6 OTHER STANDARDS EFFORTS

17.6.1 U.S. Government

The government has at least three organizations that sponsor security standards. The U.S. Department of Defense (DoD) publishes DoD standards, the Department of Commerce produces FIPS, and the General Services Administration produces federal standards [17.3].

Department of Defense Standards. The DoD sponsors both DoD standards and military standards. The department's National Computer Security Center (NCSC) has produced many recommendations, some of which have become standards. Efforts such as the Secure Data Network System (SDNS) and GOSIP are expected to produce standards in the next few years. The SDNS is a program sponsored by the NSA to develop standards for secure data networks. The program integrates security features into OSI-based networks, using the concepts of OSI security discussed above. The program will issue standards that may then be used by vendors involved in NSA's Commercial COMSEC (Communication Security) Endorsement Program.

Table 17.6
Areas of Standardization Interest

Areas of Standardization Interest
Access control
Alarm reporting
Audit
Authentication
Confidentiality
Digital signature
Electronic data interchange
Encipherment and cryptographic techniques
Integrity
Key management
Message handling
Network layer security
Nonrepudiation
Presentation layer security
Security architectures
Security management
Transport layer security
User requirements

National Computer Security Center. In 1978, the National Bureau of Standards (now called NIST) started to discuss the auditing and evaluation of computer security. In June 1981, the DoD Computer Security Center began operation. The success of the center and the need to transfer security information to the commercial sector resulted in the expansion of the DoD Computer Security Center to the NCSC [17.2]. NCSC's mission is "to develop and promulgate uniform computer security criteria and standards." The NCSC has issued the key publications listed in Table 17.7.

17.6.2 The Financial Industry

The financial community has devoted extensive resources to information security; a major concern is secure electronic transfer of money or securities. The industry works through ANSI. The ANSI groups responsible for financial security are American Standard Committees X9 and X12. ASC X9 develops financial services standards; ASC X12 develops standards for business transactions (international responsibility for financial security standards is distributed between ISO Technical Committee 68 and Technical Committee 154).

The following are the areas of concern to the financial community [17.3]:

1. *Connection Integrity* (referred to in the industry as "message integrity"). This can be provided with digital signatures. Financial standards use a message authentication code, to provide this service. The ANSI X9.9-1986 and ISO 8730 standards define the process.
2. *Connection Confidentiality*. The financial community uses the DEA (DEA X3.92-1981), which is equivalent to the DES, to protect data. Financial institutions use encryption to protect personal identification numbers (PINs)

Table 17.7
Government Security "Standards"

Orange Book	A DoD standard for evaluating stand-alone computers for security. Defines several different levels of secure computers. In order of increasing security, these levels are D, C1, C2 (all three traditional), B1, B2, B3, and A1 (all four multilevel).
Yellow Book	A guideline that states the environments ("the external circumstances, conditions, and objects that affect the development, operation, and maintenance of a system") in which each Orange class offers adequate protection.
Red Book	A standard for interpreting and applying the Orange Book requirement to the evaluation of networks.
Green Book	Guidelines that provide suggestions for the initialization, administration, and maintenance of password systems.

(ANSI X9.8-1982) and messages (ANSI X9.23); this service corresponds to the selective field confidentiality.

3. *Access Control Management.* An effort is under way in ANSI to develop a standard (ANSI X9.26-Draft) to protect computer sign-on information, such as passwords.
4. *Encipherment (Key) Management.* The financial community relies on a key management center (KMC) to distribute keys to subscribers that have a need to communicate. The center is able to audit and assign liability based on a subscriber's knowledge of the key. This is a necessary feature for the financial community. Communication among key management centers is being studied so that subscribers of different centers can communicate in a secure manner.

The last part of Appendix A provides a more detailed listing of ANSI and FIPS security standards for the financial and EDI arena.

REFERENCES

[17.1] D. Minoli, "Evolving Security Management Standards," DataPro Report NM-500-101, June 1989.

[17.2] A.J. Bayle, "Security in an Open System Network: A Tutorial Survey," *Information Age*, Vol. 10, No. 3, July 1988.

[17.3] K. Barker, L.D. Nelson, "Security Standards—Government and Commercial," *AT&T Technical Journal*, May–June 1988, pp. 9–18.

[17.4] A.L. Andreasson *et al.*, "Information Security: An Overview," *AT&T Technical Journal*, May–June 1988, pp. 2–8.

[17.5] "Defending Secrets, Sharing Data," Office of Technology Assessment, OTA-CIT-310, October 1987.

[17.6] "Information Retrieval, Transfer, and Management for OSI: OSIRM Part 4—OSI Management Framework," Revision of DIS 7498-4, ISO/IEC JTC1/SC21, October 1988.

[17.7] "Information Processing Systems: OSIRM Part 2—Security Architecture," Final Text of DIS 7498-2, ISO/IEC JTC 1/SC 21 N2890, July 19, 1988.

[17.8] "Information Retrieval, Transfer, and Management for OSI: Guide to Open System Security," Working Draft, ISO/IEC JTC1/SC21, November 1989.

[17.9] W. Diffie, "The First Ten Years of Public-Key Cryptography," *Proceedings of the IEEE*, Vol. 76, No. 5, May 1988, pp. 560–577.

Appendix A
Summary of Security Standards Activities

This appendix is based on ISO/IEC JTC1/SC21, "Information Retrieval, Transfer, and Management for OSI: Guide to Open System Security," a November 1989 working draft. It provides a detailed summary of security standardization efforts as of the beginning of 1990. (Because many of these standards will require several years to complete, this appendix is expected to be a guide to these efforts well into the early 1990s.)

International Organization for Standardization (ISO)

Project or Work Item	*Status (as of Jan. 1, 1990)*	*Documents*
Joint Technical Committee 1 (JTC1)—Information Technology		
SWG on Security Objectives	Ongoing	JTC1N531
JTC1/SC6—Telecommunications and Information Exchange between Systems		
WG3—Network Layer		
Network Layer Security	NWI proposal	
WG 4—Transport Layer		
OSI Lower Layer Security Model	Early WD	
Transport Layer Security Protocol	NWI proposal	

JTC1/SC18—Text and Office Systems

WG1—User Requirements and Management Support.

User Requirements on Security	WD	SC18WG1N497
Proposed Draft Addendum to ISO 8613 on Security	PDAD	SC18N2003

JTC1/SC17—Identification and Credit Cards

WG4—Integrated Circuit Card.

Identification cards—IC cards with contacts:

Part 1 Physical Characteristics	DIS	DIS 7816-1
Part 2 Number and Position of Contacts	DIS	DIS 7816-2
Part 3 Electronic Signals and Exchange Protocols	DIS	DIS 7816-3

JTC1/SC20—Data Cryptographic Techniques

WG1—Secret Key Algorithms and Applications

Modes of operation for 64-bit block cipher algorithm	complete	ISO 8372
Modes of operation for *n*-bit block cipher algorithms	DP	DP 10116
Data Integrity mechanism using a cryptographic check function employing an *n*-bit algorithm with truncation	IS	IS 9797
Peer Entity Authentication mechanism using an *n*-bit secret key algorithm	DP	DP 9798/2
Register of Encipherment Algorithms	DP	DP 9979
Cryptographic Mechanisms for key management		
Overview	Approved NWI	
Using secret key techniques	WD	SC20N436

WG2—Public Key Cryptosystems and Modes of Use

General Model for Peer Entity Authentication	DP	DP 9798/1
Peer Entity Authentication using a public key with two-way and three-way handshake	DP	DP 9798/3

Authentication with three-way handshake using zero-knowledge techniques	WD	
Digital Signature scheme with message recovery	DP	DP 9796
Hash Functions for digital signatures	DP	DP 10118
Cryptographic mechanisms for key management		
using public key techniques	WD	SC20N437
key management for public key register	WD	
WG3—Use of Encipherment Techniques in Communication Architectures		
Data Encipherment—Physical layer interoperability requirements	complete	ISO 9160
Transport layer cryptographic techniques	WD	SC20/3 N101
Presentation layer cryptographic techniques	WD	SC20/3 N102
Practical conditions for ACSE Authentication	WD	
Network layer cryptographic techniques	WD	SC20/3 N100

JTC1/SC21—Information Retrieval, Transfer, and Management for Open Systems Interconnection

WG1—OSI Architecture		
SC21 Security Coordination	Ongoing	SC21N2540
Overview of Open System Security (Roadmap)	WD	SC21N4258
OSI Security Architecture	IS	ISO 7498-2
Authentication Framework	DP	SC21N4207
Access Control Framework	Mature WD	SC21N4206
Non-Repudiation Framework	Early WD	SC21N4209
Integrity Framework	Outline WD	SC21N4208
Confidentiality Framework	Outline WD	SC21N3618
Framework Overview	Early WD	SC21N4210
WG4—OSI Management.		
OSI Security Management—6th draft	Mature WD	SC21N4091
OSI Systems Management—Part x: Security Audit Trail Function	WD	SC21N4092

OSI Systems Management—Part y: Security Alarm Reporting Function	DP	SC21N4064
Access Control for OSI Applications	WD	SC21N4094
Framework for Security Audit Trail	Mature WD	SC21N4093
Directory Authentication	IS	9594-8
Directory Access Control (3 parts)	DP	SC21N4041 SC21N4042 SC21N4043

WG6—OSI Session, Presentation and Common Application Services

ACSE Authentication	IS addendum	IS 8649
OSI Upper Layer Security Model	Early WD	SC21N4109

Technical Committee 68 (TC68)—Banking and Related Financial Services

TC68/SC2—Operations and Procedures

WG2—Message Authentication (Security for Wholesale Banking).

Requirements for Message Authentication	complete	ISO 8730
Approved Algorithms for Message Authentication—Part 1 DEA-1 algorithm (Same as ANSI X9.9—1982)	complete	ISO 8731/1
Approved Algorithms for Message Authentication—Part 2 Message Authentication Algorithm	complete	ISO 8731/2
Key Management (see also ANSI A9.17)	complete	ISO 8732
Procedures for Message Encipherment—Part 1 General Principles; Part 2 Algorithms (same as ANSI X9.23)	DIS	DIS 10126
Unnumbered Secure Transmission of Personnel Authentication Information and Node Authentication	WD	ANSI X9.26
Unnumbered Banking-Key Management—Multiple Centre Environment	proposed WI	ANSI X9.28
Data Security Framework for Financial Applications	WD	SC2 WG2N227

TC68/SC6—Financial Transaction Cards, Related Media and Operations

WG6—Security in Retail Banking.		
Retail Message Authentication	DIS	DIS 9807
Pin Management and Security Parts 1 and 2	DIS	DIS 9564
Retail Key Management Standard	WD	
WG7—Security Architecture of Banking Systems using the Integrated Circuit Card		
Financial Transaction Cards:		
Part 1—Card Life Cycle	DP	DP 10202
Part 2—Transaction Process	DP	DP 10202
Part 3—Cryptographic Key Relationships	DP	DP 10202
Part 4—Security Application Modules	WD	
Part 5—Use of Algorithms	WD	
Part 6—Card holder Verification	WD	

Comité Consultatif International Télégraphique et Téléphonique (CCITT)

SG VII 018—Message Handling Systems

Message Handling Systems Framework	complete	X.400 series
EDI Security	WD	EX 162 (June 89)

SG VII O19—Framework for Support of Distributed Applications

OSI Security Architecture	work item	ISO 7498/2
OSI Security Frameworks	work item	
Upper Layer Security Model	work item	
Security Model for Distributed Applications	work item	see JTC1N544

SG VII O20—Directory Systems

Authentication Framework	Complete	X.509
Access Control	Current work item	

SG VIII O28—Security in Telematic Services

Proposed Security Framework for Telematic Services	Current work item	see ISO SC18

European Computer Manufacturers Association (ECMA)

TC29/TGS—Security Aspects of Documents

Security Extensions to ODA	WD

TC32/TG6—Private Switching Networks

Integrating Cryptography in ISDN

TC32/TG9—Security in Open Systems.

Security Framework	complete	TR46
Data Elements and Service Definitions	WD	
Authentication and Security Attributes	NWI	
Secure Association Management	NWI	

American National Standards Institute and Federal Information Processing Standards

Topic	*FIPS*	*ANSI*
Data Encryption Algorithm/Standard (DES)	FIPS 46–1	X3.92–1981
DES Guidelines	FIPS 74	none
DES Modes of Operation (ISO 8372)	FIPS 81	X3.106
DES in Physical and Data Link (ISO 9160)	FIPS 139/FS1026	X3.105–1983
PIN Management and Security	none	X9.8–1982
Computer Data Authentication/ Retail Message Authentication	FIPS 113	X9.19–1985
Retail Key Management	none	X9.24
Wholesale Message Authentication (ISO 8730 and 8731)	FIPS 113	X9.9–1986

	X9.17–1985
	X9.23
tem	X9.26
	none
1028	none
	none
	X12.42
	X9.28
	X.12.58

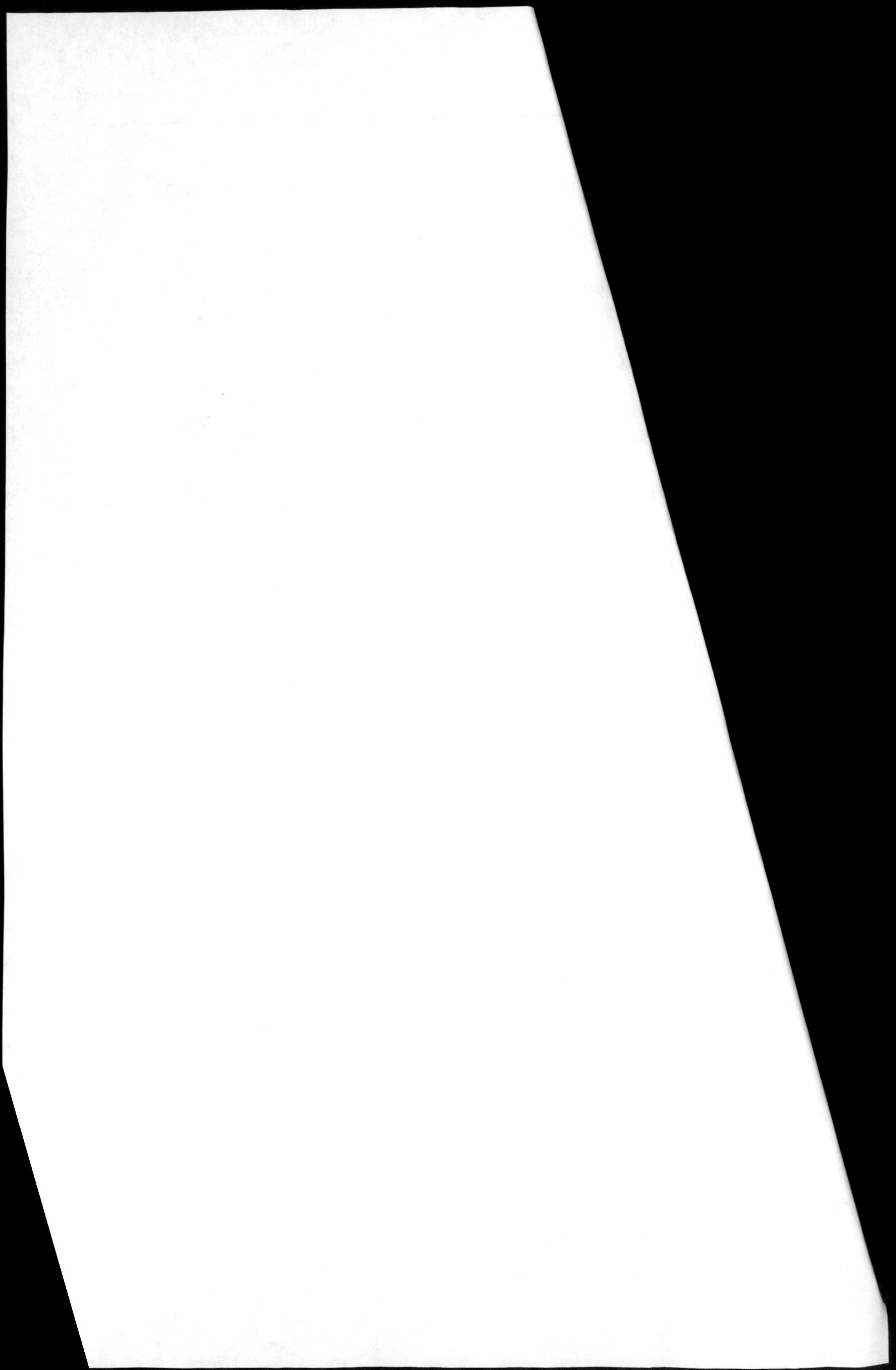

Index

The Author

Dan Minoli has 15 years of experience in the telecommunication and data communication fields. His responsibilities have included research, planning, network design, quality control, communication-related software development, and network implementation. He has worked for Bell Telephone Laboratories, ITT World Communications, Prudential-Bache Securities, and is currently working for Bell Communications Research (Bellcore) as a strategic planner. His research at Bellcore is aimed at identifying new data services that can be provided in the public network in the 1990s, using the integrated services digital network (ISDN) and the advanced intelligent network (AIN).

Mr. Minoli has published more than 150 technical and trade articles; some of his papers have been translated into German, French, and Spanish. He recently coedited an anthology on artificial intelligence applications to integrated network management (Artech House, 1989). He is a frequent speaker at industry conferences and has chaired several conference sessions. He is an advisor for DataPro Research Corporation, an adjunct associate professor at New York University's Information Technology Institute, and a lecturer at Rutgers Center for Management Development. He has acted as a columnist for *Computer World* magazine and has been a reviewer for the IEEE. In 1990, Mr. Minoli also became a contributing editor for *Network Computing* magazine.

The Artech House Telecommunications Library

Vinton G. Cerf, *Series Editor*

A Bibliography of Telecommunications and Socio-Economic Development by Heather E. Hudson

Advances in Computer Systems Security: 3 volume set, Rein Turn, ed.

Advances in Fiber Optics Communications, Henry F. Taylor, ed.

Broadband LAN Technology by Gary Y. Kim

Codes for Error Control and Synchronization by Djimitri Wiggert

Communication Satellites in the Geostationary Orbit by Donald M. Jansky and Michel C. Jeruchim

Current Advances in LANs, MANs, and ISDN, B.G. Kim, ed.

Design and Prospects for the ISDN by G. DICENET

Digital Cellular Radio by George Calhoun

Digital Image Signal Processing by Friedrich Wahl

Digital Signal Processing by Murat Kunt

Digital Switching Control Architectures by Giuseppe Fantauzzi

Disaster Recovery Planning for Telecommunications by Leo A. Wrobel

E-Mail by Stephen A. Caswell

Expert Systems Applications in Integrated Network Management, E.C. Ericson, L.T. Ericson, and D. Minoli, eds.

Handbook of Satellite Telecommunications and Broadcasting, L. Ya. Kantor, ed.

Innovations in Internetworking, Craig Partridge, ed.

Integrated Services Digital Networks by Anthony M. Rutkowski

International Telecommunications Management by Bruce R. Elbert

Introduction to Satellite Communication by Bruce R. Elbert

International Telecommunication Standards Organizations by Andrew Macpherson

Introduction to Telecommunication Electronics by A.Michael Noll

Introduction to Telephones and Telephone Systems by A. Michael Noll

Jitter in Digital Transmission Systems by Patrick R. Trischitta and Eve L. Varma

LANs to WANs: Network Management in the 1990s by Nathan J. Muller and Robert P. Davidson

Long Distance Services: A Buyer's Guide by Daniel D. Briere

Manager's Guide to CENTREX by John R. Abrahams

Mathematical Methods of Information Transmission by K. Arbenz and J.C. Martin

Measurement of Optical Fibers and Devices by G. Cancellieri and U. Ravaioli

Meteor Burst Communication by Jacob Z. Schanker

Minimum Risk Strategy for Acquiring Communications Equipment and Services by Nathan J. Muller

Mobile Information Systems by John Walker

Optical Fiber Transmission Systems by Siegried Geckeler

Optimization of Digital Transmission Systems by K. Trondle and G. Soder

Principles of Secure Communication Systems by Don J. Torrieri

Principles of Signals and Systems: Deterministic Signals by B. Picinbono

Private Telecommunication Networks by Bruce Elbert

Quality Measures and the Design of Telecommunications Systems by John H. Fennick

Radiodetermination Satellite Services and Standards by Martin Rothblatt

Setting Global Telecommunication Standards: The Stakes, The Players, and The Process by Gerd Wallenstein

Signal Theory and Processing by Frederic de Coulon

Techniques in Data Communications by Ralph Glasgal

Telecommunications: An Interdisciplinary Text, Leonard Lewin, ed.

Telecommunications in the U.S.: Trends and Policies, Leonard Lewin, ed.

Telecommunication Systems by Pierre-Girard Fontolliet

Television Technology: Fundamentals and Future Prospects by A. Michael Noll

Terrestrial Digital Microwave Communications, Ferdo Ivanek, ed.

The ITU in a Changing World by George A. Codding, Jr. and Anthony M. Rutkowski

The Law and Regulation of International Space Communication by Harold M. White, Jr. and Rita Lauria White

The Manager's Guide to the New Telecommunications Network by Lawrence D. Gasman

The Telecommunications Deregulation Sourcebook, Stuart N. Brotman, ed.

Traffic Flow in Switching Systems by G. Hebuterne

Troposcatter Radio Links by G. Roda

Voice Teletraffic System Engineering by James R. Boucher

World Atlas of Satellites, Donald M. Jansky, ed.

F